Handbook of Surface and Interface Analysis

Handbook of Surface and Interface Analysis

Methods for Problem-Solving

edited by

J. C. Rivière
AEA Technology
Oxford, England

S. Myhra
Griffith University
Nathan
Queensland, Australia

MARCEL DEKKER, INC. NEW YORK • BASEL • HONG KONG

Library of Congress Cataloging-in-Publication Data

Handbook of surface and interface analysis : methods for problem-solving / edited by J. C. Rivière, S. Myhra.
p. cm.
Includes bibliographical references and index.
ISBN 0-8247-0080-5 (acid-free paper)
1. Surfaces (Physics)—Analysis. 2. Interfaces (Physical sciences)—Analysis. 3. Surface chemistry. 4. Surfaces (Technology)—Analysis. I. Rivière, J. C. II. Myhra, S. (Sverre).
QC173.4.S94H35 1998
541.3'3—dc21 97-46944
CIP

ISBN: 0-8247-0080-5

The publisher offers discounts on this book when ordered in bulk quantities. For more information, write to Special Sales/Professional Marketing at the address below.

This book is printed on acid-free paper.

Marcel Dekker, Inc.
270 Madison Avenue, New York, New York 10016
http://www.dekker.com

Current printing (last digit):
10 9 8 7 6 5 4 3 2 1

PRINTED IN THE UNITED STATES OF AMERICA

Preface

The origins of the subject of this book go back to the late 1960s, when one of the authors (JCR) assisted at the birth of the field of surface analysis, in the process cutting his teeth (and bruising his knuckles) on some of the precursors of the present generation of surface analytical instrumentation. The other author (SM) missed out on the early years of the growth of the infant field, but came in for his share of knuckle bruising at a later stage in its life. Both of us have for many years now been involved directly in wrestling with practical problems of surfaces and interfaces, as well as contributing when possible to basic surface science. We believe, therefore, that we have credentials appropriate to the organization of a project such as this volume.

Not long ago a review publication in this field would be concerned with either new surface analytical techniques or fresh and exciting basic insights gained from ingenious use of the instrumentation. However, times change and the literature reflects such changes. Two factors dominate. One is that it is becoming increasingly difficult to justify in economic terms the maintenance of large research groups and very expensive instrumentation solely for the purpose of basic research in surface science. More and more institutes and laboratories are becoming dependent for their existence on industrial collaboration and participation in technological research and development. The other factor is that what might be called the "traditional" surface analytical techniques are now fully mature and capable of being applied as routine tools of the trade in a remarkably wide range of technological problems. Unlike most earlier volumes, therefore, the present book reflects that maturity in that the emphasis is not, as before, on the techniques but on the problems that they are required to help solve. New techniques and new insights will continue to be generated, one fervently hopes, but for an increasing number of surface scientists it is the short-term project that will provide their bread and butter.

It goes without saying that this book could not have been written without the time and effort that so many contributors have had to find out of their already overcommitted professional lives. Each chapter represents a distillation of the expertise and experience gained by the contributor(s) as a result of his or her devotion to one or more aspects of surface science and technology. We would like to set on record our appreciation of that time and effort, and to acknowledge the good-humored acceptance of our (sometimes substantial) recasting and rearranging of contributors' material.

Although success and the accumulation of expertise in science can be a purely individual achievement, for most of us it is largely a cooperative social phenomenon. For that reason we are indebted to all those co-workers, too numerous to mention, who have passed through our research groups over the years and participated in the two-way information transfer process that has added inestimably to our own expertise. We hope that they have found it equally rewarding.

We have both had longstanding connections with what used to be called AERE Harwell, but is now known as the Harwell Laboratory, AEA Technology, and we must acknowledge the logistical and informational help of that organization in compiling this volume. Our special thanks, too, to those individuals at the laboratory who have been of particular assistance to us over the last couple of years; they will know whom we mean. One of the editors (SM) also wishes to acknowledge a grant from the Australian Department of Industry, Science and Technology, which funded crucial aspects of the collaborative enterprise.

Finally, we must say how grateful we are to our immediate families. Engagement in a scholarly endeavor may be a source of satisfaction to those directly involved, but tends instead to be an ordeal to those having to live alongside the resultant upheaval.

J. C. Rivière
S. Myhra

Contents

About the Contributors

I. (Imre) Bertóti received a Ph.D. in the chemical sciences from Roland Eötvös University in Hungary in 1973. Since then he has held appointments in the United Kingdom while retaining a permanent connection with the Research Laboratory for Inorganic Chemistry of the Hungarian Academy of Sciences in Budapest, where he is currently a scientific adviser and Head of the Surface Chemistry Group. His present interests are concerned mainly with ion/atom-beam-induced surface reactions and with the characterization of surfaces.

A. (Alex) Brennenstühl worked for many years on corrosion problems of defense equipment in the Ministry of Defence in the United Kingdom. In 1981 he moved to Canada, but remained in the defense industry. In 1987 he joined the Corrosion and Tritium Technology Section of Ontario Hydro in Canada, after which his principal interest became concerned with corrosion of heat exchangers and steam generators.

R. (Ross) Davidson worked for many years at Atomic Energy of Canada in Chalk River, Ontario, gaining experience with a wide range of surface analysis equipment. In 1987 he joined Surface Science Western in London, Ontario, where he has contributed to work on surface and materials science problems for industry and academic clients.

C. (Christophe) Donnet received a Ph.D. in analytical chemistry from Lyon University in France in 1990. He has since joined École Centrale de Lyon, where he works in the Laboratoire de Tribologie et Dynamique des Systèmes. His principal interests are concerned with applications of surface-specific techniques and methodologies to tribology, with particular emphasis on solid lubrication and boundary layer phenomena.

T. (Thomas) Gross received a Ph.D. in chemistry from Humboldt University in Berlin in Germany in 1979. Subsequently he worked at the Institute of Physical Chemistry of the Academy of Sciences in Berlin until 1991. He joined the Federal Institute for Materials Research and Testing in 1992, where he is a researcher in the Surface and Thin Film Analysis Laboratory. He has major interests in methodologies and applications of electron spectroscopies to adsorption, electrochemistry, catalysis and tribology, and has been committed to the formulation of international standards for surface chemical analysis.

B. (Birgit) Hagenhoff received a Ph.D. from the University of Münster in Germany in 1993. She then joined the Centre for Manufacturing Technology at Philips in Eindhoven in the Netherlands. Her particular interests include the development of static SIMS for industrial applications, and the combination of SIMS with other surface specific techniques for problem-solving in a technical and industrial setting. From 1997 onwards she expects to launch a consulting service offering a range of surface analytical services to industrial and other customers.

S. (Steven) J. Harris received a Ph.D. from Southampton University, United Kingdom, in 1988. He joined the Materials Sciences Department at Sowerby Research Centre (BAc) in October 1988, and has been in charge of the Surface Analysis facilities at British Aerospace since 1989. His major interests are adhesion, corrosion, diffusion bonding and sealants. He is a member of ISO TC201 "Surface Chemical Analysis" and is currently the secretary of the UK ESCA Users Group.

G. (Gar) B. Hoflund received a Ph.D. in chemical engineering from the University of California at Berkeley in 1978. He then joined the faculty in the Department of Chemical Engineering at the University of Florida in Gainesville. His major interests are concerned with chemical reactions at solid surfaces, such as hyperthermal atom sources, with applications to catalysis, semiconductor materials and processing, electrochemistry, and tribology. He has contributed toward the development of instrumentation and methodologies for surface and interface analysis.

I. (Ingrid) Hyder received an ME. Sc. in engineering science from University of Western Ontario, London, Canada in 1995. She has now joined Air Products and Chemicals, Inc. in the United States. Her principal interests are concerned with the electrochemical dimension of the corrosion of metals and alloys.

N. S. (Stewart) McIntyre worked for 10 years in the nuclear industry in Canada and the United States. In 1981 he joined the University of Western Ontario in London in Canada, where he helped to found Surface Science Western. Subsequently he became its Director. He has worked on problems in the broad area of corrosion, and has special interests in the application of surface specific techniques to nuclear materials.

M. (Miklos) Menyhard received a Ph.D. from the Hungarian Academy of Sciences in Budapest. He is now the Head of the Surface Physics Group at the Research Institute for Technical Physics of the Hungarian Academy of Sciences. His principal interests are concerned with Auger depth profiling and with developing instrumentation and methodologies for profiling of surfaces.

S. (Sverre) Myhra received a Ph.D. in physics from the University of Utah in 1968. Subsequently he worked at the University of Queensland in Australia. He is now at Griffith University, also in Queensland, where he helped to establish the Brisbane Surface Analysis Facility. Recently he has become interested in the development of methodologies for SPM analysis of a wide variety of systems. He has had a longstanding professional relationship with the UK Atomic Energy Authority's laboratory at Harwell (now AEA Technology), where he has worked with the Materials Characterisation Service.

E. (Ernesto) Paparazzo received a Ph.D. in chemistry from the University of Rome in 1978. Subsequently he has worked at, or been connected with, the Consiglio Nazionale della Richerche at Frascati in Italy, the Lawrence Livermore Laboratory in Berkeley, California, and the University of Merida in Venezuela. He is now in the Istituto di Struttura della Materia at CNR in Italy. He has interests in the broad area of the study of surface chemistry of materials, in particular, semiconductors and archeomaterials, using a range of techniques such as XPS, AES/SAM, SRPS and scanning ELS.

D. (Derk) Rading has recently completed a Ph.D. in physics at the University of Münster in Germany. His principal interests are concerned with applications of static SIMS to the characterization of molecular surfaces, with particular emphasis on the interactions between incident species and chemisorbed surface species.

F. (François) Reniers received a Ph.D. in chemistry from the University of Brussels in Belgium in 1991. Since then he has worked at, or been connected with, Vrije Universiteit Brussel, University of Guelph in Canada, and the Materials Science Department of the Lawrence Livermore National Laboratory in Berkeley, California. He has recently joined the Chemistry Department at the Université Libre de Bruxelles in Belgium. His principal interests have been in the development of methodologies for depth profiling with AES and other surface specific techniques.

J. (John) C. Rivière graduated M.Sc. at the University of Western Australia in 1950, then received a Ph.D. from the University of Bristol in England in 1955. He worked in the Commonwealth Scientific Industrial Research Organisation in Melbourne before joining the Harwell Laboratory of the UKAEA (now AEA Technology) in 1958. While with the UKAEA he also held the posts of Visiting Professor in the University of Innsbruck, Austria, and the University of Cardiff, Wales, and spent much time working at those institutions. He retired officially in 1989, but has maintained a close connection with the Harwell Laboratory via consultancies and publications. His research interests over the years have always been in the realm of surface science, both basic and applied, and he and collaborators were responsible for much of the early development of the electron spectroscopies. His contributions were acknowledged by the award of the British Vacuum Council Medal in 1989 and the conferring of the degree of D.Sc. by the University of Bristol in 1995.

D. (Dorothee) M. Rück received a Dr. Phil. Nat. in physics from the University of Frankfurt in Germany. She subsequently in 1988 joined the Gesellschaft für Schwerionenforschung mbH in Darmstadt, Germany. Her interests have ranged from the development of ion sources to the application of ion beams in materials processing, characterization, and development. She is currently engaged in several projects on polymer materials and metals with industrial partners.

P. (Peter) M. A. Sherwood received a Ph.D. in chemistry from Cambridge University in the U.K. in 1970. He subsequently worked at Cambridge and at the University of Newcastle in the U.K. before joining the faculty at Kansas State University in the USA in 1985 where he is now Head of the Chemistry Department. He has headed an active and productive research program concerned with inorganic solids and surfaces with particular emphasis on electrode surfaces, corrosion, and carbon fiber surfaces. Many of the projects have been undertaken in response to problems of direct relevance to industry and emerging technologies. He has served with the U.S. National Science Foundation, and held several visiting appointments. His contributions were acknowledged by the award of a University Distinguished Professorship at Kansas State University in 1997 and the conferring of the degree of Sc.D. by the University of Cambridge in the U.K. in 1995.

R. (Roger) St. Clair Smart received a Ph.D. in chemistry from the University of East Anglia in England in 1969. He subsequently worked in Papua, New Guinea, and at Griffith University in Australia. In 1987 he moved to the University of South Australia, where he is now Deputy Director of the Ian Wark Research Institute and a leader of a surface and materials processing company. His principal research interests are in the area of surface properties of oxides, minerals, and glasses using a broad range of surface-specific techniques and methodologies. In particular the work is concerned with meeting the needs of the Australian materials and minerals processing industries.

G. (Graham) C. Smith received a Ph.D. in physics from the University of Leicester in England in 1983. After a short spell in the scientific instrument industry he joined the National Physical Laboratory to work on standards for surface analysis. In 1990 he moved to Shell Research Ltd in Chester in the U.K. and is now at Shell International Chemicals in Amsterdam in the Netherlands, returning to the U.K. in 1998. He has broad experience with techniques and methodologies for characterization of surfaces, thin films, and interfaces. Most recently his research has focused on catalysis, tribology, polymers, and other applications related to petrochemistry.

A. (András) Tóth received a Ph.D. in inorganic chemistry from Eötvös University in Budapest in 1980. He then worked in Italy before joining the Research Laboratory for Inorganic Chemistry of the Hungarian Academy of Sciences in 1982. His main interests are in surface modification of polymers by ion and atom beams and by cold plasmas, as studied by XPS.

W. (Wolfgang) E. S. Unger received a Ph.D. in physical and analytical chemistry from the Institute of Physical Chemistry of the Academy of Sciences in Berlin in Germany in 1987. He remained at this institution until 1991 where he became Head of the Electron Spectroscopy Sub-Department in 1989. Currently he is at the Federal Institute for Materials Research and Testing in Berlin, where he is the project leader of the Surface and

Thin Film Analysis Laboratory. He has major interests in the development of methodologies for electron and mass spectroscopies, which he has applied to characterization of catalysts, polymer surfaces, and polymer-metal interfaces.

J. (John) F. Watts received a Ph.D. in materials science from the University of Surrey in England in 1981 and was awarded a D.Sc. in 1997 from the same university. He subsequently returned to an academic position in Adhesion Science at Surrey, where he now heads the Surface and Interface Reaction Group. Current research interests are concerned with aspects of adhesion and polymer surface treatment, and with the applications of XPS, AES, ToF-SIMS, and SPM to those problems. He is the author of an introductory text and a software package on electron spectroscopy, and has lectured widely in Europe, the United States, and the Far East.

R. (Bob) K. Wild received a Ph.D. in physics from Reading University in England in 1966. After a brief period at the University of Virginia, Charlottesville, he returned in 1968 to the Research Laboratories of the Central Electricity Generating Board at Berkeley in England. Many of the activities were subsequently transferred in 1992 to the Interface Analysis Centre at the University of Bristol. His principal professional interests are in applying AES, XPS, and SIMS to the study of corrosion of steels, and effects of grain boundary segregation on Ni-based alloys. More recently he has focused his attention on diffusion bonding of Al and Ti alloys.

Handbook of Surface and Interface Analysis

Finally, the demographic situation within the broad field needs to be considered. Many of those scientists that grew up with the field as it came into being in the 1960s are either of retirement age or have moved into managerial/executive positions. Hence, there is a generational transition in which a new cadre is entering an already mature field but is subject to entirely different imperatives and pressures. Part of the rationale for this book is in the area of information transfer from one generation to another. The intention is to speak to members of an audience at the early stages of their careers rather than to the "founding fathers and mothers."

3. THE INTENDED AUDIENCE

The upshot of the foregoing is that there is an expanding group within the surface science community consisting of those working mostly in industrial laboratories or in analytical service centers, whose job it is to provide reliable and relevant answers to problems involving surfaces and interfaces with rapid turnaround and at relatively low cost. They will know enough about the nature of the information obtainable from a range of techniques, surface specific and otherwise, to be able to decide which ones to use, and will know how to deploy them to best effect. They will not in general have sufficient in-depth background, and certainly not the time, to be able to engage in long-term basic research. Although many might well have expertise in one or more areas of materials science, the nature of their work is such that they are liable to meet at any time a material or a problem from a totally unfamiliar area. It is the members of this group at whom this book is aimed, in an effort to provide them with a guide to the procedures to be adopted when they do come up against something unfamiliar.

Tertiary education institutions are no longer immune to the dicta of efficiency, productivity, and flexibility. Not long ago it could be assumed that senior academic staff would devote most of their time and energy to research and the two-way teaching and learning process. As a consequence, research students and postdoctoral researchers would gain insight through lengthy one-to-one interactions in the laboratory with an academic supervisor. Regrettably, most academic supervisors are now increasingly preoccupied with fundraising, administrative duties, and responding to corporate imperatives. This book is thus intended also as a complement and, in some cases, an alternative to traditional methods of learning for researchers at the formative stages of their careers.

4. THE STRUCTURE OF THE VOLUME

The general thrust of this volume will be a description of the technical, methodological, and phenomenological aspects of surface and interface analysis, in the course of which an attempt will be made to chart the most efficient path(s) from an initial question to a credible answer for a series of generic problem areas. There will be less concern with the many techniques as ends in themselves than as means to an end. Since this approach is an inversion of the more usual one of first describing a technique and then giving examples of its application, it can be thought of as a "top-down" approach, which has been shown to be effective and efficient in many other areas. In view of the ways in which an increasingly large number of workers in the field operate, the structural inversion seems logical.

From the foregoing it may be deduced that the structure of the volume will be somewhat reminiscent of that of an expert system. Starting with some of the most general materials science questions that could arise in surface/interface science and technology, the user will be guided to ever more detailed and specific levels of questions, the choice of path between them being based on the experience of experts. There will be descriptions of the principal surface specific techniques, but not in nearly such great detail as may be found elsewhere; the intention is simply to help the user toward a preliminary choice of techniques. The reader will be directed to whichever chapter (or chapters) is appropriate to the problem, and will find there much essential information, a recommendation for the most productive methodology(ies), and an indication of the likely answer, in phenomenological terms. Throughout there will be an emphasis on the multiskilling approach, by the demonstration in each chapter of how the information provided by one or more of the traditional surface–specific techniques is complemented and reinforced by that either from some of the less common techniques or from so-called bulk techniques. In particular, attention will be paid to how the recent exciting developments in scanned probe microscopy can assist in the provision of a more complete analytical answer.

A visual summary of the structure is given in Fig. 1, in which the various routes that users of the book might take are indicated. The chapters fall neatly into a set of four modules. This chapter and the next are introductory, one to the aims and objectives of the book and the other to the principles of analytical problem solving. The third chapter sets out exactly how to use the information in the book and occupies its own module. Chapters 4 to 10 describe the tools of the trade, how to use them, when and when not to use them, what information they provide, how to quantify that information, etc. Finally, Chapters 11 to 19 represent an authoritative distillation of the expertise and experience of authors each of whom is a specialist in the application of surface analytical and complementary techniques to the particular subject area.

All readers will need to consult the How-to-Use Module (Chapter 3) in order to find their way around the book in search of the information they are after. Some may wish to skip the introduction module and go straight to Chapter 3; others, perhaps of less experience, will benefit from some introduction. Those

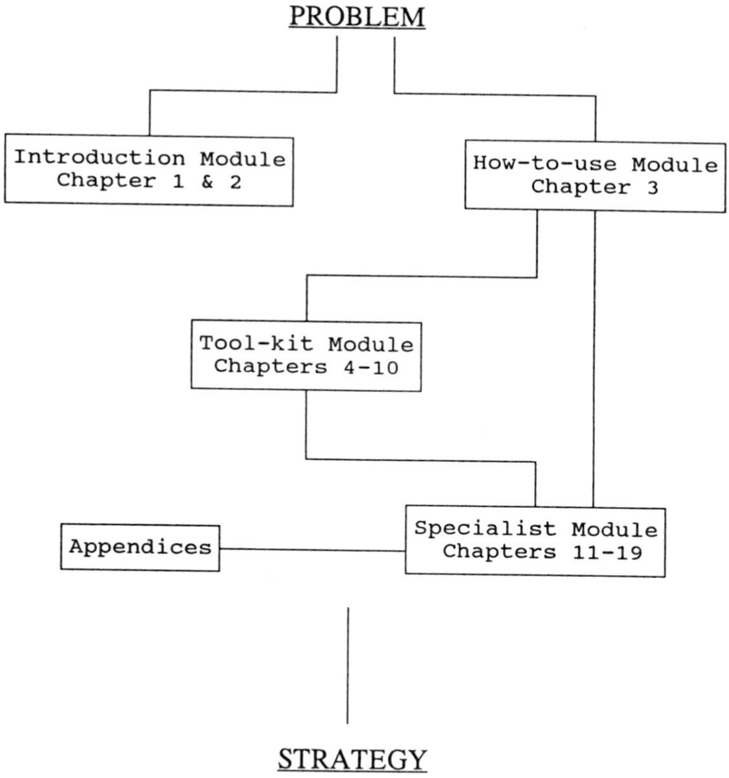

Figure 1. The structure of the volume: routes to solutions in surface and interface science and technology.

that already have significant instrumental (i.e., Tool-kit) experience will no doubt bypass the Tool-kit module and proceed directly to the specialist module. Several variations are therefore possible. Whatever the level of experience, however, and whatever the route chosen the objective is to provide a practical guide to problem-solving methods for surfaces and interfaces. The authors hope that this book will fill the need that they have perceived for such a guide.

2

Elements of Problem-Solving

S. MYHRA AEA Technology, Harwell Laboratory, Oxon, England
and Griffith University, Queensland, Australia

J. C. RIVIÈRE AEA Technology, Harwell Laboratory, Oxon, England

1. INTRODUCTION

The approach taken in this book assumes that some readers are expert in many aspects of problem solving and others are novices or at an early stage on the learning curve. For the latter readership this chapter sets out the underlying logic of problem solving, with particular reference to surface analysis, in the belief that such a discussion might be informative and useful. The successive stages that go to make up the problem-solving route will be described together with the associated thought processes, many of which will be intuitive for the more experienced practitioners. Along the route some of the more down-to-earth practicalities will be encountered and discussed.

2. SURFACE, INTERFACE AND BULK

In the interrogation of any system in any scientific discipline, to obtain information about that system at least three components are necessary: a stimulus, an interaction volume in the system that responds to the stimulus, and a means of gathering the information generated within that volume as a result of the stimulus. Since not all the information will necessarily be able to be gathered, there is also an information volume, by definition equal to or smaller than the interaction volume.

So far the description is entirely general; realization of the general components in terms of the different types of stimulus and of means of gathering information leads naturally to the many analytical techniques available. Now suppose that one of the dimensions of the information volume is limited to between one and a few atomic diameters as a result of the choice of operating parameters of the stimulus. Such a limitation has the effect that the volume becomes quasi-two-dimensional. If the system being interrogated is a solid, the information volume must then correspond to the surface of the solid, and the relevant techniques are those whose information content is specific to the surface. For "interrogation" now read "surface and interface analysis."

In some situations, mostly in basic research, it is the information generated only within the quasi-two-dimensional volume, or surface layer, itself that is of interest; that is, the analysis is concerned with conditions in that layer in isolation, as it were. When dealing with solids, however, it is rare that the properties of a surface layer can be considered without reference to the same properties of regions further away from the surface, beyond the shortest (atomic scale) dimension of the information volume. This is so because some properties of the surface depend on what is happening, or has already happened, in those remoter regions, generally termed the "bulk" of the solid. Even in homogeneous solids such a dependence exists, but of course few solids are homogeneous. They often contain internal interfaces of many types, such as phase boundaries, grain boundaries, artificially created compositional and/or structural interfaces, and so on. In principle it is clear that the presence of a surface must imply the existence of an interface, even if the latter is only with the UHV environment. The properties of interfaces are of crucial importance to the behavior of the solid from both surface and bulk aspects, and will need to be accessible for analysis. Since the interfaces are themselves effectively quasi-two-dimensional volumes, analysis of their properties should ideally be undertaken by surface-specific methods. However, they are of course internal and therefore, in current jargon, "buried"; how to gain access to them for analysis without at the same time causing perturbation leading to ambiguity has been the central question in surface and interface analysis problem solving for a long time. The question has in fact been addressed in this book by several authors using various methods, but it cannot be said that it has yet been answered in a universally satisfactory way.

It should now be clear that, except in special cases, surface analysis is not restricted merely to the "genuine" surface, i.e., the original outermost atomic layers, but is applied very frequently to internal surfaces exposed in a variety of ways. When applied thus to features that are really properties of the bulk, at least of the near-surface bulk, the information derived from the surface analysis ought to show some complementarity to the information from certain other techniques whose information volume is such that they are normally regarded as "bulk" analytical techniques. Indeed, it has been an increasing trend recently for analysis by surface-*specific* techniques to be complemented by analysis by "bulk" techniques used in surface-*sensitive* ways, a trend that is illustrated in several places in this volume. This holistic approach is essential to the complete solving of many problems, but of course the choice of complementary techniques needs to be made carefully.

3. THE PROBLEM-SOLVING SEQUENCE

The logical sequence described here is not specific to problem solving by surface analysis, but could be applied to problem solving in any branch of material science. However, it is cast in terms of surface problems.

3.1. Identification of the Problem (e.g., Teflon peeling from pan) and Formation of an Initial Hypothesis

The onset of the problem triggers the subsequent sequence of events. Normally it will be up to the person or the organization suffering the consequences of a problem to identify the nature of it and to initiate a process of remedy. To do that they will, as customers, need to commission one or more analysts to provide both relevant and reliable analytical data to enable them to reach a resolution of the problem via the building up of an overall picture. Preliminary consideration of the nature of the problem may have suggested a hypothesis as to the reasons for its occurrence, and the type(s) of analysis initially chosen will then be based on that hypothesis. On the other hand, there may have been no clues arising from the way that the problem first occurred, and a working hypothesis will in that case have to wait for the results from analysis used in "shot-in-the-dark" or intuitive manners. With increasing experience, intuition can become surprisingly accurate. Experts in problem solving develop the ability to propound successful hypotheses based either on intuition or on very scant information.

Often the analysts are merely providers of data, and it is the customer who works toward the solution. More interesting for the analyst and usually more rewarding for the customer are those instances in which the analyst is allowed to play an interactive role in the entire problem-solving sequence. Such interaction can often lead to short cuts, and therefore savings in cost, in the sequence.

3.2. Identification of the Essential Variable(s) (e.g., composition at the Teflon/pan interface)

Once the problem has been identified and a preliminary hypothesis set up, the nature of the information required, based on that hypothesis, can also be identified. From that it is a short step to deciding what can actually be measured experimentally to provide that information (i.e., the essential variable(s)). Practicalities enter here, and indeed at many other points in the sequence, in that the customer will naturally be seeking the maximum amount of information at the minimum cost.

3.3. Reduction of the Problem as Far as Possible without Losing Essential Information (e.g., disregard bulk composition of substrate and concentrate on interface composition)

Although in basic research on model systems there is usually complete freedom of experimental design, in the real world the problems are too complex to be able to "solve" in the same complete sense. Life is too short and available finance is strictly limited. In any case, the materials associated with problems, and often the problems themselves, may be ill-defined, with possible hidden variables. It is vital at this stage in the sequence to avoid the temptation to start analysis at once, since the result may simply be to end up with a mass of irrelevant or even misleading data. The most important step in the art of problem solving is pinning down what is the crucial experimental variable that must be measured, and to that end all relevant background information on the system giving rise to the problem should be used. Such information might include the bulk compositions of constituents, any chemical or physical pretreatment, dopant concentrations, ambient atmosphere, etc. It is necessary for the customer and the analyst to have full discussions so that no pieces of the information jigsaw are overlooked.

3.4. Selection of the Techniques(s) Likely to Provide the Crucial Information by the Most Reliable and Economic Route (e.g., XPS would be a good choice, but AES and SSIMS would be poor ones)

Not many institutes or organizations can offer the full range of analytical instrumentation and expertise, and most only a subset of that range. Where there are only a few techniques available in-house, there is an understandable tendency not only to try to apply them regardless of suitability or otherwise but also to push individual techniques beyond their limits of reliable and efficient function. That tendency must be resisted. In the long run uncritical application of the same few techniques to the attempted solving of every analytical problem that comes along is counterproductive, in that for many problems the information and data produced will be quite inappropriate to solution. The technique(s) must be matched to the crucial information or the whole process becomes nonsensical; if the necessary technique is not available in-house, then it must be sought elsewhere.

3.5. Choice of Methodology(ies) Consistent with the Selection of Technique(s) (e.g., use large sample area for XPS, avoid obscuring contamination, compare data from failed components with those from control specimens)

Once the choice of necessary technique(s) has been made, then details, such as the way in which the technique(s) must be used to produce the crucial information, and any precautions that must be taken to avoid ambiguity in interpretation have to be decided. In many problems vital clues can be extracted from "good versus bad" or "before versus after" analytical comparisons, which require control specimens. There is also the ever-recurring and vexed question, referred to earlier, of how to access buried information. Here practicalities enter again. The customer may decide that solution of the problem can be achieved only by such access, but then he must be made aware not only that extra time, and therefore money, are involved but that the mechanics of obtaining access, typically by ion bombardment to removal material, may change the nature of the information irretrievably. There is no firm and final answer to that situation; every sample and problem has to be considered individually, and the pros and cons weighed up.

3.6. Acquisition and Processing of Data of Adequate Quantity and Quality (e.g., energy resolution, signal-to-noise ratio, scan widths required for background subtraction, etc.)

At this point the analysis can actually be started. Unless the data are of a minimum quality necessary for correct interpretation, they will not help toward solution of the problem. By this is meant that they must be good enough to disprove a preliminary hypothesis if necessary, and lead to alternative hypotheses, rather than being of just sufficient quality to support the preconceived hypothesis. The level of quality required is a function of the nature of the crucial information sought and will vary from problem to problem.

3.7. Interpretation of the Data (e.g., are the results consistent with hypotheses and with other experimental evidence, are they credible, are they reproducible, and if so, at what confidence level?)

In an ideal situation there would be abundant data of high quality obtained from several complementary techniques, requiring fairly minimal and straightforward interpretation. In practice the situation may be constrained to be far from ideal, and interpretation must be undertaken carefully and realistically. The more direct and transparent the interpretation, and the greater the extent to which it is compatible with prior information about the problem, the more likely it is that the conclusions from it will be of use to the customer. The main hazards at this stage are that the interpretation is wrong or irrelevant to the problems.

3.8. Review and Evaluation, and Iteration if Necessary (e.g., the data have been misinterpreted, or in hindsight emphasis should have been put on other variables, or the customer asked the wrong questions in the first place)

This is one of the most important stages in the whole sequence and should be taken seriously. Both individual and institutional credibility are at stake, as is perhaps the long-term viability of the institute or organization in the problem-solving field. A lifetime of carefully nurtured reputation and customer relationships can be undone so easily by one or two shoddy jobs. It is far better, if in doubt, to perform additional measurements and analyses before a report is submitted to the customer.

3.9. Presentation (e.g., to-the-point reports, false color maps, succinct and adequately descriptive captions, and easy-to-follow conclusions are more useful to the customer than lengthy explanations, particularly if the essential conclusions have to be presented to management)

Of course a report must be accurate, factual, relevant, and complete, but its essential message(s) must be transparent to the relatively nonexpert reader without, for instance, descending into the excessive use of jargon. Nowadays, with so many excellent computer packages available, there is no excuse for submitting a report without substantial visual impact. In addition, it is perfectly possible to assemble data from several techniques for incorporation into an overall report. With the trend toward full digitization of output, including even high-resolution images, full flexibility of composition, layout, and mode of presentation can be achieved.

The sequence described suggests that the scope of problem solving is limited to support of R&D in an industrial setting. However, the ambit of the book is far wider than that. For instance, the emphasis in strategic research may be on a broad characterization of a particular class of material or on the investigation of a generic variety of surface/interface reaction. In either case the requirement will be for completeness, which can be promoted only by adopting an integrated approach exploiting complementary techniques and methodologies. Another setting in which surface/interface analysis has much to offer is that of quality control. There the emphasis will be on demonstrating that the relevant variable(s) is(are) indeed being controlled, and that the methodology remains constant over time. Also, it must be remembered that a significant fraction of surface/interface analysis is carried out within the tertiary education sector by central facilities providing a variety of services for academic customers. The sequence thus includes a component of technology transfer, in the sense that both providers and customers are engaged in a teaching and learning endeavor, as well as a problem-solving process.

4. PRACTICAL MATTERS IN PROBLEM SOLVING FOR SURFACES AND INTERFACES

Up to now the discussion and description of the logical sequence of problem solving have been couched in general terms. What follows is a more detailed series of observations on some of the practicalities that have to be taken into account before and during surface analysis. Most of the points are described, often in greater depth and breadth, in subsequent chapters, but it is useful at this stage to provide a general overview.

4.1. Specimen Handling, Preparation, and Configuration

Whether surface analysis is being used in a basic scientific experiment, in which the origin and history of the specimen are known, or whether it is applied in problem solving, where the specimen is likely to arrive in a much less well defined state, the essential first requirement in dealing with the specimen is cleanliness. This should be obvious from the extreme surface specificity of surface analytical techniques, but must be emphasised nevertheless. Even though specimens exposed to the ambient atmosphere or subjected to various pretreatments will carry surface contamination, it is vital not to add to or alter that contamination by manual contact or by any other contact that might cause transfer of material onto the surface. For the same reason the vacuum ambient in which the analysis is performed should be such that it cannot contribute to the contamination, i.e., it should be oil free and in the UHV region of pressure.

One of the reasons for avoidance of any alteration to the existing contamination layer is that there will be instances where some or all of the information needed for solving a problem may actually reside in the nature of the contamination itself. Such instances might arise, for example, in tribology, adhesion, or corrosion. It is in any case essential to carry out an analysis of the specimen in the "as-received" condition, if for no other reason than to establish the extent of the contamination and therefore to be able to decide on subsequent procedures.

Specimen preparation and treatment can take place either outside the vacuum envelope of the analysis system (ex situ), or within the system (in situ). Ex situ preparation could be in the uncontrolled laboratory ambient or in a controlled atmosphere, while in situ preparation and treatment could be in the analysis position itself or in an associated chamber.

4.1.1. Ex situ preparation

In surface and interface problem solving most specimens will have originated from an external source (i.e., the customer) and therefore by definition they will have had some ex situ preparation or treatment, if only as a result of transfer from their original environment to that of analysis. However, that is not what is really meant by ex situ preparation. Many analytical techniques in materials science, those that can be classified as providing "bulk" analysis, require preparation of the specimen by methods such as polishing, abrasion, and sectioning. These methods can also be used in the preparation of specimens for certain types of surface analysis, but of course the cleanliness requirements are much stricter than in preparation for bulk analysis. For example, lubricating or cutting liquids cannot be used because they will leave residues on the surface, while exposure to polar fluids may degrade the surface. In addition, any debris left on the surface after such ex situ methods have been used should be removable by ultrasonic cleaning in a bath of high-volatility solvent. The usual reason for employing such methods is to expose a "buried" interface located at a depth too great to be reached by in situ methods.

4.1.2. In situ preparation

Once inside the analytical system there are many types of preparation or treatment of the specimen that might need to be carried out in the course of any one problem-solving procedure. They fall into the approximate categories of cleaning, depth profiling, interface exposure, and surface treatment.

The most common method of removing surface contamination is by ion beam erosion, normally with energetic Ar^+ bombardment. By adjusting the ion energy and by using a low ion dose (roughly equivalent to about one incident ion per surface site) it is possible to remove just the first one or two atomic layers, which usually consist of contamination. In basic research such erosion is often used with specimen heating in sequential cycles in order to achieve as near perfect cleanliness and structured order on single-crystal surfaces, as possible. The heating is necessary to allow the surface to recrystallize after the disorder introduced by the ion beam. Clearly such a cleaning procedure is inappropriate to specimens involved in problem solving since their information content would be lost immediately. However, cleaning by gentle ion beam erosion (without heating) is convenient and reliable, and is the most widely used method largely because for the vast majority of specimens there is no alternative. One or more ion guns are therefore mandatory accessories on most surface analytical installations. The principal disadvantages of a technique in which relatively heavy charged particles such as Ar^+ are accelerated to impact on a surface are structural and chemical damage, mentioned in more detail later.

In a few special cases in situ cleaning can be performed by mechanical means. It is possible to fix special tools to the end of an auxiliary manipulator arm so that the operations of abrasion (with a diamond file) scraping (with scalpel or razor blades), or even cleavage (razor blade or chisel) can be carried out. When abrading or scraping, one must be careful that material from one surface is not transferred to the next.

As well as for cleaning, ion beam erosion is used very extensively for the (relatively) controlled removal of material to allow analytical access to regions of the specimen beyond the surface information volume. Erosion and analysis can be simultaneous (e.g., when using AES) or sequential (e.g., when using XPS); in both cases the plot of elemental concentration as a function of amount removed is called a depth profile. Examples of depth profiles abound in the literature and there are several in this volume. Also in this volume in various places (Chapters 6–9) are described the deleterious effects to be expected from ion bombardment of a surface. The latter arise from unavoidable knock-on structural and chemical damage, which disrupts the surface crystal structure and alters the electronic structure. In addition, for a multiatomic specimen, as most are, the removal of surface species by the ion beam will be selective, so that some species are enriched and others depleted as the eroded face progresses into the specimen. As a result there may be selective desorption, interlayer mixing, and the appearance of subvalent species and nonstoichiometry. For much more detail see the above-mentioned chapters. Unfortunately there is no realistic alternative to ion beam depth profiling at the moment, particularly in the realm of problem solving, and the best that can be done is to be aware of the effects and attempt to account for them in a semiquantitative way.

If internal interfaces are also regions of structural weakness, they can often be exposed for analysis by mechanical methods. The classical example is that of grain boundaries in polycrystalline metals that have been weakened as a result of impurity segregation to them following heat treatment. The boundaries can be exposed by fracture of the metal, but any attempt to measure the nature or level of the impurity by fracture ex situ must always be unsuccessful because of reaction of the impurity with ambient air and of the accumulation of contamination. Thus, there are many available designs of fracture stage for use in situ, mostly for impact fracture but some for tensile fracture. The stages can also be used for ceramic fracture, since there are instances in which ceramic materials also lose cohesion through the presence of grain boundary impurities.

In special cases, such as in adhesion and in thin oxide film studies, it may be possible to expose the interface of interest by peeling techniques. Devices designed to do that are usually constructed individually and are not normally available commercially, unlike fracture stages.

Finally, in situ preparation includes also all those treatments used from time to time to try and create on a specimen the same surface condition as might be found after some technological treatment or other. Since it is normally undesirable to carry out such treatments in the analysis chamber itself, many instruments have an additional chamber, communicating with but isolatable from the analysis chamber, in which the treatments can take place. According to what is required, the additional, or reaction, chamber might therefore be equipped with the means of heating or cooling the specimen, or of exposing it to a gas or mixture of gases via a metered gas-handling system, or of depositing thin films of one or more materials on its surface, or of causing alterations in its surface by ion implantation or plasma discharge. Many of these treatments will be described in various places in this book.

4.1.3. Specimen configuration

Just because in some techniques (e.g., STM) the quasi-two-dimensional interaction volume is exceedingly small does not mean that the specimen configuration can be similar. There is no strong relationship between the two. Since specimens have to be positioned and oriented with the naked eye, and handled in so doing with, for example, clean high-quality tweezers, it follows that they are all macroscopic. In addition, many of the surface analytical techniques, when used in routine fashion, are macroscopic in two of the three dimensions of the interaction volume; SSIMS and XPS in their nonimaging forms are good examples.

Despite the above statement, the analyst engaged in problem solving will not infrequently be presented with nonstandard specimen configurations, which will require ingenuity and inventiveness if a representative analysis is to be achieved. Examples of such "difficult" specimens are powders, carbon and polymer fibers, biomaterials, and microelectronic circuits; again, mention is made of these in this volume in the appropriate chapters, where ways of presenting them for successful analysis are described.

4.2. Technique Destructiveness

Basic physics says that any measurement process is irreversible; in that very fundamental sense, then, no measurement technique can claim to be nondestructive. In the world of practical problem solving, however, such a limitation can be set aside as being too rigid. Given that all the surface analytical techniques can be destructive to greater or lesser extents, a pragmatic approach must be adopted. Depending on the constraints applicable in any particular problem or to a particular type of specimen, a set of functional criteria can be set up to classify techniques and procedures in terms of their potential destructiveness, as follows:

The functionality criterion

If the intended function(s) of a specimen or material are unaffected by the problem-solving process, then the techniques and methodologies required to solve the problem can be said to be nondestructive.

The market value criterion

Using "market" here in its broadest sense, if the market value of an object is not reduced by the problem-solving process, then the process can be said to be nondestructive. There might even be a gain in value as a result of the analytical procedure, which would be offset against any loss arising from possible destructiveness. Further, if there is a set of nominally identical items, it is possible that the loss in value of those that are analyzed and perhaps irretrievably damaged would be more than made up by the gain in value of the remainder (e.g., due to demonstrable quality control).

The sequential analysis criterion

When using a multitechnique approach to problem solving, care must be taken that the application of one technique to a specimen does not jeopardize the validity of subsequent analyses by other techniques. Under this criterion if that validity is unaffected, then the technique first applied can be said to be nondestructive. The criterion puts obvious constraints on the sequence in which techniques should be used, beginning with the least destructive and ending with the most. A corollary of this criterion is that in cases where there are continuing ex situ or in situ specimen treatments requiring periodic analysis, then the analytical procedure should not itself affect the course of the treatment(s).

The information volume criterion

There is a direct relationship between the quality and quantity of information retrievable from the information volume and the fluence of the stimulus being used. The greater the fluence, the larger will be the number of interactions per unit time, and therefore the better will be the signal-to-background quality of the information. On the other hand, it is the direct, and sometimes indirect, effects of the stimulus that cause destructiveness. Among these effects, for example, might be the rate of energy deposition, or the rate of momentum transfer, or the type of interaction mechanism, or the duration of the analytical procedure. Thus a balance has to be struck between the quality of information and the level of damage. In practical terms this may mean having to take longer over an analysis than normal in order to maintain the rate of energy deposition below a certain critical level, or using an energy of the stimulating probe different from normal to avoid or minimize a particular interaction mechanism. Where damage is unavoidable, the analysis may reduce simply to having to accept whatever information can be collected before the specimen surface is altered to a predetermined extent.

There will be instances in which concerns over destructiveness override all others, as where an object is either irreplaceable and/or of great value, for example archaeological artefacts, national treasures, objets d'art, etc. At the other extreme there will be objects that can be replaced at essentially zero cost, compared with the cost of the analysis, such as mass-produced consumables, waste products, common naturally occurring products, etc. Even before analysis starts, it might be the specimen preparation itself that has to be considered because of its destructiveness, e.g., if a large item has to be sectioned because it cannot be accommodated in the analysis position.

It is clearly difficult from the above discussion to give fixed criteria for taking destructiveness into account. There will generally be a degree of uncertainty with respect to a particular specimen or problem. The most important rule is always to tell the customer beforehand what the potential effects on the specimen might be and to have complete agreement on the procedure to be adopted.

4.3. Quality Assurance, Best Practice, and Good Housekeeping

In an ideal world the quality of an analytical process, and its outcome, would be guaranteed automatically by scrupulous adherence to the "scientific method," independently of the intention of the analysis. That is, it should not matter whether the end user of the information is a research scientist or a customer, nor whether the results are to be used to generate new basic knowledge, improved techniques, better products or financial gain. In the ideal situation the quality would indeed be assured without the need for further formality.

When it comes to the type of contractual relationship that is now commonplace between the supplier of the analytical service and the customer, a more formal guarantee basis is needed. The customer will not necessarily be familiar with the "in-principle" merits of the scientific method, but will understand a published set of guidelines to which the supplier must adhere. Such a set will include rules for the "best practice" in all aspects of handling, preparing, and treating specimens, and instructions as to how an analysis must be carried out so that its results have been quantified in an approved way and are traceable back to an agreed international standard. The latter proviso is essential in order that all results from any laboratory using the same technique are intercomparable. Only by general adherence to the same set of guidelines and instructions can quality assurance in an analysis be guaranteed. In many countries there are now standards organizations that assess analytical supplier laboratories for accreditation in quality assurance, and the number so accredited, although few, is growing steadily.

No analytical service can operate efficiently without the internal procedures which can be lumped together under the heading of "good housekeeping." Such procedures are basically commonsense routines that ensure that both personnel and instruments function smoothly and reliably, that proper records are kept and kept up-to-the-minute, that specimens are not mislaid, mishandled or cross-contaminated, that reports are compiled and sent on time, that customers are always fully informed, that safety standards are being met, etc. The volume will have much to say about the esoteric aspects of surface and interface analysis. It must always be borne in mind, however, that attention to detail is at least as important to the final outcome.

ACKNOWLEDGMENTS

The chapter is a description in summary form of the activities that are taking place in many laboratories. The authors are fortunate to have worked with some of the best practitioners of problem solving. In particular we value our long-standing connections with the Harwell Laboratory of AEA Technology. We wish to acknowledge many useful comments by Dr. H. E. Bishop.

3
How to Use This Book

S. MYHRA AEA Technology, Harwell Laboratory, Oxon, England, and Griffith University, Queensland, Australia

J. C. RIVIÈRE AEA Technology, Harwell Laboratory, Oxon, England

1. INTRODUCTION

In this chapter will be found directions on how to use the information in the rest of the book. These directions consist of sets of key words and references linking most aspects of the process of problem solving using surface analytical and complementary methods. The references themselves consist of headings, subheadings, or sub-subheadings, etc., of the appropriate sections in the appropriate chapters. The sequence in which the references should be used in the attempted solution of a particular problem is based on flowcharts containing a succession of questions, or prompts, starting with general questions and progressing to ever more specific ones. Any pathway thus found through a flowchart will not necessarily be unique, since there may be more than one route to a solution, but it is intended that the information can be accessed in such a way that the most efficient route can be chosen.

2. DEFINITIONS

In constructing the flowcharts and tables of references, terms and concepts have been used that require definition here. The definitions are mostly in line with common usage, but for the sake of convenience and completeness some have been framed slightly differently. They are as follows:

Class of material: This includes the usual categories, such as metals, alloys, semiconductors, polymers, organics, etc., to which it is convenient to add some more specialized and specific ones such as composites, catalysts and adhesives.

Properties of materials: Defined as those attributes, largely physical, that are intrinsic to the class of material. They would, for instance, include melting point, electrical and thermal conductivities, hardness, density, reflectivity, elasticity, etc.

Material variables: To be differentiated from properties in that they are those aspects of materials that are usually the subjects of measurement during analysis, such as surface, interface, and bulk composition, grain size and orientation, surface and bulk crystal structure, topography, porosity, etc. Often the variables will not only reflect in some direct or indirect manner the intrinsic properties of materials but will also govern their functions and applications (see below).

Material functions: Often the terms "function" and "application" can be used interchangeably, but for the present purposes it is useful to distinguish them. Thus "function" is defined as the ability of a material to act in a certain way, for example, as a high-permeability membrane, or an impact-resistant surface, or a photochromic barrier, or an oxidation-resistant coating, etc. The exact application is not specified.

Material applications: This is the most precise and specific of the definitions, in that an application invariably involves a tangible end product. Applications might therefore include catalysts for particular reactions, soft contact lenses, carbide-tipped machine tools, CMOS-based devices, and so on.

Specimen configuration: Those aspects of a specimen presented for analysis that need to be taken into account in deciding how to handle it and which must be considered in the context of the analytical instrument and the chosen methodology. The principal aspects are dimensions, form, topography, structure, and fragility. Others such as porosity (leading to outgassing problems), toxicity, and radioactivity might need to be considered occasionally.

Interpretation: According to the nature of the information provided by each surface/interface technique, so the raw data after initial processing will require interpretation in terms of one or more of composition, electronic structure, crystalline or phase structure, topography, etc. A correct interpretation will in turn lead to a "solution" to the problem.

3. DECISION MAKING IN PROBLEM SOLVING

Before any analytical work actually starts, a series of logical decision steps have to be taken to arrive at the correct procedure or set of procedures that will produce the required information about a specimen. These steps are outlined visually in Fig. 1, which represents the thought processes used subconsciously by experienced practitioners. However, it is worth pointing out that even those most experienced cannot be expected to know everything and may well find the cross-referenced information in this book helpful.

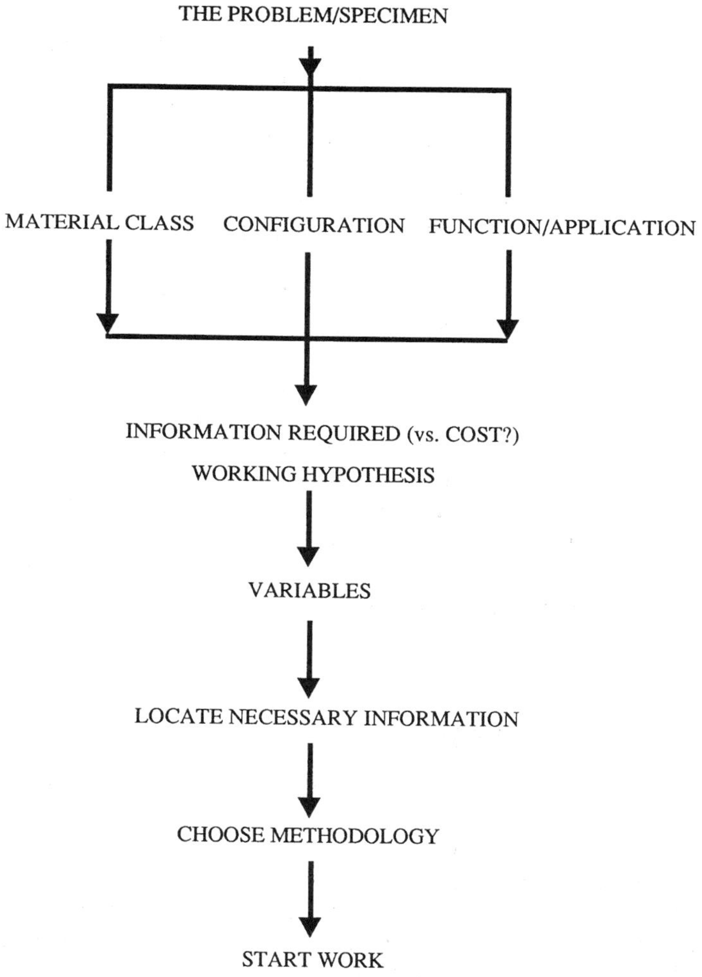

Figure 1. Outline of logical sequence used to decide on correct analysis procedure.

Before, during and after the actual analytical measurements are carried out other, more complex, decisions may have to be taken. It would, for example, become clear quite quickly if an inappropriate initial choice of methodology in terms of technique(s) and application had been made, perhaps because of the specimen configuration or due to an inability to extract the desired information. In that case a rethink and restart would be required. More difficult is the situation in which the interpretation of the results does not agree with the original hypothesis as to the nature of the problem associated with the specimen. The decision then is likely to be multibranched, in that either more of the same data or different data may be needed, from the same technique, or other techniques should be used, or specimens of different configuration, or even a fresh hypothesis. (There is a saying that if the map disagrees with the terrain, then it is usually the map that is wrong.) Here additional parallel information from other sources is invaluable in coming to a decision as to what to do. These decision steps are set out in the flowcharts in Figs. 2 and 3; again the more experienced analyst will regard many of them as intuitive. The principal message is that problem solving is an iterative process.

The first flowchart (Fig. 2) relates to the tackling of a particular problem, and the second (Fig. 3) to gaining information about a particular material. The operational decision steps are of course the same for each, but in the first the sequence terminates when the problem is judged to be "solved," whereas in the second there is greater emphasis on *completeness* of information. The iterative loops may be triggered at any stage in the sequence if it becomes obvious that an unsatisfactory decision has been made at an earlier stage. In Fig. 3 an additional iterative loop is activated if it is judged that in fact the acquired information is incomplete. At most steps in the decision sequence the reader is referred to one of several tables as an aid to making an informed decision about the next step. The tables are described and set out in the next sections.

4. ACRONYMS AND JARGON

Communications within any field of human endeavor, whether it be skateboarding or brain surgery, will sooner or later be conducted with specialized terminology, which to the uninitiated is either just so much alphabet soup or incomprehensible and confusing jargon. The field of surface and interface analysis is no different. Even to an insider the acronyms and abbreviations may sometimes present barriers to understanding rather than be a means of efficient communication. Regrettably they have to be used, if for no other reason than brevity. Other complications arise from redundant terminology where several acronyms are in common use, but refer to essentially the same technique or

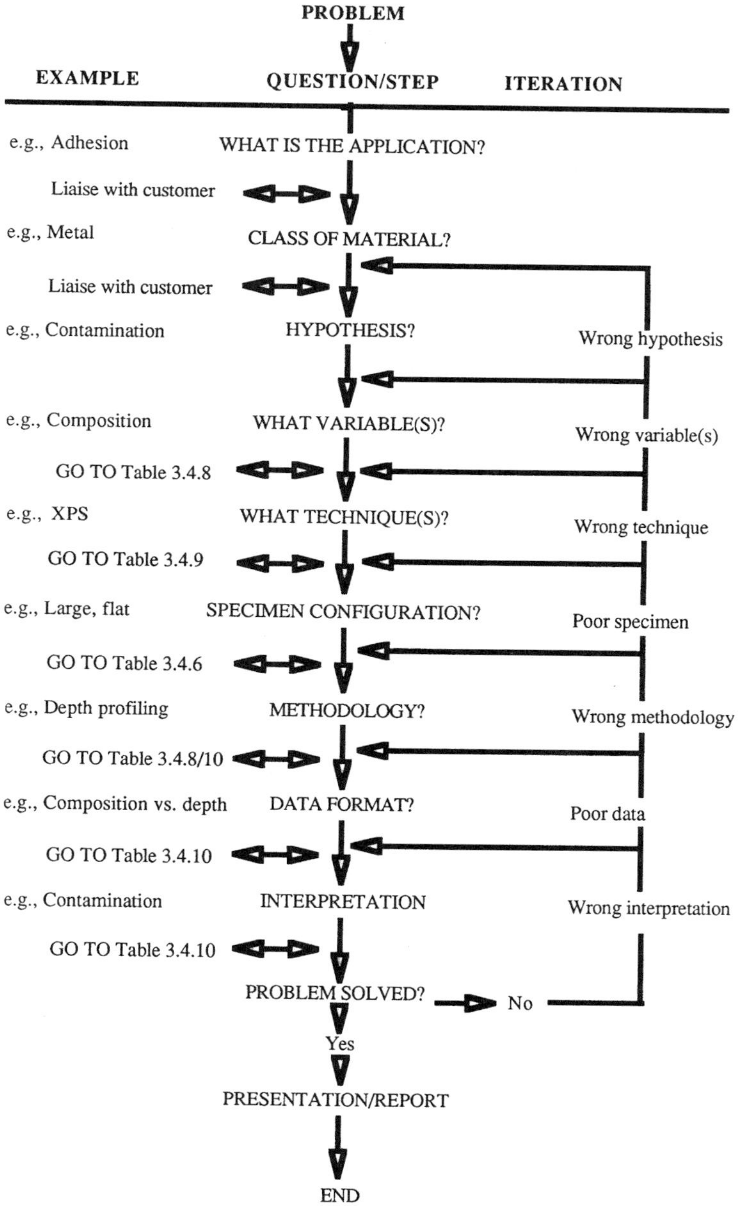

Figure 2. Flowchart showing the sequence of decision steps to be taken in the tackling of a particular analytical problem.

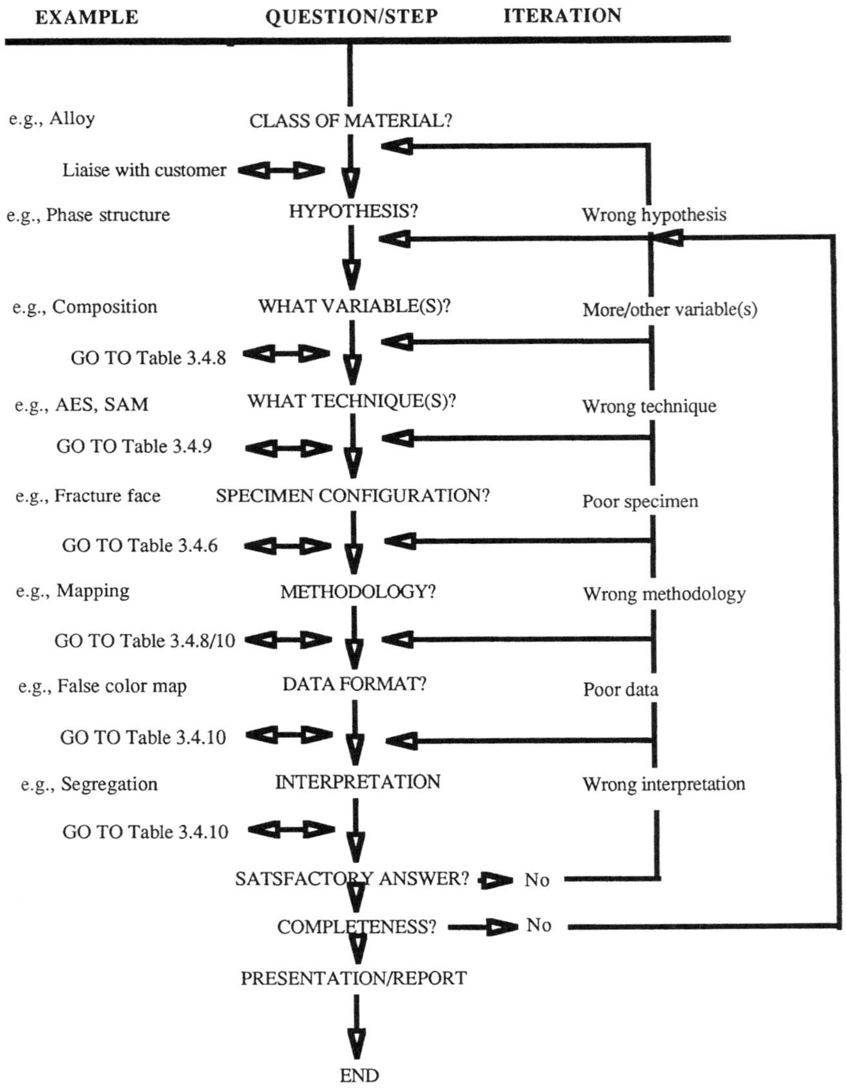

Figure 3. Flowchart showing the sequence of decision steps to be taken in gaining information about a particular material.

methodology (e.g., EDS = EDX = EDAX = EDXS). Therefore Tables 3.4.1 to 3.4.5 set out the origins and meanings of most of the acronyms, abbreviations and definitions used throughout the volume. These tables are simply vehicles for translation and have no other use.

Table 3.4.1. Acronyms: Techniques for Surfaces and Interfaces

AAS	atomic absorption spectrometry
AES	Auger electron spectroscopy (electron induced)
AFM	atomic force microscopy
ARAES	angularly resolved AES
ARXPS	angularly resolved XPS
ATEM	analytical TEM
ATR	attenuated total reflection (usually with FTIR)
CEMS	conversion electron Mössbauer spectroscopy
DCEMS	depth selective CEMS
DRIFT	diffuse reflectance FTIR
DSIMS	dynamic SIMS
EDAX	EDS
EDS	energy dispersive (x-ray) spectroscopy
EDX	EDS
EDXS	EDS
EELS	electron energy loss spectroscopy (high incident energy, cf. ELS)
EIS	electrochemical impedance spectroscopy
ELNES	electron (energy) loss near-edge spectroscopy
ELS	electron (energy) loss spectroscopy (low incident energy, cf. EELS)
EPMA	electron probe microanalysis
ERDA	elastic recoil detection analysis
ESCA	XPS
ESD	electron-stimulated desorption
ESDIAD	electron-stimulated desorption/ion-angular distribution
ESDIED	electron-stimulated desorption/ion-energy distribution
EXAFS	extended x-ray absorption fine structure
EXELFS	extended (electron) energy loss fine structure
F-d	force-distance "spectroscopy" (in AFM)
FEGSTEM	field emission gun STEM
FESEM	field emission SEM
FTIR	Fourier transform infrared spectroscopy
GDOES	glow discharge optical emission spectroscopy
GIXRD	glancing-angle incidence XRD
HEBS	high-energy backscatter spectroscopy
HREELS	high-resolution electron energy loss spectroscopy
HRTEM	high-resolution transmission electron microscopy
ICP	inductively coupled plasma emission spectrometry
ISS	ion scattering spectroscopy (= LEIS)

(*continued*)

Table 3.4.1. (*continued*)

iXPS	imaging XPS
LAED	low-angle electron diffraction
LEED	low-energy electron diffraction
LEIS	low-energy ion spectroscopy (= ISS)
L(F)FM	lateral (friction) force microscopy
LIMA	laser induced mass analysis
LIMS	laser induced mass spectrometry (= LIMA)
LRM	laser Raman microprobe spectroscopy
MALDI	matrix-assisted laser desorption
NEXAFS	near-edge x-ray absorption fine structure
NRA	NRS
NRS	nuclear reaction spectroscopy
OIM	orientation imaging microscopy
PEELS	parallel EELS (in combination with TEM)
RAIRS	reflection absorption IR spectroscopy
RBS	Rutherford backscattering spectroscropy
REELM	ELS (in scanning mode)
SA(E)D	selected area (electron) diffraction
SAM	scanning Auger microscopy
SAX	small area XPS
SEELS	surface electron energy loss spectroscopy (= ELS)
SEM	scanning electron microscopy
SERS	surface enhanced Raman spectroscopy
SEXAFS	surface extended x-ray absorption fine structure
SFA	surface force apparatus
SFM	scanning force microscopy
SIMS	secondary ion mass spectrometry
SNMS	secondary neutral mass spectrometry
SPM	scanned probe microscopy
SRPS	synchrotron radiation photoelectron spectroscopy
SSIMS	static SIMS
STEM	scanning transmission electron microscopy
STM	scanning tunnelling microscopy
STS	scanning tunnelling spectroscopy
TEM	transmission electron microscopy
ToF-SIMS	time-of-flight SIMS
TPD	temperature programmed desorption
UFI	ultrathin film interferometry
UPS	ultraviolet photoelectron spectroscopy
VBXPS	valence band XPS

Table 3.4.1. (*continued*)

WDS	wavelength dispersive (x-ray) spectroscopy
XAES	x-ray induced AES
XANES	x-ray absorption near-edge spectroscopy (= NEXAFS)
XAS	x-ray absorption spectroscopy
XPS	x-ray photoelectron spectroscopy
XRD	x-ray diffraction
XRF	x-ray fluorescence spectroscopy
XSW	x-ray standing wave
XTEM	cross-section TEM

Table 3.4.2. Acronyms: Surface and Interface Methodologies

BET	Brunauer-Emmett-Teller (surface area measurement)
BSE	backscatter electron (image)
COMPO	compositional (SEM image)
CVD	chemical vapor deposition
IBA	ion beam analysis
IBAD	ion beam assisted deposition
LAPVD	laser ablation PVD
MO	molecular orbital calculations
PECVD	plasma enhanced CVD
PIII	plasma immersion ion implantation
PVD	physical vapor deposition
SCF	self-consistent field calculations
SE	secondary electron (image)

5. FINDING THE INFORMATION

A volume of the present format, breadth and depth is not intended to be read through once, to have its content absorbed, and then to be consigned to the shelf. The success of any genuine handbook or how-to-do-it manual is dependent not only on its content but even more on the ease with which information can be accessed. Several tables have therefore been provided, any one of which can be entered through key words, and which will lead to one or more appropriate sections in the book. The tables are interlocking in the sense that different key words and different tables may lead to the same section or that the same key word may be found in more than one table. Most of the tables are concerned with providing information on which choices and decisions may be based.

Table 3.4.3. Acronyms and Trade Names: Compounds

APS	amino-propyl-trimethoxy-silane
CPO	chlorinated polyolefins
DLC	diamond-like carbon
Dupont E-120	pitch-based fiber
EGS	electrogalvanized steel
GPS	γ-glycidoxy-propyl-trimethoxy-silane
HDGS	hot dipped galvanized steel
HOPG	highly oriented pyrolytic graphite
Krytox	perfluorinated polyether lubricant
LB	Langmuir-Blodgett film
MoDDP	molybdenum dithiophosphate
MOS	metal oxide semiconductor device
Nicalon	silicon carbide
NMP	*N*-methyl pyrrolidone
PAN	polyacrylonitrile
PBD	polybutadiene
PDMS	polydimethylsiloxane
PMDA-ODA	polymethacrylate-dimethyl-amide-*ortho*-dimethyl-amide
PMMA	polymethacrylate-methyl-amide
PMMI	polymethacrylate-methyl-imide
PP	polypropylene
PS	polystyrene
PTFE	polytetrafluoroethylene
PVA	polyvinyl-acetate
PVC	polyvinyl-chloride
Toray M40	PAN-based carbon fiber
TMV	tobacco mosaic virus
ZDDP	zinc dithiophosphate

Table 3.4.6 presents key words dealing with specimen configuration and preparation for several classes of material. The configuration refers to geometry as-delivered and/or as will be required for the intended analysis. The key words then direct the reader to particular methods or techniques for preparing a particular specimen in the optimum manner for a particular method of analysis; this is then cross-referenced to the most appropriate section(s) of the book.

Table 3.4.7 covers aspects of how the techniques can best be used in order to obtain the required information and/or refers to more technical aspects of

Table 3.4.4. Acronyms: Miscellaneous

AC	alternating current
AL	attenuation length
ASTM	American Society for Testing and Materials
BE	binding energy
BP	bulk plasmon
CF	corrosion fatigue
CHA	concentric hemispherical analyzer
CMA	cylindrical mirror analyzer
DC	direct current
ECU	electronic control unit
EHD	elastohydrodynamic (lubrication)
FFT	fast Fourier transform(ation)
FWHM	full-width half-maximum (of spectral peaks)
IEPS	isoelectronic point of a surface
IMFP	inelastic mean free path
ISO	International Standards Organization
IR	infrared
KE	kinetic energy
MIC	microbiologically influenced/(induced) corrosion
NIST	National Institute of Standards and Technology (formerly NBS)
PROFILE	ranges of ions in matter (computer code)
RDF	radial distribution function
RF	radio frequency
RMS	root mean square
RSF	relative sensitivity factor
SCC	stress corrosion cracking
SCE	saturated calomel electrode
(D-)TRIM	(dynamic-)transport of ions in matter (computer modeling code)
UHV	ultrahigh vacuum
UV	ultraviolet

Table 3.4.5. Definitions: Miscellaneous

Bremsstrahlung	electron impact photon radiation
Gaussian	reference to a particular peak shape
Langmuir	1 L = 10^{-6} torr-sec = 1.33×10^{-4} Pa-sec
Lorentzian	reference to a particular peak shape
Shirley	reference to the shape of a spectral background
Temkin	referring to adsorption isotherm
Tougaard	reference to the shape of a spectral background

Table 3.4.6. Choices and Decisions: Specimen Configuration and Preparation

Form or configuration	Methodology/information	Section(s)[a]
Bulk, film (Au calibration)	XPS calibration	4.1.2.3
Bulk, film (catalyst)	analytical requirements	17.2
Bulk, etc. (catalyst)	in situ preparation	17.4
Bulk, film	ARAES requirements	7.2
Bulk, film	ARXPS requirements	7.2
Bulk, film	SIMS requirements	7.2
Bulk, film	XPS requirements	4.1.2.5
Bulk (metal, alloy)	chemical etching	11.2.2.1
Bulk (metal, alloy)	hydrogen charging	11.2.3.2
Bulk (metal, alloy)	intergranular fracture	11.2.3.1
Bulk (metal, alloy)	impact fracture	11.2.3.2
Bulk (metal, alloy)	ion etching	11.2.2.2
Bulk (metal, alloy)	polishing	11.2.2
Bulk, film (polymer)	irradiation damage	8.3.2
Bulk, film (polymer)	ion beam processing	8.3.2
Bulk (semiconductor)	AES/SAM/ELS requirem.	12.4.2
Fibers (composite)	configuration	14.2.1
Film, interface (adhesive)	ball cratering	18.4
Film, interface (adhesive)	sectioning	18.4
Film, interface (adhesive)	T-peel specimen	18.3.4
Film, interface (tribomaterial)	in/ex situ/in vivo/post mort.	16.2.3
Film, interface (tribomaterial)	configurations	16.3.x
Film (semiconductor)	XPS/AES requirements	12.3.2
Interface (semiconductor)	ARXPS/SRPS requirem.	12.5.2
Interface (semiconductor)	AES/SAM/ELS requirem.	12.6.2
Interface (semiconductor)	profiling	12.5.2
Leaded bronzes	preparation	19.4.2
Lead pipe	preparation	19.3.2
Multilayer	Auger depth profiling	8.2.1.1
Multilayer (organic)	SIMS quantification	6.2.4.3
Powders, pellets (catalyst)	analytical requirements	17.2
TEM foil	preparation	11.2.4.1
Thin film	quantification	8.2.2.3

[a] The first number refers to the relevant chapter, e.g., 11.2.2 refers to Section 2.2 in Chapter 11.

Table 3.4.7. Choices and Decisions: Instrumental Aspects

Technique(s)	Instrument elements	Function/description	Section(s)[a]
AES	energy analyzers	CMA, CHA	4.2.2.2
	depth profiling	Zalar rotation	8.2.1.2/3
			4.2.4.1
			7.4.2.6
	electron sources	characteristics	4.2.2.1
	add-on	charge neutralization	4.2.2.2
Implantation	various	instrumentation	9.3
ISS	equipment	general	4.3.2
SPM	equipment	general	10.2/3/4
		spatial positioning	10.3.2
		control loop	10.3.3
		raster	10.3.4
		noise/drift	10.3.5
	probe	spring constant	10.4.6/7
	tip shape	calibration	10.4.9/10/11
(S)SIMS/SNMS	ion source	general	6.1.2.1
		general	7.4.1
		electron impact	6.1.2.1
		Cs	6.1.2.1
		LMIG (Ga)	6.1.2.1
	mass analyzers	magnetic sector	6.1.2.2
		quadrupole	6.1.2.2
		time-of-flight	6.1.2.2
	add-on	charge neutralization	6.1.2.3
	add-on	postionization	6.1.2.3
		postionization	7.4.5.1.1
XPS	x-ray sources	bremsstrahlung	14.2.3
		choice	4.1.2.1
		monochromatization	14.2.3
	analyzer	CMA, CHA	4.1.2.2
	add-on	charge neutralization	4.1.2.5
		charge neutralization	14.5.2
iXPS, SAX		implementation	4.1.4

[a] The first number refers to the relevant chapter, e.g., 8.3.2 refers to Section 3.2 in Chapter 8.

using instrumentation. The aspects are correlated with particular instrumental functions and attributes, and again the reader is directed to the most appropriate section(s) of the book.

Table 3.4.8 is intended to correlate the types of information being sought with techniques of analysis. Choosing which variables to investigate is at the heart of the problem-solving process. Only slightly less important is the decision about which methodology(ies) and technique(s) to deploy.

A brief summary of the properties of the techniques used in this book is given in Table 3.4.9 in terms of operating characteristics, information content and principal merits/demerits of each. The summary should be useful for those readers who are unfamiliar with some or all of the techniques and need a quick guide.

Table 3.4.10 contains the key words and references to the accumulated expertise and experience found in Chapters 4 to 19 and in the Appendices. The underlying assumption is that the analyst either will have a priori information (i.e., from the customer) as to the class of material that is involved or will be able to make an accurate initial assessment. Thus, the search can be limited to entries under one main heading (class of material). In the majority of instances the customer will be able to provide additional information about the specimen (e.g., thin film on Si substrate, carbon fiber in epoxy matrix, protective organic coating on steel, etc). Thus, the analyst can continue the search for a similar material, with similar properties, used for the same application. With luck there will be a good match, in which case the appropriate section should offer much valuable information. If no match can be found, then the reader should locate the closest comparable system (e.g., if a reference to thin film on InSb cannot be found, then thin film on GaAs might be the closest alternative). Also, if the specified problem for a particular system appears remote from those listed for the same system, then entries for related systems should be scanned in order to find a better match for the problem (e.g., the actual problem might have to do with thin films on InSb, but no entry under the material appears to be relevant; instead search other semiconductor entries for a better match).

Finally, the volume contains well over 2000 references chosen by the expert contributors. If all else fails, then one or more of the tables should guide the reader to one or more references allowing entry to the specialist literature on any topic in the broad area of surfaces and interfaces.

Table 3.4.8. Surface and Interface Techniques: Information and Methods

Information sought	Technique(s)	Type of analysis	Section(s)[a]
α*-parameter	XAES	spectral	4.1.3.2/5.2.3/17.6
		interface	12.5.3.2/18.3.3
Artefacts	XPS	spectral	5.2.1.1
Binding energies	XPS	spectral	5.2.1.1/17.6
Charging effects	AES	neutralization	5.2.1.3/12.2.7
	ISS	neutralization	4.3.3.2
	XPS	neutralization	5.2.1.3/17.5/14.5.2/ 12.2.7
Chemical inform.	(P)EELS	spectral	11.2.4.2/16.2.4.1/16.3.6
	ELS	band transitions	A.3.2
		mapping	12.4.3.2
		plasmon loss	A.3.2
		spectral	12.6.3
	iXPS/SAX	mapping	4.1.4/18.3.2
	UPS	spectral	A.3.1/14.5.6.3
	XAES	spectral	12.5.3.2
	XAS	spectral	16.3.2.1/16.2.4.1
	XPS	band transitions	4.1.3.3
		core levels	19.3.2
		curve fitting	4.1.3.2/14.5.5/17.6
		date processing	4.1.2.4
		energy calibration	4.1.2.3/14.5.2
		interface	12.3.3.2/15.2.7/15.6/ 18.3.1/19.3.3
		line shape	5.2.1.2
		monochromatic	14.5.5.3
		multiplet splitting	4.1.3.3
		plasmon	4.1.3.3/5.2.1.2/17.6
		shake-up/down	4.1.3.3/5.2.1.2/17.6
		standards	14.5.2
		valence band	14.5.1
Chemical shift/state	AES	spectral	4.2.3/5.2.2.2
	XPS	background subtract.	5.3.1.3.1/4.1.2.4/14.5.5
		curve fitting	5.2.1.4/4.1.3.2/14.5.5
		spectral	5.2.1.3/4.1.2.4
Composition surface	AES	interface	12.4.3.1
		quantification	5.3
		sensitivity factors	5.3.1.1/5.3.1.3.2
		statistical errors	5.3.1.4
	ESD	mass analysis	A.3.3

(*continued*)

Table 3.4.8. (*continued*)

Information sought	Technique(s)	Type of analysis	Section(s)[a]
Composition surface	ISS	data processing	4.3.3.11
		elemental sensitivity	4.3.3.7
		surface coverage	17.9
		general	4.3.3.1/14.4.9
		quantification	4.3.3.10
	NRS	spectral	16.3.8
	SSNMS	general	6.1.3
	SSIMS	internal standards	6.2.4.1
		metal substrates	6.1.5
		monolayers	6.2.4.2
		quantification	6.1.3/6.2.4
	TPD	mass analysis	14.4.8
	XPS	depth variation	5.3.3
		imhomogeneity	5.3.2
		interface	12.5.3.2/19.3.2/19.4.2
		quantification	5.3
		sensitivity factors	5.3.1
		statistical errors	5.3.1.4
Composition bulk	EDS	general	15.2.6
	WDS	general	14.4.5
Composition depth	many	cross-sectioning	15.2.4
		depth calibration	7.4.2.3
		depth profile general	6.1.1.1/7.4.2/14.5.3
		depth resolution	7.4.2.4
		quantification	7.4.2.2
		sputtering general	6.1.1.1/8.2/8.3
	AES	curve fitting	7.4.2.7
		factor analysis	7.4.2.7
		sputtering	4.2.4.1/8.2/8.3/12.3.3.1
	ARAES	angular dependence	4.2.4.2/7.3.2.1/2
	ARXPS	angular dependence	4.1.4/7.3.2.1/2/12.5.2/ 14.5.3/18.3.6
	GDOES	quantification	7.4.3.2
		special methods	7.4.3.2
		sputtering	7.4.3.1/7.4.3.2
	ISS	sputtering	4.3.3.12
	RBS	spectral analysis	7.3/7.3.1.3
	SIMS	general	7.4.4/17.8.3
		external standards	7.4.4.2
		quantification	7.4.4.2

(*continued*)

Table 3.4.8. (*continued*)

Information sought	Technique(s)	Type of analysis	Section(s)
Composition depth	SNMS	general	7.4.5
		quantification	7.4.5.2
	SRPS	sputtering	12.5.2
	XPS	quantification	17.9
		sputtering	15.2.7/15.6/19.3.3/ 19.4.3
Composition lateral	SAM	mapping	4.2.4.3/12.4.3.2/19.4.2
	iXPS/SAX	mapping	4.1.4
	AES	various	5.1/2
	XAES	various	5.2.3
	XPS	various	5.1/2
Kinetic energies	AES	spectral	5.2.2.1
Molecular structure	ATR	spectral	A.3.4.1.1
	FTIR	spectral	13.3/14.4.2/16.3.9/ 18.6.4
	HREELS	spectral	A.3.4.1.3/13.3
	RAIRS	spectral	A.3.4.1.2/18.5
	Raman	spectral	13.3/14.4.2/16.3.10
	SERS	spectral	18.5
	SSIMS, etc.	peak assignment	6.2.3
		spectral	6.2.3/17.6
	ToF-SIMS	spectral	18.3.5/18.5
Topography	AFM	image analysis	10.2/4/13.3/4/14.4.4/ 16.2.4.2
	EPMA	image analysis	14.4.5
	(FE)SEM	image analysis	13.3/4/14.4.3
	STM	image analysis	13.3/14.4.4/16.2.4.2
	UFI	interferogram anal.	16.3.1
Phase structure	(D)CEMS	spectral	9.4.1
Structure bulk	EXAFS	spectral	13.3/16.2.4.1
	(HR)TEM	diffract./image anal.	13.3/16.3.2/6
	NEXAFS	spectral	13.3
	XRD	diffraction analysis	13.3/14.4.1/16.3.7/ 19.2.1
Structure surface	GIXRD	diffraction analysis	9.4
	ISS	angular analysis	13.3
	LEED	diffraction analysis	13.3
	STM	image analysis	10.2/4/13.4.7/16.3.4
	XSW	diffraction analysis	13.3

[a]The numbers refer to chapter and section as in previous tables.

Table 3.4.9. Surface and Interface Techniques: Characteristics and Attributes

Technique	Primary excitation	Analyzed emission	Information[a]	Depth range	Lateral resolution	Map[a] info.	Quantification[b]	Charging[c]	Destructiveness[c,d]
Photon excitation									
XPS	X-rays 1250–1500 eV	e^- 0–1500 eV	E,C	1–5 nm	≈1 mm	—	***	*	*
SAX	X-rays 1250–1500 eV	e^- 0–1500 eV	E,C	1–5 nm	≈150 μm	E,C	***	*	*
iXPS	X-rays 1250–1500 eV	e^- 0–1500 eV	E,C	1–5 nm	≈5 μm	E,C	***	*	*
XAES	X-rays 1250–1500 eV	e^- 0–1500 eV	C	1–5 nm	≈1 mm	—	—	*	*
UPS	UV 20–40 ev	e^- 0–40 eV	C	0.5–2 nm	≈1 mm	—	—	*	—
SRPS	synch. rad. 50–500 eV variable	e^- 0–500 ev variable	E,C	0.5–3 nm variable	100 μm	—	**	*	—
FTIR (in refl.)	IR 400–4000 cm^{-1} IR	IR 400–4000 cm^{-1}	V,M (S)	1 nm with AR	mm	—	*	—	—
Raman	IR	IR	V,M	1 nm	mm	—	*	—	—
LRM	Visible	Visible	(S)	>10 nm	0.5 μm	S,V	*	—	—
XAS	X-rays 100–1000 eV	X-rays 100–1000 eV	C,(E)	≈1 μm	mm	—	*	*	—
EXAFS	X-rays	X-ray struct. above abs. edge	S,(C)	≈2 μm	100 μm	—	***(S)	—	—

SEXAFS	X-rays	Surf. enhanced EXAFS	S,(C)	1–5 nm	100 μm	—	***(S)	—	—
XRD	X-rays	X-rays	S	≈2 μm	≈1 mm^2	—	***(S)	—	—
GIXRD	X-rays	X-rays	S	10–100 nm	≈5 mm^2	—	**	—	—
UFI	Visible	Visible	(T)	0.1 nm	ca. 5 mm^2	—	**	—	—
Electron excitation									
AES	e$^-$ 5–50 keV	e$^-$ 1–2000 eV	E,(C)	0.5–3 nm	5–10 μm	—	**	***	**
SAM	e$^-$ 20–50 keV	e$^-$ 1–2000 eV	E,(C)	0.5–3 nm	50–500 nm	E	*	***	***
ELS	e$^-$ 200–2000 ev variable	e$^-$ 0–100 eV loss	C	1–3.5 nm	5 μm	(E)	—	**	**
SEM	e$^-$ 10–30 keV	e$^-$ 10–30 keV	T,(E)	0–5 μm	0.5 μm	T	—	**	**
SEM/BSE	e$^-$ 10–30 keV	e$^-$ 10–30 keV	Z-contrast	0–5 μm	0.5 μm	Z	*	**	**
SEM/SE	e$^-$ 10–30 keV	e$^-$ 0–20 eV	T	100 nm	50 nm	T	—	**	**
FESEM	e$^-$ 1–5 keV	e$^-$ 0–20 eV	T	100 nm	10 nm	T	—	*	*
EDS (SEM)	e$^-$ 5–40 keV	X-rays 0–20 keV	E	0.5 μm	0.5 μm	E,T	**	**	**
EPMA	e$^-$ 5–40 keV	e$^-$ 5–40 keV	T,(E)	0.5 μm	0.5 μm	T	—	**	**
WDS	e$^-$ 10–50 keV	X-rays 0–20 keV	E	0.5 μm	0.5 μm	E,T	***	**	**

(continued)

Table 3.4.9. *(continued)*

Technique	Primary excitation	Analyzed emission	Information[a]	Depth range	Lateral resolution	Map[a] info.	Quantification[b]	Charging[c]	Destructiveness[c,d]
Electron excitation *(cont.)*									
TEM	e^- 100–300 keV	e^- 100–300 keV	S,T	0.2–0.5 μm	0.5 μm	T,(S)	**	*	**
SAD (TEM, etc)	e^- 100–300 keV	e^- 100–300 keV	S	0.05–0.5 μm	2–300 nm	S	***	*	**
HRTEM	e^- 100–200 keV	e^- 100–200 keV	S	<0.05 μm	0.2 nm	S	***	*	**
STEM	e^- 100–300 keV	e^- 100–300 keV	S,T	0.1–0.3 μm	<10 nm	S	**	*	**
FEGSTEM	e^- 100–300 keV	e^- 100–300 keV	E,(S,T)	0.1–0.3 μm	<1 nm	E,(S)	**	*	**
EDS (STEM, etc.)	e^-	X-ray 0–20 keV	E	0.1–0.3 μm	1–10 nm	E	**	*	**
EELS	e^- 100–200 keV	e^- 100–200 keV less loss	E,(C)	0.1–0.3 μm	1 nm-μm	—	*	*	**
PEELS	e^- 100–200 keV	e^- 100–200 keV less loss	E,(C)	0.1–0.3 μm	1 nm-μm	—	**	*	*
EXELFS	e^- 1–2 keV	e^- above ionization edge	S	1–2 nm	mm	—	*	*	*

Ion excitation									
SSIMS	Ar^+,Ne^+ 2–5 keV	M^+,M^- mass anal.	M,(E)	0.5–1 nm	0.5 mm	—	*	*	*
DSIMS	Ar^+, N^+ Ne^+, Xe^+ 10–30 keV	M^+,M^- mass anal.	M,(E)	10 nm	100 nm	M,(E)	**	**	***
SNMS	Ar^+,Ne^+ 2–10 keV	M^0 mass anal.	M,(E)	10 nm	100 nm	M,(E)	**	**	***
ISS	He^+,Ne^+ 1–5keV	He^+, Ne^+ energy anal.	E,S	0.5 nm	0.5 mm	—	***	*	*
ToF-SIMS	Ar^+,Ne^+	M^+,M^- ToF mass analysis	M,(E)	0.5 nm	100 nm	M	*	*	**
RBS	He^+ 0.5–3 MeV	He^+ 0–KE_{max}	E	μm/20 nm	0.5 μm	(E)	***	*	**
ERDA	Cl^{n+} 30 MeV	H to N few MeV	E	100/20 nm	0.5 μm	—	***	*	**
NRS/A	p^+, d^+ 0.2–MeV	p^+,d^+,He^{2+}	E	μm/20 nm	0.5 μm	(E)	**	*	**
Neutral particle excitation									
CEMS	γ	γ	(E),S	2 nm	mm	—	**	—	—
DECMS	γ	γ	(E),S	2 nm	mm	—	**	—	—
FABMS	Ar^0	M^+,M^- Mass analysis	M,(E)	0.5–1 nm	mm	—	*	*	*

(continued)

Table 3.4.9. *(continued)*

Technique	Primary excitation	Analyzed emission	Information[a]	Depth range	Lateral resolution	Map[a] info.	Quantification[b]	Charging[c]	Destructiveness[c,d]
Other techniques									
APFIM	E-field	M^+ ToF mass analysis	E,S	0.1 nm	0.1 nm	E	***	—	***
STM	e^- tunnel	e^- ±0–5 eV	T,S, M,C	0.01 nm	0.1 nm	S,T	**(S)	**	—
STS	e^- tunnel	e^- ±0–5 eV	C,S,M	0.01 nm	0.1 nm	C	—	***	—
AFM	force	displacement	T,S	0.05 nm	0.2 nm	T	**(S)	*	*
F-d	force	displacement	C,M	0.05 nm	0.2 nm	—	*(Force)	*	*
L(F)FM	force	torsion	Friction	0.05 nm	0.2 nm	F	*	—	*
TPD	heat	M^+,M^-	E,M	0.1 nm	—	—	*	—	***

[a]E = elemental; C = chemical/electronic; V = vibrational; M = molecular; S = structural; F = frictional; Z = atomic number contrast; T = topographical.

[b]The more stars the better.

[c]The more stars the worse.

[d]The destructiveness refers to the measurement process; it must be borne in mind that specimen preparation itself is often a destructive process, in the sense that the functionality of a device or product can be affected adversely.

Table 3.4.10. Classes, Functions and Applications of Materials: Key Words and Locations

Material	Form, function and/or applic.	Technique(s)	Location section(s)[a]	Variable(s) phenomena, information
Ceramics, oxides, minerals and glasses				
AgO	model compound	XPS	4.1.3.2	surface chemistry, oxidation, reduction
Al_2O_3/metal	friction at interface	AES	16.3.5	interface/surface composition
AlF_3	powder, model compound	XPS	17.6	binding energies
γ-alumina	powder, catalyst	XPS	17.6	chemical information
Bauxite	ore powder	XPS	13.4.8	topography, structure
	catalysis	FTIR	13.4.8	phase analysis, defects
		SIMS	13.4.8	chemical information
		SEM	13.4.8	composition
		TEM	13.4.8	surface chemistry
		XRD	13.4.8	
B_4C/B_4C	friction at interface	AES	16.3.4	interface composition
α/β-CrF_3	powder, model compound	XPS	17.6	binding energies
$CaCO_3$	growth	AFM	13.4.2	growth mechanism
$CaTiO_3$	poly-xtal waste form	TEM	13.5.4	grain boundary structure
		XPS	13.5.4	composition
		SAM	13.5.4	segregation
		SIMS	13.5.4	boundary chemistry
Carbon	CO on W(112), W_2C, graphite, diamond	AES	4.2.3	chemical signatures
Carbon	energy standard	XPS	5.1.2.1	C 1*s* shape and BE

(*continued*)

Table 3.4.10. (*continued*)

Material	Form, function and/or applic.	Technique(s)	Location section(s)[a]	Variable(s) phenomena, information
Ceria	model compound	XPS	5.2.1.2	line shape, interpretation
Cr_2O_3	powder, model compound	XPS	17.6	binding energies
Cr/O/Si	cermet film	XPS	8.3.1.3	depth profiling
				interface chemistry
		RBS	8.3.1.3	stoichiometry
		XRD/SAD	8.3.1.3	structure
E-glass	fibers, plates	XPS	13.6.1/6	surface composition
		FTIR	13.6.1	surface chemistry
	functionalization	ARXPS	13.6.4	adsorption chemistry
Feldspars	weathering	TEM	13.4.5	structure, topography
	dissolution	SEM	13.4.5	chemical information
		XPS	13.4.5	composition
		SIMS	13.4.5	
Fe_2O_3	single xtal	STM	13.4.2	surface structure
FeS_2	(100) surface	STM	13.4.2	surface structure
Fe_7S_8	surface alteration	(AR)XPS	13.4.3	surface chemistry
		SAM	13.4.3	depth dependence
GeO_x	oxide film	XPS	8.3.2.1	depth profiling
				oxygen loss
Glass	contamination	SSIMS	6.3.4	identification of
	on substrate	XPS	6.3.4	organic contaminent
		IR	6.3.4	

Kaolinite	surface modification	FTIR	13.4.8	functionalization
		XPS	13.4.8	surface chemistry
MgO	"smoke"	HRTEM	13.4.1	surface roughness structure, defects
Nb_2O_3	model compound	XPS	4.1.3.2	chemical shifts
PbS	surface alteration	STM	13.4.2	(001) structure
		STS	13.4.2/6	Au-particle sorption
		XPS	13.4.2	oxidation
		FTIR	13.4.2	U sorption
		SIMS	13.4.6	adsorption of organics
PNL glass	waste form	SIMS	13.6.5	dissolution mechanisms
		XPS	13.6.5	composition
		FTIR	13.6.5	surface chemistry
		SEM	13.6.5	precipitation products
SiC/SiC	friction at interface	AES	16.3.6	
		XPS	16.3.6	interface composition
		TEM/EELS	16.3.6	interface chemistry
		HRTEM	16.3.6	interface crystallinity
Silica	organics on silica	(AR)XPS	13.5.6	long-chain alcohols
		FTIR	13.5.6	surfactant
		SAM	13.5.6	adsorption chemistry
		SEM	13.5.6	surfactant chemistry
Pt/SnO_x	model compound	ISS	4.3.3.2	oxidation and reduction
Sn-oxide	model compound	ESD/AES	A.3.3	surface composition
		ELS	A.3.2	plasmon losses

(*continued*)

Table 3.4.10. *(continued)*

Material	Form, function and/or applic.	Technique(s)	Location section(s)[a]	Variable(s) phenomena, information
Ta_2O_5	oxide film	XPS, AES	7.4.2.4.1	depth profile
	depth calib.			composition
TiN	coating on substrate	XPS	8.3.1.1	surface/interface chemistry
				composition, structure
				depth profiling
TiO_2	single xtal	STM/STS	13.4.2	surface structure
				electronic structure
Ti_nO_{2n-1}	Magneli phases	TEM	13.5.4	grain boundary
		SAD	13.5.4	structure
W-N	model compound	AES	7.4.2.7	depth profile
				chemical information
WO_3	particle on support	ISS	17.9	depth profile
Zn-Cr-O	spinel powder	ISS	4.3.3.12	depth profile
Composites				
Ag/α-Al_2O_3	catalyst (ethylene)	XPS	4.1.3.1	surface composition
		ARAES	4.2.4.2	depth profile
		ISS	4.3.1	surface composition
Au/Al/SiO_2/Si	multilayer, for depth calib.	XPS/AES	7.4.2.6	depth profile composition
C_{60}/steel	coating on substrate	Raman/LRM	16.3.10	molecular structure
DLC/substrate	coating on substrate	FTIR	16.3.9	molecular structure
Fe/Al_2O_3	friction at interface	AES	16.3.4	interface composition

MoS_2/steel	lubricant film on substrate	XRD	16.3.7/8	structure
		NRS	16.3.7/8	O content
		AES	16.3.7/8	interface composition
		AFM	16.3.7/8	topography
		HRTEM	16.3.7/8	structure
Ni-Mo/Al_2O_3	particle on ceramic support	ISS	17.8.3	depth profile
Pt/SnO	model system	XPS	4.1.3.2	surface chemistry
				oxidation, reduction
Rh/alumina	particle on ceramic support	SSIMS	17.7	molecular fragments
Pt-Re	particle on ceramic support	XPS	17.8.2	chemical information
Pt-Sn	particle on ceramic support	XPS	17.8.2	chemical information
		SSIMS	17.8.2	molecular fragments
		EXAFS	17.8.2	atomic coordination
Pt-Sn/alumina	particle on ceramic support	FABMS	17.7	molecular fragments
SiN/carbon fiber	protective coating on fiber	XPS	14.5.2	chemical information
Metals and alloys				
Al	oxide layer	XPS	18.2.2	composition
		AES	18.2.2	composition
Alloys 600/800	heat exchanger	XPS	15.2.7	corrosion products
		SIMS	15.2.9	surface oxidation
Astroloy	sharp tip	FIM/ToF	11.3	interface composition
Au-Pd	nm-sized particles	XPS	17.6	binding energy reference
Bronze	Roman artefact	SEM	19.4.2/3	topography/bulk composition
		EDS	19.4.2/3	
		AES	19.4.2/3	composition profile
		SAM	19.4.2/3	lateral composition
		XPS/XAES	19.4.2/3	chemical information and composition

(*continued*)

Table 3.4.10. *(continued)*

Material	Form, function and/or applic.	Technique(s)	Location section(s)[a]	Variable(s) phenomena, information
Cr/substrate	implanted film	AES	4.2.4.1	N-implantation, depth profile
Cr/steel	N-implanted protective film	AES	9.5.2.1/2	depth profile
		XPS	9.5.2.1/2	surface chemistry
		ISS	9.5.2.1/2	composition
Fe/Fe/PFDE	friction at interface	AES	16.3.4	interface composition
		STM	16.3.4	morphology
Fe/Cr/Mo/V	crack propagation	SEM	11.3	topography
steel		AES	11.3	lateral mapping
Fe/3%Ni	fracture mechanism	SEM	11.2.3.1	topography
		AES	11.2.3.1	segregation
		SAM	11.2.3.1	porosity
		iXPS	11.2.3.1	segregation
Inconel 690	thin foil	FEGSTEM	11.2.4.2	segregation
		EDS	11.2.4.2	interface composition
Mg	single xtal	ISS	4.3.2	surface composition
Monel 400	heat exchanger	EIS	15.3/6	aqueous corrosion
		SIMS	15.3	characterization, B-inclusion
			15.6.1/6	profiling
		XPS	15.6.1	characterization
			15.6.3.3	composition, chemistry
		AFM	15.6.3.2	topography
		FESEM	15.6.3.2	topography

Nb/Pb	Josephson junctions	AES	12.3.3	interface composition
		XPS	12.3.3/4	interface chemistry
				oxidation, degradation
Ni(111)	clean metal	UPS	A.3.1	band structure
Ni/Cr	multilayer, for depth calib.	depth profiling	7.4	depth profile composition
Ni-Cr	alloy	AES	4.2.3	chemical information
		XPS	4.2.3	
		ISS	4.3.3	surface oxidation
Ni-Cr	alloy grains	AES	15.2.8	surface oxidation
Nimonic PE16	nuclear fuel can	AES	11.2.3	radiation segregation
Pb	Roman lead pipe	SEM	19.3.2/3	topography/bulk composition
		EDS		
		AES	19.3.2/3	lateral composition, profile
		XPS/XAES	19.3.2/3	chemical information and composition
Pd	metal hydride	ISS	4.3.3	hydriding mechanisms
Pt_3Sn	alloy	ARAES	4.2.4	depth profile
		SAM	4.2.4	compositional mapping
		ISS	4.3.2	surface composition
Sn	clean metal	ELS	A.3.2	plasmon losses
Steel	Eu-implanted foil	D/CEMS	9.4.1.2	phase analysis
				depth profile
Steel/steel	friction at interface	XPS	16.3.4	interface chemistry
		HREELS	16.3.4	
Steel 304	cooling loop	AES	15.2.8	corrosion products

(*continued*)

Table 3.4.10. *(continued)*

Material	Form, function and/or applic.	Technique(s)	Location section(s)[a]	Variable(s) phenomena, information
Ti6Al4V	prosthetic joint,	RBS	9.5.1	implantation by N, C, O
	wear,	HEBS	9.5.1	profiling
	polymer	ERDA	9.5.1	microstructure
	coating	XPS	9.5.1	
		XRD	9.5.1	
Semiconductors				
GaAlAs	B-doped	SIMS	7.4.2.6	depth profile composition
GaAs	etched surface	ARXPS	7.3.2.1	depth profile
Ge/Si/SiO_x/Si	multilayer structure	XTEM	8.2.1.1	interface structure
		XPS	8.2.1.2/3	depth profile
				Zalar rotation
In-Ga-As-P	surface oxidation	AES	12.4.2/3/4	composition
		SAM	12.4.2/3/4	chemical information
		ELS	12.4.2/3/4	oxygen uptake
InP/SiO_2	device interface	SAM	12.6.2/3/4	interface composition
		ELS	12.6.2/3/4	interface chemistry
Si	Si (100) single xtal	ARXPS	4.1.3.1	surface composition
Si	Ir on Si substrate	RBS	7.3.1.3	interface composition
Si(100)	model compound	ESDIED	A.3.3	H-desorption
SiO_2/Si	film on substrate	XPS	12.5.2/3/4	interface composition
		SRPS	12.5.2/3/4	interface chemistry

SiO_2/Si	native oxide	AES	4.2.3	chemical information
		ARAES	4.2.4.2	depth profile
Organics and polymers				
Acrylic lacquer	coating on Al-brass	XPS	18.3.1	interface composition
				chemical information
		AES	18.3.1	interface composition
Amine	adsorption on Al-oxide	XPS	18.6	interface chemistry
		SIMS	18.6	molecular fragments
Adhesive	coating on Al/Al_2O_3	AES	18.4	interface composition
		XPS	18.4	interface composition
Benzene	on Ni(111)	UPS	A.3.1	electronic structure
Car paint	surface defects	SEM	6.3.1	topography
		SSIMS	6.3.1	ion spectra
CF_x	model compound	XPS	5.2.1	curve fitting
				chemical states
C_6H_{12}	adsorbed layers	HREELS	A.3.4.2.1	physisorption
		RAIRS	A.3.4.2.1	physisorption
Cyclosporin	drug A&D versions	ToF/SSIMS	6.2.4	presence in blood
Dupont E-120	treated fiber	XRD	14.5.1	structure, VB spectra
		XPS	14.5.1	chemical information
EGS	coating and substrate	XPS	18.4	interface composition
Epoxy	coating on Al	SAX	18.3.4	interface composition
Epoxy	coating on mild steel	iXPS	18.3.2	lateral chemistry
		SAX	18.3.2	and composition
		ToF-SIMS	18.3.2	molecular fragments
Epoxy	resin on carbon fiber	XPS	14.5.5.2	C 1*s* structure

(continued)

Table 3.4.10. (*continued*)

Material	Form, function and/or applic.	Technique(s)	Location section(s)[a]	Variable(s) phenomena, information
Ethyl tri-fluoroacetate	model compound	XPS	4.1.1	chemical states (C 1*s*)
			4.1.3.2	
GPS	adhesion promoter on metal	SSIMS	18.5/7	molecular fragments interface chemistry
Hydrocarbon	contamination at interface	ARXPS	18.3.2	chemical information
		ToF-SIMS	18.3.2	molecular fragments
Hydrocarbon	contamination on steel	XPS	18.2.1	composition, profiling chemical information
		AES	18.2.1	composition, lateral
Krytox	polymer lubricant	SSIMS	6.2.3	fingerprint ion spectra
Kynol	novoloid fibre	XPS	14.5.6.1	VB information
MoDDP	adsorbed lubricant	AES	16.3.2.3	interface composition
		XPS	16.3.2.3	
MoDDP	chemisorbed lubricant	AES	16.3.2.3	interface chemistry
		Raman	16.3.2.3	and
		TEM/EELS	16.3.2.3	composition
Nomex	aramid fiber	XPS	14.5.6.1	VB information
Organosilicon	membrane	XPS	8.3.2.3	ion beam processing
PAN	carbon fiber degradation	XPS	14.5.4	surface degradation
			14.5.5.2	C 1*s*, VB information
PBD	coating on mild steel	XPS	18.3.2	interface composition
PBD/PS	polymer mixture	SSIMS	6.2.4.3	quantification of phases

PDMS	release agent on substrate	XPS	18.2.1	composition
				chemical information
		SIMS	18.2.1	molecular fragments
PDMS	monolayer on Ag	SSIMS	6.1.3.1	fingerprint spectra
PMDA-ODA	bulk polymer	XPS	8.3.2.2	ion beam processing
PMMA	polymer on oxide	ARXPS	18.6	chemical information
PMMA	ion processing for waveguides		9.5.3	He ion bombardment
				optical properties
Polymer	thin film on metal	XPS	18.4	chemical information
				interface chemistry
Polysulfones	film or membrane	XPS	8.3.2.1	ion beam processing
PP/CPO	adhesion mechanism	SSIMS	6.3.2	ion spectra
				Cl^- diffusion
PTFE	bulk model compound	XPS	5.2.1.1	chemical information
PTFE	bulk model compound	SSIMS	6.2.3	fingerprint ion spectra
PVA	sensor surface	SSIMS	6.3.3	functionalization sequence
				surface bio-organic chemistry
PVC	coating on HDGS	XPS	18.3.3	chemical information
				interface composition
Resin	coating on ceramic/Al	XPS	18.3.6	chemical information
				interface chemistry
Rubber	adhesion to brass	XPS	18.3.1	chemical information
				interface composition
		SSIMS	18.3.1	molecular fragments
Silane	adhesion promoter on glass fiber	XPS	18.5	interface chemistry
		ToF-SIMS	18.5	molecular fragments

(*continued*)

Table 3.4.10. *(continued)*

Material	Form, function and/or applic.	Technique(s)	Location section(s)[a]	Variable(s) phenomena, information
Silane	adhesion promoter on metal	XPS	18.5	interface chemistry
		SSIMS	18.5	molecular fragments
		RAIRS	18.5	vibrational modes
		SERS	18.5	vibrational modes
TMV	image artefacts	AFM	10.4.10	tip shape evaluation nanodissection
Toray M40	PAN-based carbon fiber	XPS	14.5.5	C 1*s* structure
ZDDP	adsorbed lubricant	AES	16.3.2.3	interface composition
		XPS	16.3.2.3	
ZDDP	thin-film lubricant	XAS	16.3.2.1	chemical information
		EXAFS	16.3.2.1	coordination
		ELNES	16.3.2.1	symmetry

[a]The notation refers to chapter and section, e.g., 13.5.4 means Section 5.4 in Chapter 13.

4

Spectroscopic Techniques: X-ray Photoelectron Spectroscopy (XPS), Auger Electron Spectroscopy (AES), and Ion Scattering Spectroscopy (ISS)

GAR B. HOFLUND University of Florida, Gainesville, Florida

1. X-RAY PHOTOELECTRON SPECTROSCOPY (XPS)

1.1. Introduction and History

XPS is a photoemission technique which is used widely to examine the composition and chemical state distribution of species at solid surfaces. For this reason it was originally named electron spectroscopy for chemical analysis (ESCA). In principle XPS is closely related to ultraviolet photoemission spectroscopy (UPS), and the two techniques are often considered jointly as photoelectron spectroscopy (PES). A detailed history of the development of XPS has been published by Jenkin et al. [1]. According to Sherwood [2], it was Lenard who first built an apparatus, in 1899, for measuring the kinetic energies of photoelectrons and for performing photoemission experiments. The photoemission effect was later explained by Einstein [3], for which he was awarded the Nobel prize in 1905. When irradiated, a solid absorbs discrete quanta of energy resulting in direct emission of electrons, and the kinetic energy distribution of the emitted electrons provides detailed information about the solid. Early measurements of the distribution recorded with x-ray excitation were performed at too low an energy resolution [4–7] and consequently did not exhibit the distinct spectral features required for surface analysis. In 1954, Siegbahn et al. [8] measured this distribution with a high-energy-resolution electron spectrometer and observed well-defined peaks useful for studying atomic orbital structure. Further study in 1958 on copper and copper oxides [9] demonstrated that shifts in the kinetic energies of the peaks provided chemical-state information. Most importantly, this observation proved to be general [10]. In 1966, the same group demonstrated that XPS has a high surface specificity [11]. Siegbahn was awarded the Nobel prize in 1981 for his efforts in developing XPS as a surface analytical technique.

The emission process has been described as a three-step model by Berglund and Spicer [12]. The first step involves absorption of the x-ray and promotion of an electron from its ground state to the final state above the Fermi level. The final state lies within the potential field of the solid and satisfies the Schrödinger equation for this field. The second step is transport of the electron to the surface, and the third step is escape of the electron into the vacuum. Since the electron is generated within the potential of the solid, its wavefunction contains contributions from the solid even after it has escaped into the vacuum. In XPS the kinetic energies of the emitted electrons forming the spectral peaks are measured using an electrostatic charged-particle energy analyzer, from which the binding energies of these electrons can be calculated from the following equation:

$$E_{\mathrm{b}} = h\upsilon - E_{\mathrm{k}} + \Delta\phi \qquad (1)$$

where E_{b} is the electron binding energy (BE) in the solid, $h\upsilon$ is the energy of the incident photon, E_{k} is the electron kinetic energy (KE) and $\Delta\phi$ is the difference in work functions between the sample and the detector material assuming that there is no electrical charging at the sample surface (Fig. 1a; related processes are shown in b and c).

A typical XPS survey spectrum recorded from a platinized tin oxide catalyst used for low-temperature CO oxidation [13] is shown in Fig. 2. It consists of well-defined spectral peaks on a large background of inelastically scattered electrons formed during transport of photoelectrons to the surface. The electrons comprising the spectral peaks carry most of the useful information although the fraction of these electrons in the total emission yield is very small ($\ll 1\%$). The primary peaks arise from the valence-band structure, from shallow core levels (BE < 50 eV), from deeper core levels and from Auger processes (Fig. 1b) for each of the elements present at or above the detection levels in the near-surface region of the sample. These and the other, less prominent, peaks present are discussed in detail below. The BEs and relative intensities of the peaks provide compositional information about the sample since every element, except H and He, produces core-level and Auger peaks at specific energies.

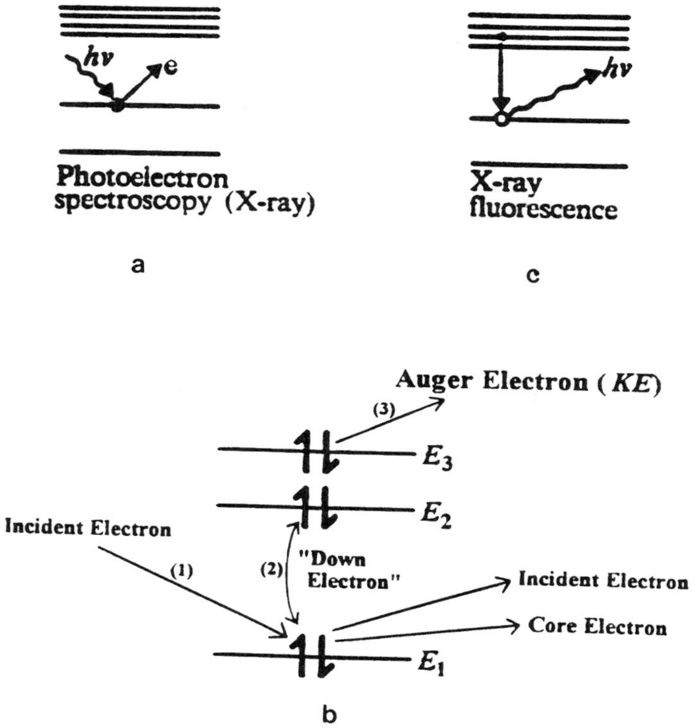

Figure 1. Schematic diagrams of the (a) photoemission, (b) Auger, and (c) fluorescence processes.

The composition of the near surface regions can be calculated from the relative intensities (areas) of the peaks assuming that matrix effects (such as the distribution of elements in the near surface region) can be taken into account. The calculation of the near-surface-region composition from XPS data is discussed in detail in Chapter 5.

The valence-band spectrum is generally difficult to understand, but it contains a large amount of information about the chemical species present and about the chemical bonding of the valence electrons which hold the solid together. The positions and shapes of the core-level features also provide detailed chemical-state information as can in principle the Auger features. The former is illustrated by the high-resolution XPS C 1*s* spectrum obtained from ethyl trifluoracetate by Siegbahn et al. [14] (Fig. 3). The four carbon atoms in this molecule each have a different chemical environment which results in four different C 1*s* binding energies. In most cases, however, the binding energy shifts

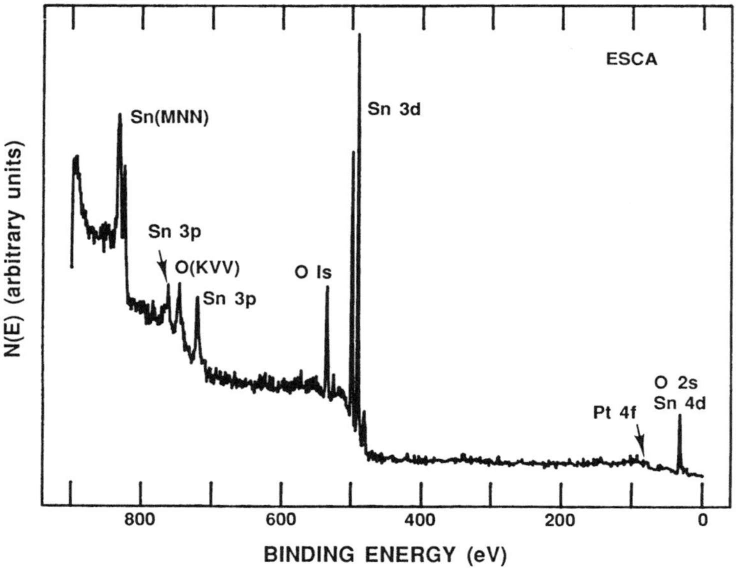

Figure 2. XPS survey spectrum recorded from a commercially available platinized tin oxide catalyst powder used for low-temperature CO oxidation. (From Ref. 13.)

are not so obvious as in Fig. 3 and can be too small to detect due to insufficient resolution. For example, Paparazzo et al. [15] have determined that the XPS Sn 3*d* binding energy shift between SnO and SnO_2 is only 0.19 eV. The data in Fig. 4 illustrate that a difference as small as that can be measured in special circumstances. The top spectrum was recorded from a polycrystalline tin oxide film annealed in air at 500°C. The Sn is present mainly in the form of SnO_2, but a small shoulder is apparent due to the presence of some SnO. After exposure to a flux of hydrogen atoms, the lower spectrum was obtained. Most of the tin oxide has been converted to metallic Sn, and what remains consists primarily of SnO. However, a slight shoulder due to SnO_2 is apparent on the high-BE side of the SnO peak. Although such small BE differences can be measured, subtle peak-shape changes can be difficult to interpret particularly when reliable data are not available from well-characterized standards. In the worst cases the binding energies of elemental species may be so closely spaced that XPS is unable to distinguish between them. This difficulty is complicated further by the presence of more than two overlapping chemical states.

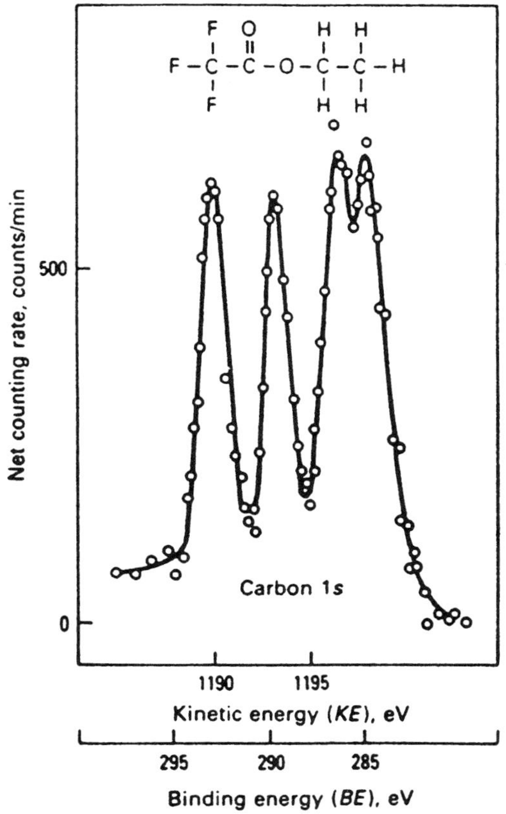

Figure 3. XPS C 1*s* spectrum recorded from ethyl trifluoracetate. (From Ref. 14.)

The surface specificity of XPS derives from the fact that the average net distance traveled in a solid by an electron before being inelastically scattered (the attenuation length, AL) is very short. As shown in Fig. 5, the AL is dependent on the electron KE and the type of material, i.e., elements, inorganic compounds or organic compounds [16]. X-rays have a low absorption cross section in solids so that most penetrate many hundreds of nm. However, most of the photoelectrons, which in XPS have KEs between 100 and 1500 eV, travel only a few nm before losing energy by suffering an inelastic collision, then becoming part of the broad background rather than contributing to the elemental peaks. If θ is the angle of emission of the photoelectron with respect to the perpendicular to the surface, then the average escape depth is given by AL ($\cos\theta$). Thus, even at normal emis-

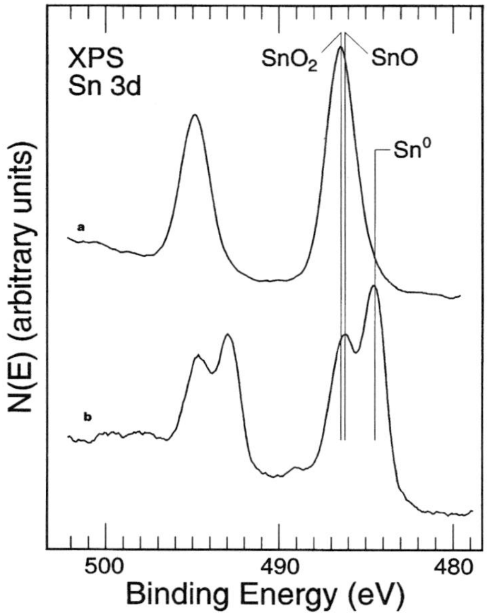

Figure 4. XPS Sn 3*d* spectra recorded (a) before and (b) after exposure of a tin oxide film to a flux of H atoms.

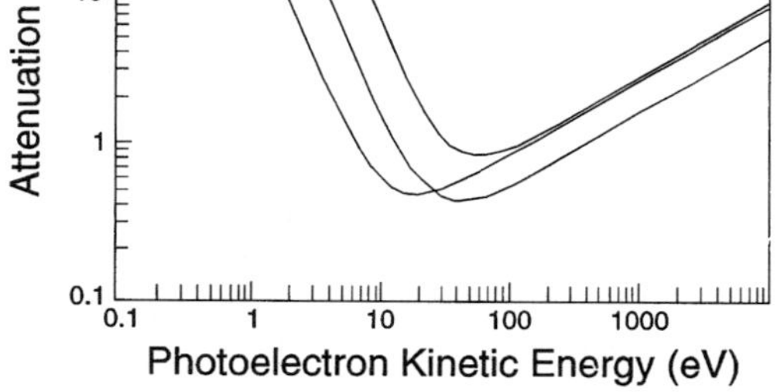

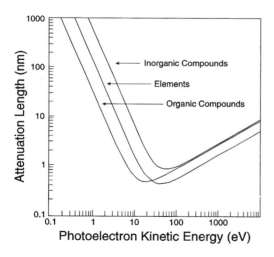

Figure 5. Dependence of attenuation length (AL) on the photoelectron kinetic energy for elements, inorganic compounds and organic compounds. (From Ref. 2 based on equations from Ref. 16.)

sion the average depth of origin of the photoelectrons is only 1–4 nm, according to Fig. 5. These points are discussed more quantitatively in Section 2.4.2.

1.2. Experimental Equipment and Data Collection

XPS instrumentation consists of an x-ray source and controller, an electrostatic charged-particle energy analyzer, a sample mount system, a UHV system and a computerized electronic system to control the electrostatic analyzer and collect and process the signal. A detailed description has been given by Rivière [17]. The energy analyzer, sample mount system and x-ray source are contained within the vacuum envelope. For studies of reactive surfaces such as those of clean metals, which rapidly adsorb background gases, the pressure should be below 10^{-8} Pa so that spectra can be collected before the accumulation of contamination from the ambient is enough to interfere with the analysis. In studies of less reactive surfaces, including metals such as Au or Ag, most oxides and most polymers, the pressure can be as high as 10^{-5} Pa. Higher pressures can cause arcing in the high-voltage region of the x-ray source, resulting in damage.

1.2.1. X-ray sources

X-rays are generated by bombarding a metallic anode with high energy electrons (10 to 15 keV) as shown in Fig. 6. The anode material determines the characteristic x-ray energy and the width of the x-ray line. The most commonly used anode materials are Mg and Al, whose principal photon energies are 1253.6 and 1486.6 eV, respectively. In addition to the primary line, small satellite lines are present, which also produce XPS spectral features as shown in Fig. 7 [18]. These smaller features are always displaced by a fixed energy from the parent peak, so that there is no difficulty in their identification unless they overlap other peaks. The satellite peak positions and relative magnitudes for Mg and Al anodes are given in Table 1.1. Less prominent spectral features can also arise either from contaminants on the anode, which produce x-rays with energies characteristic of the contaminant elements or, in the case of dual anodes, from "crossover" radiation from the second anode material. Such features are known as ghost lines. Fortunately, they are rarely observed, but when they do appear, it is a sign that the source is faulty. In order to minimize contaminants, the filaments are placed out of line-of-sight of the anode. An Al window between the anode and the sample protects the latter from stray electrons and from contamination. Anode materials other than Al and Mg can be used, usually for special purposes.

In XPS the overall energy resolution (ΔE in units of eV) is given by

$$\Delta E = (\Delta E_x^2 + \Delta E_A^2 + \Delta E_2^2)^{1/2} \tag{2}$$

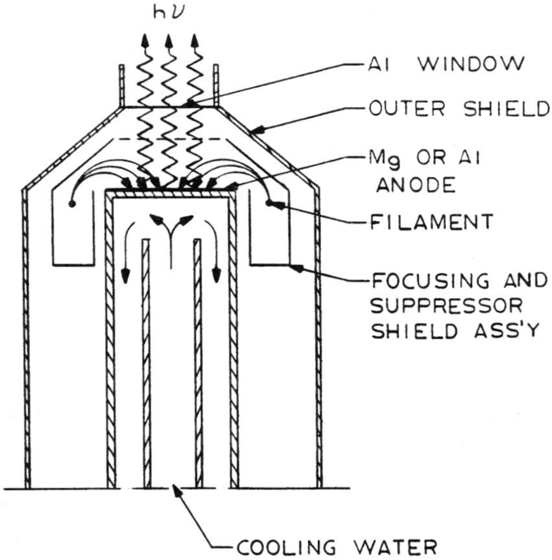

Figure 6. Schematic diagram of an x-ray generator used as a source for XPS.

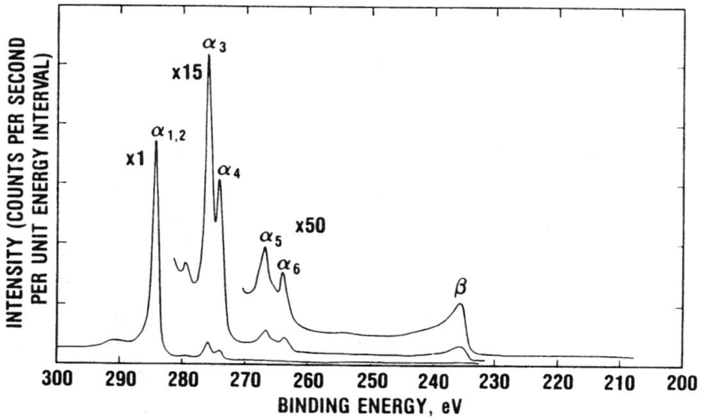

Figure 7. Mg x-ray satellites exhibited by the C 1*s* spectrum recorded from graphite. (From Ref. 18.)

Table 1.1. X-ray Satellite Energies and Intensities

		$\alpha_{1,2}$	α_3	α_4	α_5	α_6	β
Mg	Displacement (eV)	0	8.4	10.2	17.5	20.0	48.5
	Relative height	100	8.0	4.1	0.55	0.45	0.5
Al	Displacement (eV)	0	9.8	11.8	20.1	23.4	69.7
	Relative height	100	6.4	3.2	0.4	0.3	0.55

Source: Ref. 18.

where ΔE_x is the full width at half-maximum (FWHM) of the x-ray line, ΔE_A is the width of the analyzer energy window, and ΔE_2 is the natural line width of the orbital in the atom. ΔE_2 cannot be altered, while the analyzer energy window width can be set so that it does not limit resolution. Thus, it is the x-ray line width that usually limits the overall energy resolution of XPS, which is the critical factor in the unambiguous extraction of chemical-state information. Mg K_α x-rays have a natural line width of 0.75 eV while line widths from other possible anode metals can be as broad as several eV. Reduction in x-ray line width, and hence improvement in resolution, can be achieved by the use of a monochromator. For example, the Mg and Al K_α widths can be reduced to about 0.35 eV. Monochromatization also removes the satellite structure and the Bremsstrahlung background, leading to "cleaner" spectra. The disadvantage of a monochromator is that the incident x-ray flux is reduced significantly (by as much as a factor of 40), thereby reducing the photoemission signal strength too. Improvements in signal detectability are beginning to overcome this problem. When a spectrum contains several closely spaced and overlapping peaks whose separation and identification is essential for the analysis, then it is worth the loss in signal strength caused by using a monochromator. Another benefit of removing satellites and background from the x-ray flux is that radiation-sensitive materials are less adversely affected.

According to Eq. 1, only core levels with BEs less than that of the x-ray line can be observed, which with the conventional Mg and Al sources can occasionally be a limitation. Photoelectrons from levels with higher BEs can be excited by x-rays of higher energy. One source that has been found useful in such an application is Ti, whose K_α line energy is at ca. 4510 eV. However, the line width is 2.0 eV, so that the energy resolution, and hence the ability to acquire chemical-state information, is reduced or lost. The situation is not helped by the fact that chemical shifts in the core levels at higher BEs are usually small.

The cross section for photoemission of an electron in a core or molecular level is dependent on the photon energy. An important consequence of this statement relates to the valence-band electrons which have large cross sections for photoemission by UV light but very small cross sections for photoemission by x-rays. A good choice of anode for XPS valence-band photoemission is Zr, whose M_ξ line has a photon energy of 151.4 eV and a line width of ca. 0.77 eV. With Zr M_ξ radiation the valence-band cross sections are large, the spectral resolution is not significantly limited by the photon line width, and electrons in core levels with BEs between 20 and 145 eV can also be excited. The latter are not accessible with most UV photon sources and are useful because they are quite sensitive to chemical state, even if a compositional analysis cannot be made.

1.2.2. Energy analyzers

There are two types of charged-particle electrostatic analyzers commonly used for XPS, the double-pass cylindrical mirror analyzer (DPCMA) [19] and the concentric hemispherical analyzer (CHA). Schematic diagrams of them are shown in Fig. 8a and b, respectively. Most surface analytical systems sold in the last 10 years have been based on the CHA, and the DPCMA is being phased out. Both analyzers function as bandpass filters with the window width being set to any chosen value by adjusting the voltages within the analyzers. As the window width is increased, the signal strength increases but the energy resolution decreases. The two types of analyzers have different focusing properties, transmission properties and acceptance geometries, are be described below.

If the radii of the coaxial cylinders in a CMA are r_1 (inner) and r_2 (outer), the focal relationship between an electron of a particular energy E_0 passing through the analyzer and the deflecting voltage V applied to the outer cylinder, with the inner cylinder grounded, is

$$E_o = KeV[\ln\left(\frac{r_2}{r_1}\right)]^{-1} \tag{3}$$

where K is a characteristic constant and e is the elementary charge. For an entrance angle α value of 42°18.5′, the CMA becomes a second-order focusing device, for which K has a special value of 1.3099 [20]. All commercial CMA devices are built to be second-order focusing.

Electrons entering an analyzer are not restricted to a single entrance angle but have a spread $\delta\alpha$ of entrance angles about α. For a CMA, if this spread is not too great, $\delta\alpha \leq 6°$, then the base resolution ΔE_b is given by

$$\frac{\Delta E_b}{E_o} = \frac{0.6W}{r_1} + 5.50(\delta\alpha)^3 \tag{4}$$

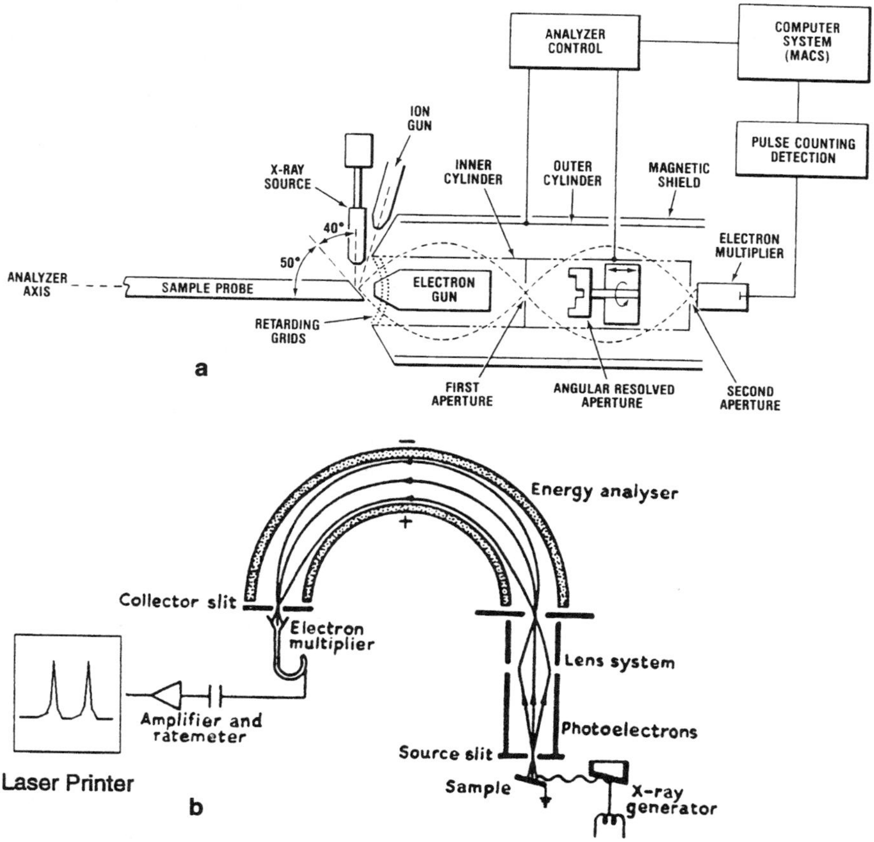

Figure 8. Schematic diagrams of (a) a double-pass cylindrical mirror analyzer (DPCMA) (from Ref. 19), and (b) a concentric hemispherical analyzer (CHA) (from Ref. 16).

where ΔE_b is the width of the base of the chosen spectral feature at median energy E_0, and W is the notional width of the entrance slit, effectively equal to the size of the source on the sample for the angular spread normally used. If the source size can be approximated by a point, as in AES, then the base resolution becomes simply

$$\frac{\Delta E_b}{E_o} = 5.50(\delta\alpha)^3 \tag{5}$$

Since the lateral extent of the area from which electrons are admitted to the analyzer is not a point source, the energy resolution available from a single-pass CMA (SPCMA) can never be adequate for XPS. If, however, a second CMA is added end-to-end, so that the exit aperture of the first stage becomes the entrance aperture of the second stage, then the source area for the second stage can be restricted to acceptable dimensions because the overall energy resolution is set by the second stage. The resultant DPCMA has been used extensively for XPS, but it has too many disadvantages in comparison with the CHA. These are that the precise energetic position of an observed peak is overly sensitive to the sample position in front of the analyzer, that since all characteristics scale with the radius of the inner cylinder improvements can be achieved only by an increase in the overall size, that there is very little space available near the specimen position for additional components, and that because a CMA cannot accommodate an electron transfer optical arrangement it is much less capable of further development than is a CHA. The importance of the SPCMA for use in AES lies in its high solid angle of acceptance, which means that the transmission is also high. The transmission T can be calculated from

$$T = 2(\sin\,\alpha)\delta\alpha = 1.346\delta\alpha \tag{6}$$

For $\delta\alpha = 6°$, $T \approx 14\%$.

For a CHA, if the radii of the inner surfaces of the two hemispheres are R_1 (inner) and R_2 (outer), then the focal relationship [21] between an electron of kinetic energy E_0 entering the analyzer and the deflecting voltage V applied between the hemispheres is

$$eV = E_o\left[\frac{R_2}{R_1} - \frac{R_1}{R_2}\right] \tag{7}$$

As with the CMA, electrons entering a CHA do so with a spread $\delta\alpha$ of angles about the correct tangential angle α. If the widths of the entrance and exit slits in a CHA are put equal to W, which is what is done in practice in most instruments, and if R_0 is the radius of the median equipotential surface (approximately equal to $(R_1 + R_2)/2$), then the base resolution is given by

$$\frac{\Delta E_b}{E_o} = \frac{W}{R_o} + (\delta\alpha)^2 \tag{8}$$

The form of Eq. 8 shows that the base resolution can be improved either by decreasing the slit width W or by increasing R_0, i.e., the size of the spectrometer. For both CHA and DPCMA designs, it is normal practice to aim at optimum relative resolution in XPS, that is $\Delta E_b/E_0$, by retarding electrons entering the analyzers to a chosen constant energy called the pass energy. The lower the pass energy, the better the relative resolution, but at the same time the transmission decreases, which means that in practice a balance has to be struck between intensity (count rate) and resolution. In XPS the signal intensity depends not on the transmission, as such, but on the so-called étendue [22], the product of transmission and slit area, which is given by

$$G = LW\left(\frac{W}{R_o}\right)^{1/2}\left(\frac{\beta}{\pi}\right) \tag{9}$$

where L is the slit length and β is the semiangular acceptance of the CHA in the plane of the slit.

1.2.3. Energy calibration

An XPS spectrum appears in the form of electron current as a function of KE of the electron. In order to extract the desired information, the KE must be transformed into BE. This step is extremely important and can be difficult. The goal is accurate assignment of the photoelectron peaks in terms of the corresponding BEs. In practice, "accurate" means to within 0.1 eV. A number of ways have been proposed in the literature for converting the KE scale into one of BE, and they all have problems. Generally, the conversion is easier for conducting samples and can be difficult, occasionally impossible, for insulating samples. Conductive samples are electrically coupled to the detector (i.e., analyzer) so that their Fermi levels are aligned. Thus, the $\Delta\phi$ term in Eq. 1 is equal to the difference in the work functions of sample and detector material. In principle, therefore, by measuring the work functions directly using conventional methods and then entering them in Eq. 1, which relates KE to BE, the conversion is simple. The work function of the detector hardly varies, so that if a work function probe were to be included in the vacuum system to monitor the sample work function, then BE could be assigned for any conducting samples. However, it is generally very inconvenient to have to measure the sample work function at frequent intervals, and other approaches must be adopted.

One popular method is to assign the energy of the C 1*s* peak due to carbon contamination (referred to as adventitious carbon) on the sample to the internationally accepted value of 284.6 eV, which then allows the BEs of all other features to be found. Although this method has serious drawbacks, it has been used more often than all others. The drawbacks include the facts that the adventitious carbon is not well characterized, that it may not be in electrical contact with the species being examined, and that it may consist of several different chemical states of carbon leading to a complex peak shape. In addition, as a result of ion erosion during depth profiling, the contaminant carbon may well be removed. In the ideal case this carbon consists only of hydrocarbons in electrical contact with all surface species. The use of adventitious carbon as a binding energy standard has been discussed in detail by Barr and Seal [23]. A second method is to measure the energy of the Fermi level of a conductor or that of the top of the valence band of an insulator and then assign all other features with respect to those values. Since the valence-band intensity in XPS is weak, it can be difficult to determine the value of the Fermi level accurately, because at the energy resolution necessary for accurate Fermi level determination the data acquisition times are very long. For an insulator it is necessary to know the value of the energy gap between the top of the valence band and the Fermi level so that the energies of all features can be referenced with respect to the Fermi level. A third method involves depositing a small amount of metal, usually Au, with a known core-level binding energy, which is then used as the reference. The difficulties with this method are that the binding energies of all metals (even Au) can be influenced by the support material and by crystallite size effects and that the deposited material may be charged differentially with respect to the substrate surface.

There is also an empirical method used by the author and many other workers, based on internal spectral self-consistency. Instead of assigning the BE of an alleged contamination peak to the C 1*s* value given above, a peak from one of the major constituents in the sample is chosen, and on the basis of the probable chemical state of that constituent (from other evidence) a BE is assigned to that peak. This BE is taken from the large body of reference literature that has accumulated over the last 25 years [18, 24–26]. With this trial assignment, the BEs of all other features can then be assigned, at which point self-consistency must be examined. If the assignment is correct, then not only will the assignment of a C 1*s* peak to adventitious carbon and of the position of the Fermi level appear reasonable, but the interpretation of all the other spectral features will be compatible with published values of BEs. If consistency is not obtained in that interpretation, then either an adjustment must be made to the trial BE or another element selected and the procedure repeated. Basically, the method involves treatment of $\Delta\phi$ in Eq. 1 as an unknown constant to be determined. For conducting materials and for insulating materials which charge uniformly over the whole surface, a value of $\Delta\phi$ will exist which will be consistent with the data. As discussed, $\Delta\phi$ is the work function difference for conducting materials. For insulating materials the detector and the sample surface are not electrically connected, in which case $\Delta\phi$ is determined by the field that builds up at the surface. The field strength is determined by the relative numbers of electrons lost by the sample due to photoemission and those picked up from the UHV chamber or any conduction paths to the sample surface. In the worst case different regions of the sample may charge by different amounts so that no single value of $\Delta\phi$ will be consistent with the data. There are experimental methods of dealing with samples which exhibit such differential charging, to be discussed later. Differential charging is not always a disadvantage. If two regions on a sample charge by sufficiently different amounts, then a separation of the spectral features may occur which could assist in the interpretation of the composition and chemistry of each region.

1.2.4. Data processing

Apart from trivial operations such as on-screen spectrum shifting, expansion, contraction, etc., there are two procedures that are available in all XPS data processing packages supplied by manufacturers, namely smoothing and background subtraction. These two are far from trivial, and it is important to be aware of their limitations.

Noise, or random electronic signals, can enter an acquired spectrum from many sources and cannot be completely eliminated even under optimum operating conditions. One of the challenges in analysis by XPS is the unambiguous identification of a very small genuine signal arising from an element present at a surface, in the presence of the spurious signal arising from noise. The signals in XPS are in fact in the form of pulses, which are counted for a preset length of time at each step in the chosen KE range. If the pulse count in the genuine signal is S and that in the noise at the same step is N, then it is the signal-to-noise ratio S/N that determines the detectability or otherwise of S. One way of improving S/N is to count for a longer time at each step, thus accumulating more pulses, because the signal S will increase directly as the number of counts n, whereas the noise N will increase only as $n^{1/2}$. The obvious disadvantages in doing that are that the sample is exposed to the x-ray flux and to any heating effects from the x-ray source for much longer, with the consequent risk of damage, and that excessive instrument time is involved. Another way of improving S/N, but with exactly the same disadvantages, is to scan the same KE region multiple times and add the spectra together. This maintains a short time per energy step.

The above disadvantages can be avoided (but others possibly introduced) by digital smoothing of spectra after acquisition, an operation that is easy with current data systems. It is usually based on the method of Savitsky and Golay [27]. The number of counts at any particular energy step are averaged with the counts at equispaced energy steps on either side of it; the number of steps to be included and the statistical weight to be given to each step can be varied in a chosen manner in order to vary the extent of smoothing. The dangers in the procedure arise because it is rarely known a priori just how much smoothing is necessary for any one spectrum. If too little were to be applied, then the S/N ratio would not be improved adequately. If too much were used spectral resolution would be lost, small features would be reduced unacceptably, and undesirable artefacts might be introduced. Momentary electrical excursions producing a sudden increase in the number of counts, usually confined to one channel, generally called "spikes," in a spectrum, can give rise to serious misinterpretations if their presence is not realized. Most data systems have routines for removing spikes before smoothing, although such routines are themselves subjective. Removal of a spike from a region of flat background is one thing, but from the tops or sides of genuine peaks is quite another and can introduce further artefacts.

It is possible to establish a set of smoothing parameters by recording high-resolution multiple-scanned spectra over a long time period from a stable surface such as that of a clean metal in UHV, and then comparing them with rapidly acquired spectra from the same surface that have been smoothed with several different choices of parameters. Provided that there are no time-dependent fluctuations in the spectra, the correct parameters will be those that yield a spectrum most similar in appearance to that recorded over the long period of time. Unfortunately, the same set of parameters will not necessarily apply to spectra recorded with different energy resolutions.

It is clear from the discussion on energy calibration that it is necessary to be able to measure accurately the BE corresponding to a peak position. For precise measurement this is complicated by the presence of the inelastic spectral background, which in essence adds a variously sloping contribution beneath a peak, as can be seen in Fig. 2. There are two current methods of dealing with this problem. The first and more difficult one is that of background subtraction using linear [28], Shirley [29], modified Shirley [30], or Tougaard [31] backgrounds, as discussed in detail in Chapter 5. Of the various backgrounds quoted, only that of Tougaard has any pretensions to physical reality. The second method is purely graphical, and consists of drawing a set of lines within a peak envelope, parallel to an assumed linear background, and then drawing a line through the midpoints of the lines, as illustrated in Fig. 5 of Ref. 18. Where the midpoint line intersects, the top of the peak is taken as the location of the BE.

1.2.5. Sample configuration

There are many forms of sample including single crystals, polycrystalline foils, polymer films, powders, porous pellets, sputter-deposited films, etc. They are all amenable to analysis by XPS, but different types of sample mount and of treatment capability are required for the various sample forms. Generally, it is important that the sample be larger than the analysis area and electrically grounded as effectively as possible even when it is believed to be an insulator. Metals and semiconductor materials are usually easy to ground with a spot weld or a mechanical clip attached to the grounded sample holder. Polymer films should be thin and on metal substrates, while powders should be pressed into a metal cup which is itself grounded. Although such materials are normally insulators, charging problems with them can often be reduced by mounting samples in ingenious ways. Very little is understood about charge buildup on insulators, so that predictions regarding charging behavior during XPS are not easy, and if a given sample charges, then several different approaches should be tried. Differential charging can make data interpretation very difficult because peaks can be artificially broadened or even split into several peaks. This and other charging problems can often be avoided or minimized by flooding the surface with low-energy electrons (several eV) from a hot filament placed near the sample. The flood electrons neutralize the positive charge built up by the loss of photoelectrons. Care must be taken to avoid contamination of the sample surface by species emitted from the filament. Heating a sample can sometimes reduce charging by increasing the surface electrical conductivity, but, on the other hand, heating can alter the chemical composition of surfaces. Charging and charge compensation techniques for insulators have been discussed by Hofmann [32], and recent experimental developments are being marketed by Physical Electronics, Inc.

1.2.6. Sample treatment

In situ sample cleaning is an important treatment for some types of sample, particularly those to be the subject of basic research. Generally speaking, no attempt should be made to clean samples of technological origin, since vital analytical information might be lost thereby. If cleaning is necessary, the method to be used depends on both the sample material and the nature of the information sought. A review of cleaning procedures for various materials has been published by Musket et al. [33]. Metals are usually cleaned by ion bombardment coupled with annealing in a reactive gas or in vacuum. Thus it is usually desirable for an XPS system to contain an ion sputter gun, a sample heater and a system for handling and delivering pure gases. Fairly volatile species such as water, or species that decompose at low temperature, can often be removed by annealing at low temperature. The surface compositional changes are monitored using XPS or another surface technique while the desorbing gas-phase species are monitored with a mass spectrometer. New methods for cleaning surfaces are being developed currently based on the use of novel, UHV-compatible H-, O- and N-atom sources [34, 35]. An example of the cleaning of a Ni/Cr alloy surface using H atoms is shown in Fig. 9. The XPS survey spectrum (a) recorded from the as-received sample exhibits very few peaks except those of contaminating O and C. During exposure to the H atoms, reaction products including H_2O and CH_4 form and desorb producing the cleaned surface (spectrum b). The O 1*s* peak is nearly eliminated, the C 1*s* peak is greatly reduced, and both photoelectron and Auger Ni peaks are now prominent. The atoms are of low energy (ca. 1 eV) so that they do not damage the surface in the same way as ion sputtering. Other methods of producing clean surfaces include evaporation of a thin film from either a heated or a sputtered source onto a specimen holder, and in situ fracture. Ex situ cleaning includes ultrasonic cleaning in various solvents, fracturing, scraping, and grinding to a

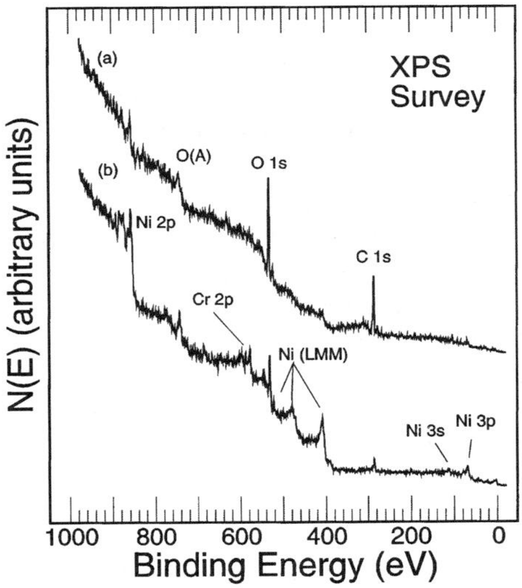

Figure 9. XPS survey spectra recorded from a polycrystalline Ni/Cr (80:20 at%) surface (a) before and (b) after exposure to a flux of H atoms.

powder. Each situation is different, and since surface contamination can accumulate during ex situ cleaning, XPS should be used to monitor the effectiveness of the cleaning process.

Sequential treatment of a sample, with the resultant surface changes being monitored by XPS after each step, is one of the ways in which data interpretation in XPS is made easier. Since the nature of the treatment is known, the progressive alteration in surface chemistry can be interpreted with little ambiguity. Air exposure invariably alters the surface chemistry. It is therefore necessary to perform the treatments in the UHV system, or in an attached chamber, so that the sample can be transferred between the chambers without air exposure. The treatment chamber might contain a film-growth system, an electrochemical cell, a chemical (or catalytic) reactor or an integral cell for annealing the sample in reactive gases at high pressures [36]. By moving the sample back and forth between the analysis and treatment chambers, successive spectra evolve in a manner which can be more easily interpreted than could a single spectrum. In some cases it may be possible to reverse the effects of a given treatment. An example is the oxidation/reduction behavior of a $TiO_2(001)$ surface [37]. Both reduction in hydrogen and ion sputtering lower the XPS O/Ti ratio and produce suboxides such as Ti_2O_3 and TiO, while annealing in oxygen restores the TiO_2 stoichiometry by reoxidation of the suboxides.

1.3. Spectral Features and Interpretation

The information sought from XPS data is normally that of composition and/or chemical state. It is also possible sometimes to obtain structural information or to study fundamental interactions between photons and matter using XPS, but the latter applications are not discussed in this chapter. A detailed description of the calculation of composition is given in Chapter 5 so only a few points on that topic will be made here.

1.3.1. Determination of composition from XPS data

The sampling depth in XPS is the outermost 1 to 4 nm, depending upon the ALs of the photoelectrons being collected. The distribution of depths of origin of photoelectrons tails exponentially into a solid, which means that the composition determined from XPS data is a weighted average over the region being examined with the near-surface layers contributing more intensity to the total signal. Since the characteristic peaks of the elements appear at different KEs, i.e., with different corresponding ALs, it follows that the elemental depth distribution functions in the near-surface region are also different. If the composition of the near-surface region were homogeneous, this effect could be taken into account. In fact, an assumption of homogeneity is nearly always made and is programmed into all the commercial software quantification packages. The problem is that a sample is rarely homogeneous in the near-surface region, and indeed it is the very nature of the geometric distributions of elements within the near-surface region that is the most sought-after information from the quantification of XPS data. The homogeneity assumption is therefore wrong in most cases, and the magnitude of the error can be quite large. This point is illustrated by the XPS data and the corresponding SEM images shown in Fig. 10, recorded from Ag/α-Al_2O_3 ethylene epoxidation catalysts as a function of time in the reactor [38]. The morphological changes with reaction time can be seen in the SEM micrographs. Initially, the Ag was in the form of a thin film which covered the α-Al_2O_3 fairly uniformly. With aging in the chemical reactor, the Ag first formed globules and then sintered to form large Ag crystallites which interacted very weakly with the alumina support. Although the actual amount of Ag on the alumina support surface did not change with aging, the Ag-to-O and Ag-to-Al peak-height ratios in the XPS spectra changed significantly. As the Ag film sintered to form large Ag crystallites, more of the alumina support became exposed and more of the Ag on the surface became effectively subsurface Ag. This matrix effect, and not a change in composition, caused the observed relative peak-height changes. If the homogeneity assumption were to be applied to the calculation of composition from the XPS data in Fig. 10, incorrect conclusions would have been drawn that the amounts of Ag, O and Al on the surface

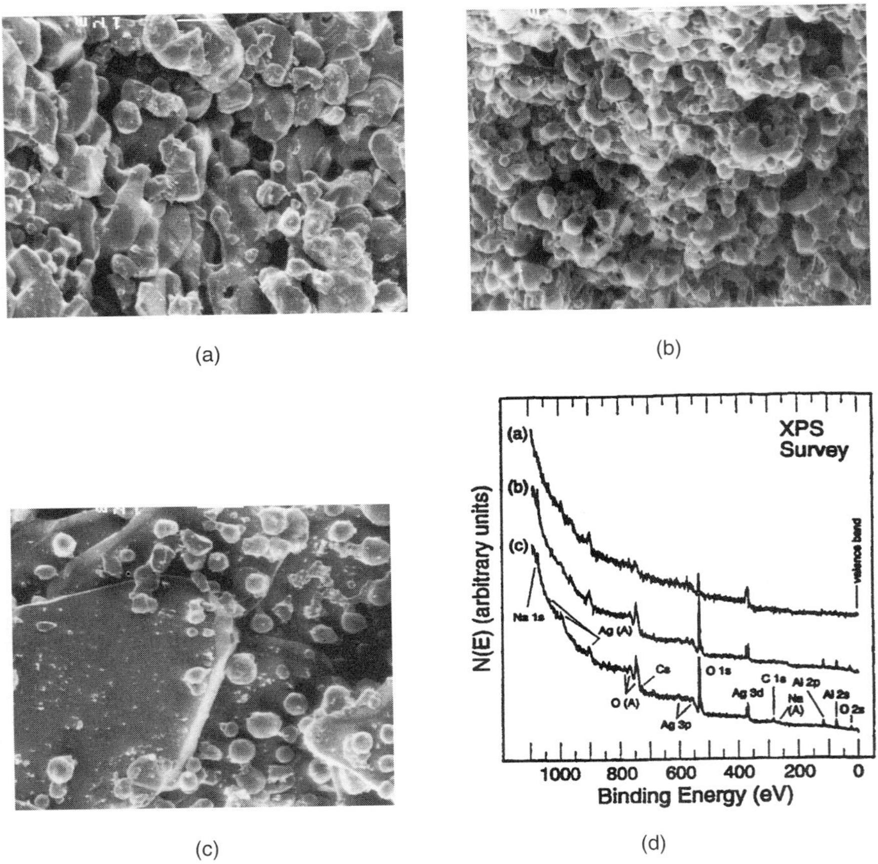

Figure 10. XPS spectra and corresponding SEM micrographs recorded from (a) fresh, (b) active and (c) aged Cs-promoted Ag/α-Al_2O_3 ethylene epoxidation catalyst. (From Ref. 38.)

had changed with time. Errors due to the neglect of matrix effects and to the assumption of homogeneity are usually much larger than others encountered in the quantification of XPS data. Unfortunately, there is no routine way presently available for taking matrix effects into account, and of course the matrix differs for each type of sample. Simple depth distribution models applied on a trial and error basis have had some success, but often the best approach is to gain information about the matrix from other techniques and then use that information to assist in the quantification.

For very flat surfaces angle-resolved XPS (ARXPS) can provide information about the spatial distribution of elements in the near-surface region. By tilting the sample with respect to the detector, so that the collection angle θ between the detector and the normal to the surface increases toward grazing exit, the average depth sampled decreases by a factor of cos θ, as illustrated in Fig. 11. The x-ray incidence angle does not affect depth sensitivity in XPS since x-rays penetrate deeply into the solid. Thus ARXPS provides a means of varying the depth sensitivity. From ARXPS data it is possible to derive compositional profiles using mathematical models published in the literature [39–41]. However, certain assumptions are required, and the solutions are not unique. As an example, the data shown in Fig. 12 [42] were collected at different exit angles from a Si(100) surface with a very thin native oxide layer, and provide clear evidence of the spatial depth distribution. The spectrum recorded at the near-grazing angle is more characteristic of the native oxide layer, as expected. When a surface is rough, it must be noted that the asperities will contribute more of the signal at grazing collection angles; if it is known beforehand that the surface is rough, then there is still useful qualitative information to be gained from the angular resolution.

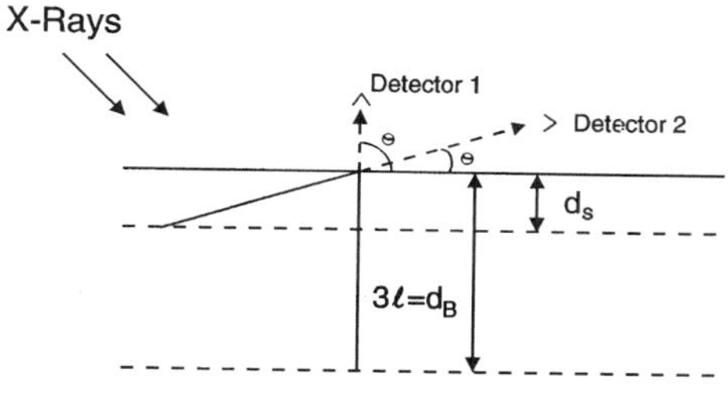

Figure 11. Schematic diagram illustrating how collection angle affects XPS depth specificity. If the AL of a photoemitted electron is l, then 99% of the photoelectrons detected by detector 1 are emitted within a distance of $3l$ (d_B) beneath the surface. However, moving the detector to position 2 results in an equivalent detection of electrons emitted within a depth d_S.

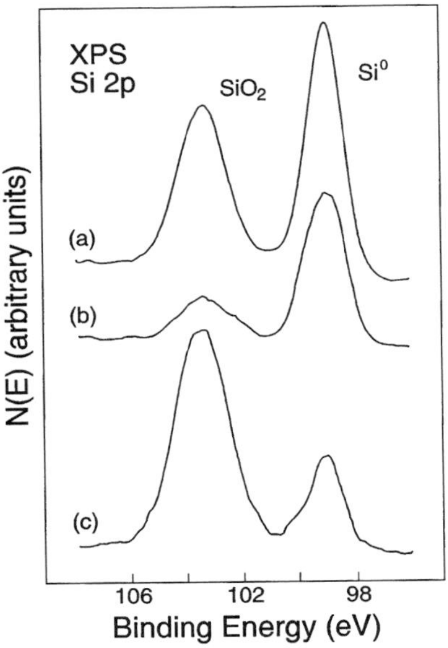

Figure 12. Si 2*p* XPS spectra recorded from the native oxide layer on the Si (111) surface with the sample tilted 52° with respect to the DPCMA normal. (a) Collection of electrons emitted over all angles into the DPCMA. (b) Collection of electrons emitted normal to the sample, in a less surface-specific mode. Elemental Si is more pronounced in this spectrum than Si bound as SiO_2. (c) collection of electrons emitted at 80° with respect to the sample normal, resulting in enhanced surface specificity. Elemental Si is less pronounced in this spectrum than is Si bound as SiO_2. (From Ref. 42.)

1.3.2. Determination of chemical state

Determination of the chemical-state distribution from XPS data is often straightforward provided that the energy axis has been calibrated properly and that differential charging does not produce distortion of the peak shapes. An excellent example is the XPS spectrum shown in Fig. 3 in which the BEs of the different species contributing to the C 1*s* feature are well separated. There are four chemically nonequivalent carbon atoms in the molecule that can be associated unambiguously and respectively with the four well-defined C 1*s* peaks in the XPS spectrum. Examples such as that are frequently given in expositions on XPS, but unfortunately they are not representative of those encountered in practice.

Because no two elements have the same set of BEs, the photoelectron spectrum from an element will be characteristic of that element and will possess several features at different BEs. However, the precise positions of the features, coupled with their peak shapes, will depend on the chemical state(s) of the element at the surface. Information on BEs and peak shapes for various chemical states can be found in the Perkin-Elmer PHI XPS handbook [18], the National Institute on Standards and Technology (NIST) database [24], a polymer handbook by Beamson and Briggs [25], a relatively new journal entitled *Surface Science Spectra* [26], and throughout the surface science literature. Chemical-state assignments are usually based on the most prominent peaks because they have the highest detectability, but some of the less prominent peaks often exhibit larger shifts in BE for a given chemical state, so it is useful in such cases to base chemical-state analysis on them. Because BE shifts are not always available for the less-prominent peaks such as those arising from shallow core levels, it may be necessary to record spectra from one or more suitable reference materials.

A difficult problem to handle in chemical-state analysis occurs when BEs are so closely spaced that separate features cannot be distinguished easily. This point was made earlier with regard to the XPS data obtained from Sn oxide before and after reduction (Fig. 4). Since it is normal for many elements to be present at a surface in more than one chemical state, resolution of the spectral features of one element from those of another and resolution of the features arising from different chemical states are essential to complete analysis.

Such resolution is normally attempted by curve-fitting techniques, of which there are many variations. The approach adopted by the author is first to smooth the data and subtract an appropriate background, then to add known reference spectra in varying relative proportions until the best fit to the experimental spectral envelope is obtained. An example is given in Fig. 13 from Ref. 43. In the upper part of the figure is the Nb 3*d* spectrum from Nb_2O_5, which can be resolved easily into the 5/2–3/2 doublet with peaks of equal FWHM centered at 210.3 and 207.6 eV, respectively. After sputtering the oxide with Ar^+ at 5 keV and at a current density of 28 mA m^{-2} for 120 min, the Nb 3*d* envelope shown in the lower part of the figure was recorded. Oxygen is sputtered preferentially from most oxides resulting in the formation of lower oxides and sometimes, as in this case, metal. The spectral envelope from the sputtered Nb_2O_5 can be fitted almost exactly by adding contributions from Nb, NbO, NbO_2 and Nb_2O_5, with the maximum contribution coming apparently from NbO.

In principle the relative concentrations of the various species found to be present by curve fitting could be determined directly from the proportionate contributions, as in the example of Fig. 13, but to do that would involve the assumption of homogeneity with depth. Such an assumption is normally made in analytical work, but it cannot be justified in isolation, as discussed earlier. A single additional measurement at another, much shallower, collection angle would yield information about the validity of the assumption.

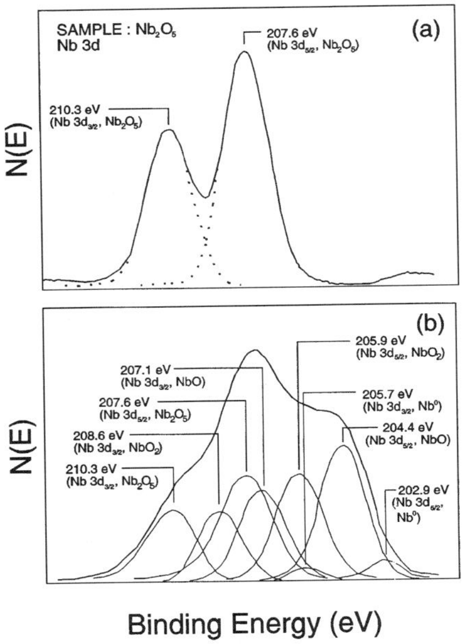

Figure 13. Curve-resolved XPS data. Nb $3d$ XPS data recorded from Nb_2O_5 (a) before and (b) after sputtering with 5 keV Ar^+. (From Ref. 43.)

Not infrequently the reference spectra cannot be added in any proportions to fit the experimental envelope, in which case other species may be present. If it is assumed that the peak arising from the unknown species has the same FWHM and, in the case of a doublet, the same separation and relative magnitudes, as for the reference materials, then it can be resolved by trial and error in a computer, but the process can be onerous. The more reference spectra that have to be used, and the greater the number of possible unknown species, the greater is the likelihood of large error occurring and therefore of misinterpretation. In any case no interpretation should be accepted that is at odds with what is known about the chemistry of the system from other sources, including information about the chemical states of other elements present from the rest of the XPS spectrum. The same stricture applies to the other popular method of curve fitting, that of using model Gausso-Lorentzian peaks of variable width and mixing to fit the experimental envelope. A computer program, if given enough latitude in choice of parameters, will always find a fit, but it may not be physically realistic. More is said on this topic in Chapter 14.

As described, spectra from standard materials are often used as reference spectra in chemical-state resolution, but even with standards problems arise in the making of unambiguous assignments to features in spectra.

A concept that may be useful in the determination of chemical state is that of the Auger parameter, originated by Wagner [44]. The so-called modified Auger parameter α^* is defined as the sum of the BE of a prominent photoelectron peak and the KE of a prominent Auger peak, for the same element in the same spectrum. α^* has the advantage that it is independent of energy calibration and therefore of any charging. In its general usage α^* is basically an empirical parameter, but if defined more accurately it can be shown to be linked theoretically with the extra-atomic relaxation energy involved in electron emission processes. See Chapter 5 for further description.

The Auger parameter may vary more with chemical state than do chemical shifts in the core level BEs, hence its usefulness. For some elements, however, there is not much to be gained by trying to use α^* rather than BEs, and Ag is a good example. Table 1.2 shows that although α^* can distinguish between metallic Ag and its oxides more easily than by examination of the BEs, there is no benefit in using it to distinguish between AgO and Ag_2O. Obviously, great care must be taken to measure both BE and Auger KE as accurately as possible in order to avoid misinterpretation, and this is difficult due to the width and complex shape of the Auger feature. The Auger parameter cannot be used when more than one chemical state is present and the features overlap.

The use of well-defined reference materials is crucial with regard to chemical-state determination by XPS. However, it is very difficult to prepare well-defined surfaces for many chemical compounds due to the presence of surface contaminants and multiple chemical states and to the fact that the surface composition is seldom similar to the bulk composition. This point is illustrated by the use of an AgO powder sample to obtain reference BEs and peak shapes for AgO and AgO_2 [45, 46]. The AgO was heated in vacuum in order to decompose first the contaminating carbonate and hydroxyl species and then the AgO itself to Ag_2O and finally to Ag metal. The XPS Ag 3*d* spectra recorded after heating the sample to various temperatures are shown in Fig. 14. The spectral changes are complex because several Ag species are present and their BEs are separated by only a few tenths of an eV (Table 1.2). Heating at 100°C results in the loss of carbonate and hydroxyl species, at 200°C in the decomposition of AgO to Ag_2O, and at 400°C in the decomposition of Ag_2O to Ag metal. Spectrum f was recorded from a clean

Table 1.2. Values of the Ag $3d_{5/2}$ and Auger Parameter α^* for Ag Compounds

	Ag $3d_{5/2}$ (eV) [45, 46]	α^* (eV) [18]
Ag^o	368.3	726.1
Ag_2O	367.8	724.5
AgO	367.4	724.0

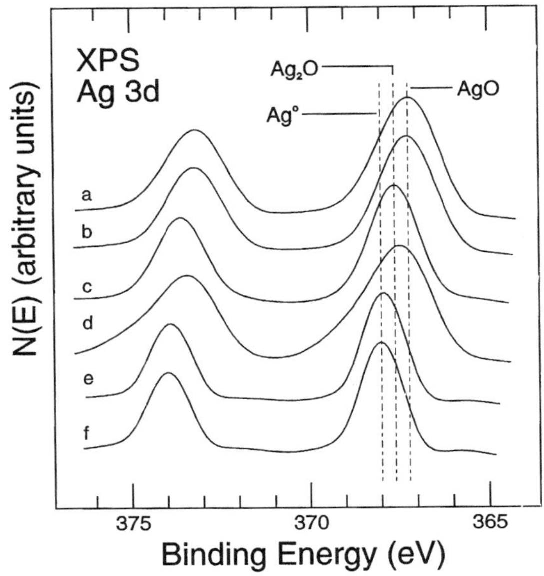

Figure 14. XPS Ag 3*d* spectra recorded (a) from the as-prepared AgO powder sample, (b) after annealing at 100°C, (c) after annealing at 200°C, (d) after annealing at 300°C, (e) after annealing at 400°C and (f) from a polycrystalline Ag foil after annealing at 250°C and sputtering with 2 keV Ar^+. (From Ref. 45.)

Ag foil and is included for comparative purposes. Although discrete peaks cannot be distinguished, the carefully controlled manner in which the experiment was performed helped the BEs and peak shapes of AgO [47], Ag_2O [48] and Ag metal [49] to be assigned unambiguously. The corresponding O 1*s* and C 1*s* spectra were also analyzed in order to assess the cleanliness of the reference surfaces and any possible influences of impurities on the Ag 3*d* features. The point is that reference surfaces must be clean, but that the cleaning procedures themselves may as often as not introduce ambiguities into the spectra so that the preparation of reference surfaces for many compounds is challenging.

The potentially complex nature of XPS features is illustrated further by the Pt 4*f* data recorded from a platinized tin oxide surface and shown in Fig. 15 [50]. It is important to understand that every Pt atom with a different chemical environment gives rise to an unique Pt 4*f* feature, but that the features are often closely spaced and have similar shapes. Vertical lines have been drawn in Fig. 15 at the BEs for Pt metal (crystallites), PtO_2, PtO, $Pt(OH)_2$ and Pt-O-Sn. The as-deposited Pt (spectrum A) consists mostly of Pt-O-Sn and $Pt(OH)_2$ with a small shoulder at 71.3 eV due to the presence of small crystallites of metallic

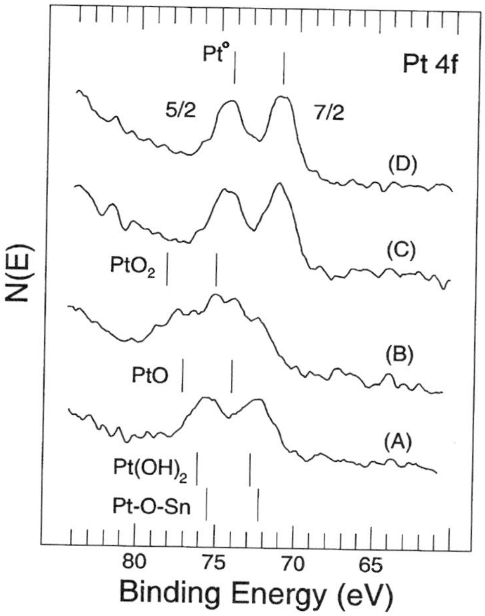

Figure 15. XPS Pt 4*f* spectra recorded from a platinized tin oxide film (A) immediately after evacuation, (B) after annealing in 150 Pa O_2 at 600°C for 30 min, (C) after annealing in 0.13 Pa H_2 at 500°C for 30 min, and (D) after annealing in vacuum at 800°C for 30 min. (From Ref. 50.)

Pt. Spectrum B was recorded after annealing at 600°C in O_2 at a pressure of 150 Pa for 30 min. Except for the Pt bonded as Pt-O-Sn, which anchors the Pt species to the tin oxide surface, most of the Pt was converted to PtO and PtO_2 during the oxidation step. The sample was then heated to 500°C in H_2 at a pressure of 0.13 Pa for 30 min. Spectrum C collected after this treatment indicates that the PtO and PtO_2 had been converted to Pt metal; however, small Pt-O-Sn and $Pt(OH)_2$ features are also apparent. After heating to 800°C in vacuum (spectrum D), the Pt remained metallic and the Pt-O-Sn and $Pt(OH_2)$ features were still present, but the Pt metal peak shapes had changed. The Pt $4f_{7/2}$ metal peak in spectrum C exhibits a shoulder on the low BE side at 70.8 eV. This shoulder is characteristic of bulk metallic Pt and indicates that large Pt crystallites were present. High-temperature annealing in vacuum caused sintering of the Pt and therefore formation of larger Pt crystallites. The broadening of the Pt 4*f* peaks in spectrum D with an enhanced contribution at 70.8 eV was therefore due to the growth of Pt crystallites. This illustrates the point that important chemical and physical processes at surfaces may manifest themselves only as subtle peak-shape changes in XPS spectra.

Core-level BE shifts often correlate with some measure of electronegativity, as shown in Fig. 16a and b [51]. Such plots can assist in determining which types of species may be contributing to a given spectral feature. The simple theoretical explanation is that as more charge is withdrawn from an atom, the atomic nucleus attracts the remaining electrons more strongly thereby increasing their binding energies. This oversimplified explanation implies that the BEs of core levels of metals are always shifted to higher values upon the formation of oxides. This is clearly not the case for Ag metal, Ag_2O and AgO (Fig. 14) so that the argument does not apply in all cases.

Further complexities are contributed by quantum mechanical effects within the atom. Because of the spin-orbit coupling [52], photoemission features based on *p, d* and *f* orbital excitation appear as doublets. Theoretically the peak-area ratios should be 2:1, 3:2 and 4:3 for the *p* 3/2 and 1/2 peaks, *d* 5/2 and 3/2 peaks and *f* 7/2 and 5/2 peaks, respectively, but the ratios are not always found

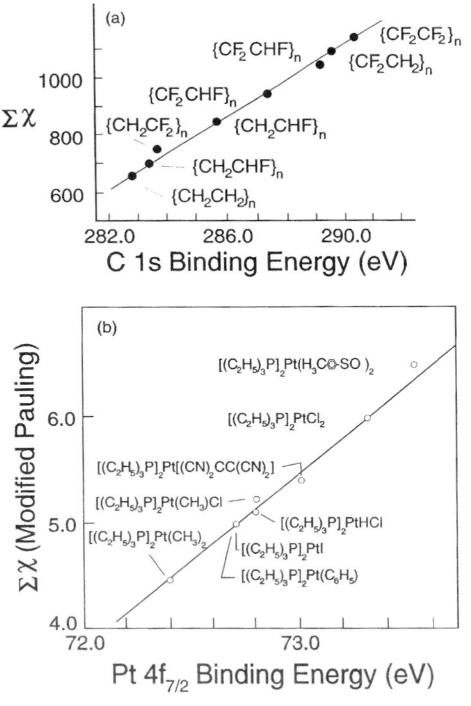

Figure 16. (a) XPS C 1*s* peak positions in fluoropolymers correlated with substituent electronegativity and (b) XPS Pt $4f_{7/2}$ peak positions in organometallic compounds correlated with substituent electronegativity. (From Ref. 51.)

in those exact proportions. During chemical-state changes such as in oxidation, the behavior of the doublets can lead to complex overlapping structures, as in Fig. 15d, where the Pt $4f_{7/2}$ peak from $Pt(OH)_2$ appears as a shoulder on the low-BE side of the Pt $4f_{5/2}$ peak from Pt metal.

1.3.3. Additional features in XPS spectra

In the interpretation of complex spectral features, the possibility of spectral interferences must always be considered. Every elemental spectrum contains as its principal features those arising from photoemission from valence and core levels and from Auger processes. The Auger features usually appear as groups of associated peaks which have originated from the decay of a core hole through the Auger process shown in Fig. 1b and are described more thoroughly in the next section on AES. Since the Auger peak energies do not depend on the photon energy, whereas the direct photoemission features do, overlapping Auger and photoemission features can be separated by changing the photon energy, i.e., by using a different anode material. In addition to these features, there are the x-ray satellite and x-ray ghost features discussed earlier. The positions and relative heights of the satellite features with respect to the principal x-ray line are known for many anode materials, while the presence of ghosts, although rare, indicates that the x-ray anode is faulty.

Not all photoemission processes are simple and direct and obey Koopman's theorem [52]. This is the "frozen-orbital" approximation in which it is assumed that there is no change in the energies of electrons in other orbitals as the photoelectron is ejected. However, there is a finite probability that the ionized atom will be left in an excited state by simultaneous transfer of energy from the ejected electron to another electron as the core hole is created. This process is referred to as "shake-up" and results in the photoemitted electron losing up to several eV of energy. Examples are shown in Figs. 17 and 18 [18]. In the C 1*s* spectrum from polystyrene in Fig. 17, a subsidiary peak lying 6.7 eV higher in BE than the C 1*s* peak can be seen; it is characteristic of a shake-up process occurring in compounds with unsaturated carbon bonds involving the energy of the $\pi \rightarrow \pi^*$ transition. For this example it is quite small, but sometimes it can be much larger. The energetic positions and peak shapes of shake-up satellites vary significantly in unsaturated compounds and are used diagnostically. With paramagnetic compounds the shake-up satellites can often be nearly as large as the main feature, as seen in Fig. 18. In some cases multiple shake-up features are present as for $CuSO_4$. Again, for such compounds the positions, relative intensities and line shapes are often useful for chemical-state identification, provided that spectra are available from reliable standard materials.

Multiplet splitting is yet another electronic process whose effect can be seen in XPS spectra. It occurs during emission of a core-level electron from an atom which has net spin due to unpaired electrons either in the valence levels or in

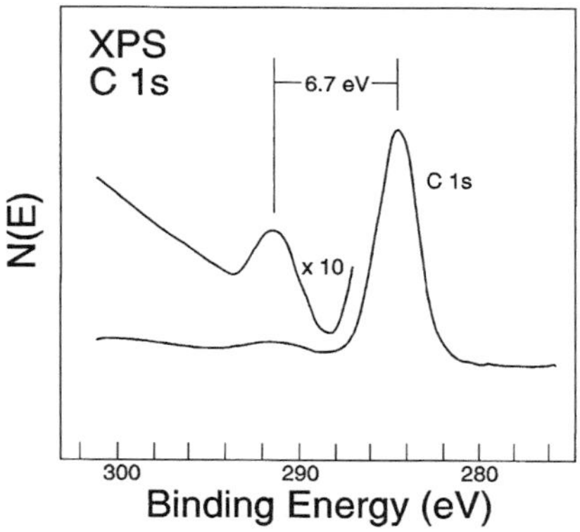

Figure 17. π-bond shake-up satellite for the XPS C 1*s* feature recorded from polystyrene. (From Ref. 18.)

shallow core levels. A variety of possible final states can be created through coupling of the new unpaired electron in an *s*-type orbital with other unpaired electrons [18]. This is illustrated in Fig. 19, where the Cr 3*s* features are shown from Cr metal and Cr_2O_3. Cr_2O_3 exhibits multiplet splitting, but Cr metal does not. The rare earth metals, in which the shallow 4*f* states are highly localized, show complex multiplet structure in their 4*f* photoelectron spectra.

As photoemitted electrons travel through a solid, they may lose discrete amounts of energy by exciting plasmons in the solid or by promoting electrons from filled to unfilled levels (intra- or interband transitions). As expected, the surface and bulk plasmon loss features are larger for metals [53, 54] while those due to electronic transitions are more prominent in nonconducting materials [55]. The features are usually quite small in XPS spectra, but they can sometimes be used for obtaining chemical-state information in situations where the photoelectron features are either ambiguous or difficult to interpret [54, 55]. The nature of the information obtainable from plasmon loss and band transition features is discussed in more detail in the description of the technique of electron loss spectroscopy in Appendix 3.

1.4. Spatially Resolved XPS

In its conventional form XPS is basically an area-averaging analytical technique in which the area analyzed can be varied from ca. 1 to 10 mm^2. Such a relatively large area (cf. 10 nm^2 to 10 μm^2 typically available from AES) has been regarded as an advantage since individual lateral compositional variations were integrated in the analysis, and in addition the signal strength was more than

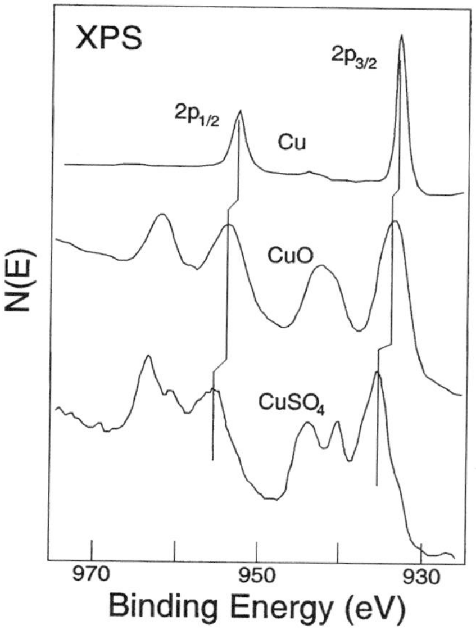

Figure 18. Shake-up features in XPS Cu 2*p* spectra recorded from Cu metal, CuO and $CuSO_4$, (From Ref. 18.)

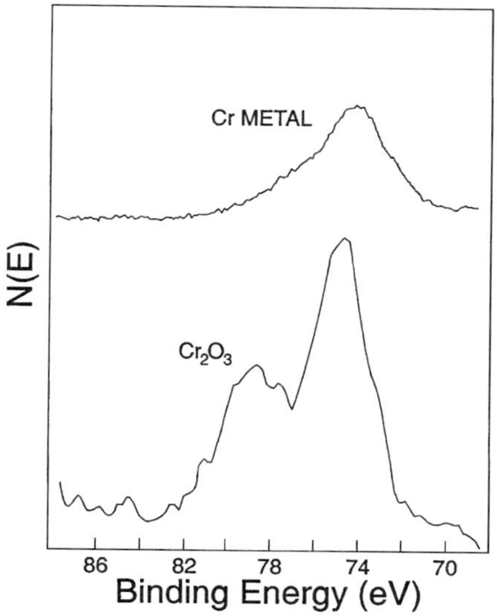

Figure 19. XPS Cr 3*s* feature recorded from Cr metal and Cr_2O_3. The Cr_2O_3 spectrum exhibits multiplet splitting. (From Ref. 18.)

adequate for the acquisition of spectra at high enough energy resolution to enable chemical-state information to be extracted. However, the ability to analyze just those lateral variations in terms of both composition and chemical state has over the past few years become increasingly important. AES can of course provide the compositional information at high spatial resolution, but there are still considerable difficulties in obtaining chemical-state information from the technique, and there is always the worry about surface degradation. In many problem-solving instances the condition of the specimen and the required information are not consistent with spectral integration over large areas. In those instances the relevant data are not being obtained. This provides the impetus to improve the spatial resolution in XPS.

Attempts to achieve spatially resolved XPS have taken many forms over the last 20 or so years and will not be described in detail here. Among the earlier work was that of (a) Cazaux [56] and Hovland [57], who used an Al anticathode behind a very thin specimen, with a rastered electron beam generating x-rays in the Al backing; (b) Keast and Downing [58], who interposed a 0.7 mm-diameter collimating tube between the specimen and the entrance to a CHA analyzer lens, in such a way that only photoelectrons passing through the tube were accepted in the analyzer; (c) Yates and West [59], who placed an aperture of diameter 150 μm between a CHA analyzer lens and the input plane of the analyzer so that the analyzer area was defined by the virtual image on the specimen; (d) Gurker et al. [60], who replaced the exit aperture and multiplier assembly in a CHA with a two-dimensional position-sensitive detector in order to make use of the imaging properties of a CHA to provide line width resolution of about 0.5 mm; and Turner et al. [61], who designed a revolutionary approach to imaging XPS by surrounding the specimen with a very high magnetic field, along whose diverging lines of field strength electrons are transported to form a magnified image at the detector. Spatial resolutions that have been achieved by such methods have varied from 700 to 5 μm.

Currently available methods for the improvement of spatial resolution in XPS fall into two general categories, rastering and imaging.

In an extension of the idea of Yates and West, above, Seah and Smith [62] introduced a scanning system inside the analyzer input lens, between the specimen and the transfer lens, to raster the virtual image of the analysis area. The spatial resolution available depends of course on the size of the aperture at the entrance to the CHA. The instrumental arrangement is shown in Fig. 20, from Ref. 63. It forms the basis for the AXIS system marketed by Kratos Analytical.

Although x-rays themselves cannot be rastered they can be focused to a fine point via a monochromator with an appropriately shaped focusing single crystal. This property is employed in different ways in two current commercial systems, one relatively new to the market.

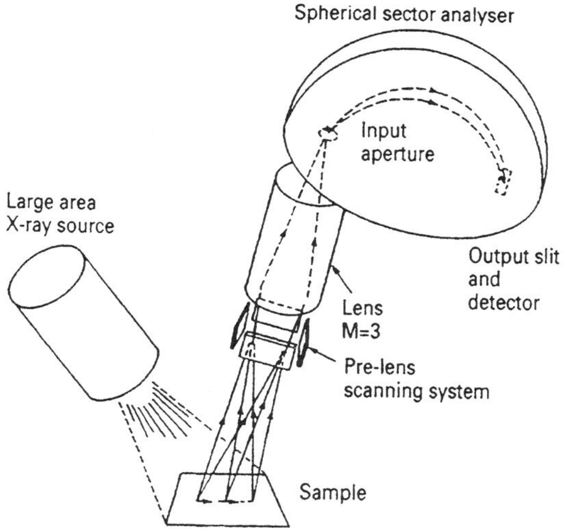

Figure 20. Instrumental arrangement for spatially resolved XPS by virtual image rastering. (From Ref. 62) A prelens scanning system, combined with a small input aperture, allows acquisition of XPS maps with a resolution of ca. 100 μm. (From Ref. 63.)

The Surface Science Instruments (SSI) (part of the Thermo group) m-probe focuses the x-rays to a spot of diameter ca. 100 μm on the specimen surface. Rastering to form a two-dimensional image is not carried out in the source or the analyzer lens, but is achieved by moving the specimen stage with stepper motors.

The newest system aimed at high spatial resolution XPS is the Physical Electronics Quantum 2000. In one particular its design derives from the idea of Cazaux in that rastering is performed by scanning the exciting electron beam in the x-ray source over the anode. The x-rays are then focused in a monochromator onto the sample surface, but there is one essential item of design. As the position of the exciting electron beam on the anode changes, so in a conventional spherical diffracting crystal would the diffraction conditions change, and therefore the selected energy of the photons would change. To circumvent that problem the diffracting crystal is ellipsoidal and not spherical, thus compensating for the movement of the electron beam. The x-ray beam diameter on the specimen surface is a function of the focussing conditions of the exciting electron beam and is claimed to be variable from 10 μm upward. Taking into account the power required for adequate count rate at a reasonable energy resolution (0.9 eV), it seems that a more practicably useful spot size might be 40 μm.

Both systems that employ imaging for improved spatial resolution make use of the dispersion properties of the CHA. The Scienta ESCA-300 design follows the ideas of Gurker et al. [60] by replacing the single detector normally found with a position-sensitive detector parallel to the nondispersive plane at the output of the analyzer. In that plane positional information, i.e., on the specimen surface, is retained. The Scienta instrument also uses a monochromator with a very high power (8 kW) rotating anode, and a large analyzer, to maximize count rate at good energy resolution. The output is an *E*:*X* map—that is, energy along one axis and distance along the other. Spatial resolution of about 25 μm is available.

In the VG Scientific ESCASCOPE the dispersive characteristics of the CHA and the electron optical properties of electron lenses are both used. The system is shown schematically in Fig. 21 from Coxon et al. [64]. Compared to the configuration of the standard CHA, two lens elements have been added, one at the input and the other at the output of the analyzer. Magnified images of the analyzed area are presented to the object plane of the extra input lens (no. 3 in the figure). The latter is so arranged that the distance between the object plane and

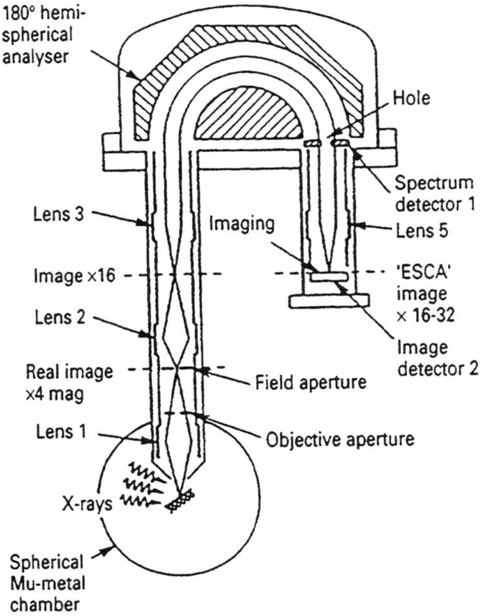

Figure 21. VG ESCASCOPE schematic. Parallel imaging XPS achieves spatial resolution of 1–2 μm. Quasi-Fourier transform optics are located in lenses 3 and 5. X-Y information is converted into angular information in lens 3, and reconverted to X-Y in lens 5. (From Ref. 64.)

the analyzer input is equal to the focal length of the lens, which means that electrons entering the analyzer do so in an approximately parallel beam. Effectively there is a Fourier-transformed image of the analyzed area at the entrance slit of the analyzer, and *x-y* positional information has been converted into angular information, which is conserved, as is energy information. After energy dispersion in the CHA in the usual way, the transformed image appears in the image plane at the output of the analyser. Photoelectrons passing through the exit slit are then collected by the additional output lens (no. 5), which is positioned similarly but inversely to lens no. 3 so that the Fourier transformation is inverted and angular is converted back into positional information. The image is that of the two-dimensional energy-filtered analyzed area. Spatial resolution down to 2 μm has been demonstrated using metallic grids, and tracks of width 5–10 μm on semiconductor surfaces have been resolved and analyzed.

The whole field of improved spatial resolution in XPS is in the throes of rapid development, and it is possible that eventually the capability of the technique in that respect will approach that of AES.

2. AUGER ELECTRON SPECTROSCOPY (AES)

2.1. Introduction and History

Auger electron spectroscopy (AES) has probably been the most widely used surface analytical technique over the last 30 years. In many surface studies AES has been used to monitor surface cleanliness and composition even if the focus of the study has been on the use of other techniques. Until now AES has been used primarily for the qualitative and semiquantitative determination of the composition of the near-surface region, but it is being used increasingly to obtain chemical-state and structural information. There are many similarities between XPS and AES in that both provide compositional and chemical-state information, but of course significant differences also exist. The two techniques provide complementary information, so that by applying them together to the analysis of a surface a more complete picture emerges.

The Auger effect was predicted on theoretical grounds by Rosseland [65] in 1923 and discovered experimentally at about the same time by Auger [66–68] while working with x-rays and cloud chambers. The Auger process in an atom results from the decay of a core-level hole created by the interaction of energetic photons or electrons or ions with an electron in the core level. The decay of this hole takes place through a many-body process in which an electron from a shallower level fills the core-level hole, causing emission of either an Auger electron by transfer of the excess kinetic energy to a third electron (Fig. 1b) or a photon (fluorescence as in Fig. 1c). The kinetic energy, E_{AE}, of the KL_1L_2 Auger electron can be written as

$$E_{AE} = E_K - E_{L_1} - E_{L_2} - \Delta \tag{10}$$

where E_K, E_{L_1} and E_{L_2} are the binding energies of electrons in the K, L_1, and L_2 energy levels and Δ is a complicated term containing both the sample and spectrometer work functions as well as many-body corrections which account for energy shifts during the Auger process and other electronic effects. The Δ term is often small (<10 eV) and varies with chemical state. The probability of decay via an Auger process is greater for light elements than that via fluorescence as shown in Fig. 22 for K-shell electrons [69]. Auger processes occur for all elements except hydrogen and helium which have no core-level electrons. As can be seen from Eq. 10, the Auger KE does not depend on primary beam energy, and the threshold energy for the transition is that required to produce the core-level hole.

The KE distribution of secondary electrons produced by bombarding a polycrystalline Ag metal target with a primary beam of 1000 eV electrons is shown in Fig. 23 [70]. The predominant feature is a large peak, referred to as the elastic peak, at the incident beam energy. Although the shapes of the Auger features do not depend on the KE distribution of the electrons in the incident beam, the shape of the 1000 eV elastic peak is in fact dependent on that distribution. Electrons which have lost small amounts of energy (<3 eV) inelastically to vibrational modes also lie within the elastic peak since the energy resolution normally used in AES is not high enough to resolve the features due to them. Small features can be found at energies just below (950–1000 eV) the elastic peak. As discussed in the section on XPS, these electron energy loss peaks are due to electrons which have lost energy in exciting plasmon or intra-/interband transitions, and they are discussed further in the description of ELS in Appen-

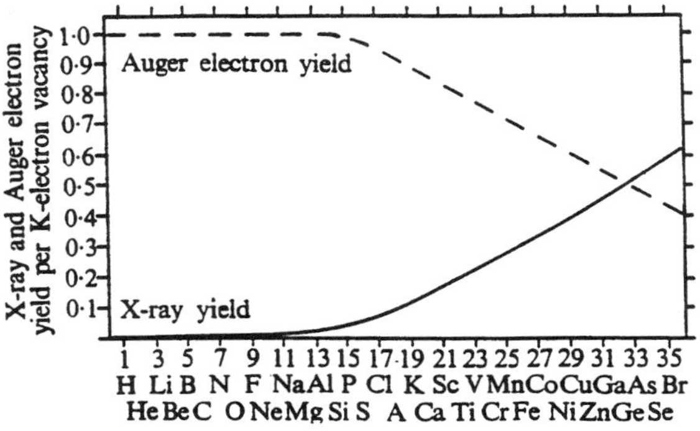

Figure 22. Auger electron emission and X-ray fluorescence yields for K-shell electron vacancies as a function of atomic number. (From Ref. 69.)

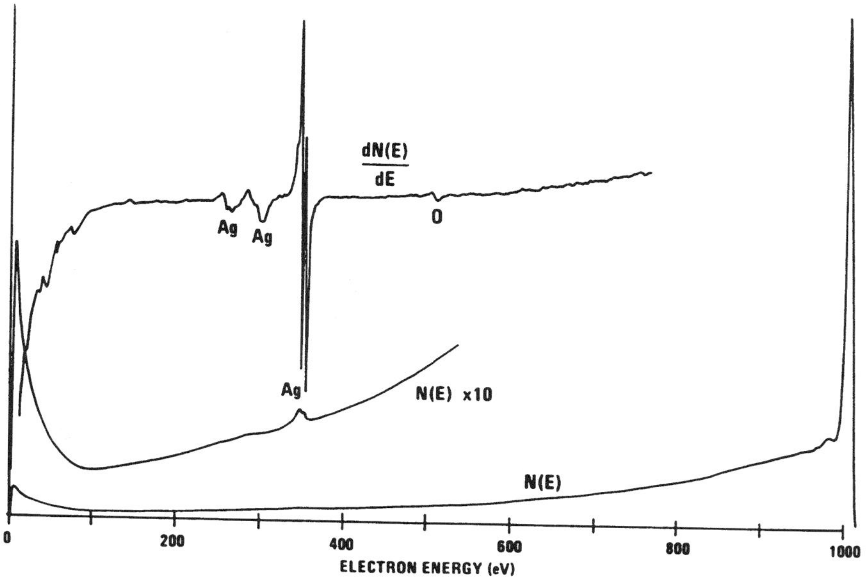

Figure 23. *N(E)* and *dN/dE* electron energy distributions recorded as a result of 1 keV primary electron beam impinging upon an Ag surface. (From Ref. 70.)

dix 3. The broad sloping background ranging from 0 eV to the elastic peak arises from secondary electrons that have been scattered inelastically by a variety of processes. The Auger features on this background are not usually apparent, but after amplification an Ag Auger feature can be observed at ca. 380 eV. Smaller Ag Auger features also appear as does a small peak due to oxygen. Even greater visibility can be achieved by differentiation of the spectrum with respect to energy, as shown in the upper part of Fig. 23. This was first demonstrated by Weber and Peria [71], who used a retarding field analyzer to record differentiated energy distributions. The differentiation process removes much of the slowly varying inelastic background and accentuates the Auger features. All elements except H and He produce Auger features at specific KEs as shown in Fig. 24. [72]. The complexity of an elemental Auger spectrum is determined not only by the number of electron energy levels in the atom, which means that light elements have intrinsically simpler Auger spectra than heavy elements, but also by the respective transition probabilities. This is illustrated by the Auger spectra from Be and Pt [69] shown in Fig 25. Only one prominent Be feature is present in the Be spectrum, whereas a large number of Pt peaks are found in the Pt spectrum.

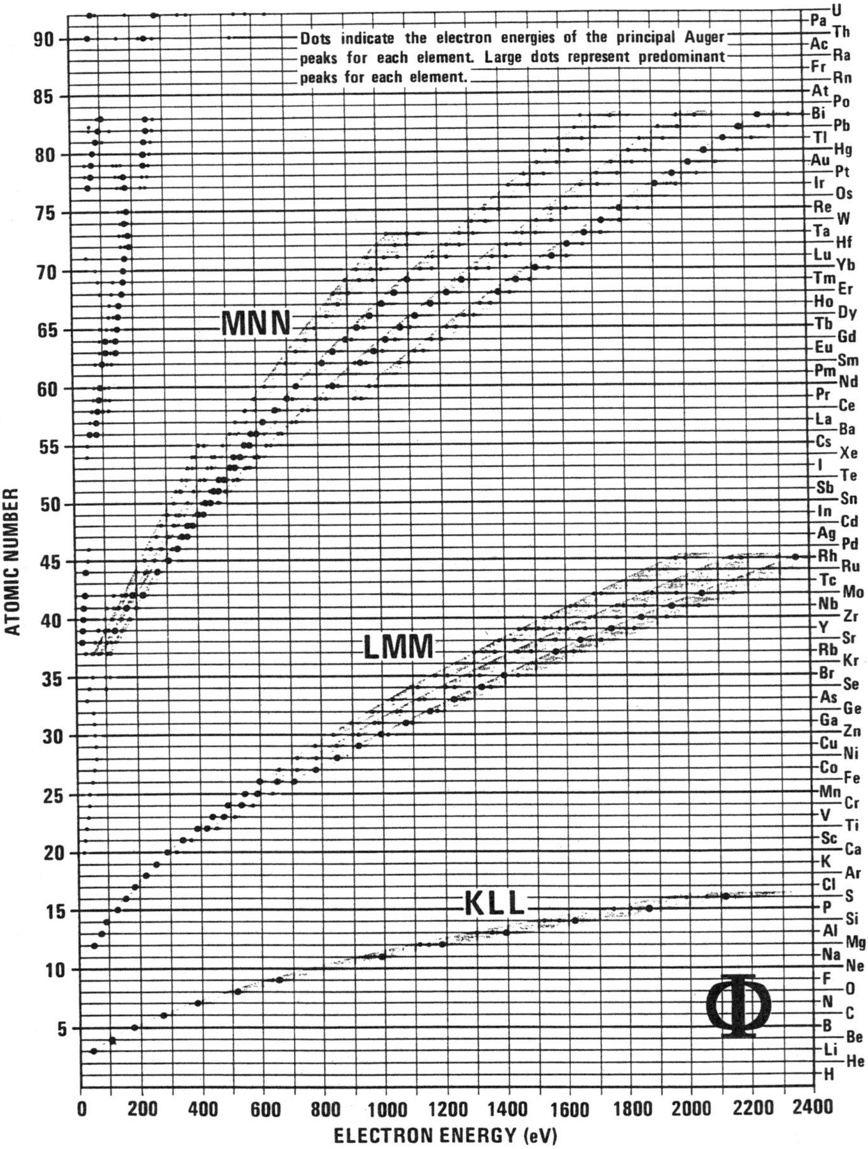

Figure 24. Principal elemental Auger energies. (From Ref. 72.)

2.2. Experimental Equipment and Data Collection

2.2.1. Electron sources

An electron source is used to provide the excitation in conventional AES. Primary beam energies usually range from 3 to 10 keV and beam currents from 2 nA to 10 μA. By focussing the primary beam to sub-μm dimensions, compositional information can be obtained from areas of similar size. Under ideal conditions such information could be obtained from areas less than 10 nm in diameter but 100 nm is more typical. If a focused beam is rastered while the Auger signal from a particular element is monitored, the result is a two-dimensional elemental map, and the technique is called scanning Auger microscopy (SAM). As discussed below, sample damage is a serious problem in AES. Unfortunately, the way in which AES is usually operated results in significant damage. The problem is that most commercially available electron sources have been designed to have optimum performance at a focussed spot size of a few tenths of a millimeter. For the normal mode of operation of a CMA, i.e., with an angular spread of ±6° (Section 1.2.2), the acceptance area is significantly greater than the electron beam spot size on the sample, and unless the beam is being rastered there is little to gain by using the optimum focusing conditions. To minimize sample damage the beam should therefore be defocused to the size of the acceptance area.

2.2.2. Energy analyzers

Several different types of energy analyzer have been used in AES. In the early days retarding field analyzers (RFA) that had been developed for LEED were later modified to perform AES [71, 73]. RFAs are still used, since they are necessary for LEED, and it is convenient to check surface cleanliness with an RFA when a vacuum system does not contain a more sophisticated analyzer. The overriding problems with RFAs are the poor signal-to-noise characteristics and the inherently high capacitance of the large spherical grids, which lead to long signal collection times. Unacceptable surface damage can result.

To date the most widely used analyzer for AES has been the CMA. Both the single-pass (SP) (Fig. 26) [72] and DPCMAs (Fig. 8a) are suitable, but high-resolution spectra cannot be obtained with a SPCMA. The SPCMA is always operated in the nonretarding mode in which the inner cylinder is grounded and the negative deflecting voltage applied to the outer cylinder is ramped. Since the transmission of the CMA is proportional to the energy E of the transmitted electron, which in turn is proportional to the voltage difference between the cylinders, the recorded energy distribution is $EN(E)$, and the differential distribution is $E\,dN(E)/dE$. With the DPCMA both cylinders are ramped while maintaining a constant voltage difference (called the pass en-

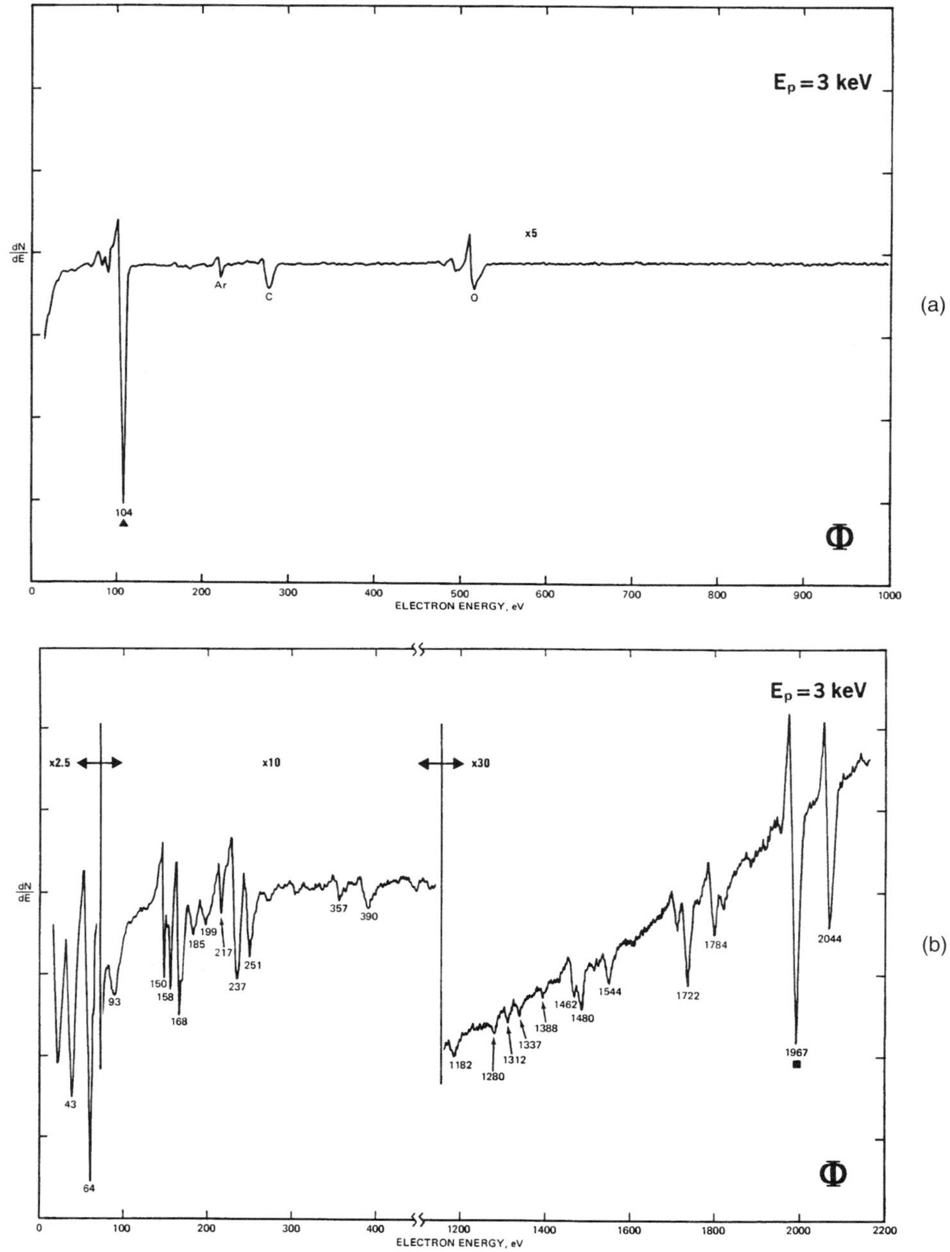

Figure 25. Auger spectra recorded from (a) Be and (b) Pt. (From Ref. 72.)

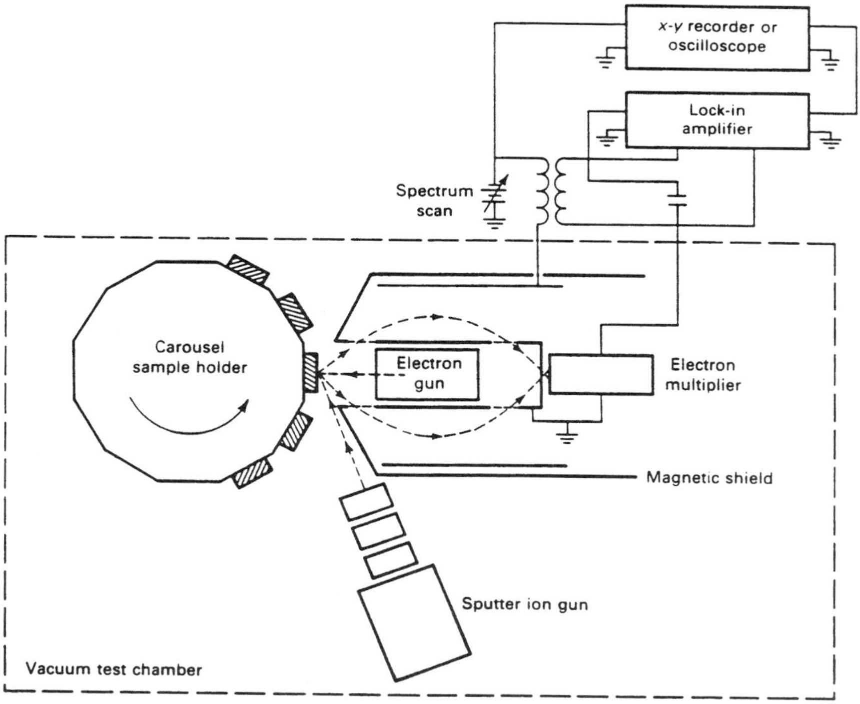

Figure 26. Schematic diagram of an Auger system based on a single-pass cylindrical mirror analyzer (SPCMA). (From Ref. 72.)

ergy) between them, which results in constant energy resolution over the whole kinetic energy range and the collection of *dN(E)/dE* spectra. The operating modes have been discussed mathematically by Gilbert et al. [74]. With sufficiently large primary beam currents, analog signal detection (e.g., phase sensitive) methods can be used to collect data in the differential or direct modes. Burrell et al. [75] have shown how systems designed to record differential spectra may be modified to acquire integral spectra, while Seah et al. [76] have described simple mathematical routines which can be used to convert integral spectra into differential spectra. If small primary beam currents are used, spectral data can be collected in digital form, thus minimizing surface damage. The principal advantage of the CMA for AES is its high transmission, but, as pointed out earlier (Section 1.2.2), it has several disadvantages, of which the principal one is the great sensitivity to precise specimen position. Changes in position of less than a millimeter can result in considerable differences in the measured electron KEs and intensities [77]. Other analyzer types are more tolerant with regard to sample position.

The CHA (Fig. 8b) can also be used for AES, with data collection in the integral mode similar to that in XPS, but primary beam currents must be kept low so that the electron multiplier and other data collection circuitry do not become saturated. With the CHA the electron gun cannot be an integral part of the analyzer as it usually is with the CMA. Since the transmission characteristics and the solid angles of acceptance of the two analyzers are different, Auger spectra recorded by them from the same surface can appear dissimilar. Such effects have to be taken into account when comparing spectra recorded in different laboratories.

The same statements regarding the vacuum system made above for XPS apply to AES, but there are two important differences. One is that contamination from the electron gun filament can be deposited onto the sample surface, particularly if the filament is in direct line of sight to the surface. This can be minimized by outgassing the filament before placing the sample in the analysis position. Another problem is that primary electrons can dissociate background gas molecules with the production of "active" species which will adsorb readily on the surface being analyzed. For the latter reason it may not be possible to attempt treatment of the surface with some gases at elevated pressures while performing AES.

Sample charging can be a serious problem in AES. As with XPS charging results from an imbalance of electrons leaving and entering the solid. In AES the surface of a solid may charge either positively or negatively. If the charging occurs uniformly across the surface, then a simple overall shift in the KE scale will be the result. In particularly bad cases an electrostatic mirror potential can build up at the surface completely preventing any recording of spectra. Charging can also produce large fluctuations in the detected signal, resulting in spectra containing many large noise spikes. There are various methods that have been tried to minimize charging, as discussed by Hofmann [32]. One is to balance the incident beam current with the overall current leaving the sample using a primary beam energy of 3 keV. Another is to change the incident beam angle so that it is closer to grazing incidence (>60° from normal). An angle near grazing incidence tends to enhance secondary electron emission, thus minimizing surface charge buildup. However, such a change in angle will almost certainly alter the spectra due to angular effects, as discussed later. The techniques described in the section on XPS can also be used, e.g., flooding with low-energy electrons.

There are various ways of recording AES data, and the choice is usually governed by the system available. Ideally the choice should be made on the basis of the type of information sought. To obtain the derivative spectrum, phase-sensitive detection is still often used, in which the ramped voltage on the cylinders is modulated with a small sinusoidal voltage and the fundamental frequency of the output signal amplified. Alternatively the intensity of the primary electron beam can be chopped, as described by Mogami and Sekine [78, 79]. Gilbert et al. [74] have developed a digital-derivative-generation method in which the function of the lock-in amplifier used in phase-sensitive detection is replaced by a computer, enabling rapid collection of high-resolution AES data in the *dN/dE* mode. Burrell et al. [75] have described an alternative AES collection circuit for obtaining *EN(E)* spectra with a SPCMA in the nonretarding mode by using an isolation amplifier directly after the electron multiplier, while Roberts and O'Neill [80] have used a voltage-to-frequency converter which converts the output signal of the CMA to optically isolated pulses that are then counted by a computer. Frank and Vasina [81] have used a high-resolution analyzer and traditional pulse-counting techniques to produce *N(E)* spectra, but the long counting times (1 h or more) require either strict control of, or compensation for, fluctuations in the primary beam current, which is not easy. Woodward et al. [82] have developed a regulator which maintains a constant primary beam current at levels as low as 10 nA. Nearly all AES systems now use computers for data collection, storage and processing. When analog data collection is used, *RC* smoothing is performed electronically before digitization of the data. Digital pulse-counting circuits provide data which can be smoothed by the method of Savitsky and Golay [27], as in XPS.

2.3. Spectral Features and Interpretation

Auger spectra can be found displayed in both derivative and integral forms, with a trend toward the latter. Unlike XPS spectra, only Auger features and a background are present in Auger spectra. The Auger energies for most elements have been tabulated in standard references [72, 83] and are shown graphically in Fig. 24. An Auger spectrum can consist of a very large number of peaks particularly if there are either many elements or some heavy elements present at the surface of the sample. The first step in identification is to assign the major peaks to elements on the basis of their KEs. This is usually straightforward but can be complicated if peaks overlap or if there are large peak shifts due to differences in chemical state. After the major peaks have been assigned tentatively to various elements, the assignments should be confirmed by checking that the smaller peaks associated with each element are also present using the elemental standard reference spectra [72, 83]. Most of the remaining features will then be due to elements present in small amounts, and these should be identified next. There may be some small peaks present arising from interatomic, or "cross," Auger transitions involving the energy levels of two different but adjacent atoms. These can be identified using either compound reference spectra [26] or calculation of their approximate kinetic energies using Eq. 10. The relative intensities of AES peaks vary considerably with element and with the type of Auger transition as shown in Fig. 27, and with primary beam energy as shown in Fig. 28 [83]. The signal strengths of the very small peaks often

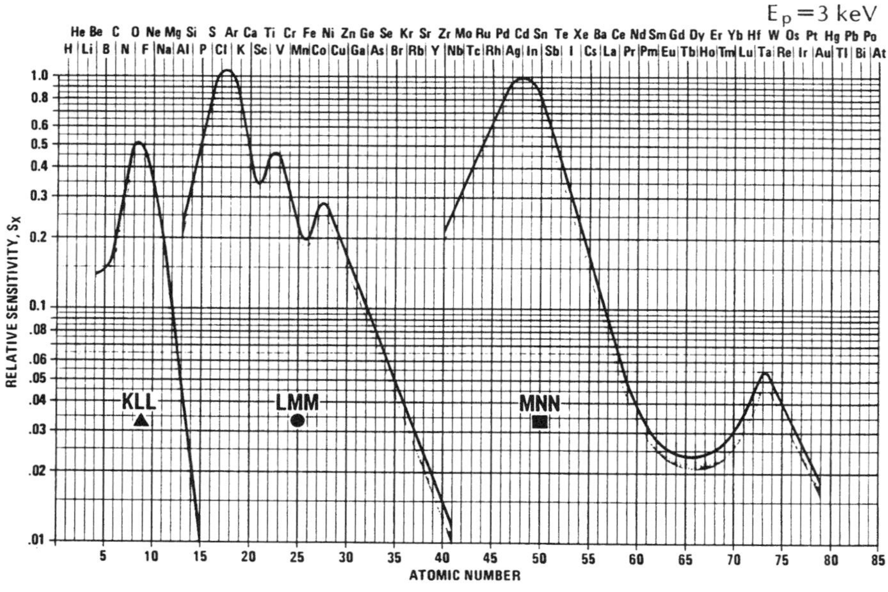

Figure 27. Relative Auger sensitivities of the elements using a 3 keV primary electron beam. (From Ref. 72.)

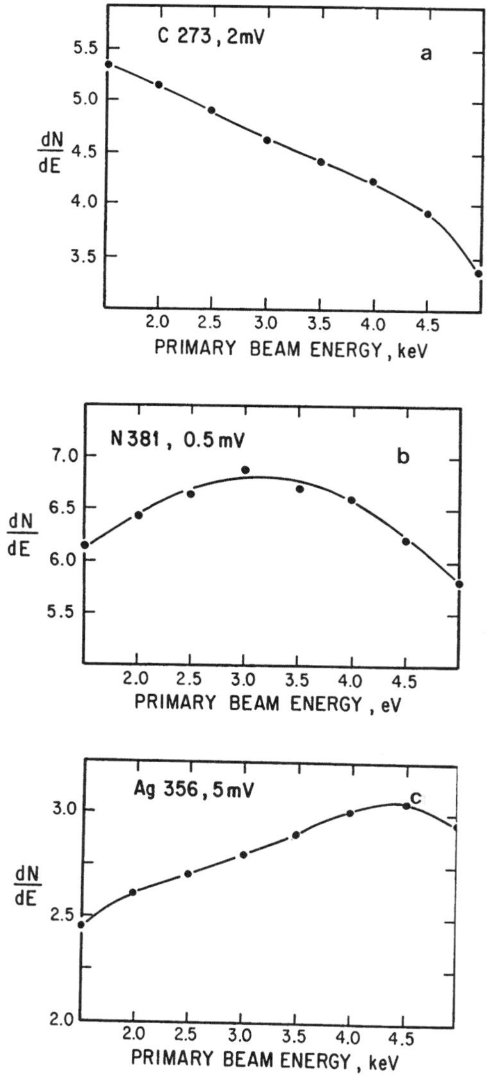

Figure 28. Variations of the Auger peak heights of selected elements with primary beam energy. (From Ref. 83.)

present can be enhanced by scanning slowly across a peak at high sensitivity (amplification), from which it can be determined if the peaks arise from the principal elements present, from interatomic transitions, or from other elements including contamination present in small quantities. Since AES is used frequently to establish the cleanliness or otherwise of a surface, the very small peaks must often be examined carefully. Operational settings need to be chosen to extract such peaks from the background. In some cases contaminant peaks overlap the Auger peaks of the materials being analyzed, e.g., S in Zr [84] and Si in Pt [85], and then it is particularly useful to use several complementary techniques to assess surface cleanliness.

Auger peak shapes and positions are often strongly influenced by chemical state. The Auger line shapes determined from either *N(E)* or *dN(E)/dE* spectra can provide sensitive probes of the local chemical and electronic environments of elements. Since the valence electronic structure strongly reflects the chemical state and local bonding environment, the Auger lines which generally contain the most chemical information are those which involve final electronic states containing one or two holes in the valence levels, i.e., core-core valence (CCV) or core-valence-valence (CVV) transitions. Houston and Rye [86] and Jennison [87] have shown that local electronic densities of states can be determined by measuring the relative Auger intensities of CVV transitions from different elements bonded to each other. The bonding structure of an adsorbed species also can be studied by comparing the Auger line shapes of the adsorbed species with those obtained from gas-phase studies [87, 88]. The differences are due to correlated behavior between the two final-state, valence-level, holes which cause them both to become localized in a particular adsorbate orbital. This phenomenon creates a functional group dependence of the line shape which can be observed [86, 89] and which allows determination of the electronic structure about an atomic site. The shape of the Auger line depends on both the energy levels involved and the transition probability density. Efforts have been made to calculate Auger line shapes from quantum mechanical principles [90–93], with reasonable progress despite the inherent difficulties.

Chemical state dependence is illustrated by the Auger spectra in Figs. 29 to 31. Carbon KLL peak shapes for CO adsorbed on W(112), for W_2C, for graphite and for diamond are shown in Fig. 29 [94]. The line shapes are significantly different, and for diamond the peak minimum is lower in KE than that of graphite by ca. 12 eV. The Auger line shape can often be used to extract chemical-state information, but the procedure becomes complex if two or more chemical states of the element are present because the Auger line shape is complex. Nevertheless, changes in the Auger peak shape are indicative of changes in chemical state, and a useful correlation can often be made between sample treatment and the extent of alteration of chemical state based on such spectral differences. Auger spectra recorded from the native oxide layer on Si(111) (shown in Fig. 30) [95]

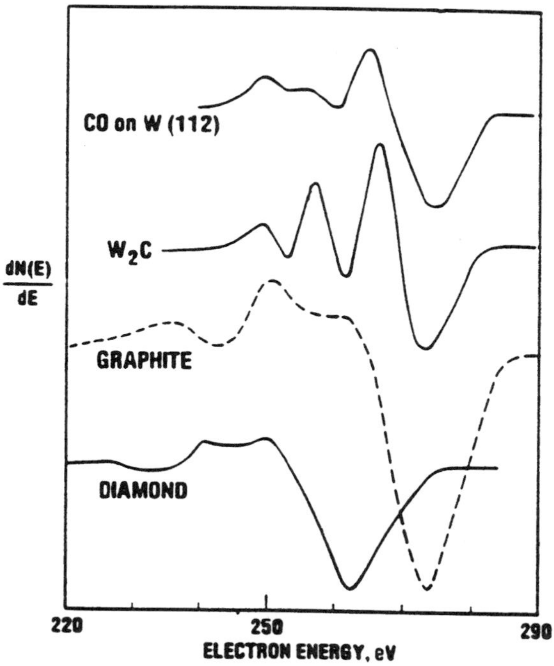

Figure 29. Carbon Auger peak shapes recorded from different chemical forms of C including CO adsorbed on W(112), W_2C, graphite and diamond. (From Ref. 94.)

illustrate an important point. Due to the thinness of the oxide layer, the spectra indicate that Si is present as both SiO_2 and elemental Si. As seen in Fig. 30a, the features due to SiO_2 lie below 86 eV with the predominant feature lying at a KE of ca. 76 eV. That due to elemental Si lies at 92 eV. In this case the two chemical states are so widely separated in energy, 16 eV, that they can be examined independently of each other. Their separation is in fact much larger than the corresponding 4.2 eV separation of the XPS Si 2*p* features. The point is that it is not uncommon for AES to provide better chemical-state information than XPS, a fact that should always be considered. The best approach is to record AES and XPS data at the same time and then carry out a chemical-state analysis using both sets of data. To obtain chemical-state information with AES alone, it is necessary to record spectral features at high energy resolution [74]. It is better to use x-ray-excited Auger spectra, as in XPS, where of course both Auger and photoelectron features appear in the same spectrum at high resolution. The other Auger spectra shown in Fig. 30 are discussed in conjunction with angle-resolved AES.

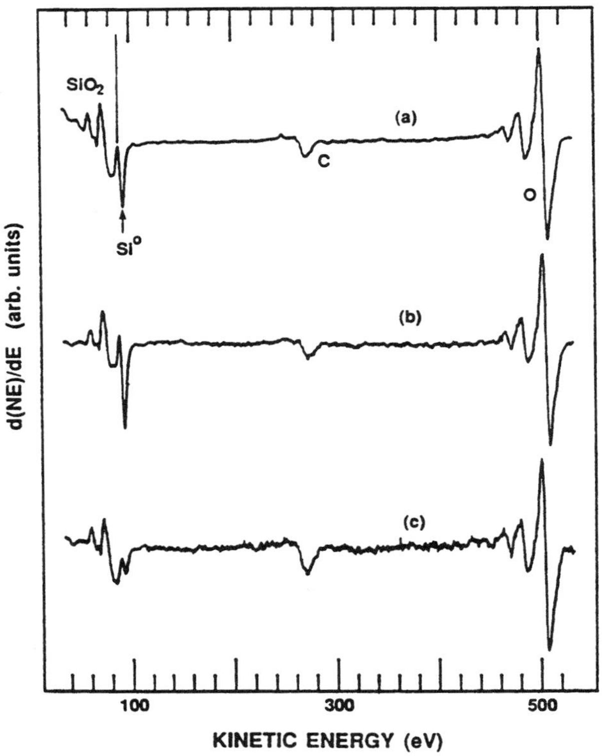

Figure 30. AES spectra recorded from the native oxide layer on an Si(111) surface using the internal coaxial CMA electron gun. (a) Sample normal to the CMA axis, electrons collected in a cone about 42° off sample normal. (b) and (c) Sample rotated 45° off CMA normal. Spectrum (b) is less surface specific since electrons were collected in the direction normal to the sample. Spectrum (c) is more surface specific because electrons were collected at 70° off the sample normal. (From Ref. 95.)

Chemical-state changes during the oxidation of alloys such as Ni/Cr (80:20 at %) can also be studied using AES [96], as illustrated by the spectra shown in Fig. 31, in which low-energy Auger features involving valence levels are shown for clean and oxidized Ni, Cr, Ni/Cr surfaces. The corresponding peak assignments are presented in Table 2.1. The readily observable differences in peak position of the various species allow chemical-state changes to be determined even in a complex system such as this. The peaks lie at KEs in a range from 25 to 65 eV, which is near the minimum in the dependence of AL on KE (Fig. 5) of about 0.4 nm, which means that ca. 78% of the Auger electrons orig-

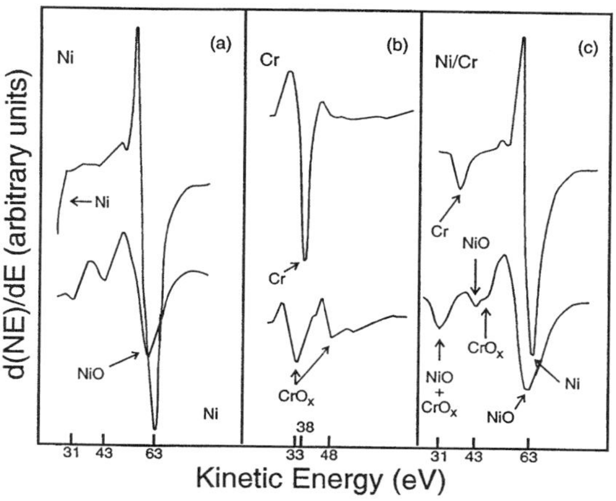

Figure 31. Low-energy Auger spectra recorded from (a) polycrystalline Ni, (b) polycrystalline Cr and (c) an Ni/Cr(110) alloy surface before and after exposure to 200 L of O_2 at room temperature. (From Ref. 96.)

inate from within about 1.2 nm of the outer surface. This is as highly surface specific as AES can achieve using a conventional geometric configuration. Chemical-state information could also have been obtained using the Cr and Ni $2p_{3/2}$ features with BEs of 574.0 and 852.5 eV, corresponding to ALs of 2.0 and 2.5 nm. Thus, if Auger features at low KE are used, XPS samples more deeply than AES in this example. Due to the complex nature of the depth distribution functions for AES and XPS, quantification of the compositional depth profile is

Table 2.1. Cr/Ni Low-Energy Auger Peak Assignments

Kinetic energy (eV)	Species
24	Ni^o
31	NiO
33	CrO_x
38	Cr^o
43	NiO
48	CrO_x
53	CrO_x
62, 63	NiO, Ni^o

Source: Ref. 96.

possible only in a limited number of cases. Nevertheless, qualitative or semi-quantitative differences between the AES and XPS spectra, corresponding to different sampling depths, can be observed and used, demonstrating one of the benefits of using a multitechnique approach in surface science.

Hydrogen cannot give rise to an Auger feature because it has no core-level electrons. Nevertheless, its presence may be detected in certain cases by its influence on the Auger peak shapes of elements bonded to it. For example, Madden [97] has found that hydrogen adsorption influences the line shapes of the Coster-Kronig transitions of amorphous silicon, while Malinowski [98, 99] has observed substantial changes in the MVV lines of Sc and Ti induced by adsorption of hydrogen. The presence of hydrogen at a surface may often be responsible for the chemical behavior of that surface. If the hydrogen concentration is large, then there is a possibility that its effects can be observed through AES line shape analysis. If the hydrogen concentration is less than a fraction of a monolayer, it may still be important chemically but not observable even indirectly by AES.

2.4. Associated Methodologies

2.4.1. Depth profiling with AES

The sampling depth in AES is typically less than 4 nm and is dependent on the KE of the Auger electron and the angles of incidence and collection. It is possible to depth profile several hundred nanometers into a surface by combining AES with inert-gas-ion sputtering (or etching). Sputtering and analysis may be performed either simultaneously or in a stepwise manner. The sputtering process itself is complex because it comprises many individual processes that can alter the composition of the sputtered region, making quantification difficult (see Chapters 7 and 8). Nevertheless, depth profiling with AES and ion sputtering currently provides the most convenient means of obtaining subsurface compositional information.

In order to understand depth profiling [100], it is necessary to consider the ion-sputtering process illustrated in Fig. 32. As a result of the various individual processes, a damage layer is produced in which the composition, the chemical states of the species present, and the morphology, will have been altered. The depth of this damage layer depends on several factors, including the type, mass, charge state, and KE of the primary ion and the incident angle. Characteristics of the damage layer can be calculated using the TRIM computer code developed by Biersack and Haggmark [101]. The layer is usually quite thick compared to the average sampling depths of the various surface analytical techniques, which means that each of the techniques analyzes mostly ion-affected material. The results of Monte Carlo calculations by Ishitani and Shimizu [102] for 4 keV Ar^+ incident on a target of randomly dis-

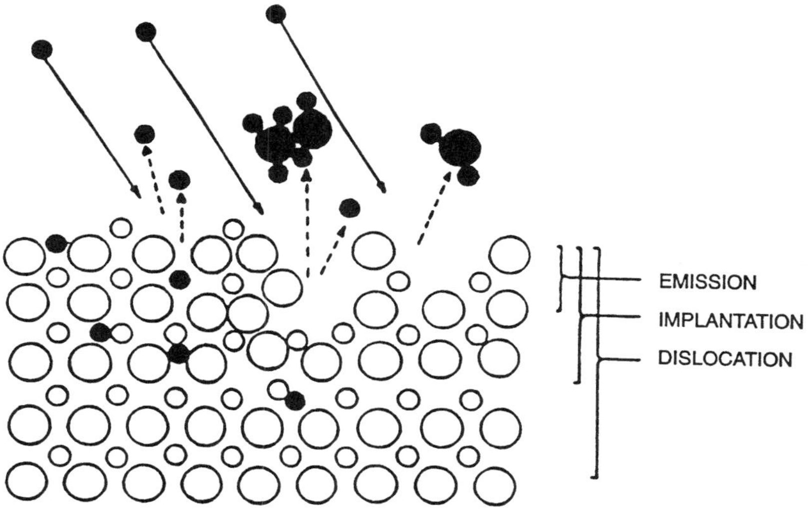

Figure 32. Schematic diagram of the ion-sputtering process. (From Ref. 100.)

tributed Cu atoms, which provide information about the damage layer, are shown in Fig. 33.

The ion-sputtering process in pure materials is described reasonably well by the linear cascade model developed by Sigmund [103]. The model predicts quantitatively sputter yields and energy distributions as functions of incident and target atom masses and of incident angle. The sputtering rates also depend on the surface binding energy. A comparison of the calculated and experimental sputtering yields for 400 eV Xe^+ for various elements is shown in Fig. 34. The agreement is excellent. Unfortunately, the depth profiling of a pure material is of little interest, and the ion-sputtering process is much more complex for multicomponent solids. Artefacts are introduced because the different constituent elements are sputtered away at different rates, causing the composition and chemical-state distribution to be altered progressively with depth (i.e., with ion dose). Taglauer et al. [104] have demonstrated that the chemical bond strength also has a strong influence on sputter yield. The difficulty is that a general correlation cannot be developed for the sputtering yield of an element because the specific nature of the chemical bonding is so important that it must be taken into account directly.

The ion-sputtering process is statistical in nature in that the second and subsequent atom layers may be depleted before the first layer is removed entirely. This fact leads to a progressive degradation in depth resolution with sputtering

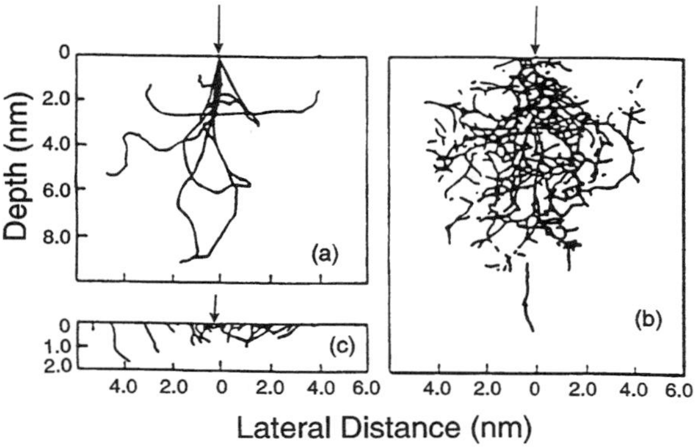

Figure 33. Results of Monte Carlo calculations for 4 keV Ar^+ incident on a target of randomly distributed Cu atoms. (a) Ten incident particle trajectories, (b) the resulting recoil distribution, and (c) the sputtering events resulting from 50 incident ions. (From Ref. 102.)

time, as shown in Fig. 35. The apparent interface thickness is greater due to loss in depth resolution when the interface lies more deeply beneath the surface. Sputtering yields also depend upon crystallographic orientation. Polycrystalline materials have randomly oriented grains, and the observed profile broadening follows the equation

$$\Delta z = \phi z^n \tag{11}$$

where z is the depth, Δz the depth resolution, ϕ depends upon roughness, sputter yield and incident angle, and n varies from 0 to 1, depending on structural factors. Another parameter is the primary-ion KE, which also affects depth resolution according to the equation

$$\Delta z \propto (KE)^{1/2} \tag{12}$$

Hofmann [105, 106] has reviewed the principles of sputter depth profiling, distortional effects in depth profiling, conditions and methods for optimized sputter profiling, and deconvolution of sputtering profiles. He claims that the method of sample rotation during sputtering developed by Zalar [107, 108] is the most effective for improving depth resolution (see also Chapter 7). This claim is illustrated by the profiling data given in Fig. 36. With sample rotation the depth resolution remains almost constant with depth, whereas without rotation it becomes progressively worse.

In depth profiling with AES, either complete Auger spectra are recorded intermittently during ion sputtering, or selected peak heights are monitored continuously as a function of sputtering time. As described in Chapter 5, the peak

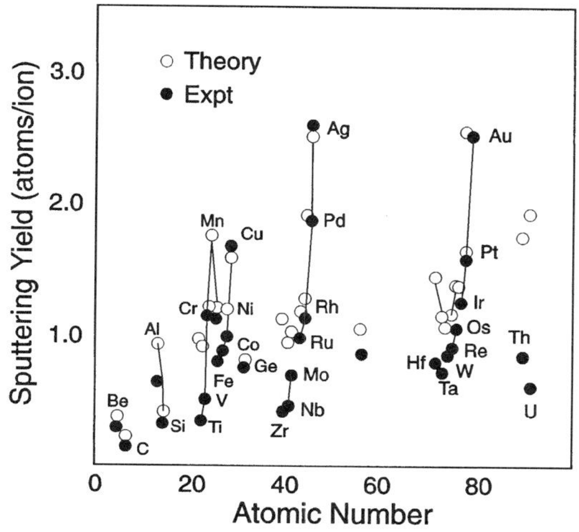

Figure 34. Comparison of calculated and experimental sputtering yields for 400 eV Xe^+ impinging on various elements. (From Ref. 103.)

heights are then converted into composition. If the sputter rate can be predetermined in some such manner as the use of standards or of marker experiments, then a plot of composition versus depth can be produced similar to the one shown in Fig. 37 [109]. This depth profile was recorded from an electrodeposited Cr film, with formic acid added to the plating bath. Such films have very high hardnesses after annealing at elevated temperatures. After annealing the film was implanted with 90 keV N^+ to increase further the hardness. In the depth profile the anticorrelation of the individual N and C profiles indicates that implanted N displaces C, suggesting that they are in competition to bond with the Cr. Such depth profiles suffer of course from all the problems described above as well as others discussed by Lam [110]. Although a reasonable, semiquantitative, understanding exists for most of the individual processes involved, Lam suggests that models that consider all the processes collectively along with their potential interactions are in the early stages of development and that a fundamental understanding of the synergistic effects of these processes is necessary to determine how artefacts due to ion sputtering alter the experimental depth profiles.

In spite of the many processes which can induce artefacts into depth profiling, the technique can often provide an excellent picture of the subsurface region. Careful selection of experimental parameters can minimize or eliminate artefacts. Generally, it is desirable to use high ion-sputter rates, low temperatures, and sample rotation. The influence of incident current density is shown in Fig. 38 [110]. Unfortunately, some problems such as that of preferential sput-

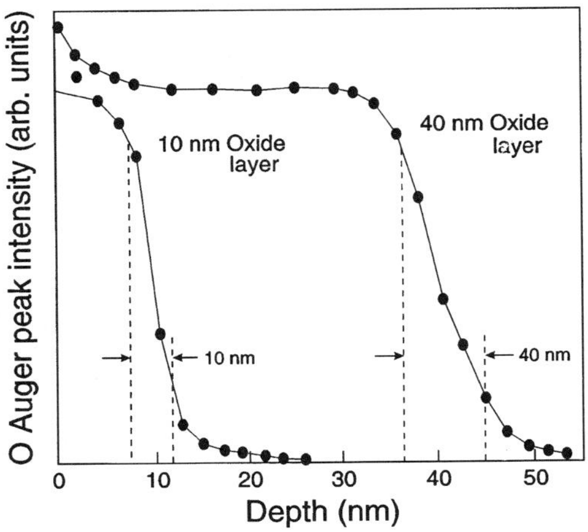

Figure 35. Depth profiles obtained by ion-sputtering two niobium oxide films 10 and 40 nm thick, respectively, on Nb metal. The interface resolution is worse for the 40 nm layer due to the statistical nature of the sputtering process. (From Ref. 100.)

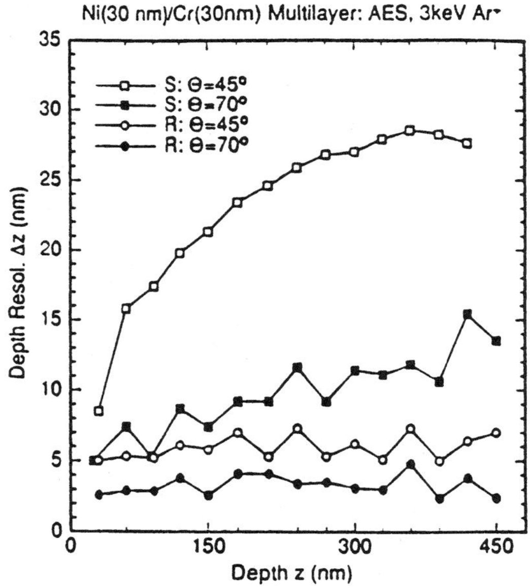

Figure 36. Depth resolution as a function of the sputtered depth for Ni (30 nm)/Cr (30 nm) multilayer samples using AES and 3 keV Ar^+, with sample rotation R (1 rpm) and without rotation S for ion incident angles of 45° and 70°. (From Ref. 108.)

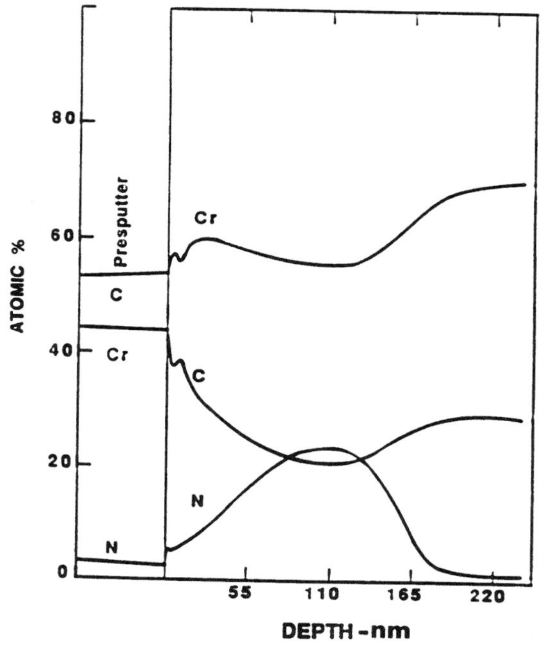

Figure 37. AES depth profile recorded from an amorphous, bright, chromium deposit annealed at 100°C and implanted with a dose of 90 keV, 4×10^{17} N^+ cm^{-2}. (From Ref. 109.)

tering cannot be avoided, and the magnitude of this effect is strongly dependent on the nature of the sample.

2.4.2. Angle-resolved AES (ARAES)

There are many parameters in AES which can be varied and two of the most important are the angles of incidence and collection. Systematic variation of them produces ARAES data. There are two reasons why angle-resolved measurements are made. First, a nondestructive alternative to sputter depth profiling of amorphous and polycrystalline samples is often needed. By variation of the incidence and collection ("take-off") angles, the signals from species at greater depths are conversely reduced or enhanced [95]. Second, by accumulating sets of data in which the Auger signal is measured as a function of collection angle for a succession of selected incident angles, the structures of single-crystal surfaces can be determined [111–116].

In ARAES (and also ARXPS) from a homogenous solid the measured signal is strongest along the surface normal and decreases cosinusoidally as the collection angle moves toward grazing emergence. An ARAES technique based

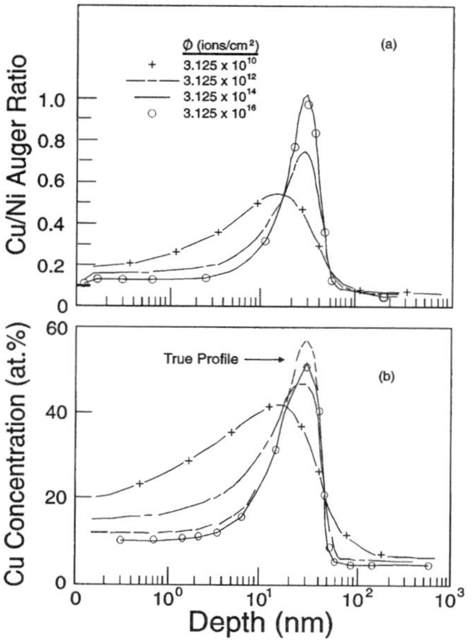

Figure 38. Calculated (a) Cu/Ni Auger ratios and (b) the corresponding Cu depth profiles obtained for various sputter ion fluxes. The true initial profile is indicated by the solid curve. (From Ref. 110.)

on a DPCMA has been developed by Hoflund et al. [95], in which both incident and collection angles can be varied independently from normal to 10° off the surface. Such a variation is generally difficult to accomplish in standard Auger systems because they are not designed for it. The geometrical configuration is shown in Fig. 39b. The DPCMA contains a movable aperture, developed by Knapp et al. [117], for selection of the collection angle. In addition, sample angle with respect to the analyzer can be varied, while two incident angles can be selected by using either the coaxial electron source in the DPCMA or an off-axis electron source. Since the signal strengths in AES are usually large, the decrease at grazing incident and collection angles does not cause problems.

The contribution from each atom layer beneath an Ag(110) surface to the total AES Ag MNN signal has been calculated for both bulk-sensitive and surface-sensitive geometric configurations [118], based on the lattice spacings, by Kuk and Feldman [119]. The calculations are based on the standard form for the probability of an electron traveling a given distance in a solid without scattering.

$$P(x) = ke^{-x/\lambda} \tag{13}$$

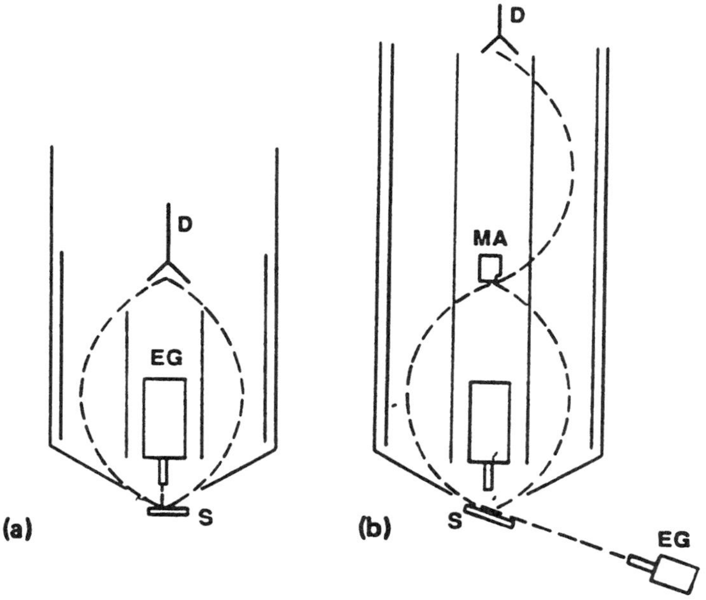

Figure 39. (a) The most common form of operation of AES using an SPCMA is to place the sample (S) normal to the CMA axis. A coaxial electron gun (EG) provides the primary electrons, and electrons emitted in a cone about 42° from the normal are collected by the detector (D). This arrangement yields AES spectra for which the bulk sensitivity of the technique is close to maximum. In the ARAES experimental arrangement shown in (b), an off-axis electron gun (EG) is used. The sample is tilted 52° from the CMA normal, and the primary electrons impinge upon the sample surface at an incident angle of 10°. The emission angle of the detected electrons can be varied from 0° to 80° from the sample normal by rotating the internal movable aperture (MA), thereby varying the surface specificity from low to high, respectively. (From Ref. 95.)

where x is the distance traveled, λ is the attenuation length, k is a normalization constant, and P is the probability. For this calculation it was assumed that the attenuation of both the electrons in the primary beam and the Auger electrons followed Eq. 13. The former assumption is not correct since primary electrons can suffer one or more inelastic collisions and still have enough energy to produce a core hole, but that does not introduce much error because the primary electrons are of high KE (3 keV) and have long attenuation lengths compared to those of the Auger electrons. The bulk-sensitive data were calculated using a normal incidence angle and a collection angle 42° off normal. The surface-sensitive data were calculated using incidence and collection angles both 70° off normal. Diffraction effects were neglected. The results are shown in Fig. 40. For the bulk-sensitive case, less than 8% of the signal originates from the outermost atomic layer and almost 1% from the 30th layer. Thus the distribution is low and broad with about 52% of the signal originating from the outermost nine layers. The atoms in the second layer contribute slightly more to the signal than do those in the outermost layer due to the outward expansion of the outermost layer. The more surface-sensitive distribution is quite different (note the change in scale on the *y*-axis). In that case about 46% of the signal originates from the outermost three atomic layers.

The ARAES data shown in Figs. 30 and 41 illustrate how the surface specificity of AES varies with incident and collection angles. Figure 30a is an Auger spectrum taken from the native oxide layer formed on an Si(111) surface, using

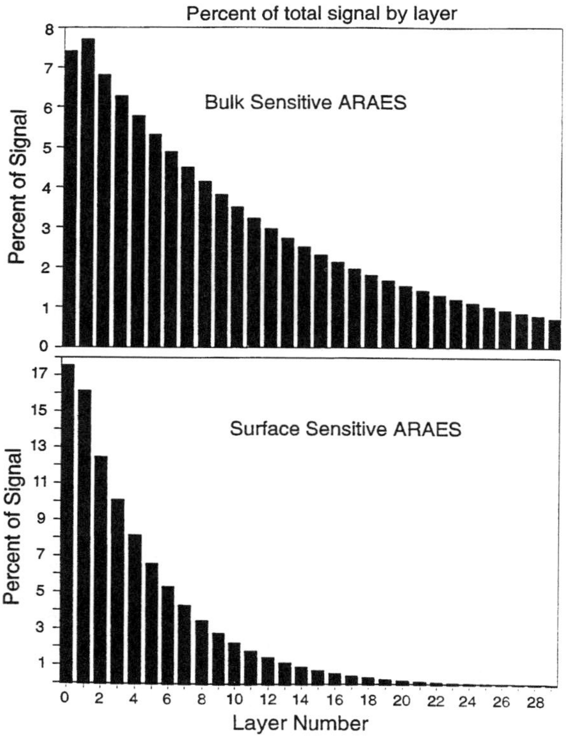

Figure 40. Calculated contributions from each layer of Ag atoms to the total Ag MNN Auger signal for bulk-sensitive and surface-sensitive geometrical configurations. (From Ref. 118.)

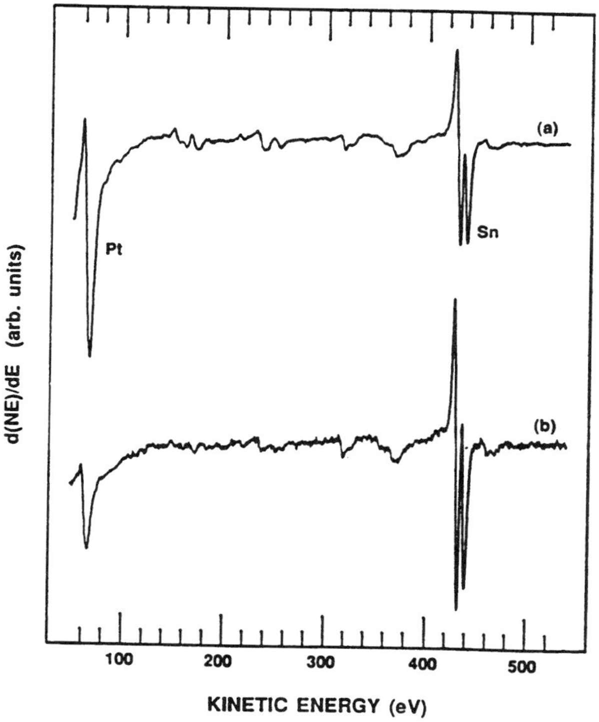

Figure 41. AES spectra recorded from a clean, sputtered Pt_3Sn surface using the experimental configuration shown in Figure 39b with a 3 keV primary electron beam at a glancing angle of 10°. An emission angle almost normal to the sample was used to collect spectrum (a), and one 80° from the sample normal for spectrum (b) (more surface specific). (From Ref. 95.)

the conventional geometry shown in Fig. 39a. Figure 30b was recorded after rotating the sample 45° away from the DPCMA normal and then collecting Auger electrons emitted along the sample normal, since the acceptance angle into the analyzer is 42° (Section 1.2.2.). Such an analyzing condition should result in a more bulk-sensitive spectrum, and this is confirmed by the observations that the Si_{ox}/Si^o, O/Si^o, and C/Si^o peak-height ratios were all reduced. The spectrum in Fig. 30c, on the other hand, was recorded using the same incident angle but with collection of Auger electrons at 70° from the normal, and should correspond to a a more surface-sensitive configuration. This was also seen from the observations since the Si_{ox}/Si^o peak-height ratio was greatly increased and the C peak became larger than the Si^o peak. Further enhancement of surface specificity could be achieved by using incident and collection angles even closer to grazing. The spectra in Fig. 41 were recorded from a sputter-cleaned polycrystalline alloy surface [95]. For both the 3 keV primary electron beam was incident at a grazing angle of 80° from the surface normal. Spectrum (a) was collected with an emission angle nearly normal to the surface, and (b) with an emission angle 80° from the sample normal (i.e., more surface specific). The Sn/Pt peak-height ratio was increased by a factor greater than 6 by using the more surface-specific emission angle. The results indicate that the Pt_3Sn surface consists of a tin-rich layer over the underlying alloy surface, consistent with the results of previous studies [120, 121].

Some of the problems encountered in the quantification of AES data are highlighted by ARAES observations. As for XPS (Section 1.3.1.) the assumption of a homogeneous distribution of species with depth when attempting to quantify data is not usually a good one. The example already used to illustrate the riskiness of the assumption, for XPS data (Fig. 10) [38] can be used again here for AES. AES spectra recorded from a Cs-promoted Ag/α-Al_2O_3 ethylene epoxidation catalyst, at the same time, and from the same surfaces, as were the XPS spectra and the SEM micrographs in Fig. 10, are shown in Fig. 42. Although the total Ag content did not vary, there were large changes in the relative magnitudes of the Ag, O and Al peaks. It is still very difficult to take these

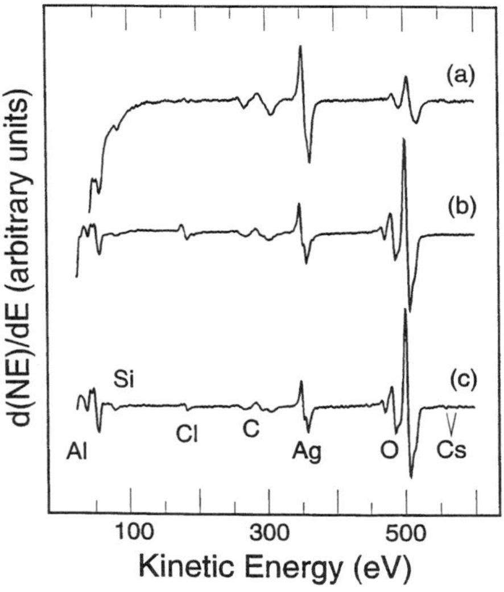

Figure 42. Auger spectra recorded from (a) fresh, (b) active and (c) aged Ag/α-Al_2O_3 ethylene epoxidation catalysts. (From Ref. 38.)

highly inhomogeneous distributions into account quantitatively even though in the foregoing example they are obvious in the SEM micrographs. Accurate quantification is seldom possible on real, complex samples. Another problem in AES quantification, revealed by ARAES, is that Auger spectra taken from the same inhomogenous surface in different systems with different analyzers will not be comparable because the angles of incidence and collection will also be different. Relative elemental peak heights would not be the same, nor therefore would the surface composition if derived on the assumption of a homogenous distribution.

Hubbard et al. have developed an angle-resolved method which they refer to as angular distribution Auger microscopy (ADAM) for structure determination [111–115]. A schematic of the experimental configuration is shown in Fig. 43. A grazing incidence electron source is used, and the Auger electrons can be collected at any emission angle via a small hemispherical analyzer capable of being positioned almost anywhere in the half-space above the sample. An example of a contour map of the 65 eV Auger emission from a Pt(111) surface is shown in Fig. 44. The center of the map corresponds to emission normal to the surface, and the edge to emission 70° from normal. From distributions such as these, the surface structure can be determined, as reviewed by Frank [116]. The technique is useful only for ordered surfaces including single crystals, layered crystals, and monolayer and bilayer structures.

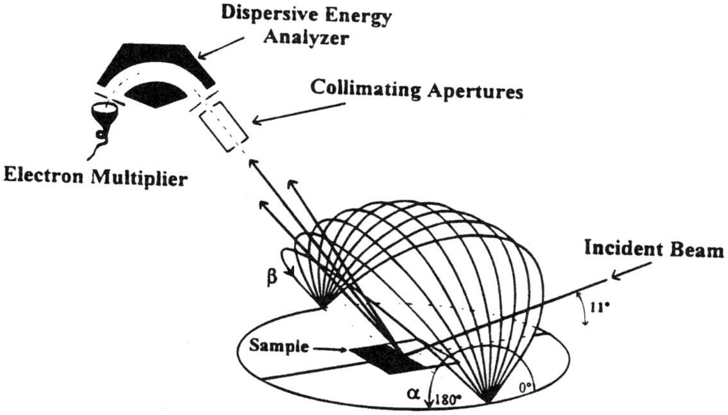

Figure 43. Angular mapping system using a tracked hemispherical analyzer. Auger electrons emitted from the surface with the selected trajectory and kinetic energy pass through angle-resolving collimating apertures and an energy-resolving dispersive analyzer before being amplified and counted. (From Ref. 116.)

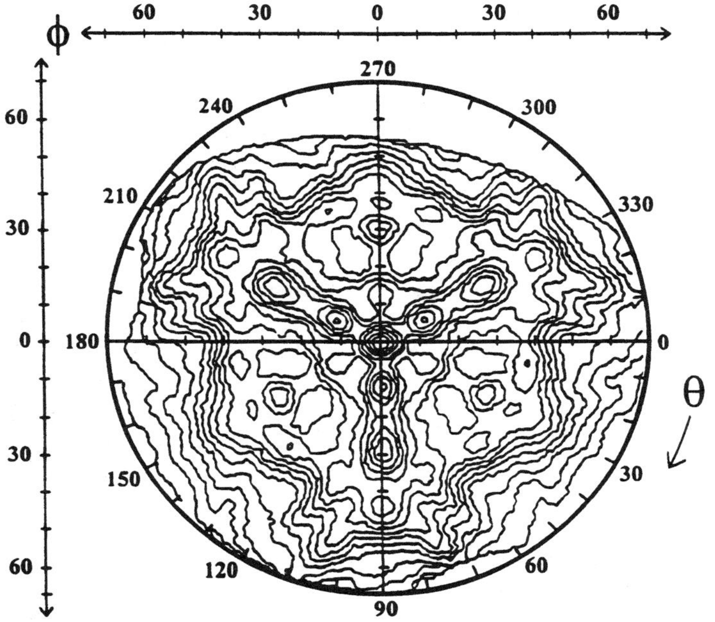

Figure 44. Contour map of the Pt(111) Auger emission intensity at 65 eV. Contours are drawn at 5% intervals. The center of the map corresponds to emission normal to the surface, and the edge to emission 70° from normal, as indicated by the scales. Coordinates are defined in Figure 43. (From Ref. 116.)

2.4.3. Scanning Auger microscopy (SAM)

In SAM the primary electron beam is focused to a small spot and rastered across the sample surface. By setting the analyzer to detect a particular Auger peak or by programming it to record peaks from several elements sequentially and cyclically (called "multiplexing"), resolved compositional information can be obtained. The spatial resolution depends on both the focusing capability of the electron gun and the nature of the surface. Electron sources which focus to spot diameters of less than 10 nm are commercially available, and there are others providing spot sizes in the range up to 1 μm. The actual resolution achievable on a given surface may be limited by surface roughness, by the charging characteristics or by the ease with which the surface is damaged by the electron beam. If a sample charges, it is particularly difficult to keep the primary electron beam from drifting due to the charge-induced fields. Through electron stimulated desorption, the surface composition can be altered significantly by the electron beam, and the problem can be so severe in SAM that the desired information is either lost or badly distorted. For the focusing conditions for high spatial resolution (small spot diameter), a very small primary beam current is necessary. The Auger signal strengths are reduced accordingly with the consequence that longer counting times must be used to obtain a reasonable signal-to-noise ratio resulting in increased damage. The effect can be reduced by increasing the spot size, i.e., degrading the spatial resolution, or sometimes by cooling the specimen.

It is important to be able to perform SEM at the same time as SAM since there is nearly always a direct correlation between the surface morphology revealed by SEM and the SAM compositional mappings. It is necessary only to incorporate a secondary electron detector in the system in which the SAM instrumentation already exists. Such a correlation is illustrated by the SEM and corresponding SAM micrographs in Fig. 45, recorded in a study of the oxidation of a polycrystalline Pt_3Sn surface by Hoflund et al. [122]. The SEM micrograph shown in (a) is from the sputter-cleaned surface, with the Sn SAM micrograph from the same surface region in (b); the magnification factor is about 250. In (b) greater brightness corresponds to higher Sn concentration, so that the brightest regions are Sn-rich. In Fig. 45a the crystallites and grain boundaries are clearly observable. Marked features include A: grain boundaries, B: flat, smooth regions, C: rough regions and D: bulk surface deposits. Correlation of micrographs (a) and (b) indicates that the regions of highest Sn concentration are at the grain boundaries. The rough areas (C) are also Sn-rich but not as much so as the grain boundary regions. Both the smooth regions and those with surface deposits have lower Sn concentrations. A Pt SAM micrograph (not shown) was simply the negative of the Sn SAM micrograph indicating that the Sn-rich regions were depleted in Pt except for D which was both O- and C-rich due to residual contamination. The micrographs demonstrate a clear correlation between the Sn and Pt distributions and the surface morphology. The concentration of Sn was highest at the grain boundaries, and decreased with distance away from a grain boundary until a Pt-rich region was reached near the center of a crystallite. This suggests that Sn segregates to the surface by diffusion along grain boundaries and then spreads out over the surfaces of the alloy crystallites. Since the thickness (or concentration) of the Sn layer was greatest near the grain boundaries and least near the center of the crystallites, sputtering exposed the Pt-rich regions by removing the thinnest Sn layers on the crystallite centers. Hence, the patterns observed in (a) and (b). The O SAM micrograph in (c) was recorded after a saturation exposure to O_2 at 1.3×10^{-4} Pa and room temperature. The brighter regions are O-rich. The pattern indicates that oxygen adsorbs at regions which are Sn-rich and, therefore, preferentially at grain boundaries. Many other examples appear in the literature, as discussed by Joshi et al. [70] and Joshi [123].

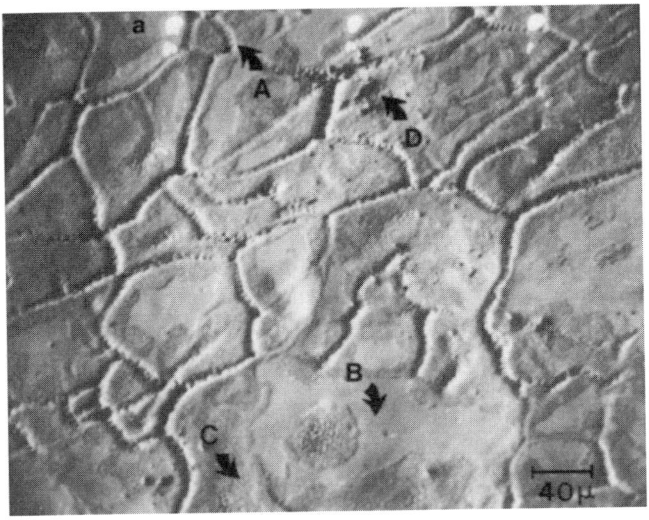

(a)

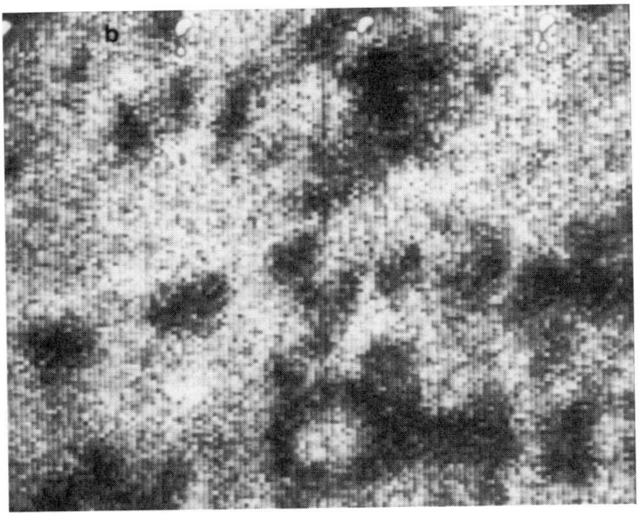

(b)

Figure 45. (a) Secondary electron micrograph of Pt_3Sn alloy surface after cleaning; the magnification is 180; (b) scanning Auger map of Sn (same region as in (a)); the lighter regions are Sn-rich; (c) scanning Auger map of O (same region as in (a)) after a saturation exposure to O_2 at 1.3×10^{-4} Pa and room temperature. The lighter regions are O-rich. (From Ref. 122.)

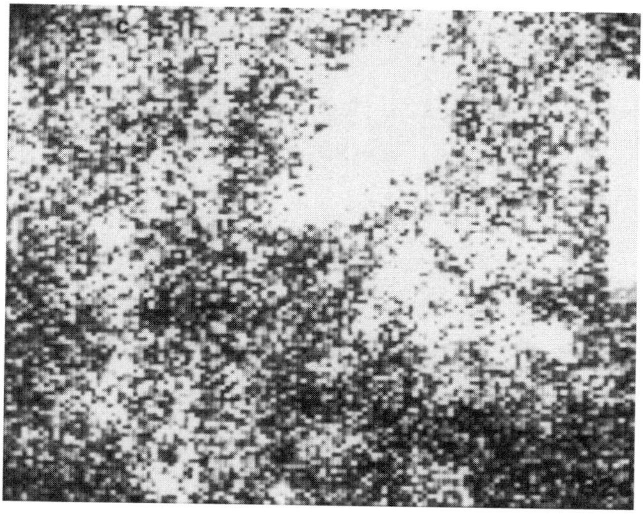

(c)

3. ION SCATTERING SPECTROSCOPY (ISS)

3.1. Introduction and History

ISS (also referred to as low-energy ion-scattering spectroscopy (LEIS)) is a technique which can yield compositional information about the outermost atomic layer of a surface and structural information about well-ordered surfaces. Its unusually high surface specificity places ISS in an unique position among surface science techniques because all surface processes are strongly dependent on the composition of the outermost atomic layer. Other surface techniques such as XPS and AES sample significantly greater depths and therefore cannot provide compositional information related only to the outermost layer. ISS, XPS and AES are thus complementary techniques in that a more complete understanding of a surface can be obtained by using them in conjunction with each other.

ISS is performed by impinging an incident flux of monochromatic, inert-gas ions on a solid surface and energy analyzing the ions which scatter from the surface at some preselected angle as shown schematically in Fig. 46. The ion-solid interactions can be approximated as elastic binary collisions between the ions and the individual atoms in the solid. This approximation works quite well because the collision times are very short (10^{-15} to 10^{-16} s) compared to the time constant of a characteristic lattice vibration (10^{-13} s). The ion therefore strikes a surface atom and leaves the surface region before the recoiling atom has time to interact with the solid. The conservation of energy and momentum in the binary scattering process can be written as

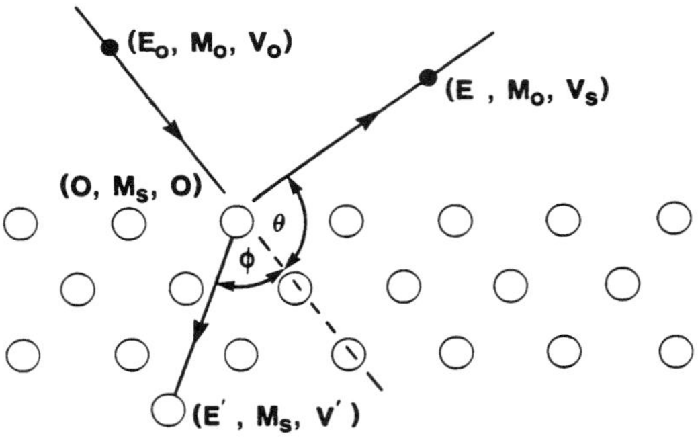

Figure 46. Schematic representation of the ISS process.

$$E_o = E + E' \tag{14}$$

$$\begin{aligned} M_oV_o &= M_oV_s \cos\theta + M_sV'\cos\phi \\ 0 &= M_oV_s \sin\theta - M_sV'\sin\phi \end{aligned} \tag{15}$$

and combined to yield

$$\frac{E_s}{E_o} = \left(\frac{M_o}{M_o + M_s}\right)^2 \left[\cos\theta \pm \left(\left(\frac{M_s}{M_o}\right)^2 - \sin^2\theta\right)^{1/2}\right]^2 \tag{16}$$

The symbols in these equations are defined in Fig. 46. If $M_s/M_o > 1$, then only the plus sign applies and each target mass gives rise to a single peak in the spectrum of scattered intensity as a function of E/E_o. If $M_s/M_o < 1$, then both signs apply subject to the constraint

$$\frac{M_s}{M_o} \geq \sin\theta \tag{17}$$

and each target mass gives peaks at two energies in the above spectrum. In Eq. 16, E_o is the KE of the incident inert-gas ion (set by the ion source), E is the kinetic energy of the scattered ion measured with an electrostatic energy analyzer, M_o is the mass of the primary ion (selected by the choice of inert gas), and θ is the scattering angle determined by the experimental geometry. The variables are all known so that M_s, the masses of the surface atoms, can be determined from the positions of the peaks in the ISS spectrum. Typical ISS spectra recorded from α-alumina-supported Ag catalysts used for ethylene epoxidation [124] are shown in Fig. 47. The spectra are complex because many elements are present on the catalyst surfaces.

The extremely high surface specificity of ISS is important in many areas of application, including heterogeneous catalysis and growth of semiconductor surfaces because the nature of the outermost atomic layers determines the

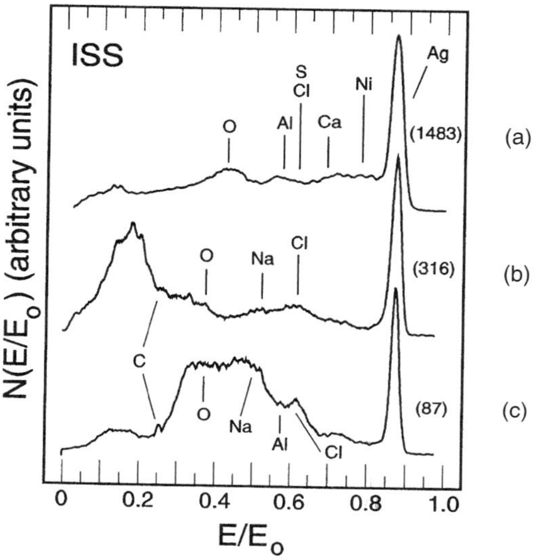

Figure 47. ISS spectra recorded from Ag/α-Al_2O_3 ethylene epoxidation catalysts: (a) fresh, (b) active and (c) aged. The spectra are normalized with respect to the height of the Ag peak. The actual Ag peak heights are listed in parentheses. (From Ref. 124.)

chemical properties of the surface. In describing ISS as being specific to the outermost atomic layer, it is necessary to consider the definition of the outermost atomic layer. Schematic diagrams of two types of outermost atomic layer are shown in Fig. 48. In Fig. 48a the outermost atomic layer is well defined and consists of both A and B atoms. In Fig. 48b the B atoms are displaced toward the bulk with respect to the A atoms. If the primary ion beam impinges on the surface from a grazing angle, then scattering occurs only from the A atoms because they shield the B atoms from the primary beam. For the surface in Fig. 48b the ISS spectrum would therefore contain a single peak due to scattering from A. If the primary ion beam impinged on the surface along the sample normal, then scattering would occur from both A and B. Provided that the incident ions scattered from B atoms were collected at angles close to the sample normal, peaks due to both A and B atoms would be present in the ISS spectrum. This effect can be illustrated using the concept of the shadow and blocking cones shown in Fig. 49. Any atoms within the shadow cone of another atom will not be observed in an ISS spectrum. This is also true for atoms in which the scattering ions are blocked by other atoms. If both the incident and collection angles are changed, the relative sizes of the peaks due to scattering from A and B will also change. These effects can make quantification in ISS diffi-

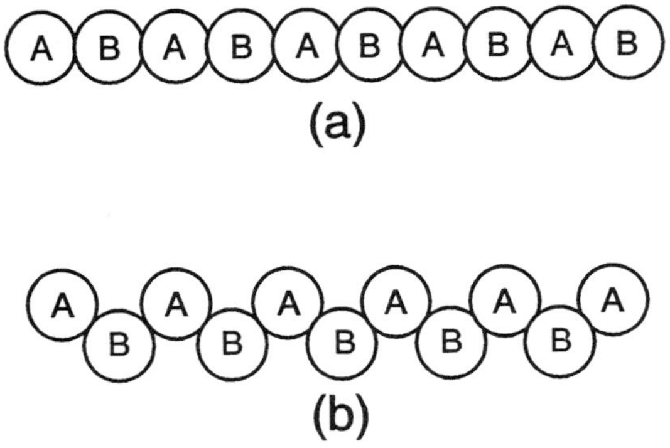

Figure 48. Two schematic models of the outermost atomic layer of a solid.

cult, so there are similarities to the matrix effects in the quantification of AES and XPS data.

Many early studies of the interaction between ions and solids contributed to the development of ISS. Brunee [125] and Panin [126, 127] both suggested the possibility of using backscattered ions to determine the surface composition of solids. Brunee [125] demonstrated that the maximum energies of 0.4–4 keV alkali ions scattered from Mo could be predicted by single binary elastic collision theory, and Panin [126, 127] found a similar result for H^+, He^+, N^+, O^+ and Ar^+ scattered from Mo and Be at high energies (7.5–80 keV). Later studies by Mashkova and Molchanov [128], Fluit et al. [129] and Datz and Snoek [130] confirmed the findings and provided evidence that the interaction potential could be described by a screened Coulomb potential. Most of these studies used high-energy ions (>5 keV), reactive ions, or both, which limited the usefulness of the technique for surface analysis, as described below.

Smith [131] was the first to use ISS as a surface analytical technique by impinging primary beams of He^+, Ne^+ and Ar^+ at energies of 0.5 to 3.0 keV on

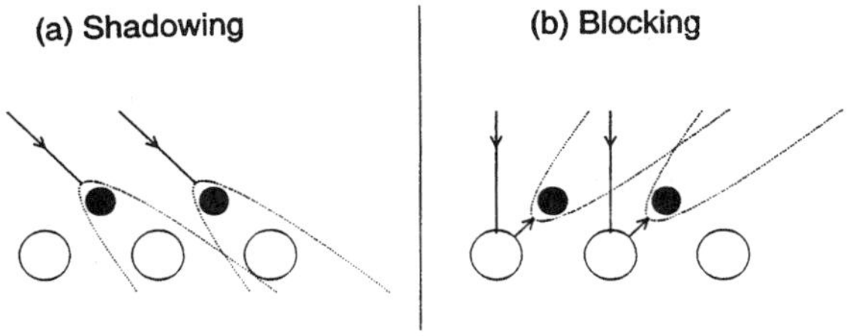

Figure 49. Schematic representation of the shadow cone and blocking effects.

polycrystalline Mo and Ni targets. Sharp peaks due to scattering from C, O and substrate atoms were found, demonstrating that multiple surface species could be detected using ISS. In contrast, if reactive ions such as H_2^+ were used, as shown in Fig. 50, the ISS spectra exhibited broad tails compared to those using inert-gas ions [132]. The tails are due to ions which penetrate beneath the surface, undergo numerous binary collisions, thereby losing energy, and escape as ions. The binary collisions are elastic so that referring to ions in the tail as inelastically scattered is incorrect. The tails associated with the use of inert-gas ions are far smaller because the probability of neutralization of such ions is much higher.

Strehlow and Smith [133] examined both of the cleaved surfaces of a CdS single crystal and were able to distinguish between the Cd-rich and S-rich sides. A similar experiment was performed by Brongersma and Mul [134] on the opposite (111) faces of a ZnS crystal, and the results are shown in Fig. 51. Smith also studied the oxidation of a polycrystalline Al surface with ISS [135], and the results are shown in Fig. 52. In those experiments Smith demonstrated that ISS was able to detect multiple elements on a surface, was highly surface specific, and yielded semiquantitative information relating to crystal structure.

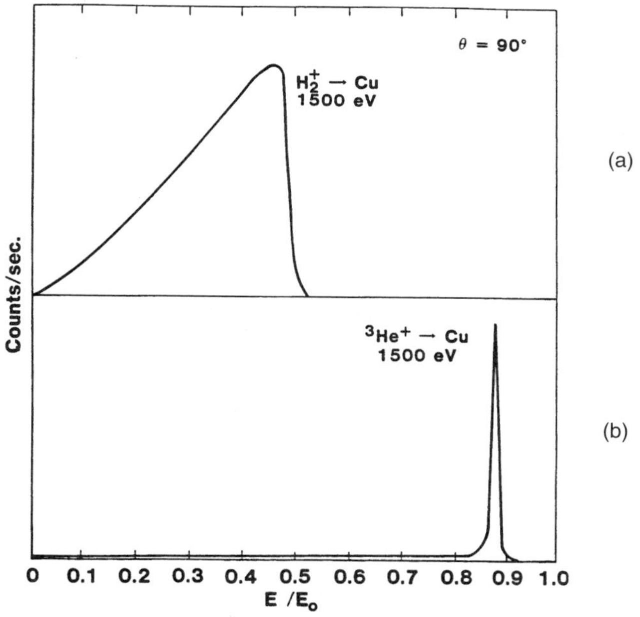

Figure 50. ISS spectra of (a) H_2^+ and (b) He^+ scattered from Cu. (From Ref. 131.)

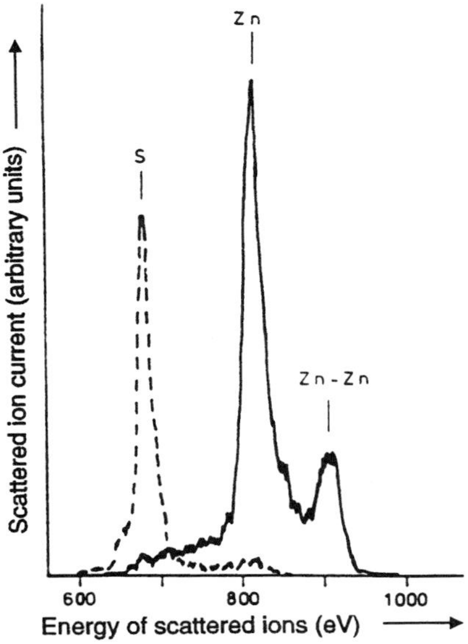

Figure 51. $^{20}Ne^+$ ISS spectra recorded from the two polar (111) faces of a ZnS crystal using 1 keV $^{20}Ne^+$ and a scattering angle of 45°. (From Ref. 134.)

3.2. Experimental Equipment and Data Collection

The information sought in ISS is the composition and/or structure of the outermost atomic layer. Since ion bombardment is a destructive process, it is crucial to perform ISS in a manner which provides the desired information without causing significant damage to the outermost layer. Unlike in XPS and AES, it is necessary to collect ISS data at the lowest resolution and signal-to-noise ratio compatible with obtaining the information required, in order to minimize ion-beam effects on the outermost atomic layer. This vital point is not generally understood by the suppliers of commercial surface science equipment, who assess incorrectly ISS performance in terms of peak widths and signal-to-noise ratio similar to XPS. Using their suggested parameters for data collection, the outermost atomic layer would in most cases be removed by sputtering before an ISS spectrum could be recorded. A reasonable guideline is that only a few percent of the atoms in the outermost atomic layer should be struck by ions during analysis. In order to collect ISS data, the ion beam current and the data collection time should both be minimized, while the sample area examined and the collection efficiency should both be maximized. As discussed below, $^3He^+$ should be used as the primary ion whenever possible because it yields the largest signals and results in the least amount of sput-

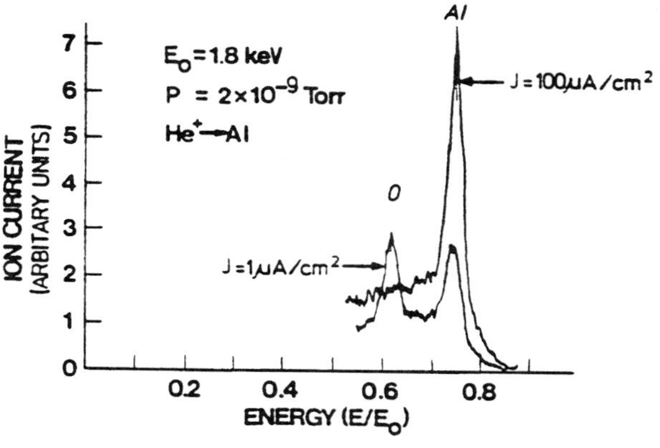

Figure 52. ISS spectra recorded from polycrystalline Al metal, and from an Al_2O_3 film of thickness 5 nm on Al metal. (From Ref. 135.)

ter damage. Due to the problem with sputter damage, ISS cannot be performed in a small-spot mode.

ISS requires a source capable of producing inert-gas ions of energies in the range 0.5 to 3 keV. Ideally the source should be capable of being defocused in order to irradiate an area equivalent to the total acceptance area of the analyzer. If the source is focused, then the beam must be rastered sufficiently quickly over a large area to meet the damage criteria. The radial ion density within a beam is typically Gaussian with FWHMs ranging from 1 μm to 3 mm in diameter. The flux must be contaminant free, and very good vacuum conditions are essential if contamination is not to accumulate on the surface at a rate at which it would influence the ISS spectrum. Sources are either differentially pumped to minimize the pressure in the sample chamber or require backfilling of the chamber with inert gas to about 1.3×10^{-3} Pa. In the latter case a Ti sublimation pump should be employed to ensure purity of the inert gas since outgassing from the walls may lower the purity. A continuous-flow system is also a good way of maintaining the purity of the inert gas.

The ion energy distribution from ion sources is also Gaussian, and its FWHM significantly affects the ISS resolution, as illustrated by the calculated ISS spectra shown in Fig. 53 [136]. The spectra are for 2 keV He^+ impinging on an Mg surface and analyzed at a scattering angle width of 1°. The primary ion beam FWHMs range from 10 to 90 eV. At a FWHM of 10 eV (a), the Mg isotopes are completely resolved. As the FWHM is increased, the resolution is degraded until at a FWHM of 90 eV (d) a single asymmetric peak is obtained.

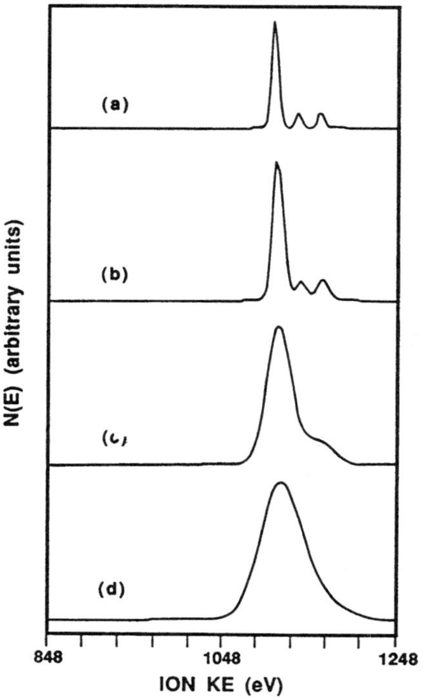

Figure 53. Calculated Mg ISS spectra as from a DPCMA for 2 keV He^+ incident on a Mg surface, using a scattering angle width of 1° and primary ion beam FWHM of (a) 10, (b) 20, (c) 50 and (d) 90 eV. (From Ref. 136.)

Electrostatic analyzers used for AES and XPS can also be used to collect ISS data in most systems by switching their polarities. As reviewed by Miller [137], the first ISS experiments were carried out with a 127° sector analyzer, but this analyzer is inefficient because its entrance solid angle is too small to collect more than a small fraction of the ions. The first commercially available ISS instrument was based on a SPCMA with a coaxial ion gun (rather than an electron gun as in AES). This analyzer had a high collection efficiency but could operate at only one scattering angle, 138°. In the early 1980s Hoflund et al. [138] modified a DPCMA to perform ISS. Since the DPCMA already contained a coaxial electron gun, the ion gun had to be positioned off-axis. Scattered ions were selected using a movable aperture developed by Knapp et al. [139]. In principle the scattering angle can be varied continuously by moving this aperture, but in practice most of the ions scatter in-plane so that only the forward-scatter and backscatter directions yield acceptably large signals.

Hemispherical analyzers can also be used for ISS, but a small aperture must be placed in front of the inlet lens to limit the width of the scattering angle. The latter is an important parameter in ISS experiments. As the width of the scattering angle is increased, the signal is essentially averaged over a broader range of scattering angles, resulting in degradation of resolution. The effect is shown in Fig. 54 [136]. In spectrum (a) the three isotopes of Mg are completely resolved. As the width of the scattering angle is increased ((a) to (d)), the resolution is degraded until at (d) only a broad asymmetrical peak is recorded. However, as the resolution is decreased, the signal strength increases, which means that a lower ion dose can be used, resulting in less surface damage. It is therefore desirable to use the largest scattering angle width possible, i.e., the largest width compatible with the peak resolution necessary for the analysis. Electrostatic analyzers, particularly CHAs, often have the capability of changing apertures to alter the width of the scattering angle.

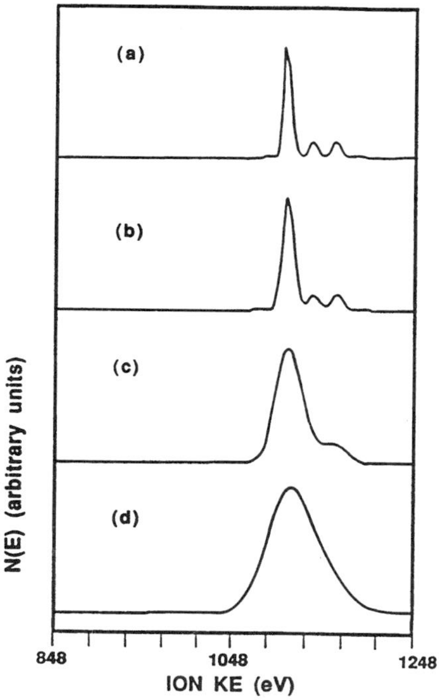

Figure 54. Calculated Mg ISS spectra as from a DPCMA using a 10 eV FWHM primary ion beam and scattering angle widths of (a) 1°, (b) 2°, (c) 6° and (d) 12°. (From Ref. 136.)

For ISS, a DPCMA can be operated either in the nonretarding mode or, if increased spectral resolution is required, in the retarding mode [140]. Hoflund and Asbury [141] compared ISS spectra from a polycrystalline Pt/Sn alloy, using 1 keV He^+, in both the nonretarding mode and the retarding mode at 50 eV pass energy. The resolution of the Pt and Sn peaks was similar in both modes, but the signal strength was 30 times larger in the nonretarding mode. This result was expected since the ISS features are quite broad. The important point for both DPCMA and CHA is that the energy resolution should be decreased as far as conveniently possible in order to increase the signal strength, decrease the data collection time and, therefore, minimize sample damage. The SPCMA is always operated in the nonretarding mode so that the energy resolution is determined only by the geometrical characteristics of the analyzer.

Brongersma and co-workers [142–145] have developed the energy and angle-resolved ion-scattering spectrometer shown in Fig. 55. In this analyzer a beam of mass-selected inert-gas ions is directed coaxially along the sample normal. The double-toroidal analyzer accepts ions that have been scattered into a solid angle of $145° + \Delta\theta$ ($0 \leq \Delta\theta \leq 3.2°$), where $\Delta\theta$ is selected using variable slits. The width of the total energy range is 10% of the analyzer pass energy. A zoom lens at the entrance of the analyzer allows selection of a section of the energy range. The analyzer images the scattered ions onto a two-dimensional position-sensitive detector which simultaneously determines the azimuthal angles and the energies of the scattered ions.

All types of sample are amenable to analysis by ISS, but insulators can sometimes cause charging problems. The latter can usually be minimized using the charge reduction techniques discussed above. Many insulating samples yield excellent ISS spectra without any charging problems. However, some samples do cause problems, and the reasons are not fully understood. Tilting the sample with respect to the incident ion beam may reduce charging or move spectral features due to charging out of the range of interest in an ISS spectrum. Tilting does not usually alter the relative ISS peak heights for rough, polycrystalline or amorphous samples, but can have a large influence on relative intensities from highly ordered surfaces, due to the shadow or blocking effects.

3.3. Spectral Features and Interpretation

3.3.1. General features

Some types of spectral features are common to nearly all ISS spectra. For example, in the spectra recorded from the Ag/α-Al_2O_3 ethylene epoxidation catalysts shown in Fig. 47, there are elemental features due to elastic scattering from atoms in the outermost atomic layer and there is a background due to multiply scattered ions which have penetrated beneath the surface. In addition there

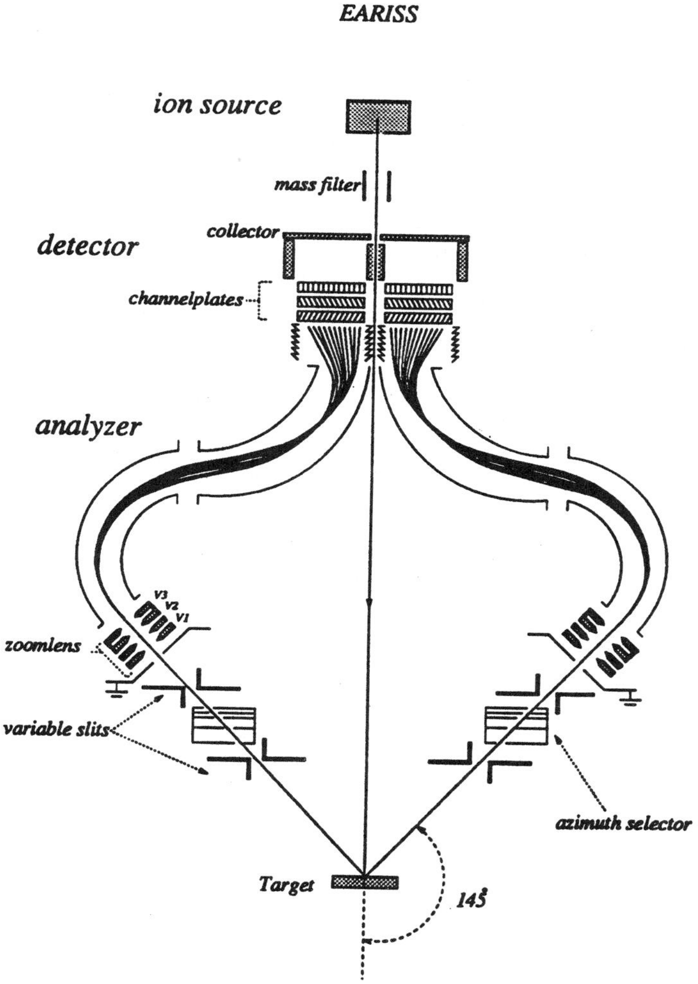

Figure 55. Schematic cross section of the EARISS system. (From Ref. 145.)

are features present below $E/E_0 = 0.25$. In spectra (a) and (c) these features arise from emission of secondary ions and consist mostly of H^+. If a narrow energy range of ions were to be selected by electrostatic filtering and the mass of the ions analyzed, the technique then becomes SSIMS. The large low-energy feature in spectrum (b) is characteristic of charging. There is no apparent reason why charging occurred in (b) but not (a) and (c). The SEM micrographs and the XPS and AES spectra obtained from the same catalysts are shown in

Figs. 10 and 41, respectively. Clearly, compositional estimates from the ISS, XPS and AES spectra yield different results due to the different sampling depths and matrix effects.

3.3.2. Background and neutralization

The background and neutralization effects are closely related. In ISS nearly all the incoming inert-gas ions are neutralized and leave as atoms, so they are not detected by the electrostatic analyzer. The point is illustrated by the data in Fig. 56, in which the ion fraction of Ar^+ scattered from polycrystalline Au is shown as a function of incident ion energy from 6 to 32 keV [146]. The extent of ion neutralization increases as the primary-ion energy decreases. At 6 keV about 5% of the primary ions scatter from the surface as ions in the elastic elemental peak. The proportion drops to less than 1% when the primary-ion energy is reduced to the normal range of 1–2 keV used for ISS. This is a disadvantage since, of course, 100% of the ions cause surface damage while only ca. 1% provide useful information. The ratio of background intensity to elemental peak intensity

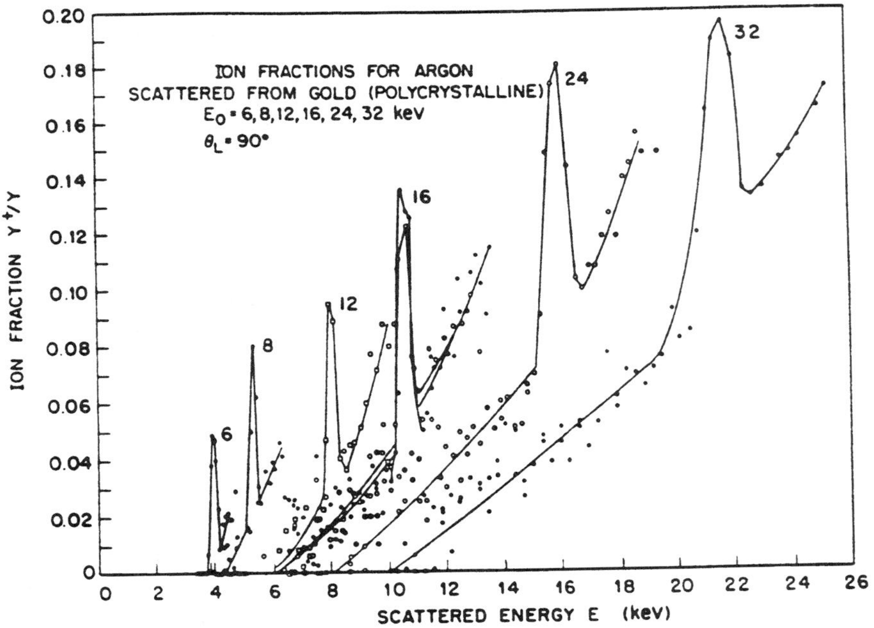

Figure 56. Ion fractions, i.e., (ions)/(ions plus neutrals), for Ar^+ scattered from polycrystalline Au as a function of incident energy in the range 6 to 32 keV. (From Ref. 146.)

also depends on the primary-ion energy as illustrated by Fig. 57 [137]. A larger fraction of the higher energy ions penetrate beneath the surface, undergo multiple collisions in which they lose energy and then leave the surface still as ions. They may even become neutralized and reionized one or more times.

The background intensity also depends on the electrical properties of the surface. Metals typically produce very low backgrounds but insulators much larger ones. The reason for this is that conducting materials are more effective in neutralizing ions in the near-surface region because there is a higher density of conduction electrons with high mobilities. This is advantageous with regard to producing a small background but disadvantageous in that high neutralization probabilities result in smaller signals in ISS. The change in the intensity of the background relative to elemental peaks is illustrated by the ISS data in Fig. 58, recorded by Gardner et al. [147] from an air-exposed platinized tin oxide surface before (spectrum a) and after reduction in vacuum by annealing at 300°C (spectrum b) and at 450°C (spectrum c). Several features appear in spectrum (a). The Sn peak is present only as a shoulder on the Pt peak, since the two species cannot easily be resolved using He^+, and the peak at $E/E_o = 0.6$ is due to Na contamination. The secondary ions desorb with a threshold of $E/E_o = 0.12$ due to charging effects. A small oxygen peak is also present at $E/E_o = 0.49$, a value which is higher than predicted due to the charging effects. The background is large compared to the elemental features, which is characteristic of a surface with low electrical conductivity. Several changes occur during anneal-

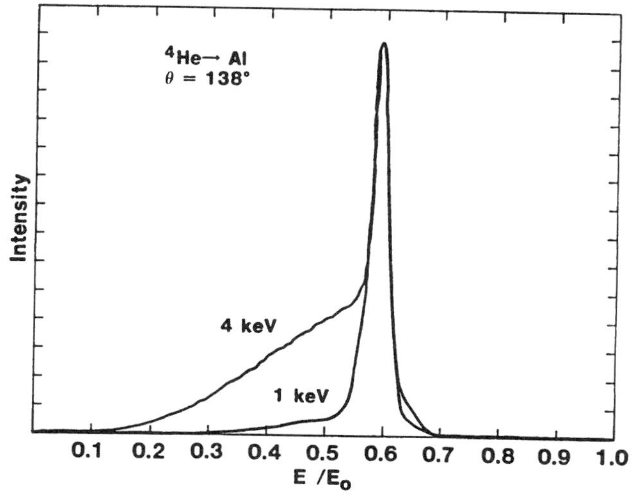

Figure 57. Normalized ISS spectra of $^4He^+$ scattered from Al at incident energies of 1 and 4 keV. (From Ref. 137.)

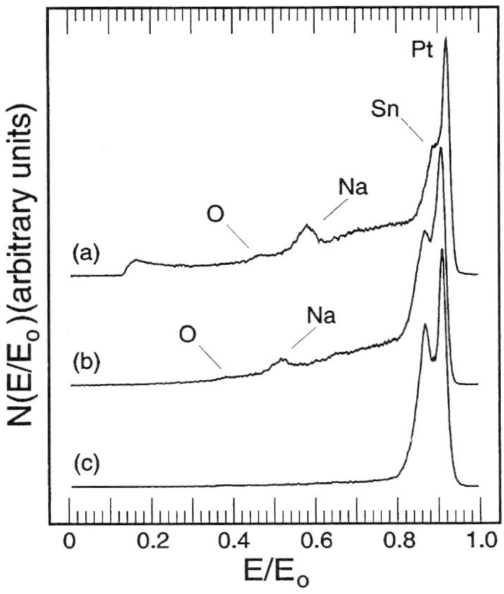

Figure 58. ISS spectra recorded from a platinized tin oxide surface (a) after insertion into the vacuum system, and after annealing in vacuum at (b) 300°C and (c) 450°C. (From Ref. 147.)

ing of the sample in vacuum at 300°C. The Sn peak is more prominent and has increased in size with respect to the Pt peak. Charging has disappeared so that the O and Na peaks now appear close to the energies predicted by Eq. 16. Also, the background has decreased relative to the elemental features. These observations are indicative of an increase in conductivity. The trends continue upon annealing in vacuum to 450°C. The Na feature has vanished due to sublimation of Na_2O, the Sn/Pt ratio has increased due to alloy formation and the O peak has diminished further. The background intensity is now very low and is characteristic of a metallic surface.

3.3.3. Multiple scattering

Another type of feature which may be present in some ISS spectra is that due to multiple scattering as shown in Fig. 51. The Zn-Zn feature arises from primary ions which have scattered from one Zn atom, hit another Zn atom and then left the surface as ions. The geometry of the multiple event can be modeled by estimating the two scattering angles and applying Eq. 16 twice. If the two scattering angle estimates are accurate, then the correct E/E_o of the multiple scattering feature will be obtained.

3.3.4. Multiply charged ion scattering

Sometimes small peaks may appear at energies higher than that of the principal scattering peak of an ion from an element, due to the scattering of multiply charged ions. Ion sources produce doubly and even triply charged ions in addition to singly charged ones. These multiply charged ions can be filtered out, but often are not and they produce distinct peaks in the ISS spectrum as shown in Fig. 59 [148]. The multiply charged ions scatter according to Eq. 16, but due to the characteristics of the electrostatic analyzer they appear at energies which are two and three times, respectively, the energy of the elastically scattered, singly charged ions. The flux of multiply charged ions from an ion source can be reduced by lowering the ionization voltage.

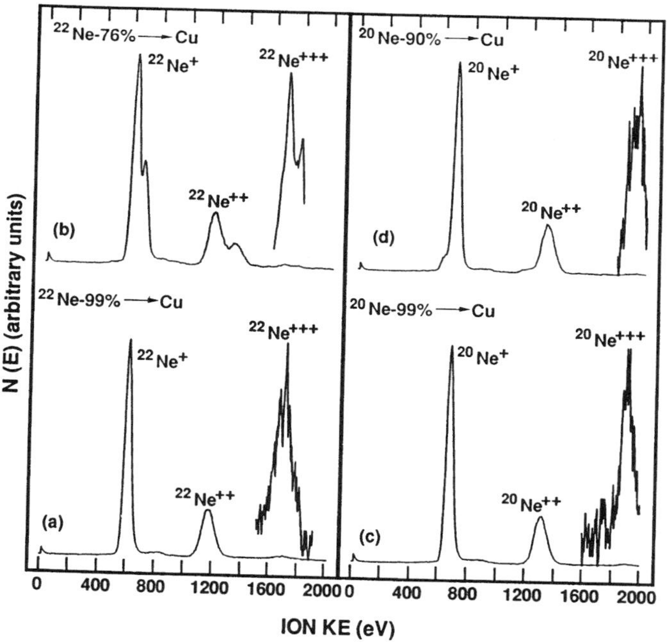

Figure 59. ISS spectra recorded by scattering 2 keV $^{20}Ne^+$, 99.95%; $^{20}Ne^+$, 24%/$^{22}Ne^+$, 76%; $^{20}Ne^+$, 90%/$^{22}Ne^+$, 10% and $^{22}Ne^+$, 99.95% from polycrystalline Cu showing contributions from Ne^+, Ne^{2+} and Ne^{3+}. (From Ref. 148.)

3.3.5. Choice of primary ion

There are only a limited number of inert gases to select for the primary ion, i.e., ^{3}He, ^{4}He, ^{20}Ne, ^{22}Ne and ^{40}Ar. ISS studies using Xe, Kr or Rn have not been reported for reasons discussed later. With Ne it is important to use isotopically pure gas rather than the naturally abundant mixture because with the latter multiple peaks will be obtained, as demonstrated with $^{20}Ne/^{22}Ne$ mixtures in Fig. 60 [148].

As discussed, the neutralization efficiency for a given ion is greater at lower primary energies. This is due to the fact that ions which have lower energies move more slowly and therefore spend more time in the vicinity of the solid, thereby increasing their probability of neutralization. The same argument also applies to an increase in the mass of the ion at a fixed primary energy. The effect is quite large as shown by the data in Fig. 61 [148]. The corollary is that

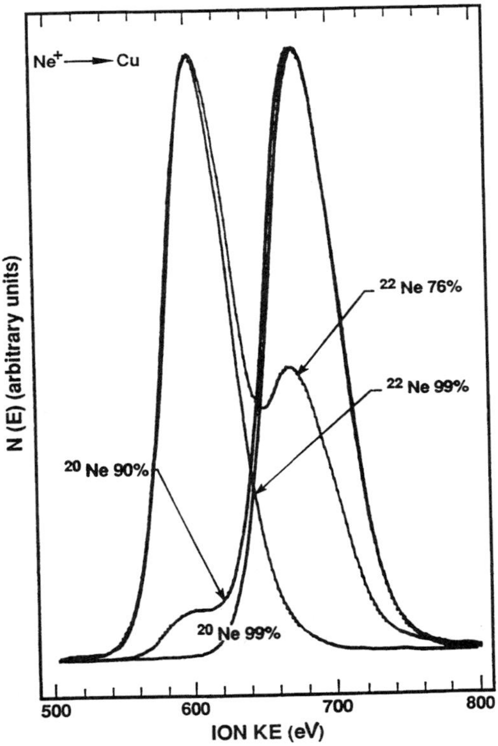

Figure 60. ISS spectra recorded by scattering 2 keV $^{20}Ne^{+}$, 99.95%; $^{20}Ne^{+}$, 24%/$^{22}Ne^{+}$, 76%; $^{20}Ne^{+}$, 90%/$^{22}Ne^{+}$, 10%, and $^{22}Ne^{+}$, 99.95% from Cu. (From Ref. 148.)

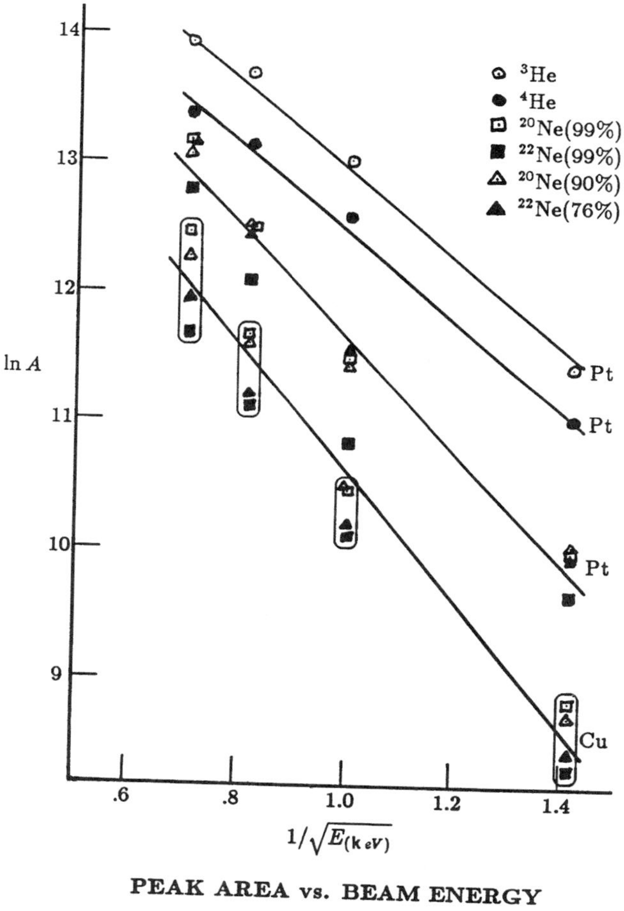

Figure 61. Plot of the ISS logarithmic peak areas as functions of $E^{-1/2}$ for various inert gas ions scattered from polycrystalline Pt and Cu. The slope is the neutralization decay rate (From Ref. 148.)

the use of lighter ions yields larger signals in ISS. Damage is also minimized by using lighter ions because a lower dose is required to achieve a given signal-to-noise ratio, and lighter ions may create less surface damage per ion than heavier ions.

Another advantage of using light primary ions is that all elements heavier than the chosen ion can be observed in the spectrum, according to Eq. 16 and the associated condition given in Eq. 17. For example, if $^{20}Ne^{+}$ is used as the primary ion, then only elements heavier than Ne will be observed in the ISS spectrum unless the scattering angle can be altered to satisfy Eq. 17 for a lighter element. ISS data obtained using 1 keV Ne^+ during the low-pressure oxidation of a polycrystalline Ni/Cr sample are shown in Fig. 62 [149]. The near-surface oxygen content was high after exposure to 100 L of oxygen, and decreased with annealing as the oxygen penetrated into the bulk by a place exchange mechanism. There is no feature specific to oxygen in the ISS spectra shown in the figure because O is lighter than Ne and Eq. 17 is not satisfied. However, there is a broad background below $E/E_O = 0.48$ in spectra (b)–(e), which correlates with a change to an insulating or semi-insulating character of the surface as a

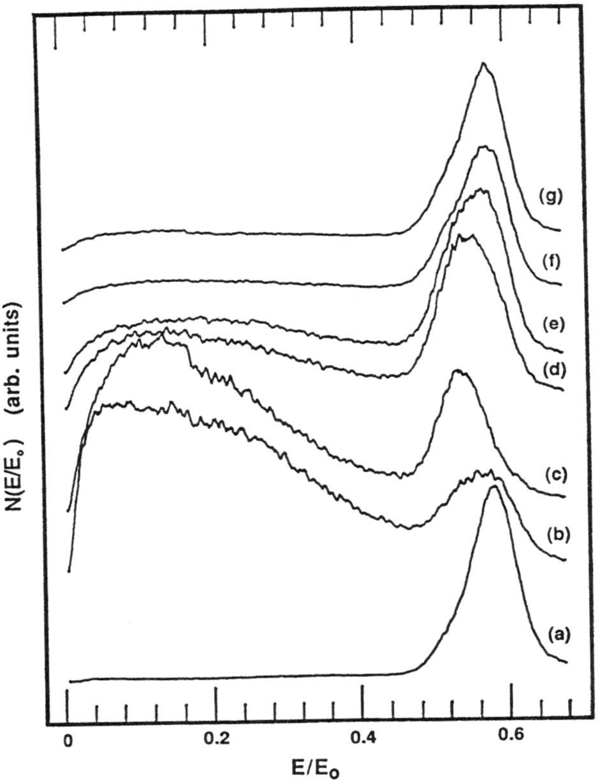

Figure 62. ISS spectra recorded from a polycrystalline Ni/Cr alloy surface using 1 keV Ne^+ (a) after cleaning by sputtering and annealing, (b) after exposing the cleaned surface to 100 L of O_2 at room temperature, (c) immediately after heating the O-exposed sample to 500°C, and after annealing the O-exposed sample at 500°C for (d) 0.5 h, (e) 1 h, (f) 2 h and (g) 4 h. (From Ref. 149.)

result of oxidation of the outermost atomic layer. As the oxygen penetrates on annealing, and the surface becomes conducting again, the broad background disappears.

3.3.6. Hydrogen and carbon

Detection of hydrogen and carbon pose interesting and important problems for ISS. Even though a feature due to surface hydrogen cannot be observed directly under usual conditions, hydrogen influences the ISS spectrum by shielding other atoms. The effect is illustrated by the ISS data in Fig. 63 recorded from Pd before and after exposure to H_2 (3×10^2 Pa for 30 min) [150]. The chemisorbed hydrogen blocks almost completely the underlying Pd from the He^+. Hydrogen can be observed directly in two other ways, however. The first, as discussed, is by mass-analyzing the secondary ions (SSIMS). The second is to use a very small scattering angle which satisfies Eq. 17 so that H can be observed as a distinct spectral feature using $^3He^+$. Such scattering angles of only a few degrees are rarely used in ISS because they result in very poor mass resolution. The possibility of detecting hydrogen in that way is illustrated by the data shown in Fig. 64, recorded by Shoji et al. [151].

Another challenge for ISS is the detection of surface carbon. Although both $^3He^+$ and $^4He^+$ are lighter than carbon, carbon has a low scattering cross section because it is light and has a high neutralization probability. Portions of two ISS spectra recorded from carbon using 2 keV $^3He^+$ and 2 keV $^4He^+$, respectively [152], are shown in Fig. 65. They illustrate the importance of using a lighter scattering ion to reduce the neutralization probability and increase the sensitivity to lower-mass elements. An ISS spectrum recorded from an oxide surface heavily contaminated by hydrocarbons, using 2 keV $^3He^+$ at a scatter-

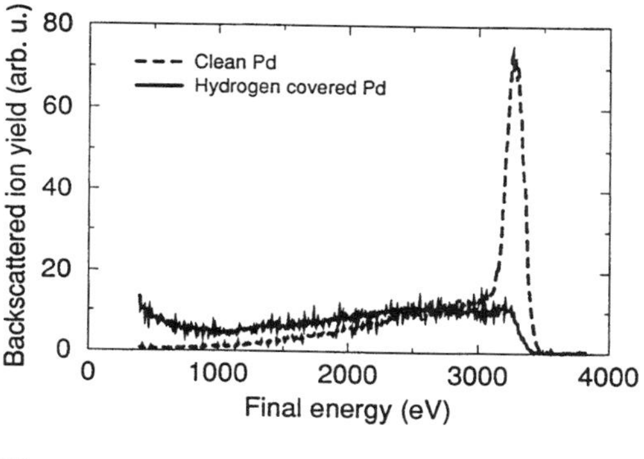

Figure 63. ISS spectra recorded using 4 keV He^+ scattered from clean Pd (a) before and (b) after a 30 min treatment in 3×10^2 Pa of H_2. (From Ref. 150.)

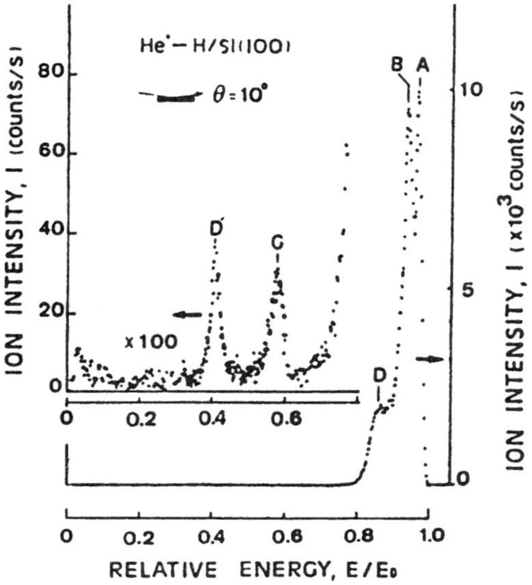

Figure 64. $^3He^+$ ISS spectra recorded from an H/Si (100)-(1 × 1) surface prepared by exposing Si(100) to atomic hydrogen at room temperature. The peaks labeled D and D′ are the scattering peaks from surface hydrogen, A is from Si, B is from O or C and G is recoil from H or O. (From Ref. 151.)

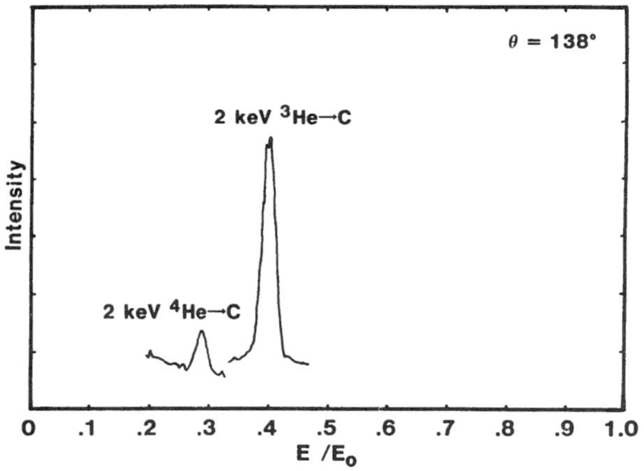

Figure 65. ISS spectra recorded from carbon at a scattering angle of 138° using 2 keV $^3He^+$ and $^4He^+$. (From Ref. 152.)

ing angle of 138° [137], is shown in Fig. 66. The large low-energy feature is due to secondary ions, presumably H^+.

3.3.7. Elemental sensitivity

The sensitivity of ISS is a strong function of mass and of the experimental parameters used. So many factors are involved in determining the sensitivity of an element in a particular case that it is not appropriate to state specific numbers unless all the parameters involved are completely specified. Nevertheless, guidelines have been provided in the literature. Leys [152] has suggested that absolute sensitivities range from about 0.3 to 10^{-4} monolayers for Li to Au, and relative sensitivity factors for 2 keV $^3He^+$ and 2 keV $^{20}Ne^+$ are shown in Fig. 67 [137]. The curve for $^3He^+$ indicates that the relative sensitivity varies by a factor of more than 500 over the periodic table and that elements with $Z < 10$ have low relative sensitivities. The relative sensitivity is proportional to the scattering cross section and inversely proportional to the neutralization probability. The detection of higher atomic number elements is favored because the cross section increases, and the neutralization probability decreases with Z.

The experimental geometry also has a strong influence on the elemental sensitivities. ISS spectra obtained from polycrystalline Ag after numerous sputtering and annealing treatments [153] are shown in Fig. 68. Forward-scattered and backscattered ion spectra obtained from the same surface are compared. In-

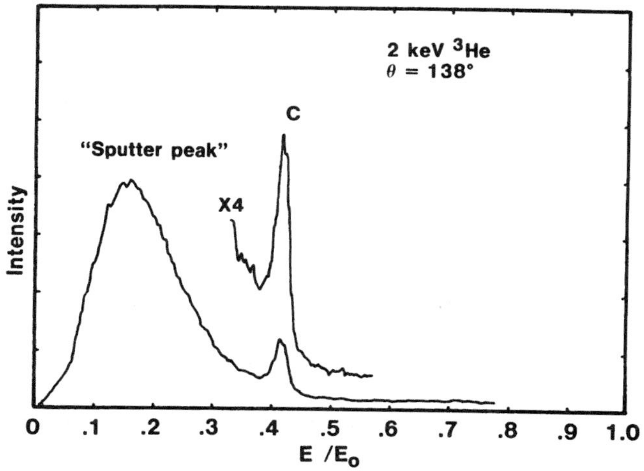

Figure 66. ISS spectrum recorded from an oxide surface heavily contaminated by hydrocarbons. (From Ref. 137.)

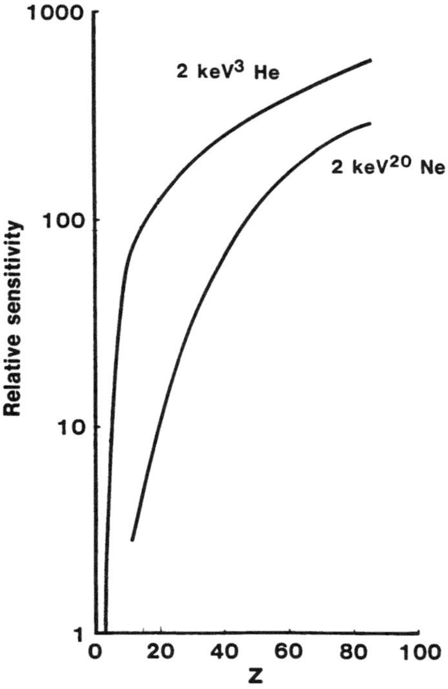

Figure 67. Relative elemental sensitivity curves for 2 keV $^{3}He^{+}$ and $^{20}Ne^{+}$, as functions of atomic number *Z*. (From Ref. 137.)

plane scattering yields large signals compared to out-of-plane scattering. The surface was 'clean' according to AES, and the backscattered ISS spectrum is in agreement with this observation even after amplification by a factor of 200. However, the forward-scattered spectrum exhibits a significant O peak, and large C and O peaks after amplification by a factor of 43, demonstrating that ISS in the forward-scattered mode is much more sensitive to light elements. A minor disadvantage for forward-scattering is that resolution is lost for the heavier elements, as observed in the rather close spacing of the C, O and Ag peaks. This is relatively unimportant since the objective in performing forward-scattered ISS is to identify the presence of light contaminating species.

The above data prompt the question as to whether the routinely used standard condition "clean according to AES" is really adequate for assessing surface cleanliness. Many surface processes are dependent on, or sensitive to, very low levels of surface species. ISS can generally provide greater sensitivity to surface species than either AES or XPS and is probably a better technique for

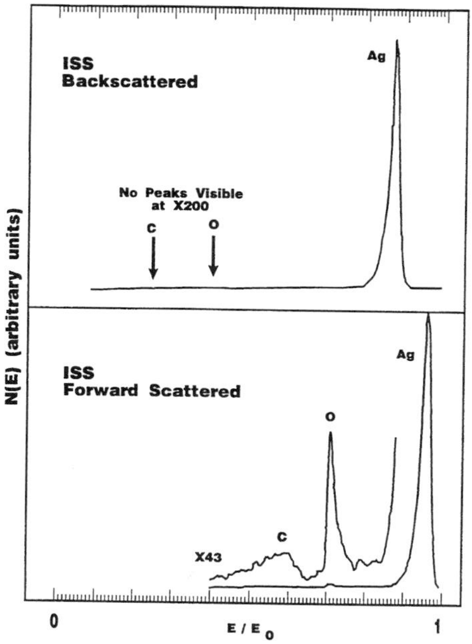

Figure 68. Forward- and backscattered ISS spectra recorded from the same Ag surface using a DPCMA after numerous sputtering and annealing treatments of the sample. (From Ref. 153.)

assessing surface cleanliness when the contaminants are restricted to the outermost atomic layer. As always, the best approach is to use several surface techniques such as ISS, AES, XPS and SSIMS to assess surface cleanliness because each technique has a different depth specificity and different elemental sensitivity. The differences in depth specificities are shown in Fig. 69 [153]. ISS, more surface-specific ARAES, and less surface-specific ARAES, data were recorded from cleaned (sputtered and annealed) polycrystalline Ag. In the "clean" state the oxygen concentration in the outermost atomic layer was below the detection level of ISS, and no oxygen was observed using less surface-specific ARAES. However, more surface-specific ARAES indicated that a few percent of oxygen remained just beneath the surface regardless of how many cleaning cycles were repeated. The subsurface oxygen was not observed in the ISS spectrum because it lay beneath the outermost atomic layer and was not observed either in the less surface-specific ARAES spectrum presumably because the average concentration over the sampling depth was too low. The Ag was then exposed to 75 Torr of O_2 for 1 h at (b) 75, (c) 150, (d) 200 and (e) 250°C. The oxygen content of the near-surface region increased with temperature of

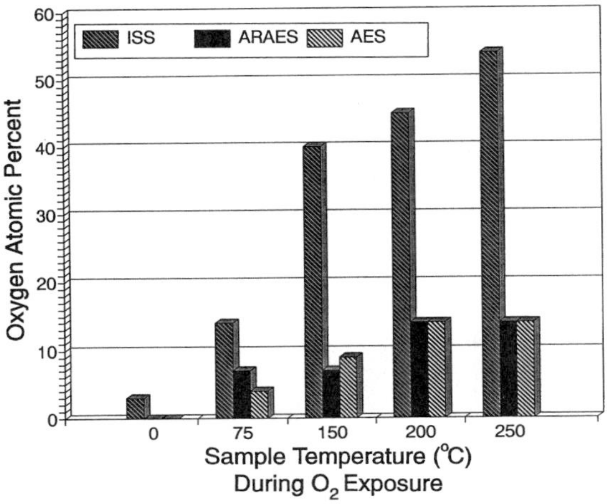

Figure 69. Oxygen concentrations (at.%) calculated from ISS, more surface-specific (ARAES), and less surface-specific (AES), data recorded from (a) clean polycrystalline Ag after exposure to 75 torr of O_2 for 1 h at (b) 75, (c) 150, (d) 200 and (e) 250°C. (From Ref. 153.)

exposure, and in the outermost atomic layer increased monotonically to ca. 50% coverage. The subsurface oxygen concentration also increased with temperature of exposure. The interpretation is that the near-subsurface region filled first with oxygen at lower temperatures and then the subsurface oxygen content became uniform at 14 at.% at higher temperatures. The data demonstrate how ISS and ARAES can be used in complementary roles to obtain depth-specific information.

3.3.8. Energy resolution

Spectral resolution is usually defined either as the width of a chosen peak in a spectrum or as the extent of overlap between closely spaced peaks. According to Eq. 16, elemental peaks are more closely spaced for higher mass elements for a given scattering angle. Using $^3He^+$ or $^4He^+$, peaks from the lower mass elements will be widely separated, but elements of much greater mass such as Sn and Pt, which are widely separated in mass (119 and 195 amu), are nevertheless very closely spaced in He^+ ISS spectra, as illustrated in Fig. 58. Greater separation can be achieved by using either a larger scattering angle and/or a higher mass primary ion. These effects have been discussed quantitatively by Miller [137], who calculated mass resolution from the expression

$$\frac{M_s}{\Delta M_s} = \frac{E}{\Delta E}\,\frac{2M_s/M_o}{1+M_s/M_o}\,\frac{M_s/M_o+\sin^2\theta-\cos\theta(M_s^2/M_o^2-\sin^2\theta)^{1/2}}{M_s^2/M_o^2-\sin^2\theta-\cos\theta(M_2^2/M_o^2-\sin^2\theta)^{1/2}} \qquad (18)$$

The results are shown in Fig. 70 for $^{3}He^{+}$ and $^{20}Ne^{+}$ at scattering angles of 90° and 137° and $E/\Delta E$ = 60. The mass resolution is very high when the masses of the elements detected are close to that of the primary ion. Note that the curves approach the mass of the scattering ion for the scattering angles shown. Even when using $^{20}Ne^{+}$, the mass resolution is still quite poor for heavy masses. Where a peak cannot be identified in an ISS spectrum due to resolution problems, another technique such as AES, XPS or SSIMS can be useful in assigning the ISS peak. Young and Hoflund [149] have calculated the energy resolution for Na/Pb and Cr/Ni as a function of scattering angle for 2 keV $^{20}Ne^{+}$. The results are shown in Fig. 71. The energy resolution is quite different for Na and Pb peaks except at Θ = 95° where the curves cross. If a third element were present, then it would not be possible to choose a scattering angle at which the energy resolutions of all three peaks were equal. When the masses are similar, the resolution will be similar over the whole scattering angle range. For Cr and Ni they cross at Θ = 105°. The curves all have similar shapes with

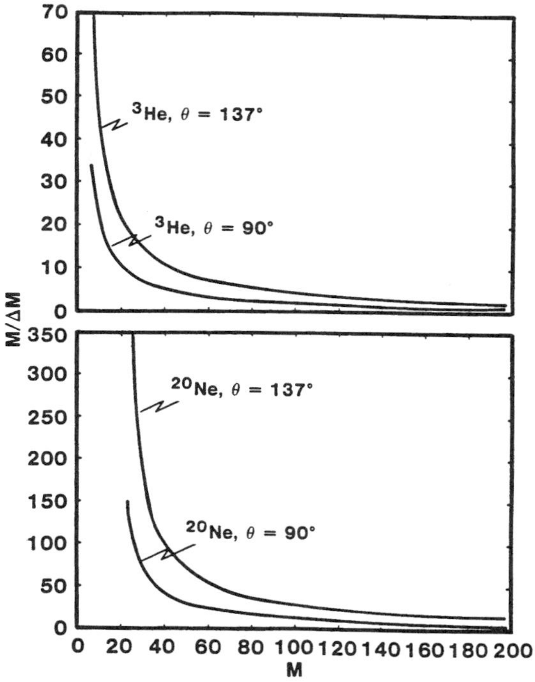

Figure 70. Mass resolution for $^{3}He^{+}$ and $^{20}Ne^{+}$ at scattering angles of 90° and 137° and an $E/\Delta E$ of 60. (From Ref. 137.)

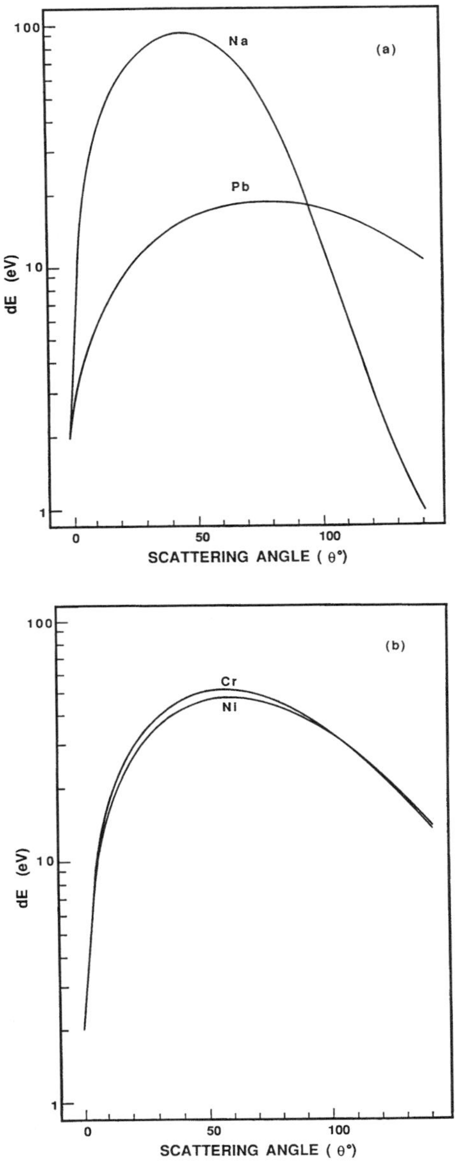

Figure 71. Calculated energy resolution *dE* as a function of scattering angle Θ for 2 keV $^{20}Ne^{+}$ scattered from surfaces containing (a) Na and Pb and (b) Ni and Cr. (From Ref. 149.)

high resolution at both small and large scattering angles. The effects of other factors such as the energy spread of the primary-ion beam and the width of the scattering angle have been discussed in the experimental section.

3.3.9. Peak shape

Elemental ISS peak shapes are important not only for gaining a fundamental understanding of the interaction between low-energy ions and solids but also in practical applications involving ISS as an analytical tool. In the development of Eq. 16, a hard-sphere potential was used to model the collision between the primary ion and a single atom in the outermost atomic layer of the solid. In essence, the presence of the rest of the solid plays no role in this model so that the atoms in the outermost layer can be considered as isolated from each other in space. The model does not account for neutralization effects or the presence of a background. Although the model is surprisingly accurate for predicting the ISS elemental peak energies, it is not completely precise. The actual peak positions are usually shifted by 20–30 eV from those predicted by elastic binary scattering theory. This shift is due to the fact that the interaction potential is not perfectly modeled by a hard-sphere potential. Potentials which can be used to model this interaction include the Bohr screened-Coulomb potential, the Born-Mayer potential, the Thomas-Fermi-Firsov potential, the Molière potential and the Ziegler-Biersack-Littmark (ZBL) potential. They have been discussed by Miller [137], Young et al. [154] and Niehus et al. [155]. The ZBL potential generally yields the best agreement between theory and experiment. Young et al. [154] have shown that the ISS elemental peak shape is dependent on the nature of the interaction potential but that ISS experiments must be performed at high resolution if any differences between the various potentials are to be discerned. In principle, it should be possible to determine the nature of the interaction potential from high-resolution ISS data. The peak-energy shift may also be due to the form of the interaction potential, but the results of such calculations have not yet been published. Studies by Sonda et al. [156, 157] and Thomas et al. [158] claim that the ion-scattering peak consists of two components. One component is due to elastic binary scattering of an ion which remains as an ion throughout the whole scattering event. The other component results from neutralization of the primary ion during collision followed by reionization of the atom as it leaves. During the latter process the KE of the scattered species is reduced by an amount similar to the first ionization potential of the incident species. These two components make up the overall envelope, and its shape depends upon the relative magnitudes of them, which is a property of the target material. The model provides a partial explanation for the asymmetry, which is often apparent on the low-energy side of ISS peaks, and for the small peak position shifts described before.

As shown by Young and Hoflund [149], the isotopic distribution of an element also influences its ISS peak shape. In this initial study the peak shapes of Ni and Cr were fitted by taking the elemental isotopic distributions into account. In a later study [136] they examined how the isotopic distribution of Mg influences ISS peak shape. As shown in Figs. 53 and 54, it has a strong influence. In another isotopic study Melendez et al. [148] examined the effects on ISS peak shapes of using different isotopic mixtures of Ne^+ scattered from Pt and Cu. Based on the results of these studies, the isotope effect is clearly significant and must be taken into account in analyzing ISS data.

3.3.10. Quantification

Quantification of ISS is a challenging problem because there are a number of important variables which can be estimated in only a rather crude manner. The intensity of an ISS peak can be written as

$$I_t = I_p CTD\, \Delta\Omega\, \sigma_{t,\mathrm{ave}} P_t N_t R G_t \tag{19}$$

where I_t is the ion intensity for primaries scattered from atoms of mass t, I_p is the primary-ion flux, N_t is the surface density of atoms of mass t, R is a surface roughness factor, T is the analyzer transmission, D is the detector efficiency, $\Delta\Omega$ is the solid angle of collection, $\sigma_{t,\mathrm{ave}}$ is the differential scattering cross section for element t, P_t is the ion survival probability for element t, G_t is a geometrical factor which quantifies the shadowing and blocking effects and C is a constant dependent on the quantity being used as the measure of intensity. As discussed by Young and Hoflund [149], peak areas provide the best measure of intensities since the peak shapes differ for each element due to differences in the isotopic distributions and neutralization/reionization effects. I_p can be measured in an ISS experiment and I_t can be taken as the area under the elemental peak after some form of background subtraction is applied as discussed later. $\Delta\Omega$, D and T may be supplied by the manufacturer of the analyzer. $\sigma_{t,\mathrm{ave}}$ can be estimated from the results of calculations based on a chosen interaction potential. Examples of such calculations are shown in Fig. 72a [159] and b [137]. The different potentials yield similar variations with ion energy and primary-ion and elemental mass, but the numerical values for the cross sections can differ considerably. Jacobs et al. [160] have suggested that smooth surfaces yield ISS features which are about twice as great as similar features for rough surfaces, but beyond this estimate there is no method for determining R quantitatively. The importance of G has been discussed earlier, but there is no general way of determining its value unless the surface is well ordered and careful structural studies are carried out. If all the factors in Eq. 19 were known except for N_t and C, then calibration experiments could be performed to determine C. In such experiments N_t could be varied in some known manner

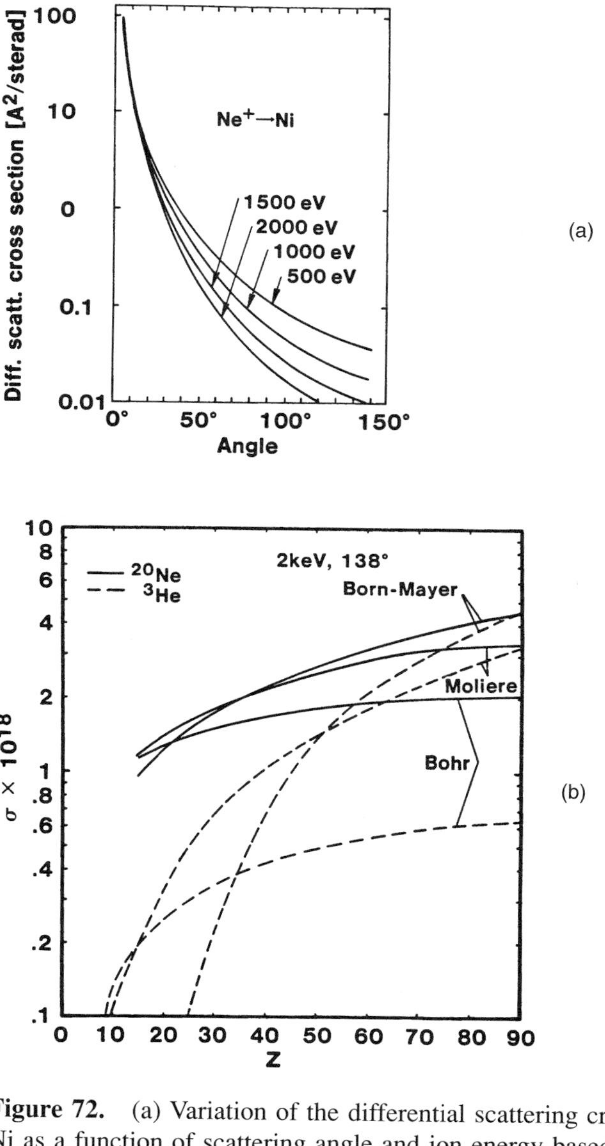

Figure 72. (a) Variation of the differential scattering cross section for $Ne^+ \rightarrow$ Ni as a function of scattering angle and ion energy based on calculations using the Born-Mayer potential. (From Ref. 159.) (b) Differential scattering cross sections for 2 keV $^3He^+$ and $^{20}Ne^+$ at a scattering angle of 138° as a function of atomic number Z calculated using the Born-Mayer, Molière and Bohr potentials. (From Ref. 137.)

perhaps based on AES results, and I_t could be measured, thereby yielding the value of C. P_t is a very important quantity, and unfortunately almost nothing is known about it. To illustrate its importance, consider a hypothetical ISS experiment in which an element t covers 50% of a surface and its neutralization probability is 99% so that the ion survival probability is 1%. Now suppose that the elemental coverage remains unchanged, but the chemical state is changed by some process and that the neutralization probability changes by only a very small amount to 98%. Now the ion survival probability is 2%, i.e., doubled, so that the intensity measure has doubled even though the surface coverage has remained unchanged. Currently, there is no way to determine the neutralization probability in ISS experiments, but its importance cannot be neglected in the interpretation of ISS data.

3.3.11. Data processing

The general approach in the processing of ISS data is to fit the background and peaks, following which they can be separated and the peak areas determined. The simplest way to remove the background from the peaks is to assume that it is linear. This is obviously incorrect, but the magnitude of the error introduced is probably small compared to those errors introduced by other factors involved in the quantification. Nelson [161] has developed a semiempirical model for generating ISS spectra numerically which allows the neutralization decay rate to be determined. His model assumes that each element i yields a peak for which the intensity as a function of energy, I_i(E), can be described as a combined Gaussian and Lorentzian function, e.g.,

$$I_i(E) = \frac{A\ \exp[-(1-R)\times 2.77259\left(\dfrac{(E-E_o)}{\mathrm{FWHM}}\right)\times 2]}{R(E-E_o)^2 + (FWHM/2)^2} \tag{20}$$

where R is the Gaussian/Lorentzian factor (1 for a pure Lorentzian and 0 for a pure Gaussian), FWHM is the full width at half-maximum, E is the energy variable, and A is the peak amplitude. The inelastic background intensity $I_B(E)$ is

$$I_B(E) = B\left\{\pi - 2\ \tan^{-1}\left[\frac{2(E-E_o)}{\mathrm{FWHM}}\right]\right\} \tag{21}$$

and the intensity of the tail is

$$I_T(E) = \exp\left(\frac{-K}{E^{1/2}}\right) \tag{22}$$

where B is the intensity of the tail and K is the neutralization decay rate. The total intensity as a function of energy for a single element is given by adding the products of Eqs. 21 and 22 to Eq. 20; i.e., $I_i(E) + I_B(E)I_T(E)$. Thus, there are six adjustable constants for the fit: A, B, R, FWHM, E_o, and K. Young et al. [136] combined the peak-shape model of Young and Hoflund [149] with the model of Nelson to fit experimental ISS spectra. A typical result is shown in Fig. 73.

3.3.12. Depth profiling

Depth profiling with ISS is a relatively simple process experimentally since the same primary-ion beam used to generate the ISS data also can be used for sputtering surface layers. All that is required is to collect ISS spectra in a stepwise fashion. Although ISS is capable of producing high-resolution depth profiles because of its high surface specificity, it is still susceptible to all the artefacts encountered in the sputtering process such as ion beam mixing and preferential sputtering. The ISS spectra shown in Fig. 74 [162] were recorded from a pressed powder sample of a Zn/Cr/O spinel promoted with K, which catalyzes the production of isobutanol and methanol from a mixture of CO and H_2. The spectra were obtained by depth-profiling the near-surface region of an aged catalyst. In the outermost atomic layer of the catalyst, spectrum (a), the predominant component is K with lesser amounts of Si and Zn. Si is a contaminant which accumulates during reaction. Spectrum (b) was obtained after sputtering for 5 min with 1 keV He^+. The Zn feature has increased with respect to those for Si and K. After sputtering the sample with 1 keV Ar^+ for 10 min, spectrum (c) was recorded. Zn is now the predominant feature and a Cr peak has become observable. K and Si peaks are still present, but they are now quite small.

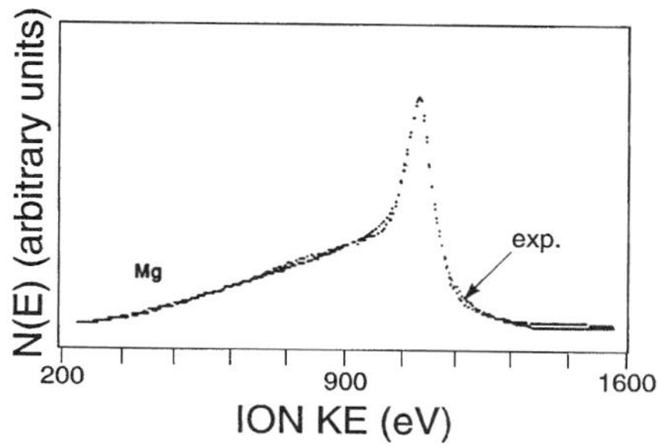

Figure 73. An experimental ISS spectrum recorded from polycrystalline Mg, with the calculated best-fit spectrum. (From Ref. 136.)

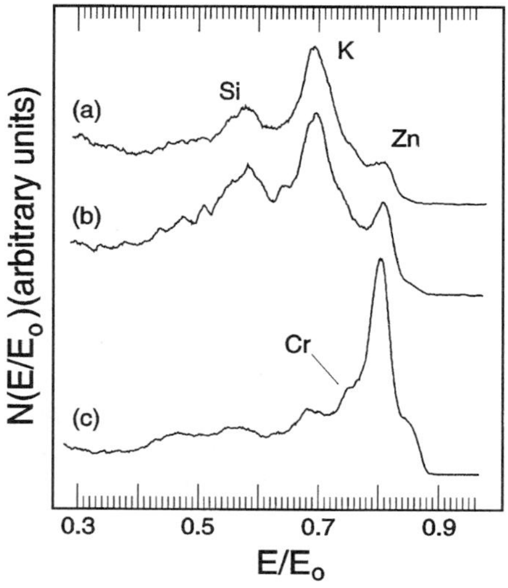

Figure 74. ISS spectra recorded from an aged isobutanol production catalyst after (a) entry into the analysis chamber, (b) 5 min sputtering with 1 keV He^+ and (c) an additional 10 min of sputtering with 1 keV Ar^+. The catalyst consisted of a Zn/Cr spinel support promoted with 5 wt.% K. (From Ref. 162.)

ACKNOWLEDGMENTS

Financial support for this effort was received from the National Science Foundation through grant No. CTS-9122575. The efforts of William A. Epling and John T. Wolan with regard to figure and manuscript preparation are appreciated. Numerous comments and contributions by John C. Rivière are deeply appreciated, particularly with regard to small-spot XPS.

REFERENCES

1. J. G. Jenkin, R. C. G. Leckey and J. Liesegang, *J. Electron Spectrosc. Relat. Phenom, 12*(1977), 1.
2. P. M. A. Sherwood, X-ray photoelectron spectroscopy, in *The Handbook of Surface Imaging and Visualization* (A. T. Hubbard, ed.), CRC Press, Boca Raton, FL, 1995, Chap. 63, p. 875.
3. A. Einstein, *Ann. Phys. 17*(1905), 132.
4. H. Robinson and W. F. Rawlinson, *Phil. Mag. 28*(1914), 277.
5. H. Robinson, *Proc. Roy. Soc., Ser. A, 104*(1923), 455.

6. H. Robinson, *Phil. Mag. 50*(1925), 241.
7. M. de Broglie, *Compt. Rend. 172*(1921), 274.
8. K. Siegbahn, *Alpha, Beta and Gamma-Ray Spectroscopy*, North-Holland, Amsterdam, 1954.
9. C. Nordling, E. Sokolowski and K. Siegbahn, *Ark. Fys. 13*(1958), 483.
10. S. Hagström, C. Nordling and K. Siegbahn, *Phys. Lett. 9*(1964), 235.
11. K. Larsson, C. Nordling, K. Siegbahn and E. Stenhagen, *Acta Chem. Scand. 20*(1966), 2880.
12. C. N. Berglund and W. E. Spicer, *Phys. Rev. A 136*(1964), 1030.
13. J. E. Drawdy, G. B. Hoflund, S. D. Gardner, E. Yngvadottir and D. R. Schryer, *Surf. Interface Anal. 16*(1990), 369.
14. K. Siegbahn, C. Nordling, A. Fahlman, R. Nordberg, K. Hamrin, J. Hedman, G. Johansson, T. Bergman, S. Karlson, I. Lindgren and B. Lindberg, *Electron Spectroscopy for Chemical Analysis—Atomic, Molecular, and Solid-State Structure Studies by Means of Electron Spectroscopy*, Almquist and Wiksells, Stockholm, 1967.
15. E. Paparazzo, G. Fierro, G. M. Ingo and N. Zacchetti, *Surf. Interface Anal. 12*(1988), 438.
16. M. P. Seah and W. A. Dench, *Surf. Interface Anal. 1*(1979), 2.
17. J. C. Rivière, in *Practical Surface Analysis: Auger and X-ray Photoelectron Spectroscopy*, 2nd ed. (D. Briggs and M. P. Seah, ed.), Wiley, New York, 1990, Chap. 2, pp. 19–83.
18. C. D. Wagner, W. M. Riggs, L. E. Davis, J. F. Moulder and G. E. Muilenberg (eds.), *Handbook of X-ray Photoelectron Spectroscopy*, Perkin-Elmer Corp., Physical Electronics Division, Eden Prairie, MN, 1979.
19. P. W. Palmberg, *J. Vac. Sci. Technol. 12*(1975), 379.
20. V. V. Zashkvara, M. I. Korsunskii and O. S. Kosmachev, *Sov. Phys. Tech. Phys. 11*(1966), 96.
21. E. M. Purcell, *Phys. Rev. 54*(1938), 818.
22. M. P. Seah, Electron and Ion Energy Analysis, in *Methods of Surface Analysis* (J.M. Walls, ed.), Cambridge University Press, Cambridge, 1989, Chap. 3.
23. T. L. Barr and S. Seal, *J. Vac. Sci. Technol. A 13*(1995), 1239.
24. NIST X-Ray Photoelectron Spectroscopy Database, data compiled by C. D. Wagner, Surfex Co, Program written by D. M. Bickham, Distributed by Standard Reference Data, NIST, Gaithersburg, MD, 1989.
25. G. Beamson and D. Briggs, *High Resolution XPS of Organic Polymers: The Scienta ESCA 300 Database*, Wiley, New York, 1992.
26. *Surface Science Spectra*, an official journal of the American Vacuum Society published by the American Institute of Physics.
27. A. Savitsky and M. J. E. Golay, *Anal. Chem. 36*(1964), 1627

28. C. D. Wagner L. E. Davis, M. V. Zeller, J. A. Taylor, R. H. Raymond and L. H. Gale, *Surf. Interface Anal. 3*(1981), 211.
29. D. A. Shirley, *Phys. Rev. B 5*(1972), 4709.
30. D. Briggs and M. P. Seah (eds.), *Practical Surface Analysis: Auger and X-ray Photoelectron Spectroscopy*, 2nd ed., Wiley, New York, 1990.
31. G. Tougaard, *Surf. Sci. 216*(1989), 343.
32. S. Hofmann, *J. Electron Spectrosc. Relat. Phenom. 59*(1992), 15.
33. R. G. Musket, W. McLean, C. A. Colmenares, D. M. Makowiecki and W. J. Siekhaus, *Appl. Surf. Sci. 10*(1982), 143.
34. G. B. Hoflund and J. F. Weaver, *Meas. Sci. Technol. 5*(1994), 201.
35. G. B. Hoflund, in *Encyclopaedia of Scientific Instrumentation* (A. Levin, L. G. Rubin and F. A. Settle, Jr., eds.), American Chemical Society, 1998, p. 00.
36. G. B. Hoflund, M. R. Davidson and G. R. Corallo, *J. Vac. Sci. Technol. A 9*(1991), 2412.
37. G. B. Hoflund, H.-L. Yin, A. L. Grogan, Jr., D. A. Asbury, H. Yoneyama, O. Ikeda and H. Tamura, *Langmuir 4*(1988), 346.
38. G. B. Hoflund and D. M. Minahan, *J. Catal. 162*(1996), 48.
39. C. S. Fadley, R. J. Baird, W. Siekhaus, T. Novakov and S. Bergström, *J. Electron Spectrosc. Relat. Phenom. 4*(1974), 93.
40. C. S. Fadley, *Prog. Surf. Sci. 16*(1985), 275.
41. P. C. McCaslin and V. Y. Young, *Scan. Electron Microsc. 1*(1987), 1545.
42. G. B. Hoflund, D. A. Asbury, C. F. Corallo and G. R. Corallo, *J. Vac. Sci. Technol. A 6*(1988), 70.
43. J. L. Sullivan, S. O. Saied and T. Choudhury, *Vacuum 43*(1992), 89.
44. C. D. Wagner, *Farad. Discuss. Chem. Soc. 60*(1975), 291; C. D. Wagner, L. H. Gale and R. H. Raymond, *Anal. Chem. 51*(1979), 466.
45. J. F. Weaver and G. B. Hoflund, *J. Phys. Chem. 98*(1994), 8519.
46. J. F. Weaver and G. B. Hoflund, *Chem. Mater. 6*(1994), 1693.
47. G. B. Hoflund, J. F. Weaver and W. S. Epling, *Surf. Sci. Spectra 3*(1995), 151.
48. G. B. Hoflund, J. F. Weaver and W. S. Epling, *Surf. Sci. Spectra 3*(1995), 157.
49. G. B. Hoflund, J. F. Weaver and W. S. Epling, *Surf. Sci. Spectra 3*(1995), 163.
50. D. F. Cox, G. B. Hoflund and H. A. Laitinen, *Langmuir 1*(1985), 269.
51. W. M. Riggs and M. J. Parker, in *Methods of Surface Analysis* (A. W. Czanderna, ed.), Elsevier, Amsterdam, 1975, p. 157.
52. M. Tinkham, *Group Theory and Quantum Mechanics*, McGraw-Hill, New York, 1964.
53. G. B. Hoflund and G. R. Corallo, *Surf. Interface Anal. 13*(1988), 33.
54. G. B. Hoflund and G. R. Corallo, *Phys. Rev. B 46*(1992), 7110.

55. D. F. Cox and G. B. Hoflund, *Surf. Sci. 151*(1985), 202.
56. J. Cazaux, *Rev. Phys. Appl. 10*(1975), 263.
57. C. J. Hovland, *Appl. Phys. Lett. 30*(1977), 274.
58. D. J. Keast and K. S. Downing, *Surf. Interface Anal. 3*(1981), 99.
59. K. Yates and R. H. West, *Surf. Interface Anal. 5*(1983), 217.
60. N. Gurker, M. F. Ebel and H. Ebel, *Surf. Interface Anal. 5*(1983), 13.
61. D. W. Turner, I. R. Plummer and H. Q. Porter, *J. Microsc. 136*(1984), 259.
62. M. P. Seah and G. C. Smith, *Surf. Interface Anal. 11*(1988), 69
63. J. F. Watts, *Vacuum 45*(1994), 653.
64. P. Coxon, J. Krizek, M. Humpherson and I. R. M. Wardell, *J. Electron Spectrosc. 52*(1990), 821.
65. A. Rosseland, *Z. Phys. 14*(1923), 172.
66. P. Auger, *Comp. Rend. 177*(1923), 169.
67. P. Auger, *Comp. Rend. 180*(1923), 65.
68. P. Auger, *Ann. Phys. 6*(1926), 183.
69. G. A. Somorjai, *Principles of Surface Chemistry*, Prentice Hall, Englewood Cliffs, NJ, 1972.
70. A. Joshi, L. E. Davis and P. W. Palmberg, Auger electron spectroscopy in *Methods of Surface Analysis* (A. W. Czanderna, ed.), Elsevier, Amsterdam, 1975, Chap. 5, p. 159.
71. R. E. Weber and W. T. Peria, *J. Appl. Phys. 38*(1967), 4355.
72. L. E. Davis, N. C. MacDonald, P. W. Palmberg, G. E. Riach and R. E. Weber, *Handbook of Auger Electron Spectroscopy*, Perkin-Elmer Corporation, Eden Prairie, MN, 1976.
73. C. C. Chang, in *Characterization of Solid Surfaces* (P. F. Kane and G. R. Larrabee, eds.), Plenum, New York, 1974, Chap. 20.
74. R. E. Gilbert, G. B. Hoflund, D. A. Asbury and M. R. Davidson, *J. Vac. Sci. Technol. A 6*(1988), 2280.
75. M. C. Burrell, R. S. Kaller and N. R. Armstrong, *Anal. Chem. 54*(1982), 2511.
76. M. P. Seah, M. T. Anthony and W. A. Dench, *J. Phys. E 16*(1983), 848.
77. E. N. Sickafus and D. M. Holloway, *Surf. Sci. 51*(1975), 131.
78. A. Mogami and T. Sekine in Proc. 6th European Congr. Electron Microscopy, TAL International, Jerusalem, 1976, p. 422.
79. T. Sekine, Y. Nagasawa, M. Kudoh, Y. Sakai, A. S. Parkes, J. D. Geller, A. Mogami and K. Hirata, *Handbook of Auger Electron Spectroscopy*, JEOL Ltd., Tokyo, 1982.
80. R. H. Roberts and M. K. O'Neill, *J. Phys. E 19*(1982), 97.
81. L. Frank and P. Vasina, *Comput. Physics Comm. 26*(1982), 113.
82. W. S. Woodward, D. P. Griffis and R. W. Linton, *Anal. Chem. 54*(1982), 1229.

83. G. E. McGuire, *Auger Electron Spectroscopy Reference Manual*, Plenum, New York, 1979.
84. J. S. Foord, P. J. Goddard and R. M. Lambert, *Surf. Sci. 94*(1980), 339.
85. H. P. Bonzel, A. M. Franken and G. Pirug, *Surf. Sci. 104*(1981), 625.
86. J. E. Houston and R. R. Rye, *Comments Sol. St. Phys. 10*(1983), 233.
87. D. R. Jennison, *J. Vac. Sci. Technol. 20*(1982), 548.
88. C. T. Campbell, J. W. Rogers, Jr., R. L. Hance and J. M. White, *Chem. Phys. Lett. 69*(1980), 430.
89. D. E. Ramaker, *Appl. Surf. Sci. 21*(1985), 243.
90. H. D. Hagstrum, *Phys. Rev. 150*(1966), 495.
91. G. E. Becker and H. S. Hagstrum, *Surf. Sci. 30*(1972), 125.
92. E. N. Sickafus, *J. Vac. Sci. Technol. 11*(1974), 308.
93. G. Cubiotti, G. Mondio and K. Wandelt (eds), *Auger Spectroscopy and Electronic Structure*, Springer-Verlag, Berlin, 1989.
94. T. W. Haas, J. T. Grant and G. J. Dooley, in *Adsorption-Desorption Phenomena* (F. Ricca, ed.), Academic Press, New York, 1972, p. 359.
95. G. B. Hoflund, D. A. Asbury, C. F. Corallo and G. R. Corallo, *J. Vac. Sci. Technol. A 6*(1988), 70.
96. S.-P. Jeng, P. H. Holloway, D. A. Asbury and G. B. Hoflund, *Surf. Sci. 235*(1990), 175.
97. H. H. Madden, *J. Vac. Sci. Technol. A 2*(1984), 961.
98. M. E. Malinowski, *J. Vac. Sci. Technol. 15*(1978), 39.
99. M. E. Malinowski, *J. Vac. Sci. Technol. 19*(1981), 120.
100. G. B. Hoflund, Depth profiling, in *The Handbook of Surface Imaging and Visualization* (A. T. Hubbard, ed.), CRC Press, Boca Raton, FL, 1995, Chap. 6, p. 63.
101. J. P. Biersack and G. R. Haggmark, *Nucl. Inst. Meth. 174*(1980), 257.
102. I. Ishitani and R. Shimizu, *Phys. Lett. 46A*(1974), 487.
103. P. Sigmund, *Phys. Rev. 184*(1969), 383.
104. E. Taglauer, W. Heiland and V. Beitat, *Surf. Sci. 89*(1979), 710.
105. S. Hofmann, Prog. *Surf. Sci. 36*(1991), 35.
106. S. Hofmann, *Appl. Surf. Sci. 70/71*(1993), 9.
107. A. Zalar, *Thin Solid Films 124*(1985), 223.
108. A. Zalar, *Surf. Interface Anal. 9*(1986), 41.
109. H. Ferber, G. B. Hoflund, C. K. Mount and S. Hoshino, *Surf. Interface Anal. 16*(1990), 488.
110. N. Q. Lam, *Surf. Interface Anal. 12*(1988), 65.
111. D. G. Frank, N. Batina, T. Golden, F. Lee and A. T. Hubbard, *Science 247*(1990), 182.
112. D. G. Frank, N. Batina, J. W. McCargar and A. T. Hubbard, *Langmuir 5*(1989), 1141.

113. D. G. Frank, T. Golden, O. M. R. Chyan and A. T. Hubbard, *J. Vac. Sci. Technol. A. 9*(1991), 1254.
114. O. M. R. Chyan, D. G. Frank, C. A. Doyle and A. T. Hubbard, *J. Vac. Sci. Technol. A 11*(1993), 2659.
115. D. G. Frank, O. M. R. Chyan, T. Golden and A. T. Hubbard, *J. Phys. Chem. 98*(1994), 1895.
116. D. G. Frank, Angle-resolved Auger electron spectroscopy, in *The Handbook of Surface Imaging and Visualization* (A. T. Hubbard, ed.), CRC Press, Boca Raton, FL, 1995, Chap. 1, pp. 1–22.
117. J. A. Knapp, G. J. Lapeyre, N. V. Smith and M. M. Traum, *Rev. Sci. Instrum. 53*(1982), 781.
118. M. R. Davidson, G. B. Hoflund and R. A. Outlaw, *J. Vac. Sci. Technol. A9*(1991), 1344.
119. Y. Kuk and L. C. Feldman, *Phys. Rev. B. 30*(1984), 5811.
120. D. A. Asbury and G. B. Hoflund, *Surf. Sci. 199*(1988), 552.
121. S. D. Gardner, G. B. Hoflund and D. R. Schryer, *J. Catal. 119*(1989), 179.
122. G. B. Hoflund, D. A. Asbury, P. Kirszensztejn and H. A. Laitinen, *Surf. Interface Anal. 9*(1986), 169.
123. A. Joshi, Auger electron spectroscopy, in *Metals Handbook*, Vol. 10, *Materials Characterization*, American Society for Metals, Metals Park, Ohio, 1986.
124. G. B. Hoflund and D. M. Minahan, *J. Catal. 162*(1996), 48.
125. C. Brunee, *Z. Phys. 147*(1957), 161.
126. B. V. Panin, *Sov. Phys. JETP 42*(1962), 313.
127. B. V. Panin, *Sov. Phys. JETP 45*(1962), 215.
128. E. S. Mashkova and V. A. Molchanov, *Sov. Phys. Dokl. 7*(1963), 828.
129. J. M. Fluit, J. Kistemaker and C. Snoek, *Physica 30*(1964), 870.
130. S. Datz and C. Snoek, *Phys. Rev. A 134*(1964), 347.
131. D. P. Smith, *J. Appl. Phys. 38*(1967), 340.
132. G. C. Nelson, *J. Coll. Interface Sci. 55*(1976), 289.
133. W. H. Strehlow and D. P. Smith, *Appl. Phys. Lett. 13*(1968), 34.
134. H. H. Brongersma and P. M. Mul, *Chem. Phys. Lett. 19*(1973), 217.
135. D. P. Smith, *Surf. Sci. 25*(1971), 171.
136. V. Y. Young, G. B. Hoflund and A. C. Miller, *Surf. Sci. 235*(1990), 60.
137. A. C. Miller, in *Treatise on Analytical Chemistry*, Part 1, Vol. 11 (J. D. Winefordner, ed.), Wiley, New York, 1989, p. 253.
138. G. B. Hoflund, D. A. Asbury, P. Kirszensztejn and H. A. Laitinen, *Surf. Sci. 161*(1985), L583.
139. J. A. Knapp, G. J. Lapeyre, N. V. Smith and M. M. Traum, *Rev. Sci. Instrum. 53*(1982), 781.
140. R. E. Gilbert, G. B. Hoflund, D. A. Asbury and M. R. Davidson, *J. Vac. Sci. Tech. A 6*(1988), 2280.

141. G. B. Hoflund and D. A. Asbury, unpublished results.
142. G. J. A. Hellings, Ph.D. thesis, Eindhoven University of Technology, 1986.
143. P. A. J. Ackermans, Ph.D. thesis, Eindhoven University of Technology, 1990.
144. R. Bergmans, Ph.D. thesis, Eindhoven University of Technology, 1991.
145. G. J. A. Hellings, H. Ottevanger, S. W. Boelens, C. L. C. M. Knibbeler and H. H. Brongersma, *Surf. Sci. 162*(1985), 913.
146. T. M. Buck, W. F. van der Weg, Y.-S. Chen and G. H. Wheatley, *Surf. Sci. 47*(1975) 244.
147. S. D. Gardner, G. B. Hoflund, M. R. Davidson and D. R. Schryer, *J. Catal. 115*(1989), 132.
148. O. Melendez, G. B. Hoflund, R. E. Gilbert and V. Y. Young, *Surf. Sci. 251/252*(1991), 228.
149. V. Y. Young and G. B. Hoflund, *Anal. Chem. 60*(1988), 269.
150. R. Germans, Ph.D. thesis, Eindhoven University of Technology, 1996.
151. F. Shoji, K. Kashihara, K. Oura and T. Hanawa, *Surf. Sci. 220*(1989), L719.
152. J. A. Leys, Proc. 1973 Pittsburgh Conf. Analytical Chemistry and Applied Spectroscopy, Cleveland, 1973.
153. M. R. Davidson, G. B. Hoflund and R. A. Outlaw, *J. Vac. Sci. Technol. A 9*(1991), 1344.
154. V. Y. Young, N. Welcome and G. B. Hoflund, *Phys. Rev. B 48*(1993), 2891.
155. H. Niehus, W. Heiland and E. Taglauer, *Surf. Sci. Rep. 17*(1993), 213.
156. R. Sonda, M. Aono, C. Oshima, S. Otani and Y. Ishizawa, *Surf. Sci. 150*(1985), L59.
157. R. Sonda and M. Aono, *Nucl. Instrum. Meth. Phys. Res. B 15*(1986), 114.
158. T. M. Thomas, H. Newmann, A. W. Czanderna and J. R. Pitts, *Surf. Sci. 175*(1986), L737.
159. E. Taglauer and W. Heiland, *Appl. Phys. 9*(1976), 261.
160. J.-P. Jacobs, S. Reijne, R. J. M. Elfrink, S. N. Mikhailov and H. H. Brongersma, *J. Vac. Sci. Technol. A 12*(1994), 2308.
161. G. C. Nelson, *J. Vac. Sci. Technol. A 4*(1986), 1567.
162. G. B. Hoflund, W. S. Epling and D. M. Minahan, unpublished data.

5

Compositional Analysis by Auger Electron and X-ray Photoelectron Spectroscopy

GRAHAM C. SMITH Shell Research and Technology Centre Amsterdam, Amsterdam, The Netherlands

1. INTRODUCTION

Auger electron spectroscopy (AES) and x-ray photoelectron spectroscopy (XPS) are the two most useful and widely applicable electron spectroscopic techniques for problem solving in surface analysis. The underlying physical principles of the techniques have been described in the introductory chapters of this book; here practical methods for their use in surface compositional analysis are discussed. Although the important equations are provided, the aim is not to give a rigorous technical treatment of the theories involved as that is available elsewhere [1, 2]; rather it is to give sufficient background to ensure the fundamental aspects are sufficiently understood for their successful application, and to provide a guide, or route map, to their use for compositional analysis.

Surface compositional analysis may be thought of as a cascade of processes in which the data interpretation is taken to increasing degrees of complexity, depending on the level of detail required to solve the particular problem under consideration. At the initial level, it may simply be sufficient to identify which elements are present near the surface of the specimen under examination. This may be adequate in simple cases where the requirement is, for example, to confirm the presence of a film or to assist in identification of a source of surface contamination. However, most problems will need additional information for their solution. The second level in the hierarchy of interpretation of AES and XPS data is the derivation of chemical state identification. Both techniques are essentially probes of the local electronic environment around the photoemitting or Auger electron-emitting atoms. The local electronic structure is intimately bound up with the nature of the chemical bonding, which may therefore be determined, at least in part, by careful examination of spectral line shapes and energy shifts. This may allow particular chemical species to be identified at the surface. For example, sulfur may have been identified at the first, elemental, level of data analysis. However, measurement of its XPS line energy to indicate its presence in SO_4^{2-} groups may be much more relevant to solving a practical problem. Chemical-state identification is a much more advanced science in XPS than AES. Nevertheless, useful complementary information can be gained by examination of Auger peaks, and this will be discussed further. At the third level in the hierarchy of data interpretation, the analyst is concerned with quantitative aspects of the specimen structure. This may include the determination of the relative atomic ratios of the elements present, the measurement of the overall composition of the specimen within the sampling depth of the technique, or the composition and likely thickness of any overlayers on the specimen surface. To continue the example, if the SO_4^{2-} groups were found to be confined to the outermost nanometer of the specimen, at a level corresponding to 20 at.% of the material present to this depth, and if it were known that the specimen came from an environment in which sulfuric acid might have been generated under certain conditions, then the analyst is well on the way to providing a true insight into the nature of the specimen and its provenance.

It is important to emphasize that surface compositional analysis by electron spectroscopy should not be carried out in isolation, and that wherever possible use should be made of evidence from other techniques. For example, a surface-specific infrared (IR) spectroscopy measurement can give bonding information which is complementary to that obtained from XPS. This complementarity may be exploited in studies of organic overlayers on metallic substrates where grazing-angle IR spectroscopy using reflection geometry can have monolayer sensitivity and can give evidence of, for example, carbon-carbon double bonds that are not distinguishable from carbon-carbon single bonds in the XPS photoelectron signal. For near-surface or thin-film analyses, a micro-ATR (attenuated total reflectance) IR accessory can prove very useful. As a counterpoint, XPS can access bonding configurations whose vibrational spectra are out of the usual range of conventional IR spectroscopy, and has the advantage that it is a quantitative technique (although the same claim may be made for IR spectroscopy when internal standards are used). Another popular and fruitful combination of bulk and surface techniques is that of electron probe microanalysis (EPMA), or energy/wavelength dispersive x-ray spectroscopy (EDS/WDS) as it is sometimes known, with AES. The electron beam in the AES spectrometer excites atoms which may decay nonradiatively, to give Auger electrons, or radiatively, to give characteristic x-rays. The Auger electrons are detected only if they originate from the top few nanometers of the specimen, whereas the x-rays can be emitted from depths up to a micron. Therefore, examination of the AES and EDS/WDS data from the same point on the specimen surface gives an immediate comparison of the difference between the outer surface and the more bulklike composition of the specimen. This can be very important in many technological applications, and for this reason an EDS detector may frequently be found fitted to an Auger electron spectrometer designed for practical problem-solving analyses.

Information is extracted from electron spectra from measurements of the energies and shapes of spectral features and from their intensities. In general, it is necessary to understand the nature of the spectral features before using their relative intensities to deduce structural or quantitative information. For this reason the remainder of this chapter is split into two parts. First, measurements of spectral energies and line shapes are discussed with a view to their use in giving elemental and chemical information. The second part of the chapter covers the determination of quantitative compositional or structural data from peak intensities.

2. SPECTRAL INTERPRETATION

Correct spectral interpretation relies to a great extent on accurate measurement of peak energies. A fundamental requirement for both XPS and AES is therefore an accurate calibration of the energy scale of the electron spectrometer, and, in an increasingly quality-conscious world, a traceable measurement through which the energy scale is validated. This requirement may be met by making use of the work of Seah [3] at the UK National Physical Laboratory (NPL). The NPL has published measurements of copper, silver and gold Auger and photoelectron peak energies which are traceable to atomic standards and which may be used with confidence. Essentially, measurements made of the appropriate lines on the spectrometer to be calibrated are compared with the reference values and the spectrometer is then adjusted until the difference between the measured and the reference values is less than an acceptable minimum across the energy range of interest. For XPS this is likely to become the subject of an ISO standard.

2.1. XPS Spectra

2.1.1. Elemental line energies

The principles of the XPS technique have been explained in Chapter 4. In summary, a soft x-ray, usually Al K_α at 1486.6 eV or Mg K_α at 1253.6 eV, gives all its energy to a core-level electron in an atom in the surface region of the specimen under examination. This electron is then photoemitted with a kinetic energy corresponding to the difference between the photon energy and the binding energy of the core level from which it originated, after allowing for the work functions of the specimen and the spectrometer. The spectrum of electron intensity versus energy acts as a fingerprint of the emitting atom type, allowing the identification of the elements present in the sample to a depth within which the photoelectrons can escape without losing energy. Usually, in modern spectrometers, the computer data system contains a database of peak positions and can provide reasonably reliable identification of the main spectral features. These may be verified by reference to handbooks of standard XPS spectra [4].

However, in practice, spectra are found to show a number of minor peaks which may or may not be analytically significant but need to be understood for a full interpretation. As a simple example, Fig. 1 shows an Mg K_α XPS spectrum for a reasonably clean piece of polytetrafluoroethylene (PTFE). This material contains only two elements, carbon and fluorine; nevertheless the spectrum shows a wealth of detail. Apart from the fluorine and carbon 1*s* photoelectron lines, the most obvious spectral features are the background intensity, which increases toward the high-binding-energy side (left-hand side, in this case) of each photoelectron peak, and the strong fluorine KVV Auger peaks. The background originates from electrons generated by the x-ray beam in the specimen, but which have suffered one or more inelastic collisions before being emitted from the surface. The shape of the background may be used to give information on the structure of the sample as a function of depth, using methods described in Section 3. In the case of the spectrum of Fig. 1, the sample is believed to be of composition approximately uniform with depth. The fluorine Auger peaks arise from decay of the photoemitting atoms from the excited state with a hole in the 1*s* level back to the ground state via the Auger process. Radiative decay to give a characteristic x-ray, as for EDS carried out in conjunction with AES, is also a possibility, but it would be very unusual to find an x-ray detector fitted to an XPS instrument for this purpose. The use of the x-ray excited Auger peaks to give additional chemical state information is of fundamental importance in practical XPS analyses and is treated separately in Section 2.3.

There are a number of minor features in the spectrum of Fig. 1 that are worthy of attention. As well as the fluorine Auger peaks, there is also the carbon KVV Auger feature weakly visible at around 990 eV apparent binding energy. Near the zero of binding energy there is a peak produced by valence-band electrons. In problem-solving applications the valence-band structure (or the conduction band in the case of a metallic or semiconducting sample) is usually ignored. However, the newer instruments have the sensitivity and resolution to enable the spectra of these relatively weak features to be acquired at

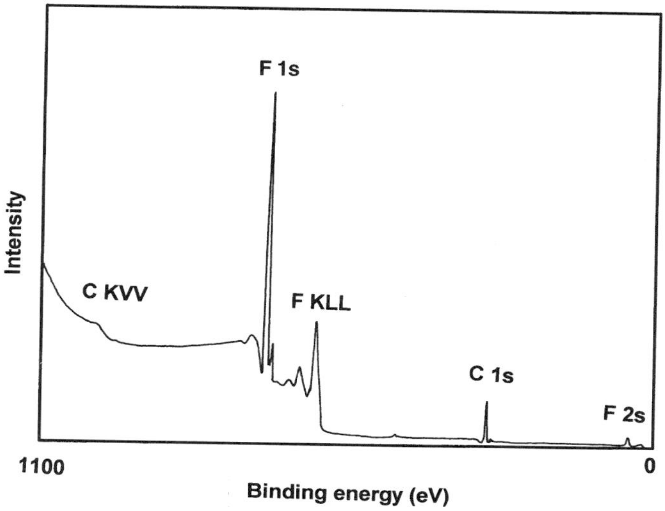

Figure 1. XPS survey spectrum of polytetrafluoroethylene obtained using Mg K_α x-radiation.

useful levels of signal-to-noise ratio without subjecting the sample to unacceptable levels of x-ray or thermal damage. In such cases, the study of valence-band structure can give information of immediate use in a practical analysis. The question of whether the sample is a conductor can be addressed by examining the electron density near the Fermi level. For polymeric samples, useful reference data exist for valence-band spectra with which the unknown may be compared [5].

The remainder of the features in the spectrum shown in Fig. 1 are, to a greater or lesser extent, artefacts that must be understood and categorized to avoid the risk of identifying them falsely as additional elemental photoelectron peaks (which an automated spectral-labeling computer program almost certainly will do, if given the opportunity). Each of the main photoelectron peaks (the C 1*s* and F 1*s* in this case) has a minor peak associated with it at approximately 8–10 eV lower binding energy and with approximately 5% of the intensity of the main peak intensity. The spectrum was recorded using nonmonochromatic Mg K_α radiation, which has a number of satellites associated with it, each of which produces its own photoelectron spectrum. They are generally so weak that only the most intense lines in the main spectrum, produced with the $K_{\alpha1,2}$ line, have a visible satellite peak from the associated $K_{\alpha3,4}$ and other lines. The satellite peaks mentioned originate from the $K_{\alpha3,4}$ lines; occasionally K_β satellites are also observed for particularly strong photoelectron peaks. Nonmonochromatic Al radiation also produces satellite peaks, but at different energies. Some of these may be very inconvenient. For example, the Mg K_β satellite of the C 1*s* line overlaps neatly with the expected position for Mo 3*d* electrons. In such cases, there is no substitute for the use of a monochromatic source. As an aid to the interpretation of XPS data from nonmonochromatic x-radiation, Table 2.1 gives a compilation of the principal satellite energies and relative intensities.

The spectrum of Fig. 1 was generated using a twin-anode x-ray source which produces either Mg or Al radiation, depending on which filament is energized. Unfortunately, the mutual exclusion of Mg and Al in such a source is never perfect and there is a small amount of crossover of intensity. The small feature at a binding energy of approximately 456 eV in the spectrum is, in this case, due to a low level of Al K_α radiation exciting the F 1*s* peak. Such features always occur at a separation of 233 eV from the main peak, that being the energy difference between the Mg and Al K_α lines. Other sources of unwanted radiation may also occur with the conventional twin-anode Mg/Al design. After extended use, the Mg anode may become oxidized, particularly if it has been used at high power in less than ideal vacuum conditions. The oxide coating on the anode results in the emission of weak oxygen x-rays which also contribute to the spectrum of photoelectrons emitted from the specimen. Generally these are not a serious problem except for a few critical peak overlaps. The most inconvenient of these is probably the overlap of the Mg K_α-excited carbon KVV Auger peak, which contains important chemical information, with the O K_α-excited carbon 1*s* line. The handbook of Wagner et al. [4] gives a table of the common spurious peaks and the means by which they may be identified. In practice, it is worthwhile running a spectrum from a simple material, such as PTFE, to assess the extent of artefact generation before embarking on an im-

Table 2.1. X-ray Satellite Energies and Relative Intensities for Mg and Al Radiation

	Mg		Al	
Satellite	Displacement (eV)	Relative height	Displacement (eV)	Relative height
$\alpha_{1,2}$	0	100	0.4	100
α_3	8.4	8.0	9.8	6.4
α_4	10.2	4.1	11.8	3.2
α_5	17.5	0.55	20.1	0.4
α_6	20.0	0.45	23.4	0.3
β	48.5	0.5	69.7	0.55

portant analysis of a series of complex materials, particularly where it is thought that a number of less common elements may occur at low intensity.

As though to compensate for the production of a number of satellite peaks, the twin-anode Mg/Al source has the important advantage that running the same sample with the two different x-ray energies can assist greatly in the correct identification of the minor spectral components. In essence, the Auger peaks change their apparent binding energies by 233 eV, whereas the photoelectron peaks do not. The satellite structure associated with the more intense photoelectron peaks also changes, again giving a valuable aid to interpretation. In extreme cases, where several overlaps occur, this may be an essential prerequisite for an analysis.

2.1.2. Photoelectron line shapes

The shape of a photoelectron line in an XPS spectrum is a convolution of three factors: the true shape of the excited peak, the line shape of the x-ray source, and the energy spread function (the resolution) of the electron spectrometer. The true line shape itself may be modified by processes occurring in the specimen as a result of the excitation and emission from the solid of the photoelectron. The line shape is a strong function of the local electronic environment around the photoemitting atom; therefore an understanding and appreciation of the factors which govern this important parameter are essential if the maximum information on chemical bonding is to be extracted from the spectrum.

Taking the instrumental factors first, the electron energy analyzer resolution is the easiest with which to deal. For many instruments this can be approximated quite well by a simple Gaussian line shape. However, it is essential for any kind of detailed interpretation that the spectra be acquired with the highest practical resolution. There is always a trade-off between resolution and signal strength, and increasing the analyzer resolution is subject to a law of diminishing returns, as other terms, such as the x-ray source line width, then become of comparable magnitude. Manufacturers of instruments typically quote the count rate on the unmonochromatized Ag $3d_{5/2}$ peak at a resolution of around 1.0 eV as a figure of merit, and without a monochromator there is little point in trying to operate at better resolution. The shape of the exciting x-ray line is frequently approximated by a Lorentzian function, which is why curve fitting is often carried out using a mixed Gaussian-Lorentzian line shape function (see Section 2.1.4), and it is thus, in principle, possible to deconvolute the line shape from the spectra. However, experience suggests that this can result in the generation of artefacts and should perhaps be avoided except with spectra of all but the very highest signal-to-noise ratio. If a monochromator is available, its use will greatly facilitate the interpretation of chemical information from the spectra. Earlier designs suffered from a severe loss of signal strength, but the best modern instruments retain good sensitivity even at an overall system resolution of around 0.5 eV. While resolution at this level can be very valuable, it should be remembered that much good work can also be done with the conventional twin-anode systems typically used in practical problem-solving analyses.

The simplest line shapes are often produced by photoemission from the 1*s* levels of the lighter elements of the periodic table. Of these, the most frequently met is probably the carbon 1*s* line at a binding energy of approximately 285 eV. Even nominally carbon-free samples will exhibit this line as a result of exposure to atmospheric contaminants, including residual hydrocarbons in the UHV atmosphere of the spectrometer. For saturated hydrocarbons, the line is approximately symmetric, although at the highest energy resolution broadening on the high-binding-energy side can be discerned and is attributable to excitation of vibrational levels. In graphitic carbon, the line shows a distinct asymmetry. The two shapes are shown in Fig. 2, recorded using nonmonochromatic Mg K_α radiation.

The asymmetry in the graphitic carbon 1*s* line shape arises from the presence in the overall peak envelope of those electrons emitted from the 1*s* level which have lost small amounts of energy to the delocalized π electrons characteristic of the planar atomic arrangement and hexagonal symmetry of the graphite structure. These delocalized electrons may also undergo collective oscillations, known as plasmons, on passage of a photoelectron. The photoelectron then loses the appropriate amount of energy required to excite such an oscillation and appears in the spectrum as a distinct peak separated from the main peak by an amount known as the plasmon loss energy. For graphite, this is around 7 eV, and a weak feature can indeed be observed at this energy in the spectrum of Fig. 2. In aromatic compounds, π-bonding occurs but does not result in the formation of a conduction band as there is no overlap between parallel atomic planes. However, the photoelectron can excite transitions from the filled π orbitals to unfilled π* orbitals, resulting in a small loss of energy and giving rise to a subsidiary peak at approximately the same energy difference from the main 1*s* peak as the plasmon peak in graphite. Polystyrene and similar materials show this effect most strongly, and the intensity of the π-π* feature can approach 10% of that of the principal peak.

Free-electron metals such as aluminum show plasmon effects strongly, and occasionally double or even triple plasmon loss peaks may be observed. Inspection of the *Handbook of XPS* [4] reveals several examples, particularly from aluminum and similar materials. Simple free-electron theory gives the plasmon energy, ω_p, as [6]

$$\omega_p = \left(\frac{ne^2}{\varepsilon_0 m} \right)^{1/2} \tag{1}$$

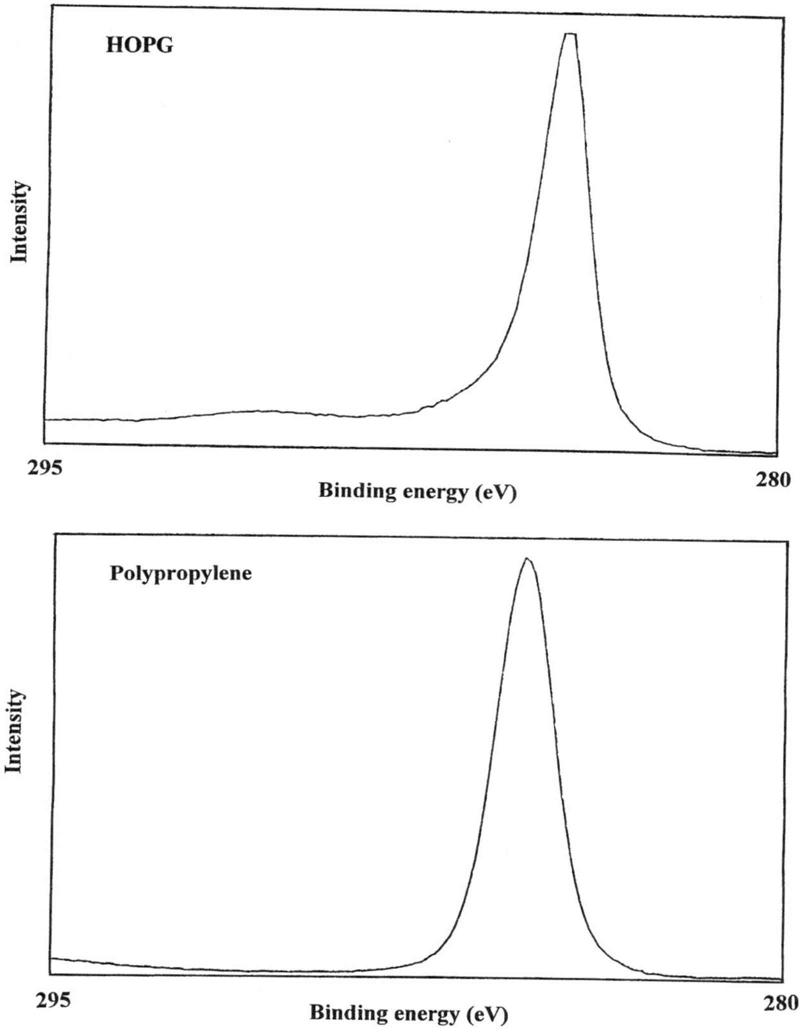

Figure 2. Carbon 1*s* XPS spectra for graphite and polypropylene.

where n is the electron density, e is the electronic charge, ε_0 is the permittivity of free space and m is the mass of the electron. Where applicable, good agreement between theory and experiment is found. At a flat surface the theory predicts an additional surface plasmon at an energy of $\omega_p/\sqrt{2}$, but this is rarely observed in applied surface analytical work.

In transition metals, other factors come into play and tend to give added structure to the shapes of the primary photoelectron lines. These metals typically have a high density of filled and unfilled electron states near the Fermi level. Photoelectron peak broadening may occur through small energy losses caused by the excitation of electron-hole pairs by the primary photoelectron. The effect is stronger for metals with the highest densities of states at the Fermi level, giving rise to a marked asymmetry, which must be taken into account in peak fitting (see Section 2.1.4). The resulting peak shape is accurately described by the Doniach-Sunjic equation [7]

$$I(\varepsilon) = \frac{\Gamma(1-\alpha)\cos[\pi\alpha/2 + \theta(\varepsilon)]}{(\varepsilon^2 + \gamma^2)^{(1-\alpha)/2}} \tag{2}$$

where Γ is the gamma function, ε is the energy, γ is the lifetime width of the core hole, and

$$\theta(\varepsilon) = (1-\alpha)\tan^{-1}\left(\frac{\varepsilon}{\gamma}\right) \tag{3}$$

$$\alpha = 2\Sigma_l(2l+1)\left(\frac{\delta_l}{\pi}\right)^2 \tag{4}$$

l is the orbital angular momentum quantum number, and δ_l are the appropriate partial wave phase shifts. Where there are no phase shifts (i,e, no losses), α goes to zero and the line shape goes to a Lorentzian form. Although it is important in the understanding of the origin of this source of asymmetry, the equation for the Doniach-Sunjic line shape is not often used in practical work. Instead it is typical to model the line shape with a mixed Gaussian-Lorentzian function and to treat asymmetry in a pragmatic way by, for example, adding an exponentially decaying tail to the peak.

As well as causing changes to the shape of the primary line, electronic interactions during the photoemission process can also give rise to additional minor spectral features. In the first row of transition elements the $2p$ photoelectron peaks show well-separated $2p_{1/2}$ and $2p_{3/2}$ doublets with increasing asymmetry on going from Sc to Mn for the metals and the oxides. From Fe to Cu, subsidiary peaks become evident on the high-binding-energy sides of the principal components. At the end of the first row of transition elements, after Zn, the subsidiary peaks lose their intensity. The subsidiary peaks are known as shake-up features and are due to electronic rearrangements that occur during the photoemission process. In principle, photoemission can be considered as either a sudden or an adiabatic process. In the sudden approximation the electronic structure of the emitting atom is frozen and does not change during the emission of a photoelectron. The kinetic energy of the photoelectron then accurately reflects the binding energy of the level from which it was emitted. In the adiabatic approximation the atom is allowed to relax from the excited state in which it finds itself after absorption of an x-ray photon and creation of a core-level hole, before the photoelectron is emitted. The photoelectron energy in that case represents the excess energy left after any internal electronic rearrangement. The relaxation may involve sufficient energy to promote a valence electron to an unfilled higher level, representing a fixed energy loss suffered by the photoelectron which appears as a satellite, or shake-up, peak on the higher-binding-energy side of the main peak. The process is most common where there exist unpaired electrons in $3d$ or $4f$ levels, and in such cases its presence is a good indicator of the valence state of the emitting atom. CuO, NiO and CoO are good examples. Cu metal, for example, has the electronic configuration of Ar + $4s^2 3d^9$, with overlap of $3d$ levels to form a well-defined conduction band. In the Cu^{2+} state two electrons are lost to the ionic bond, and two electrons are promoted to give a filled $5s$ level, leaving a half-filled $3d$ level and the configuration Ar + $4s^2 3d^5 5s^2$. The $5s$ electrons readily take part in shake-up processes, leading to strong satellite peaks from, for example, CuO. In Cu_2O, on the other hand, Cu^+ has a $3d^8$ configuration, and does not show shake-up features.

To show that complex shake-up features are of more than academic interest, consider the Ce $3d$ photoelectron spectrum in Fig. 3. Ceria is widely used industrially, particularly in catalysis, where it can act in alcohol synthesis, heavy-oil cracking, and automotive exhaust gas catalytic converters. In the automotive application, the reversible $CeO_2 \Leftrightarrow Ce_2O_3$ reaction is used to store oxygen from the reduction of NO_x and release it for the oxidation of CO and hydrocarbons. XPS analysis is often used to determine the oxidation state of the ceria after a range of operation cycles, to act as a diagnostic, for example, for catalyst formulation optimization. In CeO_2 the Ce is believed to be in the $4f^0$ configuration, having lost two $5s$ electrons and two $4f$ electrons to give Ce^{4+}. However, overlap of the O $2p$ orbitals with the Ce $4f$ levels gives a strong f-like character to the valence band. In Ce_2O_3 there is less overlap and the Ce is in the $5s^0 4f^1$ configuration, giving Ce^{3+}. Photoelectron excitation of either of these materials will result in shake-up effects, lowering the $4f$ levels in Ce_2O_3 and CeO_2 to a point where strong overlap with the O $2p$ levels can occur. Possible final states for CeO_2 include Ce $4f^2$ O $2p^4$, Ce $4f^1$ O $2p^5$ and Ce $4f^0$ O $2p^6$, whereas for Ce_2O_3 only the Ce $4f^2$ O $2p^4$ and Ce $4f^1$ O $2p^5$ occur. Therefore the complete Ce $3d$ spectrum for a partially reduced specimen of CeO_2 contains a total of 10 features, taking into account the $3d_{5/2}$ and $3d_{3/2}$ contributions. These are indicated in Fig. 3, from Romeo et al. [8], where the CeO_2 states in the $3d_{5/2}$ line are labeled v, v'' and v''', and the Ce_2O_3 states are labeled v_o and v'. The corresponding $3d_{3/2}$ states are labeled u, according to the same scheme. The intensity of the u''' peak around 918 eV binding energy is frequently taken as a measure of the ability of the specimen to undergo reversible oxygen storage,

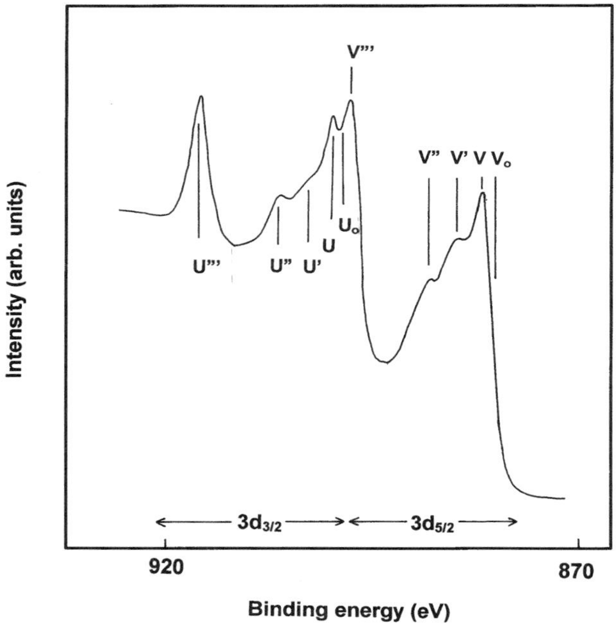

Figure 3. XPS spectrum of the 3*d* photoelectron lines of ceria. (After Ref. 8.)

although the relative intensity of the peak is not directly proportional to the amount of CeO_2 because of the effects of hybridization on the bonds.

2.1.3. Chemical shifts

XPS is a powerful technique when used for the determination of surface composition, but to exploit it to the full, use should be made of its sensitivity to chemical bonding. As discussed above, XPS is a probe of both atom type and local electronic environment. The shake-up features, plasmon losses and line shape changes all give information on the electronic structure of the photoemitting atom. However, for general problem-solving work, much more use is made of the so-called chemical shifts that occur as a result of changes in chemical bonding. These shifts in peak position may be quite large, up to 10 eV in some cases, and are readily interpreted with the aid of reference materials or a suitable database.

The chemical shifts originate from the changes in core-level binding energies resulting from bonding differences. In the simplest model, due to Siegbahn et al. [9], the change in binding energy of a particular core level a, due to bonding to neighboring atoms b, ΔE_a, is given by

$$\Delta E_a = k\Delta q_a + \Delta \sum_b \frac{q_b}{r_{ab}} + L \quad (5)$$

where q_a and q_b are the charges on atoms a and b, separated by a distance r_{ab}. The constants L and k are determined by experiment. This model has been validated for simple gaseous compounds [10]. In practical work, measured core-level shifts are compared to reference data [11] or, potentially more reliably, to data acquired in the same laboratory from a range of known compounds or materials similar to the specimen under analysis.

Different elements show great variations in chemical shifts. The shifts tend to be larger for elements near the beginning of the periodic table, where the core levels giving the photoelectrons of interest are less screened from the bonding orbitals than in heavier elements, and for elements participating in the formation of anions rather than cations. For example, the 2*p* photoelectron peak from sulfur shows energies ranging from 161 to 162 eV for metallic sulfides, through 163 to 165 eV for organosulfur compounds, and up to 168 to 169 eV for inorganic sulfates. The phosphorus 2*p* peaks show a similar range, although in this case the wide range of possible valencies makes it difficult to distinguish between, for example, ortho-, meta- or pyrophosphates. Nitrogen 1*s* binding energies also vary over a large range, from typically 397 eV for nitrides through approximately 399 eV for amines, 400–401 eV for amides, 401–402 eV for quaternary nitrogen compounds, up to 406–407 eV for nitrates. However, other elements do not necessarily act as such good chemical state indicators. The oxygen 1*s* line, for example, is seen at around 530 eV from metal oxides but generally gives a rather broad spectral feature in the range 531–533 eV for a range of organic bonding configurations. Many metals give distinct shifts which enable identification of their valency states, but are not particularly anion selective. For example, Ca metal can be readily distinguished from Ca^{2+}, but it would be difficult to differentiate with certainty between Ca^{2+} in $CaCO_3$ or $CaSO_4$. Similarly, Mo, Mo^{4+} and Mo^{6+} are clearly separable, but it is not so easy to be confident about the separate identification of MoS_2 or MoO_2. Of course, in both these examples, the additional information available from the C, O or S lines would enable a distinction to be made. This illustrates the importance, particularly in problem-solving applications, of always checking for self-consistency in chemical-state assignments. Put simply, if, for example, S is identified as a sulfide, is there a metal ion of the appropriate valency with which it could form a compound?

The carbon 1*s* line deserves special mention in any discussion of the practical use of chemical shifts. It is found on the surface of almost any specimen that has not been specially created or treated within the ultrahigh vacuum system, and it forms the foundation for a number of practical applications areas including polymers and carbon fibers. Polymers have been extensively researched by Briggs and Beamson [5, 12, 13]. Nevertheless, it is worthwhile covering at least part of this work here in overview form.

Carbon in the amorphous hydrocarbon form typical of the contaminant layers encountered in practice usually gives a 1*s* photoelectron peak around 285.0 eV binding energy. If it is a few eV or more from this energy, then electrostatic charging of the sample is indicated. Indeed, it has been normal practice to reference the binding energies on insulating samples to this line, conventionally defined to be at 285.0 eV. Small variations in the energy of the C 1*s* line do occur, even for C—C bonded carbon. The graphitic type of carbon found on the surfaces of carbon fibers, for example, gives a peak at 284.6 eV, whereas pure saturated aliphatic hydrocarbons may give peaks in the range 285.1–285.3 eV [14]. Oxygen-bonded carbon is frequently found in polymers, organic coatings, surface release agents and plasticizers, among other materials, with the different types of carbon-oxygen bond giving distinct chemical shifts in the carbon 1*s* line. The shift depends to a certain extent on the nature of the carbonaceous material (for example, whether it is primarily graphitic or organic), so it can be important to identify the type of carbon structure present before proceeding with the interpretation. This may be accomplished with the aid of the carbon KVV x-ray-excited Auger line shape, as discussed later in Section 2.3. Shifts in the oxygen-induced carbon 1*s* line, and the accompanying oxygen 1*s* and 2*s* shifts, have been studied for oxygen-containing polymers by Beamson and Briggs [12]. Similar shifts for graphitic carbon surfaces have been investigated by Desimoni and coworkers [14]. The results from the two sets of investigations are summarized in Table 2.2. In cases where carbon-oxygen bonding is present but other species also occur, the different chemically shifted components may overlap and it can prove very difficult to obtain a definitive chemical-state identification. A typical example may be in oxygen- and nitrogen-containing polymers where the C—O and C—N contributions to the C 1*s* line overlap and are not usually separately identifiable unless prior knowledge is available. This is not usually a problem if the other element is fluorine, as carbon-fluorine shifts are very large and there is in that case little room for doubt when assigning the chemical states.

When assigning chemical states it is important to ensure consistency between the different elements present. In the example of the oxygen-containing polymers discussed above, if there is, say, evidence for ester bonding in the C 1*s* line, then the O 1*s* line should be checked for the presence of the equivalent state. Table 2.3. shows data taken from various sources in the literature for nitrogen in a variety of chemical environments. Where carbon-nitrogen bonding is involved, the carbon binding energy for the same state is also shown. By making comparisons in this way, it is possible to ensure a self-consistent interpretation of chemical-state data for all elements present in an unknown spec-

Table 2.2. Oxygen-Induced C 1*s* Shifts for Polymers and Carbon Fibers[a]

Carbon chemical environment	C 1*s* binding energy in polymers (eV)	C 1*s* binding energy on carbon fibers (eV)
Graphite, aromatics		284.6
Saturated hydrocarbon	285.0	285.1–285.3
Alcohol, phenol, ether	286.5	286.1
C–O in ester	286.7	
Carbonyl	287.9	287.6
C=O in ester	289.1	
COO	289.3	289.1
Carbonate	290.4	290.6
Plasmon		291.3
π-π^*	291.7	

[a]The values for polymers have been averaged and rounded to one decimal place.
Source: From Refs. 12 and 14.

imen. If it is also possible to make this agreement quantitative (see Section 3), then a good level of confidence may be placed on the result.

It should be mentioned that so-called secondary shifts can be of importance, particularly when working at high resolution and using monochromated radiation. All the chemical shifts discussed so far have been primary shifts. That is, the binding energy for the core level of the atom in the particular electronic en-

Table 2.3. Binding Energies for Carbon and Nitrogen in Various Chemical Environments

Chemical environment	N 1*s* binding energy	C 1*s* binding energy
Nitride	397.3–398.5	—
Nitrile C≡N	399.6	286.8
Amine	399.3 (primary)	285.8–286.2
	400.2 (secondary)	
Amide	399.5–400.6	287.5–288.0
Urea, NH_2CONH_2	400.6	289.0
Carbamate, NH_2COO^-	398.7	289.5
Quaternary salt, NH_4^+	401.3	—
NO	403.0	—
NO_2^-	405.9	—
NO_3^-	407.5	—

vironment in question is the one showing the shift. However, in some cases, adjacent species can also be affected. An example is shown in Fig. 4, where the CH_2 component of the fluoropolymer illustrated is not at the usual hydrocarbon value of approximately 285.0 eV. Instead, it appears at a higher binding energy due to the effect of the adjacent CF_2 group; i.e., it has suffered a secondary shift. This is important for polymer studies, where secondary shifts due to a range of bonding configurations may be detected. The subject is discussed in detail by Briggs and Beamson [12].

2.1.4. Curve fitting

Frequently the elements in a specimen undergoing XPS analysis are found to exist in more than one chemical state. Unless the chemical shifts are very large, for example sulfide to sulfate in the S 2*p* line or CH_2 to CF_2 in fluoropolymers, it is necessary to apply some procedure to separate the corresponding peaks. This is usually accomplished through curve fitting. In curve fitting the objective is to find a set of model XPS line shape functions which, when correctly combined over the energy range of interest, accurately simulate the measured experimental data. Usually, guesses are made at the forms of the functions required, based on prior knowledge of the specimen chemistry or on a combination of experience and intuition, which are then iterated until the desired degree

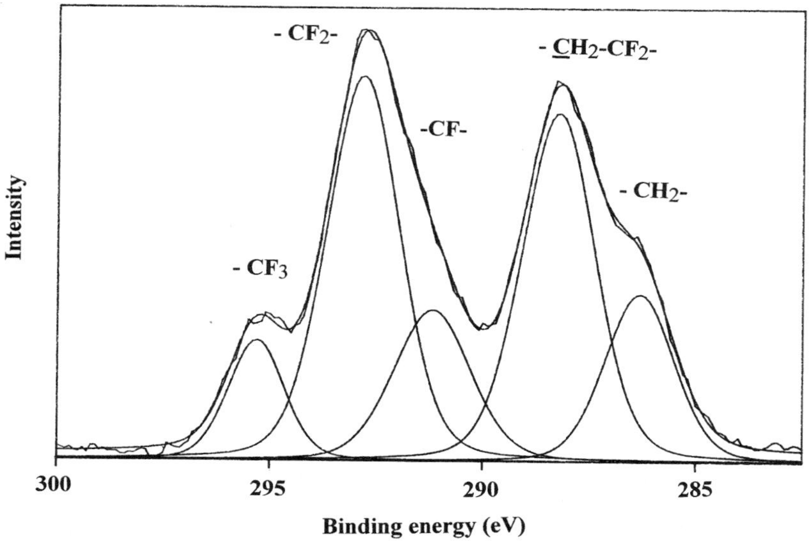

Figure 4. Carbon 1*s* photoelectron spectrum for the commercially important elastomer hexafluoropropylene vinylidene fluoride.

of convergence between the model and the experimental data is achieved. The process is different from deconvolution, which attempts to remove instrumental and other terms from the measured data leaving only the "true" data, and the two should not be confused.

Curve fitting may be carried out either in a "free" manner or using constraints. Generally, the use of physically realistic contraints results in a fit that is more physically acceptable. Typical examples of the use of constraints would be the fixing of the $2p_{3/2}$ to $2p_{1/2}$ ratio in an overlapping doublet from a simple element such as sulfur or phosphorus, at least initially, and to set equal the linewidths of the two components. Alternatively, where different chemical forms of the same element are known to be present in certain ratios, that information could be included in the fit. For example, in the curve-fitted carbon 1*s* spectrum of the fluoropolymer hexafluoropropylene-vinylidene fluoride shown in Fig. 4, the areas of the three components of the hexafluoropylene block, namely CF_3, CF_2 and CF, are equal. The CF_2 and CH_2 components of the vinylidene fluoride monomer are also of equal intensity, and a very satisfying fit is found.

The components in Fig. 4 were fitted using a mixed Gaussian/Lorentzian line shape. The rationale behind such a model line shape is that, generally, the finite energy resolution of the spectrometer will contribute a Gaussian form to the line, and other factors such as the x-ray source line shape and the intrinsic width of the photoelectron line will tend to be of Lorentzian form. In practice, either a sum or a product of Gaussian and Lorentzian functions may be used. The sum and product functions respectively are given by Evans [15] as

$$f^{\text{sum}}(x) = H_{\text{max}}\left[\frac{m}{1+F} + (1-m)\exp(-F\ln 2)\right] \tag{6}$$

and

$$f^{\text{product}}(x) = \frac{H_{\text{max}}}{(1+mF)\exp[1-m)F\ \ln 2]} \tag{7}$$

where, in each case, H_{max} is the peak height, m is the mixing ratio and

$$F = \frac{(x-x_0)^2}{\omega^2} \tag{8}$$

where x_0 is the center value of the peak and ω is the peak width parameter, and m has a value of zero for a pure Gaussian curve and unity for a pure Lorentzian. The parameter ω is the half-width at half the maximum height in the case of the sum function, and a value very close to this for the product function. Calculation shows that the sum function gives a closer representation of theoretical line shapes; however fits to real experimental data usually give lower residuals (i.e., less residual unfitted data) if the product function is used

[15]. This probably occurs because, in practice, spectra are not measured over a sufficiently wide energy range to allow accurate reproduction of the long tails of the Lorentzian component, and a truncated Lorentzian function is very much closer to the shape of the product function than to the sum function.

Once the line shape has been decided, the user of the curve-fitting program will wish to specify the number of components, together with their estimated widths, energies and heights. These may or may not be constrained according to any prior knowledge of the specimen, depending on the sophistication of the software, following which the computer attempts to minimize iteratively the difference between the measured data and the calculated fitting curves until some minimum is reached. This minimization involves nonlinear methods outside the scope of this chapter, with different computer software suppliers using a range of alternative procedures. The damped nonlinear least-squares method of Hughes and Sexton [16] seems particularly well suited to XPS data.

For practical purposes, convergence of the fitting algorithm at the level of 1 part in 10^4 is probably satisfactory, but it can prove difficult to decide when an optimum fit has been achieved. Generally, the technique is to vary the constrained parameters of the fit while monitoring the weighted sum of least squares, χ^2, given by

$$\chi^2 = \sum_i \left[\frac{c_i - c_i^{\text{calc}}}{\sigma} \right]^2 \tag{9}$$

where the summation is from 1 to N, c_i and c_i^{calc} are the measured and calculated values of the counts in channel i, and σ is the standard deviation of the number of counts in that channel. Alternatively, the reduced χ^2 may be used, where

$$\text{reduced } \chi^2 = \frac{\chi^2}{N - M} \tag{10}$$

with N the number of channels in the spectrum to be fitted and M the number of independently adjustable parameters, for example, peak positions or widths.

Unfortunately for the practicing electron spectroscopist, it is very easy to produce an apparently good fit to the experimental data which is actually physically meaningless. Recognition of this has led to research on the nature of the errors involved in the curve-fitting process. Cumpson and Seah [17] point out that the probability of finishing a good fit with a value of $\chi^2_{\min}$ (denoted by t) greater than some value τ purely by chance is $Q(\tau \mid \nu)$, where

$$Q(\tau \mid \nu) = \int_\tau^\infty f(t, \nu)\,dt \tag{11}$$

Here, ν is the number of degrees of freedom given by $N - M$, and $f(t,\nu)$ is the probability density function

$$f(t,\nu)\ dt = [2^{\nu/2}\Gamma\left(\frac{\nu}{2}\right)]^{-1} t^{\nu/2-1}\ \exp\left(\frac{-t}{2}\right)\ dt \tag{12}$$

$Q(\tau \mid \nu)$ closely resembles a complementary error function for values of $\nu \geq 30$. An approximate method of calculating $Q\ (\tau \mid \nu)$ is given by Cumpson and Seah [17]. For a good fit, $Q(\tau \mid \nu)$ should be around 0.5. A very small value of $Q(\tau \mid \nu)$, say ≤ 0.005, indicates the presence of systematic uncertainties in the fit, which should be modified to bring $Q(\tau \mid \nu)$ within the range $0.005 \leq Q(\tau \mid \nu) \leq 0.995$. Within this range, Cumpson and Seah [17] give a method for estimating the uncertainties in the fitted parameters. However, the method is not yet available on commercial software. In practical cases, where an estimate of the uncertainty in a fitted parameter is required, then the approach of Evans [15] may be more appropriate. The method involves monitoring the change in χ^2 as each parameter in the fit is individually varied. It does not give an estimate of the precision of the fit, but does give an indication of the sensitivity of the fit to the parameters. Not surprisingly, in practical examples the fit is generally more sensitive to peak positions, heights and widths than to the Gaussian-Lorentzian mixing ratio or the degree of asymmetry.

2.2. Auger Electron Spectra

The fundamentals of Auger electron spectroscopy (AES) have been described in Chapter 4. The Auger effect is a nonradiative process whereby an atom in an excited state, with a core-level hole, can relax to its ground state, losing an electron as it does so. The Auger process competes with the alternative radiative decay route, x-ray fluorescence. Generally the higher-atomic-number elements tend to favor x-ray emission while the lighter elements favor Auger electron production, although for core-level binding energies of less than approximately 2 keV Auger emission is the dominant process. The initial excitation to create the core-level hole may be through an electron beam, an x-ray beam or (rather less commonly) an ion beam. The latter tends to be a curiosity, suitable for very specific purposes only and is not in any real sense an analytical spectroscopic technique. The term "Auger electron spectroscopy" is reserved for electron beam excitation, this being the preferred method offering the advantages of focused beams with their high flux density and ability to form images. The use of an x-ray beam is an extension of XPS and has already been mentioned; it is discussed further in Section 2.3.

2.2.1. Elemental line energies

Consider an atom in which a vacancy has been created in core level x. In the Auger process an electron drops from a higher level, y, to fill the hole, and a further electron is emitted from level *z*, as an *xyz* Auger electron, taking the excess energy as its kinetic energy. X-ray notation is used rather than the spectroscopic notation conventionally employed in XPS. Thus, the strong Auger peaks arising as a result of a hole in the 2*s* level of first-row transition elements may typically be LMM lines, whereas second-row elements show stronger MNN lines due to 3*s* holes. If valence electrons are involved, for example in emission from the first row of elements in the periodic table, then the notation V may be used as a shorthand. The carbon or oxygen KLL peaks at kinetic energies of approximately 270 or 500 eV, respectively, may be designated KVV in this notation. In the preceding hypothetical example given, the kinetic energy of the Auger electron, E_{xyz} is

$$E_{xyz} = E_x - E_y - E_z \tag{13}$$

This is oversimplified, as it ignores the relaxation of the ionized atom between excitation and emission. In practice, the measured kinetic energies are always lower than would be calculated from such a simple formula. As a first step toward improving the calculation of E_{xyz} the change in energy levels between the initial and the final states of the emitting atom of atomic number *Z* can be approximated by replacing the final term with the equivalent energy for the element of atomic number *Z* + 1 to give

$$E_{xyz}^{Z} = E_x^{Z} - E_y^{Z} - E_z^{Z+1} \tag{14}$$

Recognizing that an *xyz* transition is indistinguishable from an *xzy* transition leads by a logical extension to

$$E_{xyz}^{Z} = E_x^{Z} - \frac{E_y^{Z} + E_y^{Z+1}}{2} - \frac{E_z^{Z} + E_z^{Z+1}}{2} \tag{15}$$

Use of this equation, due to Chung and Jenkins [18], gives reasonably accurate results. However, the use of such equations has now largely been rendered unnecessary for practical analytical purposes by the availability of good reference data for elemental line energies [19–22]. Of these, only the *Physical Electronics* handbook of Davis et al. [19] seems to be widely available in surface analytical laboratories; nevertheless, tabulations exist in other texts [2].

Calculations of Auger electron energies from first principles give, of course, the peak positions in the direct spectrum. However, many electron spectrometers, particularly the earlier designs, give spectra in the differentiated form. For differential spectra the convention is to take the position of the most intense negative excursion as the energy of the peak. This of course does not correspond to the true position, and the difference between the two should be borne in mind when using tabulated values. Modern spectrometers, particularly of the concentric hemisphere design, tend to produce direct, undifferentiated spectra but such is the strength of historical precedent that the spectra are often subsequently differentiated numerically for presentation to the user. Differentiation has the advantage of eliminating, or at least greatly reducing, the sloping spectral background and making the peaks more visible. However, it is accompanied by smoothing, so fine structure in the peak may be lost. It is up to the spectroscopist to make an informed choice of the type of data presentation required for solving a particular problem.

2.2.2. Chemical shifts

The basic equation for Auger electron kinetic energies given above may be modified in a physically realistic way to give the correct expression by the inclusion of an interaction energy, U, such that

$$E_{xyz} = E_x - E_y - E_z - U \tag{16}$$

U can be calculated and combined with experimental values of the binding-energy terms to give accurate values for E_{xyz}, for which tabulations have been published [23]. Shifts in energies of Auger lines as a result of changes in the local electronic environment of the emitting atom may occur as a result of either changes in the binding energies of the levels concerned (as in XPS) or of changes in the interaction term, U.

Generally, the chemical effects in AES are not so extensively exploited as in XPS. This is partly historical, as the early instruments tended to give broad differentiated peaks from which it was difficult to distinguish chemical shift information. This is not a problem with modern equipment. The most commonly used shifts are those occurring between metals and their oxides and in the carbon KVV line. There is no equivalent of the XPS handbook or NIST database for Auger chemical shifts, although there is a body of literature from which data can be extracted for comparison. Some examples of chemical differences in Auger line shapes and energies can be found in the *Physical Electronics* handbook [19]. The most striking of these are probably the differences between the metallic and oxide states of Mg and Al, due to the presence of strong plasmon loss features in the metallic spectra that are not present in the oxide spectra. The low-kinetic-energy peak due to Si in silicon nitride is seen to be similar to that for pure Si at 92 eV, but in SiO_2 the peak shifts to lower kinetic energy and splits to form a doublet. However, other series of related materials show less prominent effects with, for example, very little difference in the K LMM peak shape or position between KCl, KBr and KI, or in the I MNN peak between KI, RbI or CsI. Some data on Auger peak-energy shifts and on the Auger parameter (see Section 2.3) are available in the NIST XPS database (for direct rather than derivative spectra), but, in general, experience suggests that the analyst may fare better by performing analyses of reference materials of interest rather than relying on published data.

The only element for which good Auger chemical shift and line shape data are readily available in the literature is carbon, for which the KVV line at approximately 272 eV kinetic energy (negative peak, differential spectrum) has been the subject of a number of studies. Most practicing surface analysts would recognize immediately the signature in the Auger spectrum of carbon as graphitic carbon or carbon in the form of a metallic carbide. The use of the carbon KVV peak to obtain further information on the nature of the chemical bonding in the specimen in discussed in Section 2.3.

Recently, some impetus has been given to the use of Auger chemical shift data by the availability of factor analysis or principal component analysis software on commercial instruments. This has been especially valuable in the interpretation of series of data such as are acquired during a depth profile.

2.3. X-ray-Excited Auger Electron Spectra

Auger peaks in an electron spectrum may be produced by any incident radiation that is sufficiently energetic to cause core level ionisation. Auger peaks are therefore found frequently in XPS spectra. Sometimes they merely act as a nuisance factor. For example, in first-row transition metals the LMM peaks occupy a wide energy range and frequently overlap with photoelectron peaks from minor species, thereby giving problems of identification and peak-area measurement. On other occasions they may be particularly valuable. Their presence may act to confirm the presence of a species whose identification by XPS alone was in doubt. The use of a twin-anode x-ray source to give peak identification through the observation of 233 eV shifts between Mg K_α and Al K_α radiation has been mentioned. Of greater importance is the use of the Auger peaks to give additional chemical information. This may be accomplished through measurement of the line energy and shape or by measurement of the difference in energy between the Auger line and a principal photoelectron line. This latter measurement is known as the Auger parameter, α. First described by Wagner [24], it has proved a very useful tool in the determination of chemical states in surface analysis by electron spectroscopy.

The chemical-state sensitivity arises as follows. In XPS, the change in binding energy due to a change in the local electronic environment is [25]

$$\Delta E_b = \Delta V - \Delta R \tag{17}$$

where ΔV is an initial-state contribution due to the dependence of the core potential on the local electronic environment, and ΔR is a final-state contribution arising from the response of the local electronic structure to the creation of the core hole (i.e., a relaxation term). The equivalent expression for the change of kinetic energy in the Auger case is [25]

$$\Delta E_k \approx -\Delta V + 3\Delta R \tag{18}$$

Defining a modified Auger parameter, α^*, as the sum of the kinetic energy of the Auger peak and the binding energy of the photoelectron peak allows the change in α^* to be written as

$$\Delta\alpha^* = \Delta E_b + \Delta E_k \sim 2\Delta R \qquad (19)$$

simply by combining Eqs. (17) and (18).

Using x-ray excitation and measuring ΔE_b and ΔE_k in the same spectrum eliminates the problem of electrostatic charging which occurs when a single measurement is made on an insulating sample. Incorrect referencing of the spectrometer energy scale is also eliminated, so long as the spectrometer error is constant across the energy range. These practical advantages are of great value when problem solving on difficult samples by electron spectroscopy.

$\Delta\alpha^*$ involves only the final-state contribution, and for this reason it is sometimes known as the final-state Auger parameter to distinguish it from an alternative measure, $\Delta\beta$, known as the initial-state Auger parameter, which is [26]

$$\Delta\beta = \Delta E_k + 3\Delta E_b \sim 2\Delta V \qquad (20)$$

Strictly, the expressions for $\Delta\alpha^*$ and $\Delta\beta$ apply only to Auger transitions which do not involve valence electrons. Where valence electrons participate in the Auger process, a more sophisticated analysis is required, beyond the scope of this book. At present, the understanding of the factors contributing to initial- and final-state effects in the Auger process is primarily of theoretical interest. As the theoretical underpinning becomes stronger, the use of the two parameters $\Delta\alpha^*$ and $\Delta\beta$ together will become increasingly common in practical applications of XPS and XAES. For further details the reader is referred to the review by Weightman [27]. In 1979, Wagner [28] suggested that the Auger parameter could best be used through the medium of two-dimensional chemical-state plots in which the kinetic energy of the Auger peak and the binding energy of the photoelectron peak are plotted as ordinate and abscissa respectively. Measurement of the two energies then gives a point on the plot, which may be compared with the positions on the plots of the photoelectron and Auger lines for the element in a range of known chemical environments. This is probably still the best approach to take. As an example, Wagner's plot for Na is shown in Fig. 5 [65].

An interesting and underexploited aspect of XAES is the use of Auger peaks at kinetic energies above the photon energy of the x-ray source. These are excited by the background bremsstrahlung radiation extending to the incident electron beam energy of the conventional twin-anode source design. This renders accessible the Si, P and S KLL lines, the LMM lines of the second-row transition elements and the MNN lines of the heavy metals from Ta to Bi, in addition to those lines normally seen below the Fermi energy in

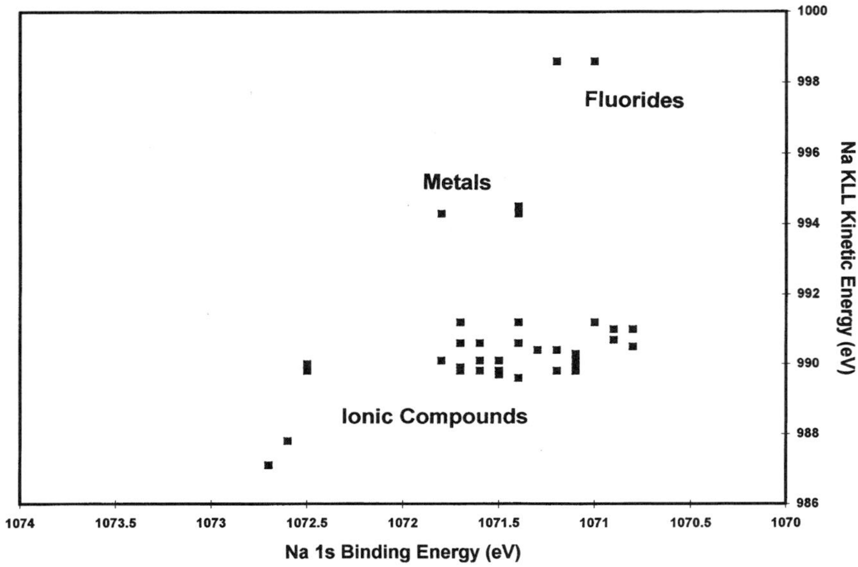

Figure 5. Auger parameter plot for Na. (From Ref. 65.)

conventional Mg or Al K_α excitation. The study of Auger lines by this means, rather than by electron excitation, has the important advantage that the background is generally very low, enabling measurements to be made at a good signal-to-noise ratio in the peak even of weak spectral features. It is, of course, not possible if a monochromator provides the only source of x-rays.

3. QUANTITATIVE AND STRUCTURAL ANALYSIS

When the spectral interpretation is complete and the chemical nature of the species present has been identified, then attention will turn to the structure and composition of the sample. The methods used in the determination of such information from XPS and AES data are described in this section. The emphasis is on providing a route map, or guide, to the various possible approaches, to a depth of mathematical complexity that is sufficient for practical problem solving without necessarily being rigorous in all respects. The scheme follows the logic of the flowchart of Fig. 6. The quantification of data from specimens that are homogeneous both laterally and in depth is discussed in Section 3.1. The modifications necessary to include the effects of lateral and depth variations are covered in Sections 3.2 and 3.3, respectively.

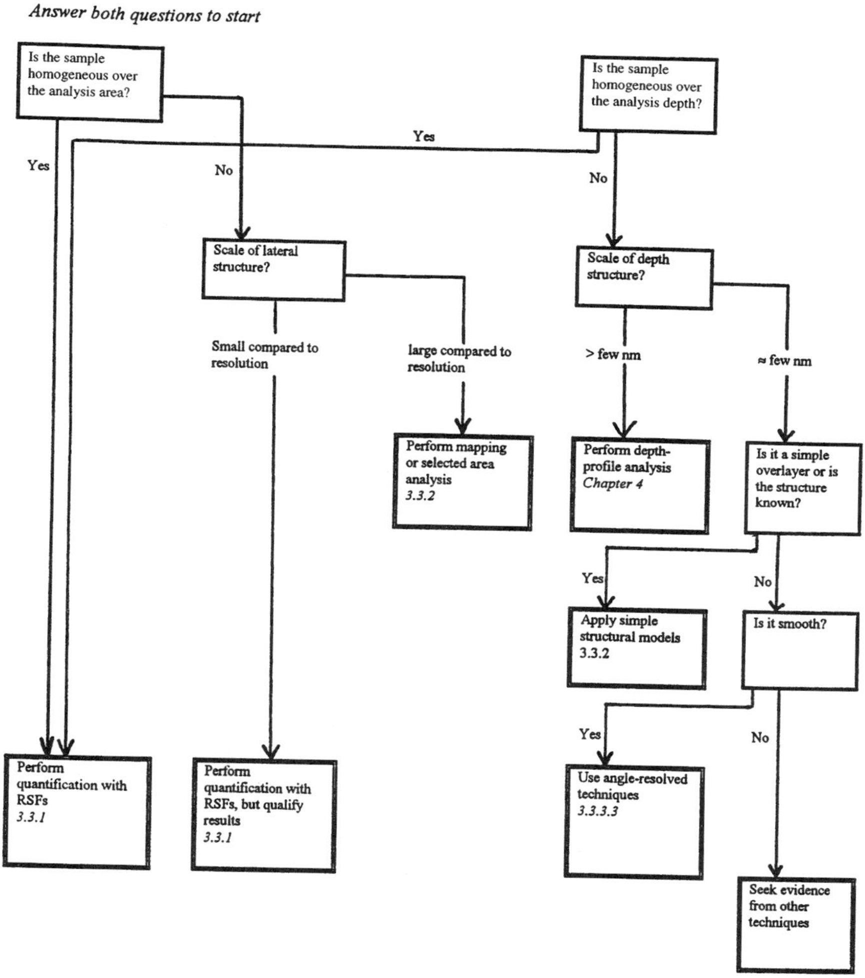

Figure 6. Flowchart for quantitative and structural analysis. Numbers in italics refer to the section headings in which the topic is covered. Boxes with a double outline represent end-points of the decision process.

3.1. Quantification of Homogeneous Samples

3.1.1. Use of sensitivity factors

The objective of quantification is to relate the intensities measured in the Auger or photoelectron spectrum to the composition of the specimen. Intensities in electron spectra are governed by sample- and instrument-dependent factors. The instrumental factors include the electron or x-ray beam flux and area of irradiation, and the analyzer transmission and detector sensitivity as a function of energy. In a well-organized laboratory these will be known and controlled and, for the moment, may be ignored. The principal sample-dependent factors affecting the relative intensities of spectral peaks are the different probabilities of Auger or photoelectron emission for the different elements that may be present. As different elements give different peak intensities, even when present at the same concentration, it is not sufficient to attempt to relate directly the peak intensities to the sample composition.

If the specimen is uniform both laterally and in depth, then quantification can be accomplished by comparing the measured signal intensities with the intensities that would be measured from pure elemental samples of the materials found in the specimen under investigation. The approach used is that of relative sensitivity factors (RSFs). It is the most commonly used method of quantitative analysis in AES and XPS, and in practice is frequently applied to specimens which do not fulfill the criteria of homogeneity. As we shall see, it is useful when comparisons between similar samples are made, but its overall accuracy may not be as good as when more rigorous methods are used.

Various reviews of the RSF method have been published, with the most authoritative probably being that of Seah [29]. The nomenclature used by Seah [29] is widely accepted throughout the surface analysis community and will be followed here, with minor modifications where appropriate.

In the RSF method the analyst compares the measured intensity I_j from element j, for example, with the intensity that would be measured from a specimen of pure element j, I_j^0. Similar ratios for all elements present are then normalized to give the atom fraction composition X_j for element j in an n-element sample according to

$$X_j = \frac{I_j / I_j^{\mathrm{o}}}{\Sigma_i (I_i / I_i^{o})} \tag{21}$$

The summation is from 1 to n. The I^0 values are typically normalized to either the C 1s or F 1s line in XPS or to the Ag MNN line in AES. The use of Eq. (21) immediately raises two questions: namely how should the I_j values be measured, and where do the I^0 values come from? The answers to these questions depend on the technique and the purpose to which the results of the quantitative analysis will be put. They are addressed in the following sections.

3.1.2. Measurement of intensity

In XPS the I_j may be peak heights or peak areas. All XPS instruments in laboratories equipped for practical problem solving are fitted with computer data processing systems. These make it easy to calculate peak areas once a suitable background has been subtracted. Consequently peak-height measurements, with their associated lack of certainty arising from resolution variations and statistical effects, are rarely used. The choice of background may be that of a simple straight line, or of an integral or Shirley background [30], or of a Tougaard background [31]. The straight line is simple but physically unrealistic. The Tougaard background is physically realistic but requires a rather large energy range on either side of the peak and may not work very well in practice with complex specimens containing many elements. The Shirley background is neither physically totally correct nor particularly simple. Nevertheless, it gives results which look attractive and is very widely used.

In XPS the photoelectron peaks are frequently observed to have a higher background on the high-binding-energy side than on the low-binding-energy side. The reason for this is that photoelectrons generated in the peak, but at too great a depth in the material to escape without losing energy, simply act as an extra source of background, which is seen only at lower kinetic energies (i.e., higher apparent binding energies in XPS) than the parent peak. The Shirley background attempts to use this phenomenon to determine a background correction. If each electron acts as a source for scattered electrons, the background at any kinetic energy in the spectrum is proportional to the total number of electrons in the spectrum higher than that energy. This is realized in the Shirley method by subtracting a constant (horizontal) background equal to the intensity on the low-binding-energy side of the peak and then subtracting a background correction Δn_i from each successive spectral channel going from low to high binding energy such that

$$\Delta n_i = b \sum_j n_j \qquad (22)$$

where the summation is from $j = i+1$ to N, and b is a fitting parameter determined iteratively. The Shirley background is useful in cases where there is a step change in background intensity on the high-binding-energy side of the peak, but the background remains approximately horizontal. Unfortunately it is often found that the background slopes behind the peak, in which case the Shirley expression is not satisfactory. For example, this is typically seen for the $2p$ photoelectron peaks of transition metals. Modifications to the background can be made to account for this [32], but the need for empirically determined fitting parameters reduces the flexibility of the method.

It has been demonstrated that subtraction of a Shirley background can result in a significant underestimate of the true intensity of a peak [33]. This is avoided if the Tougaard method is used [33, 34]. The latter method extracts the "true" electron spectrum (the primary excitation function, $F(E)$) from the measured spectrum $j(E)$, after correcting for instrumental effects, taking into account inelastic electron scattering in a realistic manner. The equation given by Tougaard [34] for $F(E)$ is

$$F(E) = j(E) - B_1 \int_E^{\infty} \frac{E' - E}{[C + (E' - E)^2]^2} j(E')\ dE' \qquad (23)$$

where B_1 and C are fitting parameters. An important outcome of the use of this form of background correction is that there often appears a region of the spectrum extending to perhaps 50 eV toward the higher-binding-energy side of the peak which contains a significant contribution from primary electrons. This intensity is not included in the straight-line or Shirley background methods, which consequently may give a large underestimate of the true intensity.

On current data systems available for commercial electron spectroscopy instruments the Tougaard background subtraction method is usually available as an option. However, it is not often used in problem-solving applications as it requires a wider energy range of data than the simpler alternatives, and it may have a worse precision (even though the accuracy should be higher). The situation may change as experience is gained in the wider surface analysis community and the benefits of the approach are realized. The Tougaard approach also allows information on the distribution of elements with depth to be obtained from the data. Further discussion of this aspect is given in Section 3.3.

In AES, spectra may be measured in direct or derivative modes, thereby placing an extra level of complexity on the intensity measurement decision. If derivative spectra are used, the conventional measure of peak intensity is the peak-to-peak height between the maximum positive and negative excursions of the peak. However, this may be subject to a large number of errors, both instrumental and related to the nature of the specimen. For example, for an element whose principal Auger line is sensitive to the local chemical environment, such as carbon, the peaks can show large variations in shape between different specimens, even if the amount of the element in each specimen is nominally the same. The effect is generally stronger on the positive excursion of the peak, so it has been suggested that the signal level between the mean (zero) level and the minimum of the negative excursion should be taken as a more representative measure of the intensity.

Although chemical effects on the Auger line shape are important, instrumental factors can be much greater. The line shapes, and consequently the peak-to-peak or peak-to-zero signal levels, are strongly dependent on the resolution of the spectrometer and the width and form of the differentiating function used. This was recognized in the early stages of the development of the technique, and the *Physical Electronics* handbook of AES [19] gives examples of the effects of increasing the modulation amplitude on the Au 69 eV and 2024 eV peaks. The 2024 eV peak is relatively broad, and as the modulation amplitude is increased the peak-to-peak signal increases approximately linearly until a modulation of about 5–6 V, after which the rate of increase of the peak height falls off. The 69 eV peak is narrow, and consequently the saturation effect is seen at lower modulation amplitudes, above approximately 2 V. In general, as the resolution is degraded or the modulation amplitude increased, the heights of narrower peaks are attenuated more strongly than those of broader peaks. The problem has been considered by Anthony and Seah [35]. They show that the measure hW_m^2 is a good measure of intensity for differentiated peaks. Here, h is the (negative) height and $W_m/2$ is the measured width of the negative excursion of the differentiated peak. The value of hW_m^2 is much less dependent on the analyzer resolution or the modulation amplitude than is the height h. It is obviously important in quantification that the signals from the sample to be quantified are obtained under precisely the same experimental conditions of modulation and resolution as those from the pure specimens with which they are to be compared. If this cannot be accomplished, then the measure suggested by Anthony and Seah [35] should be used instead of a simple measurement of peak height.

Where Auger spectra are acquired in the direct mode, it is probably better to use the direct spectra rather than apply numerical differentiation and measure peak excursions. The measurement of a peak height over a background immediately above the peak in kinetic energy should give a result which is less sensitive to instrumental factors than the peak-to-peak height or the negative peak-to-background height in the differential spectra, and relative sensitivity factors for AES based on this measure have been published [36]. However, it is more attractive to measure the peak area, as this should be independent of instrumental and chemical-state factors. As for XPS, this means that the precise form of the background to be subtracted must be carefully considered.

A schematic Auger spectrum is shown in the direct mode in Fig. 7, over the energy range from zero kinetic energy to the primary beam energy, E_p. In principle, subtraction of the underlying contributions to the background from the backscattered primary electrons and from the secondary electrons should leave the true Auger peak intensity, provided there are no peak overlaps and there is a reasonable amount of background available from which to extrapolate.

Over the energy range 0.2 to 0.8 E_p the background contribution due to the backscattered primary electrons is described accurately by a simple exponential of the form [37]

$$B(E) \propto \exp(E_p - E) \tag{24}$$

With a typical value of E_p = 5 keV, this form of background may be found under, for example, the higher-energy Si or P KLL lines, the LMM lines of Mo or the MNN lines of heavy metals from Ta upward. For many elements it is common to use the lower-energy lines in the range approximately 50–1000 eV. These can be stronger, and have the advantage of giving increased surface sensitivity. The background under these peaks consists primarily of secondary electrons. Provided a good correction has been made for the energy dependence of

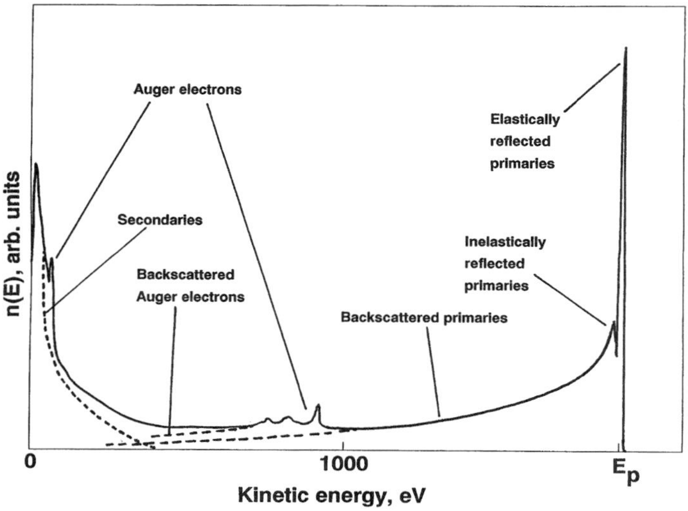

Figure 7. Schematic electron-beam excited Auger spectrum in the direct mode.

the spectrometer transmission, it is observed that, on plotting the low-energy region of the electron-beam-excited Auger spectrum on logarithmic axes, the spectrum falls into a series of straight-line sections separated by the Auger peaks. As a consequence of this observation, the secondary electron contribution to the background may be fitted by a function of the form [38]

$$S(E) \propto E^{-m} \tag{25}$$

Using Eqs. (24) and (25) on properly corrected experimental data should allow the true Auger peak areas to be determined. However, the method is not yet well established and it suffers from the disadvantage that data must be acquired over a very wide spectral range before the extrapolations can be done accurately. An advantage is that, after normalization, the sensitivity factors required for use in Eq. (21) should be exactly equal to the theoretical values, thereby reducing a source of uncertainty in quantification.

In present practice the full background correction described tends not to be made in AES; instead, an approach similar to that for XPS is used. The spectrum is acquired over a relatively narrow range around the peak of interest and a choice of straight-line, Shirley or Tougaard background is subtracted to give an estimate of the peak area. The straight-line and Shirley methods suffer from the same problems as they do when applied in XPS. In particular, it is frequently not clear where to put the lower kinetic energy limit for a straight-line background. The Tougaard background requires a wider energy range than the other two methods. As a result, the Shirley method is quite commonly employed where peak areas are to be used for quantification of Auger spectra. There is no databank of relative sensitivity factors readily available for peak areas in AES, so, at least for the more frequently met elements, the analyst may well be faced with the task of constructing a personalized set of RSFs for in-house use. If this is the case, then the precise choice of background method may not be so critical, the important factor being that it be used consistently. Quantification of Auger spectra is an area of current scientific research, and RSFs based on peak-area measurements may become available in future.

3.1.3. Modified sensitivity factors for improved quantification

The previous section, "Use of Sensitivity Factors," introduced the use of RSFs for quantitative analysis by electron spectroscopy. The choice of RSF needs to be considered carefully if realistic numbers are to obtained from their use. The factors affecting the correct choice and method of use of RSFs are different for AES and XPS, so they are covered separately.

3.1.3.1. Quantification of XPS data

As defined, the RSF or $I°$ values are the intensities that would be measured from an infinitely thick pure elemental specimen after some suitable normalization. This immediately implies that the RSFs should be obtained experimentally. However, experimental data may suffer from problems related to statistical uncertainties, the purity of the material and the nature of any surface treatment it may have undergone, the effect of contamination from the residual vacuum, and sensitivity variations between spectrometers. For these reasons, the use of theoretical sensitivity factors has been popular, but then it is important to ensure that the intensity measure used on the data to be quantified accurately reflects the calculation used in the derivation of the theoretical sensitivity factor. That is, the background subtracted should include all the intensity, not just the amount that appears to be in the main part of the photoelectron peak. Probably the most useful set of RSFs available in the literature is that of Wagner et al. [39], where a combination of approaches has been used. The data of Wagner et al. [39] should not be applied directly unless an instrument of characteristics identical to their original is used. To understand the correct application of RSFs it is worthwhile considering the fundamental factors that contribute to the intensity of an XPS photoelectron peak. An overview is given here; for further details the work of Seah [29] and references therein should be consulted.

The intensity of a photoelectron line depends primarily on the intensity of the incident radiation, J_o, the cross section for photoemission for the particular photoelectron line of interest from element A at the photon energy $h\nu$, $\sigma_A(h\nu)$, the probability of the photoelectron traveling to the surface and being emitted without losing its initial energy, and the sensitivity of the electron energy analyzer and detector system. These factors may be combined with other terms to give, for a pure reference sample uniformly irradiated by x-rays,

$$I_A^o = \sigma_A(h\nu)J_o L_A(\gamma)Q(E_A)N_A^o \lambda_A(E_A)\cos\theta \tag{26}$$

The additional terms are the angular asymmetry parameter $L_A(\gamma)$ for an included angle γ between the x-ray source and the electron energy analyzer, the transfer characteristic $Q(E_A)$ of the spectrometer, the number density N_A^o of atoms in the pure reference sample, the mean free path or attenuation length $\lambda_A(E_A)$ for photoelectrons of kinetic energy E_A traveling in material A, and the angle of emission, θ, of the photoelectrons from the specimen surface relative to the surface normal. If θ is not restricted to a narrow range of angles about a single value, the appropriate integral must be included in Eq. (26). The term $\lambda_A(E_A)\cos\theta$ represents the integral of the probability of escape for electrons generated at depth d, over the limit $0 < d < \infty$, assuming an exponential form for the attenuation of signal with depth.

Taking the terms in Eq. (26) in order, the photoemission cross sections have been calculated by Scofield [40] for all lines of interest in XPS for Mg and Al K_α radiation. Values for the important lines in XPS using Al Kα radiation are shown graphically in Fig. 8. Several of the earlier attempts to compare theoretical and experimental cross sections concluded that the Scofield values could be in error by large amounts. However, it now seems likely that the lack of

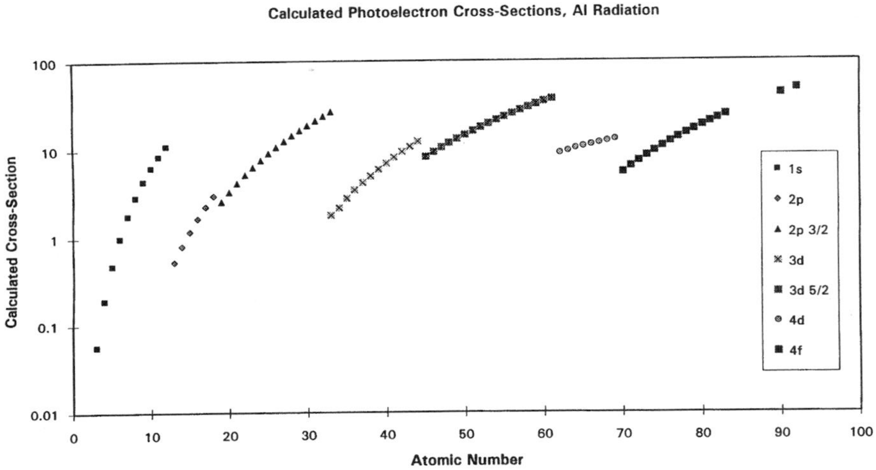

Figure 8. Scofield cross sections, calculated for Al K_α radiation, shown for the principal lines of interest for quantification in XPS.

agreement was caused at least in part by a failure to measure the full line intensity, and the Scofield cross sections are now believed to be correct. A comparison with intensities properly corrected for instrumental effects and calculated using the Tougaard background would be valuable in this respect, but this does not yet seem to have been done.

The angular asymmetry parameter $L_A(\gamma)$ describes how the intensity of a particular type of core-level line varies with the angle γ between the incoming x-rays and the outgoing photoelectrons.

$$L_A(\gamma) = 1 - \frac{\beta_A(3\ \cos^2\gamma - 1)}{4} \tag{27}$$

β_A is a constant given by the calculations of Reilman et al. [41] for the the photoelectron peaks of importance in XPS. At an angle $\gamma = \cos^{-1}(1/\sqrt{3})$, i.e., approximately 54.7°, $L_A(\gamma)$ is unity and the asymmetry parameter has no effect on the quantification. This angle, the "magic angle," is close to that used on many XPS systems, so the effect of $L_A(\gamma)$ on the quantification tends to be less important than the other factors. The variation of $L_A(\gamma)$ is further reduced by elastic scattering of the photoelectrons, which acts to smooth out the initial angular distribution of photoelectrons as they travel to the surface prior to photoemission. For these reasons the angular asymmetry parameter is usually neglected in practical applications, although for careful work, particularly if geometries far from the magic angle are involved, the effect should be considered.

The transfer characteristic $Q(E_A)$ describes the energy dependence of the transmission and detection sensitivities of the spectrometer. For XPS instruments operated at a constant energy resolution ΔE, calculations show that $Q(E)$ should typically be $E^{-0.5}$ or E^{-1} [42]. In practice, individual instruments deviate from this behavior and the true energy dependence should be derived from traceable reference spectra. At present, the only source of such data is the UK National Physical Laboratory (NPL), which also provides a methodology for its use [43].

The number density of atoms in the sample, N_A^o in Eq. (26), is given by the reciprocal of the atomic volume, a_A, calculated from

$$a_A = \left(\frac{A_M}{N_A \rho_A}\right)^{1/3} \tag{28}$$

where A_M is the atomic weight, N_A is Avogadro's number and ρ_A is the density.

The final term in Eq. (26), $\lambda_A(E_A)$, is the attenuation length for electrons at energy E passing through element A. The search for accurate values of this parameter has been a recurring theme throughout the recent history of electron spectroscopy. Valuable early work was performed by Seah and Dench [44], who used a compilation of all the available experimental data at the time to show that values for elements, inorganic compounds and organic compounds could be obtained from the following parametric equations:

for elements:

$$\lambda(E) = 538aE^{-2} + 0.41a^{3/2}E^{1/2} \quad \text{nm} \tag{29}$$

for inorganic compounds:

$$\lambda(E) = 2170aE^{-2} + 0.72a^{3/2}E^{1/2} \quad \text{nm} \tag{30}$$

for organic compounds:

$$\lambda(E) = 49E^{-2} + 0.11E^{1/2} \quad \text{mgm}^{-2} \tag{31}$$

In these equations, a is the average monolayer thickness, in nm, given by

$$a = 10^{9}\left(\frac{Z}{\rho n N_{\mathrm{A}}}\right)^{1/3} \quad \text{nm} \tag{32}$$

The mean scatters between the experimental values and the results given by Eqs. (29)–(31) were typically in the region of 40%. The parameter given by these equations was described in the original work as the inelastic mean free path. However, the parameter measured in an experiment includes the effects of elastic scattering and is known as the attenuation length. The measurements of attenuation in the compilation of Seah and Dench [44] were subject to systematic error, principally from assumptions made about the structures of the thin films used in the experiments. Although Eqs. (29) to (31) gave the best fit to the data available at the time, more recent work has tended to show a rather stronger energy dependence, with an exponent that depends on the material. For example, Wagner et al. [45] found energy exponents ranging from 0.54 for gold to 0.81 for silicon. Mean calculated exponents of 0.80 ± 0.04 for attenuation lengths and 0.76 ± 0.04 for inelastic mean free paths of 28 elements have been found using Monte Carlo methods [46]. Tanuma, Powell and Penn [47–49] give calculations of the inelastic mean free paths of electrons in selected elements, inorganic compounds and organic compounds from 50 to 2000 eV kinetic energy. The calculations employ a semiempirical approach in that experimental optical constants and materials parameters are fitted to well-established theory. They are believed to be accurate, and represent a good source of numerical values for use in quantitative electron spectroscopy. Unfortunately, they show no simple energy dependence and cannot be approximated by simple analytical functions. The user must refer to the original publications and correct for elastic scattering where possible.

From this discussion it would appear to be possible to calculate sensitivity factors for XPS. In practice, this is not normally attempted. However, simplified forms of Eq. (26) are used in practical XPS quantification routines. In one approach the RSFs and the experimental data are taken to apply to the same set of analysis conditions, and the data are normalized to the C 1*s* line. The J_o and cos θ terms therefore cancel out. The variation in $L_A(\gamma)$ is assumed negligible, and the number density term is ignored. Under such conditions the intensity from element A relative to that from C 1*s* is

$$\frac{I_A^o}{I_{C1s}^o} = \frac{\sigma_A Q(E_A)\lambda_A(E_A)}{\sigma_{C\,1s} Q(E_{C\,1s})\lambda_A(E_{C\,1s})} \tag{33}$$

As the data are to be defined relative to the C 1*s* line, all terms in the denominator on the right-hand side may be set to unity. Assuming the energy dependence of the attenuation lengths may be approximated by an exponent *p* such that $\lambda(E) \propto E^p$ and, further, that the energy dependence of the analyzer response function may be represented in a similar manner as $Q(E) \propto E^q$, then Eq. (33) simplifies to

$$\frac{I_A^o}{I_{C1s}^o} = \sigma_A \left[\frac{E_A - E_{C1s}}{E_{C1s}}\right]^{p+q} \tag{34}$$

Although not stated explicity by the authors, Eq. (34) is the method by which the sensitivity factors given in the handbook of XPS [4] are derived. Assuming that reasonable choices of *p* and *q* can be made, this approach looks attractive. It has the advantages of simplicity and of not being reliant on measured data with all the associated problems of sample contamination, statistical noise and measurement uncertainty. However, its reliability depends critically on the extent to which the calculated intensity reflects that which is measured from the sample to be quantified. Figure 9 shows the relative sensitivity factors for the principal XPS lines calculated using Eq. (34), for Al K_α radiation. An analyzer transmission energy exponent of –0.5 and an attenuation-length energy dependence exponent of 0.75 have been used in the calculations.

The approach of Eq. (34) has been popular and is implemented on some commercial instrument data systems. When tested using samples of known composition, it can give good results. This is surprising, as it is well known that the simple background subtraction methods used in the measurement of intensity often fail to include a significant proportion of the true intensity calculated by Scofield [40]. The agreement may be, at least in part, attributed to fortuitous cancellation of errors in the ratios of intensities used to derive the quantification. Because of the difference between that which is calculated and that which is measured, it may be more appropriate to use relative sensitivity factors based on measurements of known materials. Many such sets of data exist in the literature, some dating from the earliest days of the development of the technique. They are of interest, but are not used for practical problem-solving work and so are not discussed further. For more details the interested reader should consult Seah [29]. However, the data set of Wagner et al. [39] is of particular im-

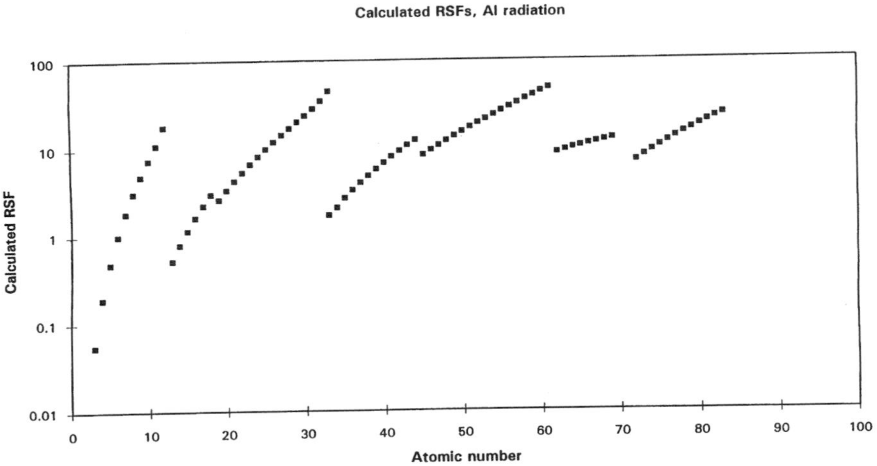

Figure 9. Calculated RSFs for Al K_{α} radiation using Eq. (34).

portance as a combined semiempirical approach was used in which measured data for series of elements or photoelectron lines were fitted to curves similar to those given by Eq. (34). This hybrid approach appears to make good sense, and the data of Wagner et al. [39] are probably the most generally useful of those available. The compilation of Wagner et al. [39] is, strictly speaking, only valid for an instrument with the same combined energy dependence of transmission and detection as that used for the original work. Two different spectrometers were used for this work, both of the cylindrical mirror type, and the energy dependence of each was assumed to be $E^{-0.5}$. For a different energy dependence, calibrated using the method of Seah [43], the appropriate correction must be made. To give some visualization of the potential errors involved in quantification using different sets of relative sensitivity factors, Fig. 10 shows the ratio between the values of Wagner et al. [39] and those calculated using Eq. (34), normalized to C 1s = 1. The two sets show an rms scatter of approximately 20%.

An alternative method of deriving relative sensitivity factors would be to use the calculated intensities of Scofield [40], after correction for the energy dependences of the attenuation length and analyzer function, as in Eq. (34), but to use an accurate measure of the full intensity for the experimental data, such as would be obtained from the Tougaard background method [31]. A full evaluation of such an approach is not yet available. Preliminary work seems to suggest that the overall results may be more accurate, but possibly with lower precision. That is, there may be benefits in using this method if the analysis is

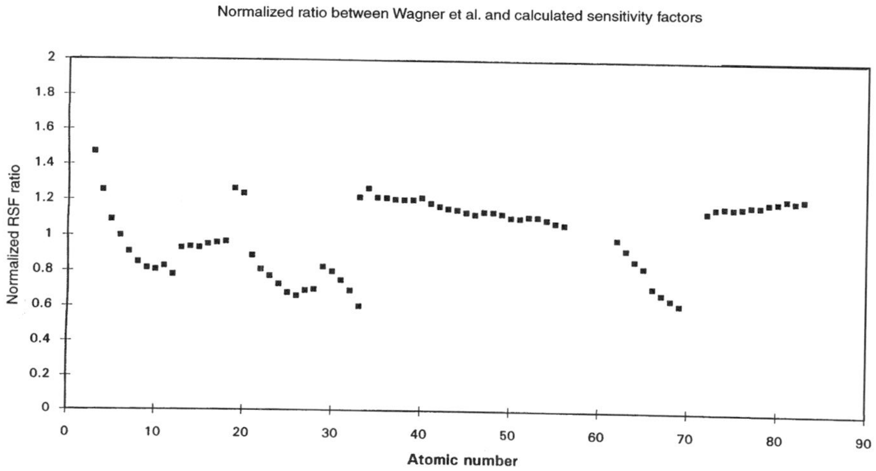

Figure 10. Ratios between RSFs given by Wagner et al. [39] and those calculated using Eq. (34).

for a one-off sample and it is important to get an answer as near the truth as possible, but if small differences between samples in a series with similar composition are important, then a simpler approach may be more appropriate. As noted, the Tougaard background does suffer from disadvantages if there are a large number of peaks present in a narrow energy window, and it requires the accumulation of data over a relatively wide range around the peak of interest.

3.1.3.2. Quantification of AES data

As was the case for XPS, to understand the correct use of sensitivity factors in AES it is necessary to consider briefly the origin of the various terms that give rise to the Auger intensity in the electron spectrum. The intensity of emission of Auger electrons originating from a particular electronic transition in a uniformly illuminated specimen of element a is

$$I_a = I_o \rho \gamma \sigma [1 + r_M(E, \alpha)] Q(E) N_a \lambda \cos\theta \tag{35}$$

where I_0 is the electron beam current, ρ is a factor to allow for the increase in effective surface area of the specimen due to surface roughness, γ is the probability of the particular Auger transition of interest occurring, σ is the cross section for ionization of the core hole by the electron beam of energy E_p, $r_M(E, \alpha)$ is a factor allowing for the additional intensity due to primary electrons incident at angle α relative to the surface normal that have been backscattered in the specimen but still retain sufficient energy to cause ionization of the core level at energy E, $Q(E)$ is the overall spectrometer efficiency, N_a is the number density of atoms of type a in the specimen, and θ is the angle of emission of the Auger electrons relative to the surface normal. The most important parameters in Eq. (35) are probably the cross section for ionization, σ, and the backscattering factor r_M. Various workers have given expressions for the ionization cross section, among them Bethe [50], Gryzinski [51] and Lotz [52]. These are shown in Fig. 11. They all show a rapid rise up to a peak in the range U = 2–5, followed by a gradual reduction. The most satisfactory of these has generally been that due to Gryzinski, who gives the following approximate expression for σ:

$$\sigma = \left[\frac{\pi e^4 n_x}{E_x{}^2 U}\right]\left[\frac{U-1}{U+1}\right]^{3/2}\left\{1+\left[\frac{2(1-1/2U)}{3}\right]\ln\left(2.7+(U-1)^{1/2}\right)\right\} \tag{36}$$

where n_x is the number of electrons in the shell, x, to be ionized, and U is the overvoltage given by the ratio between the primary beam energy E_p and the energy of the shell to be ionized, E_x. Once the core level is ionized, the atom decays from the excited state to give an Auger electron, with a probability γ which depends on the atomic number of the element and the binding energy of the particular core level. Auger emission competes with x-ray emission as a decay process. For binding energies less than approximately 2 keV, Auger emission is the dominant process, but its probability falls off with increasing atomic number.

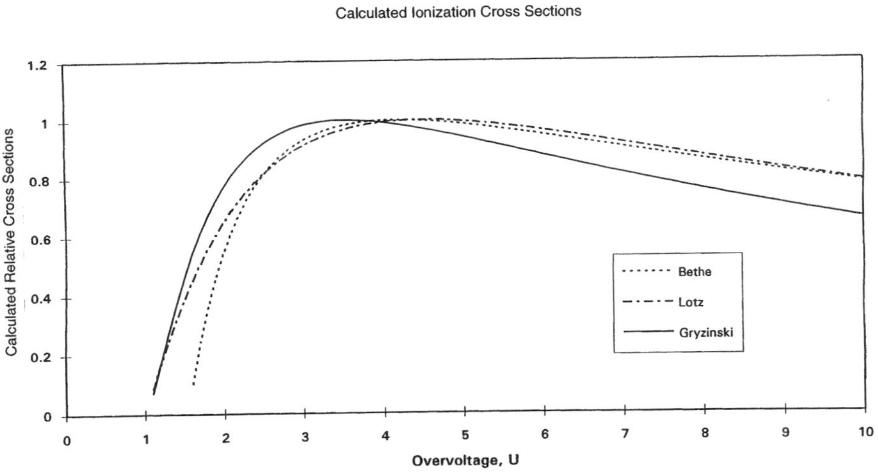

Figure 11. Comparison of AES cross sections using the methods of Bethe [50], Gryzinski [51] and Lotz [52].

The backscattering factor, r_M, is obtained from the parametric fits to the results of Monte Carlo calculations by Ichimura and Shimizu [53] given by Shimizu [54] for the angles of incidence of 0° (normal incidence), 30° and 45°:

$$r = (2.34 - 2.10Z^{0.14})U^{-0.35} + (2.58Z^{0.14} - 2.98) \quad (\alpha = 0°) \tag{37}$$

$$r = (0.462 - 0.777Z^{0.20})U^{-0.32} + (1.15Z^{0.20} - 1.05) \quad (\alpha = 30°) \tag{38}$$

$$r = (1.21 - 1.39Z^{0.13})U^{-0.33} + (1.94Z^{0.13} - 1.88) \quad (\alpha = 45°) \tag{39}$$

where Z is the atomic number and U is the overvoltage, E_p/E. The behavior of the backscattering factor calculated using Eqs. (37)–(39) for a range of atomic numbers and energies is shown in Fig. 12.

The presence of this backscattering factor results in a modification to Eq. (21) for quantification using relative sensitivity factors to include a matrix factor, F, such that the atomic fraction

$$X_j = \frac{F_{j,M}(I_j / I_j^o)}{\sum_i F_{ij}(I_i / I_i^o)} \tag{40}$$

where the summation is $i = 1$ to N. The matrix factor $F_{j,M}$ refers to Auger electrons originating from atoms of element j and passing through the sample matrix M, containing n elements in total, which accounts for the difference in intensity expected between a pure sample of element j and atoms of element

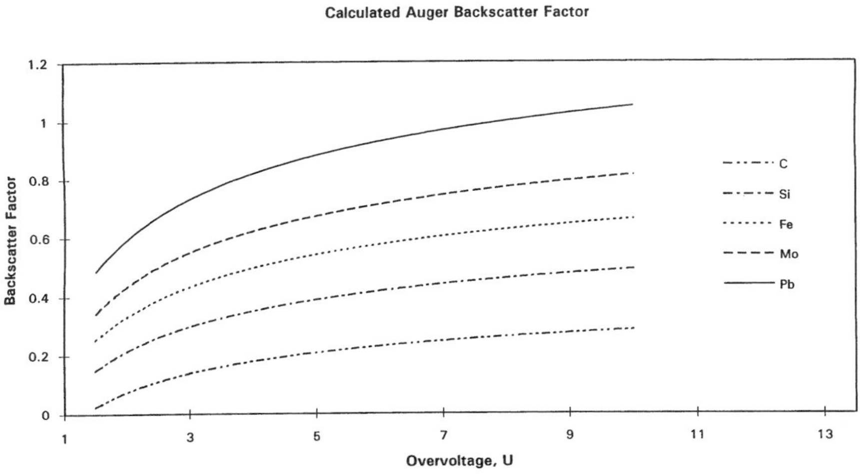

Figure 12. Backscattering term in AES for various elements as a function of overvoltage, U, calculated using the equations of Shimizu [54] for an angle of incidence of 30°.

j embedded in the sample containing n elements. Using Eq. (35) for the Auger intensity, together with the empirical equations for the attenuation lengths in elements given in Eq. (29), it can be shown [29] that for a binary alloy system containing elements A and B, the matrix factors are given approximately by

$$F_{\mathrm{AB}}^{\mathrm{A}}(X_{\mathrm{A}} \to 0) = \left\{\frac{1 + r_{\mathrm{A}}(E_{\mathrm{A}})}{1 + r_{\mathrm{B}}(E_{\mathrm{A}})}\right\}\left(\frac{\mathrm{a}_B}{\mathrm{a}_A}\right)^{3/2} \tag{41}$$

and

$$F_{\mathrm{AB}}^{\mathrm{A}}(X_{\mathrm{A}} \to 1) = \left\{\frac{1 + r_{\mathrm{A}}(E_{\mathrm{B}})}{1 + r_{\mathrm{B}}(E_{\mathrm{B}})}\right\}\left(\frac{a_{\mathrm{B}}}{a_{\mathrm{A}}}\right)^{3/2} \tag{42}$$

In general, the ratio $r_{\mathrm{A}}/r_{\mathrm{B}}$ is not strongly dependent on energy and Eqs. (41) and (42) give approximately equivalent results. In the case of a multielement sample, average values for the atomic numbers must be used to estimate the backscattering factors. This appears to raise the problem that prior knowledge of the composition needs to be available before the quantification can be carried out. This may be overcome using bulk data from EDS or WDS analysis, for example, or by using an uncorrected estimate of the composition from Eq. (21). Where the Auger intensities are known to be influenced by backscattering from an underlying substrate of different composition, the correction of Barkshire et al. [55] may be used. The effect of backscattering on the results of quantification can be quite significant with, for example, light elements in a heavier matrix having an apparently enhanced signal intensity. Given the relative ease with which matrix factors can be used to correct for this effect, it is perhaps surprising that they are not often seen in practical use.

3.1.4. Statistical errors in quantification

As with any other analytical data, electron spectra are subject to statistical noise. In general, it is likely that for any particular spectrum the errors in quantification are dominated by systematic effects such as the uncertainty in the relative sensitivity factor or in the attenuation length. However, if comparisons are to be made within series of similar spectra, the random errors arising from the statistical noise in the data become important. Similarly, if a particular element is present in the specimen at a very low level it becomes necessary to consider the statistical nature of the detection limit of the technique.

Electron spectroscopic data acquired using pulse counting should have a count rate which shows a Poissonian distribution. For the count rates typically found in practical electron spectroscopy, this approximates well to a Gaussian distribution. The following discussion assumes a Gaussian distribution of random noise in the spectrum.

A typical peak in either XPS or direct-mode AES will appear generally as shown in Fig. 13. Here, data are shown acquired with a channel width ΔE, and a background is drawn between two energies E_1 and E_2. For such a peak, the measured intensity to be used in the quantification by Eq. (21) is given by the peak area [56]:

$$A = \frac{\Delta E}{t}\left[\sum_i N_i - \frac{(n-2)(N_1 + N_n)}{2}\right] \tag{43}$$

where the data are acquired for t seconds per channel. The summation is from $i = 2$ to $n - 1$. N_1 is the count rate of the background at energy E_1 at one side of the peak, and N_n is the count rate of the background at the other side of the peak, for a peak containing n channels. The random error associated with the peak area, A, is given by Harrison and Hazell [56] as

$$\sigma(A) = \frac{\Delta E}{t}\left[\sum_i N_i + \frac{(n-2)^2(N_1 + N_n)}{4}\right]^{0.5} \tag{44}$$

where the summation is from $i = 2$ to $n - 1$. From this, the random error in the quantification due to statistical noise is estimated as [56]

$$\sigma_R(X_i) = [(1-2X_i)^2\sigma_R^2(A_i) + \sum_{j\neq i} X_j^2\sigma_R^2(A_j)]^{0.5} \tag{45}$$

In Eq. (45), $\sigma_R(X_i)$ is the relative precision in the concentration of element i, given by $\sigma_R(X_i) = \sigma(X_i)/X_i$, and $\sigma_R(A_i)$ is the relative precision in the peak area for element i, given by a similar expression. These equations are relatively

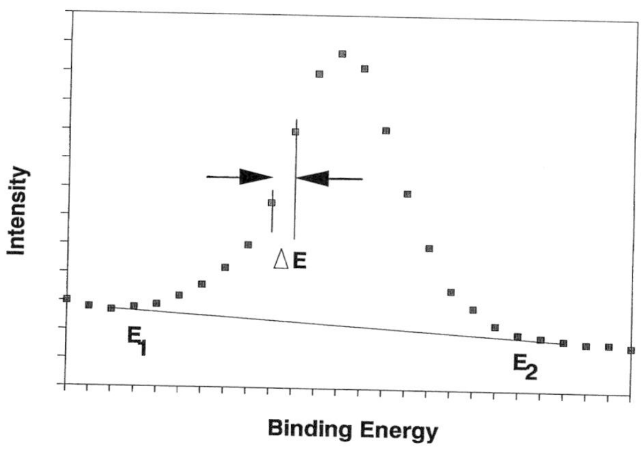

Figure 13. Schematic diagram of a generalized computer-acquired peak in AES or XPS.

straightforward to implement in a computer quantification routine, and their use can lead to insights into the certainty with which compositions derived from XPS or AES data may be quoted.

3.2. Analysis of Specimens with Spatially Varying Compositions

The discussion of the interpretation of relative intensities in electron spectra has so far concentrated on specimens where the concentration is uniform over the analyzed depth and the analyzed area. For many materials encountered in the laboratory for problem-solving applications, these approximations do not hold. The derivation of information concerning variations in concentration with depth is dealt with in the following section; here, laterally inhomogeneous surfaces are considered.

There are two cases of lateral inhomogeneity to be considered. In the first, the spatial variation across the surface to be analyzed is on a smaller scale than the smallest probe size available, and in the second the analyzing technique has sufficient spatial resolution to be able to address the local variation of the surface composition. Historically, the first case has been generally typical of XPS and the second of AES. With recent instrumentation, XPS is now able, under favorable circumstances, to image with a resolution down to 1 μm, although, depending on the instrument type, it is generally not possible to acquire an individual spectrum from an area as small as this. As a result, many surface analysis applications previously addressable only by AES may now be tackled with XPS, with the advantage of better chemical-state differentiation and, probably, more reliable quantification. However, AES data acquisition is usually rather quicker than for small-area XPS. The spatial resolution in AES has also improved over the years, and spot sizes in the 50–100 nm range are commercially available.

Where the lateral resolution of the probing technique is lower than the structure on the specimen surface, all that can be achieved is an average of the surface composition. If this is the case, then there is no advantage in using a small analysis area; instead, a larger area should be analyzed to give a good average. For example, in XPS analysis of a specimen surface where the lateral variation was suspected to be on the 50 μm scale, there would be no benefit in using a small-area XPS instrument with an analyzed area of diameter 100 μm. An average over a 1-mm-diameter area would give a much more statistically significant result, as well as a higher count rate and a consequent saving in analysis time.

More interesting is the case where the instrument is able to resolve the detail on the specimen surface. Scanning the electron beam (in scanning Auger microscopy (SAM)) or the sample position (in small-spot XPS) then allows the acquisition of maps of signal intensity versus position. In the newer XPS instruments maps of signal may be acquired either in parallel, using a position-sensitive detector, or by scanning the position of the x-ray spot on the specimen. Normally the image is built up by measuring the spectral intensity at a particular energy, with the usual measure being the peak height. To avoid unwanted contrast arising from topographic variations or changes in the spectral background, it is normal to apply some kind of correction procedure. The simplest is to measure a second image at an energy in the spectral background close to the peak of interest and to subtract this image from the peak intensity image. Denoting the peak intensity as P and the background intensity as B, this simply involves the calculation of $P–B$ at each point in the image. Advantages are claimed for more sophisticated approaches involving, for example, normalization to give $(P - B)/(P + B)$ [57] and these are normally available as an option on commercial instruments. Occasionally, such methods can give rise to errors as a result of changes in the slope of the spectral background, even without a true change in signal intensity. This may be overcome by recording the background at two points, either on both sides of the peak, in which case an interpolation may be used to determine the background to be subtracted, or on the same side of the peak, in which case an extrapolation is used. Such methods are rarely encountered in a practical problem-solving environment.

Images are valuable in the analysis of specimens with difficult or complex structures. Their value is increased if images representing several elements or, in the case of XPS, chemical states, are acquired with pixel-to-pixel registration over the same area of the specimen surface. Combinations of images can then be used to great effect to show such features as, for example, the correlation or anticorrelation between different elements on the surface. The use of scatter diagrams, pioneered for multispectral Auger microscopy [58], is becoming popular for this purpose and can show hitherto unrecognized surface phases on complex specimens.

3.3. Analysis of Specimens with Compositional Variations in Depth

Surface analysis is performed because scientifically and technologically interesting specimens show compositions at their surfaces that are different from the bulk. The discussion so far has been concerned with the elucidation of the nature of the surface composition and chemistry of the specimen to within the information depth of the technique. In many cases it is also necessary to determine the manner in which the composition varies with depth into the specimen. The traditional approach to this is by sputter-ion erosion of the specimen surface using, for example, a beam of argon ions. This method is extremely effective and is the subject of Chapters 7 and 8. However, there are other methods, particularly when the structure of interest lies close to the surface, within the probe depth of the AES or XPS technique, and these are discussed here. Some methods have the advantage that they are essentially free; that is, the data required are already present in the spectrum and simply need interpretation. Others require further measurements, but avoid some of the problems associated with argon-ion erosion of very thin layers or of materials sensitive to degradation by the ion beam.

A simple method of generating a compositional depth profile is by using either the AES electron beam or the analysis area of a selected-area or small-spot XPS instrument to scan across a taper section of the specimen of interest. Creation of a taper section at a sufficiently shallow angle results in the conversion of a variation in composition with depth into a lateral variation, often with a considerable magnification factor. The geometry is shown in Fig. 14. When the taper section is cut at an angle α to the horizontal, a spot size or analysis area of diameter d is able to give a depth resolution of $d \tan \alpha$. The method is surprisingly successful on plastic, polymer and elastomer specimens, provided a very sharp clean blade is used to produce the section. Metallic specimens may be prepared by careful polishing, but often a damaged layer is left that must be removed by argon-ion sputtering. This will have the effect of disrupting the local chemistry across the section; nevertheless useful elemental information may still be obtained. The method has the advantage that relatively large depths may be accessed, compared to those over which ion-beam-sputtered profiles are normally performed. In a variation of the technique, a focused programmable ion beam may be used to differentially erode the specimen and produce the taper section in situ over a small area of the surface identified for this treatment.

An attractive means of gaining some knowledge of the distribution of elements with depth into a sample is by making use of information contained within the spectral background. This is usually possible only in XPS, where the peaks are sharp and there are no redistributed or backscattered primary electrons. As mentioned, in XPS each photoelectron peak acts as a source of inelastically scattered electrons which contribute to the background. A simple steplike change in the background level between the high- and low-kinetic-energy sides of the peak is an indication that the photoemitting element has a

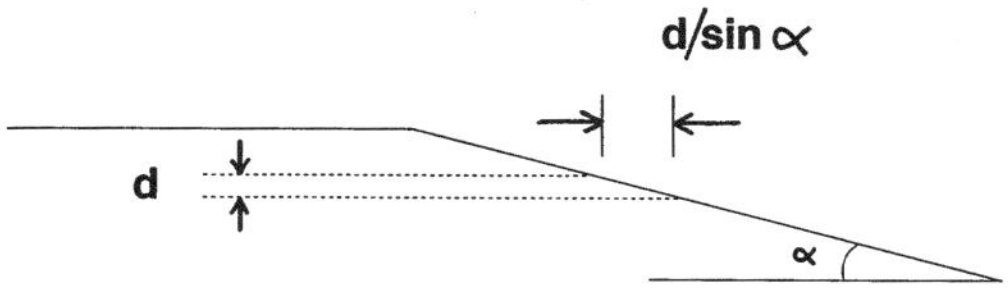

Figure 14. Geometry of taper section for revealing depth information.

reasonably uniform distribution with depth, as all possible loss energies appear equally. However, if some losses are favored over others, this may indicate a nonuniform distribution with depth. In one limiting case, there may be no losses, and consequently no change in background either side of the peak, if the element is present only in a thin layer on the surface. This arises because the electrons may not have traveled sufficiently far within the material to suffer a significant number of inelastic collisions before being photoemitted. Alternatively, if the background rises only gradually behind a peak before reaching a constant level, it may indicate that the emitting element is buried beneath an overlayer, and electrons that have suffered only small losses have a reduced probability of traveling to the surface and escaping.

The method of using the shape of the background behind a photoelectron peak to give information on the elemental depth distribution has been formalized by Tougaard [59]. The method relies on the measurement of a parameter D, given by A_p/B, where A_p is the peak area and B is the increase in background at a defined energy behind the peak. For XPS peaks of transition metals, D is found, to a good approximation, to be independent of concentration, peak shape and peak energy. However, it is strongly dependent on the distance traveled by the photoelectron before being emitted from the specimen. Measurement of D can therefore be related to the depth of the emitting species [59]. In further work, the method was extended to include data from the whole spectral background [60]. The method has now become available as a stand-alone PC-compatible computer software package [61].

An alternative and conceptually simple method of gaining information on the distribution of species with depth in a specimen is to examine the relative intensities of peaks at different electron kinetic energies. Generally, the electron attenuation lengths show an energy dependence which has a minimum around 100 eV kinetic energy and then an increase approximately as some power of the energy at higher kinetic energies. Therefore, within a particular spectrum, peaks from the same element at different kinetic energies will show differing degrees of surface sensitivity. For example, in the Auger electron spectrum of gold, the low-energy peak at approximately 69 eV will contain electrons that have originated much closer to the surface than those in the higher-energy peak at approximately 2025 eV kinetic energy. Similarly in XPS spectra of transition metals, the 2*p* photoelectron peaks show more surface sensitivity than the lower-binding-energy (higher kinetic energy) 3*s* and 3*p* peaks. To use this effect in practical cases, the spectrum from the specimen under examination should be compared with elemental spectra acquired from homogeneous samples. An apparent relative intensity enhancement in the higher-kinetic-energy peak compared to the homogeneous sample indicates preferential attenuation of the more surface sensitive low-kinetic-energy peak and is good evidence that the emitting element may be buried below an overlayer. Alternatively, an apparent increase in the more-surface-sensitive peak is consistent with the element being present as a surface layer.

In many cases it may be known in advance that the specimen consists of a thin overlayer of element A on a substrate of element B. Typically, this would apply, for example, to an adventitious oxide layer on a clean Al or Si substrate or to a thin lubricant film on a sample of magnetic recording media. If the overlayer completely covers the specimen and is known to be uniform in thickness, the thickness may be estimated using an equation derived from the expression for exponential attenuation of signal with overlayer thickness, originally given by Fadley [62]:

$$d_A(\theta) = \lambda_A \cos\theta \ \ln\left[1 + \frac{I_A I_B^o}{I_B I_A^o}\right] \tag{46}$$

where $d_A(\theta)$ is the thickness of the overlayer measured at an emission angle θ, λ_A is the attenuation length in the overlayer, I_A and I_B are the measured intensities from the overlayer and substrate, and I_A^o and I_B^o are the intensities that would be measured from pure samples of elements A and B. A useful check is to take measurements at two or more emission angles and check that the calculated apparent thickness is not a strong function of angle. The accuracy of the method is limited by the uniformity of the overlayer, the roughness of the substrate, and knowledge of the attenuation length.

Provided the structure of the sample is known, equations such as (46) can be written for the relative intensities of the spectral lines from the elements present. It is a simple matter to derive similar equations for multilayer structures or partial overlayers, for example. However, in many cases there is only limited prior knowledge of the structure of the sample. Changing the angle of emission at which the electrons are detected varies the surface sensitivity of the measurement, so it appears an attractive proposition to attempt to deduce the concentration variation as a function of depth from a series of measurements of apparent concentration at different emission angles. For XPS the intensity for a particular element measured at an angle of emission θ is related to the depth dependence of the concentration, $X(z)$, by

$$I(\theta) = K\int_o^\infty X(z) \exp\left(\frac{-z}{\lambda \cos\theta}\right) dz \tag{47}$$

where K is some constant, including instrument-dependent terms and the photoelectron cross section. The objective then is to deduce $X(z)$ from measurements of $I(\theta)$. By making the change of variable $p(\theta)=1/(\lambda \cos\theta)$, Eq. (47) can be rewritten as a Laplace transform:

$$I(p) = \int_o^\infty X(z) \exp(-pz)\, dz \tag{48}$$

and the problem then appears to be reduced to one of solving the inverse transform:

$$X(z) = L^{-1}[I(p);z] \tag{49}$$

where L^{-1} is the inverse Laplace transform. In recent years much research effort has been devoted to the solution of this problem, and a variety of methods suggested. However, as pointed out by Cumpson [63] the information content of angular-dependent XPS data is actually rather low, and any solution to Eqs. (47)–(49) is extremely sensitive to the noise in the data; a severe compromise between depth resolution and compositional certainty becomes necessary. In view of this, reconstruction of depth profiles from angular-dependent XPS may not be of practical use for problem solving on real, nonideal, specimens with rough surfaces, varying overlayer thicknesses, or partial, incomplete overlayers. However, if some information about the structure of the specimen is available, such as the composition of the substrate for a simple overlayer system, then it may well be worthwhile performing calculations of the expected data from such a structure and iterating the overlayer thickness or composition until a good agreement between the measured and calculated data is obtained. Although sophisticated methods such as regularization and maximum entropy have been applied to the problem, Cumpson [63] recommends a simple least-squares fit to a layer-by-layer model of the sample structure. If there is no prior knowledge about the sample structure, then, rather than use a sophisticated reconstruction algorithm with its associated uncertainty, the best approach is probably to attempt to obtain some other information by, for example, comparison with electron microprobe measurements or by argon-ion depth profiling.

If a simple stratification of the specimen within the escape depth of the photoelectrons is required, rather than a complete compositional depth profile, then the method of Seah et al. [64] has advantages. They simply take the ratio of two spectra recorded at different emission angles, and from the result deduce the layering of the elements within the sample, i.e., which elements are nearer the surface and which are buried. This is simple, reliable and straightforward to perform, and is well suited to the solution of practical surface analysis problems involving elements in layers near surfaces. It can also be applied to the individual chemically shifted components in a photoelectron line.

REFERENCES

1. J. C. Rivière, *Surface Analysis Techniques*, Monographs on the Physics and Chemistry of Materials, Oxford Science Publishers, Clarendon Press, Oxford, 1990.
2. D. Briggs and M. P. Seah, (eds.,) *Practical Surface Analysis*, 2nd ed., Vol. 1, *Auger and Photoelectron Spectroscopy*, Wiley, Chichester, 1990.
3. M. P. Seah and G. C. Smith, Appendix 1, in *Practical Surface Analysis*, 2nd ed., Vol. 1, *Auger and Photoelectron Spectroscopy* (D. Briggs and M. P. Seah, eds.), Wiley, Chichester, 1990, pp. 531–540.

4. C. D. Wagner, W. M. Riggs, L. E. Davis, J. F. Moulder and G. E. Muilenberg, *Handbook of X-ray Photoelectron Spectroscopy*, Perkin Elmer Corporation, Eden Prairie, 1978; J. F. Moulder, W. F. Stickle, P. E. Sobol and K.D. Bomben, *Handbook of X-ray Photoelectron Spectroscopy*, Perkin Elmer Corporation, Eden Prairie, 1992.
5. G. Beamson and D. Briggs, *High Resolution XPS of Organic Polymers: The Scienta ESCA300 Database*, Wiley, Chichester, 1992.
6. H. Raether, in *Excitation of Plasmons and Interband Transitions by Electrons*, Springer Tracts in Modern Physics, vol. 88, Springer, Heidelberg, 1980.
7. S. Doniach and M. Sunjic, *J. Phys. C: Solid State Phys., 3*, 285 (1970).
8. M. Romeo, K. Bak, J. El Fallah, F. Le Normand and L. Hilaire, *Surf. Interface Anal., 20*, 508–512 (1993).
9. K. Siegbahn, C. Nordling, G. Johansson, J. Hedman, P. F. Heden, K. Hamrin, U. Gelius, T. Bermark, L. O. Werne, R. Manne and Y. Baer, *ESCA Applied to Free Molecules*, North-Holland, Amsterdam, 1969.
10. U. Gelius, *Phys. Scr., 9*, 133 (1974).
11. C. D. Wagner and D. M. Bickham, *NIST Standard Reference Database 20: NIST X-ray Photoelectron Spectroscopy Database*, National Institute of Standards and Technology, Gaithersburg, 1989.
12. D. Briggs and G. Beamson, *Anal. Chem., 64*, 1729–1736 (1992).
13. D. Briggs and G. Beamson, *Anal. Chem., 65*, 1617–1623 (1993).
14. E. Desimoni, G. I. Casella, A. Morone and A. M. Salvi, *Surf. Interface Anal., 15*, 627–634 (1990).
15. S. Evans, *Surf. Interface Anal., 17*, 85–93 (1991).
16. A. E. Hughes and B. A. Sexton, *J. Electron Spectrosc. Relat. Phenom., 46*, 31 (1988).
17. P. J. Cumpson and M. P. Seah, *Surf. Interface Anal., 18*, 345–360 (1992).
18. M. F. Chung and L. H. Jenkins, *Surf. Sci., 22*, 479 (1970)
19. L. E. Davis, N. C. MacDonald, P, W. Palmberg, G. E. Riach and R. E. Weber, *Handbook of Auger Electron Spectroscopy*, 2nd ed., Physical Electronics Industries Inc., Eden Prairie, 1976.
20. G. E. McGuire, *Auger Electron Spectroscopy Reference Manual*, Plenum Press, New York, 1979.
21. T. Sekine, Y. Nagasawa, M. Kudoh, Y. Sakai, A. S. Parkes, J. D. Geller, A. Mogami and K. Kirata, *Handbook of Auger Electron Spectroscopy*, JEOL, Tokyo, 1982.
22. Y. Shiokawa, T. Isida and Y. Hayashi, *Auger Electron Spectra Catalogue—A Data Collection of the Elements*, Anelva Corporation, Tokyo, 1979.
23. F. P. Larkins, *Atomic Data and Nuclear Tables, 20*, 311 (1977)
24. C. D. Wagner, *Faraday Disc. Chem. Soc., 60*, 291 (1975).

25. T. D. Thomas, *J. Elec. Spectrosc. Relat. Phenom., 20*, 117 (1980).
26. J. A. Evans, A. D. Laine, P. Weightman, J. A. D. Matthew, D. A. Woolf, D. I. Westwood and R. H. Williams, *Phys. Rev., B46*, 1513 (1992).
27. P. Weightman, *Microsc. Microanal. Microstruct., 6*, 263 (1995).
28. C. D. Wagner, *Anal. Chem., 51*, 468 (1979).
29. M. P. Seah, Quantification of AES and XPS, in *Practical Surface Analysis*, 2nd ed., Vol. 1, *Auger and Photoelectron Spectroscopy* (D. Briggs and M. P. Seah, eds)., Wiley, Chichester, 1990, chap. 5, pp. 201–255.
30. D. A. Shirley, *Phys. Rev., B5*, 4707 (1972).
31. S. Tougaard, *Surf. Sci., 216*, 343 (1989).
32. H. E. Bishop, *Surf. Interface Anal., 3*, 272 (1981).
33. S. Tougaard and B. Jorgensen, *Surf. Interface Anal., 7*, 17 (1985).
34. S. Tougaard, *Surf. Sci., 216*, 343 (1989).
35. M. T. Anthony and M. P. Seah, *J. Electron Spectrosc. Relat. Phenom, 32*, 73 (1983).
36. T. Sato, Y. Nagasawa, T. Sekine, Y. Sakai and A. D. Buonaquisti, *Surf. Interface Anal., 14*, 787 (1989).
37. D. Jousset and J. P. Langeron, *J. Vac. Sci. Technol., A5*, 989 (1987).
38. R. Browning, D. C. Peacock and M. Prutton, *Appl. Surf. Sci., 22/23*, 145 (1985).
39. C. D. Wagner, L. E. Davis, M. V. Zeller, J. A. Taylor, R. H. Raymond and L. H. Gale, *Surf. Interface Anal., 3*, 211 (1981).
40. J. H. Scofield, *J. Electron Spectrosc. Relat. Phenom., 8*, 129 (1976).
41. R. F. Reilman, A. Msezane and S. T. Manson, *J. Electron Spectrosc. Relat. Phenom., 8*, 389 (1976).
42. M. P. Seah, *Surf Interface Anal., 2*, 222 (1980).
43. M. P. Seah, *J. Electron Spectrosc. Relat. Phenom, 71*, 191 (1995).
44. M. P. Seah and W. A. Dench, *Surf. Interface Anal., 1*, 1 (1979)
45. C. D. Wagner, L. E. Davis and W. M. Riggs, *Surf. Interface Anal., 2*, 53 (1980).
46. H. Ebel, M. F. Ebel, P. Baldauf and A. Jablonski, *Surf. Interface Anal., 12*, 172 (1988).
47. S. Tanuma, C. J. Powell and D. R. Penn, *Surf. Interface Anal., 17*, 911 (1991).
48. S. Tanuma, C. J. Powell and D. R. Penn, *Surf. Interface Anal., 17*, 927 (1991).
49. S. Tanuma, C. J. Powell and D. R. Penn, *Surf. Interface Anal., 21*, 165 (1993)
50. H. Bethe, *Ann. Phys. (Leipzig), 5*, 325 (1930).
51. M. Gryzinski, *Phys. Rev., A336*, 138 (1965).
52. W. Lotz, *Z. Phys., 232*, 101 (1970).
53. S. Ichimura and R. Shimizu, *Surf. Sci., 112*, 386 (1981).

54. R. Shimizu, *Jap. J. Appl. Phys., 22*, 1631 (1983).
55. I. R. Barkshire, M. Prutton and D. K. Skinner, *Surf. Interface Anal., 17*, 213 (1991).
56. K. Harrison and L. B. Hazell, *Surf. Interface Anal., 18*, 368 (1992).
57. M. Prutton, L. A. Larson and H. Poppa, *J. Appl. Phys., 54*, 374 (1983).
58. M. Prutton, C. G. H. Walker, J. C. Greenwood, P. G. Kenny, J. C. Dee, I. R. Barkshire, R. H. Roberts and M. M. El-Gomati, *Surf. Interface Anal., 17*, 71 (1991).
59. S. Tougaard, Surf. Interface Anal., *8*, 257 (1986).
60. S. Tougaard, *Appl. Surf. Sci., 32*, 332 (1988).
61. *Quantitative Analysis of Surface Electron Spectra*, V1.2, QUASES Tougaard Aps., Odense, Denmark, 1994.
62. C. S. Fadley, *Prog. Surf. Sci., 16*, 3 1984.
63. P. J. Cumpson, *J. Electron Spectrosc. Relat. Phenom., 73*, 25 (1995).
64. M. P. Seah, J. H. Qiu, P. J. Cumpson and J. E. Castle, *Surf. Interface Anal., 21*, 336 (1994).
65. C. D. Wagner, L. H. Gale and R. H. Raymond, *Anal. Chem., 51*, 468 (1979).

6

Ion Beam Techniques: Surface Mass Spectrometry

BIRGIT HAGENHOFF Philips Centre for Manufacturing Technology, Eindhoven, The Netherlands

DERK RADING Physikalisches Institut der Universität Münster, Germany

1. PRINCIPLES

In 1910 Thomson [1] discovered that metals emitted secondary particles when bombarded by his so-called Kanalstrahlen. A small fraction of these "sputtered" particles was found to be charged. Nowadays, the twin techniques of secondary ion mass spectrometry (SIMS) and secondary neutral mass spectrometry (SNMS), both of which are based on Thomson's discovery, belong among the most powerful surface analytical techniques for the compositional characterization of surfaces. In both the surface to be analyzed is bombarded by ions of several keV energy, the so-called primary ions (PI). Depending on whether the masses of the PI-induced secondary ions or neutrals are analyzed the terms SIMS or SNMS are used. They can be included under the heading surface mass spectrometry since it must be emphasized here that it is a surface and not, e.g., the gas phase, that is the target for the mass analytical characterization [2].

1.1. Physical Effects of Ion-Induced Sputtering

The physical effects that lead to the eventual desorption of secondary particles are not yet completely understood. Whereas the desorption of secondary neutrals from elemental targets can be described fairly well, an explanation of the emission of charged particles and, in particular, the emission of molecular species, is still being sought. Nevertheless, there is consensus that sputtering is based primarily on the formation of a collision cascade in the target caused by the impinging primary ion [3, 4].

1.1.1. Sputtering

When a primary ion hits a solid, it loses its energy and momentum by elastic and inelastic collisions with target atoms, which are then displaced. These moving atoms (i.e., primary recoils) themselves induce the movement of further target atoms (i.e., secondary recoils) so that the paths of the primary ion and of the highly energetic primary recoil atoms are surrounded by a cloud of recoil particles of lower energies (i.e., collision cascade; see Fig. 1). Recoil atoms can leave the solid if their momenta are directed at least partially toward the surface and if their energies are sufficient to overcome the surface binding energy.

The greatest number of recoils is produced toward the end of the collision cascade and their average energy is consequently comparatively low. Only re-

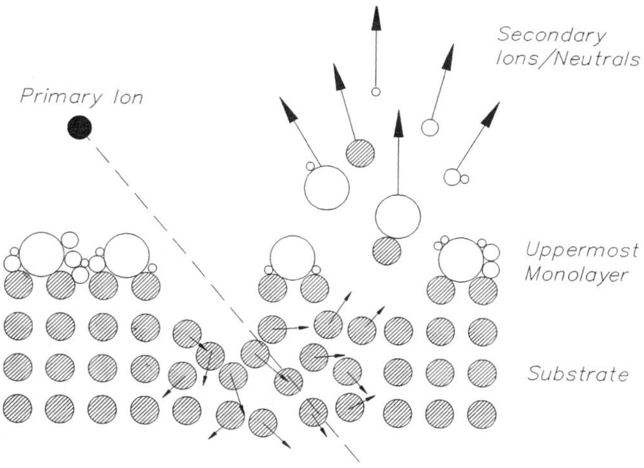

Figure 1. Schematic of the sputtering process.

coils originating in the outermost monolayer of the solid can overcome the surface barrier and contribute to the sputtered flux. The low energy of the late recoils is sufficient not only to cause the ejection of atoms into the gas phase but also intact molecules as well (see Section 1.1.3). The total sputtered flux is therefore characteristic of both elemental and molecular composition of the uppermost monolayer of the bombarded surface (see [3–8] for detailed descriptions). If the primary ions are focused into a small spot and the sputtered species are detected as a function of the primary-ion spot position, then it is even possible to generate elemental and chemical maps of the analyzed surface (i.e., the "imaging mode").

1.1.2. Ionization

By far the greatest proportion of the sputtered flux has zero charge, i.e., consists of neutrals. The fraction desorbed as ions is only 10^{-1}–10^{-6}. The number of sputtered ions per incident primary ion (i.e., the secondary-ion yield terms often used in SIMS and SNMS and explained in Section 1.4) can also vary tremendously with the chemical environment in the near-surface layers, a phenomenon known as the matrix effect. Examples of the secondary-ion yield enhancement, which in some cases can reach several orders of magnitude, are described in [9, 10] for positive metal ions in oxidic environments.

With such yield variations direct quantification of surface species based on the number of desorbed secondary ions (i.e., from the SIMS data) is in general impossible (see Section 2.4 for analytically important exceptions). Sputtered neutrals suffer much less from changes in the chemical matrix. However, SNMS has so far been applied far less often than SIMS for the simple reason that mass analysis can be achieved only if the species is charged. Sputtered neutrals must therefore be ionized prior to mass analysis (i.e., postionization). This process not only complicates the experimental setup but also reduces the number of surface species that can be detected to such an extent that only rarely does SNMS approach the levels of sensitivity (in terms of minimum detectable surface species) of SIMS.

1.1.3. Formation of molecular species

Whereas the desorption of atoms and small molecular clusters can be described in a satisfactory way by the formation of collision cascades, it is surprising that large molecular species and even molecular constituents that are thermally unstable and cannot be vaporized are also emitted from the first monolayer of a surface as intact neutral particles and as positive or negative secondary ions. This phenomenon provides the basis for the mass spectrometric analysis of molecular surfaces [11, 12]. It is, however, beyond the scope of this chapter to discuss in detail the emission of large particles. For possible explanations of the phenomenon the reader is referred to the literature on molecular dynamics models [6, 13, 14] and precursor models [5, 15, 16]. A summary of most of the models currently under consideration for elemental and molecular ion emission can be found in [17].

As mentioned, the entire energy of the impinging primary ion is transported in the collision cascade. Those areas affected by the cascade therefore undergo dramatic changes in their composition, including the breaking and reformation of chemical bonds. If information on the molecular structure of a surface is required, collection of data from already damaged areas must be avoided. From a statistical point of view this can be achieved if the primary-ion dose density (i.e., the number of primary ions per bombarded area) is reduced to such an extent that a maximum of only 1% of the surface is bombarded. For such an operational mode the terms static SIMS (SSIMS) and static SNMS are used in contrast to dynamic SIMS (DSIMS) and dynamic SNMS. In the dynamic versions operational conditions are such that successive atom layers are removed completely and rapidly under the primary-ion bombardment, and the sputtered flux is used to obtain a "depth profile" of the elements present [5]. The static limit, i.e., the highest primary-ion dose density that can be tolerated without causing significant surface damage, is of the order of 10^{11}–10^{13} ions-cm^{-2}, depending on the size of the species analyzed (see also Section 1.4). This chapter focuses on surface analysis and therefore concentrates on the static operational mode. Aspects of other modes are dealt with in Chapters 7–9.

Surface analysis under static conditions implies that the actual amount of material available for characterization is very limited—in fact only about 1% of the first monolayer may be consumed. Recalling again that secondary-ion yields in SIMS and useful yields in SNMS (i.e., after postionization) can be extremely small (10^{-1}–10^{-9}), it becomes clear that very efficient use must be made of the ions in order to achieve acceptable sensitivities, a fact which has considerable influence on the choice of the most suitable instrumentation.

1.2. Instrumentation

In general the instrumentation for SSIMS and SNMS can be divided into two parts, that of the primary-ion column in which the primary ions are generated, focused and transported toward the target and that in which desorbed secondary species are extracted, postionized (where appropriate), mass separated and detected (see Fig. 2).

1.2.1. Primary-ion bombardment

Among the variety of ion guns that have been developed so far, two types are particularly suitable for static applications. If reasonably high primary-ion currents and maximum mass resolution at moderate spot sizes are required,

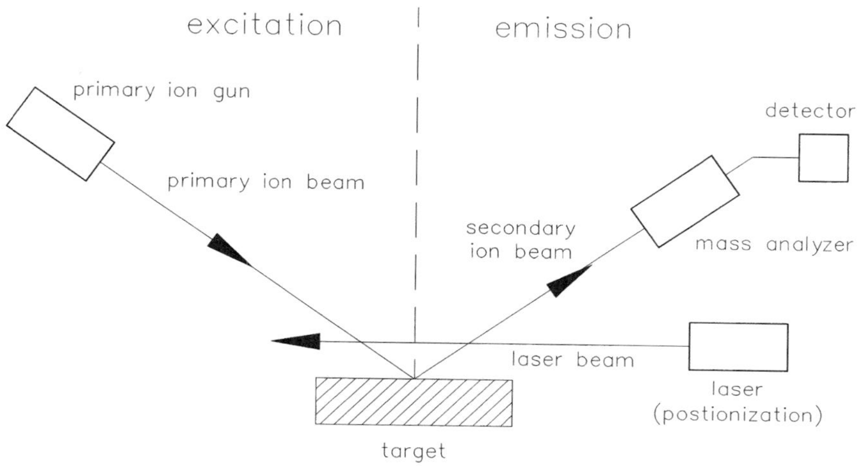

Figure 2. Schematic of an SIMS/SNMS apparatus.

then electron impact (EI) ion sources can be used. In an EI gun gas atoms are ionized by collisions with electrons of some 10 eV energy (i.e., five to six times the ionization potential) drawn to an anode from a heated filament. The gas ions so generated are accelerated toward an extraction electrode and can then be focused onto the target [5, 18–20, 20]. For SSIMS and SNMS applications noble gases are generally used which offer the advantage that there are no chemical reactions between primary-ion and target species which could affect the correct determination of the surface composition. Caesium ions may also be used, with the advantage that Cs^+ bombardment enhances the secondary yield of negative elemental ions as well as those of some polymer fragments [5]. (Cs^+ is produced by surface ionization, with extraction similar to that in EI.)

For the production of elemental and chemical surface maps with high lateral resolution (imaging) spot sizes available from EI guns are inadequate, and liquid metal ion guns (LMIG) are used in which the primary-ion beam can be finely focused, but at the cost of smaller ion currents and lower mass resolutions when used with time-of-flight (ToF) analyzers (see Section 1.2.2). In an LMIG a small needle is wetted by a liquid metal (usually Ga or In). A positive voltage of some 10 keV is applied to the needle and a spray process occurs at the needle tip. The desorbed stream of ions (jet) is used as the primary-ion beam [21–23]. Because the ion production volume at the tip is very small, LMIG sources have high brightness, and consequently the ion beam can be focused to a small area having high current density. The mandatory use of liquid metals, (Ga, In) means that chemical reactions at the surface cannot be completely excluded, although they have not been reported so far. Table 1.1 compares the performance data of both ion gun types [24].

1.2.2. Mass analyzers

In principle, four types of mass analyzers could be used for SSIMS and SNMS applications:

1. Double-focusing magnetic sector field analyzers
2. Quadrupoles
3. Fourier transformation ion cyclotron resonance (FT-ICR) mass analyzers
4. Time-of-flight analyzers

Table 1.2 compares the most important performance data for these analyzers.

For the choice of a mass analyzer suitable for static applications the following factors must be considered:

- The available number of desorbed ions and postionized neutrals is very limited. High sensitivities can therefore be achieved only if the analyzer has a high transmission and/or can detect different masses simultaneously.
- In order to be able to separate elemental from molecular species and molecular species from each other, the mass resolving power (usually given as separable mass Δm at mass m) should exceed 10^4.
- Desorbed molecular species are mostly metastable, i.e., they decompose on a time scale of µs, and therefore mass separation and detection should occur on a similar or shorter time scale.

As can be seen from Table 1.2 the ToF analyzer is the one of choice under these circumstances and ToF-SIMS (i.e., SIMS using a ToF analyzer) has almost become a synonym for SSIMS with high sensitivity. The ToF analyzer is therefore described in more detail in the following. For a description of the other analyzer types the reader is referred to [18, 19].

In a ToF analyzer (Fig. 3) all desorbed ions (or postionized neutrals) are accelerated to a common energy qU (q = charge of the ions, U = applied voltage), and the time t taken by the ions to reach a detector after traveling along a field-free drift path of given length s is measured. According to

$$\frac{ms^2}{2t^2} = qU \Rightarrow t^2 = \frac{ms^2}{2q\mathrm{U}} \propto \frac{m}{q} \qquad (1)$$

The q/m ratio can be calculated from this flight time. Because a very well defined start time is required for the flight time measurement, the primary-ion gun has to be operated in a pulsed mode in order to be able to deliver discrete primary-ion packages [20, 25, 26]. Electric fields (e.g., ion mirrors [26, 27] or

Table 1.1. Typical Performance Data of Ion Guns Used in SSIMS and SNMS

	EI	LMIG
Ion	Xe^+, Ar^+	Ga^+, In^+
Ions per pulse	100/1000	3/40/100
Size of source	1 mm	50 nm
Brilliance (A-sr^{-1}-cm^{-2})	1	10^6
Mass resolution	$>10^4$/$>10^4$	1/$>5 \times 10^3$/$>10^4$
Min. spot size on target[a]	4 μm/30 μm	80 nm/200 nm/4 μm

[a]In combination with ToF mass analyzers.

electrical sectors [28, 29]) are introduced into the drift path in order to compensate for different incident energies and angular distributions of the ions (see also Section 1.4.1). ToF analyzers are able to provide simultaneous detection of all masses of the same polarity. Furthermore, the pulsed operation makes it particularly suitable for combination with laser systems for the postionization of sputtered neutrals.

1.2.3. Add-Ons

Charge compensation

The primary-ion bombardment leads not only to the desorption of sputtered elements and molecules but also to the emission of further secondary species, among them electrons. The number of emitted electrons per primary ion (the so called ion-induced secondary-electron coefficient) can be as large as 10 according to the particular material. Ion bombardment of insulating surfaces therefore leads to the buildup of positive charge at the surface which can se-

Table 1.2. Typical Performance Data of Mass Analyzers[a]

	Transmission	Mass res. $m/\Delta m$	Accessible mass range	Simultaneous ion detection
Double focusing	**10^{-1}**–10^{-3}	300–10^3	10^3–10^4	Few masses
Magnetic analyzer	10^{-2}–10^{-4}	10^3–**10^4**		
Quadrupole	10^{-3}–10^{-5}	100–500	1000	None
FT-ICR	$<10^{-3}$	**10^4–10^7**[b]	10000	**All masses**
ToF	**1–10^{-1}**	10^3–**10^4**	**Unlimited**	**All masses**

[a]The requirements for the static operational mode are boldface.
[b]Minimum lifetime of ions: some ms, as required for reliable mass determination.

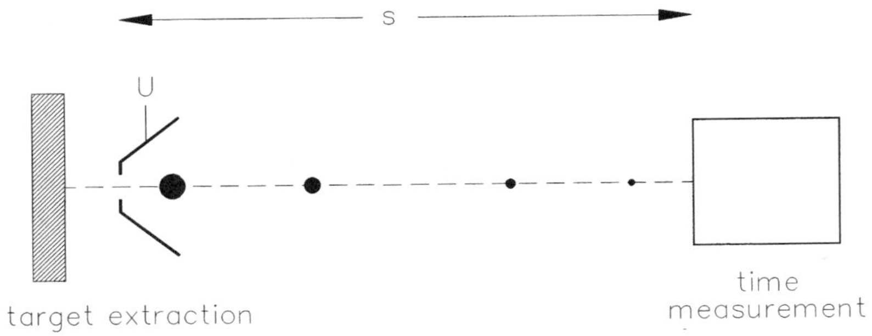

Figure 3. Schematic of a ToF analyzer. The flight path length is S and the extraction voltage is U.

verely disturb the secondary-ion extraction. The charge can be compensated by auxiliary electron bombardment. Low-energy electrons are generally used (<20 eV) because, first, at those energies the electron-induced secondary-electron coefficient is <1 (if the electron-induced secondary-electron coefficient is less than unity, then every incoming electron leads to less than one outgoing electron and an existing positive charge can be compensated), and, second, because sensitive organic molecules are hardly affected by the additional bombardment. For the latter reason high-energy electrons (e.g., >3 keV, for which the electron-induced electron yield coefficient would also be below unity) are less suitable for charge compensation purposes.

If ToF analyzers are used for mass determination, low-energy electrons can easily be injected into the target zone in the period between secondary-particle extraction and the next primary-ion pulse. During that period all high voltages near the target zone, which might deflect low-energy electrons, can be switched off. With this arrangement stable spectra of both polarities can be obtained even for extremely insulating materials, including powders. For details see [30].

Postionization

As mentioned, sputtered neutrals must be postionized prior to mass analysis. In general, electrons beams, plasma environments and laser beams are suitable for this task, but only laser beams allow the postionization of both elements and molecules. For this reason only postionization by a laser beam is described here in detail.

Desorbed neutrals can drift away from the surface due to the energy they have acquired during sputtering (see Section 1.1.2) irrespective of any electric field present. When this cloud of neutrals reaches the ionization volume some μm above the target (desorbed ions are rejected by a retarding field), the laser is fired. Excimer systems operating at 193 and 248 nm are usually used since they make nonresonant ionization of almost all elements and many molecules in a two-photon process (first photon: excitation to a virtual state; second photon: ionization into the continuum) possible. Elemental postionization probabilities (see Section 1.4.3) of nearly unity can be achieved in the center of the ionization volume. For molecules the situation is much more complex because resonances can occur and fragmentation can take place during and shortly after the excitation. Therefore the wavelength and laser power density used must be adjusted carefully in order to match the individual molecule. Due to the geometric (space-time distribution) and postionization conditions the useful yields that can be achieved are only occasionally higher than those obtained when analyzing secondary ions under optimum conditions. For a detailed description see [31].

1.3. Typical Spectra

1.3.1. Typical characteristics of SSIMS spectra

Figure 4 shows the positive secondary-ion mass spectrum ("positive" refers to the detection of positively charged secondary ions) of an isolated oligomer of polydimethylsiloxane (PDMS) prepared as a monolayer on an etched silver substrate (solution: 0.1 mg/mL in toluene, 1 μL spread onto 80 mm^2). This particular molecule is used here as an example since it allows most of the typical spectral features of SSIMS spectra to be described. In contrast to polymer preparations containing more than one oligomer, spectra of isolated oligomers, though technically irrelevant, do not suffer from peak interferences with other oligomers and the fragments can arise from only one type of molecule instead of from the whole molecular weight distribution. The chosen molecule was kindly provided by Wacker Chemie AG, München. Furthermore, PDMS molecules, known also as silicone oils or rubbers, are among the most troublesome contaminants in adhesion problems (see Chapter 18). The reason for the choice of the silver substrate is described in detail in Section 1.5.

The spectrum can be divided into three general sections: the fragment area (1–500 amu), the area of quasimolecular ions (1150–1350 amu) and an intermediate area, where almost no secondary ions can be detected.

The term "quasimolecular ions" refers to those ions which are formed either by the attachment of low-molecular-weight cations and/or anions (e.g., salt and metal ions) to the parent mass or by the loss of small fragments (mostly functional groups such as CH_3, OH, etc.). The most intense group of peaks among the quasimolecular ions can be attributed to a silver attachment to the intact molecule ($(M + Ag)^+$, confirmed by a comparison of the calculated and measured isotopic patterns). Almost the same holds for the peak at mass 701.1 amu. Here, cationization has been achieved by the attachment of two Ag ions. Consequently, the ion is doubly charged

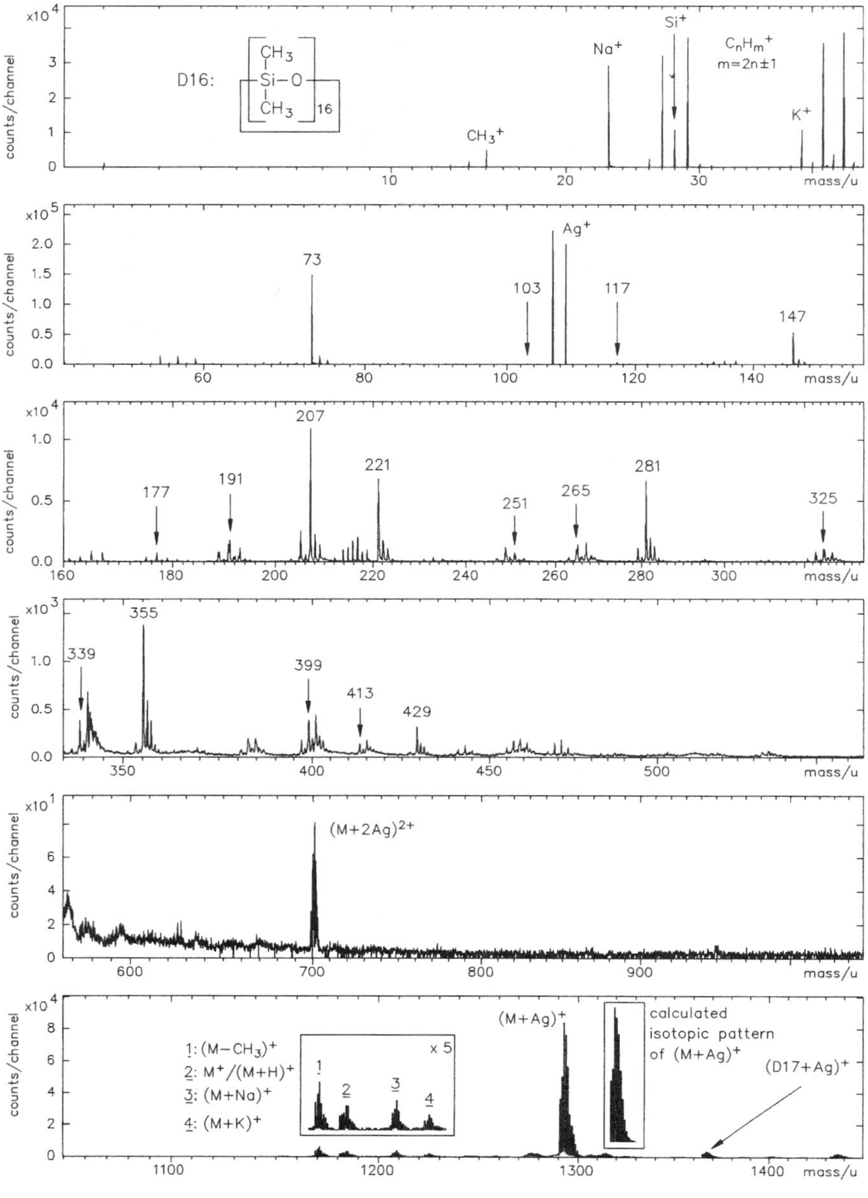

Figure 4. Positive secondary-ion mass spectrum of D16 (polydimethylsiloxane oligomer consisting of 16 repeat units), prepared as a monolayer on silver.

and appears at half the mass of the parent molecule (1402.2 amu) (detection of *m/q*, Section 1.2.2). (The low intensity of this peak shows that formation of doubly charged species is not a dominant process in SSIMS.) Further quasimolecular ions can be attributed to $(M–CH_3)^+$, M^+, and $(M+H)^+$ (the isotopic patterns of M^+ and $(M+H^+)$ overlap), as well as $(M+Na)^+$, and $(M+K)^+$. The differences in the intensities of these ions compared to that of the $(M+Ag)^+$ peak demonstrate the high formation probability of the cations (see Section 6.1.5 for details). If a complete oligomer weight distribution were to be analyzed, the form of the distribution could be derived directly from the distribution of the respective quasimolecular ions.

The mass range up to 500 amu is dominated by fragment ions. Four fragment series can be identified starting at masses 73, 103, 117, and 133 amu, respectively. All further fragments can be explained by the addition of the PDMS repeat unit (74 amu). All the ion structures are given in Fig. 5. With the exception of the primary ionization process, fragments in SSIMS are nearly all formed according to the conventional rules of electron impact mass spectrometry (α and β cleavages, rearrangement processes [32]). As other polymers give rise to other fragment patterns, the type of polymer can be determined by evaluation of the low-mass fragments alone.

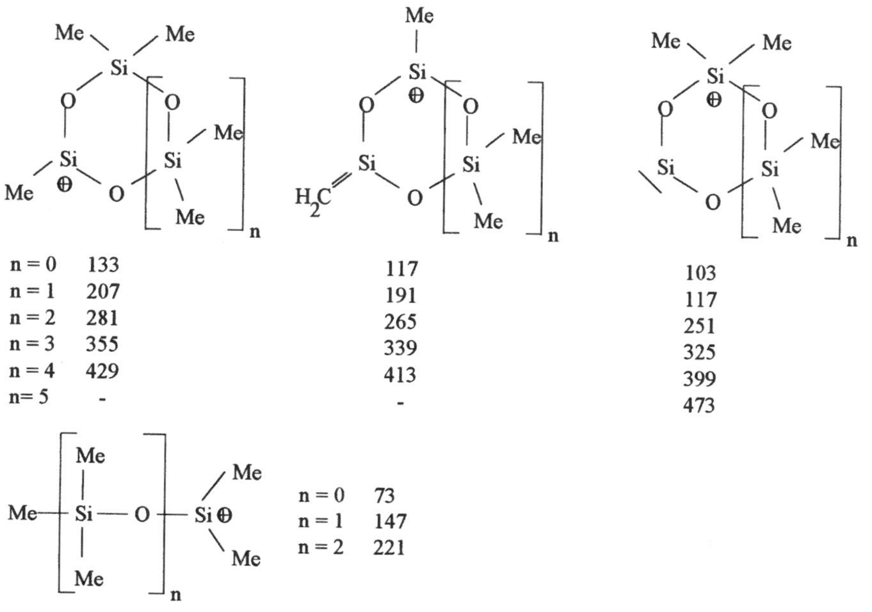

Figure 5. Fragment structures of PDMS obtained under primary-ion bombardment.

In the intermediate-mass range, where the PDMS does not emit any characteristic secondary ions, large fragments can be observed from other types of polymers originating from single or multiple-main-chain scissions. In many cases these fragments are neutral and can therefore be observed only as $(F + Ag)^+$ (F is an appropriate fragment) from monolayer preparations on silver. From the mass distance between two fragment peaks the mass of the polymer repeat unit can be derived.

Of course, negatively charged secondary ions can also be formed, as quasimolecular ions (e.g., $(M–H)^-$) as well as fragments. Small negatively charged fragments are often more characteristic of a molecule than those positively charged because charge stabilization is often related to oxygen, a heteroatom having a profound influence on chemical behavior. From PDMS, for example, no negative quasimolecular ions are formed, but a strong fragment series starting at mass 75 amu ($CH_3SiO_2^-$) is apparent.

In summary, SSIMS spectra provide not only evidence of all the elements present (see, e.g., the Na^+, Si^+, Ag^+ peaks) but also detailed insight into the molecular composition. Quasimolecular ions can be desorbed intact up to 3500–15000 amu, depending on the particular molecule [33] and whether an effective ionization mechanism is present. Larger molecules tend to fragment under the SSIMS excitation conditions due to stronger intra- and intermolecular bonds.

If no quasimolecular ions are formed, due to hindered desorption (large molecules) or hindered ionization (e.g., thick molecular overlayers without contact with cations), small-fragment ions still remain visible, because they can be formed even if strong intra- and intermolecular bonds exist and because they carry an intrinsic charge. The observed fragment peak patterns are characteristic of the particular molecules and for the mass range up to 500 amu the term “fingerprint region” is thus used. Whereas quasimolecular ions are comparatively easy to identify, the interpretation of the fingerprint region requires much experience and should best be left to the expert at least until reliable libraries are published [34].

To date no material has been reported from which SSIMS spectra could not be obtained, provided that it is stable in a vacuum environment, making SSIMS a universal analysis technique.

1.3.2. Typical characteristics of SNMS spectra

In general, SNMS spectra do not differ much from SSIMS spectra; i.e., quasimolecular ions, fragment ions and elemental ions can be observed [31]. Of course the primary ionization process is different and results always in positively charged ions. Typical quasimolecular ions are therefore M^+. Quasimolecular ions can easily be generated from aromatic ring systems (due to resonance excitation).

Fragments are similar to those generated in SSIMS, although the relative intensities can be different and additional fragments can be formed by the energy uptake from the laser beam. The extent of fragmentation therefore depends strongly on the laser parameters (power density, flux, laser pulse length).

Most elements can easily be postionized by a two-photon laser scheme [35]. SNMS is thus also a suitable tool for elemental as well as molecular analysis, although the identification of molecules requires at least some information on the expected structures in order to adjust the laser parameters. In particular, for elemental analysis SNMS offers the advantage that the results can easily be quantified (determination of surface concentrations).

1.4. Useful Definitions in SSIMS and SNMS

1.4.1. General

Disappearance yield $Y_D(M)$

$$Y_D(M) = \frac{\text{No. of surface species } M \text{ having disappeared}}{\text{No. of primary ions}} \tag{2}$$

This quantity refers to those surface species that have disappeared under the influence of the primary-ion beam independent of their fate (i.e., due to sputtering, fragmentation, thermal desorption, etc.)

Disappearance cross section σ: Defined as the area from which surface species have disappeared under PI bombardment, irrespective of their fate; the magnitude of σ depends on the species fragment analyzed (i.e., $\sigma = \sigma(M)$). Typical values are around 10^{-13}–10^{-14} cm^2 (see Section 1.4.2).

Static limit: In general, the probability P of collecting data from an already damaged area must be maintained very small; that is,

$$P = \frac{\sigma_{\text{Tot}}}{A} \ll 1 \tag{3}$$

where A is the bombarded area and σ_{Tot} is the total damaged area. If P is sufficiently small, then the bombardment processes can be assumed to be independent of each other; i.e.,

$$\sigma_{\text{Tot}} = \langle z \rangle \sigma = \text{PID} \times \sigma \tag{4}$$

where σ is the disappearance cross section (see above) and PID is the primary-ion dose. Then

$$P = \frac{\text{PID} \times \sigma}{A} = \text{PIDD} \times \sigma \ll 1 \tag{5}$$

where PIDD (= PID/A) is the primary-ion dose density. For $P = 0.01$ and $\sigma = 10^{-13}$–10^{-14} cm^2, the corollary is that $(\text{PIDD})_{\max} = 10^{11}$–$10^{12}$ ions-cm^{-2}, while for $P = 0.1$, $(\text{PIDD})_{\max} = 10^{12}$–$10^{13}$ ions-cm^{-2}. It must be emphasized that the static limit $(PIDD)_{max}$ depends on σ, which itself depends on the particular species.

Sputter yield $Y_s(M)$

$$Y_S(M) = \frac{\text{No. of sputtered species M}}{\text{No. of primary ions}} \quad (6)$$

Energy distribution: Elements and molecules are desorbed with different energy distributions. For elements the maximum of the distribution is about 5–10 eV with a slow decrease toward higher energies (proportional to $E/(E + E_B)^3$; E = energy, E_B = surface binding energy [36]). The maximum for molecules is shifted toward lower energies, from 5 to 10 eV to ca. 1 eV, and the distribution decreases exponentially with increasing energy (proportional to $\exp(-E/E_m)$; E_m = maximum in the distribution [5, 37]). This spread in energy must be compensated for by mass analyzers in order to achieve high mass resolution.

Angular distribution: For primary-ion bombardment perpendicular to the target surface the sputtered material is ejected in a cosine distribution according to [38]

$$\frac{dY}{d\Omega} \propto \cos \nu \quad (7)$$

where ν is the emission angle with respect to the surface normal. For other angles of incidence the cosine distribution shows an anisotropy in the direction of the reflected primary ions [39]. As for the desorbed particle energy distribution, the angular distribution must also be taken into account in the development of suitable mass analyzers.

1.4.2. SSIMS

Secondary ion yield Y_i

$$Y_i(X^q) = \frac{\text{No. of secondary ions } X^q}{\text{No. of primary ions}} \quad (8)$$

where Y_i is also dependent on the particular instrumental conditions (e.g., transmission, detection efficiency). For measurements performed with the same instrument it is the easiest way to compare different spectra quantitatively.

Number of detectable ions $N_d(t)$: The number of detectable ions at time t is

$$N_d(t) = N_d(0)\exp\left(\frac{-\sigma\ It}{Ae}\right) \quad (9)$$

with $N_d(0)$ being the number of detectable ions at $t = 0$, I the primary-ion current, A the bombarded area and σ the disappearance cross section, i.e., the area on the surface from which the particular species undergoing primary-ion bombardment has disappeared completely (independent of its fate, i.e., by sputtering, fragmentation or evaporation). From a plot of $N_d(t)$ versus t, σ can be found. $N_d(0)$ itself is given by

$$N_d(0) = N(M)P(\mathrm{M} \rightarrow \mathrm{X}^q)TD \tag{10}$$

with N(M) the number of molecules M present at the surface, T the transmission and D the detection efficiency. $P(\mathrm{M} \rightarrow \mathrm{X}^q)$ is the so-called transformation probability (see below).

Transformation probability $P(M \rightarrow X^q)$

$$P(\mathrm{M} \rightarrow \mathrm{X}^q) = \frac{\text{No. of emitted secondary ions } X^q}{\text{No. of sputtered surface species M}} = \frac{Y_i(X^q)}{Y_\mathrm{D}(\mathrm{M})} \tag{11}$$

$P(\mathrm{M} \rightarrow \mathrm{X}^q)$ is the probability of a surface species M being desorbed as an ion X in a charge state q. The transformation probability can be determined by evaluation of the curve $N_d(t)$ versus t if N(M) is known. $P(\mathrm{M} \rightarrow X^q)$ is used mainly for the description of monolayers. It is assumed that desorbed ions are stable. For metastable ions measured values of $P(\mathrm{M} \rightarrow \mathrm{X}^q)$ are too low.

Useful yield Y_u

$$Y_\mathrm{u} = \frac{\text{No. of detected secondary ions } X^q}{\text{No. of sputtered surface species M}} \tag{12}$$

Y_u is still dependent on instrumental parameters (i.e., transmission, detection efficiency). For a determination of Y_u the total number of sputtered species has to be determined by independent techniques.

Matrix effect: The matrix effect in SSIMS is defined as the dependence of $P(\mathrm{M} \rightarrow \mathrm{X}^q)$, Y_i, and Y_u on the chemical composition of the analyzed surface.

1.4.3. SNMS

Geometrical yield $Y_G(M, X_I^Q)$

$$Y_\mathrm{G}(\mathrm{M}, \mathrm{X}_\mathrm{I}^\mathrm{Q}) = \frac{\text{No. of species } \mathrm{X}_\mathrm{I}^\mathrm{Q} \text{ that have drifted into the ionization volume}}{\text{No. of sputtered species M}} \tag{13}$$

where the generalized superscript Q encompasses all charge states of X_I (10). The geometrical yield takes into account that only those neutrals can be ionized which have reached the ionization volume at the time of postionization.

Postionization probability $\alpha(X_I^Q \rightarrow X_i^q)$

$$\alpha(X_I^Q \rightarrow X_i^q) = \frac{\text{No. } X_i^q \text{ of ions formed}}{X_I^Q} \tag{14}$$

$\alpha(X_I^Q \rightarrow X_i^q)$ is the fraction of postionized neutrals in the ionization volume.

Transformation probability $P(M \rightarrow X_i^q)$

$$P(M \rightarrow X_i^q) = \sum_{I,Q} Y_G(M, X_I^Q)\alpha(X_I^Q \rightarrow X_i^q) \tag{15}$$

$P(M \rightarrow X_i^q)$ is the probability that a surface species M is transformed into a postionized ion.

Useful yield $Y_u(M, X_i^q)$

$$Y_u(M, X_i^q) = \frac{\text{No. of detected postionized ions } X_i^q}{\text{No. of sputtered surfaces species M}} = P(M \rightarrow X_i^q)TD \tag{16}$$

where T is transmission and D is the detection efficiency.

1.5. Use of Noble Metal Substrates

Although noble metals do not number among the more frequently used materials in industry (reactive metals, metal oxides and semiconductors being found more often in that context), it is nevertheless worth considering them as substrates for SSIMS and SNMS analyses of organic materials. The materials to be analyzed should be prepared as (sub)monolayers on noble metal substrates either by preparation from solution or simply by physical contact of the substrate with the particular material (e.g., by rubbing). The noble metal surfaces must be freshly prepared by etch processes (e.g., in dilute HNO_3, or *aqua regia* and subsequent rinsing in water), by evaporation of noble metals onto other substrates, by cleaning in an Ar plasma, or by galvanic deposition.

Preparation of monolayers of organic materials on noble metal substrates leads to high secondary-ion yields and can even allow the detection of organic materials at levels that would otherwise remain below the detection limits. Three properties can be identified that make noble metals such exceptional SSIMS and SNMS substrates: (1) enhanced desorption efficiency, (2) cationization and (3) stability and nonreactivity.

1. Enhanced desorption efficiency. The sputter yields of metals are distinctly higher than those of organic materials. If the collision cascade takes place principally in the metal substrate (i.e., after monolayer preparation), the sputter yields of the organic species are up to factors of 3 higher than those

obtained from the bulk materials [40]. Furthermore the surface area excited by the collision cascade is larger. Independently of ionization mechanisms metal substrates should therefore be used in SSIMS as well as in SNMS for the analyses of organic species. Noble metals should also be preferred to silicon wafers or to glass substrates.

2. Cationization. Noble metals provide a very efficient ionization mechanism in SSIMS resulting in the formation of $(M + NMe)^+$ ions (NMe = noble metal). These quasimolecular ions are formed almost independently of chemical structure, the noble metal cation attachment being therefore an universal ionization process. (Note that cagelike structures may not accommodate the size of the metal cation, e.g., polyethyleneglycol.) If silver is used, the isotopic pattern of Ag is an additional help in identifying the $(M + Ag)^+$ isotopic peak patterns. In other cases the differentiation between $(M + H)^+$, $(M + Na)^+$, $(M + K)^+$, and $(M + NMe)^+$ requires careful evaluation, in particular if only one peak group is observed.

3. Stability and nonreactivity. The $(M + NMe)^+$ quasimolecular ion is distinctly more stable than all the other secondary ions formed, which becomes important in those cases where only a very limited amount of analyte is available. This fact together with the higher sputter yield and the efficient ionization means that sometimes the use of noble metal substrates is the only method of detecting some organic materials in SSIMS. Furthermore, noble metals are chemically less reactive than other suitable metal substrates (e.g., Cu and Al) and dissociation of organic materials upon contact with the metal is prevented.

In summary, noble metals are the substrates of choice if

- Monolayers can easily be deposited.
- The amount of analyte available is limited.
- Information on intact molecules is required and no protonatable/deprotonatable groups are present.

For the detection of contaminants a noble metal preparation can easily be obtained by either rubbing suspect surfaces over the substrate or subjecting the substrate to the suspect environment for some time. For more details see [41].

1.6. Performance Summary

Table 1.3 lists the most relevant performance characteristics of SSIMS and SNMS, correct at the time of writing. Instrumental performance (e.g., mass resolution, lateral resolution, etc.) is always likely to improve progressively with time.

Table 1.3. Performance Data of SSIMS and SNMS

	SSIMS	SNMS
Excitation	Ions 5–30 keV	
Detection	Secondary ions	Secondary neutrals
Suitable materials	All, including insulators and powders, with the caveat that they must be compatible with a vacuum environment	
Available information	All elements including H, molecules; molecular structure	All elements (molecules)
Information depth	First monolayer	
Detection limits	ppb (elements), fmol (molecules)	
Quantification	See Section 6.2.4	Yes
Mass resolution	>10,000	>3,000
Lateral resolution	<100 nm (LMIG)	<100 nm
Max. field of view	700 × 700 μm² (EI) and 100 × 100 μm² (LMIG)	
Maximum depth	Some μm by sputtering	
Screening for unknown		
elements	Yes	Yes
molecules	Yes	No

2. OPERATIONAL METHODOLOGY

The following sections concentrate on the ways which SSIMS has been operated and the spectra quantified. Similar considerations for SNMS have not been included because SNMS with laser postionization is still in an experimental stage and is not widely used in industrial laboratories. For future expectations see Section 4.

2.1. The Analytical Question

A universal analysis technique should be able to (a) locate points and areas of interest on the surface as well as in deeper layers, (b) identify elements and molecules found, and (c) determine the concentrations of those species with an overall sensitvity of at least ppm for surfaces and ppb for the bulk (see Fig. 6). Unfortunately, no known analytical technique can satisfy all these requirements. Nevertheless, the performance of a technique in the three areas of location, identification, and quantification can be used as a measure of its usefulness in routine analysis. In the following these three areas are evaluated for surface analysis by ToF-SIMS.

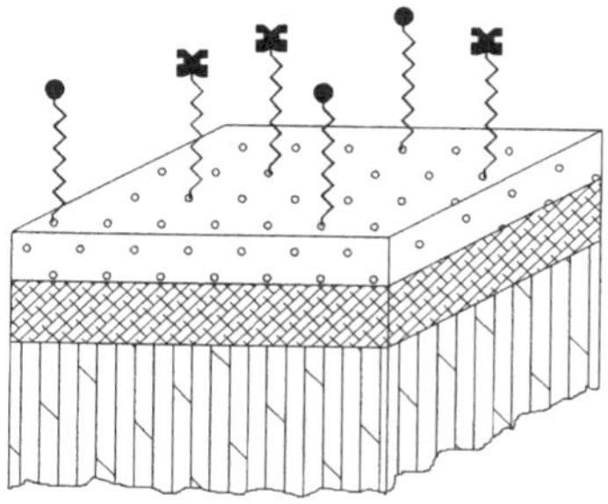

Figure 6. The analytical question.

2.2. Spatial Location

In general, three factors are essential for successful location of surface species, and they therefore govern the overall performance (i.e., the level of spatial resolution) of a technique. They are

1. Physics of the analysis process
2. Lateral resolution provided by the instrumentation
3. Achievable sensitivity of the particular surface analytical technique

In SSIMS, secondary ions are formed in a collision cascade following the impact of the primary ion. The diameter of this collision cascade therefore determines the ultimate limit of the accuracy with which the original position of a surface species can be located. Depending on the type of primary ion, on the bombardment energy and on the particular surface species, the collision cascade diameter is about 2–5 nm [42, 43].

The lateral resolution offered by the instrumentation is determined mainly by the spot diameter of the primary-ion beam. When liquid metal ion guns are operated in a static mode, i.e., not pulsed, a focus diameter of 10 nm can be achieved [44]. However, when modern, sensitive ToF analyzers are used for mass determination and ion detection, the static operational mode of the LMIG allows the acquisition either of total secondary-ion images or of total ion-induced secondary electron images only. Mass information cannot be obtained. The total secondary-ion and secondary-electron images nevertheless do give valuable topographic information. The image quality nowadays is close to that of dedicated SAM instruments. If mass information is required, then the LMIG must be pulsed, resulting in an inferior focus diameter. With nominal mass resolution, however, 80 nm can still be achieved. A focus diameter of 200 nm allows the acquisition of spectra with a reasonable mass resolution of $m/\Delta m = 5000$, which is sufficient to differentiate elements from organic species reliably. For maximum mass resolution an LMIG focused diameter of ca. 4 μm has to be used (see Table 1.1).

A small focus diameter, however, is not of much help if not even one secondary ion can be detected from the the bombarded area due to insufficient sensitivity. As described in Section 1.4.2 the number of detectable ions is given by Eq. (10):

$$N_d = N(\mathrm{M})P(\mathrm{M} \rightarrow \mathrm{X}^q)TD$$

For ToF analyzers the product of transmission T and detection efficiency D can be estimated to be about 0.1. One monolayer (N(M)) consists of about 10^{15} atoms-cm^{-2} or 10^{14} molecules-cm^{-2}. Transformation probabilities ($P(\mathrm{M} \rightarrow \mathrm{X}^q)$) can be as large as 10^{-1} for elements but less than 10^{-7} for molecules. The consequences for the number of detectable ions N_d are listed in Table 2.1. Assuming that at least one detectable secondary ion is required for a successful detection, Table 2.1 shows that for elements ($P(\mathrm{M} \rightarrow \mathrm{X}^q) = 10^{-1}$–$10^{-2}$) the number of emitted secondary ions is sufficient even for primary-ion-beam spot diameters smaller than 100 nm. Organic fragments in the fingerprint region ($P(\mathrm{M} \rightarrow \mathrm{X}^q) = 10^{-1}$–$10^{-3}$) can also be imaged with sub-μm resolution. The situation is not as good for larger fragments and quasimolecular ions ($P(\mathrm{M} \rightarrow \mathrm{X}^q) = 10^{-2} \leq 10^{-7}$). Here, the useful lateral resolution can be as poor as 10 μm even though the primary-ion gun is capable of much better spot diameters. This simple calculation explains why elements and small fragments are so important for sub-μm imaging. If, for example, a spot size of 200 nm were to be chosen for an LMIG in combination with a ToF analyzer, the mass resolution of about 5000 would still answer the question as to whether or not an imaged organic fragment contained a heteroatom, an important clue in the identification of the species under investigation.

As an example of the lateral resolution obtainable with a ToF-SIMS instrument, Fig. 7 shows negative secondary-ion images of tabular silver bromide crystals coated with iodine. The field of view is 6 × 6 μm^2, and the elemental

Table 2.1. Number of Detectable Secondary Ions as a Function of Bombarded Area and Transformation Probability[a]

		$P(\mathrm{M} \rightarrow \mathrm{X}^q)$			
Sampled area	N(M)	10^{-1}	10^{-3}	10^{-5}	10^{-7}
0.1 mm^2	10^{10}	10^8	10^6	10^4	10^2
10 μm^2	10^8	10^6	10^4	10^2	1
1 μm^2	10^6	10^4	10^2	1	10^{-2}
100 nm^2	10^4	10^2	1	10^{-2}	10^{-4}

[a]TD = 0.1; a monolayer coverage of 10^{14} cm^{-2} is assumed

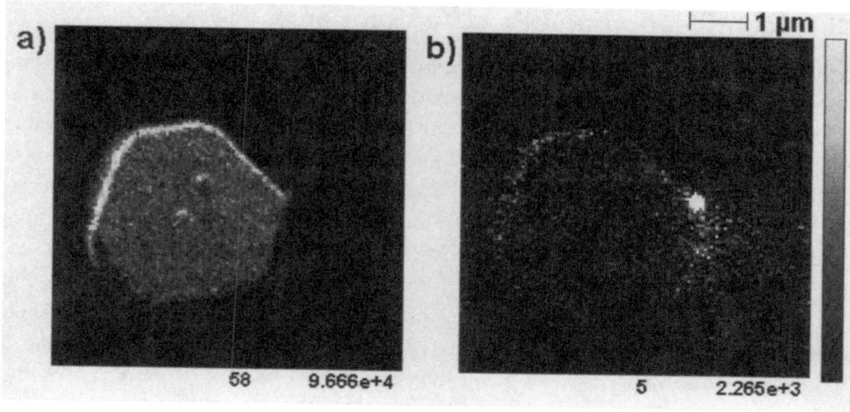

Figure 7. Negative mass resolved ToF-SIMS images of $AgBr_{0.9}I_{0.1}$ crystallites; field of view was 6 × 6 μm²; (a) Br^- map; (b) I^- map; the sample was provided by G. Janssens, University of Antwerpen, Belgium.

maps of Br^-(Fig. 7a) and I^- (Fig. 7b) were acquired with a primary-ion dose density of 3×10^{14} cm^{-2} (the sample consumption was about one monolayer). The image resolution is approximately 75 nm.

In summary it can be stated that the performance of LMIGs has improved continuously over the last few years and that the lateral resolution for elemental detection has moved toward its physical limit. The gap between dedicated electron microscopes and ToF-based instruments is thus closing. For molecular analysis the lower transformation probabilities reduce the useful lateral resolution to about 100 nm for small fragments and about 5 μm for larger fragments and intact molecules. As atomic force microscopy (AFM), with its power to resolve single molecules, has not yet been able to identify molecules directly, the combination of SSIMS and AFM might be interesting if used with overlapping fields of view.

Although it is not the subject of this chapter, it is important to note that SSIMS can also be used to locate elements in subsurface layers. To do so, sputtering is increased to remove surface material gradually (sputter depth profiling). For details see Chapter 7 and 8. Due to the collision cascade which destroys most chemical information it is not possible to monitor organic molecules during depth profiling; in that case it is advisable to make mechanical cross sections of the material and to image the resulting sample. As SSIMS receives its information from the outermost monolayer of the particularly surface, it is, however, necessary to ensure that the section is not contaminated by embedding agents, knife contaminants, etc.

2.3. Identification and Peak Assignment

SSIMS is a mass spectrometric technique; i.e., the analyte is identified by its precise mass. Identification is straightforward for elements, small inorganic molecules (e.g., phosphates, sulphates, nitrates, etc.) and inorganic compounds (e.g., oxides) because the number of constituent elements is limited, their mass is known accurately and the characteristic isotopic pattern of each element facilitates the identification process. The masses of elements and inorganic molecules are generally found at values lower than the respective nominal masses. This is caused by the specific mass deficiency of nuclear particles bound in the atomic nuclei in contrast to free protons and neutrons. On the atomic mass scale the mass of ^{12}C is by definition set to 12.000, which leads to an atomic mass of H of 1.007825, whereas ^{16}O has a mass of 15.9949 and ^{56}Fe of 55.9349. Inorganic compounds containing the latter elements therefore have masses below nominal, whereas the high amount of H in organic materials leads to masses above nominal.

Identification is much more complicated when it comes to organic materials. Although organic molecules consist mainly of C, H, and some heteroatoms such as O, S, N, the various combinations of these elements lead to an immense variety of organic species. Whereas for nominal mass 13 only two possible assignments exist, namely ^{12}CH and ^{13}C, this number increases enormously with mass; at nominal mass 50 more than 1000 different organic species are possible, all differing only slightly in mass. A high mass resolution is therefore mandatory in order to be able to identify the correct sum formula of a peak. Organic species occur mostly at masses above nominal.

The situation becomes even more complex if mixtures of different materials exist at the surface. Fortunately in most cases the spectra of mixtures can be explained by superposition of the spectra of the individual compounds. Unfortunately such a superposition cannot necessarily be used for quantification since the matrix effect in SSIMS has to be taken into account, but is useful for qualitative identification; i.e., the peaks present in the spectra of the individual compounds appear also in the spectrum of the mixture. Mixed ions, i.e., those formed by reaction between the individual compounds, are rarely observed. Spectra libraries can therefore be used successfully to identify the individual compounds.

Completely new spectra should be evaluated using the following three guidelines.

1. Principal elements from the matrix or from a substrate. In most cases some minimum information about a sample is available: e.g., it might be known to be a glass, or a Si wafer, or a ceramic, or a polymer, or an organic material on Ag, etc. The peaks belonging to these materials, if not known from previous experience, can often be found in spectra libraries and may account for

most of the observed peaks. Any remaining peaks must then belong to other compounds or to possible contaminants.

2. Identification of quasimolecular ions (monolayer preparation on Ag only). For unknown materials prepared as (sub)monolayers on Ag it is advisable first to screen for quasimolecular ions. If, for example, a twin peak were observed at masses 497 and 499 amu, it would probably be due to a molecule M of mass 390 amu to which a silver ion was attached (107, 109 amu). If it had protonatable groups, then an $(M + H)^+$ peak should occur at mass 391 amu in the positive spectrum, and an acidic group would give rise to an $(M - H)^-$ peak at mass 389 amu in the negative spectrum. The next question would then be: are there similar twin peaks nearby? If so, from the mass separation it would be possible to obtain further insight into the molecular structure. A mass separation of $\delta m = 14$ amu would hint, for example, at the existence of aliphatic chains (δm corresponding to CH_2). A detailed look at the lower mass range might then reveal the existence of special functional groups or fragments. A large peak at mass 57 amu in the positive spectrum would indicate the existence of tertiary butyl groups ($(CH_3)_3C^+$). The complete interpretation of the spectrum from an unknown material may require a lot of patience and experience. Nevertheless, with the recent publication of standard spectra many materials can be identified even by inexperienced users.

3. Evaluation of the fingerprint region. In most cases only the fingerprint region will be available for the identification of a material. In addition to the identification of the substrate and principal matrix peaks (see above), it is advisable to start by screening for elements in the mass range below 150 amu. Elements and their oxides (the outer layers of most surfaces are oxidized due to contact with ambient air) give rise to spectral features at masses below nominal and in modern ToF instruments such features can be distinguished easily from those of organic fragments. The second step is to look for overall peak patterns: what are the masses of discrete peaks, and what are their separations? Is there an obvious repeat separation which could hint at the mass of a polymeric repeat unit? Fortunately, SSIMS spectra are very reproducible; i.e., the same material always leads to the same spectrum. This holds in a qualitative sense even for different types of instruments, e.g., ToF analyzers and quadrupoles. The characteristic peak pattern in the fingerprint region will show up in both instruments, although with slightly different relative intensities. (In a quadrupole the transmission drops with increasing mass, whereas in a ToF analyzer it does not.) This fact allows standard spectra to be used for reference purposes independently of the particular instrumentation.

In summary it can be stated that SSIMS belongs among the most useful techniques with respect to the identification of material present at surfaces. In contrast to most other techniques it gives not only elemental information but also detailed information on molecules present. In that respect it is a very sensitive probe of chemical structure. This is demonstrated in Fig. 8, which shows the positive spectra from the perfluorinated polyether Krytox, a lubricant, and polytetrafluorethylene (Teflon), both of which are polymers based on fluorocarbons. As can be seen, both materials emit the same characteristic peak, based on $C_xF_y^+$ clusters, but the relative intensities are quite different. In the Teflon spectrum, for example, the fragment series $(CF_2)_nCF^+$ (at 31, 81, 131, 181 amu, etc.) is much more prominent than in the Krytox spectrum due to the fact that solitary CF_2 chains in Teflon are much longer than in Krytox, in which O is incorporated into the polymeric backbone. Peak intensities in this respect are often a measure of the formation probability and stability of an ion, and the chemical structure thus reveals itself in the spectrum. In the negative spectra of these two materials this is even more evident because O-containing species occur in the Krytox spectrum which are not present in the Teflon spectrum. The positive spectra are presented here in order to demonstrate that chemical structure can be eludicated even by comparison of the same sets of peaks.

For SSIMS it is also an easy task to distinguish between molecules of the same substance class but of different masses. For example, it is possible not only to establish the presence of fatty acid residues as a whole but also to show precisely the distribution of the various fatty acids present. Thus it is also possible to discriminate naturally occurring fatty acids from synthetic ones.

2.4. Quantification

In Section 1.1 the matrix effect in SSIMS was introduced, i.e., the dependence of the number of detected secondary ions on the particular chemical environ-

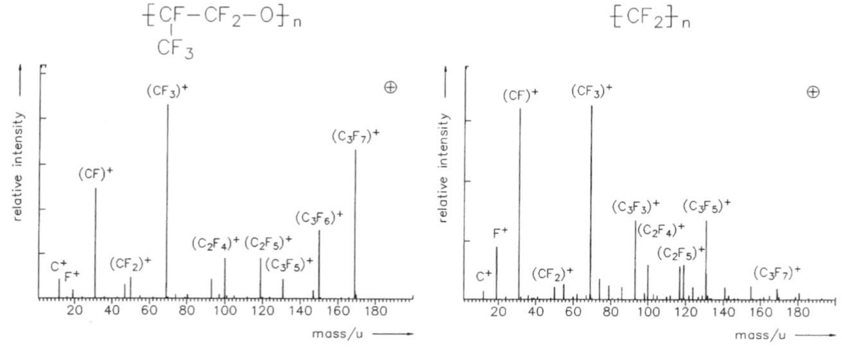

Figure 8. Positive ToF-SIMS spectra of the fluorocarbons Krytox and Teflon.

ment. In general, the effect prevents a direct quantification of SIMS data because no direct functional correlation exists between the number of surface species and the number of detected secondary ions characteristic of those species. As stated in Section 1.4.2 the number of detectable secondary ions is given by Eq. (10):

$$N_d = N(M)P(M \rightarrow X^q)TD$$

with $P(M \rightarrow X^q)$ the transformation probability (i.e., the probability that a given species M at a surface will be transformed into an ion X of charge state q), which can also be written as

$$P(M \rightarrow X^q) = \frac{Y_i(X^q)}{Y_D(M)} = \frac{\text{Secondary-ion yield}}{\text{Disappearance yield}} \qquad (17)$$

showing that both the sputtering event (represented in Y_D) and the ionization process (represented in Y_i) influence the transformation probability. It is the variation of $P(M \rightarrow X^q)$ with chemical environment which causes the matrix effect and prevents quantification. For example, consider a monolayer of organic molecules deposited on a metal substrate. A certain number of these molecules can be transformed into secondary ions and be detected in an SSIMS experiment ($N_{d,\ mono}$). If the layer thickness increases, say to about five monolayers, a distinct drop in the number of detected secondary ions ($N_{d,\ multi}$) is normally observed, even though the outermost monolayer, from which all secondary ions originate, still contains the same number of molecules as for the monolayer coverage. The conclusion that fewer detected secondary ions corresponded to fewer surface molecules would be wrong. The decrease in the number of detected secondary ions is, on the contrary, caused by a decreased sputter yield from the thicker organic layer (thus influencing Y_D). In yet other cases an *increase* in the number of detected secondary ions can be observed with increasing layer thickness. This is normally caused by an increase in the secondary-ion yield, which in fact exceeds the above-mentioned effect of decreasing sputter yields. The situation becomes even more complex if mixtures are analyzed when chemical interactions can also influence the various transformation probabilities.

Is quantification of SSIMS therefore completely impossible or unreliable? Fortunately, three areas of application exist where quantification is indeed possible. Quantification here is used in the sense that the change in the number of detected secondary ions does mirror truly the change in surface concentration.

2.4.1. Use of internal standards

Internal standards are elements or molecules added to the analyte in known concentrations. If chemically identical (i.e., isotopically labeled) or chemically similar (i.e., same substance class) materials are used, analyte and standard react in the same way to changes in the chemical environment, which means that the respective transformation probabilities change in the same ratios. The evaluation of *I*(analyte)/*I*(standard), with *I* being the peak areas of the corresponding characteristic peaks, therefore allows quantification of the amount of analyte material if a calibration curve had previously been constructed from which the (fixed) ratios of the transformation probabilities could be derived. Figure 9 demonstrates this approach for the quantification of cyclosporin A (CsA), a powerful immunesuppressive drug used in transplant surgery. The upper part shows a spectrum of a mixture of CsA (analyte) and its physiologically inactive version CsD (stan-

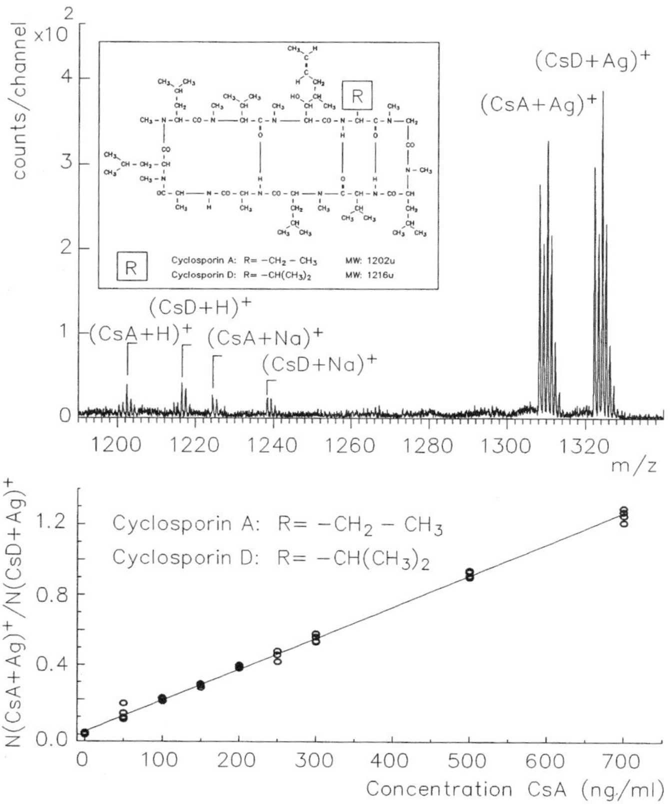

Figure 9. (upper) Positive ToF-SIMS spectrum of a mixture of cyclosporin A and cyclosporin D extracted from a whole blood sample and deposited as a monolayer on a silver substrate. (lower) The calibration curve derived from the results.

dard). Their molecular weights differ by 14 amu (CH_2, see formula). For preparation a known amount of CsD was added to a whole blood sample. A solid-phase extraction of this sample was then prepared as a monolayer on a silver substrate. The most intense peaks can be assigned to intact molecules of CsA and CsD cationized by a silver ion from the substrate. The lower part of the figure shows a calibration curve obtained from several whole blood samples. By this approach an accuracy of better than 10% could be achieved for the absolute quantification of CsA in blood [45, 46]. (ToF-SIMS was chosen as the quantification method because of its high sensitivity. Physiological concentrations of CsA are of the order of ng/mL. Alternative techniques are high-pressure liquid chromatography (HPLC) and immunoassays, both with inherent problems. ToF-SIMS furthermore allows the detection of physiologically relevant metabolism products.)

Although the use of internal standards can be a very powerful approach for absolute quantification, there are several limitations to its use in SSIMS.

- Choice of a suitable standard: The standard (S) must mix homogeneously with the analyte (A). Both S and A must show the same behavior with respect to segregation or crystallization. If a sample has to be specially prepared prior to analysis, both materials must show the same behavior in all preparational steps. All these requirements can be fulfilled for isotopically labeled materials, which are of course chemically identical, but problems can occur when using chemically similar materials. On the other hand, the latter are generally more readily available, more stable and cheaper compared to isotopically labeled materials.
- As analyte and standard must be mixed prior to the measurement, the use of internal standards will usually be limited to solutions and subsequent droplet deposition onto substrates.
- Best results can be achieved when analyte and standard are present in approximately equal amounts; i.e., prior knowledge is required about the expected amount of analyte.

2.4.2. (Sub)Monolayer coverages

If the analyte is present on a substrate at submonolayer or single-monolayer coverage, quantification can be achieved by normalizing an analyte peak area to a substrate peak area. In this approach linear relationships can be found between the surface coverage and the analyte signal normalized in this way. Absolute quantification is possible if at least one coverage value can be determined by independent quantitative techniques.

As an example, Fig. 10 compares normalized analyte signals with known coverages of sputter-deposited standards [47]. The quantification refers to Ni on a Si wafer. Ni can cause severe problems in semiconductor production and must be quantified even for extreme submonolayer coverages. The advantage of ToF-SIMS lies in the possibility of analyzing small areas with good detection limits.

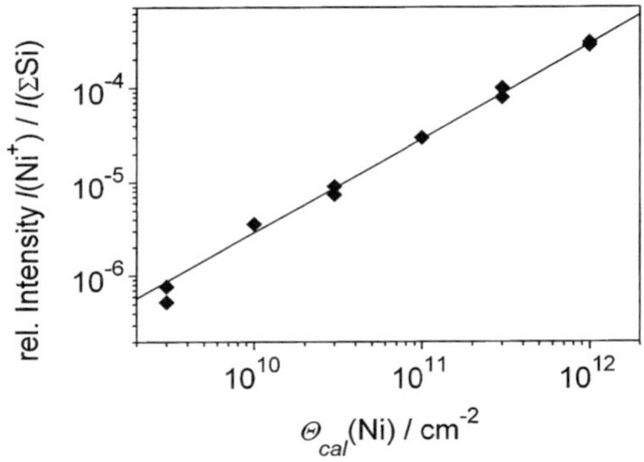

Figure 10. Plot of Ni peak areas normalized to the Si substrate peak areas as a function of the Ni surface coverage. From Ref. 47.

For submonolayer coverages quantification is possible because in such dilute systems the matrix of an analyte species always corresponds to the neighboring substrate and therefore $P(M \rightarrow X^q)$ is stable. Nevertheless, it would be expected that deviations from linearity would occur when approaching monolayer coverage. The surprising finding that the linear behavior extends even to monolayer coverage can be explained in a simplistic way by the fact that desorption in SSIMS is caused by the collision cascade, which in this case takes place entirely in the substrate and "pushes" the species from below the surface. Interactions in the vertical direction (between the layers) are thus much more important for desorption and ionization than interactions in the horizontal direction (within a layer). As long as the maximum coverage does not exceed one monolayer, a stable ion formation process (i.e., stable $P(M \rightarrow X^q)$) can be expected. This argument implies that severe problems could occur in quantification during the transition from monolayer to multilayer. In that case the collision cascade could take place partially in the substrate and partially in the analyte which would affect principally the sputter yield. In addition, the ion formation process could also change during the transition. Coverage effects on the secondary-ion yield can even be observed for those cases where a second monolayer has just started to be formed (i.e., submonolayer coverage of the second monolayer on a complete first monolayer); the effect on $(M + H)^+$ formation is an example. For the same reason quantification is also unreliable for isolated multilayer clusters ("island" nucleation) on an otherwise uncovered substrate.

2.4.3. Organic multilayers

Quantification of organic multilayer systems is surprisingly successful in view of the enormous chemical variety of organic materials. For quantification purposes a typical analyte molecule A_f (mostly a fragment from the fingerprint area) is first chosen and its intensity then normalized to either

- That of a fragment of a second compound or that of the sum of both compounds (in binary mixtures)
- The total spectral intensity
- A sum over the intensities of peaks (including all relevant compounds)
- That of an uncharacteristic hydrocarbon fragment $C_xH_y^{\pm}$

An example is given in Fig. 11, which shows the quantification of polybutadiene (PBD) in a binary system of PBD and polystyrene (PS). Both polymers are

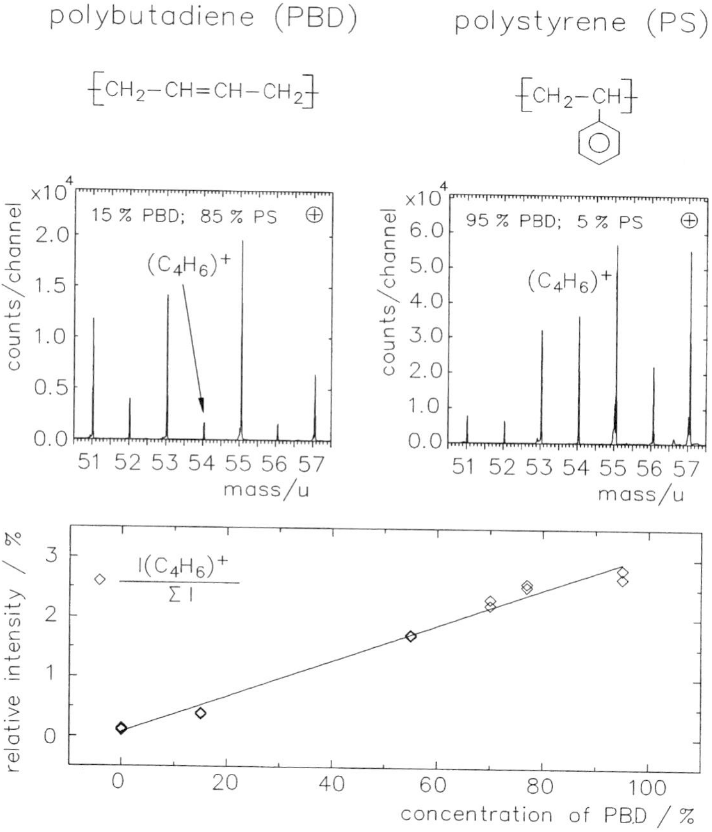

Figure 11. Positive ToF-SIMS spectra of binary mixtures of polybutadiene and polystyrene. The lower part shows the calibration curve derived from the results.

unsaturated hydrocarbons and quantification by other techniques is not easy. A peak typical of PBD is the $C_4H_6^+$ fragment at mass 54 amu. A linear relationship can be obtained if the peak area of this fragment, normalized to the total spectral intensity, is plotted against the PBD concentration in the particular sample. The quantification is achieved directly at the surface allowing the monitoring of surface phenomena such as segregation, preferential adsorption and diffusion.

Quantification in such a system is possible mainly because the chemical environment consists only of hydrocarbons, which from a physical (in contrast to chemical) point of view do not vary much (only C, H and few heteroatoms present). If fragment ions are used, then the matrix is formed partially by the (larger) molecule itself, which helps to stabilize $P(\mathrm{M} \rightarrow \mathrm{X}^q)$ further.

However, this quantification approach requires considerable experience with SSIMS. First of all, the choice of a fragment is not always straightforward and not all characteristic peaks are suitable. In particular, the most intense peaks are not necessarily the most successful candidates for quantification purposes. Second, a suitable normalization procedure must be chosen. In reality it might prove necessary to try several combinations of fragments and normalization procedures while using any additional foreknowledge of the sample. The selection should then, in principle, be tested on a model system of known composition in order to establish adequate reliability. Fortunately, statistical procedures based on multivariate statistical modeling can help to select suitable normalization candidates. In particular, the use of partial least-squares fitting shows promising results [48, 49]. Furthermore, standard spectra can be consulted in order to identify characteristic fingerprint peaks. Absolute quantification in multilayer systems is only rarely achieved, but semiquantitative information can be obtained in many cases.

In summary it can be concluded that quantification of surface coverages is possible much more often than might be expected given the matrix effect in SSIMS. Internal standards should be used whenever possible, since they give the most reliable results. When working with (sub)monolayer coverages, quantification by normalization to a substrate-related peak is always worth the effort. If the matrix is genuinely stable, then information on the surface coverage can be derived. If the matrix is varying (e.g., due to oxygen treatment in a plasma or from UV/ozone exposure), at least some new insight into the matrix effect itself might be gained. In this approach even a change in oxidation state can be monitored. For organic multilayers it is well worth while attempting quantification. If there are no initial clues as to which peaks and normalization procedures might be suitable, then a start could be made by normalizing identified fragment ion intensities to those of uncharacteristic hydrocarbon fragments (e.g., $C_3H_5^+$, $C_3H_2^-$) which are always present on samples that have been in contact with ambient air. This is surprisingly successful in many cases and easy to apply. Further progress could then be made from spectra libraries and eventually multivariate statistical modeling. Quantification remains difficult for systems where there are only a few monolayers (one to five) covering the substrate.

It should be emphasized that the determination of surface concentrations and coverages is only one aspect of quantification. SSIMS can also be used successfully for the determination of molecular weight distributions and average molecular weights in polymers, for the determination of diffusion constants (vertical and horizontal diffusion) and for the determination of surface coverages by evaluation of ion images. Some examples will be shown in the next section.

3. PROBLEM SOLVING

It will be shown here, taking four selected examples, how SSIMS with ToF analyzers can be used to solve surface-related problems. All examples have originated directly from a production site or are related closely to development and production problems.

3.1. Defects in Car Paint

Car paints generally consist of several layers on the underlying metal. As one of the last steps in car production visual checks are performed with respect to the quality of the paint layers. If defects are observed, then that car cannot be processed further and paint repair is necessary. For quality and cost reasons it is therefore essential to find the cause of the paint defects occurring in the various stages of the production process, even if that means cutting samples from the (rather expensive) cars themselves.

One example of such a defect is described in Figs. 12 and 13. The upper sequence of Fig. 12 shows a series of optical images of the defect, which is a crater in a red paint coat. In the center of the crater some paint remnants can be seen. The images differ only by their field of view (that of the right-hand image is about $150 \times 150\ \mu m^2$). In the second row of Fig. 12 there are three secondary electron images showing clearly the crater topography. The images resemble closely those obtained by SEM. However, instead of excitation by an electron beam the finely focused ion beam of the SIMS instrument has been used. A detector for ion-induced secondary electrons enabling topographical information about a sample to be obtained is a routine addition to modern ToF-SIMS instruments. The lateral resolution in such images is limited to the spot size of an LMIG in the static mode (10–40 nm).

As the next step in the analysis, secondary-ion spectra were recorded from within the defect and from its surroundings. The results are shown in Fig. 13.

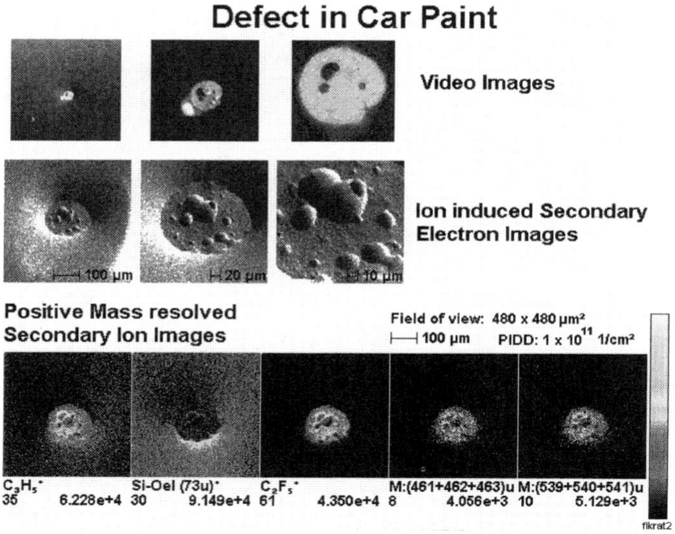

Figure 12. Video images, ion-induced secondary electron images and mass-resolved ion images of a defect in car paint. For the mass-resolved secondary ion images a thermal color scale is used to represent intensities (white: high intensities; black: low intensities).

Whereas all peaks in the spectrum from its surroundings could be assigned to the paint, the spectrum from the defect shows clearly the presence of a perfluorinated polyether (peaks due to $C_xF_y^+$ and the peak at mass 47 amu (CFO^+) are diagnostic of the polyether structure; see also Fig. 7). Such materials are commonly used lubricants. The lower sequence in Fig. 12 shows mass-resolved secondary-ion images recorded from the crater. That on the far left shows the lateral distribution of the uncharacteristic hydrocarbon $C_3H_5^+$. Since low-mass hydrocarbons are present on almost every sample which has been in contact with ambient air and are usually distributed homogeneously, the images of such hydrocarbons can give valuable topographical information, if sputter yield and ion formation are constant over the field of view. In the present example the sputter yield from the crater bottom is higher (collision cascade mainly taking place in the underlying metal) than from the neighboring paint areas (collision cascade in the organic paint material), which explains why the crater bottom seems brighter than the paint. Within the crater bottom (uniform sputter and secondary-ion yields) the image does indeed contain topographical information (compare the corresponding ion-induced secondary-electron images). The next image shows the lateral distribution of the silicone oil, which is included in the

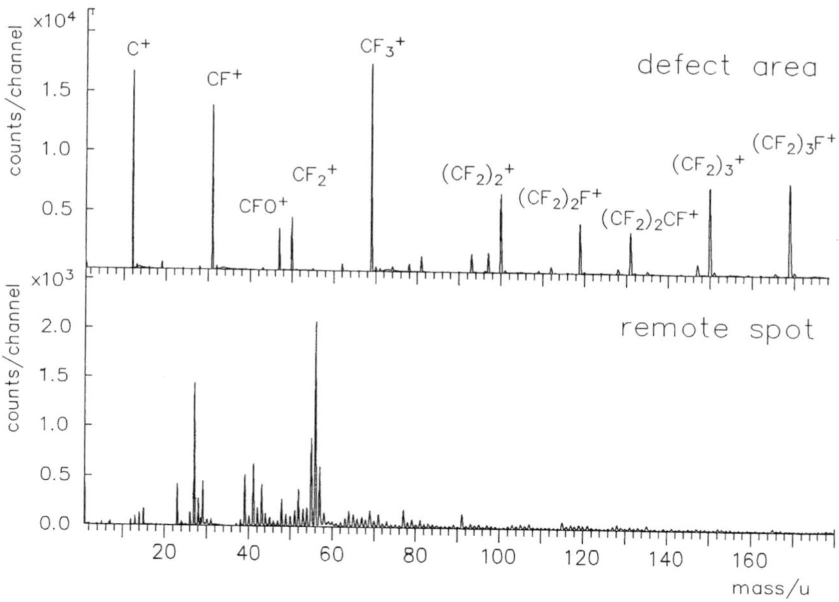

Figure 13. Positive ToF-SIMS spectra from the defect area and from a spot remote from the defect (see Fig. 12).

paint as a smoothing agent. Silicone oil could be detected only from the paint itself. This observation is important in that silicone oils can also cause paint defects if they are present in excessive concentrations. Silicone oil can therefore be excluded as the cause of the defect. The last three images show the lateral distributions of several $C_xF_y^+$ fragments, indicating clearly that the perfluorinated polyether was found exclusively in the crater.

Intensive discussions with the line engineers responsible for the painting process revealed that the lubricant originated from the transport belts used in the production. Lubricant droplets can fall into the paint bath and prevent adhesion of the paint to the metal. Tests in which the metal was deliberately contaminated by the lubricant proved that typical craters were indeed produced.

ToF-SIMS was decisive in solving the crater problem because it gave organic information about the various species present and because its sensitivity was sufficient to detect submonolayer coverages of the lubricant at the crater bottom. The lubricant concentration was so low that neither EPMA nor SAM, imaging techniques normally applied to these kinds of problems, was able to detect the fluorine in the crater.

3.2. Cl Diffusion in Polymer Materials

Nowadays, products ranging from car bumpers to CD players to shavers consist of plastics covered by (colored) organic lacquers. One of the plastics used most often is polypropylene (PP). Unfortunately, not every lacquer adheres well to PP-based substrates. In order to overcome the problem either the PP surface can be modified (e.g., by corona or plasma treatment) or adhesion promoters can be used. For PP, chlorinated polyolefines (CPO) have proved to be very useful for adhesion improvement. However, not much is known about the underlying adhesion mechanism. Furthermore, experience shows that the improvement is also dependent on the nature of the PP substrate.

One mechanism proposed is the diffusion of CPO into the PP substrate [50]. Differences in adhesion behavior in this model should arise from different diffusion constants for CPO in the various PP substrates. In order to test this hypothesis a 50-μm CPO layer was deposited onto two different PP substrates (the PP-ethylene-propylene-diene terpolymers Keltan and Hifax). From the samples prepared in that way cryosections were cut and the PP-CPO interfaces (as revealed by the surfaces of the sections) were analyzed with imaging ToF-SIMS. The chlorine distribution (originating from the CPO molecules) in the interface was monitored as a function of time in order to observe possible diffusion processes. Figure 14 shows the Cl images of the CPO-treated Hifax interface after 8 days (left) and 57 days (right). Broadening of the interface after 57 days can be seen clearly, a first hint of the occurrence of diffusion. Linescans across these images, shown in Fig. 15, also demonstrate the broadening of the inter-

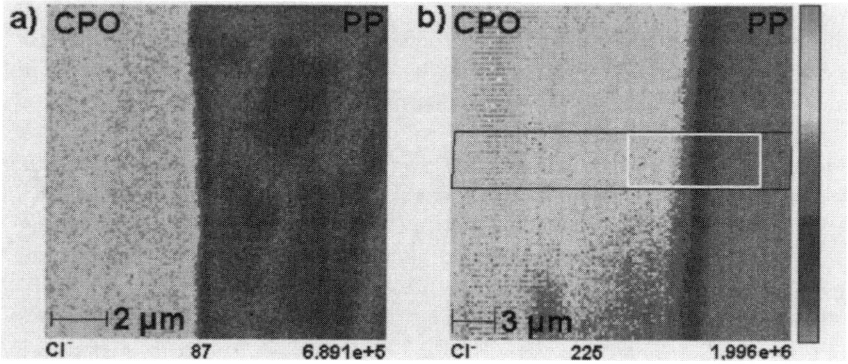

Figure 14. Cl^- ToF-SIMS images of cryosections of a CPO layer deposited on the PP substrate Hifax (a) 8 days after preparation and (b) 57 days after preparation. For the Cl^- images a gray scale is used to represent intensities (dark: high intensities; bright: low intensities). (From Ref. 50.)

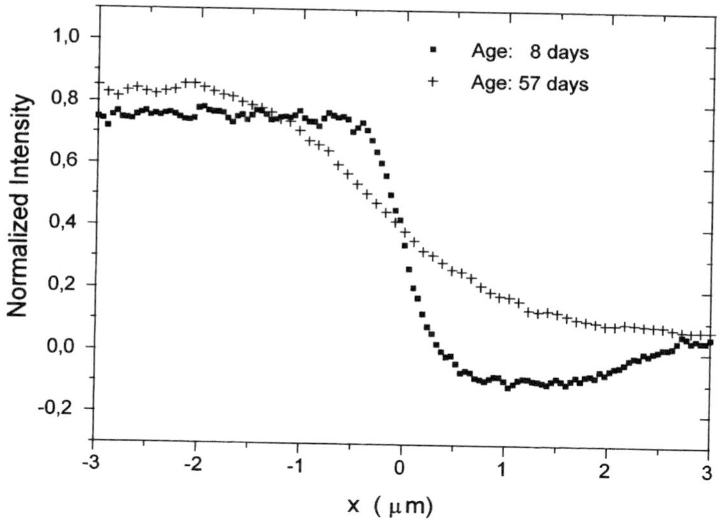

Figure 15. Cl^- line scans across the images of Fig. 14, recorded by adding up all ion intensities in the white marked box of Fig. 14. (From Ref. 49.)

face with storage time. A mathematical evaluation of the interface broadening based on diffusion theory (determination of the distance at which the original Cl concentration in CPO had decreased to one-tenth of its value as a function of the storage time, according to Fick's second law) for the two substrates (Fig. 16) shows not only that diffusion of CPO into the PP had indeed occurred but also that the diffusion constant in Keltan was significantly greater than in Hifax. This was in very good agreement with the empirical finding that lacquer adhesion was better on Keltan than on Hifax [50].

Imaging ToF-SIMS was used to investigate the adhesion mechanism because the values of the diffusion constants were not known in advance and it was feared that collisional mixing, as occurs unavoidably in sputter depth profiling, could have destroyed the effect to be measured. Sectioning of samples instead of sputter depth profiling is a very useful methodology for the determination of depth distributions in those cases where the collision cascade would destroy sensitive material (e.g., organic layer systems).

3.3. Monitoring of Surface Modifications

The fabrication of biosensing interfaces must include the transition from inorganic substrates (glasses, semiconductors, etc.) or polymers to large biomolecules necessary for the sensing itself. Biological recognition mainly involves

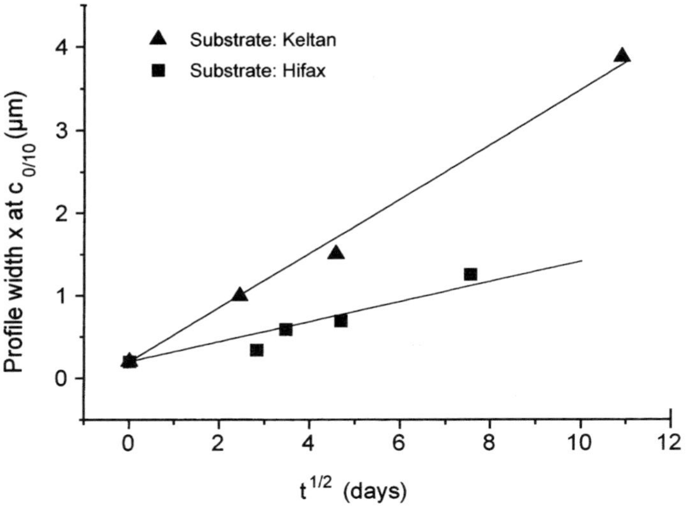

Figure 16. Cl diffusion profile width (10% definition) as a function of sample storage time (room temperature) for the PP substrates Keltan and Hifax. From the slopes of the curves the diffusion constants can be determined to be 4.3×10^{-15} cm^2/s for Keltan and 5.1×10^{-16} cm^2/s for Hifax. (From Ref. 50.)

complex stereochemical interactions, which means that the immobilization of a biomolecule must be arranged in such a way that the molecules are still able to change their conformation. Immobilization is therefore a multistep process in the specific tailoring of the inorganic surface so that its outer monolayer consists of long spacer molecules with specific recognizable endgroups distributed in such a way that biomolecules can be covalently bound to the end groups without being inactivated.

Figure 17 shows how such a multistep surface modification can be monitored by ToF-SIMS. The substrate is polyvinylacetate. This is confirmed in the negative SSIMS spectrum (Fig. 17a) in which at mass 59 amu the characteristic side-chain fragment $(CH_3–CO–O)^-$ is observed. The larger peaks in the lower mass range can be assigned to C^- ··· CH_2^-, O^-, OH, C_2^- and C_2H^-, fragments that are less characteristic of the substrate, because they can be observed from almost any organic surface. The first step is saponification of the side chain, resulting in a polyvinylalcohol. The success of this step is revealed in the SSIMS spectrum (Fig. 17b) by the obvious quenching of the peak at mass 59 amu. Additionally, typical peaks of the alcohol formed were found in the positive spectrum (not shown here). The transformation was not complete because a small peak at mass 59 amu still remains visible (for information about the

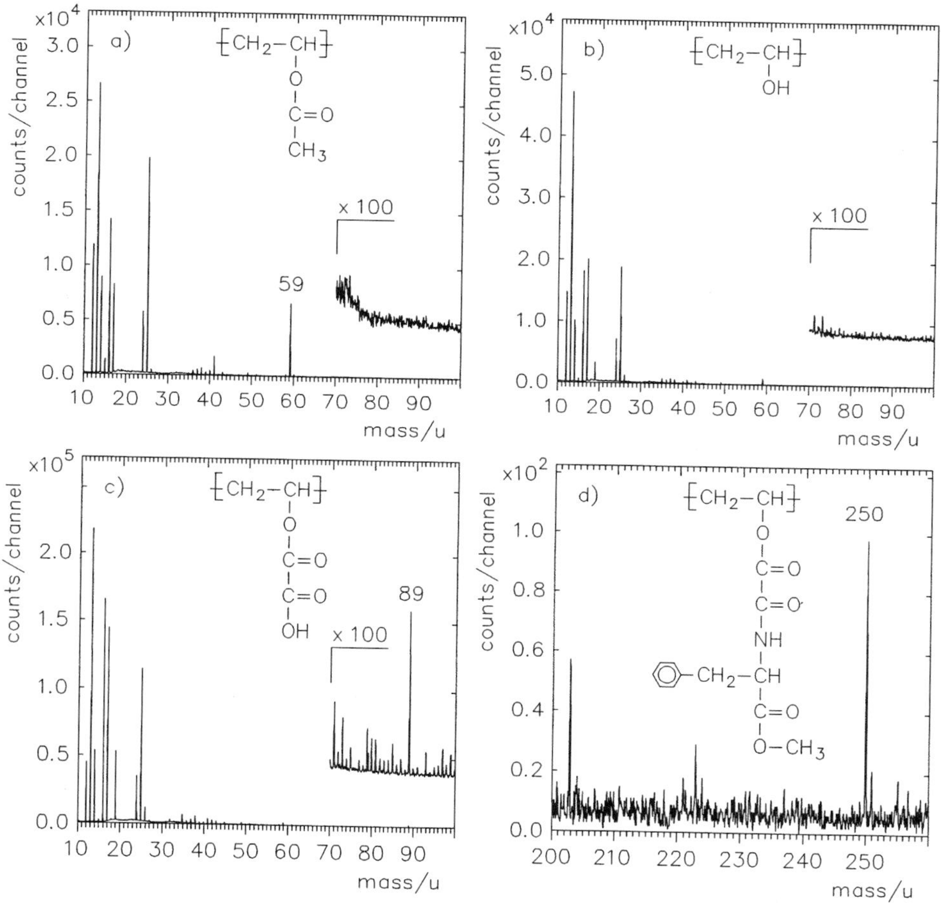

Figure 17. Negative ToF-SIMS spectra from a modified polymer surface used as model substrate for the formation of a biosensing interface: (a) unmodified, (b) after saponification, (c) after spacer addition, (d) after amino acid addition. (From Ref. 51.)

possibility of quantification in organic systems see Section 2.4). The saponification is followed by the chemisorption of a spacer molecule (oxalic acid). The expected polymeric structure is shown in Fig. 17c along with the corresponding spectrum. That the modification was at least partially successful can be seen from the fact that at mass 89 amu a new peak occurs originating from the new side chain (OH–CO–CO–O^-). The last step in the modification of this model system is the covalent attachment of a biomolecule, here the amino acid phenylalanine. Successful immobilization can be confirmed from the presence of the peak at mass 250 amu in the negative spectrum (Fig. 17d), corresponding to the complete new side chain CH_3–O–CO–CH(CH_2–C_6H_5)–NH–CO–CO–O^-); this fragment includes the aminoacid and the spacer. Such a fragment could not have been formed if the aminoacid had only physisorbed on the spacered substrate. More details are described in [51].

SSIMS is an uniquely suitable technique for the monitoring of such surface modifications because it offers detailed molecular information exclusively from the outermost monolayer (i.e., the only layer affected by the modification). Without the help of SSIMS only indirect information could have been obtained (e.g., by contact angle measurements, XPS, IR spectroscopy), which could not have distinguished so precisely between chemi- and physisorption. Additionally, SSIMS allows the sensitive detection of surface contamination (silicone oils, fatty acids, etc.), which could severely disturb or even prevent the successful surface modification. However, SSIMS cannot unfortunately observe the immobilization of physiologically more important large biomolecules (mass $>$ 3500 amu) due to hindered desorption (see Section 1.3.1) and because of its very high surface specificity (biomolecules can be as large as a μm in diameter). Future progress in this respect may be expected from matrix-assisted laser desorption (MALDI) and atomic force microscopy (AFM). Nevertheless, SSIMS can help to ensure the success of the important first immobilization steps.

3.4. Residues on Glass

The last example in this section is taken directly from a Philips production site. The product in question consists mainly of plastic materials and glass attached by various glues. The product also contains moving and rotating metal parts which need to be lubricated. Unfortunately, the process technology of the product is confidential and cannot therefore be shown, but the example demonstrates how ToF-SIMS can be used to solve real production problems.

By optical inspection of the glass parts of products that had failed in a lifetime test a contamination layer was detected, which in some areas was as thick as a few μm. In order to characterize the contaminant, IR and XPS measurements were performed. ToF-SIMS was not chosen in the first place because the problem seemed neither to require very sensitive analysis nor to be related in any way to the outer monolayer of the glass. The analyses showed that the material must be organic in origin, most probably based on aliphatic hydrocarbons (single C 1*s* peak in the XPS spectrum, CH vibrations in the IR spectrum), but no further information could be obtained (no loss features or chemical shifts in the XPS spectra, no typical bonds in the IR spectra). The diagnosis "aliphatic hydrocarbon" still included several glues and fats as possible contamination sources.

More information on the organic structure was therefore required. With the prior information from XPS and IR spectroscopies it was clear that to take a ToF-SIMS spectrum directly from the material on the underlying glass would not be advisable. First of all the large amount of contamination present could pollute the vacuum system of the SSIMS instrument (SSIMS is sensitive to readsorbed submonolayer coverages and can therefore suffer severely from memory effects), and, second, aliphatic hydrocarbons need an effective ionization mechanism in order to be visible in a SSIMS spectrum. Preparation was therefore carried out by rubbing a freshly etched silver substrate over the contamination, thus transferring about a monolayer of the material to the silver. For reference purposes the procedure was repeated for a component that had not failed (for the use of silver substrates see Section 1.5). The respective spectra are compared in Fig. 18a and b. Whereas the spectrum from the contaminated product showed a distinct peak pattern above 500 amu, this pattern was completely absent from the spectrum of the properly working product (note the enlarged section in the latter spectrum). The peaks in the spectrum of the contamination have a mass separation of 140 amu indicating some kind of polymer-like material. The isotopic pattern of the peaks is mainly that of silver, indicating that the spectra originate from neutral molecules cationized by silver from the substrate (i.e., $(M + Ag)^+$ formation). From the evaluation of the absolute masses it could be concluded that the repeat unit of the material must have the base composition $C_{10}H_{20}$. The first peak of the distribution would then correspond to a trimer of the repeat unit with hydrogen end groups, thus being a purely aliphatic hydrocarbon.

After further analysis of six glues and three fats used in the production (all dissolved in suitable solvents and deposited as monolayers on silver substrates) one particular fat was found to have the same characteristic peak pattern as the contamination. The fat formula is based on polyalphaolefines (Fig. 18c shows the spectrum of the extracted base oil). Intact molecules cationized by silver produced the characteristic peak pattern in the spectrum. By comparison of Figs. 18a and c it can be concluded that the contamination must have occurred via the gas phase because the contamination and the fat base oil have significantly different molecular weight distributions. Low-mass molecules ($n = 3, 4$) are more pronounced in the contaminant compared to the base oil. This difference is caused by the fact that low-molecular-weight molecules can desorb more easily into the gas phase than those of higher molecular weights, and this shifted distribution is found after readsorption of the desorbed molecules on the glass. The mechanism was confirmed by storing some fat material together with a clean silver substrate in a closed environment for some hours. The base oil indeed desorbed into the gas phase and readsorbed on the silver where it could be detected by ToF-SIMS, showing the shifted molecular weight distribution. After the fat was replaced, the number of products failing the lifetime test was

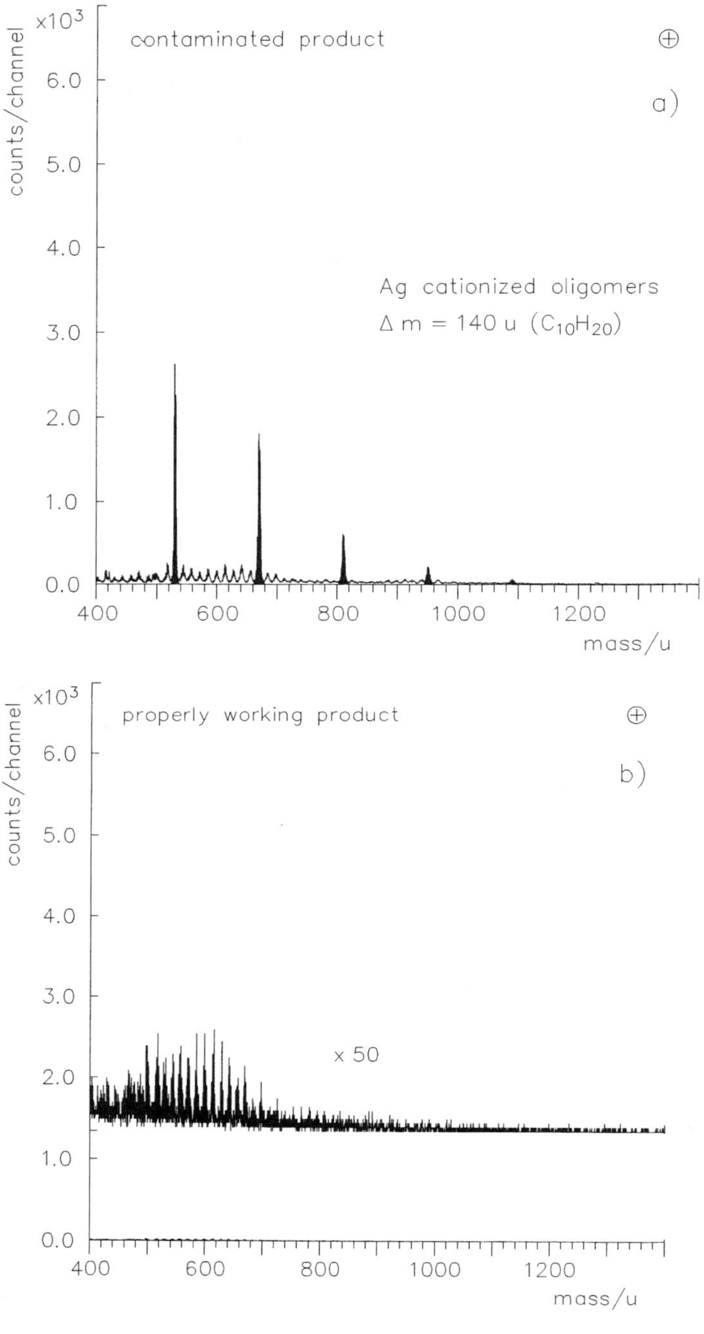
x10³
counts/channel
contaminated product
⊕
a)
Ag cationized oligomers
Δ m = 140 u (C₁₀H₂₀)
mass/u
x10³
counts/channel
properly working product
⊕
b)
x 50
mass/u

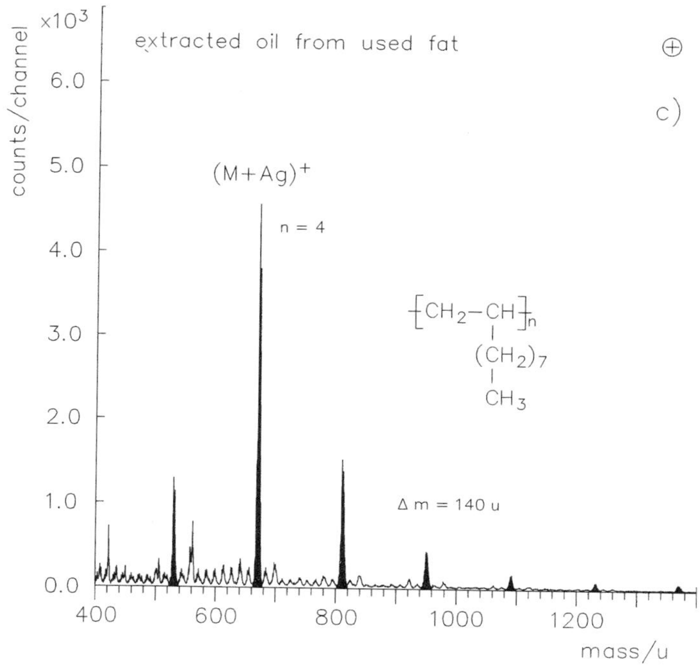

Figure 18. Positive ToF-SIMS spectra from (a) contaminated product, (b) a properly working product, and (c) the base oil extracted from one of the fats employed. Spectra (a) and (b) were recorded after rubbing freshly etched silver substrates over the respective products. The oil was prepared from a 1 mg/mL solution in hexane (about 3μL deposited onto 80 mm^2 of an etched silver substrate).

significantly reduced and no further optically visible contamination of the glass surfaces was observed.

4. SUMMARY AND OUTLOOK

Surface mass spectrometry (SSIMS and SNMS) is a very powerful tool for the chemical characterization of surfaces. It is able to identify elements and molecules present, it is sensitive down to the ppb and fmol levels, it is very surface specific (information depth = the outermost monolayer) and allows the determination of the lateral distributions of elements and molecules with sub-μm resolution. Furthermore it can be applied to almost any material in general without any special preparation (i.e., as-received). It therefore should be the method of choice in those cases where molecular information is requested or where the material available for analysis is very limited. If ToF analyzers are used for mass detection, screening of unknown samples is possible; i.e., information on all elements and molecules present on the surface can be obtained simultaneously without any need for prior sample knowledge.

On the negative side it must be said that SSIMS is inherently a nonquantitative technique, although in some analytically important cases at least semiquantitative information can be obtained. SNMS might well be considered as a quantitative technique but industrial application of this version of surface mass spectrometry is still in its infancy. Furthermore the analysis of organic materials is more complex in SNMS compared to SSIMS. The high sensitivity of surface mass spectrometry can sometimes be troublesome when dealing with production problems. Submonolayer coverages of possible contaminants are present on virtually all surfaces originating from production sites, and careful thought must always be given as to whether a molecule that has been detected is really the one causing the problem. Working with surface mass spectrometry still requires considerable experience, although the technique currently stands on the threshold of becoming routinely applicable. A German idiom says: "If the only tool you have is a hammer you tend to think of every problem as a nail." In this sense surface mass spectrometry is not always the most suitable analytical tool for solving surface-related problems. Sometimes cheaper techniques (e.g., measurement of contact angle) can give information faster and more easily. Nevertheless, surface mass spectrometry is currently the only one available for the unambiguous identification of those organic molecules present at surfaces which so often cause problems in adhesion, wetting, corrosion, etc. The surface mass spectrometric techniques must therefore be regarded as powerful tools that contribute several large and important pieces to the general analytical jigsaw puzzle that has to be solved by a combination of several techniques when dealing with complex surface-related problems.

Surface mass spectrometry is still developing very fast. Here are some brief reports on recent progress.

Temperature-programmed SSIMS: If the target temperature is controlled, then temperature-dependent surface processes (e.g., lateral diffusion) can be monitored in real time. Cooling the sample also allows the study of volatile species, while the evaluation of secondary-ion intensities as a function of target temperature can offer new insight into binding processes as well as into secondary-ion formation.

Depth profiling: Depth profiling in ToF-SIMS/SNMS instruments has become possible by the introduction of a second ion gun specifically designed and optimized for sputtering. A depth profile can then be recorded by alternately sputtering (using the sputter gun) and analyzing (using the analysis gun). Although the available sensitivity is still rather lower than that of dedicated

DSIMS instruments, ToF instruments already offer better depth resolution, with atomic mixing effects reduced to almost a monolayer. Furthermore all elements of one polarity can be profiled simultaneously.

SNMS: In SNMS most progress can be expected from a combination of laser postionization and sputter depth profiling. In contrast to the case of secondary ions where the yields can fall dramatically with depth due to depletion of oxygen, the ion yields of postionized secondary neutrals remain stable with depth resulting in much easier quantification and enhanced sensitivies. In combination with imaging, information on the three-dimensional distribution of elements becomes easily accessible.

REFERENCES

1. J. J. Thomson, *Phil. Mag. 20*, 752 (1910).
2. A. Benninghoven, B. Hagenhoff and E. Niehuis, *Anal. Chem. 65*, 630A (1993).
3. P. Sigmund, *Phys. Rev. 184*, 383 (1969); *187*, 768 (1969).
4. P. Sigmund, in Sputtering by Particle Bombardment I, (R. Behrisch, ed.), Springer, Berlin, 1981, p. 9.
5. A. Benninghoven, F. G. Rüdenauer and H. W. Werner, *Secondary Ion Mass Spectrometry*, Wiley, New York, 1987.
6. B. Garrison and N. Winograd, *Science 216*, 805 (1982).
7. P. Williams, *Surf. Sci. 90*, 588 (1979).
8. R. Behrisch (ed.), *Sputtering by Particle Bombardment I*, Springer, Berlin, 1981.
9. A. Benninghoven, E. Löbach and N. Treitz, *J. Vac. Sci. Technol. 9*, 600 (1972).
10. K. H. Müller, P. Beckman, M. Schemmer and A. Benninghoven, *Surf. Sci. 80*, 325 (1979).
11. A. Benninghoven, D. Jaspers and W. Sichtermann, *Appl. Phys. 11*, 35 (1976).
12. A. Benninghoven and W. Sichtermann, *Anal. Chem. 50*, 1180, (1978).
13. B. Garrison, *J. Am. Chem. Soc. 104*, 6211 (1982).
14. B. Garrison, *Int. J. Mass Spectrom. Ion Phys. 53*, 243 (1983).
15. A. Benninghoven, in *Secondary Ion Mass Spectrometry II*, (A. Benninghoven, C. A. Evans, R. A. Powell, R. Shimizu and H. A. Storms, eds.), Springer, Berlin, 1979, p. 116.
16. A. Benninghoven, in *Secondary Ion Mass Spectrometry III*, (A. Benninghoven, J. Giber, J. Laszlo, M. Riedel and H. W. Werner, eds.), Springer, Berlin, 1982, p. 438.
17. S. J. Pachuta and R. G. Cooks, *Chem. Rev. 87*, 647 (1987).

18. F. A. White and G. M. Wood, *Mass Spectrometry—Applications in Science and Engineering*, Wiley, New York, 1986.
19. H. E. Duckworth, R. C. Barber and V. S. Venkatasubramanian, *Mass Spectroscopy*, 2nd ed., Cambridge University Press, Cambridge, 1986.
20. H. Feld, Ph.D. thesis, Münster, 1991.
21. R. Levi-Setti, G. Crow, and Y. L. Wang, in *Secondary Ion Mass Spectrometry V*, (A. Benninghoven, R. J. Colton, D. S. Simons and H. W. Werner, eds.), Springer, Berlin, 1986.
22. J. Orloff, *Rev. Sci. Instrum. 64*, 1105 (1993).
23. L. W. Swanson, *Nucl. Instrum. Meth. 218*, 347 (1983).
24. J. Zehnpfenning, Ph.D. thesis, Münster, 1994.
25. C. Bendel, diploma work, Münster, 1994.
26. E. Niehuis, T. Heller, H. Feld and A. Benninghoven, *J. Vac. Sci. Technol. A5*, 1243 (1987).
27. V. I. Karataev, B. A. Mamyrin and D. V. Shmikk, *Sov. Phys. Techn. Phys. 16*, 1177 (1972).
28. B. W. Schueler, *Microsc. Microanal. Microstruct. 3*, 119 (1992).
29. T. Sakurai, T. Matsuo and H. Matsuda, *Int. J. Mass. Spectrom. Ion Phys. 63*, 273 (1985).
30. B. Hagenhoff, D. van Leyen, E. Niehuis and A. Benninghoven, *J. Vac. Sci. Technol. A7*, 3056 (1989).
31. C. Becker, Laser Resonant and Nonresonant Postionization of Sputtered Neutrals, in *Ion Spectroscopies for Surface Analysis* (A. W. Czanderna and D. M. Hercules, eds.), Plenum Press, New York, 1991, p. 273.
32. F. W. McLafferty and F. Tureček, *Interpretation of Mass Spectra*, 4th ed., University Science Books, Mill Valley, 1993.
33. D. van Leyen, B. Hagenhoff, E. Niehuis, A. Benninghoven, I. V. Bletsos and D. M. Hercules, *J. Vac. Sci. Technol. A7*, 1790 (1989).
34. For polymers two spectra libraries have been published. However, in both cases quadrupole mass analyzers were used limiting the mass resolution to unity. D. Briggs, A. Brown and J. C. Vickerman, *Handbook of Static Secondary Ion Mass Spectrometry* (SIMS), Wiley, Chichester, 1989 and J. G. Newmanm, B. A. Carlson, R. S. Michael and J. F. Moulder, *Static SIMS Handbook of Polymer Analysis* (T. A. Hohlt, ed.), Perkin-Elmer Corporation, Physical Electronics Division, Eden Prairie, MN, 1991); John Wiley & Sons will introduce a larger library in 1996, and most manufacturers offer their customers library packages which are integrated into their evaluation software.
35. C. H. Becker and K. T. Gillen, *Anal. Chem. 56*, 1671 (1984).
36. M. W. Thompson, *Phil. Mag. 18*, 377 (1968).
37. R. D. Macfarlane, *Nucl. Instrum. Meth. 198*, 75 (1982).
38. E. Dennins and R. J. MacDonald, *Radiat. Eff. 13*, 243 (1972).

39. J. F. Hennequin, *J. Phys. 29*, 957 (1968).
40. B. Hagenhoff, *Sekundärionenmassenspektrometrie an molekularen Oberflächenstrukturen,* Deutscher Universitätsverlag, Wiesbaden, 1994.
41. B. Hagenhoff, in *The Wiley Static SIMS Library*, Wiley, Chichester, 1996, introduction.
42. H. J. Whitlow, M. Hautala and B. U. R. Sundqvist, *Int. J. Mass Spectrom. Ion Proc. 78*, 329 (1987).
43. G. Betz and F. Rüdenauer, *Appl. Surf. Sci. 51*, 103 (1991).
44. R. L. Kubena, J. W. Ward, F. P. Statton, R. J. Joyce and G. M. Atkinson, *J. Vac. Sci. Technol. B9*, 3079 (1991).
45. B. Hagenhoff, R. Kock, M. Deimel, A. Benninghoven and H.-J. Bauch, in *Secondary Ion Mass Spectrometry VIII* (A. Benninghoven, K. T. F. Janssen, J. Tümpner and H. W. Werner, eds.), Wiley, Chichester, 1992, p. 831.
46. K. Meyer, B. Hagenhoff, M. Deimel and A. Benninghoven, *Org. Mass Spectrom. 27*, 1148 (1992).
47. A. Schneiders, R. Möllers, M. Terhorst, H.-G. Cramer, E. Niehuis and A. Benninghoven, *J. Vac. Sci. Technol. B14,* 2712 (1996).
48. K. R. Beebe and B. R. Kowalski, *Anal. Chem. 59*, 1007A (1987).
49. A. Chilkoti, B. D. Ratner and D. Briggs, *Anal. Chem. 65*, 1736 (1993).
50. H. Rulle, Ph.D. thesis, Münster, 1996.
51. A. Leute, Ph.D. thesis, Münster, 1994.

7

In-depth Analysis: Methods for Depth Profiling

F. RENIERS Université Libre de Bruxelles, Brussels, Belgium

1. INTRODUCTION

This chapter deals with the in-depth analysis of materials. Many solid samples contain within them in-depth inhomogeneities which are the result of changes in composition, chemical bonding, interfacial characteristics, etc.

The questions often put to the analyst are

What is the bulk composition of the sample?
Is the bulk composition the same at every depth?
What is the chemical state of element X at a given depth?
Is there a film-substrate interface, and where is it?
What is the composition at the interface?

The goal of the analysis is thus the determination of the in-depth distribution of the different chemical species in a sample. The solution to this simple problem should be a profile as shown in Fig. 1.

Many techniques can be used to analyze a sample: AES, SIMS, SNMS, XPS, XRD, RBS, etc. All these physical techniques need the excitation of a sample

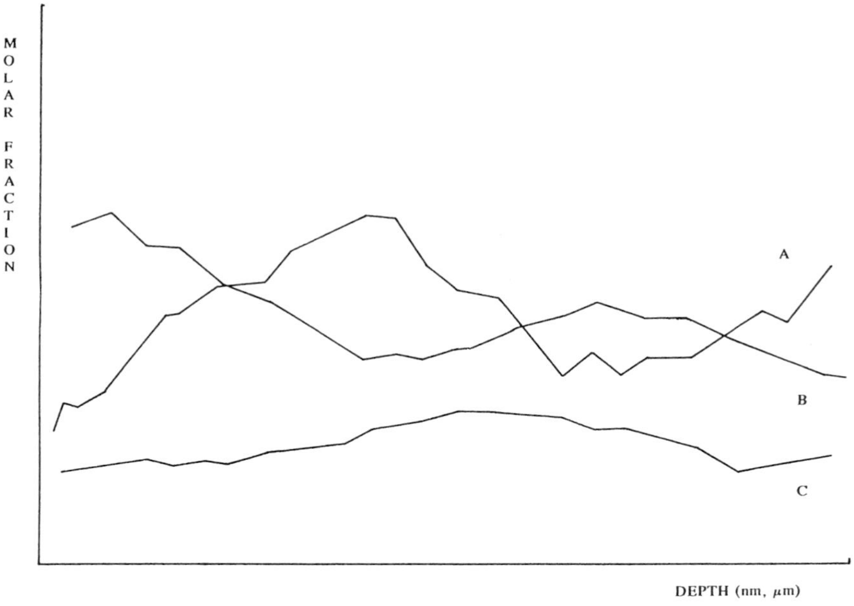

Figure 1. Absolute depth profile resulting from the in-depth analysis of a sample. A,B,C are the chemical species sought.

and the collection of a signal which may consist of ions, electrons, neutrals or photons. The signal is usually characterized by its intensity (*I*) and by the kinetic energy (KE) of the emitted particles. In-depth information is usually obtained by varying the angle of analysis (θ, defined in Fig. 2) or by ion beam etching of the sample. The results of the investigations are often presented as profiles, as in Fig. 2.

The general decisions confronting the analyst can be summarized as follows:

1. Which experimental technique(s) is (are) best suited to the given problem?
2. How can a profile of composition versus depth be extracted from the experimental data?

More specifically:

1. How can either the angle θ, or sputter time, be translated into depth?
2. What is the relationship between X, Y, Z of Fig. 2 and A, B, C of Fig. 1?
3. What is the depth resolution?
4. How can the signal intensities be translated into the true composition of the sample?
5. What are the effects of etching on the signal and on the specimen?

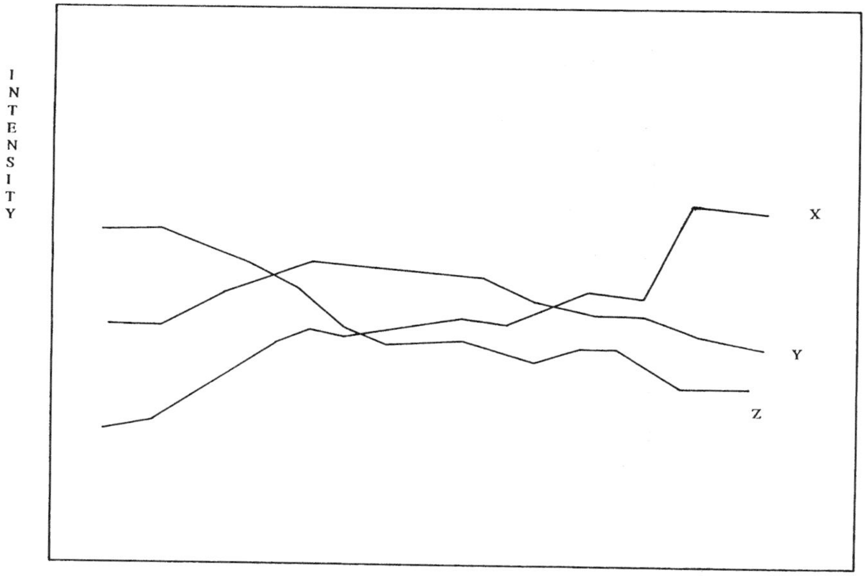

Figure 2. Typical experimental depth profile of a sample. θ is the angle of analysis; X, Y, Z are the elements detected by the method.

Many techniques are available today and this chapter will discuss only a few of them. This subjective selection is based on the following observations:

Only those techniques that are widely available and in common use will be considered, which means, in practice, that they must be available for both academic and industrial applications/users.

Only 'mature' techniques, from which quantitative information can be obtained routinely, will be included.

On this basis, the discussion will be limited to the following techniques/methodologies: Auger electron spectroscopy (AES) depth profiling; x-ray photoelectron spectroscopy (XPS) depth profiling; angle-resolved AES and XPS (ARAES, ARXPS); secondary-ion mass spectrometry (SIMS), but only in the dynamic (DSIMS) mode; secondary neutral mass spectrometry (SNMS); Rutherford backscattering spectrometry (RBS); and glow discharge optical emission spectrometry (GDOES).

The general structure of the chapter is as follows. After a general introduction, each technique will be described separately. Each subsection will contain the basic principles, the equations for quantitative analysis, the advantages and

Table 1. Principal Methods Used for In-Depth Analysis (not exhaustive)

Method	Incident beam	Analyzed particle	ambient	In-depth technique	Form of experimental data
AES	electrons	Auger electrons	UHV	ion gun	intensity/sputter time
X AES	x-rays	Auger electrons	UHV	ion gun	intensity/sputter time
ARAES	electrons	Auger electrons	UHV	sample rotation	intensity/angle
X ARAES	x-rays	Auger electrons	UHV	sample rotation	intensity/angle
XPS	x-rays	photoelectrons	UHV	ion gun	intensity/sputter time
ARXPS	x-rays	photoelectrons	UHV	sample rotation	intensity/angle
DSIMS	ions	ions	UHV	ion gun	intensity/sputter time
SNMS	ions	neutrals	UHV	ion gun	intensity/sputter time
XRD	x-rays	x-rays	air	sample rotation	intensity/angle
GDOES	Ar^+	photons	vacuum	plasma	intensity/sputter time
RBS	He^+	He^+	UHV	He^+	intensity/energy of backscattered ion
ISS	He^+, Ne^+	He^+, Ne^+	UHV	ion gun	intensity/energy of backscattered ion

disadvantages, the applications, and the particular demerits. Finally, there will be a general discussion of in-depth analysis. Table 1 presents a list of the main techniques used generally for in-depth analysis. The characteristics of the incident and detected beams are listed as well as the main technical requirements for each method. Each method, due to its individual characteristics, has advantages and disadvantages. For more detailed information, the reader is referred to Chapters 5, 6, 8 and 9. Many books have been published in which these techniques are described; including

D. Briggs and M. P. Seah (eds.), *Practical Surface Analysis*, Vols. 1, and 2, 2nd ed, Wiley, Chichester, 1992 [1, 2].

J. C. Rivière, *Surface Analytical Techniques*, Oxford Science Publications, Oxford, 1990 [3].

A. W. Czanderna (ed.), *Methods of Surface Analysis*, Vol. 1, Elsevier, Amsterdam, 1975 [4].

Thin Film and Depth Profile Analysis, in Topics in Current Physics, Vol. 37 (H. Oechsner, ed.), Springer-Verlag, Berlin, 1984 [5].

A recent review of the literature in the field of surface characterization, including in-depth analysis, has been published in *Analytical Chemistry*, Vol. 65 (1993) [6].

The researcher who has to perform an in-depth study of a given sample must choose the appropriate technique(s) according to the information that is being sought. Table 2. lists the nature of the information that can, or cannot, be extracted directly from each technique.

Table 2. Principal Information That Can Be Extracted from Each Method of Depth Profiling

Method[a]	Elements detected	Isotopic analysis	Sensitivity	Chemical information	Structure	Sampling depth (monolayers)
AES	Z≥3	no	0.3 at.%	not easily	no	3
XPS	Z≥3	no	0.3 at.%	yes	no	3
DSIMS	all	yes	<1 ppm	no	no	10
XRD	no	no	1 at.%	no	yes	
GDOES	all	no	1 ppm	no	no	100
RBS	Z>10	no	1 at.%	no	no	100
SNMS	all	yes	<1 ppm	no	no	10
ISS	all	yes	1 at.%	no	no	1
SSIMS	all	yes	0.01 at.%	yes	no	1–2

[a]The last two methods are not generally used for in-depth analysis or profiling.

2. SAMPLE PREPARATION

The aim of in-depth analysis of a sample is to obtain analytical information as a function of depth. When using AES or XPS the principal methods of depth analysis involve variation of the detection angle (ARXPS or ARAES) or, as in SIMS, removal of the surface by sputtering. For the least ambiguous depth information there is one overriding requirement and that is that the surface of the sample must be flat and as smooth as possible in the analyzed area; this is because all the theories governing ion etching and angle-resolved spectroscopies assume that the roughness is minimized. Real samples, however, have surfaces which depend on their "history." Sample preparation before analysis is therefore often necessary. The following procedures are normally used for solid samples:

1. Mechanical polishing on metallographic paper of decreasing grit size in order to eliminate macroscopic roughness. However, such a procedure often introduces SiC or WC particles into the sample which might interfere with the analysis.
2. Chemical, or better electrochemical, polishing of the sample in an appropriate bath. For metals and alloys, there are many published recipes for polishing baths [7, 8].
3. Washing, rinsing and drying the sample.
4. Examination of the surface roughness and homogeneity by optical microscopy or by SEM (usually available with SAM in most surface analysis systems), followed by selection of a flat area for analysis.

3. NONDESTRUCTIVE IN-DEPTH ANALYSIS

Basically, in nondestructive analysis, information is obtained without etching the sample. Although grazing-incidence XRD allows structural information to be obtained as a function of depth, it will not be considered here. Recent developments in that technique can be found in the literature [9–12]. The three main physical techniques described here are ARAES, ARXPS and RBS. Whereas the first two techniques are merely variations on AES and XPS, already described in Chapters 4 and 5, the technique of RBS requires a brief description and discussion.

3.1. Rutherford Backscattering Spectrometry (RBS)

3.1.1. Basic principles

The principle of RBS is as follows. A beam of energetic He^+ ions strikes a sample. Due to nuclear interactions (elastic collisions between the nuclei) the He^+ are backscattered and collected by a nuclear particle detector (Fig. 3). Extensive descriptions of RBS can be found in the literature [13, 14]. Analysis of the

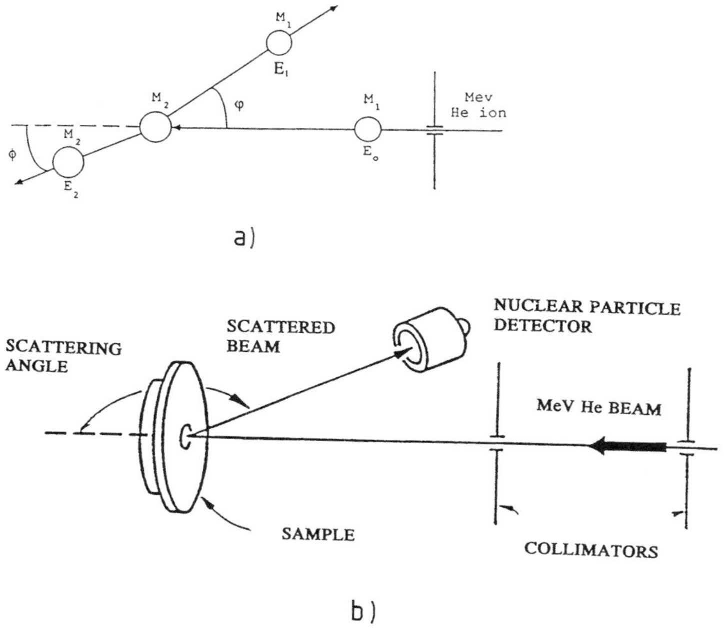

Figure 3. (a) Principle of Rutherford backscattering spectrometry; the incident particle M_1 strikes the surface atom M_2 and is backscattered with a given angle and energy; (b) diagram of an experimental setup.

energies of the scattered He^+ ions after the collision provides information on mass identification, concentration, and in-depth distribution.

3.1.2. Quantitative analysis

Due to the principle of the conservation of energy and momentum, a kinematic factor K can be extracted, which is the ratio between the energy of the detected particle (E_1) and the energy of the incident particle (E_0), namely

$$K = \frac{E_1}{E_0} = \left[\frac{(M_2^2 - M_1^2 \sin^2 \theta)^{1/2} + M_1 \cos \theta}{M_2 + M_1} \right]^2 \tag{1}$$

where M_1 is the mass of the incident particle, M_2 the mass of the target particle and θ the scattering angle. The kinematic factor thus gives direct access to the masses of the target atoms, i.e., their elemental identification. The number of backscattered particles, or the signal strength H_M, is

$$H_M = \frac{C_M Q \Delta\Omega \xi \sigma}{[\varepsilon] \cos \theta} \tag{2}$$

where Q is the flux of incident particles, C_M is the concentration of element of mass M in the sample, $\Delta\Omega$ is the aperture of the detector, ε is the stopping cross section and σ is the scattering cross section, calculated exactly from the repulsive Coulomb forces between the unscreened nuclei of the incident and target atoms. The energy loss $\Delta\varepsilon$ of the particle is related directly to the depth ΔX where the elastic collision occurred.

$$\Delta\varepsilon = \varepsilon N \Delta X \tag{3}$$

where ε is the stopping cross section (in eV/atom/cm^2) and N the particle density.

Thus in RBS, the energy loss provides the depth resolution and each element has its own depth scale. Figure 4 shows a simulated RBS energy spectrum from a 200-nm-thick W film on Si recorded with 2 MeV He^+ [15].

3.1.3. Applications of RBS

RBS is particularly suitable for following interdiffusion processes [16–18]. Figure 5 shows the RBS spectra taken at various stages in the diffusion and solid-state reaction of Ir metal with Si during heat treatment at 450°C [19]. The full

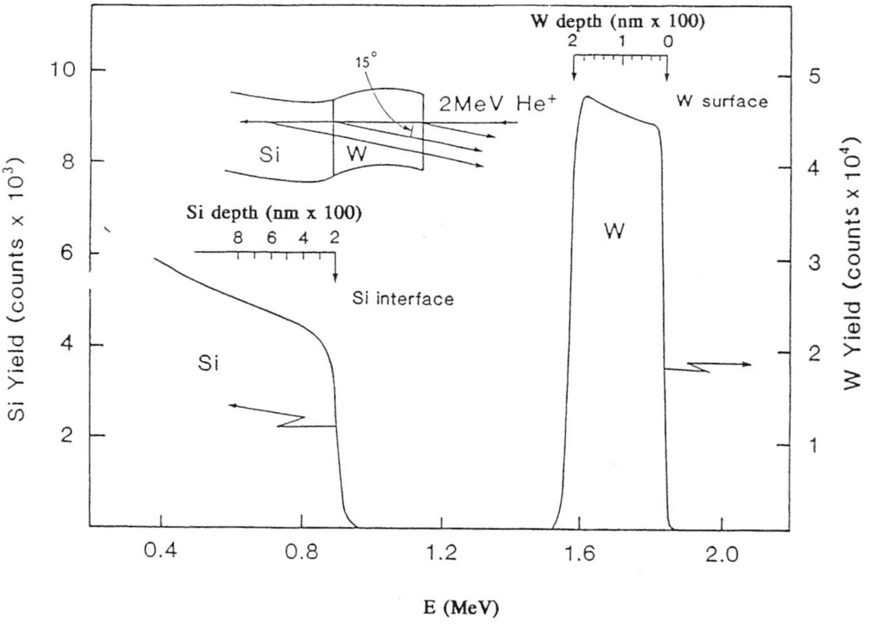

Figure 4. Simulated RBS energy spectrum from a 200-nm-thick W film on Si obtained with 2 MeV He^+ [15].

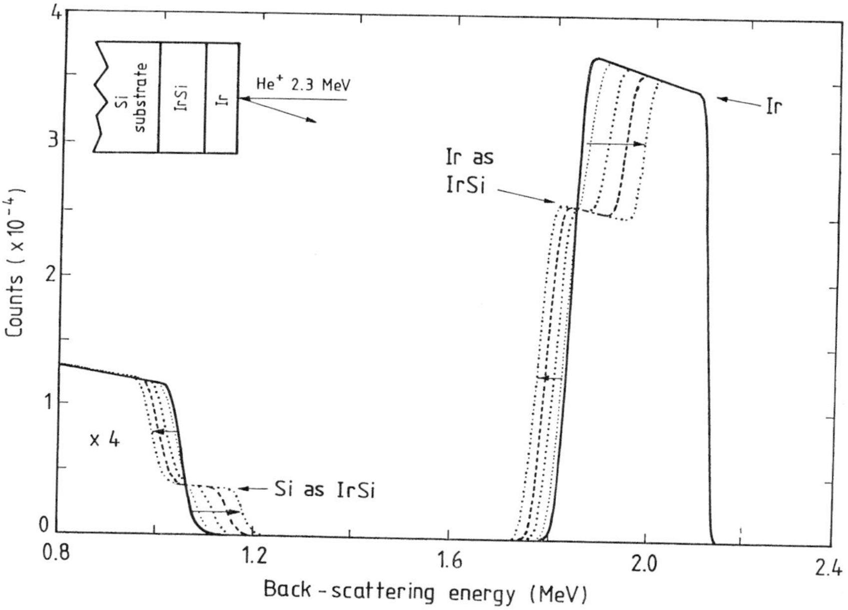

Figure 5. RBS spectra recorded at various stages in the diffusion and solid-state reaction of Ir metal with Si during heat treatment at 450°C [19].

curve is the RBS spectrum for the deposited Ir and the underlying Si substrate prior to heat treatment. The broken curves indicate the changes in the profile at different stages during the interdiffusion process.

3.2. Angle-resolved AES and XPS

3.2.1. Basic principles

AR AES and XPS provide in-depth information close to the outer surface of a sample. In normal AES or XPS, the analyzer entrance aperture is positioned normal to the surface. In the AR mode, either the analyzer is tilted or the sample is rotated. The probing depth will depend on the exit (takeoff) angle of the emitted electrons according to the relation

$$d = \lambda \sin\theta \tag{4}$$

where d is the analysis depth, λ is the attenuation length and θ is the exit angle of the ejected electrons. Conversion from signal intensity to atomic concentration is given by the usual AES or XPS relationships (see Eqs. (5) and (6)). The maximum depth analyzable by this method is ca. 3λ, which corresponds to

Table 3. Strengths and Weaknesses of RBS

Strengths
Quantitative analysis (1–2 at.% precision)
Depth information without sputtering
Easy observation of interaction and migration of species, provided $M_{deposit} \gg M_{substrate}$
Good near-surface depth resolution (5–20 nm depending on backscattering angle, material and total layer thickness)
With microbeam full three-dimensional profile determination is possible with 1-µm spatial resolution
Weaknesses
Insensitive to light elements ($Z<10$)
Overlap of light elements with substrate signal
Overlap for thick layers
Limited resolution for thick layers
Sampling thickness limited to 1 µm

about 95% of the electrons emitted. More information, as well as theoretical developments of the method, can be found in the literature [20–23] and in Chapters 4 and 5. Table 4 shows the analyzed depth (d) as a function of the analyzer angle (θ) for etched GaAs studied by XPS [24]. The photoelectron signals used were those from the Ga $3d(E_b = 20$ eV) and the As $3d(E_b = 40$ eV) levels. For both signals the escape depth (λ) is ca. 3.6 nm. See Fig. 6.

Table 4. Analyzed Depth (d) as a Function of Analyzer Angle (θ) for Etched GaAs, Studied by XPS[a]

θ (°)	d (nm)	θ (°)	d (nm)
5	0.31	35	2.06
10	0.63	40	2.31
15	0.93	45	2.55
20	1.23	50	2.76
25	1.52	55	2.95
30	1.80	60	3.12

[a]The photoelectron signals used are those from the Ga $3d(E_b = 20$ eV) and the As $3d(E_b = 40$ eV). For both signals the escape depth (λ) is ca. 3.6 nm.

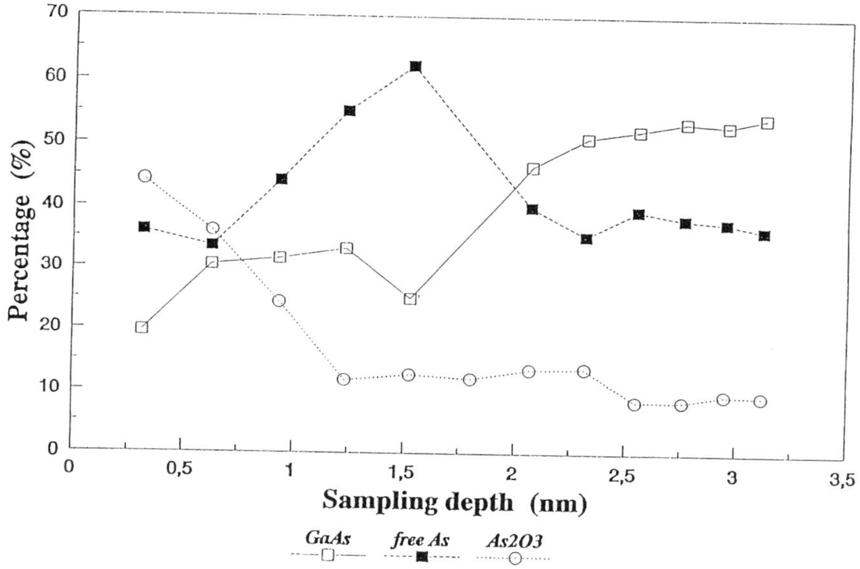

Figure 6. ARXPS depth profile using the As 3d signal from a GaAs single crystal etched in H_2SO_4–H_2O_2 solution [24]. The conversion from angle to depth is given in Table 4.

3.2.2. Applications

The two methods are particularly useful when the region of interest is close to the surface of the sample, e.g., for ultrathin films on substrates and for surface interdiffusion processes. ARXPS is probably the best tool for the study of surface oxides, as no reduction processes due to ion sputtering occur. Grafted polymers have also been widely studied by ARXPS.

3.2.3. Summary of ARXPS and ARAES capabilities

	ARXPS	ARAES
Maximum depth analyzed	3λ	3λ
Sensitivity (at. %)	0.3	0.3
Elements analyzed	$Z \geq 3$	$Z \geq 3$
Chemical information	yes	not easily
Sample requirements	flat surface	flat surface, conducting

The main advantages of AR electron spectroscopies are

1. Very good depth resolution
2. Nondestructive

The main demerits are

1. Limited to the depth corresponding to the escape depth of the electrons
2. Very sensitive to surface roughness; the sample surface must be sufficiently smooth

4. DESTRUCTIVE DEPTH PROFILING

The easiest method, in principle, for gaining access to bulk information when using spectroscopic or spectrometric techniques (AES, XPS, SIMS, SNMS) is to remove the surface layer. Historically, mechanical tools have been used to peel or to fracture the sample in order to reach new surfaces. Chemical etching has also had considerable success. However, such techniques are used less frequently today because (a) they provide very poor depth resolution (although electrochemical polishing can achieve 10 nm depth resolution), and (b) mechanical fracture occurs preferentially at grain boundaries where the composition is not the same as in the bulk. In what follows, emphasis will therefore be placed on the most widely used erosion tool, namely, ion sputtering. In sputter depth profiling either XPS or AES is used to analyze the residual surface left after a certain sputtering time, whereas SIMS and SNMS analyze the particles ejected from the sample during bombardment. Descriptions of SIMS and SNMS methods can therefore be separated from those for AES and XPS.

4.1. Ion Guns

The essential tool for depth profiling is invariably an ion gun. In Table 5 are listed the most widely used ion sources and their characteristics. For a particular application it is advisable to choose the ion source carefully.

4.2. AES and XPS

Historically, depth profiling and three-dimensional analysis have been associated with AES. It is easier to focus an exciting electron beam onto the center of a sputter crater than it is to combine a broad x-ray beam with an ion gun. For this reason, most of the depth profiling studies have employed AES. Today, however, the technological improvements in XPS (e.g., imaging, fine focus and high-intensity sources, parallel detection) allow mapping and depth profiling to be performed in ways similar to those in AES.

Table 5. Characteristics of the Most Widely Used Ion Sources[a]

Type	I_p (A)	d (m)	r (μm/h)	Application	j_p (mA-cm^{-2})
Electron impact	10^{-11}	10^{-3}	10^{-4}	SSIMS	10^{-6}
	to 3×10^{-6}	10^{-3}	4	DSIMS	0.4
Penning discharge	5×10^{-5}	4×10^{-3}	4	AES	0.4
Duoplasmatron	1×10^{-9}	2×10^{-6}	1×10^{-3}	SIMS/AES	1
	to 10^{-5}	3×10^{-4}	360		30
Liquid metal	1×10^{-9}	10^{-8}		SIMS	1

[a] r is the rate of erosion; d is the beam diameter; I_p is the bombarding current; j_p is the current density.

4.2.1. Basic principles

In XPS, x-ray excitation of an atom leads to the direct emission of a photoelectron of characteristic energy; in AES, excitation leads to the emission of Auger electrons, also at characteristic energies, produced during the deexcitation process. The theory behind the excitation processes is given in Chapter 4. AES or XPS depth profiling combines the analysis of electron energies with simultaneous ion sputtering of the sample.

The spectra obtained in either case are functions of intensity, I, versus sputtering time, t, (Fig. 2). The challenge in sputter depth profiling is the transformation of I into atomic fraction, X, and of t into depth, z.

4.2.2. Quantitative analysis

It has been shown [1–3] and in Chapter 5 that the atomic fraction can be linked to the Auger signal intensity by

$$X_i = \frac{I_i / I_i^\infty}{\Sigma_j F_{ji}^i \dfrac{I_j}{I_i^\infty}} \tag{5}$$

where F^i_{ji} is a matrix factor, X_i is the atomic fraction of element i, I_i is the signal strength of element i and I_i^∞ is the signal strength of pure element i. The ratio of signal strengths for pure elements gives a sensitivity factor. Relative sensitivity factors (RSF) are nowadays either included in surface analytical software or are taken from one or other handbooks [25–28]. However, the best method is either to measure the RSF on well-known standard samples or to use "absolute" reference spectra which are now available.

Equation (5) hides several conceptual difficulties; the sensitivity factor contains the inelastic mean free path, the backscattering factor, and the atomic density, as well as instrumental terms. In particular, matrix corrections, F, must be taken into account in most cases. Extensive discussions of each of these problems can be found in the literature [29–31] and are not the subject of this chapter. Nevertheless, the analyst should keep in mind that all these factors affecting the linear relationship between the signal strength and the composition of a sample will modify the final result.

An equation similar to Eq. (5) can be used for the quantification of XPS spectra. For a diatomic sample.

$$\frac{X_A}{X_B} = F_{AB}^x \frac{I_A / I_A^\infty}{I_B / I_B^\infty} \tag{6}$$

where F_{AB}^x is a matrix factor for XPS. Here, the matrix effects will be smaller than in AES depth profiling, as the backscattering factor for primary electrons is absent.

The above formulae assume that the sample is homogeneous over the depth analyzed, which is rarely the case in depth profiling, especially at interfaces. A mathematical model must then be used to decompose the Auger signal into layer by layer contributions [32–34].

A general equation for an arbitrary AES or XPS depth profile is

$$X_i(z) = \left(\frac{I_i}{I_i^0}\right)_z - \frac{d(I_i / I_i^0)}{dz}\lambda_i \tag{7}$$

In this equation, the composition at a given depth, $X(z)$, is linked to the signal intensity I originating at depth z corrected by the effective escape depth of the electrons λ.

The effect of the sputtering process itself on the true concentration is usually neglected. However, since the elements in a sample will have various sputtering yields, some of them will be sputtered preferentially. Analysis of the resulting surface will thus give a composition different from the true one. If the sputtering yields Y of the components are independent of their bulk concentration X_b, the surface composition X_s is inversely proportional to the respective sputtering yields, i.e.,

$$\frac{X_{sA}}{X_{sB}} = \frac{Y_B}{Y_A}\frac{X_{bA}}{X_{bB}} \tag{8}$$

where the subscripts A and B denote the two components (at steady state). Equation (8) indicates that the composition of the first layer only will be modified, the element with the lower sputtering yield being enriched at the surface.

The resulting Auger or XPS sputter profile must then be corrected by a sputtering yield ratio in order to give the true composition of the sample. Such a correction has already been developed for AgPd alloys [35].

If a steady state is not reached, a transient in surface composition is found [36], where

$$X_{sA}(t) = (X_{bA} - X_{sA})\exp\left(-\frac{t}{\tau}\right) + X_{sA} \tag{9}$$

$X_{sA}(t)$ is the instantaneous surface composition of A between the start of sputtering ($t = 0$) and the time at which secular equilibrium is obtained ($t = \infty$).

As the compositions of only the first (and second) atomic layers are modified by the sputtering process and, since in AES and XPS signals come from a few atomic layers, the higher-energy peaks (i.e., those corresponding to greater IMFP) which select more of the bulk material will be less affected by this problem.

4.2.3. Depth determination—conversion

Converting the sputtering time into depth is probably one of the most difficult tasks in AES/XPS depth profiling. Basically, the depth z is connected with the time t by the sputtering rate r:

$$z = \int_0^t r \; dt \tag{10}$$

$$r = \frac{M}{\rho N_A e} Y j_p \tag{11}$$

where M is the molar mass (kg-mole^{-1}), ρ is the density (kg-m^{-3}), N_A is Avogadro's number (6.02×10^{26} kg-mol)$^{-1}$, e is the electron charge (1.6×10^{-19} C), Y is the sputtering yield (atoms-ion^{-1}) and j_p is the primary-ion current density (A-m^{-2}). If r is constant, then $z = rt$.

The most efficient method is to measure the crater depth z_0 after sputtering time t_0, (e.g., by optical interferometry, laser measurements [37] or stylus profilometry), so as to derive directly $r = z_0/t_0$. If a multilayer system is studied, the sputtering rate in each layer must be determined. The usual method included in most software, consisting of the conversion of the sputtering time into "equivalent Ta_2O_5 thickness," is not suitable.

4.2.4. Depth resolution

Improvement in AES sputter depth profiling

One of the major problems in sputter depth profiling is the progressive decrease in the depth resolution during sputtering. The depth resolution, Δz, is defined as the width of the sputtering profile required for a signal to decrease from 84% intensity to 16% intensity at a monoatomic interface, and is usually measured on model materials in which sharp interfaces are present, such as Ta_2O_5 and

Ni/Cr multilayers. Factors known to limit the depth resolution include instrumental factors, sample characteristics and radiation-induced effects.

(1) Instrumental factors

- The residual atmosphere may contaminate the surface of the sample and the sputtering rate may change with time. Only noble gases with very low sticking coefficients are acceptable.
- Ion mass and energy; most instruments use Ar^+ of 1–10 keV energy. However, the ion-beam-induced effect on depth resolution is minimal for low-energy ions (<1 keV) and high ion mass (Xe, Kr).
- Ion beam and electron beam position; there must be a perfect match between the center of the ion beam and of the analysis spot. Moreover the ion beam intensity must be constant with time. The diameter of the electron beam must be lower (ideally 100 times lower) than that of the ion beam.

(2) Sample characteristics

- The original roughness of the surface plays a critical role in depth resolution [38–43]. This contribution to depth resolution, often written Δz_s, increases with the ion incidence angle θ [44]. It has been demonstrated that the angular distribution of the microplanes is the important parameter [41, 45]. If the original surface is perfectly smooth, on the other hand, a glancing-incidence ion beam will give better depth resolution.
- Crystalline structure and defects; since the sputtering yield depends on the crystalline orientation [46–47], a polycrystalline material will produce characteristic steps [48] and the depth resolution will be reduced. The best materials for yield measurements are either single crystals or totally amorphous materials. It has been shown that the introduction of reactive ions leads to a smoother sputtered surface [49–51] than that resulting from noble-gas ion sputtering. This is probably due to the formation of an amorphous surface oxide layer, which then sputters homogeneously [52].
- Second-phase alloys and compounds: all interfaces inside a sample will induce changes in sputtering yields, which will increase the roughness.
- Insulators: positive ion implantation; e.g., noble gas bombardment of insulators can lead to positive charging and hence to a gradual decrease of effective ion energy and of sputtering yield. This happens in XPS and SIMS and a common way to avoid it is to flood the sample with low-energy electrons [53]. The problem does not occur in AES due to the presence of secondary electrons.

In conclusion, the best materials are either amorphous, or single crystals which become amorphous during sputtering (semiconductors and oxides), rather than polycrystalline metals [54].

(3) Radiation-induced effects

The basic limitation in depth resolution is due to the sputtering process itself:

- Microtopography induced by sputtering: a $z^{1/2}$ dependence of Δz is often observed in polycrystalline metals [43, 55–58]. The principal causes are crystalline orientation and the dependence of sputtering yield on ion incidence angle [46]. If the sample is smooth, the effect can be reduced by lower ion energies, lower sputtering yields and high incidence angle.
- Atomic mixing; due to the bombardment process, target atoms are displaced into the material leading to atomic mixing and to the broadening of the interface. At steady state the atomic mixing contribution to Δz is constant with z and proportional to $E_{\mathrm{I}}^{1/2}$, the primary-ion energy, and is less for higher sputtering yields. A way to reduce atomic mixing is by the use of heavy ions (e.g., Xe^{+}), of low ion energies and/or glancing incidence.
- Preferential sputtering; already discussed in Section 4.2.2.
- Decomposition of compounds: one of the most crucial problems in ion sputtering is the decomposition of chemical compounds due to preferential sputtering. For instance, many oxides are reduced to lower oxidation states [59–66]. Table 6 lists oxides for which this reduction has been observed. In general, no reduction is detected for oxides of light elements. For some other oxides the behavior depends on the sputtering conditions.
- Radiation-enhanced induced segregation and diffusion: the phenomenon of sputter-induced segregation is well known [73–76] and is common in alloys with high elemental diffusion coefficients. Ways of reducing this problem are either to sputter at a rate significantly faster than the rate of the diffusion process or to reduce the rate of diffusion by decreasing the temperature.
- Electron beam effects in AES: the heating effect of the electron beam causes evaporation, diffusion and segregation, and thus an incorrect estimate of the composition, particularly for thin films on glass substrates [77]. The effect can be reduced if the electron beam current density is lowered.

4.2.5. Summary of optimized depth profiling conditions for AES/XPS (based on Ref. 78)

Sample characteristics

Smooth, polished surface (see Section 2, "Sample Preparation")
Noncrystalline, no second phases
Components with similar sputtering yields
Good electrical and thermal conductivities
Low interdiffusion, low Gibbsian segregation

Table 6. Sputter-Induced Reduction of Oxides

No reduction	Ref.	Reduction	Ref.
Al_2O_3	67,70,72	Ag_2O	67
Cr_2O_3	67	Au_2O_3	67
FeO	64	Co_3O_4	64,72
HfO_2	64	CdO	67
MnO	64	Cr_2O_3	64
MoO_2	64	CuO, Cu_2O	67
SiO_2	67	Fe_2O_3, Fe_3O_4	69,67,64,72
SnO, SnO_2	64	HfO_2	71
Ta_2O_5	67	IrO_2	67
Ti_2O_3	67	MoO_3	68
V_2O_3	64	Nb_2O_5	68,71
ZnO	64	NiO	67
ZrO_2	64	$Ni(OH)_2$	67
		PbO	64
		PdO	67
		SiO_2	68
		Ta_2O_5	68,71
		TiO_2	68,64,71
		WO_2	67
		WO_3	68
		ZrO_2	71

Source: From Ref. 56.

Instrumental factors

1. Low residual atmosphere ($<10^{-8}$ Pa)
2. Ion beam
    - no impurities (pure noble gases)
    - no neutrals (ion beam deflected)
    - rastered beam of constant intensity
    - low beam energy (1 keV)
    - high-mass ion species
    - large incidence angle for smooth sample (>60°)
    - low incidence angle for rough sample (<60°)
    - sample rotation, if available
2. Electron beam (AES)
    - centered within sputtered area

finely focused or rastered over small area
low current density
4. Analyzing conditions
sputtered area large compared to the analyzed area
small-spot analysis (XPS) centered on sputtered area
selection of low-kinetic-energy signal (shallow information depth).

The last condition promotes good depth resolution, but the resultant signals will be strongly affected by the change in surface composition due to sputtering (see Section 4.2.2).

4.2.6. Improvement of depth resolution by sample rotation

Since the major limitation on depth resolution arises from the sputtering process itself and from the sputter-induced microtopography, many attempts have been made to minimize such effects. The main aim is to suppress both the crystal orientation effects and the increase in roughness by rotating the sample during sputtering, as suggested by Zalar. If that is done in the correct way, then the sputter crater can be produced with a flat bottom, in the center of which is the analyzed area [79, 80].

In Figure 7 is a schematic of the principle underlying rotation and of the shape of the sputtering crater; Fig. 8 shows sputter profiles of an Au/Al/SiO_2/Si multilayer sample without (a) and with (b) sample rotation. The interface, shown by the broken lines, is sharper with sample rotation.

In some recently published papers [82], it has been shown that an increase in the rotation speed, and in the Ar^+ incident angle, lead to better depth resolution in the profile. The rotational device was designed originally for Auger spectrometers but has recently been adapted to SIMS depth profiling with considerable success [83]. Figure 9 shows the SIMS depth profile through a GaA2As Be-spiked sample without (a) and with (b) sample rotation. When there is no rotation, the depth resolution decreases progressively during sputtering, but it remains constant when the sample is rotated at 3 rpm. Provision for such rotation should be included in all modern sputter profile instruments as it seems to improve the depth resolution significantly, especially for studies of film-substrate interfaces or indeed anywhere where interfaces occur.

4.2.7. Chemical depth profiles using AES

Table 2 shows that AES sputter depth profiles provide quantitative information on elemental composition as a function of depth but are not normally used to extract chemical information. Until recently, due to the better energy resolution of the electron analyzer and to easier interpretation of spectra, chemical analysis was performed mainly by XPS. However, since the Auger transitions recorded in AES

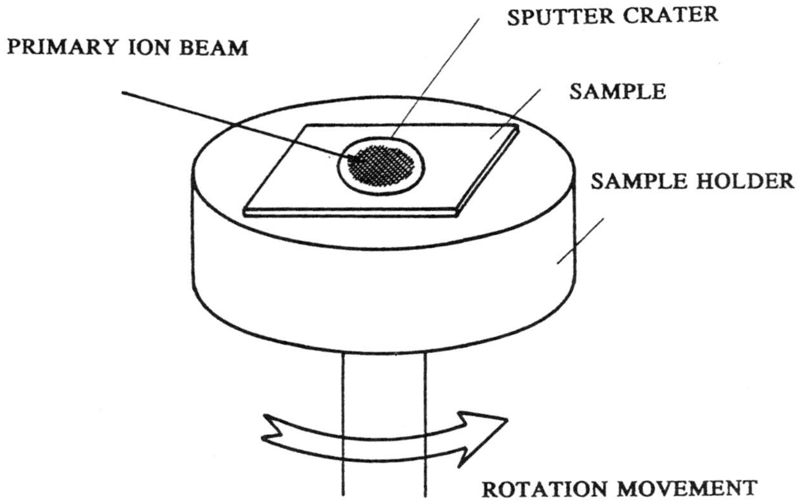

a)

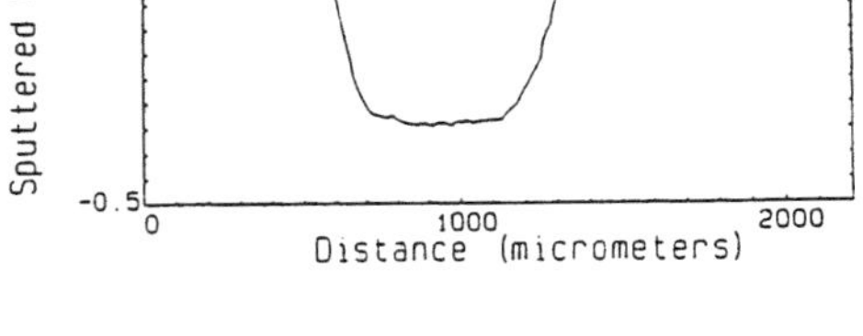

b)

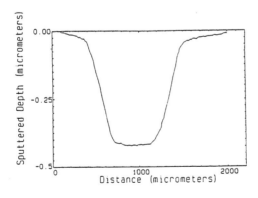

Figure 7. (a) Principle of Zalar rotation in AES; (b) typical shape of the sputter crater obtained by Zalar rotation.

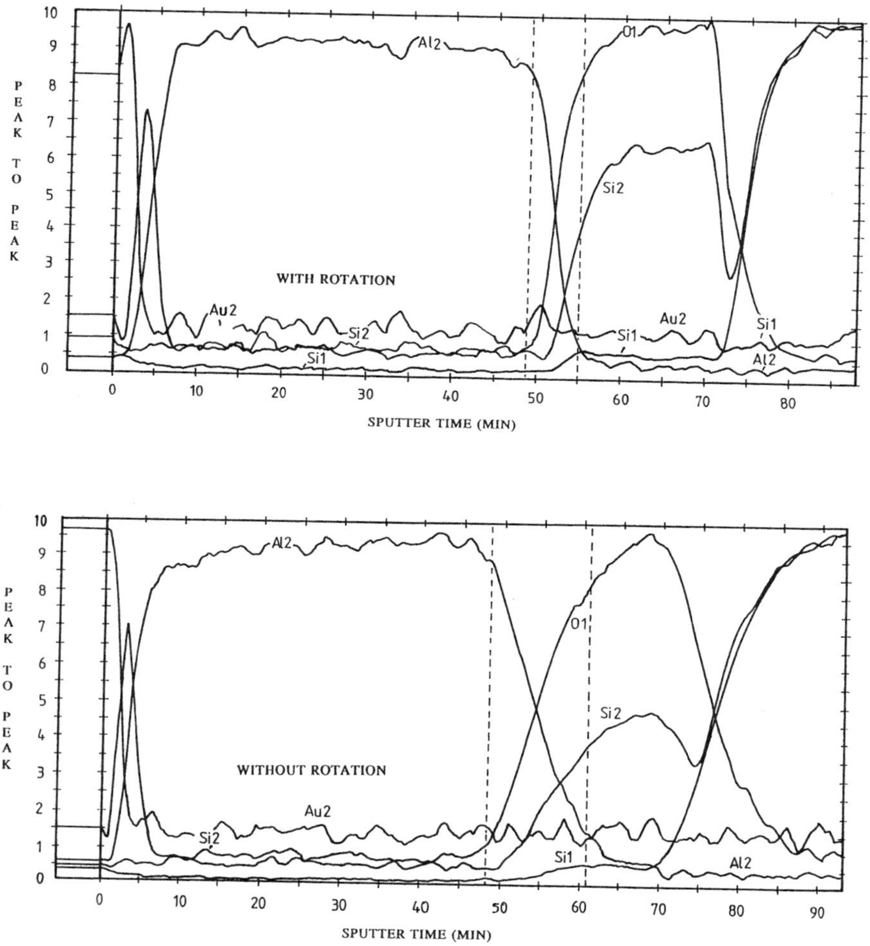

Figure 8. Effect of the rotation of the sample during sputtering on a typical Auger depth profile [81].

always involve at least one or two valence levels, chemical information is also present in Auger spectra, although analysis of peak shape is a much more complex process than in XPS. For instance, the different AES peak shapes of carbon deriving from various chemical environments are well known [84].

In depth profiles, analysis is usually confined to measurements of peak heights (i.e., the "counts"), although the peak shape also contains information. With the advent of fully computerized AES instruments, the complete treat-

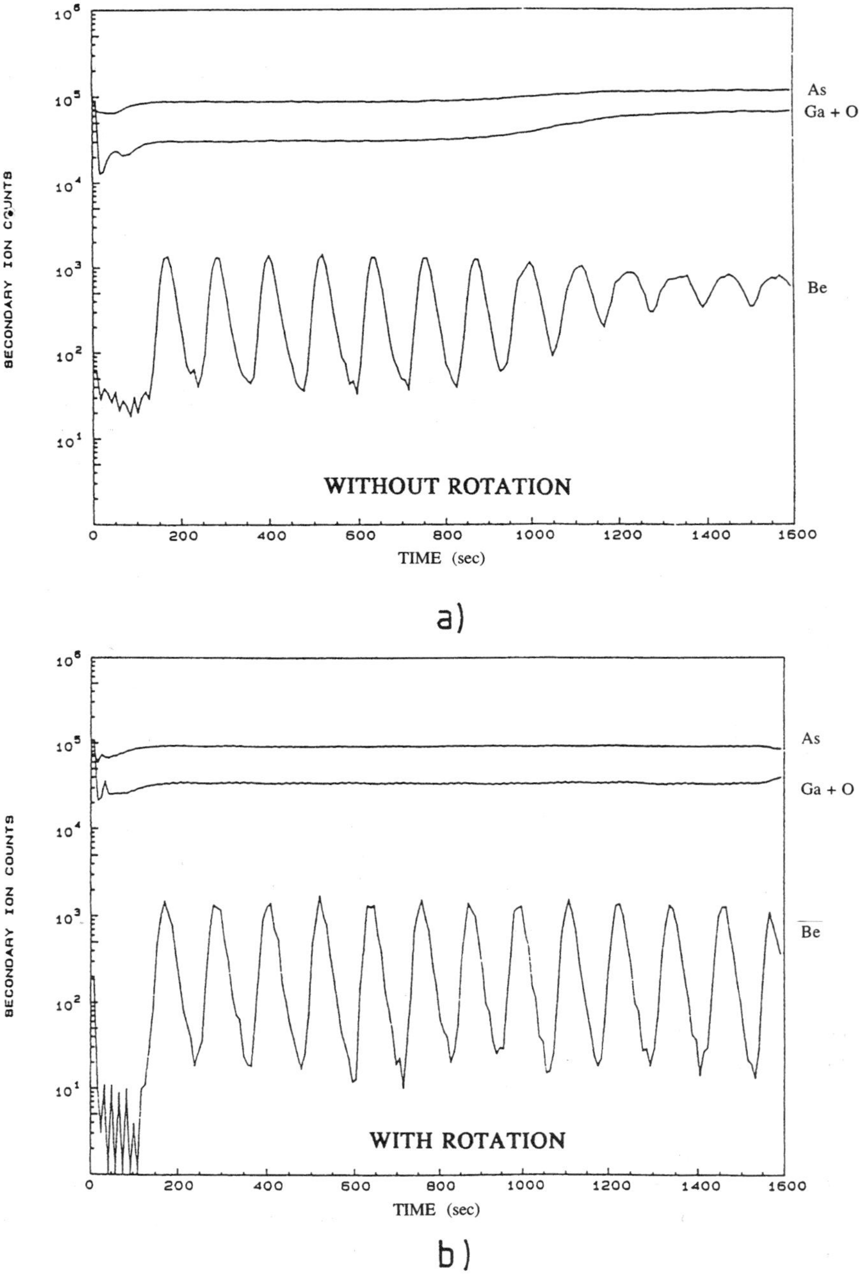

Figure 9. SIMS depth profile through a GaAlAs Be-spiked sample (a) with and (b) without sample rotation.

ment of such information has become much easier. Following Malinowski's book [85], factor analysis methods have been used increasingly frequently to resolve complex Auger peak shapes into their constituent components; the literature in this field is now extensive. Most of it deals with either unwanted overlapping of peaks (as in the classic case of Ti-N materials [86]) or with the chemical effects arising from changes in chemical environment during depth profiling [87–98].

The procedure in general use is the following:

1. Acquisition of spectra of the desired elements in selected energy windows during sputter profiling
2. Storage of the spectra in a file
3. Calculation of the peak-to-peak height in each energy window in order to obtain the usual AES depth profile
4. Analysis of the peak shapes by one of the multivariate analysis methods

A typical example of the power of this method is provided by the W-N system. In the case of the interaction between tungsten and nitrogen, the effect of nitrogen on the metal valence electrons is much weaker than that in oxides leading to a chemical effect on the metal peak shape or a shift which is considerably less obvious than in oxides and often neglected.

Figure 10a shows the AES depth profile of a W_2N film deposited on W by dc reactive sputtering [89]. Figures 10b and c show the factor analysis of the W and N peaks, respectively. The left-hand sides of (b) and (c) show the pure spectra, whereas the right-hand sides show the reconstructed concentration profile for each element. Two nitrogen components can be resolved, due to surface and to nitride species. Three tungsten components are identified, namely tungsten nitride, metallic tungsten and interfacial tungsten, the latter being present as tungsten carbide, as suggested by the carbidic shape of the carbon peak at the interface (insert in Fig. 10a). The chemical state of tungsten at the interface could not have been detected by the usual AES depth profile.

There are now several curve-fitting methods available on most AES/XPS systems. Among these may be mentioned principal component analysis (factor analysis), linear least-squares fitting, the maximum entropy method, the pattern recognition method (derived from factor analysis) and difference spectra.

It should be stressed that the methods are only mathematical tools which must be used carefully, since they will always generate a mathematical solution to a problem, even if it has no physical significance. In addition, the spectral pretreatments available (smoothing, differentiation, background subtraction, etc.) will affect the shape of a peak and may thus alter the results. Although such operations are now a standard part of all AES/XPS software, they should be used only by qualified workers aware of the physical implications.

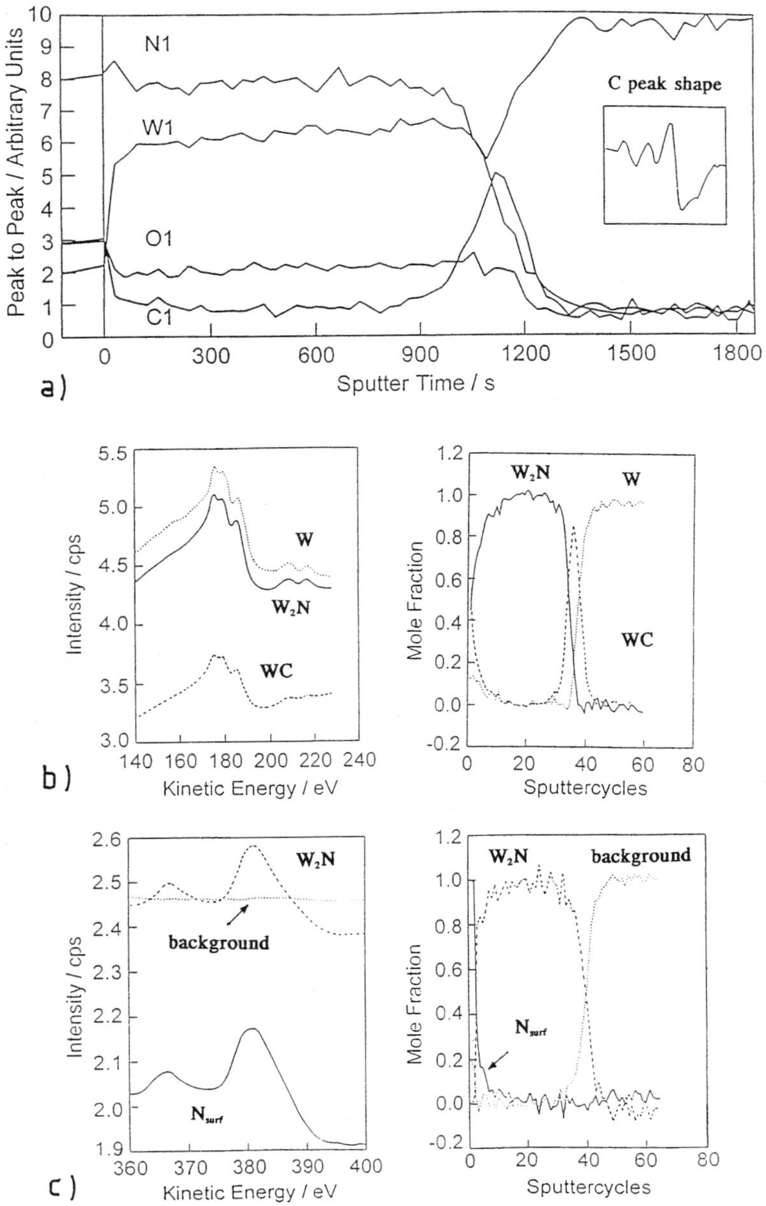

Figure 10. Factor analysis applied to the Auger sputter depth profile of a W_2N film deposited on W: (a) usual AES depth profile; (b) factor analysis of the W peak structure; (c) factor analysis of the N peak structure.

4.3. Glow Discharge Optical Emission Spectroscopy (GDOES)

4.3.1. Basic principles

GDOES had already been applied in metallurgy many years before the coming of the more sophisticated AES, XPS and SIMS techniques. It is not normally used as a "surface analysis technique" but it is certainly a very suitable "in-depth analysis technique." The principle of GDOES is the analysis of the light emitted by excited atoms after ejection from a sample by bombardment with Ar^+ created in a glow discharge. The basic device of GDOES is the glow discharge lamp designed by Grimm [99], a drawing of which is given in Fig. 11. The sample (1), located at the cathode (2), is bombarded by Ar^+. The gas is introduced via (8) to a pressure between 100 and 1000 Pa. The glow discharge is established by application of a dc voltage of ca. 1000 V between the anode (4) and the cathode. The mode of operation is in the abnormal region of the current-voltage characteristic of a discharge. The sputtered neutral atoms are excited in the plasma and the characteristic light emitted by deexcitation is collected by a spectrophotometer located behind the quartz window (5). The intensity of the light is related to the number of emitting atoms, and its wavelength is characteristic of their nature. The sputtering rate is typically from 30 nm/min to 1 μm/min, which is much faster than in AES or SIMS, and with ions of lower energy.

4.3.2. Quantitative analysis

For exact quantitative analysis of a sample M in GDOES, analysis of a reference material R is required. The intensity of an emission line is related to the

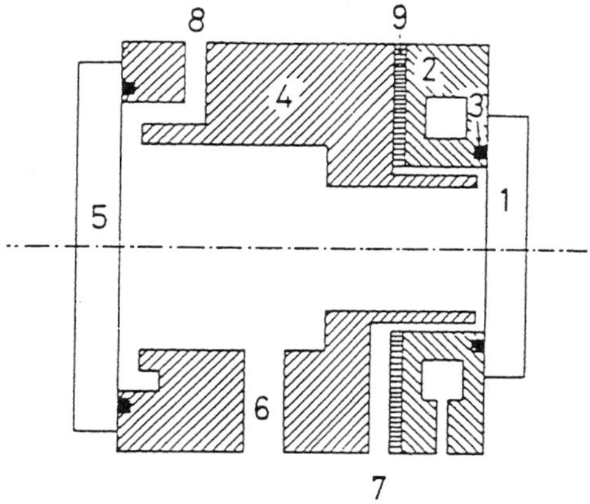

Figure 11. The Grimm lamp used in GDOES [99]: (1) sample, (2) cathode, (3) O-ring, (4) anode, (5) quartz window, (6), (7) evacuation ports, (8) gas inlet, (9) Teflon insulator.

atomic concentration of the emitting element in the solid through complex parameters such as the applied voltage, the matrix of the sample, spectral line constants, diffusion processes, etc. All these must be normalized, so that the use of a reference material is required for quantitative analysis. A good review of all the requirements has been given by Rivière [3]. If the discharge current during analysis of the reference and of the unknown sample is kept constant, the concentration is given by

$$c_{AM} = c_{AR} \frac{C_R}{C_M} \frac{I_{\lambda M}}{I_{\lambda R}} \left(\frac{V_R - V_{0R}}{V_M - V_{0M}} \right) x(\lambda, A) \tag{12}$$

where c_{AM} is the unknown concentration of A in the matrix M, c_{AR} is the concentration of A in the reference material R, $C_{(M \text{ or } R)}$ is a constant dependent on the sample and the sputtering ion, $V\text{-}V_0$ is the voltage above the threshold for the sample M and for the reference R (V_0 being the threshold voltage), $I_{\lambda(M \text{ of } R)}$ are the intensities of the emission lines from the unknown sample and from the reference material, respectively, and $X\lambda A$ is a spectral constant. Figure 12 shows quantified GDOES depth profiles from three hot-dip zinc-coated steel samples [100]. Only the Zn/Fe interface region is shown; the Zn and Fe concentrations vary from 0 to 100 at.% while the Al concentration varies from 0 to 1 at.%. This example shows two advantages of GDOES over, for instance, AES: (a) the deep location of the interface (of depth more than 10 μm) is reached quickly; and (b) the wide range of concentration recorded simultaneously (0 to 100 at.% and 0 to 1 at.%) is not available in AES. However, as the authors stated, the depth resolution was very much degraded due to the sputtering process and to grain orientation. The conversion from intensities to concentrations and from sputtering times to depths was performed by software developed by Bengtson [101–103].

4.3.3. Recent improvements in GDOES

Technical modification of the Grimm lamp (a hollow-cathode lamp operating in the microwave-boosted mode) has given increased sensitivity compared to the conventional hollow-cathode lamp [104]. Argon line interferences have been removed by introducing a third electrode to modulate the intensities [105]. A low-pressure device can be introduced in order to decrease atmospheric contamination of the sample [106]. Fast Fourier transform analysis of emitted light has provided increased sensitivity (e.g., Mo in steel down to 50 ppm [107]).

In special cases GDOES can operate at much lower erosion rates and then becomes a surface depth profiling technique with the same depth sensitivity as, for instance, AES [108]. However, the effectiveness of GDOES lies in its high-sputter-rate mode. Other emission spectroscopies from surfaces are available, such as spark or laser ablation spectroscopies, but they are less well adapted to quantitative depth profiling than GDOES.

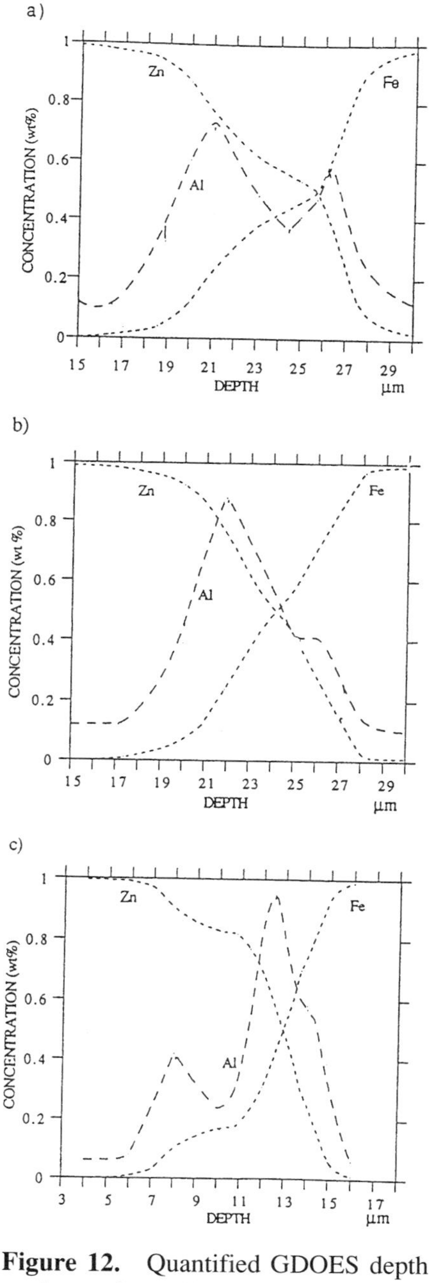

Figure 12. Quantified GDOES depth profiles from three hot-dip zinc-coated steel samples [100].

Table 7. Strengths and Weaknesses of GDOES

Strengths
Samples of thickness from 1 to 15 μm can be analyzed.
The analysis is rapid (a few minutes).
All elements can be detected (including H).
The technique is inexpensive (in comparison with AES and XPS).
A vacuum chamber is not required (except to detect O, N and H lines in extreme UV).
The technique provides elemental information.
It has high sensitivity (ppm).
Weaknesses
The depth resolution is poor (in comparison with AES, XPS or SIMS).
It is not normally a surface analysis technique (i.e., not in the nm range).
The lateral resolution is poor.
No chemical information is available.
The sample must be conducting.

4.4. SIMS

4.4.1. Basic principles

SIMS is based on the emission of atomic and molecular particles (secondary ions) from the surface of a solid under bombardment with primary particles (ions). Under DSIMS conditions, the erosion rate is of the order of monolayer(s) per second and concentration depth profiling can be performed with high spectrometric sensitivity, in favorable cases down to the ppb level. The sensitivity of SIMS is element dependent and varies from 1 ppb to 100 ppm, and the technique is therefore used for trace and ultratrace analysis.

The main problem in SIMS is quantification, because of the dependence of relative and absolute secondary-ion yields on matrix effects, on surface coverage by reactive elements (oxygen for instance), on background pressure in the sample chamber, on the effect of crystal orientation with respect to the directions of the primary- and secondary-ion beams, singular effects, etc. (see also Chapter 6).

4.4.2. Quantitative analysis

Some basic considerations are summarized here briefly. For both SIMS and SNMS the ion signal I_i of an isotope i (at a given mass) in a sample j is related to the isotopic concentration X_i in the sample by

$$I_i = (AX_iY_j\varepsilon_{ij}\Gamma_{i,j}\beta_{ij}T_{ij}D_i)I_p \tag{13}$$

where I_p = primary-ion flux (ions-sec^{-1}); A = fractional abundance of the isotope sampled (usually calibrated and neglected in SIMS); i,j = combined indices indicate that a particular parameter may depend on both the nature of the sputtered ion and on the sample matrix; X_i = average concentration of the element taken over the sputtered atom escape depth (under normal conditions virtually all the sputtered atoms come from the first layer); Y = overall sputtering yield of the sample; ε = fraction of i which is sputtered in the monitored form; Γ = preferential sputtering term (relative sputtering yield efficiency of i in the matrix j; β = ionization probability of the sputtered entity. This last parameter causes the greatest difficulty in SIMS analysis, because β depends strongly not only on the nature of the sputtered entity but also on the chemical state of the sample and on the nature of the primary ions. In SNMS neutrals are postionized (i.e., away from the sample), and there is therefore no hidden dependence of β on the sample in that technique. The other factors are

$Y^{\pm} = Y\beta$ = ion yield
τ^{*} = useful ion yield = $\beta\ T$
T_i = mass spectrometer transmission for species i
D_i = efficiency of detector for mass species i

Equation (13) is usually simplified to

$$I_i = K_i X_i I_p \tag{14}$$

where K_i is an elemental sensitivity factor that contains all the other variables of (13), including β, which may vary by several orders of magnitude from element to element but also, for any one element, from sample to sample. The latter effect is called the "matrix effect." Thus, if truly quantitative SIMS analysis is required, the use of external standards, as identical as possible to the sample under study are absolutely essential. Such standards are rare and expensive, since they must be prepared by ion implantation. Quantitative analysis at interfaces is even more difficult, since the matrix changes abruptly. However, due to the high sensitivities of mass spectrometers, detection limits in SIMS are very low (about ppm and in favorable cases 1 ppb) which allows traces and ultratraces to be detected. This explains the success of SIMS in the microelectronics industry.

4.4.3. Applications

Many problems in metallurgy, corrosion and microelectronics involve knowledge of the in-depth composition of a sample. Due to strong matrix effects and to experimental parameters, SIMS is certainly not the best technique for quantification, although it is possible to quantify accurately, if standard materials that match the sample are used. AES, XPS and SNMS, on the other hand, give signal intensities which are more or less directly related to the atomic concentration. Matrix effects also occur in AES and XPS but not to the same extent as in SIMS. Mass spectrometric techniques do provide very low detection limits, in comparison with electron detection systems. For this reason, the average sensitivity available in SSIMS, DSIMS and SNMS is much greater than in AES and XPS. The three ion excitation methods are thus particularly suitable for semiconductor analysis where small amounts of dopants modify considerably the properties of the materials, and DSIMS is indeed used widely in the microelectronics industry. Detection of molecular ions also makes SIMS useful in the polymer industry, although there only qualitative analysis is used.

4.4.4. Optimum conditions for performing SIMS depth profiling

4.4.4.1. Bombarding conditions: ions

Ar^+, O_2^+, O^+, O^- at energies from 1 to 10 keV
Cs^+ at energies from 15 to 30 keV
Ga^+, In^+ at energies up to 40–60 keV (for small high-brightness ion probes <100 nm in diameter); for depth profiling the beam is usually of 10 to 1000 nm diameter, with ion current density of 1 to 10 μA-mm^{-2}.

4.4.4.2. Angle of incidence

There are two mutually exclusive requirements in SIMS; there is a need for high sensitivity, which is optimum at normal incidence, and as in AES and XPS, a need for good depth resolution which is optimum at glancing incidence. The best solution seems as before to involve rotation of the sample during sputtering, which produces a depth resolution relatively independent of the incidence angle so that high sensitivity can still be achieved.

4.4.4.3. Effect of the choice of gas in SIMS

DSIMS depth profiling is usually performed with O_2^+, Cs^+ and sometimes N_2^+. Bombardment with these species increases the secondary-ion yield by several orders of magnitude, particularly at normal and near-normal ion beam incidence [109–111]. The introduction of reactive gases is nevertheless risky because adverse effects also exist.

O_2^+ bombardment at normal incidence prevents roughening of the surface of polycrystalline metals such as Ti, Cr, Mn, Fe, Mo by the formation of an amorphous surface oxide layer [112, 113], and the depth resolution is thereby increased.
N_2^+ has been found to improve the depth resolution in Cu/Ni, Cu/Nb, Nb/Ge and Mo/Ge multilayers [114, 115].
N_2^+ and O_2^+ have been found to increase the depth resolution in a study of $^{28}Si/^{30}Si$ multilayers [116].

The use of O_2^+ is not recommended when noble metals and alkali metal impurities are to be studied [117–122]

4.4.4.4. *Choice of ion beam energy*

Sputtering yield has an hourglass-shaped dependence on ion energy with a maximum between 1 and 10 keV [116, 123, 124] for Ar^+, N_2^+, O_2^+. Tables of sputtering yields of metals as a function of ion energy have been published, and the analyst must choose the ion type and energy taking into account the following requirements:

The sensitivity must be as high as possible, which requires a high-energy beam.
The depth resolution must be sufficient, which is best achieved with low-energy ions (<2 keV).
The sputtering yields of the different elements in the sample must be as similar as possible.

A detailed discussion of all these parameters can be found in the literature [125].

4.4.4.5. *Interferences in SIMS depth profiling*

The mass spectra often contain peaks of adventitious ions such as Na^+, K^+, Al^+, C^-, O^-, OH^-, Cl^- [126, 127].

4.5. SNMS

The principal rationale for SNMS is the elimination of the matrix effects which are present in SIMS by decoupling the emission and the ionization processes, so the ionization probability no longer depends on the sample characteristics and composition but becomes an equipment constant. Thus, quantitative analysis may be performed even at nondilute concentrations. Moreover, SNMS can be operated with low-energy incident ions (a few hundred eV), which eliminates atomic mixing. It is also worth remembering that in the sputtering process only about 1% of the ejected particles are ionized, the remaining 99% being neutrals. Finally, the artefacts due to preferential sputtering, which are present in the electron spectroscopies, are avoided in SNMS.

4.5.1. Basic principles

When bombarded by an incident ion beam, most of the ejected particles are neutrals. However, the mass analysis of neutrals requires postionization. Many techniques have been developed and they are summarized in Table 8. The most common techniques are those using electron beam and electron gas postionization. Typical operating conditions for an electron gas plasma are detailed in Table 9.

Table 8. Postionization Techniques

Variable	Thermal ionization	Electron beam	Electron gas	Penning ionization
Elements ionized	low ionization potential	all	all	all, not noble gases
Ionizing process	surface ionization	electron impact	electron impact	collisions with noble gas atoms and electron impact.
Ionizing medium	heated chamber	electron beam plasma	low pressure plasma	medium pressure
Residual gas	mass resolution discrimination	energy discrimination	energy discrimination	mass resolution
Efficiency	poor	satisfactory	satisfactory	satisfactory
Mass spectrometer	quadrupole	quadrupole	quadrupole	quadrupole

Source: Adapted from Ref. 128.

Table 9. Typical Operating Conditions for Electron Gas SNMS, Using Argon

Pressure	ca. 0.1 Pa
Ion current density	0.8–3 mA cm^{-2}
Magnetic field	2–3 mT
RF frequency	27.12 MHz
RF power	80–200 W
Plasma potential	45–55 V
Dimensions	150–nm diameter 150–mm length

4.5.2. Quantification in SNMS

The basic relationship between the ion current $I^0(x)$ of a particular mass x collected and the concentration of the element of that mass in the sample is the same as in SIMS (see Eq. (13)) and can be written

$$I^0(X) = Y_{tot}C(X)D(X)I_p \tag{15}$$

where I_p is the primary current, Y_{tot} is the total sputtering yield of the sample per primary ion, $C(X)$ is the atomic concentration of element or isotope X, and $D(X)$ is the useful yield (the ratio of the number of detected particles to the number of sputtered particles). Equation (15) assumes, contrary to SIMS, that $D(X)$ does not depend on (a) the concentration of X and (b) the concentrations of the other constituents. Equation (15) also allows the use of relative sensitivity factors (as in AES or XPS).

For two elements i and j, one can write

$$\frac{I^0(X_i)}{I^0(X_j)} D_j(X_i) = \frac{C(X_i)}{C(X_j)} \tag{16}$$

where $D_j(X_i)$ is the relative sensitivity factor (RSF) between i and j. This RSF must be determined from a standard, which may be rather different from the sample under investigation. In electron impact SNMS, RSFs are usually referred to Fe. The general equation is then

$$C(X_i) = \frac{I_0(X_i^{mi})\eta^{mi} D_{Fe}(X_i)}{\sum_j I^0(X_j^{mj})/\eta^{mj} D_{Fe}(X_j)} \tag{17}$$

where $I^0(X_i^{mi})$ is the SNMS signal of the isotope with mass mi of element X_i, η^{mj} being the abundancy of the isotope. The RSFs are relatively independent of the matrix of the sample but a dependence, especially for light elements, has been shown on ion energy and ejection angle. Tables of RSFs can be found in the literature [128].

The best way to perform quantitative SNMS with reliable standards is the following:

1. Choose the standard as having similar microstructure and texture to the sample to be analyzed in order to avoid topographical and angular effects (use a microscope for the comparison, if necessary). Similar composition is of course necessary.
2. Calibrate the instrument using the standard and fix the instrumental parameters (i.e., ion energy, plasma or e-beam characteristics, spectrometer position, etc.).
3. Measure the elemental RSF.
4. Analyze the unknown sample under the same instrumental conditions.

Steels are often used as standards because they have good bulk homogeneity, and they contain a large number of elements with a wide range of the concentrations usually sought in SNMS (e.g. from 10 at.% to 1 ppm) [129–134]; brass and glass substrates have also been used [135].

4.5.3. Applications

SNMS is particularly suitable for the analysis of alloys in which a wide range of concentrations must be measured. Semiconductors can also be analyzed, because the detection limits of SNMS are of the same order of magnitude as in SIMS. Finally, as the depth resolution is better in SNMS than in SIMS or AES, SNMS is particularly suitable for the study of interfaces [136].

5. DISCUSSION AND GENERAL CONCLUSION

Each analytical technique discussed in this chapter has its own advantages and disadvantages, arising from both the physical processes involved and technical requirements. The theory behind each of the techniques would in itself fill a book, whereas the present purpose was to set out briefly the depth profiling approach as it related to each. The theories of sputtering and of depth resolution have also been the subject of many publications. Individual applications (corrosion, semiconductors, catalysis, organic materials, etc.) will be described separately in this volume.

To complete the brief overview, a selection is given below of some of the typical problems in depth profiling, with the solution proposed. Finally, some basic questions are posed to help the analyst in the choice of analytical method.

5.1. Typical Problems That Might Be Encountered in Specific Samples When Sputter Profiling, and Their Solutions

1. The surface of the sample is rough; the effects of roughness can be eliminated by
   mechanical polishing
   electrochemical polishing, if possible
   normal ion beam incidence, noble gas ions
2. The surface of the sample is already smooth; depth resolution is enhanced by glancing beam incidence, low ion energy, noble gas ions.
3. The sample is polycrystalline and/or metallic; then use reactive gas ion beam.
4. Heavy-element oxide sample; watch out for wrong chemical states. Use ARAES or (better) ARXPS (see Table 6).
5. The sample is an insulator; there will probably be positive charging of the sample. If XPS or SIMS analysis is to be carried out, then use an electron flood gun; it may be better to use AES sputter profiling.
6. The sample is a thin film and/or has poor thermal conductivity; then use XPS (best), or SIMS or AES with a low-current electron beam.
7. The sample is a metallic alloy with high diffusion coefficients: use high sputtering rate and (if available) a cooled sample holder.
8. The sample is an organic compound: use XPS or SIMS. (If the sample contains C–F bonds, bond breaking will occur for XPS and F contamination of the vacuum chamber will be the result)
9. If boron analysis is required, use AES, but do not use Ar^+ for sputtering (there will be spectral interferences with the B 179-eV peak), instead use Xe^+.
10. If analyses of Cr, O, I are required, then if AES is used, do not use Xe^+ due to spectral interferences.
11. In the case of oxide superconductors: decomposition to lower oxidation states will occur; use ARAES, or better ARXPS.

5.2. Key Parameters/Considerations for Choice of the Appropriate Analysis Method

1. Can the sample be damaged by the analysis?
   Yes: AES, XPS, SIMS, SNMS, and GDOES are the preferred techniques.
   No: ARAES, ARXPS, and RBS should be used.
2. Is the sample conducting?
   Yes: all the techniques may be used.
   No: do not use (AR)AES, or any other electron beam technique.

3. Type of information required:
   Quick depth profile, quantitative: GDOES can be used.
   Elemental information: GDOES, XPS, AES, SIMS, SNMS, and RBS are suitable.
   Chemical and elemental information: XPS is best, in some individual cases SIMS will be suitable, and AES if careful spectral analysis is carried out.
   Trace impurities: SIMS and SNMS will provide the answers.
   Interdiffusion reactions (in situ): RBS is the preferred choice.
   Interface analysis: AES, XPS, SIMS, and (RBS) should be used.
   Very shallow surface layer to be analyzed in depth: ARAES and ARXPS are preferred.
   Quantitative depth profile analysis: AES, XPS, GDOES and SNMS
   Good lateral resolution: SAM, microspot RBS, small-area XPS, and imaging SIMS.
   Organic sample: SIMS, SNMS and XPS
   Isotopic detection: SIMS and SNMS
   Surface analysis: SSIMS, ISS, AES and XPS

ACKNOWLEDGMENTS

I wish to thank Professor Fiermans from the University of Ghent for helpful discussion about depth information in ARXPS, Professor Terryn from the University of Brussels (VUB) for his advice on AES depth profiling, Professor Vandervost from IMEC (Leuven, Belgium) for having inspired the section on RBS, Professor Bouillon (University of Brussels, ULB) for general discussion of this text, and Professor Buess for her constant interest in the work. I wish also to thank Mr. L. Binst for his technical help.

REFERENCES

1. D. Briggs and M. P. Seah (eds.), *Practical Surface Analysis*, Vol. 1, 2nd ed.
2. D. Briggs and M. P. Seah (eds.), *Practical Surface Analysis*, Vol. 2, 2nd ed.
3. J. C. Rivière, *Surface Analytical Techniques*, Oxford Science, Oxford, 1990.
4. A. W. Czanderna, (ed.), *Methods of Surface Analysis* Vol. 1, Elsevier, Amsterdam, 1975.
5. Thin film and depth profile analysis, in *Topics in Current Physics*, Vol. 37 (H. Oechsner, ed.), Springer-Verlag, Berlin, 1984.

6. *Anal. Chem. 65* (1993).
7. P. Dettner, *Electrolytic and Chemical Polishing of Metals*, Ordentlich, Tel Aviv, 1988.
8. W. J. MacTegart, *The Electrolytic and Chemical Polishing of Metals in Research and Industry*, Pergamon Press, London, 1959.
9. Z. Tan and S. M. Heald, *J. Appl. Phys. 71*, 3766 (1992).
10. P. Zschack, J. B. Cohen and Y. W. Chung, *Surf. Sci. 262*, 395 (1992).
11. X. Wang, G. L. Zhou, C. Sheng and M. R. Yu, *Mater. Res. Soc. Symp. Proc. 220*, 235 (1991).
12. H. Chen and S. M. Heald, *Phys. Rev. B 42*, 4913 (1990).
13. W. K. Chu, J. W. Mayer and M. A. Nicolet, *Backscattering Spectrometry*, Academic Press, New York, 1978.
14. L. C. Feldman, J. W. Mayer and S. T. Picraux, *Materials Analysis by Ion Channeling*, Academic Press, New York, 1982.
15. J. E. E. Baglin and J. S. Williams, in *Ion Beams for Materials Analysis*, (J. R. Bird and J. S. Williams, eds.) Academic Press, New York, 1989, chap. 3.
16. J. Li et al., *Nucl. Instr. Meth. Phys. Res. B59–60*, 989 (1991).
17. Q. C. Zhang, P. McMillan and J. C. Kelly, *Radiat. Eff. Defects Solids 114*, 115 (1990).
18. D. Y. Shih, C. A. Chang, J. Paraszczak, S. Nunes and J. Cataldo, *J. Appl. Phys. 70*, 3052 (1991).
19. S. Petersson et al., *J. Appl. Phys. 50*, 3357 (1979).
20. H. P. Bonzel, U. Breuer and O. Knauff, *Surf. Sci. Lett. 237*, L398 (1990).
21. A. Jablonski, *Surf. Interface Anal. 15*, 559 (1990).
22. A. Jablonski, *Surf. Interface Anal. 21*, 758 (1994).
23. W. S. M. Werner, W. H. Gries and H. Störi, *Surf. Interface Anal. 17*, 693 (1991).
24. R. van de Walle, R. L. Van Meirhaeghe, W. H. Laflère and F. Cardon, *J. Appl. Phys. 74*, 1885 (1993).
25. L. E. Davis, N. C. MacDonald, P. W. Palmberg, G. E. Riach and R. E. Weber, *Handbook of Auger Electron Spectroscopy*, 2nd ed., Physical Electronics Industries Inc., Minnesota, 1976.
26. Y. Shiokawa, T. Isida and Y. Hayashi, *Auger Electron Spectra Catalogue—A Data Collection of Elements*, Anelva Corp., Tokyo, 1979.
27. T. Sekine et al., *Handbook of Auger Electron Spectroscopy*, JEOL, Tokyo, 1982.
28. C. D. Wagner, W. M. Riggs, L. E. Davis, J. F. Moulder and G. E. Muilenberg, *Handbook of X-ray Photoelectron Spectroscopy*, Perkin-Elmer Corp., Minnesota, 1979.
29. T. Sekine, K. Hirata and A. Mogami, *Surf. Sci. 125*, 565 (1983).
30. S. Tanuma et al., *Surf. Interface Anal. 15*, 466 (1990).

31. F. Reniers, *Surf. Interface Anal. 23*, 374 (1995).
32. F. Pons, J. Le Hèricy and J. P. Langeron, *Surf. Sci. 69*, 565 (1977).
33. T. E. Gallon, *Surf. Sci. 17*, 486 (1969).
34. F. Reniers, M. Jardinier-Offergeld and F. Bouillon, *Surf. Interface Anal. 17*, 343 (1991).
35. H. J. Mathieu and D. Landolt, *Appl. Surf. Sci. 10*, 455 (1982).
36. P. S. Ho, J. E. Lewis, H. S. Wildman and J. K. Howard, *Surf. Sci. 57*, 393 (1976).
37. J. E. Kempf and H. H. Wagner, *Thin Film and Depth Profile Analysis*, Topics in Current Physics, Vol. 37 (H. Oechsner, ed.), Springer-Verlag, Berlin, 1984, Chap. 5.
38. S. Hofmann and A. Zalar, *Surf. Interface Anal. 10*, 7 (1987).
39. S. Hofmann, J. Erlewein and A. Zalar, *Thin Solid Films 43*, 275 (1977).
40. A. Zalar and S. Hofmann, *Nucl. Instr. Meth. Phys. Res. B18*, 655 (1987).
41. A. Zalar and S. Hofmann, *Vacuum 37*, 169 (1987).
42. A. Zalar and S. Hoffman, *J. Vac. Sci. Technol. A5*, 1209 (1987).
43. H. J. Mathieu, D. E. Mc Lure and D. Landolt, *Thin Solid Films, 38*, 281 (1976).
44. M. P. Seah and C. Lea, *Thin Solid Films 81*, 257 (1981).
45. S. Hofmann, *Surf. Interface Anal. 8*, 87 (1986).
46. H. Oechsner, *Appl. Phys. 8*, 185 (1975).
47. H. E. Rosendaal, in *Sputtering by Ion Bombardment* (R. Behrisch, ed.), Topics in Applied Physics, Vol. 1, Springer-Verlag, Heidelberg, 1981, p. 219.
48. W. Hauffe, in *Proceedings of 3rd International Conference on SIMS (III)* (A. Benninghoven, J. Giber, J. Laszlo, M. Riedel and H. W. Werner, eds.), Springer-Verlag, Berlin, 1982, p. 206.
49. S. Hoffmann and A. Zalar, *Thin Solid Films 60*, 201 (1979).
50. R. J. Blattner, S. Nadel, C. A. Evans, A. J. Braundmeier and C. W. Magee, *Surf. Interface Anal. 1*, 32 (1979).
51. M. P. Seah and M. Kühlein, *Surf. Sci. 150*, 273 (1985).
52. K. Tsunoyama, T. Suzuki, Y. Okahasbi and H. Kishikada, *Surf. Interface Anal. 2*, 212 (1980).
53. C. P. Hunt, C. T. Hoddard and M. P. Seah, *Surf. Interface Anal. 5*, 157 (1983).
54. M. P. Seah and C. P. Hunt, *Surf. Interface Anal. 5*, 33 (1983).
55. S. Hofmann, in *Wilson and Wilson's Comprehensive Analytical Chemistry*, Vol. IX (G. Svehla, ed.) Elsevier, Amsterdam, 1979, p. 89–172.
56. H. J. Mathieu, in *Thin Film and Depth Profile Analysis* (H. Oechsner, ed.) Topics in Current Physics, Vol. 37, Springer, Berlin, 1984, Chap. 3.
57. S. Hofmann, *Appl. Phys. 13*, 205 (1977).
58. M. P. Seah and M. E. Jones, *Thin Solid Films 115*, 203 (1984).
59. S. Hofmann and J. M. Sanz, *Frez. Z. Anal. Chem. 314*, 215 (1983).

60. S. Hofmann and J. M. Sanz, *J. Trace Microprobe Technol. 1*, 213 (1982–1983).
61. G. Betz and G. K. Wehner, in *Sputtering by Ion Bombardment* (R. Behrisch, ed.), Topics in Applied Physics, Vol. II, Springer-Verlag, Heidelberg, 1987, Chap. 2, p. 11.
62. H. J. Mathieu and D. Landolt, *Appl. Surf. Sci. 10*, 100 (1982).
63. S. Hofmann and J. M. Sanz, *Surf. Interface Anal. 6*, 78 (1984).
64. R. Kelly, *Nucl. Instr, Meth. 149*, 553 (1978).
65. S. Storp and R. Holm, *J. Electron Spectr. Rel. Phenom. 16*, 183 (1979).
66. J. B. Malherbe, J. M. Sanz and S. Hofmann, *Appl. Surf. Sci. 27*, 355 (1986).
67. K. S. Kim, W. E. Baitinger, J. W. Amy and N. Winograd, *J. Electr. Spectr. Rel. Phenom. 5*, 19 (1978).
68. R. Holm and S. Storp, *Appl. Phys. 12*, 101 (1977).
69. K. Konno and M. Nagayanna, in *Passivity of Metals* (R. P. Frankenthal and J. Kruger, eds.), The Electrochemical Society, Princeton, NJ, 1978.
70. T. S. Sun, D. K. McNamara, J. S. Ahearn, J. M. Chen, D. Ditchek and J. D. Venables, *Appl. Surf. Sci. 5*, 406 (1980).
71. J. M. Sanz, *Dissertation*, University of Stuttgart, 1982.
72. T. J. Chuang, C. R. Brundle and K. Wandelt, *Thin Solid Films 53*, 19 (1978).
73. N. Q. Lam, *Surf. Interface Anal. 12*, 65 (1988).
74. R. Shimizu, *Nucl. Instr. Meth. Phys. Res. B18*, 486 (1987).
75. S. Hofmann, *Mater. Sci. Eng. 42*, 55 (1980).
76. R. Kelly, *Surf. Interface Anal. 7*, 1 (1985).
77. K. Röll, *Appl. Surf. Sci. 5*, 388 (1980).
78. S. Hofmann, in *Practical Surface Analysis*, Vol. 1, (D. Briggs and M. P. Seah, eds.), 1990, p. 177.
79. A. Zalar, *Thin Solid Films 124*, 223 (1985).
80. S. Hofmann, A. Zalar, E.-H. Cirlin, J. J. Vajo, H. J. Mathieu and P. Panjan, *Surf. Interface Anal. 20*, 621 (1993).
81. K. De Boeck, Thesis, Vrije Universiteit Brussel, Belgium, 1982.
82. S. Hofmann and A. Zalar, *Surf. Interface Anal. 21*, 231 (1994).
83. D. E. Sykes and A. Chew, *Surf. Interface Anal. 21*, 231 (1994).
84. T. W. Haas, J. T. Grant and G. J. Dooley, *J. Appl. Phys. 43*, 4 (1972).
85. E. R. Malinowski and D. G. Howery, *Factor Analysis in Chemistry*, Wiley-Interscience, New York, 1980.
86. D. G. Watson, W. F. Stickle and A. C. Diebold, *Thin Solid Films 193/194*, 305 (1990).
87. P. De Volder, R. Hoogewijs, R. De Gryse, L. Fiermans and J. Vennik, *Surf. Interface Anal. 17*, 363 (1991).
88. N. Roose, M. Ye, J. Vereecken and E. W. Seibt, *Surf. Interface Anal. 21*, 474 (1994).

89. F. Reniers, A. Hubin, H. Terryn and J. Vereecken, *Surf. Interface Anal. 21*, 483 (1994).
90. S. W. Gaarenstrom, *Appl. Surf. Sci. 7*, 7 (1981).
91. S. W. Gaarenstrom, *J. Vac. Sci. Technol. 20*, 458 (1982).
92. V. Atzrodt and H. Lange, *Phys. Stat. Sol.* A *79*, 489 (1983).
93. J. S. Solomon, *Thin Solid Films 154*, 11 (1987).
94. R. Vidal and J. Ferron, *Appl. Surf. Sci. 31*, 263 (1988).
95. S. Hoffman and J. Steffen, *Surf. Interface Anal. 14*, 59 (1989).
96. N. H. Turner, J. H. Wandass and F. L. Hutson, *J. Vac. Sci. Technol. A8*, 6, 4033 (1990).
97. C. Jansson and P. Morgen, *Surf. Interface Anal. 15*, 1 (1990).
98. J. S. Solomon, *Surf. Interface Anal. 10*, 75 (1987).
99. Y. Ohashi, Y. Yamamoto, K. Tsuyonama and H. Kishidaka, *Surf. Interface Anal. 1*, 53 (1979).
100. J. Karlsson, S. E. Höronström, H. Klang and J. O. Nilsson, *Surf. Interface Anal. 21*, 365 (1994).
101. A. Bengtson, *Spectrochimica Acta B40*, 631 (1984).
102. A. Bengston and M. Lundholm, *J. Anal. Atom. Spectrom 2*, 537 (1988).
103. A. Bengtson, A. Eklund, M. Lundholm and A. Saric, *J. Anal. Atom Spectrom 5*, 563 (1990).
104. O. Senofonte, R. Tomellini, M. Cillia, M. G. Del Monte Tamba and S. Caroll, *Acta Chim. Hung. 128*, 455 (1991).
105. K. Hirokawa and K. Wagatsuma, *Tetsu to Hagane 77*, 1823 (1991).
106. G. Ehrlich, U. Stahiberg and V. Hoffman, *Spectrochim. Acta B45*, 115 (1991).
107. J.A. Broekaert, K. R. Brushwyler, C. A. Monnig and G. M. Hieftje, *Spectrochim. Acta B45*, 768 (1991).
108. R. Berneron and J. C. Charbonnier *Surf. Interface Anal. 3*, 134 (1981).
109. K. Wittmaack, *Surf. Sci. 112*, 168 (1981).
110. K. Wittmaack, *Surf. Sci. 126*, 573 (1983).
111. K. Wittmaack, *Int. J. Mass Spectrom. Ion Phys. 17*, 39 (1975).
112. K. Tsunoyama, Y. Ohashi, T. Suzuki and K. Tsuruoka, *Jap. J. Appl. Phys. 13*, 1683 (1974).
113. K. Tsunoyama, T. Suzuki, Y. Ohashi and H. Kishidaka, *Surf. Interface Anal. 2*, 212 (1980).
114. W. O. Hofer and H. Liebl, *Appl. Phys. 8*, 359 (1975).
115. R. J. Blattner, S. Nadel, C. A. Evans Jr, A. J. Brandmeier Jr. and C. W. Magee, *Surf. Interface Anal. 1*, 32 (1979).
116. K. Wittmaack and D. B. Poker, *Nucl. Instrum. Meth. B47*, 224 (1990).
117. P. Williams and J. E. Baker, *Nucl. Instrum. Meth. 182/183*, 15 (1981).
118. P. R. Boudewijn, H. W. P. Akerboom and M. N. C. Kempeneers, *Spectrochim Acta B39*, 1567 (1984).

119. S. M. Hues and P. Williams, *Nucl. Instrum. Meth. B15*, 206 (1986).
120. V. R. Deline, W. Reuter and R. Kelley, in *Secondary Ion Mass Spectrometry, SIMS V*, (A. Benninghoven, R. J. Colton, D. S. Simons and H. W. Werner, eds.); Springer Series in Chemical Physics, Vol. 44, Berlin, 1986, p. 299.
121. K. Wittmaack, *App. Phys. Lett. 48*, 1400 (1986).
122. P. R. Boudewijn and C. J. Vriezema, in *Secondary Ion Mass Spectrometry, SIMS VI* (A. Berninghoven, A. M. Huber and H. W. Werner, eds.) Wiley, Chichester, 1988, p. 499.
123. P. C. Zalm, *J. Appl. Phys. 54*, 2660 (1983).
124. E. Hechtl and J. Bohdansky, *J. Nucl. Mater. 133–134*, 301 (1985).
125. K. Wittmaack, *Surf. Interface Anal. 21*, 323 (1994).
126. A. Benninghoven, H. W. Werner and F. Rüdenauer, *Secondary Ion Mass Spectrometry*, Wiley, Chichester, 1987, Secs. 5.1.4.1, 5.2.6. Vol. 55.
127. J. B. Clegg, in *Secondary Ion Mass Spectrometry, SIMS VI* (A. Benninghoven, A. M. Huber and H. W. Werner, eds.) Wiley, Chichester, 1987, p. 689.
128. R. Jede, O. Ganschow and U. Kaiser, Sputtered neutral mass spectrometry, in *Practical Surface Analysis* (D. Briggs and M. P. Seah, eds.), Wiley, Chichester, 1992.
129. D. Lipinski, R. Jede, O. Ganschow and A. Benninghoven, *J. Vac. Sci. Technol. A3*, 2007 (1985).
130. R. Jede, H. Peters, G. Dünnebier, O. Ganschow, U. Kaiser and K. Seifert, *J. Vac. Sci. Technol. A6*, 2271 (1988).
131. K. H. Mueller and H. Oeschner, *Microchim. Acta 10*, 51 (1983).
132. A. Wucher, F. Novak and W. Reuter, *J. Vac. Sci. Technol. A6*, 2265 (1988).
133. K. H. Müller, K. Seifert and M. Willmers *J. Vac. Sci. Technol. A3*, 1367 (1985).
134. J. Tümpner, R. Wilsch and A. Benninghoven, *J. Vac. Sci. Technol. A5*, 1186 (1987).
135. H. Schoof and H. Oeschner, *Proceedings of the 4th International Conference on Solid Surfaces and 3rd ECOSS* (D. A. Degas and M. Costa, eds.), 1981, Vol. II, 1291.

8

Ion Beam Effects in Thin Surface Films and Interfaces

I. BERTÓTI and A. TÓTH Research Laboratory for Inorganic Chemistry of the Hungarian Academy of Sciences, Budapest, Hungary

M. MENYHARD Research Institute for Technical Physics of the Hungarian Academy of Sciences, Budapest, Hungary

1. INTRODUCTION

Over the last 30 years there has been a continuous increase in the range of application of thin films, surface coatings and modified surfaces. They have played a decisive role in nearly all areas of advanced technology, from heavy-duty machinery (aircraft rotors, fuselage, etc.) to wear-resistant coatings and to sub-μm layer structures, e.g., in semiconductor devices, sensing elements and optical waveguides. Surface modification for aesthetic and decorative purposes, for biomedical use, or for food packaging, should also be mentioned among the new and fast-growing areas of applications.

The widespread application of thin films can be understood for several reasons. First, only small quantities of material are necessary for the enhancement of such physical parameters as hardness, mechanical and adhesive strength, wear and chemical resistance, etc. Second, optimum property-performance combinations can be achieved which could not be obtained for the substrate bulk material alone. Last, surface films can be prepared in metastable states with unusual combinations of properties not achievable in equilibrium bulk-phase materials.

Thin films can be prepared in crystalline or amorphous states, in single layer or in multilayers, or in multicomponent form. In their applications an enormous range of materials has been used, e.g., metals, intermetallics, oxides, nitrides, carbides, elemental or compound semiconductors, glasses, and inorganic and organic polymers. For their preparation for the various applications, numerous techniques have been and are being developed.

Thin films are usually prepared from the vapor phase, but they can also be obtained from liquid or solid phases by applying physical or chemical driving forces, or various combinations of the latter [1–3]. In principle, two types of film may be identified from the point of view of surface analysis: one in which the film is deposited on a substrate (workpiece), and the other in which the film is created as a result of the modification of the surface atomic layers of a substrate. In the latter case there is a strong involvement of the constituents of the solid substrate. A brief survey of the most frequently applied deposition and implantation techniques is presented below.

Physical vapor deposition (PVD) includes the following methods [1]: *Evaporation* in its simplest form involves heating the source material in vacuum to a temperature at which its partial pressure reaches about 10^{-2} mbar. The precise control of composition using multiple sources for multicomponent films is difficult. A sophisticated version of this method is molecular beam epitaxy (MBE), (requiring UHV conditions, precise control of the temperature of the sources, computer controlled shutters, etc.) capable of producing layers from multi- to monoatomic thicknesses. Laser-ablation-assisted deposition (LAPVD) and cluster evaporation with ionized cluster beam deposition are among the new developments used to grow films by evaporation.

Sputtering is a process in which the target atoms or molecules are ejected by ion/atom bombardment with the most frequently used ion source gas being argon. In the case of conducting targets a large negative potential (dc version) is applied between the target and the grounded substrate holder in order to accelerate Ar^+ from the plasma that develops. Nonconducting ceramic or polymeric materials can be sputtered by applying RF power. The ionizing effect of electrons in the plasma is increased by confinement of their trajectories with a magnetic field (magnetron-type sources).

Ion plating is a combination of evaporation with energetic neutral and ion bombardment of the growing film on the negatively biased substrate. The chemically inert (Ar^+, Ar^0) or reactive (e.g., N_2^+, N_2^0) ion and neutral species originate from the plasma developed around the substrate.

Ion-beam-assisted deposition (IBAD), which uses ion sources separate from the plasma may help in improving adhesion, densification, texture, grain size and morphology of the growing films [3].

Chemical vapor deposition (CVD) is a general term for film-growth processes initiated by chemical reactions between gas-phase molecules carrying the elements to be deposited. Chemical reaction involving the substrate can also be included in CVD. In addition to thermal activation of the reactions, photon (UV, laser) or plasma activation are frequently applied variants [2].

Ion implantation applied at high fluences can produce surface films that include the constituents of the substrate. Implantation performed at low primary-ion energies (0.1–10 keV) may cause ion-induced defects, mixing and grain growth, and chemical interactions, as well as either congruent or preferential sputtering. Chain scission and crosslinking are some additional processes taking place in polymeric materials. Subsurface (buried) layers can be produced by high-energy implantation (in the 100 keV range), where sputtering is less pronounced. A particular feature of this method is that new types of surface film, consisting of new materials nonexistent in the equilibrium state, can be created with an unusual combination of properties.

Analysis of thin films by surface techniques (e.g., XPS, AES, etc.) does not differ very much from the application of those techniques to thick, bulklike or dispersed phases, and the analyst should be encouraged to utilize the general knowledge and skill gained during the analysis of different samples. Obviously, it should be borne in mind that the sampling depth may be greater than the thickness of the film being analyzed. (For quantification under these conditions see Chapter 5.)

There are at least two features, however, peculiar to thin films which must be taken into account during the analysis and the data interpretation:

(1) Thin films are bordered by an interface which may be vanishingly narrow, i.e., confined to a two-dimensional network of atoms on both sides of the matching phases. In general, the bonding states of these atoms will differ considerably from those associated with some atomic layers away from the interface. Due to the dissimilarity or mismatch between the atomic arrangements of the adjacent phases even an "ideal" interface contains defects such as misfit dislocations, along with unsaturated, or so-called dangling bonds. More often the interface is smeared out over several (or several tens of) atomic layers. Surface spectroscopies applied with due care are capable of characterizing the interface, i.e., of determining the depth distribution of constituents across the interface (AES, XPS, SIMS), and of evaluating the chemical states of the elements present in the interface (XPS).

(2) Thin surface films are generally obtained by applying various means of activation (ion beams, plasmas, etc.) at relatively low temperatures, i.e., the films are formed (or deposited) under conditions usually far removed from chemical and thermal equilibria. This leads, sometimes intentionally, to non-equilibrium phases in the morphological, crystalline and chemical structural senses. Here surface analysis techniques are directed to the determination of the chemical states of the constituents of the films, for which XPS is the straightforward choice. The crystal structure of thin surface layers, on the other hand,

can be determined by x-ray diffraction (XRD), low-angle electron diffraction (LAED), and, more precisely, by selected-area electron diffraction coupled with transmission electron microscopy (SAED-TEM).

Both preparation and analysis of thin films involve at some stage ion beam or plasma-assisted processes with primary-ion energies ranging from 1 eV to 10 keV [3]. For this reason it is of prime importance to be aware of the effects of ion beams on the thin surface films and on the interfacial regions to be analyzed. Beam-induced physical phenomena, mainly atomic mixing at the interface, and chemical alterations such as compositional and chemical state changes in the surface layers will be addressed in this chapter. Examples involving metals, metal oxides, nitrides, and semiconductors as well as organic and organosilicon polymeric materials will be considered.

2. LOW-ENERGY ATOMIC MIXING

This section will review the application of Auger depth profiling to the analysis of extremely thin layers. To accomplish this, a sputtering arrangement providing an extremely good depth resolution is needed. Having achieved good depth resolution, problems that can be ignored during measurement at much poorer depth resolution must be solved. Once they are solved, experimental information can be obtained about process such as atomic mixing, that are normally not readily accessible.

The term "atomic mixing" stands for a process in which an originally abrupt interface becomes broadened due to ion bombardment. Ion mixing may be either an unwanted side effect or a technologically important tool. During ion implantation, i.e., when using high-energy ions, the accompanying ion mixing is generally undesirable. On the other hand, high-energy ion mixing is a fast-developing new technique for the modification of interfaces [4]. Exactly the same is true for the low-ion-energy case. Depth profiling techniques suffer from inevitable ion mixing, while the various IBAD methods are based on the advantageous effects of ion bombardment during thin-film deposition.

In the case of low-energy ion bombardment, the penetration depth of the ions, i.e., the depths at which mixing might occur, is in the range of some nanometers (depending on energy and angle of incidence). This means that surface effects cannot be ignored, and their separation from "pure" mixing is far from straightforward. Nevertheless, it will be shown that sputter depth profiling with low-energy ions can give information about ion mixing.

All sputter depth profiling techniques (XPS, SIMS, AES) are based on the removal of atomic layers by sputtering the surface with low-energy (inert or reactive) ions. Ion sputtering, however, does not remove the material in a consecutive layer-by-layer fashion, but instead, surface morphological changes may initiate and develop during sputtering; in addition, sputtering may alter the original concentration distribution with depth by radiation-induced segregation, by preferential sputtering and by ion mixing. Since all these processes occur simultaneously, the raw depth profiling data do not usually provide information about any of them, including ion mixing. However, all the above effects except mixing can be minimized, and thus information concerning the ion mixing can be deduced from the depth profile as demonstrated below for the Ge/Si system.

2.1. Auger Depth Profiling

2.1.1. Multilayer systems

Layered structures with characteristic dimensions of some nanometers are frequently produced nowadays for use as soft x-ray and cold neutron mirrors, etc. All experimental data presented in this section have been collected from multilayer structures containing amorphous germanium and silicon. They were grown in a dual-target dc magnetron sputtering system (Thin Film Physics Division, Linköping University) with a base pressure of about 10^{-7} mbar. The films were deposited on an unheated silicon substrate with a native oxide, at a rate of 0.2 $nm \cdot s^{-1}$ and an Ar sputtering gas pressure of 5 mbar. To prevent columnar structure growth a negative substrate bias of approximately 140 V was applied [5]. For detailed study of the ion mixing process, the following structures were depth profiled:

(A) 20 nm Si–20 nm Ge–10 × (3 nm Si–3 nm Ge)–20 nm Si–20 nm Ge
(B) 8 × [10 nm Si–15 × (2 nm Ge–2 nm Si)]

The quality of a multilayer structure can be checked in several ways, but the most straightforward is cross-sectional TEM (XTEM). On an XTEM image the flatness of the layers and also the sharpness of the interfaces can be visualized. Figure 1 shows the XTEM image of the sample containing sets of 15 alternate layers of 2-nm-thick germanium and silicon, separated by layers of 10-nm-thick silicon. The interfaces are sharp and the layers are flat. The interface thickness can be estimated to be about 0.8 nm.

2.1.2. High-resolution depth profiling equipment

The concept of the design of a dedicated depth profiling equipment is based on experience gained from ion-milling experiments over more than four decades, in the preparation of thin specimens for TEM [6].

It has been established that hillocks, terraces, pits and cones are the most common surface features generated by ion sputtering. The abundance of these features, their growth rate and the general morphological formation depend strongly on the conditions of sputtering and on the nature of the bombarded material as well [7,8]. Based on TEM experience, it can be concluded that low ion

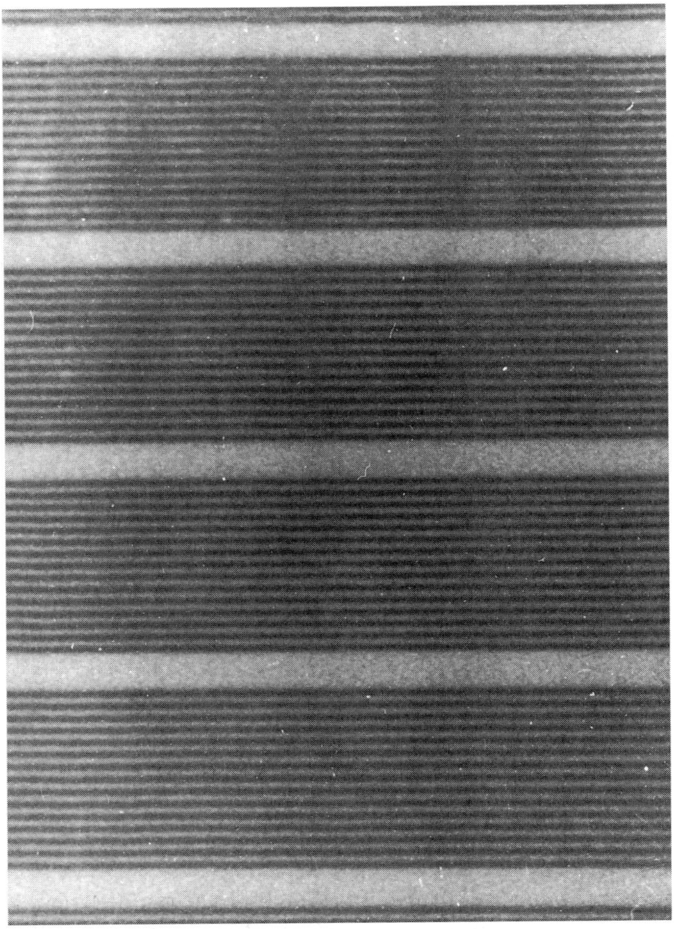

Figure 1. XTEM image of the Ge–Si multilayer structure. The bright and dark layers are silicon and germanium, respectively. The thickness of the thin layers is 2 nm, while that of the thicker ones is 10 nm.

energy and grazing angle of incidence are the necessary requirements for efficient reduction of surface roughening. It should be added that specimen rotation during sputtering was introduced as early as the 1960s [9], having been shown to be essential for the preparation of smooth surfaces. Thus the state of the art in ion milling is the simultaneous realization of low-energy ions, grazing angle of incidence, and specimen rotation [10]. The well-established observations are supported by phenomenological theories [10–12].

It is important to recognize that the same requirements should be met for the preparation of good-quality thin films as for the production of well-resolved depth profiles. In both cases the atomic planes should be removed in a layer-by-layer fashion, i.e., surface roughening should be reduced as much as possible. Good-quality depth profiling started with the introduction of specimen rotation during sputtering [13]. From the first experiment of Zalar there was agreement on the advantages of the rotation of the specimen during depth profiling [14–17].

A diagram of the dedicated depth profiling equipment used here [6, 18] is shown in Fig. 2. It consists of an ion gun, a rotating specimen holder and a cylindrical mirror analyzer (CMA) for AES, positioned in the usual UHV chamber. The holder can be translated along three directions, allowing the sample to be located at the focal point of the CMA. Wobbling of the specimen during rotation is kept within 5 μm. The ion gun has two translational degrees of free-

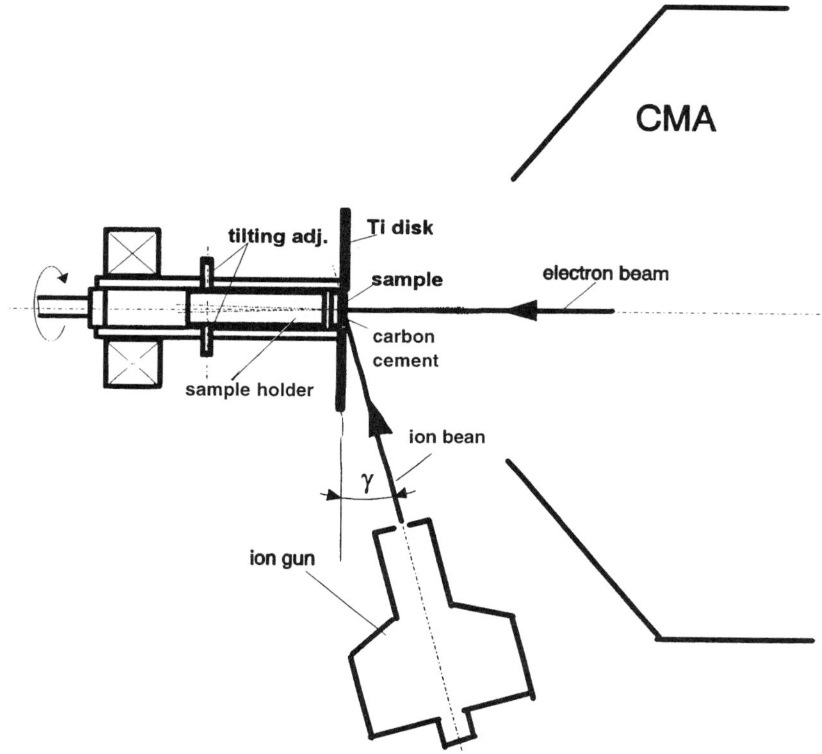

Figure 2. Sketch of a dedicated depth profiling device.

dom, allowing the ion beam to be positioned accurately; its center should be at the center of rotation of the specimen. It has a tilting capability as well.

The ion gun is a water-cooled modified version of the TELETWIN gun, originally constructed for ion thinning [10, 19]. To produce an intense ion beam at low energies, the gun was modified by adding a hot cathode and focusing lenses [18]. The half-widths of the ion beam at 500 eV, at a distance of 50 mm from the front of the gun, were 1 and 6 mm for focused and unfocused beams, respectively. The sputtering rate measured on a pure silicon wafer was 26 μm/h for 3 KeV Ar^+ energy and 60° angle of incidence. This great sputtering rate allowed a grazing angle of incidence to be used. The energy of the ion beam could be varied in the range of 0.2–10 keV, while the angle of incidence was variable from 76° to 89° (with respect to the surface normal). At low ion energies a lower angle of incidence was used. This had little effect on the depth resolution since, in this angular range, it is only weakly dependent on the angle of incidence [20]. On the other hand, the density of the ion current increases with decreasing angle of incidence, resulting in a useful increase in sputtering rate.

The sputtering gas (Xe, or He, but most frequently Ar) is introduced directly into the gun, which is enclosed apart from the ion exit aperture. The pressure in the ion gun is about two orders of magnitude higher than that in the analyzing chamber, where it is lower than 10^{-6} mbar.

2.1.3. Characteristic depth profiles

Figures 3 and 4 show two sections of the depth profile recorded from the specimen containing 2 nm thick Ge and Si layers [21]. The depth scales are measured from the original surface. The specimen was rotated, and the ion energy and angle of incidence were 1 keV and 84°, respectively, as indicated. The depth profile is an oscillating function which never reaches 100% or 0%. This is always the case when the layer thickness and the characteristic length of bombardment-induced broadening, appropriate to the chosen sputtering conditions, are comparable. In this case the amplitude of oscillation changes strongly if the sputtering-induced broadening changes. Obviously, the amplitude is closely related to the depth resolution; the higher the amplitude the better is the depth resolution.

Figure 5 demonstrates the ion-energy dependence of the amplitude of oscillation [21]. The depth profiling started with 1 keV ion energy, was changed to 218 eV and then reverted to 1 keV. It is important to note that the amplitudes of oscillation recorded during the first sputtering at 1 keV were reproduced during the second 1 keV run, and that the amplitude of oscillation increased when the ion energy was decreased. The other important feature is that the new amplitude was reached in a very short time—that is, after the removal of a layer only 1–2 nm thick [20, 21].

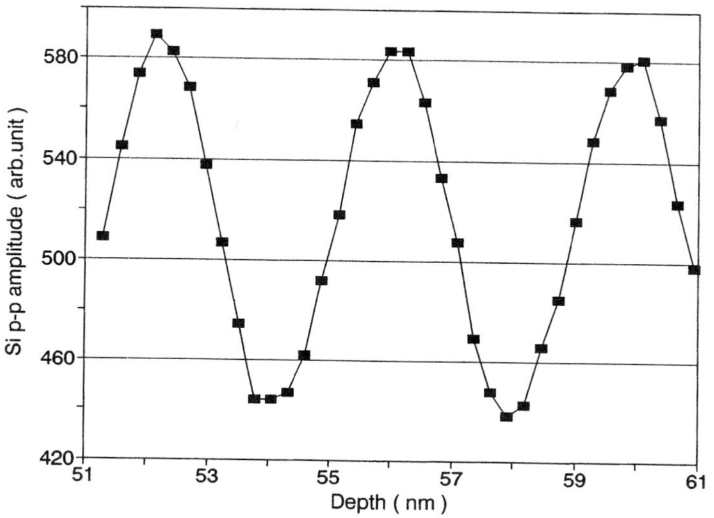

Figure 3. Part of the depth profile recorded from the Ge–Si specimen in Fig. 1. The ion energy and angle of incidence were 1 keV and 84°, respectively.

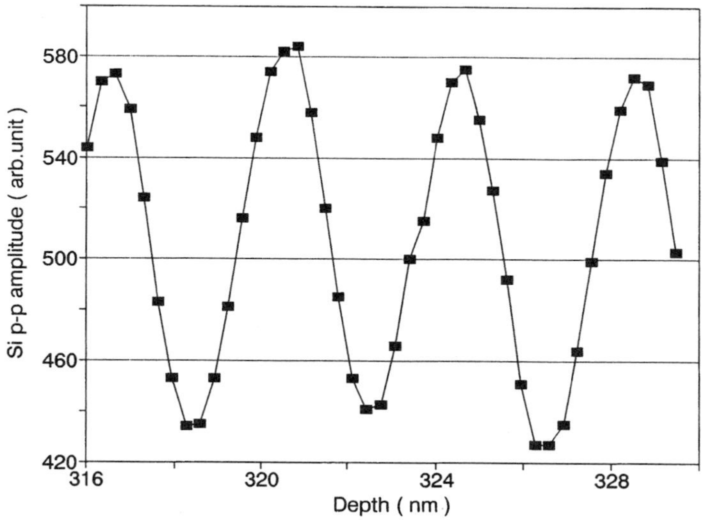

Figure 4. The same as Fig. 3 but after removal of an additional thickness of about 260 nm.

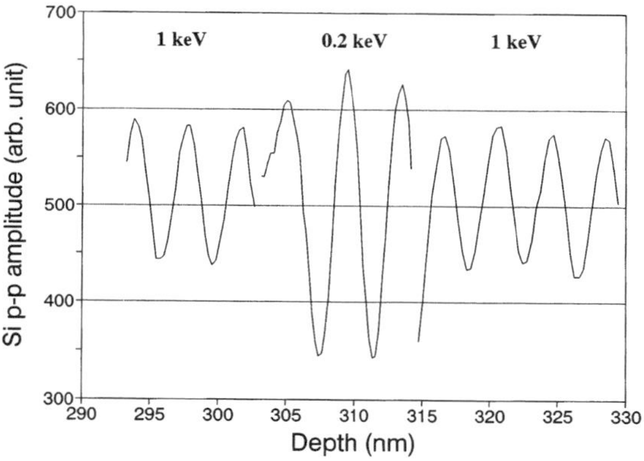

Figure 5. The effect of the variation of ion energy on the amplitude of oscillation. The depth profile was recorded from the specimen in Fig. 1. Angle of incidence is 84°. The relevant ion energies are given in the figure.

The depth profiling of a single interface between two thick layers is of special interest, since there is then no overlap between mixed regions and such a structure is therefore useful for the determination of depth resolution defined as the depth range over which a signal increases (or decreases) by a specified value, say from 16% to 84%, when profiling an ideally sharp interface between two media. Figure 6 shows the depth profile of the transition from thick germanium to thick silicon layers (sputtering conditions: ion energy 218 eV, angle of incidence 84°) [21]. The depth resolution in Fig. 6 can be measured easily to be 2.4 nm. It should be emphasized, however, that such a determination has rather a limited usefulness, unless the mixing can be described by Gaussian broadening.

2.2. Evaluation of Auger Depth Profiles

2.2.1. Sputtering-induced surface roughness

Surface roughening is known to cause alterations in the amplitude of profile oscillation, and its effect should therefore be checked for each individual case [22]. If the amplitude oscillation decreases under constant sputtering conditions, the surface roughening could be expected to determine the depth resolution. On the other hand, a constant amplitude of oscillation with depth would indicate a constant surface roughness. Although the actual roughness was not known, Figs. 3 and 4 show that the amplitude of oscillation did not change when an additional thickness of ≈260 nm was removed, which is encouraging.

The amplitude of oscillation also depends strongly on the energy of the sputtering ions, as illustrated in Fig. 5. The answer to the question as to whether

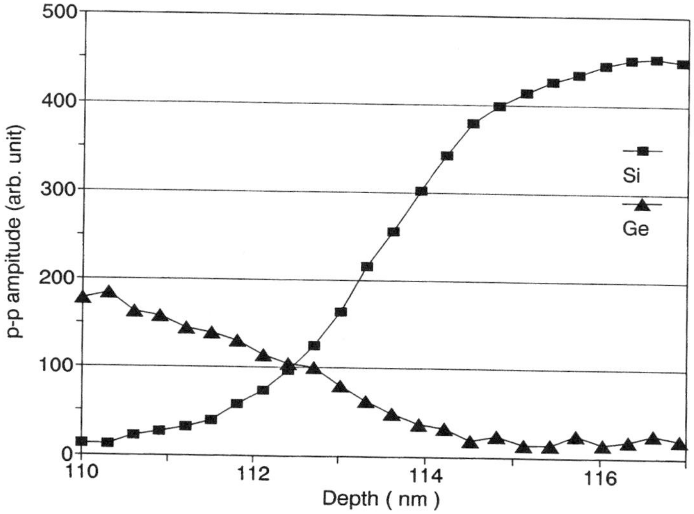

Figure 6. Depth profile of an interface between thick Ge and Si layers. Ion energy was 218 eV and angle of incidence was 84°.

the amplitude of oscillation decreased because of the increased roughening is thus *no*. It is easy to accept this answer if it is realized that, in order to change the degree of roughening, material must be removed from, or added to, the bombarded surface. Since resputtering is negligible in the configuration used, only removal need be discussed. The new amplitude of oscillation was established after removal of a very thin layer of thickness only about 1.5 nm. To develop the degree of roughness which would have been necessary to explain the decrease in the amplitude of oscillation, strongly energy dependent removal of the material from the surface would have to be assumed. Since such a process has not been reported so far [23], this explanation can be disregarded.

In the phenomenological model of Barna [10] roughness is described in terms of surface steps. Within this model an extremely strong dependence of step velocity on energy has to be assumed, in order to explain the rapid growth of roughening features. Since such a dependence was not established here the conclusion is the same as before; a rapid change in surface roughening is not a physically reasonable explanation.

Thus it can be concluded that, at least at ion energies higher than 218 eV (the lowest ion energy used), the contribution of surface roughening is negligible compared to that of ion mixing. It has been shown that by simple depth profiling definite statements on the cause of the bombardment-induced interface broadening can be made. For the case of the Ge–Si specimen analyzed here it can be concluded that under the sputtering conditions used in these experiments surface roughening was negligible compared with other effects that lead to interface broadening.

2.2.2. Intrinsic surface roughness of interfaces

Taking into account the results of the above observations, the intrinsic surface roughness of the interfaces can be estimated; it cannot be greater than the lowest broadening value obtained at the lowest ion energy (i.e., 2.4 nm). This also means that when using higher ion energies, where the amplitude of oscillation is lower (i.e., the broadening is larger), the effect of the intrinsic surface roughness can be neglected compared to the sputter-induced broadening.

2.2.3. Calculation of the surface concentration

In depth profiling it is necessary to measure surface concentrations, which can in principle be calculated from the recorded peak-to-peak heights (p-p) in the differential Auger spectrum. In the evaluation of the Auger spectra the difficulty is that the analyzed volume, which is determined by the electron beam diameter and the escape depth of the Auger electrons, is not of uniform composition. The well-known expressions for the calculation of concentrations from the intensity of the Auger peaks should be applied only for known concentration distributions. Thus the Palmberg expression [24] could be used if the analyzed volume were homogeneous. Another expression has been proposed by Gallon [25] for the case of a homogeneous matrix covered with homogeneous overlayers. An example of the application of the Gallon approach is the evaluation of the Auger spectra during depth profiling of Ge–Si multilayers [21].

For the latter case the silicon Auger peak at 92 eV may be used. The average escape depth of these Auger electrons at that energy is about 0.4 nm [26]. On the other hand, the low-energy germanium peaks are very weak; the doublet at 52 eV is about 15 times less intense than the Si 92 eV peak, while there is an even weaker peak at 89 eV which overlaps with the intense silicon peak; thus neither can be used. The high-energy 1154 eV germanium Auger peak can be measured easily. However, the average escape depth for Auger electrons at the latter energy is about 2.4 nm [26], which is greater than the layer thickness of the multilayer structure. Obviously, the high-energy germanium Auger signal corresponds to a very rough average of the germanium concentration, which rules out that peak too. The surface concentration was therefore determined by comparison of the silicon 92 eV Auger peak measured during depth profiling with the same peak in pure silicon; the ratio of the two peaks, corrected for backscattering [27], gave the concentration. This somewhat unusual concentration calculation procedure was possible here because the layer system contained thick, pure silicon layers in which the germanium Auger signal was nonexistent.

As will be shown, using the D-TRIM code, the measured depth profile can be simulated reasonably well. The simulation gives automatically the in-depth distribution of the elements and thus the surface concentration as well. Having the in-depth distribution, the Auger signal can be simulated easily and then entered in the expressions for calculation of the concentration, as shown in Fig. 7. From the above it can be concluded that by using only Auger peaks of low energy the usual concentration calculation methods can be applied.

2.3. Atomic Mixing

In addition to the surface roughening there are other processes, such as bombardment-induced segregation, preferential sputtering, etc., occurring during sputter depth profiling, which may contribute to the sputter-induced broadening of an originally sharp interface. In order to study any one of them separately experimental conditions must be chosen in which the broadening effects of all the others may be neglected. We now consider the above-mentioned processes in the case of the Ge–Si multilayer system.

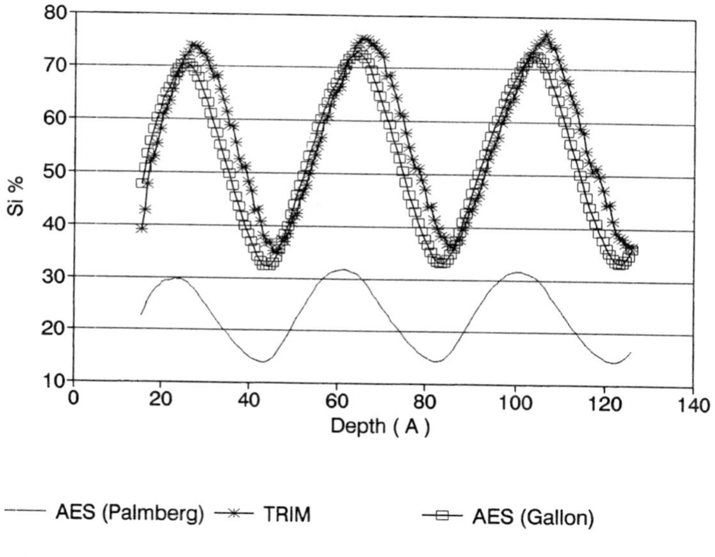

Figure 7. Simulated depth profiles in terms of concentration versus sputter depths, for the Ge (2 nm)–Si (2 nm) specimen. (*) the real depth profile, that is the concentration of the outer layer, (□) and (–) concentrations calculated from the simulated Auger intensities using the Gallon and Palmberg formulae, respectively. Ion energy was 500 eV and angle of incidence was 84°.

Preferential sputtering. The experimentally determined sputtering yields of pure Ge and Si for normal incidence and 1 keV Ar^+ are ~1.5 and ~0.6, respectively [23]. This large difference suggests the possibility of preferential sputtering in a Ge and Si mixture. The ratio of the sputtering yields was determined by using Ge and Si layers of known (20 nm) thickness at an angle of incidence of 83°. The ratio of the sputtering times necessary to remove the layers was found to be 0.96 ± 0.05, which was independent of energy in the range 0.3–1.5 keV. This means that the sputtering yields were almost equalized, and thus no preferential sputtering is expected when using grazing angles of incidence.

Bombardment-induced segregation. To check the possibility of segregation the high- and low-energy Auger peaks of both Si and Ge, can be used, since they probe different depths in the sample. Changes in the relative height ratios of high- and low-energy peaks during sputtering were not observed. Thus sputter-induced segregation could be ruled out.

Sputter-induced surface roughening has also been ruled out based on the discussion in Section 2.2.1. As a consequence, sputter-induced broadening of the interfaces during depth profiling, at grazing angle of incidence and low ion energy, and with a rotating specimen, is governed entirely by ion mixing. Analysis of the depth profiles under these conditions can therefore provide information about the ion-mixing process.

2.3.1. Energy dependence of ion mixing

The dependence of the depth profiles, related directly to ion mixing as deduced above, on the ion energy and angle of incidence in Ge–Si specimens of various structures has also been studied, but since the angular dependence is rather weak, the results given are concerned only with the energy dependence of the mixing [20, 21].

Figure 8 shows the amplitude of oscillation versus the square root of the ion energy for two Ge–Si multilayer systems, consisting of alternate Ge and Si layers 2 nm and 3 nm thick, respectively. It is clear that by decreasing the sputtering energy the amplitude of oscillation increases. The slopes of the curves are different; the thinner the layer the steeper the slope. This can be explained by the greater overlap of the mixed regions in the case of the specimen with thinner layers. Similar dependences have been established for other multilayer (Ni–Cr, Ta–Ni, Mo–Si) systems [28]. Comparable results on ion-energy dependence could be deduced by depth-profiling single interfaces between either thick layers or layers of the delta-function type.

2.3.2. Interpretation of the depth profiles

It is important to remember that a depth profile is an instantaneous function of surface concentration versus sputtering time. The thickness of the instantaneous

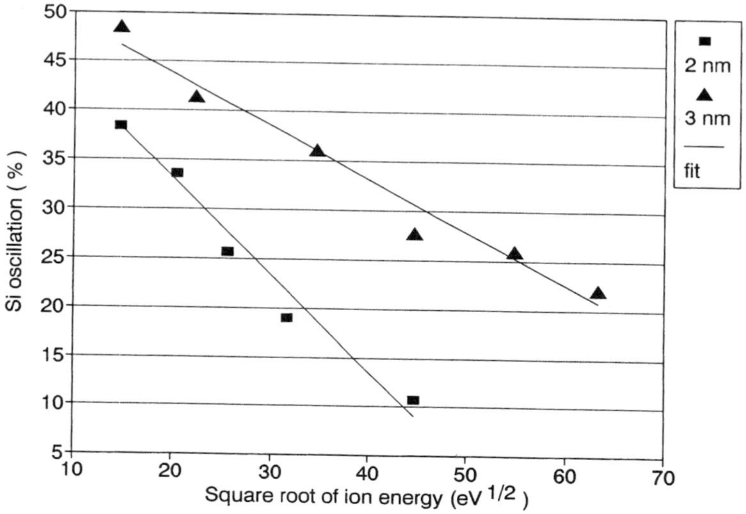

Figure 8. The amplitude of oscillation versus square root of ion-energy dependence for two multilayer systems consisting of silicon and germanium layers 2 nm and 3 nm thick, respectively.

surface layer, Δz, actually measured is different for the various depth profiling techniques. It has been seen that for the Ge–Si system, if only the low-energy Si Auger peak is used in the concentration calculation, Δz is estimated to be about 1.5 nm. With XPS Δz is larger since in that technique electrons with greater kinetic energy are used for analysis. When SIMS is used, Δz is at its lowest since the majority of the emitted fragments originate from the very first layer [29].

If the sputtering rate is known, the sputtering time can be converted to depth. In this way the function of the thickness of the amount removed versus instantaneous surface concentration function may be obtained. This conversion does not solve the original problem, because it is not the actual concentration distribution that is measured but a modified one, i.e., affected by sputtering.

To study atomic mixing, the in-depth distribution of concentration at a given time, $c(z,t_0)$, should be established after a given fluence of ions. Obviously, with depth profiling techniques one cannot determine the function $c(z,t_0)$. If a theoretical model that describes the mixing process is applied, the depth profile may be simulated. Should there be reasonable agreement between simulated and measured depth profiles, the model is acceptable.

In Auger depth profiling there are two analytical models that are used frequently to describe the broadening which, in this case, is caused by ion mixing. One of them supposes that the mixing can be described by Gaussian broadening (the parameter is the half-width, σ, of the Gaussian function) [22], while the Liau model [30, 31] assumes a constant and homogeneous mixed layer thickness. The applicability of the models separately and then combined was checked, but no agreement with experiment could be reached.

The ballistic model developed by Biersack [32], in which simulation of depth profiles is carried out using the dynamic TRIM code (T-DYN (1991), version 4.0) gave better agreement [33]. Several features of the depth profiles, e.g., the dynamics of the changes resulting from variation of the ion energy, as well as the slope of the dependence of amplitude of oscillation on the square root of ion energy, were successfully reproduced [20, 21]. The absolute value of the amplitude of oscillation was different, however, as demonstrated in Fig. 9. Since the ratio of the intercepts does not depend on the particular structure studied, a "normalization" procedure is possible. Knowing the experimentally determined slope of the oscillation versus square root of energy curve one can calculate the ion energy for the whole energy range at which the simulation gives the same amplitude of oscillation as the experimental value. Using the same normalization procedure the TRIM simulation is then also in good agreement with the measurements carried out on single interfaces and delta-function layers. Thus it may be concluded that the TRIM simulation, which considers only ballistic ef-

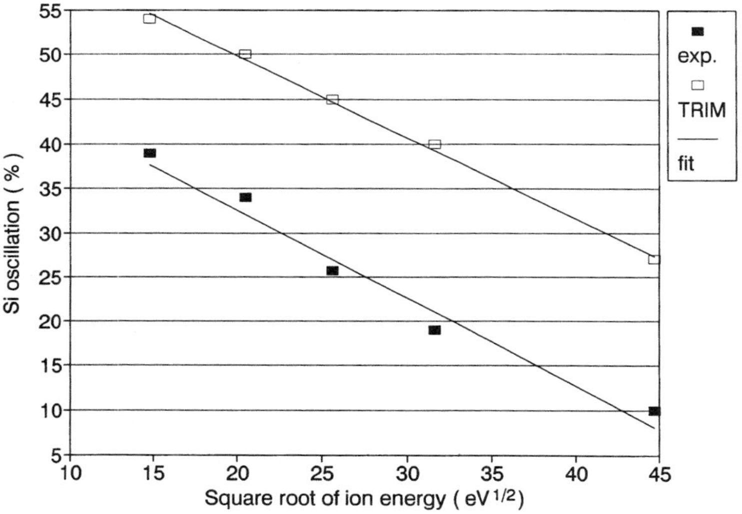

Figure 9. Comparison of the measured and simulated (using D-TRIM simulation) amplitude of oscillation versus square root of ion-energy dependences for specimens containing 2 nm-thick Ge and Si layers.

fects, seems to give a reasonable description of the sputter-induced interface broadening.

Summarizing this section it should be emphasized that, by using the sputtering arrangement described above, well-resolved depth profiles can be recorded. The depth profiling of a variety of structures is a reliable approach for the checking of the applicability of various models of ion-induced interface broadening.

3. PARTICLE-BEAM-INDUCED CHEMICAL ALTERATIONS

In this section the attention of the reader is called to some particular problems the surface analyst may encounter in analyzing thin surface films treated by various particle beams, essentially in the sputtering regime (from 1 to 5 keV). In the following subsections some representative studies will be presented in which it is demonstrated how the effects of particle bombardment on surface chemistry can be evaluated, mainly by XPS, for various inorganic compounds (e.g., TiN layers, metal oxides, Cr–O–Si cermet films) and polymeric materials (e.g., polysulfone, polyimide and poly(organosiloxanes)).

A minor problem specific to the analysis of particle-beam-treated thin surface films may arise if the conditions of the film preparation or modification do not coincide with those of the analysis. In particular, the sample position that is optimal for analysis may not always be optimal for the in situ ion beam treatment. If ignored, this circumstance may lead to the preparation of laterally inhomogeneous samples and, as a consequence, to erroneous surface analytical results.

Another specific problem may arise when the thickness of the modified layer is less than the sampling depth of the analysis. This results in a vertical inhomogeneity, which, if neglected, can lead to distorted results.

Spurious effects can also occur due to space charge, as when insulating materials are treated by charged-particle beams, and may include electric-field-enhanced diffusion, sample degradation by surface flashover, and peak shifting and broadening during XPS analysis [34, 35].

The presence of adventitious carbon contamination can have a drastic influence on the quantitative evaluation of XP spectra. The amount of the contamination will vary with the different treatments, and both removal and buildup can occur, the latter usually being faster on the freshly treated (active) surface than on the untreated one.

It is advisable, therefore, to take the following measures:

1. Ensure the lateral homogeneity of the surface treatment over the area to be analyzed. To check that, an ion-beam-induced secondary-electron imaging technique may be useful. A simple alternative may be that of a preliminary

"blank" experiment, in which separate dots of a fluorescent material (e.g., a phosphor) are located on the sample holder, instead of the real sample, in a "five-points-of-the-dice" arrangement. After switching on the ion beam, the position of the irradiated area can be adjusted by visual observation of the variations in intensity of fluorescence of the dots.
2. To minimize the unwanted in-depth inhomogeneity of the surface layer to be analyzed, prepare a layer with a thickness greater than the sampling depth of the analysis. To do that, choose an appropriate particle energy characterized by a mean projected range (calculated by the TRIM-code [36, 32]) greater than the sampling depth of the analytical method applied.
3. To diminish the buildup of a strong space charge on the sample to be bombarded or to be analyzed, the use of an electron flood gun during both ion bombardment and XPS analysis [35, 37] should be considered. The use of atom rather than ion beams may either solve or alleviate the problem, since then the surface charging is greatly reduced. (Any positive charging of dielectric surfaces in that case arises essentially from secondary-electron emission.)
4. To eliminate the problem of the varying amount of carbon contamination, use an XPS quantification method that employs variable correction factors (e.g., [38]).

3.1. Thin Surface Films of Inorganic Compounds

3.1.1. TiN layers

Outline of the problem

TiN, owing to its unique combination of properties, is used extensively as a coating on structural materials to improve their resistance to heat, wear and corrosion. It is also widely used as a decorative coating on various surfaces. Another important field of application of TiN is in microelectronics, as diffusion barriers in VLSI-MOS integrated circuits.

Recent extensive investigations of the chemistry and structure of TiN layers and coatings have revealed that their physical, electrical, optical as well as protective properties depend not only on their crystalline structure but also on small deviations from stoichiometry, i.e., on the concentration of vacancies and of interstitials of the constituents [39]. Variations in the stoichiometry are denoted by the formula $TiN_{1\pm x}$, which indicates that the N content may be smaller or greater than unity. The variation in the color of the films with increasing N content (from silvery through golden to dark red) also reflects the changes in the chemical structure and composition. Furthermore, the variability of the valence state of titanium provides a large degree of freedom for the formation of a range of compounds or phases in the Ti-N systems [40]. The almost unavoidable presence of oxygen and carbon contamination further extends this range by allowing $TiN_xO_yC_z$ type solid-solution formation [41].

Thus, detailed analysis of the chemical structure and composition of TiN layers is of prime importance. The obvious choice for this analysis is XPS, as reflected in the numerous papers published recently [42–48].

The bulk of the literature shows that the energy of the Ti 2*p* doublet exhibits large chemical shifts (up to 4 eV) and considerable alterations of peak shape due to changes in the chemical environment around the Ti atoms, e.g., due to partial oxidation. This has led to the erroneous conclusion that chemical-state determination can be performed in a quantitative manner with relative ease by using a simple peak-synthesis procedure on the Ti 2*p* envelope. As will be shown, a complex evaluation procedure should be applied to the quantitative analysis and the assignment of the chemical states of Ti, making the characterization of TiN more reliable.

It will also be shown that bombardment by Ar^+ and N_2^+ leads to significant changes in composition and short-range chemical structure which are reflected in peak-shape changes of the Ti 2*p* and N 1*s* lines. The examples to be cited provide a practical guide to the analysis of TiN layers, with attention drawn to special features easily overlooked or misinterpreted when applying the usual XPS analysis approach.

Samples

One TiN layer, polycrystalline PVD, was prepared by dc bias reactive sputtering onto (100) oriented Si substrates at 570 K [42], while another was obtained by dc plasma nitriding of chemically pure titanium at 970 K [43]. Both layers had a columnar structure with preferred (111) orientation. The PVD layer proved to be nearly stoichiometric when examined by RBS, while TiN, Ti_2N, α-Ti and traces of TiO_2 phases were revealed by XRD in the sample prepared by the dc plasma method.

For reference purposes a stoichiometric single-crystalline TiN layer was prepared by reactive ($Ar^+ + N_2^+$) sputtering onto a single-crystal MgO substrate as described in [49]. The composition and the chemical state of this reference sample after wet chemical etching (HF:H_2O=1:20, 5 min), followed by a moderate (1 keV, <1 μA·cm^{-2}, 5 min) Ar^+ cleaning were taken to be representative of stoichiometric TiN virtually unaffected by ion sputtering.

XPS measurements

The position of the Ti $2p_{3/2}$ line, determined from the stoichiometric single-crystal TiN layer was found to be 454.7 ± 0.1 eV [42]. The latter value was derived by referencing the spectra to the CH_x type C 1*s* line at BE=284.6 eV, corresponding also to the position of the Au $4f_{7/2}$ line at BE=84.0 eV. The position of the Ar $2p_{3/2}$ line of the implanted argon was found to be 242.6 ± 0.2 eV, which can also be used for energy referencing after Ar^+ cleaning or during depth profiling. For this sample the energy of the well-defined N 1*s* signal (FWHM=1.6 eV) was 396.7 ± 0.1 eV and the separation between Ti $2p_{3/2}$ and N 1*s* was 58.0 ± 0.1 eV, in agreement with an earlier finding [46]. The area ratio of the 1/2 and 3/2 components of the Ti 2*p* peaks turned out to be less than the expected 1:2 value when a Shirley-type background was subtracted [48]. For this reason only the 3/2 components were used for assessing the ratios of the various chemical states.

Synthesis of the Ti 2p envelope

It has been shown recently that the observed shoulders on the high-binding-energy sides of the Ti $2p_{3/2}$ and Ti $2p_{1/2}$ peaks are inherent features characteristic of stoichiometric TiN [46, 48–53]; they were previously often mistaken for higher-valence (e.g., oxidized) states of Ti, which affected the chemical-state assignment and the quantitative analysis.

The experimentally established existence of loss peaks in Ti 2*p* spectra [50–53] provides a straightforward explanation for the observed high-BE shoulders on the Ti 2*p* doublets that have been interpreted differently in a number of earlier works (e.g., [44, 47, 54]). In peak synthesis, the method of Strydom and Hofmann was followed [50], but applied in a slightly different way. They included only the nearest loss peak (L-1) with a separation of 1.6 eV, whereas the second nearest loss peak (L-2) at a separation of 3 eV from the main Ti-N component at 454.7 eV was also included. The energy separation between the 3/2 and 1/2 components of the doublet and between the associated loss peaks was set at 5.9 eV [44]. In order to obtain an adequate fit a third peak with a separation of about 4 eV had also to be included. Such a position does in fact correspond to that of the TiO_2-type oxide peak (at 458.8 eV), the presence of which could not be completely ruled out, but it would hardly occur in significant quantities in the single-crystal layer studied here. As a consequence, the major part of this peak must arise also from a loss process and it is denoted by L-3.

Synthesis of the N 1s peak

Strydom and Hofmann [50] have also suggested that loss peaks associated with the N 1*s* transition should be expected, with energy separations similar to those of Ti 2*p*. Accordingly, two loss features, LN-1 and LN-2, with energy separations of 1.6 eV and 3.0 eV, respectively, were included in the peak synthesis of the N 1*s* lines. The line component energies used for the peak synthesis are summarized in Table 1. The loss peak intensities determined experimentally from the stoichiometric single-crystal reference sample are also included.

Effects of Ar^+ and N_2^+ bombardment on the PVD TiN layer

This layer was subjected to sequential Ar^+ and N_2^+ bombardment, following which characteristic changes in the shapes and the intensities of the Ti 2*p* and

Table 3.1. Binding Energies (BE, eV), Doublet Separations (eV) and Relative Intensities (%) of the Main Components and Loss Peaks Used for Peak Synthesis

Line	State	BE	Doublet separation	Relative intensity
Ti 2*p*	TiN	454.7	5.9	100
	L-1	456.3	5.9	50
	L-2	457.7 ± 0.1	5.9	15
	L-3	458.8 ± 0.15	6.0	~ 6
N 1*s*	N-1	396.7 ± 0.1		
	N-2	395.8 ± 0.2		
	LN-1	398.0 ± 0.2		~ 10
	LN-2	399.7 ± 0.2		~ 5
O 1*s*	O_{Ti}	529.8 ± 0.2		
	O_C	533.0 ± 0.3		
C 1*s*	C_{Ti}	~ 281.8		
	C_H	284.6		
	C_O	~ 286.1		

Source: Ref. 43.

N 1*s* lines were observed. In Fig. 10 are shown the Ti 2*p* lines after 20 min Ar^+ and then after a further 20-min N_2^+ bombardment, with normalized intensities in (a), and with the difference spectrum in (b). It is clear that a new doublet developed, with Ti $2p_{3/2}$ and Ti $2p_{1/2}$ components at about 456.3 and 462.2 eV, respectively. Figure 11 shows the result of peak synthesis of the Ti 2*p* envelope after (a) Ar^+ and (b) N_2^+ bombardment achieved by the fitting of four doublets to the envelope; that at 454.7 and 463.6 eV corresponds to stoichiometric and also substoichiometric TiN. The other three are the loss peaks given in Table 3.1. It was observed that the intensity of the peak at 456.3 eV corresponding in energy to L-1 was always greater than 50% of the main Ti-N component for all N_2^+ bombarded samples. This surplus intensity was related to the appearance of an additional peak coinciding in energy with the loss peak and assigned to Ti in the superstoichiometric TiN_{1+x} state [43, 45].

Figure 12 shows the change in the N 1*s* line shape of the PVD sample after 20-min Ar^+ bombardment as compared to that after a subsequent 20-min dose of N_2^+. The broadening of the N 1*s* peak after N_2^+ bombardment is obvious. Figure 13 shows the results of the peak synthesis of the N 1*s* lines after (a) Ar^+ and (b) N_2^+ bombardment. The main peak after Ar^+ bombardment was at 396.7 eV. After N_2^+ bombardment, in addition to the latter peak, a new high-intensity peak at 395.8 eV appeared. The major difference between the two spectra is the

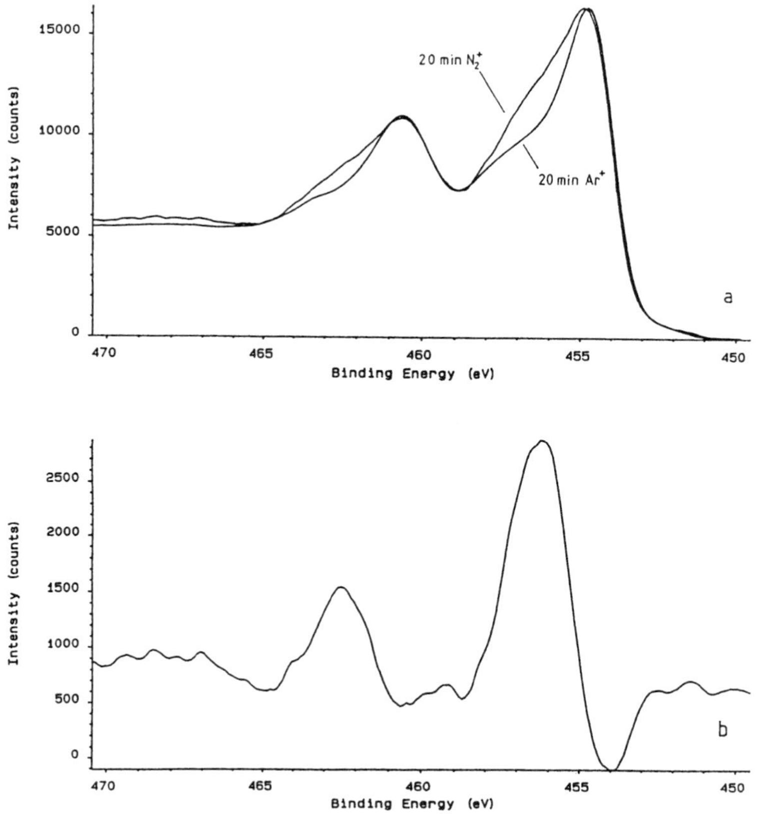

Figure 10. Comparison of the Ti 2*p* lines developed after 20 min Ar^+ and then after a further 20 min N_2^+ bombardment; (a) normalized intensities and (b) the difference spectrum. (From Ref. 42.)

high intensity of the N-2 component for the N_2^+-bombarded sample. This latter peak is almost absent in the Ar^+-bombarded state. Table 3.2 shows the surface chemical composition of the PVD TiN sample after Ar^+ and N_2^+ bombardments. Angularly resolved experiments showed that both depletion and increase, respectively, of N were more pronounced in the outer atomic layers.

Oxygen and carbon are the most common bulk and surface contaminants of titanium and TiN. The peak at 529.9 ± 0.2 eV corresponding to oxygen chemically bonded to Ti [54, 55] usually represents less than half of the O 1*s* envelope. Two other peaks found at about 531.4 and 533.0 eV can be assigned to

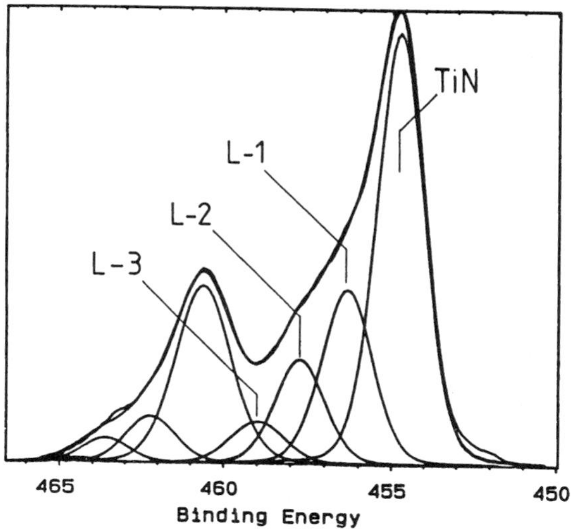

a

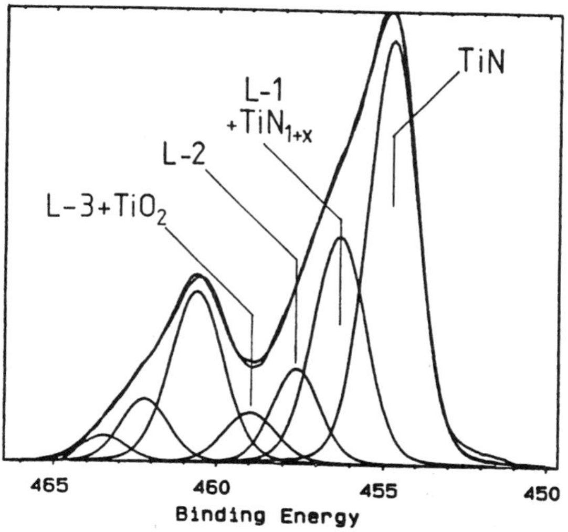

b

Figure 11. Comparison of experimental and synthesized Ti 2*p* envelopes (a) after Ar^+ and (b) after N_2^+ bombardment of the PVD TiN layer. The fit was achieved by using four doublets with the line energies from Table 3.1. (From Ref. 42.)

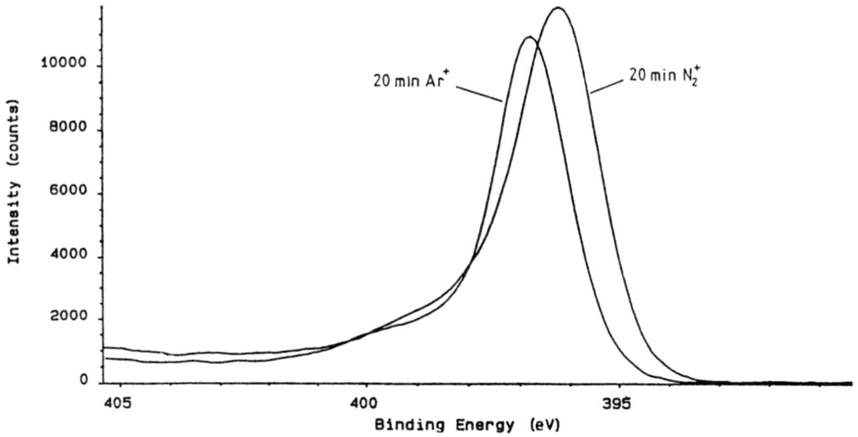

Figure 12. The change in the N 1*s* line shape of the PVD TiN sample after 20 min Ar^+ bombardment compared to that after a subsequent 20 min N_2^+. (From Ref. 42.)

adsorbed water and to C–O type surface species, respectively. In agreement with this, part of the surface carbon impurity was of the C–O–C type, and in some cases carbide type C 1*s* at 281.8 eV could also be detected. The substantial (twofold or even higher) increase in the surface concentrations of these impurities observed in angularly resolved experiments implies that they originated mainly from the residual Ar or N_2 atmosphere and were adsorbed on the freshly ion etched surfaces during the acquisition of spectra.

The N/Ti ratios during Ar^+ bombardment are shown in Fig. 14. Although there is a scatter in the experimental points, it is clear that N depletion increased slightly with ion dose. A considerably higher depletion in N observed at a 60° takeoff angle shows that N was lost preferentially from an outer surface layer of about 1 λ thickness.

Analysis of TiN layers obtained by plasma nitriding

Quantitative analysis of the plasma-nitrided titanium showed that the TiN layers so produced were characterized by a higher oxygen content [43, 45]. If the chemical states of Ti in these layers are to be determined by analysis of the chemical shifts of the Ti $2p_{3/2}$ components, then the data compiled in Table 3.3 will be needed. From the cited data it is obvious that TiO cannot be separated from TiN, and that there is overlap between the TiN_xO_{1-x}, TiN_xO_y and TiN_{1+x} states and the L-1 type, between Ti_2O_3 and the L-2 type, and between TiO_2 and the L-3 type, loss peaks (Table 3.1). An attempt to estimate the relative amounts

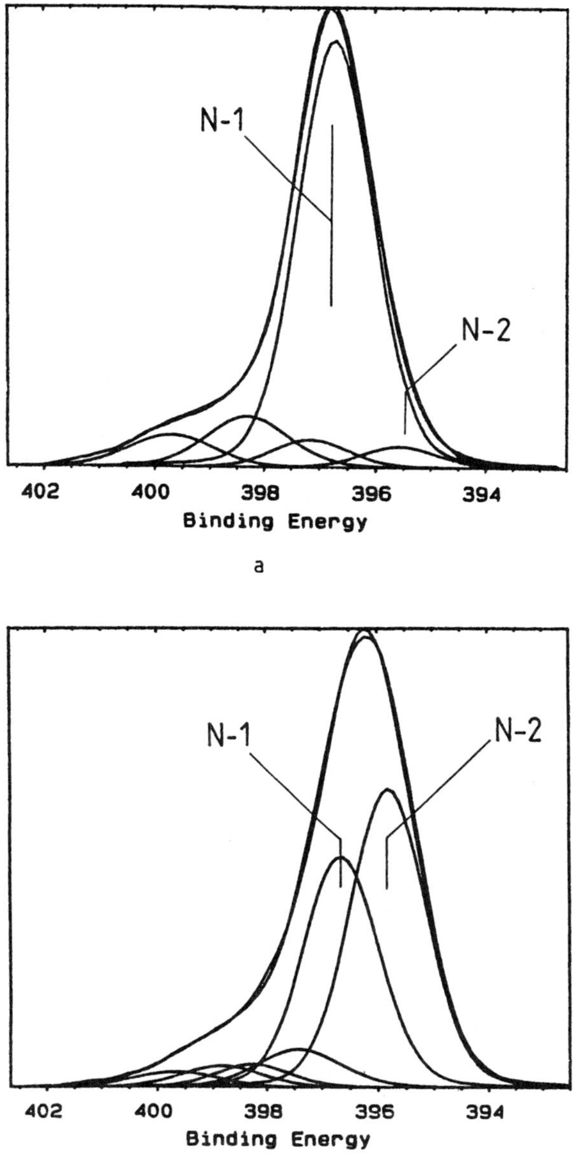

Figure 13. Synthesized N 1*s* lines from the PVD TiN sample (a) after 20 min Ar^+ and (b) after a subsequent 20 min N_2^+ bombardment. (From Ref. 42.)

Table 3.2. Chemical Composition (atomic ratio) of Ion-Bombarded PVD TiN Layers

Ion	Energy (keV)	Time (min)	Tilt (°)	Ti	N
Ar^+	2.5	10	0	1.00	0.98
			60	1.00	0.80
N_2^+	2.5	20	0	1.00	1.21
			45	1.00	1.48
			60	1.00	1.84

Source: Ref. 42

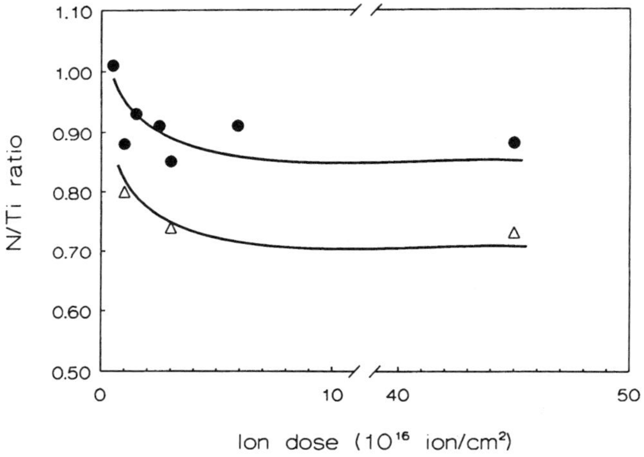

Figure 14. The change in the N/Ti ratios versus Ar^+ dose from the PVD TiN sample at 0° and at 60° takeoff angles. (From Ref. 42.)

of various chemical states in such TiN layers, based on the peak synthesis procedure described, using the Ti 2*p* and N 1*s* envelopes and the overall Ti:N:O ratio, has been described for a series of samples in [43]. The presence of relatively large amounts of N^{2-}-type nitrogen in the surface layers of all plasma-nitrided samples is another characteristic feature of this type of layer formed under high-current-density N_2^+ bombardment of the dc plasma [43].

Substoichiometry in TiN

The effect of preferential sputtering of N by Ar^+ has not been fully recognized so far, although suggestive reports in the literature can be traced back to early 1980s [44, 49]. In the majority of cases the findings were either ignored or non-

Table 3.3. Binding Energies (BE, eV) of the Ti $2p_{3/2}$ Line in Different Chemical States of Ti

State	BE	Reference
Ti^0	453.9–454.3	44, 56, 48, 57, 58, 51, 59
	453.8 ± 0.2	43
	453.6	44
TiN	454.5–455.3	44–46, 48, 57, 50]
	454.7	47
	454.7 ± 0.1	42, 43
Loss peak	456.3	50, 52
TiN_xO_{1-x}	455.4	58
TiN_xO_y	456.9–457.3	58
TiN_{1+x}	456.9	45
TiO	454.9	56
	455.2	58
Ti_2O_3	457.0	58
	457.1	55
	457.5–457.7	51, 52
TiO_2	459.3	51
	458.9	58
	458.8	55
	458.7	52
	458.0	50
	457.9	47

Source: Ref. 43.

preferential sputtering was assumed [60]. It must be admitted that the amount of N lost is rather small and can be easily overlooked, especially since it is not directly reflected in changes in the Ti $2p$ peak shape. The present findings appear to agree with those of Sundgren [49] and the effect is convincingly demonstrated by data obtained at a 60° takeoff angle, which reveal a more pronounced N loss in the outer atomic layers (Table 3.2).

Superstoichiometry in TiN

After N_2^+ bombardment N/Ti ratios much greater than unity were observed. The existence of low superstoichiometry (N/Ti=1.1–1.2) may be treated as a deviation within the limits of the original rock salt TiN structure [39], but for high superstoichiometry (N/Ti>1.3) the existence of a new TiN_2-type phase has been predicted [61, 62]. New nitride phases with suggested compositions of Ti_3N_4, Zr_3N_4 and Hf_3N_4 have recently been synthesized and characterized [63, 64]. The high N/Ti ratios (Table 3.2) obtained here by N_2^+ bombardment are in agreement with the latter findings. The appearance of N^{2-}-type N can also be assigned to such superstoichiometry, which is related to a new chemical structure with short-range order around the Ti atoms, as indicated by a new Ti $2p_{3/2}$ peak at 456.6 eV. These shifts of Ti 2*p* to higher BE and the related N^{2-}-type N 1*s* to lower BE, can be interpreted as an increased charge transfer from Ti to N, i.e., as the formation of a compound with ionic-type bonds. This is also in agreement with the measured optical band gap of 0.93 eV and the insulating properties found for Ti_3N_4 [64].

There are only a few reports in which the appearance of a second nitrogen peak on the lower BE side of the N 1*s* line, with an energy separation between 0.9–1.4 eV is mentioned [45, 47, 54]. In [54] the peak found at 395.8 eV correlates with the Ti 2*p* component for TiN_xO_y. In [47] the peak at 395.4 eV was assigned to a nonstoichiometric nitride. Only Miyagi et al. [45] explicitly related the low-BE shoulder to super stoichiometric TiN_{1+x}, in agreement with findings just described.

It has been established repeatedly that Ar^+ bombardment results in preferential loss of nitrogen from stoichiometric TiN layers. On the other hand, N_2^+ bombardment considerably increases the surface N content, with the excess N being found only in the outer atomic layers. Lastly, it has been demonstrated that the chemical states of Ti in various $TiN_{1\pm x}$ samples with low O content can be determined with confidence by using the peak synthesis procedure described here, including the proposed set of loss peaks.

3.1.2. Metal oxides

Metal oxide films for diverse applications can be produced by various PVD, CVD, PECVD or combined methods. (Oxide films unintentionally formed on metal surfaces are addressed in other chapters.) With the increase in the sophistication of optical devices for military, as well as in commercial, applications highly elaborate PVD techniques have been developed for growing oxide films. Evaporation or sputtering in combination with ion-assisted deposition and reactive ion plating on biased substrates can produce high-density single or multilayer coatings of predetermined optical properties. Among the latter control of the refractive index is of prime importance since it defines reflective or antireflective properties. The index profile may also determine the efficiency in optical waveguide applications. Highly reflective coatings are used in car mirrors or sun glasses, while antireflective coatings are increasingly used for ophthalmic lenses. A fast-developing field of application of thin oxide layers (mostly SiO_x) on polymer films is in the food packaging industry. Plasma-enhanced CVD has proven to be an efficient method for the continuous production of coatings for large areas [65].

Hence, it is obvious that energetic particle beams play an important rôle in the preparation of oxide films. On the other hand, it is well known that the consequent particle-surface interactions may alter the stoichiometry and the optical properties of the oxides. As a result, as-grown oxide films deposited from stoichiometric source materials are usually oxygen deficient unless excess oxygen is introduced. This is why the effect of ion bombardment on the composition and chemical states of the metal constituents are of major concern in the characterization of the oxides. For such characterization XPS-XAES is an obvious choice. Electron-excited AES, although widely applied, is less reliable, since it is known that use of energetic electron beams may lead to the decomposition of oxides (e.g., SiO_2, TiO_2 [66, 67]). The other drawback of AES is that the extraction of chemical information on the changing valence state of the metallic component is not straightforward.

Ion-beam-induced decomposition of oxides has been the subject of extensive investigation for decades [68–72]. Conclusions concerning the extent of oxygen loss and also the development of corresponding reduced states are, however, rather controversial. This is due partly to the lack of reliability of the applied measuring techniques, but to a greater extent to the differences in composition and chemical structure of the oxides prepared by various PVD and CVD techniques, or by anodic oxidation [69, 72]. In some cases the oxides are claimed to be stoichiometric *a priori* without any convincing evidence [72].

The intrinsic behavior of oxides under ion irradiation with the usual sputtering conditions (e.g., 2–5 keV Ar^+) has been addressed recently by studying stoichiometric single-crystal targets or well-defined bulk-phase materials [73–76]. The extent of the reduction, as established by the oxygen/metal ratio from XPS measurements, was confirmed by that calculated from the initial and reduced state(s) of the metallic components developed during ion bombardment [75]. Reliable chemical shift data for reduced components are available for a number of multivalent oxide-forming metals [77, 78], and can be used in peak synthesis.

Figure 15 (top) shows the Ge 3*d* in GeO_2 peak before and after Ar^+ bombardment (2.5 keV, 30 min, i.e., in steady state), while in Fig. 15 (lower) the results of peak synthesis of the same peak after bombardment are given. The synthesis was carried out using the method described in the literature [79], in which all Ge oxidation states Ge^{4+}, Ge^{3+}, Ge^{2+} and Ge^+ were used. A binding energy of 32.8 eV was taken for Ge^{4+}, with shifts of 0.9 eV between each successive reduced state. The stoichiometric amount of oxygen so derived corresponds to $GeO_{1.37}$, in good agreement with the value $GeO_{1.34}$ derived from quantification based on the O 1*s* and Ge 3*d* areas shown in Table 3.4 [75]. The table sets out the compositional changes observed after both Ar^+ and N_2^+ bombardment. Note that the oxides of B, Al, Si, Ge, etc., reported to be fairly stable (reported in [72]) are all reduced to a measurable extent. The remarkable similarity in the anion-to-cation ratios (O/M after Ar^+ ion treatment and (O + N)/M

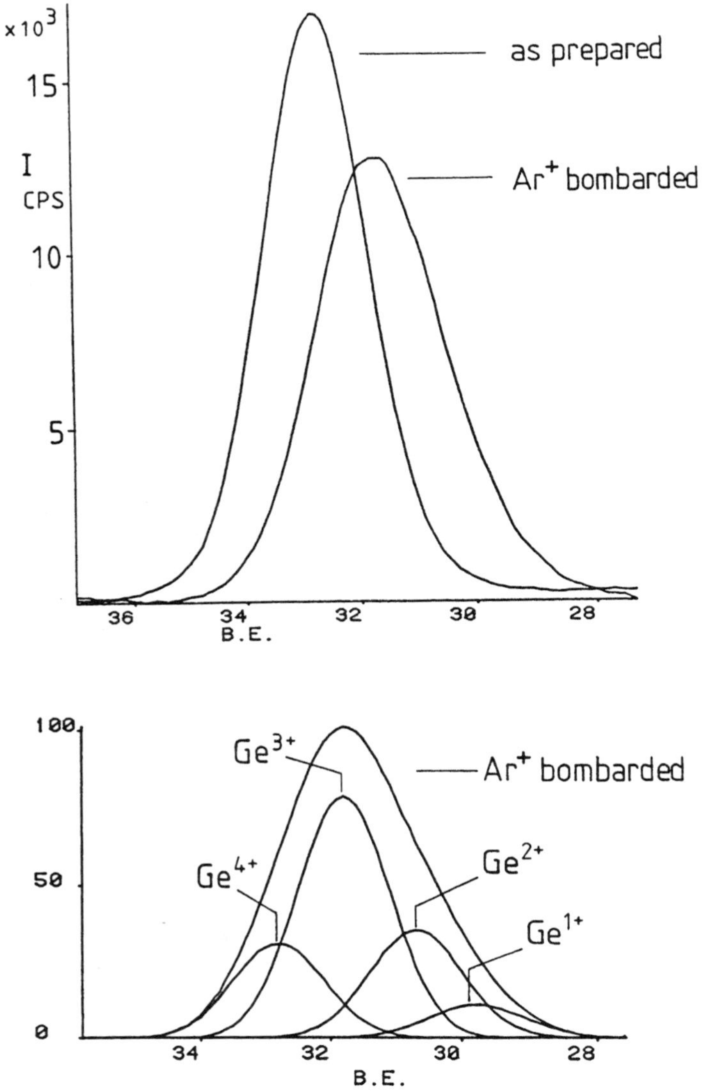

Figure 15. The Ge 3*d* peak before and after Ar^+ bombardment (2.5 keV, 30 min, i.e., in the steady state) of pelletted GeO_2 powder (top), and the result of the peak synthesis for the Ar^+-bombarded state (bottom). (From Ref. 74.)

Table 3.4. Surface Composition of Oxides after Bombardment by Ar^+ and N_2^+

Oxide	State[a]	Composition	
		Ar^+	N_2^+
B_2O_3	am.	$B_2O_{2.76}$	$B_2O_{2.47}N_{0.26}$
Al_2O_3	s.c.	$Al_2O_{2.74}$	$Al_2O_{2.2}N_{0.48}$
SiO_2	s.c.	$SiO_{1.73}$	$SiO_{1.57}N_{0.14}$
SiO_2	am.	$SiO_{1.77}$	$SiO_{1.65}N_{0.14}$
GeO_2	p.c.	$GeO_{1.34}$	$GeO_{1.22}N_{0.11}$
TiO_2	s.c.	$TiO_{1.85}$	$TiO_{1.34}N_{0.56}$
TiO_2	s.c.	$TiO_{1.66}$	$TiO_{1.12}N_{0.60}$
ZrO_2	s.c.	$ZrO_{1.63}$	$ZrO_{1.17}N_{0.42}$
V_2O_5	s.c.	$V_2O_{3.74}$	$V_2O_{2.70}N_{1.10}$
Nb_2O_5	p.c.	$Nb_2O_{3.63}$	$Nb_2O_{2.50}N_{1.17}$

[a]s.c. = single crystalline, p.c. = polycrystallnie, am. = amorphous
Source: Ref. 74.

after N_2^+ ion treatment) indicate that the nature of the defect structure formed is independent of the nature of the projectiles used [73–75].

In conclusion, it has been established that oxides, with few exceptions, tend to lose oxygen during ion bombardment in the sputtering régime. Deviations from stoichiometry can be determined by quantitative XPS analysis. In many cases the results can be confirmed by determining the relative amounts of the reduced component(s) from an appropriate peak synthesis procedure for the peak envelope of the metal component.

3.1.3. Cr–O–Si cermet films

Cr–O–Si cermet films possess high electrical resistivity, together with low values of the temperature coefficient of resistivity and with good long-term stability. They can therefore serve as precision thin-film resistors in integrated circuits, and in thermal printing heads, etc. The properties of these films depend on various factors, including chemical composition and bonding of the constituent elements. Heat treatment leads to the development of silicide phases, such as Cr_3Si, Cr_5Si_3, CrSi, $CrSi_2$ [80, 81].

While the effects of noble-gas ion beams on simple oxides have been documented, as discussed in the previous section, the corresponding response of complex oxides with two cations (e.g., $MgAl_2O_4$, $NiTiO_3$, $NiSiO_3$) to ion bombardment has received much less attention [81–85]. No general rules have been developed for predicting compositional and chemical-state changes in such complex oxides.

The effects of Ar^+ (2 keV), He^+ (2 keV) and N_2^+ (4 keV) bombardment on Cr–O–Si have been studied recently by XPS [86–89]. In order to produce experimental evidence for the rôle of Cr in the target, the results obtained from the Cr–O–Si system were compared with those obtained for an amorphous substoichiometric SiO_x layer having an O/Si ratio similar to that in the Cr–O–Si sample [88, 89].

Cr–O–Si cermet layers were prepared by RF-sputtering of a Cr:O:Si target (Leybold-Heraeus Co.) with a nominal atomic ratio of 1:1:1, onto thermally oxidized silicon wafers in a Leybold-Heraeus Z 801 load lock system as described in [86]. The thicknesses of the cermet layers were in the range of 100–200 nm. The $SiO_{1.3}$ layer was deposited in a Balzers-BAK 550 box coater by electron beam evaporation of an SiO_x ($x \approx 1$) source material at 10^{-3} Pa, onto an oxidized Al substrate at 300°C. Typical atomic ratios of Cr:O:Si = (0.8–1):(1.1–1.2):(1) for the cermet and of O:Si = 1.3:1 for the SiO_x layer were measured by RBS for the as-deposited layers, with homogeneous in-depth distributions of the constituents. The samples proved to be completely amorphous in the as-grown state as shown by both XRD and selected arc electron diffraction (SAED). Trace amounts of a Cr_3Si phase were found only after heat treatment at 600°C.

The changes in the chemical states of the constituents caused by Ar^+ bombardment can be followed in Fig. 16. Cr in the as-prepared sample was in both the elemental and oxidized states as manifested by Cr $2p_{3/2}$ components at around 574.3 eV and 577.6 eV BE, respectively [77, 78]. The majority of the Si was bound in several oxide states, as indicated by a broad $2p$ peak centered at about 102.9 eV. A very small fraction of the Si was found also in a state close to elemental, as judged by low-intensity peak at 99.3 eV. The same chemical states could also be deduced for Si by recording the Si KLL Auger lines [77, 78]. The main component of the O $1s$ line at 533.0 eV and the shoulder at about 531 eV represent O bound to Si and Cr, respectively [78]. These states were also found after removal of the outermost oxidized surface layer by wet chemical etching (in dilute HF) [86].

Ar^+ bombardment resulted in an almost complete reduction of Cr as shown by the decrease of the Cr^{3+}–O component and the predominance of Cr^0. In agreement with this, the intensity of the O $1s$ line decreased significantly, and in particular the component at about 531 eV, related to the Cr^{3+}–O bond, almost disappeared. At the same time a significant increase in the intensity of the Si^0 component took place. Applying a curve-fitting procedure using an asymmetric line shape composed of two Gaussian (with 15% Lorentzian) components with a 0.6 eV separation and a 1:2 area ratio [89], the relative amounts of Si^0 (i.e., $Si^0/\Sigma Si$) could be determined for the Ar^+-bombarded series, and are shown versus bombardment time in Fig. 17. The proportion of Si^0 obtained by curve-fitting the Si KLL envelope with experimental line shapes (of Si^0 and SiO_2) is also shown.

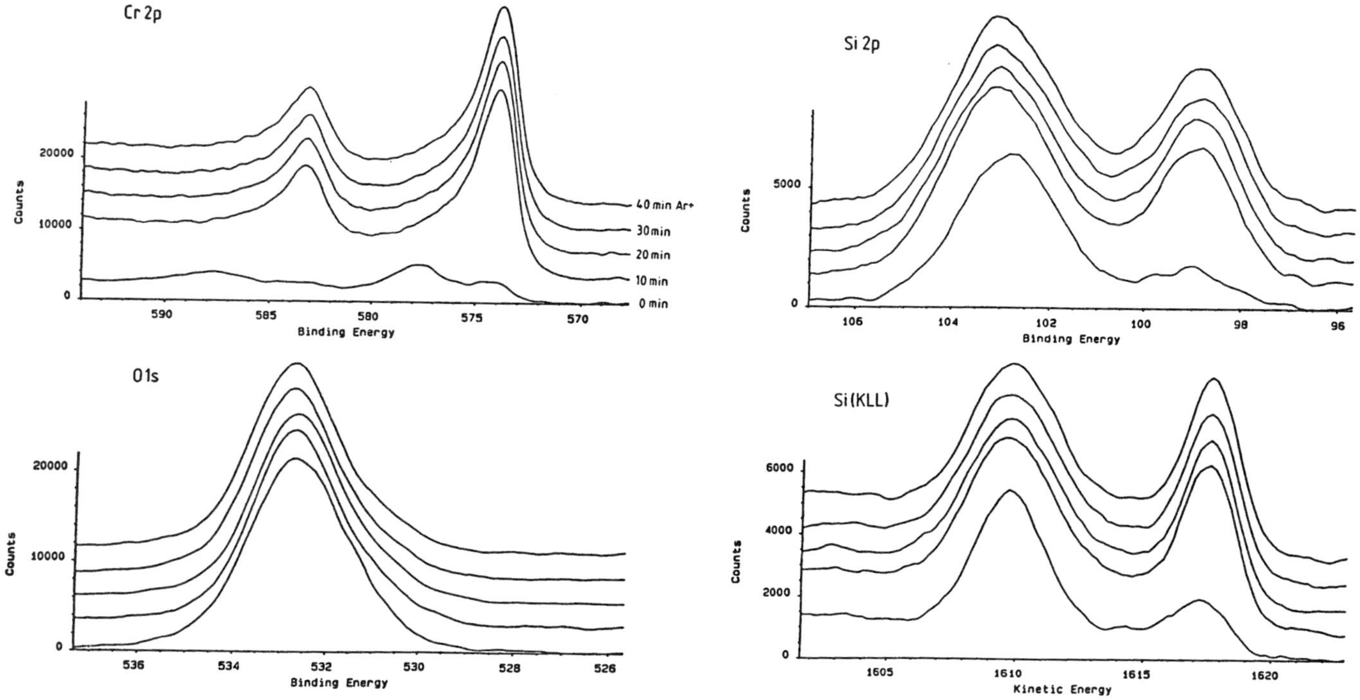

Figure 16. Changes in the line shapes of the Cr 2*p*, O 1*s*, Si 2*p* and Si KLL peaks of the Cr–O–Si sample in the as-received state and after Ar^+ bombardment steps. (From Ref. 88.)

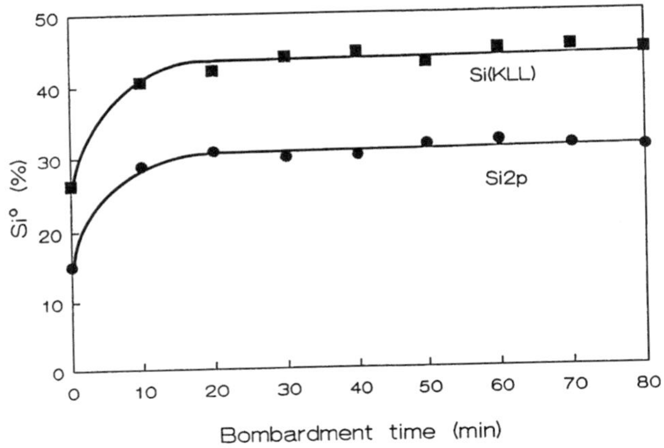

Figure 17. The relative amounts of the Si^0 component at different times of Ar^+ bombardment evaluated by peak synthesis of the Si 2*p* and Si KLL lines.

At this point it is interesting to compare the reducing effect of ion bombardment on Cr–O–Si with that observed for a SiO_x phase. According to the literature, SiO_2 can be reduced considerably by ion impact [73, 74, 90–93], but formation of elemental silicon (Si^0) has been reported only in one study [92]. Since no data were found on the effect of ion bombardment of substoichiometric silicon oxides, e.g., SiO_x with $x \approx 1$, it was decided to check as to whether a considerable amount of Si^0 could be created in a sample possessing an O/Si ratio similar to that of the investigated Cr–O–Si samples.

In Fig. 18 the Si 2*p* line shapes for the Cr–O–Si and $SiO_{1.3}$ layers are shown after 20 min Ar^+ bombardment. In the case of the Cr–O–Si sample a large Si^0 component appeared. Conversely, no Si^0 could be detected in SiO_2 and only a minor amount in the $SiO_{1.3}$ sample, when they were bombarded under conditions identical to those applied to Cr–O–Si.

In order to obtain a deeper insight into the chemical and structural transformations of the Si atoms in the Cr–O–Si and the $SiO_{1.3}$ samples, the curve-fitting procedure described was applied to the Si 2*p* envelope. The BE values for Si^{4+} (SiO_2) and the reduced states (Si^{3+}, Si^{2+}, Si^{1+} and Si^0), were taken from the literature [94]. The results of the peak synthesis obtained for the Ar^+ bombarded $SiO_{1.3}$ and Cr–O–Si samples are shown in Figs. 18b and c, respectively. The distributions of the five oxidation states were different for the two targets. For the $SiO_{1.3}$ samples the Si^{3+} peak dominated, while for their Cr–O–Si equivalent the Si^{4+} state was the most prominent. The reliability of the peak synthesis in the case of $SiO_{1.3}$ was established by relating the appropriate "stoichiometric" amount of oxygen with each oxidation state. The result was 1.33 oxygens per silicon atom, in good agreement with the RBS data.

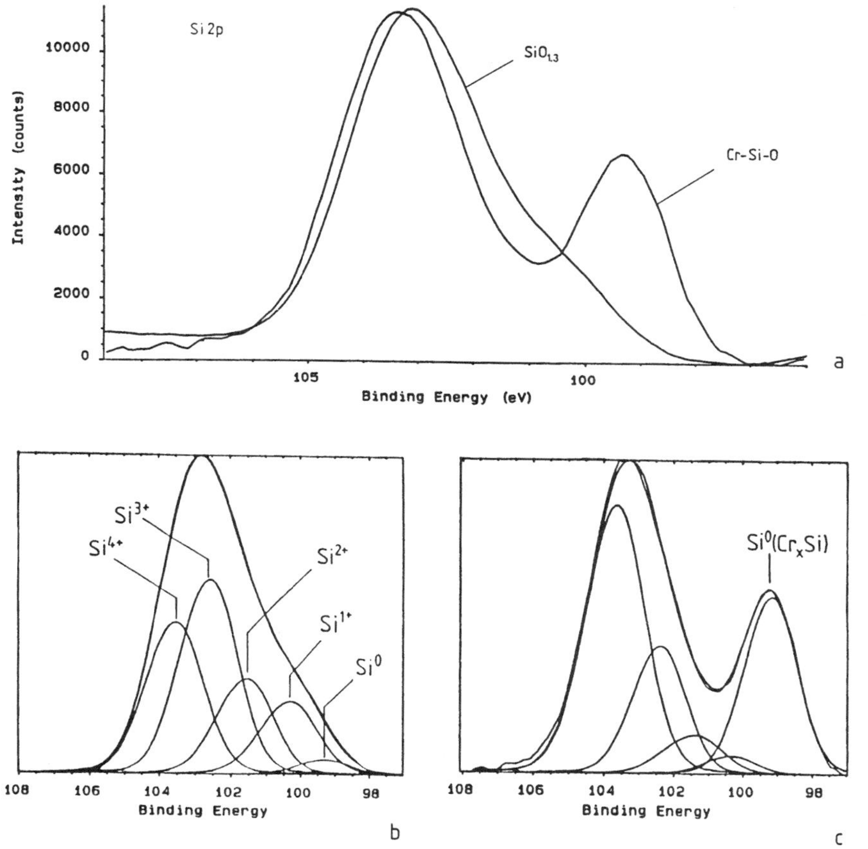

Figure 18. (a) Comparison of the Si 2*p* envelopes in the Ar^+-bombarded $SiO_{1.3}$ (5.3×10^{16} ions cm^{-2}) and Cr–O–Si samples (4.0×10^{16} ions cm^{-2}) (b) and the components used for synthesis of the same peak in $SiO_{1.3}$ and in Cr–O–Si (c). (From Ref. 88.)

The modified Auger parameter (α^*) for the Si^0 state is defined as $\alpha^* =$ Si 2*p*(BE) + Si KLL(KE) [77, 88]. By evaluating this parameter, further conclusions can be drawn on ion-induced transformations. Since α^* values are independent of any surface charging, they can be evaluated with high accuracy. It is known [95, 96] that α^* is a measure of the interatomic relaxation energy; higher values of α^* correspond to greater screening of the core hole created in the Auger transition. The present data recorded from Ar^+ bombarded Cr–O–Si produced values of α^* slightly higher (1716.4 eV) than those obtained from bulk Si, and corresponded more closely to Cr_3Si or other silicides [77, 88]. Thus, the formation of

Cr–Si bonds can be established [86] even in the case of small clusters, in which they would be undetectable by diffraction techniques. This conclusion is supported by the fact that Cr_3Si-type silicides were detectable by XRD in these layers after moderate heat treatment at 400–600°C [86]. The formation of various silicides (Cr_3Si, Cr_5Si_3, $CrSi_2$) by heat treatment has also been observed by other authors [81]. The results of this study show that Ar^+ bombardment acts in a way similar to heat treatment, but the effect is obviously restricted to the penetration depth of the ions. N_2^+ bombardment does not create Si–Cr bonds. On the contrary, it causes disruption of Si–Cr bonds formed by the preceding Ar^+ bombardment, and the implanted nitrogen then forms O–Si–N, N–Si–N and, to a lesser extent, Cr–N bonds. The intensity of the Si^0 signal also decreases and the main Si 2*p* envelope shifts toward lower BE [87]. Bombardment by He^+ after that by Ar^+ resulted in Cr/Si and O/Si ratios of 0.9–1.0 and 3.26, respectively. The relative amounts of oxidized Cr and Si species increased, at the expense of Cr^0 and Si^0, with a simultaneous increase in the O 1*s* intensity [89].

In conclusion, the effect of ion bombardment on amorphous Cr–O–Si layers is manifested in major surface chemical and short-range structural changes. Ar^+ bombardment can be used to create chromium silicide clusters. To avoid such ion-beam-induced transformations, depth profiling by wet chemical etching may be preferable.

3.2. Thin Surface Films of Polymers

Thin and ultrathin polymeric films are of great interest in the multi- and interdisciplinary field of materials science and in (surface) engineering, e.g., in micro- and nanolithography and other areas of microelectronics, optics, optoelectronics, biotechnology, membrane and sensor technology, controlled-release delivery systems, automobile industry, and packaging [97–102], etc. The interest arises from the fact that thin polymeric films may impart to the substrate–thin-film systems an impressive range of electrical, optical, dielectric, or mass transport properties, as well as good adhesion, wettability, and biocompatibility, etc., all of which increase performance and cost effectiveness.

Thin or ultrathin films of polymers can now be prepared on a variety of substrates by many types of deposition technique including plasma polymerization or plasma-enhanced chemical vapor deposition, PECVD [103], interfacial polycondensation [104], Langmuir-Blodgett techniques [97, 99], and electrodeposition [99], etc. They can also be created directly on the surfaces of bulk polymers by modification using wet or dry chemical treatments (sulfonation, halogenation, ozonation, grafting, etc.), electron beam, corona discharge, or plasma treatments [98], and particle (ion or atom) beam treatments [105–108], etc. Particle beam treatment, in particular, is a new, versatile and promising method, since very high densities of energy can be deposited in the thin surface films, and various particles can be used (both inert and reactive). Beam conditions such as energy, density and dose can be varied over wide ranges and in a precisely controllable way, allowing good reproducibility. For dielectric substrates the effects of atom beams may differ from those of ion beams [55, 109, 110], even if in the energy range being used the ions are known to undergo charge neutralization by either Auger or resonance processes within distances of a few atomic radii from the surface [111]. In effect the ions impinge on the surface as *neutral* particles. The difference between the response to ion beams and to atom beams is thus due to the difference in charging. However, the exact mechanism of the charge-induced processes is not yet known.

Knowledge of particle-beam-induced chemical structural changes can be of key importance in the applications of the various thin-film–substrate systems, since it is central to either of the preparation–structure–property or the preparation–structure–performance type relationships.

Some problems may arise during the analysis of thin polymeric films. The first is obviously the damage that may be caused by the ionizing radiation used for the analysis. Although this is a general problem in the surface analysis of polymers, it is of sufficient importance to be mentioned again. It is known that the ionization cross section of incident x-ray photons increases with the binding energy of the electrons, while for the incident electrons the situation is the opposite [112]. Thus, for an x-ray beam the excitation of valence-band electrons—i.e., those taking part in chemical alterations—is less probable than for electrons in core levels, whereas for an electron beam the case is the reverse. This is one of the reasons why the use of AES in the analysis of organic polymeric films (which are in general much more susceptible to ionizing radiation than inorganic films) is severely restricted and why XPS is usually the technique of choice. However, x-ray-induced damage may also constitute a problem during analysis, especially for halogen-containing polymers. (For more details of radiation damage including those observed during SIMS analysis, see Chapters 6 and 7.)

Time-dependent or relaxation phenomena [113] may also occur on the surfaces of polymers treated by energetic particles. A typical process is the so-called hydrophobic recovery, during which hydrophilic surfaces created by surface treatments become hydrophobic with elapsed time after treatment [98, 114, 115].

If the particle beam treatment of an organic polymer is performed ex situ, a post-treatment type oxidation of the surface layer may occur upon exposure to air due to the existence in the layer of long-lived radicals and excited species. It is therefore advisable to follow the subsequent series of steps when analyzing thin surface films of polymers prepared by particle beam modification of the bulk polymers.

1. Check the rate of radiation damage of the sample to be analyzed by consultation of published data [116], or perform analysis of the time dependence of exposure. In the case of radiation-sensitive samples either reduce the dose of exposure or take into account quantitatively the radiation dam-

age (for instance, by extrapolation of the results of the time-dependent exposure analysis to zero exposure time).

2. Check the chemical changes occurring due to elapsed time-dependent surface dynamic phenomena by performing elapsed time-dependent surface analysis. Hydrophobic recovery, for instance, may sometimes be reduced by using solvent-extracted, oligomer-free polymers as a starting material [115].
3. Minimize the post-treatment type oxidation effect in the ex situ treated samples by allowing the treated sample to relax under UHV conditions for a sufficiently long period of time.

Some representative case studies are now given in which it is demonstrated how the problem of the determination of surface chemical changes induced by energetic particles in thin surface films of polymers can be solved by surface analysis.

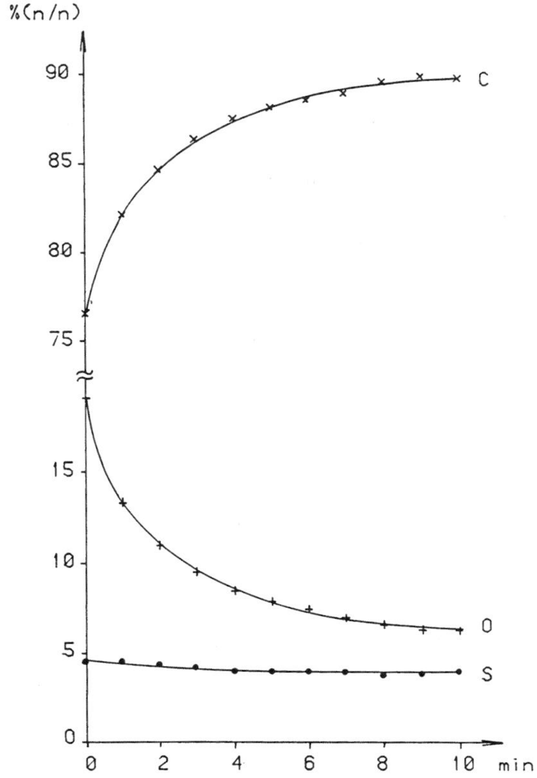

Figure 19. Ar^+ beam-induced alteration in the surface composition of PES. (From Ref. 120.)

3.2.1. Aromatic poly(ether sulfone)

Polysulfones are materials of great technological interest due to their applications as polymeric membranes, resists for microelectronics, heat-resistant materials in aerospace applications, matrices for composites, etc. It has been shown that ion beam treatment applied in the low-keV range is a suitable way to modify some relevant properties of polysulfones, such as solubility or electrical conductivity [117, 118]. XPS is probably the most suitable analytical method for the characterization of ion-beam-induced changes in the composition and chemical state of polysulfones, e.g., of aromatic poly(ether sulfone) (PES) [119, 120].

For example, a 1 keV Ar^+ beam treatment carried out at fluences up to 9×10^{12} ions cm^{-2} led to a small depletion of sulfur, a strong depletion of oxygen, and a strong enrichment of carbon in the surface layer (Fig. 19). Simultaneously, the sulfone group $-SO_2$, became partially reduced to a sulfide-like state –S– (Fig. 20). Furthermore, the shakeup satellite of the C 1*s* peak decreased drastically (Fig. 21), implying that the phenyl rings had become heavily damaged.

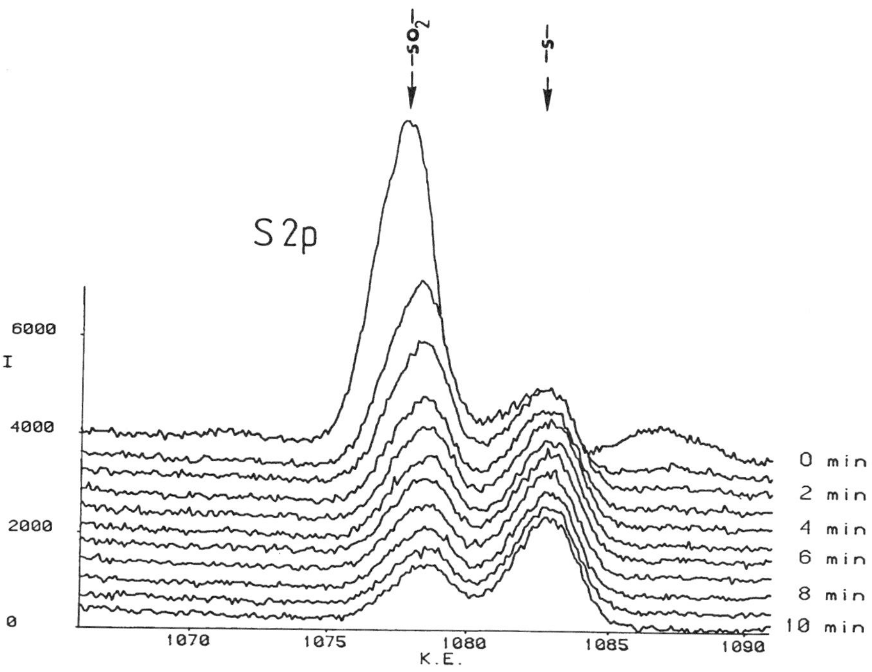

Figure 20. Ar^+ beam-induced reduction of the sulfone group of PES. (From Ref. 120.)

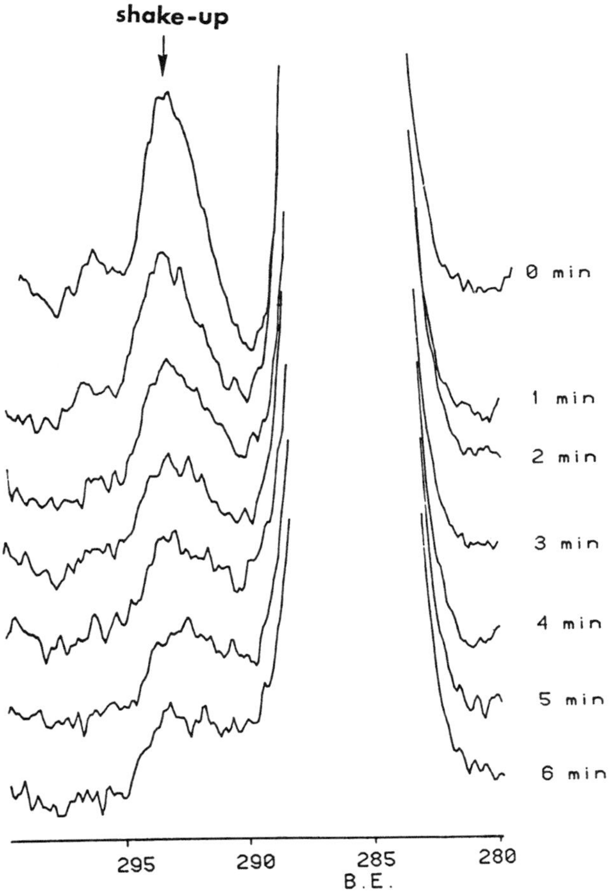

Figure 21. Evolution of the shakeup satellite of the C 1*s* peak of PES during Ar^+ beam treatment. (From Ref. 120.)

3.2.2. Aromatic polyimide

Aromatic polyimides are characterized by high-temperature stability and good mechanical and dielectric properties, which render them reliable materials for a variety of applications in a number of different advanced technologies [121]. Their ion beam modification leading to changes in conductivity, hardness, surface chemical properties, etc., has been studied intensively in the past few years (e.g., [34, 122–134]).

Characteristic surface chemical changes induced by Ar atoms can be revealed by XPS [130]. Figure 22 shows the evolution of the N 1*s* peak of the

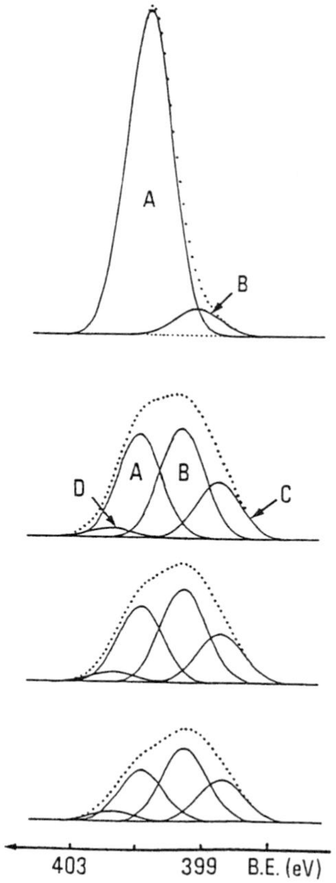

Figure 22. Evolution of the N 1*s* peak of PMDA-ODA during Ar atom beam treatment. (From Ref. 130.)

polyimide PMDA-ODA during bombardment by 5 keV Ar atoms up to a fluence of 1.4×10^{16} cm^{-2}. The N 1*s* peak for the pristine polymer is dominated by the imidic nitrogen (component A at BE = 400.8 eV). A small component B at BE = 399.3 eV is also present, probably representing a small quantity of isoimide always present after thermal curing in polyimides. For the Ar-irradiated samples the imide band decreased, but was still present in the steady state. Component B became the most intense, and two more peaks appeared, C and D at 398.6 and 401.4 eV, respectively. After bombardment, imine (–C=C–) and nitrile (–sC≡N) groups are expected to be the main contributors to component B, whereas components C and D at BEs = 398.6 and 401.4 eV can be assigned to pyridine-type and to tertiary amine groups, respectively [130, 135].

Figure 23 shows that the oxygen-content (O_{tot}) of the polyimide decreases on bombardment by Ar atoms. At the same time, the concentration of the carbonyl-like O atoms (BE = 532.3 eV) decreases monotonically, while that of the etherlike (i.e., ether and hydroxyl) O atoms (BE = 533.8 eV) passes over a local maximum at an ion dose of around 10^{14} cm^{-2}. The trend of the curve of the carbonyl-like oxygen suggests that essentially an elimination of the carbonyl groups is taking place. The transient increase of the etherlike groups at the expense of the carbonyl groups, however, indicates transformation of a portion of the C=O to C–O type bonds, probably via a pathway involving the collisional breaking of C=O and a subsequent reaction of the free–oxygen species with a suitable carbon site in the matrix [130]. Interestingly, some authors find a transient improvement of adhesion in the metal/ion beam treated PMDA-ODA system in the same fluence range of about 10^{14} cm^{-2} [136]. Thus, the transient increase in the concentration of the etherlike groups in Fig. 23 and the cited transient improvement in the adhesion appear to constitute a structure-performance-type relationship [130].

3.2.3. Organosilicon polymers

Particle beam treatment of organosilicon polymers changes their chemical composition and structure [137–140], offering a possible way of transforming them

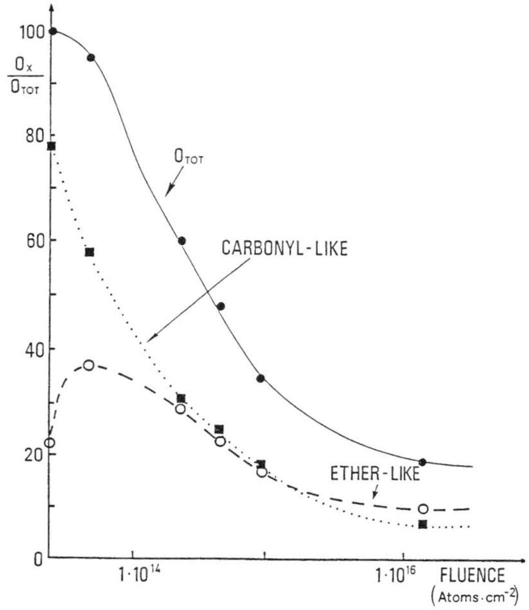

Figure 23. Percent variation of total oxygen and of related functional groups during Ar atom beam treatment of PMDA-ODA. (From Ref. 130.)

into a ceramic-like state [141] and of changing their properties, such as biocompatibility [142, 143] and mass transport properties [144–147].

For example, Ar-atom beam treatment (1 keV, 10^{15} atoms cm^{-2}) of a three-layered composite-type gas separation membrane, having a working layer containing poly(dimethyl siloxane) and poly(methyl silsesquioxane), with a ratio of the corresponding repeat units of 200:5, generally decreases the transmembrane flux of gases and increases the selectivity toward various gas pairs (Fig. 24). One reason for such changes is obviously the negative-resist-type behavior of poly(organosiloxanes), i.e., that crosslinking predominates over chain scission in these polymers when exposed to ionizing radiation [148]. However, a comparative XPS analysis performed on the pristine, and on the fast-Ar-atom beam-treated samples, revealed as well the simultaneous occurrence of significant

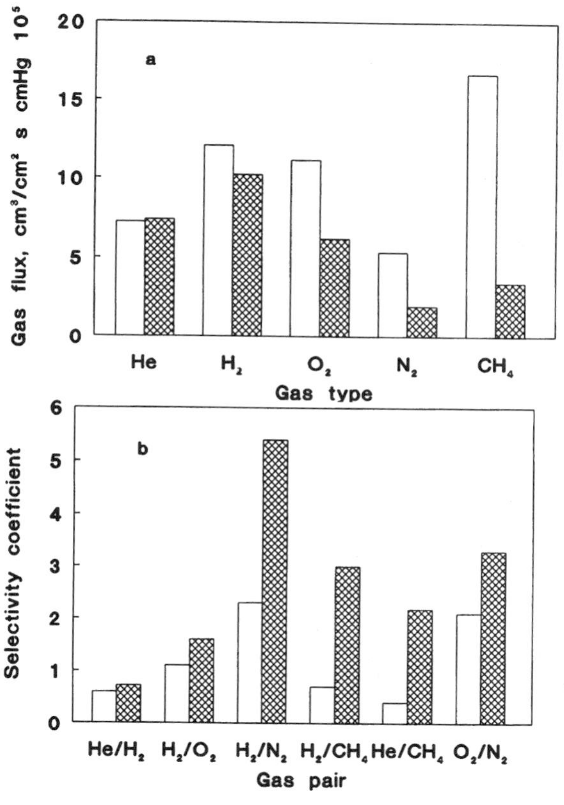

Figure 24. Changes in the transmembrane gas fluxes (a) and in the selectivity coefficients (b) of a poly(dimethyl siloxane)-based gas separation membranes, as a result of fast-Ar-atom beam treatment. (Open bars: pristine membranes, crosshatched bars: treated membranes.)

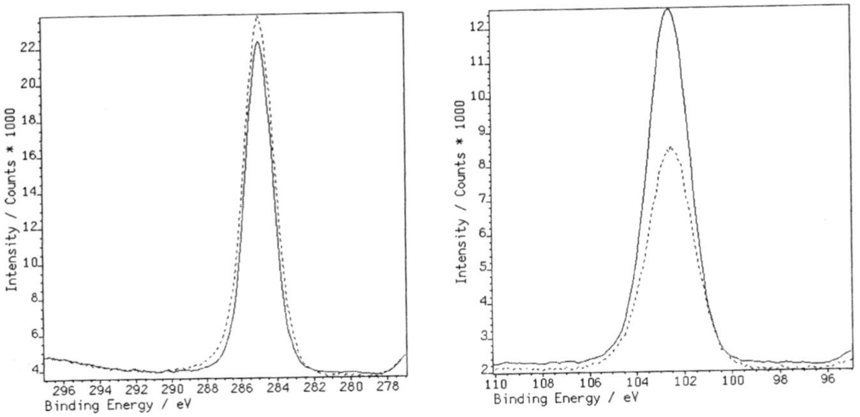

Figure 25. The C 1*s* (left) and the Si 2*p* (right) photoelectron peaks of the pristine (solid line) and of the fast Ar atom-beam-treated (dashed line) poly(dimethyl siloxane)-based gas separation membrane.

compositional changes. Figure 25 demonstrates a depletion of Si in the surface layer of the Ar-atom beam-treated membranes, probably due to preferential elimination. This gives an important indication of the mechanism of the alteration of the permeability of the penetrant gases: not only is their diffusivity affected (due to crosslinking), but their solubility must also be changed in the modified layer (owing to the strong compositioneal change realed by XPS).

REFERENCES

1. M. J. O'Keefe and J. M. Rigsbee, in *Materials and Processes for Surface and Interface Engineering* (Y. Pauleau, ed.), NATO ASI Series E, Vol. 290, Kluwer, Dordrecht, 1995, p. 151.
2. G. Whal, W. Decker, L. Klippe, A. Nürnberg, M. Pulver and R. Stolle, in *Materials and Processes for Surface and Interface Engineering* (Y. Pauleau, ed.), NATO ASI Series E, Vol. 290, Kluwer, Dordrecht, 1995, p. 185.
3. A. J. K. Hirvonen, in *Materials and Processes for Surface and Interface Engineering* (Y. Pauleau, ed.), NATO ASI Series E, Vol. 290, Kluwer, Dordrecht, 1995, p. 307.
4. W. Bose, *Mater. Sci. Eng. R12*, 53 (1994).
5. K. Jarrendahl, J. Birch, L. Hultman, L. R. Wallenberg, G. Radnoczi, H. Arwin and J-E. Sundgren, *Proc. Mat. Res. Symp. 258*, 571 (1992).
6. A. Barna, A. Sulyok and M. Menyhard, *Surf. Interface Anal. 19*, 77 (1992).
7. G. H. Wehner, *J. Vac. Sci. Technol. A 3*, 1821 (1985).
8. D. J. Barber, *Ultramicroscopy 52*, 101 (1993).

9. M. Paulus and F. Reverchon, *J. Phys. Radium Appl. 22*, 103/A (1961).
10. A. Barna, *Mat. Res. Soc. Symp. Proc., 254*, 3 (1992).
11. W. Pamler and K. Wangemann, *Surf. Interface Anal. 18*, 52 (1992).
12. G. Carter and M. J. Nobes, *Surf. Interface Anal. 19*, 39 (1992).
13. A. Zalar, *Thin Solid Films 124*, 223 (1985).
14. J. D. Geller and N. Veisfeld, *Surf. Interface Anal. 14*, 95 (1989).
15. E.-H. Cirlin, Y. Cheng, P. Ireland and B. Clemens, *Surf. Interface Anal. 15*, 337 (1990).
16. A. Barna, P. B. Barna and A. Zalar, *Surf. Interface Anal. 12*, 144 (1988).
17. S. Hofmann, A. Zalar, E.-H. Cirlin, J. J. Vajo, H. J. Mathieu and P. Panjan, *Surf. Interface Anal. 20*, 621 (1993).
18. A. Barna and M. Menyhard, *Phys. Stat. Sol. (a) 145*, 263 (1994).
19. Patents: Hungarian 205814, 194421, 190855, USA 4739463, 3847780.
20. M. Menyhard, A. Barna and H. J. Biersack, *J. Vac. Sci. Technol. A12*, 2368 (1994).
21. M. Menyhard, A. Barna, J. P. Biersack, K. Järrendahl and J-E. Sundgren, *J. Vac. Sci. Technol. A13*, 1999 (1995).
22. S. Hofmann, in *Practical Surface Analysis, Auger and X-ray Photoelectron Spectroscopy* (D. Briggs and M. P. Seah, eds.), 2nd ed., Wiley, Chichester, 1990, Vol. 1, p. 164.
23. N. Matsunami, Y. Yamamura, Y. Itikawa, N. Itoh, Y. Kazumata, S. Miyagawa, K. Morita, R. Shimizu and H. Tawara, *Atomic Data and Nuclear Data Tables 31*, 1 (1984).
24. M. P. Seah, in *Practical Surface Analysis* (D. Briggs and M. P. Seah, eds.), 2nd. ed., Wiley, Chichester, 1990, Vol. 1, p. 331.
25. T. E. Gallon, *Surf. Sci. 17*, 486 (1969).
26. S. Tanuma, C. J. Powell and D. R. Penn, *Surf. Interface Anal. 17*, 911 (1991).
27. R. Shimizu, *Jap. J. Appl. Phys. 22*, 1631 (1983).
28. A. Barna, A. Konkol, A. Sulyok and M. Menyhard, *Vacuum 45*, 333 (1994).
29. M. J. Pellin and J. W. Burnet, *Pure Appl. Chem. 65*, 2361 (1993).
30. Z. L. Liau, B. Y. Tsaur and J. W. Mayer, *J. Vac. Sci. Technol. 16*, 121 (1979).
31. S. Hofmann, *Surf. Interface Anal. 21*, 673 (1994).
32. J. F. Ziegler, J. P. Biersack and U. Littmark, in *The Stopping and Range of Ions in Solids*, Vol. 1, Pergamon Press, New York, 1985.
33. J. P. Biersack, *Nucl. Instr. Meth. B27*, 21 (1987).
34. J. M. Costantini, J. P. Salvetat, F. Brisard, M. P. Cals, J. P. Marque and F. Issac, *Nucl. Instr. Meth.* B (in press).
35. J. Cazaux, in *Ionization of Solids* (R. A. Baragiola, ed.), NATO ASI Series B, vol. 306, Plenum Press, New York, 1992, p. 325.
36. J. P. Biersack and J. F. Ziegler, in *Transport of Ions in Matter, TRIM*, Version 91. 14, IBM-Research, Yorktown, NY.

37. G. Barth, R. Linder and C. Bryson, *Surf. Interface Anal. 11*, 307 (1988).
38. M. Mohai and I. Bertóti, *ECASIA'95* (Eds. H. J. Mathieu, B. Reihl, D. Briggs), John Wiley & Sons, Chichester-New York-Brisbane-Toronto-Singapore (1995), p. 675.
39. A. J. Perry, *J. Vac. Sci. Technol. A6*, 2140 (1988).
40. H. A. Wriedt and J. L. Murray, in *Phase Diagrams of Binary Titanium Alloys* (J. L. Murray, ed.), ASM Inst., Metals Park, Ohio, 1987, pp. 183, 227.
41. S. I. Alyamovskii, Yu. G. Zaynulin and G. P. Shveykin, in *Oxycarbides and Oxynitrides of IVA and VA Metals* (Russian), Nauka, Moscow, 1981.
42. I. Bertóti, M. Mohai, J. L. Sullivan and S. O. Saied, *Surf. Interface Anal. 21*, 467 (1994).
43. I. Bertóti, M. Mohai, J. L. Sullivan and S. O. Saied, *Appl. Surf. Sci. 84*, 357 (1995).
44. H. Höchst, R. D. Bringans, P. Steiner and Th. Wolf, *Phys. Rev. B25* 7183 (1982).
45. M. Miyagi, Y. Sato, T. Mizuno and S. Sawada, in Proc. 4th Int. Conf. TITANIUM 80, The Metallurgical Society of AIME, Kyoto, 1980, p. 2867.
46. L. Porte, L. Roux and J. Hanus, *Phys. Rev. B28*, 3214 (1983).
47. K. S. Robinson and P. M. A. Sherwood, *Surf. Interface Anal. 6*, 261 (1984).
48. B. J. Burrow, A. E. Morgan and R. C. Ellwanger, *J. Vac. Sci. Technol. A4*, 2463 (1986).
49. J-E. Sundgren, Doctoral Thesis, #79, Linköping University, 1982.
50. I. le R. Strydom and S. Hofmann, *J. Electron Spectrosc. Relat. Phenom. 56*, 85 (1991).
51. R. Bertoncello, A. Casagrande, M. Casarin, A. Gilsenti, E. Lanszoni, L. Mirenghi and E. Tondello, *Surf. Interface Anal. 18*, 525 (1992).
52. C. W. Louw, I. le R. Strydom, K. van den Heever and M. J. van Staden, *Surf. Coat. Technol. 49*, 348 (1991).
53. I. le R. Strydom and S. Hofmann, *Vacuum 41*, 1619 (1990).
54. C. Ernsberger, J. Nickerson, A. E. Miller and J. Moulder, *J. Vac. Sci. Technol. A3*, 2415 (1985).
55. J. L. Sullivan, S. O. Saied and I. Bertóti, *Vacuum 42*, 1203 (1991).
56. L. Ramqvist, K. Hamrin, G. Johansson, A. Fahlman and C. Nordling, *J. Phys. Chem. Solids 30*, 1835 (1969).
57. A. Ermolieff, P. Bernard, S. Marthon and P. Wittmer, *Surf. Interface Anal. 11*, 563 (1988).
58. M. Wolf, J. W. Schultze and H-H. Strehblow, *Surf. Interface Anal. 17*, 726 (1991).
59. E. Rolinski, *Mater. Sci. Eng. 100*, 193 (1988).
60. E. Bruninx, A. F. P. M. Van Eenbergen, P. van der Werf and J. Haisma, *J. Mater. Sci. 21*, 541 (1986).
61. R. R. Manory, *Surf. Eng. 3*, 233 (1987).

62. R. R. Manory and G. Kimmel, *Thin Solid Films 150*, 277 (1987).
63. K. Schwarz, A. R. Williams, J. J. Cuomo, J. M. E. Harper and H. T. G. Hentzell, *Phys. Rev. B 32*, 8312 (1985).
64. D. S. Yee, J. J. Cuomo, A. M. Frisch and D. P. E. Smith, *J. Vac. Sci. Technol. A4*, 381 (1986).
65. P. Fayet, J. Laurent and A. Rime, in ECASIA'95 Abstracts (H. J. Mathieu, ed.), Montreux, TC-13, 1995.
66. M. Menyhard and G. Gergely, in Proc. 7th Int. Vac. Congr. & 3d Int. Conf. Solid Surf, Vienna, 1977, p. 2165.
67. L-Q. Wang, D. R. Baer and M. H. Engelhard, *Surf. Sci. 320*, 295 (1994).
68. G. Betz and G. K. Wehner, in *Sputtering by Particle Bombardment II* (R. Behrisch, ed.), Topics in Applied Physics, Vol. 52., Springer, Berlin, 1983, p. 11 (and references cited therein).
69. S. Hofmann and J. M. Sanz, *J. Trace Microprobe Techn. 1*, 213 (1982–83).
70. R. Kelly, *Mater. Sci. Eng. A115*, 11–24 (1989).
71. R. Kelly, *Nucl. Instr. Meth. B18*, 388 (1987).
72. D. F. Mitchell, G. I. Sproule and M. J. Graham, *Surf. Interface Anal. 15*, 487 (1990).
73. I. Bertóti, R. Kelly, M. Mohai and A. Tóth, *Surf. Interface Anal 19*, 291 (1992).
74. I. Bertóti, R. Kelly, M. Mohai and A. Tóth, *Nucl. Instr. Meth. B80/81*, 1219 (1993).
75. R. Kelly, I. Bertóti and A. Miotello, *Nucl. Instr. Meth. B 80/81*, 1154 (1993).
76. J. L. Sullivan, S. O. Saied and I. Bertóti, *Vacuum 42*, 1203 (1991).
77. C. D. Wagner, W. M. Riggs, L. E. Davis, J. F. Moulder and G. E. Muilenberg, in *Handbook of X-ray Photoelectron Spectroscopy*, Physical Electronics Division, Perkin-Elmer Corporation, Eden Prairie, MN, 1979.
78. J. F. Moulder, W. F. Stickle, P. E. Sobol and K. D. Bomben, in *Handbook of X-ray Photoelectron Spectroscopy*, Physical Electronics Division, Perkin-Elmer Corporation, Eden Prairie, MN, 1992.
79. Y. Takano Y. Tandoh, H. Ozaki and N. Mori, *Phys. Stat. Sol (b) 130*, 431 (1985).
80. E. Schabowska and T. Pisakiewicz, *Thin Solid Films 72*, L7 (1980).
81. G. Sobe, H. D. Bauer, J. Henke, A. Heinrich, H. Schreiber and R. Grötzschel, *J. Less-Common Metals 169*, 331 (1991).
82. G. Marletta, F. Iacona and R. Kelly, *Nucl. Instr. Meth. B65*, 97 (1992).
83. A. R. González-Elipe, R. Alvarez, J. P. Holgado, J. P. Espinos and G. Munuera, Appl. *Surf. Sci. 51*, 19 (1991).
84. A. R. González-Elipe, G. Munuera, J. P. Espinos and J. M. Sanz, *Surf. Sci. 220*, 368 (1989).

85. A. R. González-Elipe, A. Fernandez, J. P. Holgado, A. Caballero and G. Munuera, *J. Vac. Sci. Technol A11*, 58 (1993).
86. I. Bertóti, A. Tóth, M. Czermann, M. Menyhárd, A. Sulyok and E. Zsoldos, *Surf. Interface Anal 19*, 453 (1992).
87. I. Bertóti, A. Tóth, M. Mohai, R. Kelly and G. Marletta, *Thin Solid Films 241*, 211 (1994).
88. I. Bertóti, A. Tóth, M. Mohai and M. Révész, *Acta Chim. Hung. Models Chem. 130*, 837 (1993).
89. I. Bertóti, A. Tóth, M. Mohai, R. Kelly and G. Marletta, *Nucl. Instr. Meth. B 116*, 200 (1996).
90. U. Gelius, L. Asplund, E. Basilier, K. Helenelund and K. Siegbahn, *Nucl. Instr. Meth. B1*, 85 (1984).
91. J. R. Pitts and A. W. Czanderna, *Nucl. Instr. Meth. B13*, 245 (1986).
92. E. Paparazzo, *J. Phys. D: Appl. Phys. 20*, 1091 (1987).
93. J. H. Thomas and S. Hofmann, *J. Vac. Sci. Technol. A3*, 1921 (1985).
94. O. Benkherourou and J. P. Deville, *J. Vac. Sci. Technol. A6*, 3125 (1988).
95. J. C. Rivière and J. A. A. Crossley, *Surf. Interface Anal. 8*, 173 (1986).
96. J. C. Rivière, J. A. A. Crossley and G. Moretti, *Surf. Interface Anal. 14*, 257 (1989).
97. M. C. Petty, Molecular engineering using the Langmuir-Blodgett technique, in *Polymer Surfaces and Interfaces* (W.J. Feast and H. S. Munro, eds.), Wiley, Chichester, 1987, pp. 163–187.
98. F. Garbassi, M. Morra and E. Occhiello, in *Polymer Surfaces. From Physics to Technology*, Wiley, Chichester, 1994.
99. S. T. Kowel, R. Selfridge, Ch. Eldering, N. Matloff, P. Stroeve, B. G. Higgins, M. P. Srinivasan and L. B. Coleman, *Thin Solid Films 152*, 377 (1987).
100. H. K. Lonsdale, *J. Membrane Sci. 33*, 121 (1987).
101. R. W. Baker and L. M. Sanders, in *Synthetic Membranes: Science, Engineering and Applications* (P. M. Bungay, H. K. Lonsdale and M. N. de Pinho, eds.), Reidel, Dordrecht, 1986.
102. Y. Taga, *Appl. Optics 32*, 5519 (1993).
103. H. K. Yasuda, *Plasma Polymerization*, Academic Press, San Diego, 1985.
104. R. E. Kesting, Phase inversion membranes, in *Materials Science of Synthetic Membranes* (D. R. Lloyd, ed.), ACS Symposium Series 269, American Chemical Society, Washington, DC, 1985, pp. 131–164.
105. J. W. Rabalais, Interaction of ion beams with organic surfaces studied by XPS, UPS, and SIMS, in *Photon, Electron, and Ion Probes of Polymer Structure and Properties* (D. W. Dwight, T. J. Fabish and H. R. Thomas, eds.), ACS Symposium Series 162, American Chemical Society, Washington, DC, 1981, pp. 237–246.
106. T. Venkatesan, L. Calcagno, B. S. Elman and G. Foti, Ion beam effects in organic molecular solids and polymers, in *Ion Beam Modification of*

Insulators (P. Mazzoldi and G. W. Arnold, eds.), Elsevier, Amsterdam, 1987, pp. 301–379.
107. S. Tagawa, *Adv. Polym. Sci. 105*, 100 (1993).
108. G. Marletta and F. Iacona, Chemical and physical property modifications induced by ion irradiation in polymers, in *Materials and Processes for Surface and Interface Engineering* (Y. Pauleau, ed.), Kluwer, The Netherlands, 1995, pp. 597–640.
109. A. Licciardello, O. Puglisi and S. Pignataro, *J. Chem. Soc. Faraday Trans. 2, 81*, 985 (1985).
110. D. Briggs, in *Practical Surface Analysis, Ion and Neutral Spectroscopy* (D. Briggs and M. P. Seah, eds.), 2nd ed., Wiley, Chichester, Vol. 2, pp. 367–423.
111. H. D. Hagstum, *Phys. Rev. 96*, 336 (1954).
112. C. D. Wagner, The role of Auger lines in photoelectron spectroscopy, in *Handbook of X-ray Photoelectron Spectroscopy* (D. Briggs, ed.), Heyden, London, 1977, pp. 249–272.
113. J. D. Andrade, D. E. Gregonis and L. M. Smith, in *Surface and Interfacial Aspects of Biomedical Polymers. 1. Chemistry and Physics* (J. D. Andrade, ed.), Plenum Press, New York, 1985, pp. 15–41.
114. M. Morra, E. Occhiello, R. Marola, F. Garbassi, P. Humphrey and D. Johnson, *J. Colloid Interface Sci. 137*, 11 (1990).
115. A. Tóth, I. Bertóti, M. Blazsó, G. Bánhegyi, A. Bognár and P. Szaplonczay, *J. Appl. Polym. Sci. 52*, 1293 (1994).
116. G. Beamson and D. Briggs, in *High Resolution XPS of Organic Polymers. The Scienta ESCA300 Database*, Wiley, Chichester, 1992.
117. G. Marletta, S. Pignataro and C. Oliveri, *Nucl. Instr. Meth. B 39*, 773 (1989).
118. Y. Wang, S. S. Mohite, L. B. Bridwell, R. E. Giedd and C. J. Sofield, *J. Mater. Res. 8*, 388 (1993).
119. A. Tóth, I. Bertóti, T. Székely, G. Marletta, S. Pignataro and B. Keszler, *Surf. Interface Anal. 12*, 470 (1988).
120. G. Marletta, S. Pignataro, A. Tóth, I. Bertóti, T. Székely, and B. Keszler, *Macromolecules 24*, 99 (1991).
121. Several papers in *Polyimides: Synthesis, Characterization, and Applications* (K. L. Mittal, ed.), Vols. 1–2, Plenum Press, New York, 1984.
122. S. Contarini, J. A. Schultz, S. Tachi, Y. S. Jo and J. W. Rabalais, *Appl. Surf. Sci. 28*, 291 (1987).
123. G. Marletta, C. Oliveri, G. Ferla and S. Pignataro, *Surf. Interface Anal. 12*, 447 (1988).
124. B. J. Bachman and M. J. Vasile, *J. Vac. Sci. Technol. A 7*, 2709 (1989).
125. G. Marletta, S. Pignataro and C. Oliveri, *Nucl. Inst. Meth. B 39*, 792 (1989).
126. J. Davenas and G. Boiteux, *Adv. Mater. 2*, 521 (1990).
127. J. Davenas and G. Boiteux, *Synthetic Metals 35*, 195 (1990).

128. S. Pignataro and G. Marletta, Ion beam induced chemical reactions at polymer surfaces, in *Metallized Plastics*, Vol. 2 (K. L. Mittal, ed.), Plenum Press, New York, 1991, pp. 269–281.
129. S. Pignataro, *Surf, Interface Anal. 19* 275 (1992).
130. G. Marletta, F. Iacona and A. Tóth, *Macromolecules 25*, 3190 (1992).
131. D. Xu, X. Xu and S. Zou, *Rev. Sci. Instrum. 63*, 202 (1992).
132. F. Iacona and G. Marletta, *Nucl. Instr. Meth. B 65*, 50 (1992).
133. J. Davenas and P. Thevenard, *Nucl. Instr. Meth. B 80/81*, 1021 (1993).
134. E. H. Lee, Y. Lee, W. C. Oliver and L. K. Mansur, *J. Mater. Res. 8*, 377 (1993).
135. A. Tóth, I. Bertóti, T. Székely, J. N. Sazanov, T. A. Antonova, A. V. Shchukarev and A. V. Gribanov, *Surf. Interface Anal. 8*, 261 (1986).
136. T. Flottman and W. Lohmann, in *Metallized Plastics*, Vol. 2 (K. L. Mittal, ed.), Plenum Press, New York, 1991, p. 97.
137 A. Tóth, I. Bertóti, T. Székely, and M. Mohai, *Surf. Interface Anal. 7*, 282 (1985).
138 A. Tóth, I. Bertóti and V. S. Khotimsky, *Surf. Interface. Anal. 22*, 551 (1994).
139. A. Tóth, I. Bertóti, G. Marletta, G. G. Ferenczy and M. Mohai, *Nucl. Instr. Meth. B 116*, 299 (1996).
140. Y. Suzuki, M. Kusakabe, M. Iwaki and M. Suzuki, *Nucl. Instr. Meth. B 32*, 120 (1988).
141. T. Venkatesan, T. Wolf, D. Allara, B. J. Wilkens and G. N. Taylor, *Appl. Phys. Lett. 43*, 934 (1983).
142. Y. Suzuki, M. Kusakabe, H. Akiba, K. Kusakabe and M. Iwaki, *Nucl. Instr. Meth. B 59/60*, 698 (1991).
143. Y. Suzuki, M. Kusakabe and M. Iwaki, *Nucl. Instr. Meth. B 59/60*, 1300 (1991).
144. A. Tóth, V. S. Khotimsky, I. Bertóti, N. N. Fateev, T. Székely and E. G. Litvinova, in *Abstracts of the 1993 Int. Congr. on Membranes and Membrane Processes, (ICOM'93)*, Vol. 2, Heidelberg, 1993, p. 2.48.
145. A. Tóth, V. S. Khotimsky, N. N. Fateiev, I. Bertóti, and T. Székely, Hungarian Patent 211 184 (1995).
146. A. Tóth, V. S. Khotimsky, N. N. Fateiev, I. Bertóti and T. Székely, Russian Patent 93–00-66-03 (1995).
147. A. Tóth, V. S. Khotimsky, I. Bertóti and G. Marletta, *J. Appl. Polym. Sci. 60,* 1883 (1996).
148. E. Reichmanis, A. E. Novembre, R. G. Tarascon, A. Shugard and L. F. Thompson, Organosilicon polymers for microlithograpic applications, in *Silicon-Based Polymer Science. A Comprehensive Resource* (J. M. Zeigler and F. W. G. Fearon, eds.), American Chemical Society, Washington, DC, 1990, pp. 265–281.

9

Surface Modification by Ion Implantation

DOROTHEE M. RÜCK Gesellschaft für Schwerionenforschung mbH,
Darmstadt, Germany

1. INTRODUCTION

Ion implantation is a technique that involves the introduction of ionized atoms with high energies into the surfaces of metals, ceramics, semiconductors and polymers in order to modify their surface properties. Due to the high energies of the ions practically any element can be implanted in a near-surface substrate layer, and limited solubilities are not a problem. New compounds and alloys can be formed, leading to changes in the properties of a surface that can be divided arbitrarily into chemical (e.g., etchability and corrosion), physical (e.g., optical, conductivity and wettability), and mechanical (e.g., hardness, wear and strength), properties.

Although the method of ion implantation has been introduced into industrial processes, especially in the semiconductor industry and for the treatments of tools, there is a prevailing need to improve the understanding of the underlying physics.

Three international conferences exist in which the latest fundamental results and the developments in the technique are presented and discussed, namely "Ion Beam Modification of Materials" (IBMM), "Surface Modification of Metals by Ion Beams" (SM^2IB), and "Radiation Effects in Insulators" (REI). An annual conference entitled "Plasma Surface Engineering" (PSE) concentrates on the use of ion beams and plasmas as surface treatment methods, and a wide range of applications is usually presented. The "Ion Beam Analysis" (IBA) conference deals with the use of ion beams as an analytical tool. A brief overview of the main topics and results presented at these conferences over the last two decades follows.

- The main fundamental aspect that has been discussed is the ion-solid interaction, where questions of transport processes [1] and the role of chemical driving forces leading to the formation of compounds [2, 3] are discussed.
- Modeling of ion-induced damage, grain growth [4] and atom-surface interactions [5] have been covered with different approaches such as those of molecular dynamics calculations and thermal-spike-related models. High-dose ion implantation cannot be treated by the classic Monte Carlo program TRIM, therefore several other approaches have been proposed instead, taking into account the change in the modified surface layer during the implantation process [6, 7]. Monte Carlo simulation programs have been modified to try to predict ion beam mixing effects [8, 9].
- The most important class of material investigated has been that of semiconducting materials. Compound formation by ion implantation into semiconductors such as silicon has been an important issue. The formation of silicon compounds such as SiC, SiN, SiO as insulating and conducting layers in semiconductor devices has been a growing field of interest during the conferences [10–13]. An important part of this field is the use of ion beams as tools to investigate the implantation profiles and the stoichiometry of the implanted layers [14–16]. The effects of ion implantation into other semiconductors, such as GaAs [17] and InP [18], have also been discussed and some results of interest to the production of sensors have been published [19].
- Studies of metal substrates, mainly those of steel and Ti and its alloys, have been described in numerous publications, and many promising effects that it is hoped will lead to applications have been presented. When the activity in this area is analyzed, the following topics have attracted most of the interest and been investigated most intensively. Phase formation, caused by implantation of N, C and O ions [20–22], has been studied in order to clarify the role of temperature and implantation dose [23–25]. With the aim of developing applications in the machine tool and medical industries, improvements in the macroscopic properties of steel and Ti alloys by ion implantation and also by ion-beam-assisted deposition (IBAD) have been given extensive coverage [26–29]. The link between modified microstructure and improved tribological properties has been covered by only a few groups, and many results concerning the wear properties are still of a pre-

liminary nature [30–33]. Nevertheless, the number of publications which report actual industrial field tests with ion-implanted tools has increased recently [34–38]. Corrosion protection, mainly by implantation of refractory metals and noble metals or with the help of protective layers performed by IBAD [39–42], has also been reported.

- Since the mechanical properties of ceramic materials are strongly dependent on surface phenomena, such as pre-existing surface flaws, the materials should be ideal candidates for surface modification by ion implantation. Most of the investigations reported in the field of ceramics have been concerned with single crystals of Al_2O_3, MgO and TiO_2, but commercial-grade SiC and Si_3N_4 materials have also been studied [43]. The most promising result is the possibility of increasing the flexural strength and improving wear resistance to lateral cracking [44], by metal ion implantation of ceramics. The process seems, however, to be rather complicated and the transfer of the technology to ceramics used in industrial applications does not seem possible in the near future.
- Polymers are sensitive to ionizing radiation and can be modified with low ion doses. In addition to the fundamental aspects of structural changes, there are several application-related investigations under way. In the electronics packaging industry and for the protective coating of metals [45, 46], the problem of adhesion between metal and polymer may be solved by ion irradiation [47]. Electrical properties such as conductivity as well as the optical properties of polymers can be modified, and as a result investigations in the field of passive optical devices are in progress [48–50].
- The use of ion beams for ion beam mixing and IBAD [51–54] has increased rapidly over the last decade. The understanding of fundamental processes such as the role of collision cascades in the formation of intermixed layers [55], as well as the deposition of tailored layers as protective coatings on various substrates have been investigated. In particular, the development of deposition techniques has attracted increasing interest in the field of hard coatings for improved wear resistance [56–58] and for protective coatings against corrosion attack [59–61]. In Japan protective layers produced by IBAD are now used for consumer products (e.g., razor blades) [62]. The trend in the field of IBAD coatings is toward the formation of multilayers, with attempts to adapt the parameters of the crystal structure (e.g., lattice size) of the substrate to that of the layer.
- The combination of plasma and ion beams in the process of plasma immersion ion implantation (PIII) has been developed during the last few years [63–66], and it seems that this new technique shows promise in bringing down the costs of ion implantation [67]. For homogeneous implantation of workpieces with complicated geometries the process has the particular advantage that it is not a line of sight process as in conventional ion implan-

tation [68]. PIII could probably be developed into a new method for low-cost surface modification [69]. Developments are under way to combine the advantages of PIII with those of CVD and PVD coatings to create surfaces with improved tribological properties [70, 71].

- In order to commercialize ion beam methods outside the semiconductor field the development of new implanters with high-current ion sources [72–75] has been continued in recent years, and operation of such machines has been simplified by computer control [76, 77]. In the meantime the Danfysik machine is the only commercially available implanter with computer control on the market in Europe [78].

Although contributions to the above-mentioned international conferences give a fairly complete picture of the state of the art of ion implantation, especially in commercial applications, there are two areas (in addition to that of better understanding of basic physical and chemical effects) on which successful use of ion beam modification of materials depends:

1. Development of ion beam processes for dedicated applications
2. Improvement of the available equipment (implanters and target handling)

Only in these two fields can efforts transform ion beam techniques into industrially acceptable processes.

2. PHYSICAL PROCESSES

The process of ion implantation is shown schematically in Fig. 1. The ions are accelerated up to several hundred keV and penetrate into a solid, losing their energy due to the interaction with the target atoms, and in so doing modify the layer beneath the surface of the solid. Having lost all their energy, the ions come to rest at a certain distance from the surface of the target. This distance is called the projected range, R_p. Due to the fact that energy loss is a statistical process, the distribution of the ions at a certain depth in the solid has a near-Gaussian profile, as shown in Fig. 1. The surface can be eroded by sputtering, in which ions are removed from the solid surface into the vacuum, if the surface binding energy is lower than the transferred energy in a knock-on process. The value of the sputter yield, Y, which is the number of ions removed divided by the number of incident ions, is strongly dependent on the mass of the incident ions, the ion energy and the mass of the target atoms.

The energy loss of the penetrating ions is caused either by electronic interaction with the outer shell electrons or by collisions with the target atoms. If the energy transferred to an atom is above the threshold energy for atomic displacement (>5–50 eV), the atom will be displaced from its initial lattice site. Both the displaced as well as the incident ion can suffer additional collisions with other atoms in the solid, resulting in a collision cascade which causes structural and/or electronic damage. The fact that every ion initiates a different

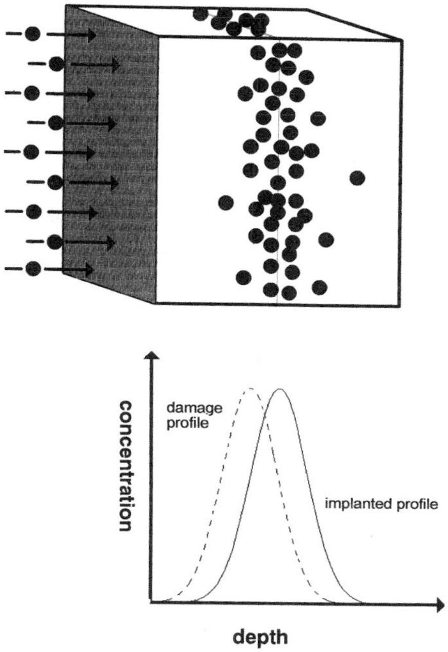

Figure 1. Ion implantation process and resulting alterations in the near-surface layers: ion profile and damage profile.

collision cascade leads to a statistical distribution of the implanted ions. In metals and semiconductors atomic displacements due to nuclear collisions are the main reason for the damage, whereas in insulators, especially polymers, the electronic interaction that leads to changes in the chemical composition (breaking and reformation of bonds) is the dominant reason for the introduced damage and therefore for the modification of the material. The actual electronic and nuclear energy losses and their relative magnitudes are dependent mainly on the properties of the target material, especially *Z*, on the ion mass, and on the ion energy. Both kinds of energy loss are shown in Fig. 2 as a function of the incident ion energy.

Figure 3 shows the energy losses for two extreme situations, the irradiation of polymeric material (low *Z*) with a light ion (He) and a heavy ion (Xe), respectively, demonstrating that in the case of light ions the electronic energy loss is dominant, whereas for Xe the amounts of the electronic and nuclear energy losses are of the same order of magnitude. The other important fact which can be seen from this figure is that the value of the energy loss increases dramatically with ion mass. In the case of the irradiation of metals

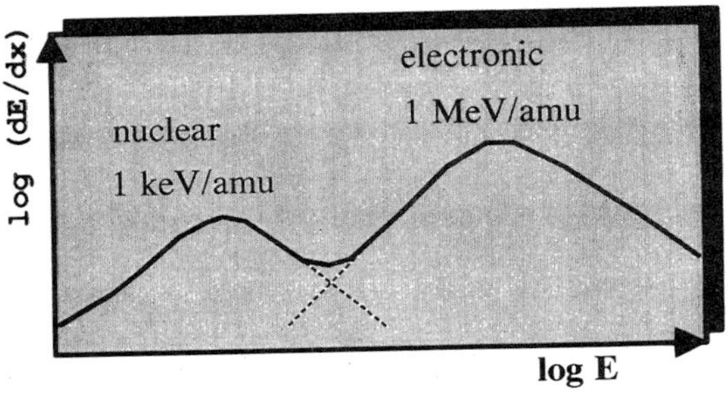

Figure 2. Energy loss as a function of the incident ion energy.

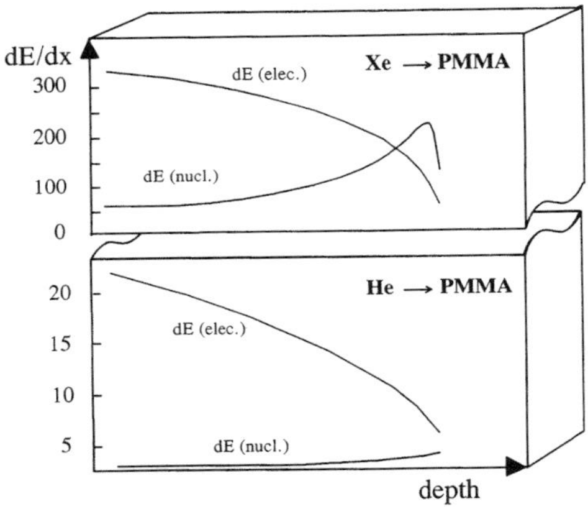

Figure 3. Electronic energy loss in a polymeric material (PMMA) for light ions (He) and heavy ions (Xe), calculated with TRIM [82].

with ions, the calculated energy loss is mainly nuclear. The range according to the nuclear energy loss is shown in Fig. 4, obtained with the program PROFILE. It is important to appreciate these facts in order to understand the ion-beam-induced changes in different materials that lead to changes in macroscopic properties.

Computer simulation programs have been developed in order to calculate and investigate the structure of cascades and the projected ranges of ions in solids [78–81]. These calculations are based on the assumptions that the collisions are of a two-body type and that energetic atoms collide only with stationary ones. The most popular program is the Monte Carlo code named TRIM (transport of ions in matter) [82]. This program does not take into account changes such as displacements and phase formation or the amorphization in-

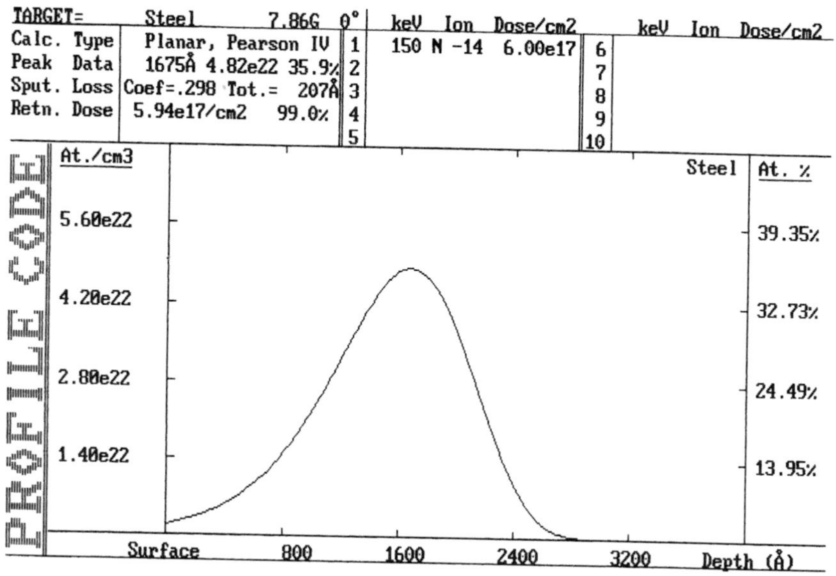

Figure 4. Calculated range of N in steel, at an ion energy of 150 keV and a dose of 6×10^{17} cm^{-2}. The profile maximum is at 167.5 nm. The dominant energy loss is nuclear.

troduced by the primary incident ions, which means that subsequent ions "see" a different structure.

Another commercially available computer code that calculates ranges, including sputtering, is PROFILE [83]. A high-dose module is included, as well as a diffusion module, which allows the calculation of diffusion after high-temperature treatment, if the diffusion coefficients are known.

An analytical simulation code has been developed at the TU Darmstadt [84] and includes the following processes:

- Stepwise creation of Gaussian implantation profiles, where the projected ranges are calculated with the stopping powers given by Kalbitzer [85] and convoluted with a width due to straggling.
- Ion-induced mixing due to recoil implantation, or cascade mixing leading to an additional broadening of the profile.
- Sputtering of the surface, taken into account [85] according to the ballistic theory of Sigmund.
- Dynamic change of the mean projected range and sputtering yield due to the changes in composition with increasing implantation dose using Bragg's rule.
- Diffusion and/or changes in composition are not included.

3. ION IMPLANTATION: INSTRUMENTATION AND PROCEDURES

The machines used to perform the process of ion implantation can be grouped into different categories. In general, an ion implanter consists of an ion generator, called a source, an acceleration unit, the ion transport system and the target chamber with target supports.

The simplest ion implanters are systems with an ion source, but without mass filtering, an acceleration unit, and a target chamber attached directly to the ion source. In current commercially available systems the ion source is often a "bucket" type delivering beams of gaseous elements [86] such as nitrogen or noble gases. A new development is the MEVVA ion source which delivers pure metal-ion beams. The advantage of these systems lies in relatively low price and relatively simple operation. The disadvantages are that

1. The ion beam is broad and cannot be scanned.
2. All ions produced in the ion source reach the target.

Larger targets or targets with complicated geometries therefore require movement of the target using expensive target manipulation systems.

Mass-analyzed ion implanters can be divided into low-, medium- and high-current machines which are used for different applications. Low-current implanters are used mostly in the semiconductor industry for *p*- and *n*-doping. The ion sources are of the Freeman type [87]. These machines are usually equipped with a mass-analyzing magnet in order to separate the ions according to their e/m ratio, thus allowing implantation of a single isotope in a defined charge state. Electrostatic focusing lenses and a scanning system are usually also included. The beam line is offset by 7° in order to avoid channeling. Target-handling facilities include electrostatic beam scanning systems to guarantee homogeneous implantation doses across the whole area of a wafer [88].

Ion implanters delivering higher beam currents have also been developed during the last 10 years. The ion sources are either of a Freeman or the bucket type CHORDIS [89, 90]. Both types exist in versions for gaseous elements, as oven versions for elements with higher vapor pressure, and as sputter versions for metallic elements. In Freeman sources a chlorination technique is often used to produce a gaseous compound from the solid feed material. The mass spectrum is then much more complex, because the chlorine compounds are present in the extracted beam prior to mass separation.

Beam focusing and scanning are usually performed with magnets. Most of the high-current implanters have two or three beamlines equipped with different target chambers and target systems (e.g., for heating or cooling, for mobile large-area targets, etc.). The target stage is usually water cooled in order to dissipate the power introduced by the beam, which can be up to several hundred watts. Large areas can be implanted using magnetic scanning systems, which in some cases cover an area of 70×70 cm^2 [91]. In this way either large tools or arrays of several tens of smaller components can be implanted. The two machines on the European market (Danfysik and until recently Wickham) are both computer controlled, which can reduce start-up times to less than 1 h for metal implantation. Figure 5 shows a schematic of the most recently developed high-current ion implanter (by Danfysik).

The latest development in the field of ion implantation is that of PIII [92–94], shown schematically in Fig. 6. The target to be implanted is positioned directly in a plasma source chamber at a pressure of ca. 2.6×10^{-2} Pa and then biased at a negative high voltage of 10–100 kV. The pulse duration and repetition rate determine the average current to the target and the efficiency of the implantation process. Target manipulation is unnecessary, because the ions are accelerated through the thick plasma sheet formed around the target, and bombard the target from all sides simultaneously. Careful plasma diagnostics are necessary in order to control the plasma parameters (e.g., plasma density), and target temperature must be monitored. Compared to conventional ion implantation the new PIII process minimizes problems of shadowing and time-consuming target manipulation. It has good prospects of becoming a low-cost type of ion implantation with respect to both capital and running costs. Nevertheless there are still several problems to be overcome, such as secondary-electron effects, cooling the target, etc., before the PIII method can be a viable commercial alternative process.

4. METHODS FOR CHARACTERIZATION OF IMPLANTED LAYERS

A modified surface layer created by ion implantation in a solid target has the following characteristics:

- The implanted layer is thin, typically between 0.1 and 1 μm in depth.
- Impurities such as C and H often follow the implantation profile.
- The concentrations of implanted elements lie between 1 and 50 at. % in the modified layer.
- The layer has a profile which is near-Gaussian.
- Phase changes can occur, during and after implantation, resulting in the creation of new phases.
- Novel chemical compounds can be formed.
- Changes in crystal structure (e.g., amorphization, changes in grain size, etc.) can occur.
- Macroscopic changes may occur due to the implantation (e.g., swelling, blistering).

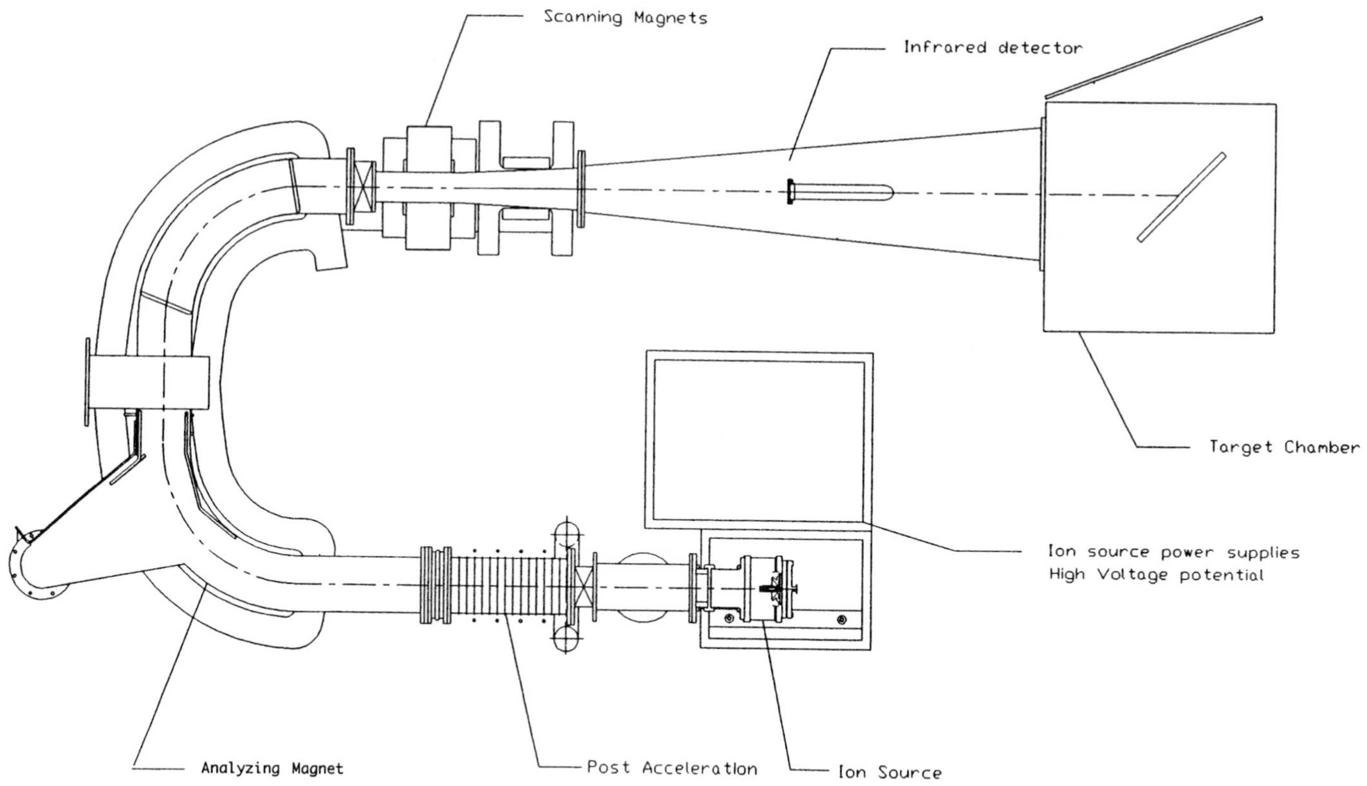

Figure 5. Schematic of the high-current industrial ion implanter, Model 1100, Danfysik.

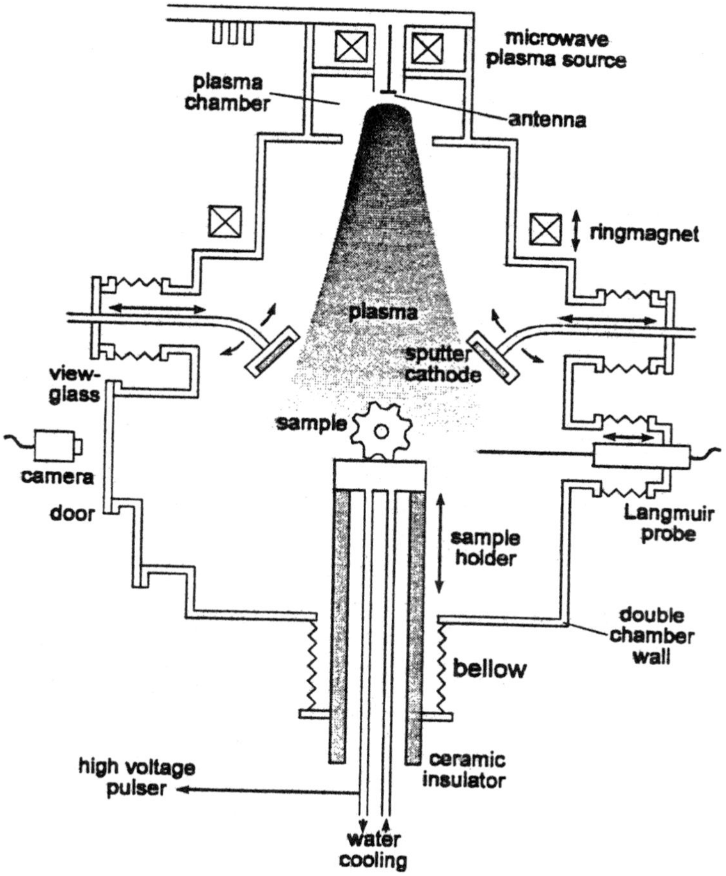

Figure 6. Schematic of plasma immersion ion implantation apparatus [70].

In order to obtain complete information about the modified microstructure of an ion-implanted layer, any characterization method should fulfill the following requirements, in relation to its detection capabilities:

- Measurement of absolute elemental concentration as a function of the depth beneath the surface (i.e., the implantation profile)
- Detection of phase changes as a function of depth and in relationship to the implantation profile
- Detection of different chemical species as a function of depth
- Detection of small quantities of new crystalline phases in a bulk matrix

A short list of the relevant characterization methods is given in the following survey. Only a few methods are discussed in detail where they are not described in Chapters 5–7 and in Appendix 3.

Nondestructive depth profiling of the elemental concentration can be carried out by high-energy ion beam analysis methods in general, including Rutherford backscattering spectroscopy (RBS), elastic recoil detection analysis (ERDA), and nuclear reaction analysis (NRA). An overview of these methods has been given recently in the *Handbook of Modern Ion Beam Materials Analysis* [95]. In general, RBS is suitable for the detection of heavy elements in a light matrix, because heavy elements have large cross sections compared to light elements. In addition, signals of the probe (He) scattered from heavy elements in a light matrix are well separated from those from the light elements, whereas signals from light elements in a heavy matrix are superimposed on the background of heavy bulk elements and have to be deconvoluted using computer codes such as RUMP [96]. In the case where resonances in the scattering cross section of the detected element occur and are larger than the scattering cross section of the matrix element, this general rule is not valid and light elements in a heavy matrix can then be analyzed. Generally if light elements in a heavy matrix need to be detected, ERDA is the better choice.

NRA gives information about the depth distribution of elements where a nuclear reaction can occur. One reaction often used for the profiling of hydrogen is N^{15} $(p, \alpha\gamma)C^{12}$ at a resonance energy of 6.385 MeV [97]. The detection limit for the absolute elemental concentration is 0.1 to 1 at. %. In special cases the detection limit can be extended to lower concentrations using standards of known composition. The detection limit for hydrogen using trace analysis techniques is currently 10 ppma [98].

The depth resolution in profiling the distribution with depth of implanted elements is a function of several parameters. The most important is the energy width of the incident ion beam. The depth resolution at a particular depth is additionally dependent on energy straggling in the sample, and scales with the mass of the matrix elements and the energy loss of the particles in the sample. In general it is 10 nm at the surface and between 100 and several hundred nm beneath the surface, and increases with depth. When using resonance methods for depth profiling the resolution is additionally dependent on the width of the resonance.

In order to detect amorphization of crystalline samples after ion beam irradiation, RBS under channeling conditions is a very useful tool [95], often applied to investigations of semiconductors and single-crystal ceramic samples or of diamond.

Secondary ion mass spectrometry (SIMS), secondary neutral mass spectrometry (SNMS), Auger electron spectroscopy (AES) and x-ray photoelectron spectroscopy (XPS) can provide depth-selective elemental profiles, but the method can be destructive. The destructiveness arises because the depth information is obtained by sputtering, starting from the original surface and progressing through the implanted profile into the depth of the solid while simultaneously detecting the emitted particles (atoms, ions or electrons). With SIMS and SNMS the ejected secondary ions and neutrals, respectively, are detected quantitatively as functions of the sputtering time. These methods are described in detail in Chapter 7. For the analysis of ion-implanted layers low sputter current densities and short time intervals must be used to obtain good depth resolution of the thin layer. Problems may arise due to inappropriate sputter and measurement times, while sputter artefacts can influence adversely the resulting profile [99].

During AES the sample is irradiated with electrons in the energy range 1–10 keV, which excite core electrons in the atoms of the solid being investigated. The emitted Auger electrons, produced by a two-step decay of core holes, are detected with kinetic energies that are specific to each of the elements present. The elemental peak intensities can provide information about the composition, and the peak shapes about the chemical environment. Due to inelastic scattering processes, the information depth is typically only 3 nm. The chemical states of the implanted elements and the chemical compositions due to ion implantation can also be studied using XPS. Here the excitation of the elements is by soft x-ray irradiation and the kinetic energies of the emitted photoelectrons are usually measured with an electrostatic concentric hemispherical analyzer (CHA). From measurement of the KEs, the BEs of the electrons can be determined, providing information about the chemical-state distributions of the elements present. For example, the kinetic energies of photoelectrons from Ti in the metallic state are shifted by several eV from that of Ti in TiO_2. To obtain good energy resolution, necessary to be able to distinguish between different chemical states, the measurement is carried out for each element over a relatively small energy range of about 10–20 eV. The depth specificity of XPS is typically less than 6 nm. As with AES, SIMS, and SNMS, depth profiling is performed by sputtering layer by layer and recording the appropriate XPS spectra after each sputtering cycle. In this way the chemical states of all elements present can be obtained as a function of the sputtering time. If the sputter rate of the material is known, the sputter time can be converted to a depth profile, but such a conversion can be complicated. Additional information can be found in Chapter 7 on depth profiling.

The microstructural changes to be discussed explicitly are the phase structural changes introduced as a result of implantation. There are two nondestructive methods which can give information about the above phenomena: Depth-selective conversion electron Mössbauer spectroscopy (DCEMS) and glancing-angle incidence x-ray diffraction (GIXRD), which must be considered as complementary methods. Destructive methods that can give information about grain sizes and precipitates are TEM and high-resolution TEM (HRTEM).

A brief description of the GIXRD method will be given, but only aspects related to the characterization of the very thin layers produced by ion implantation, since there are numerous publications dealing with the method in general. XRD techniques have been described in several review articles. The most important point is that the angle between the incident x-ray beam and the sample must be variable between 2° and 0.5° in order to be surface sensitive in the depth range of the implantation profile. The x-rays, which are scattered according to Bragg's criterion, are detected by an energy dispersive detector with a θ-2θ scan. Information about the phases is obtained by comparison of the intensity values at a specific θ with known values from a database (ASTM).

4.1. Phase Analysis by Mössbauer Spectroscopy

For nondestructive characterization of the microstructure of implanted layers conversion electron Mössbauer spectroscopy (CEMS) is a powerful tool [100–102], which can distinguish between different phases in solids and can in many cases provide chemical information. The Mössbauer effect itself describes the recoilless resonant emission and absorption of γ-quanta by identical nuclei due to the fact that the emitting and absorbing nuclei are incorporated in a crystal lattice, or more generally in a solid-state phase. If the conversion and Auger electrons emitted after the Mössbauer excitation are detected, a depth profile of phases can be obtained, similar to a depth profile of elements obtained from RBS analysis. This variation of CEMS is called DCEMS. The interaction volume has a depth of typically 300 nm with a resolution at the surface of 2 nm. As a result of recent progress in instrumentation [103, 104] that has reduced the measuring time needed and of an improved method of analysis [103–105], DCEMS has become a new powerful and practical tool in Mössbauer methodology and materials science. A depth profile of nonenriched samples can be obtained by measurements in a few days in the case of ^{57}Fe. Although the method has been developed for the Mössbauer isotope ^{57}Fe, other isotopes such as ^{119}Sn, ^{151}Eu and ^{180}W can be used. In general, Mössbauer isotopes can be implanted as probes in materials to atomic concentrations of $\approx 1\%$.

4.1.1. General aspects

The Mössbauer effect can be measured in two basic experimental arrangements, which are discussed later in more detail: transmission geometry and scattering geometry. In all versions there are a source and an absorber moving relative to each other, the Doppler effect being used to tune the energy of the γ-quanta and so to measure the energy of the resonance with a typical resolution of 10^{-13}.

Phase analysis by Mössbauer spectroscopy is possible due to the hyperfine interactions of the Mössbauer probe nucleus as follows:

1. *Isomer shift*: Electric monopole interaction of the nucleus with the *s*-electron density, which is affected by the valence electrons via shielding effects.
2. *Quadrupole splitting*: Interaction of the electric quadrupole moment of the nucleus with an electric field gradient produced by a noncubic crystal symmetry and/or by a nonsymmetric valence electron distribution.
3. *Magnetic splitting*: Interaction of the magnetic moment of the nucleus with a magnetic field.

The temperature dependence of these interactions is of particular interest for the interpretation of Mössbauer data. In the case of superparamagnetic phases the grain sizes of nanoparticles can be obtained by measurement at liquid He temperature.

The most important isotope for Mössbauer spectroscopy is ^{57}Fe (2 at.% natural abundance), using the 14.4-keV γ-transition between the 3/2 excited and the 1/2 ground states of the nucleus.

In principle two different measuring geometries, shown schematically in Fig. 7, can be distinguished:

1. *Transmission geometry (TMS)*: The resonant absorption of γ-quanta is measured. The signal comes almost exclusively from the bulk of the sample.
2. *Backscattering geometry*: (a) The detection of emitted conversion, Auger and secondary electrons results in phase sensitivity within the topmost ≈300 nm (CEMS) [105]. (b) The detection of emitted x-ray or γ-quanta will reveal again the bulk properties of the sample.

4.1.2. Depth-selective CEMS

As high-energy electrons of some keV in a solid suffer strong elastic and inelastic scattering, the monoenergetic Auger and conversion electrons are emitted from the sample surface with characteristic energies and angular distributions. With current computational capabilities the modeling of electron transport in matter can be performed by Monte Carlo simulations [105–107]. This gives the information necessary for deconvoluting the depth information from Mössbauer spectra that, by the use of an electron spectrometer, have been recorded at different electron energy settings. As an example, the resulting phase and depth information that is revealed by this special kind of Mössbauer technique (DCEMS) is shown in Fig. 8 for an implanted sample [108, 109]. For iron the depth of the interaction volume of the method ranges from 0 to ca. 300 nm.

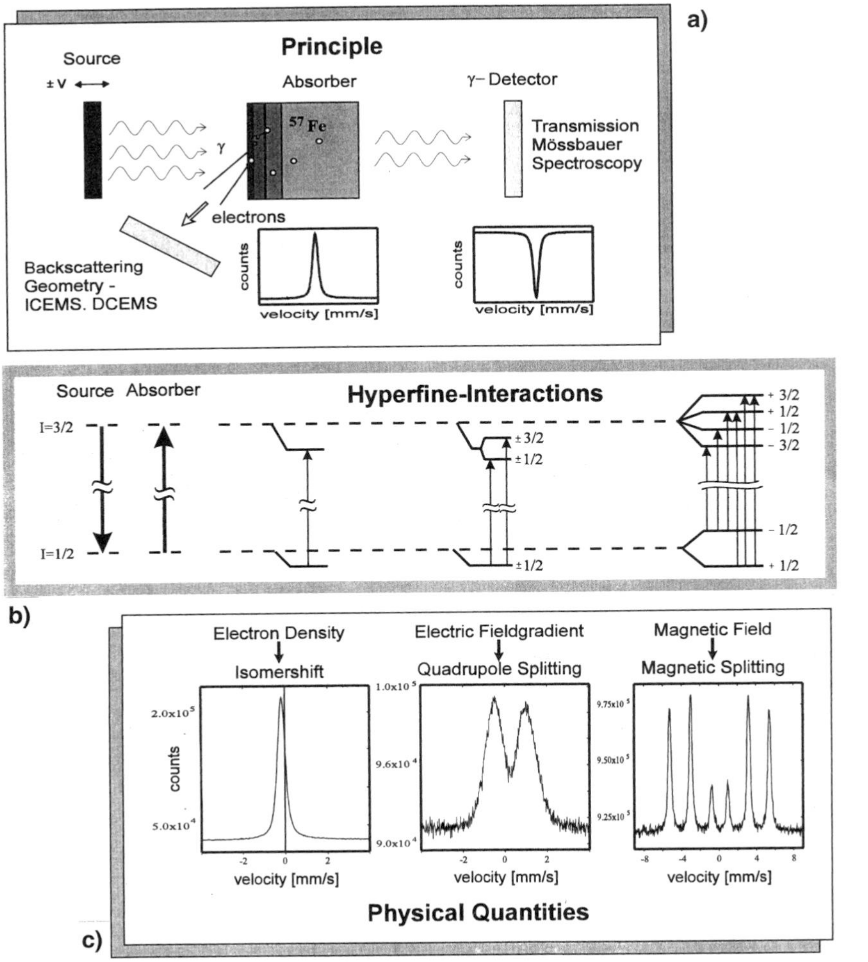

Figure 7. Principles of the Mössbauer effect: (a) measurement geometry; (b) hyperfine interactions for the ^{57}Fe isotope; (c) examples of Mössbauer spectra.

Eu was implanted into a stainless steel foil ($^{57}Fe_{62}Ni_{20}Cr_{18}$) at 400 keV and doses of (a) $1.2 \times 10^{17}/cm^2$ and (b) $6 \times 10^{16}/cm^2$, respectively. Analysis of the Mössbauer spectra for dose (a) led to the assumption of four phases: the *stainless steel* matrix, a *martensitic phase* with a broad hyperfine field distribution, a *magnetic phase* with a well-defined hyperfine field of 33.6 T and an *oxide phase* similar to FeO. Since a set of 10 Mössbauer spectra was recorded at dif-

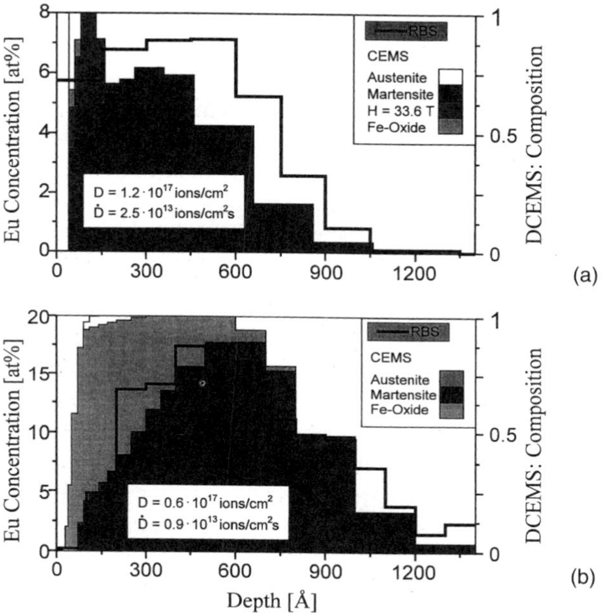

Figure 8. Depth profiles of Eu^+ implanted into stainless steel, at doses of (a) 1.2×10^{17} cm^{-2} and (b) 6×10^{17} cm^{-2}.

ferent electron energies, the depth distributions of the phases could be obtained by a least-squares analysis while assuming Gaussian-shaped profiles. They are plotted in Fig. 8 in which it can be seen that the magnetic phase at the very surface of the sample (0–15 nm) is separated clearly from the bulk phases. This is possible only in the depth-selective mode of Mössbauer spectroscopy. In integral-mode CEMS the contribution of this layer to the Mössbauer spectrum would have been too small to be observed. Further information can be obtained about the martensitic phase. The shape of the depth profile of that phase agrees very well with the RBS depth profile of elemental Eu. This implies at once that the implanted Eu ions have caused a stress-induced transformation of austenite into martensite [110]. Yet more can be gleaned about the oxide phase. The iron oxide is found in only a very thin layer of at most 4 nm thickness and at a maximum concentration of 10 at.%. This demonstrates that the implantation was performed under very clean conditions. In contrast to this, Fig. 8 shows the profile of a sample where the iron oxide phase extends across nearly the whole implantation profile. Thus the quality of the implantation procedure could be checked by depth-selective Mössbauer analysis. Finally, information also revealed by the depth analysis concerns the presence of a nonresonant adsorbate layer. This arose from the residual gas at a pressure of 2×10^{-4} Pa in the vacuum chamber of the Moll/Kankeleit version of the Orange spectrometer (see below).

The DCEMS analyses of the samples mentioned have demonstrated that this type of Mössbauer technique can be extremely valuable for the characterization and determination of phases in inhomogeneous samples, typical in the implantation field. Recently, the group of Prof. Kankeleit at TU Darmstadt has examined the applicability of DCEMS as a general method of analysis in the field of material science. Two goals have been attained, thereby establishing DCEMS as a powerful tool in fundamental and applied research.

1. In order to overcome the normally very low counting rates in a DCEMS experiment, a new UHV (10^{-7} Pa) magnetic spectrometer of the Orange type (so-called because of its geometry) has been designed and built by Stahl and Kankeleit [103, 104], based on 30 years' experience [100–112] with an earlier, poorer vacuum, version. This variation on the Orange spectrometer was invented by Kankeleit and Moll [111]. The principal feature of the new instrument is, compared to electrostatic spectrometers, an exceptionally high transmission due to the large acceptance angle, typically 20% of the 4π solid angle, at a relative energy resolution of 0.5–2%. A second feature is a bandwidth that covers 8% of the electron energy setting E, allowing the use of a multichannel detector at the focus of the spectrometer, so that Mössbauer spectra can be recorded in parallel for five electron energies. The effective luminosity of the system is increased thereby by an additional factor of 5. The improvements allow routine analyses to be carried out on nonenriched samples and permit measurements at low and high temperatures (for details see [103, 104]), avoiding systematic errors caused by possible drifts in experimental parameters during long measuring times.
2. Second, progress has been made in the theoretical approach to the analysis of DCEMS measurements. The underlying theory of resonance excitation (→ **excitation matrix**) in the sample by γ-quanta including secondary absorption and emission processes, electron transport (→ **transport tensor**), and detection response (→ **response function**) have been included in a least-squares fit routine [103, 105]. Adjustable parameters in this fitting are, on the one hand, the hyperfine interaction and line shape variables, and, on the other hand, the variables that give a parametric representation of the depth profiles. The response parameters are also included to allow energy calibration of the experimental apparatus.

The transport tensor is generated by a Monte Carlo simulation that includes differential cross sections for elastic scattering (screened Rutherford potential), inner-shell ionization, and dielectric losses by plasmon and electron hole excitation [103–107]. In Fig. 9 a reduced representation of the tensor is plotted for the experimental situation in the Orange spectrometer (acceptance angle

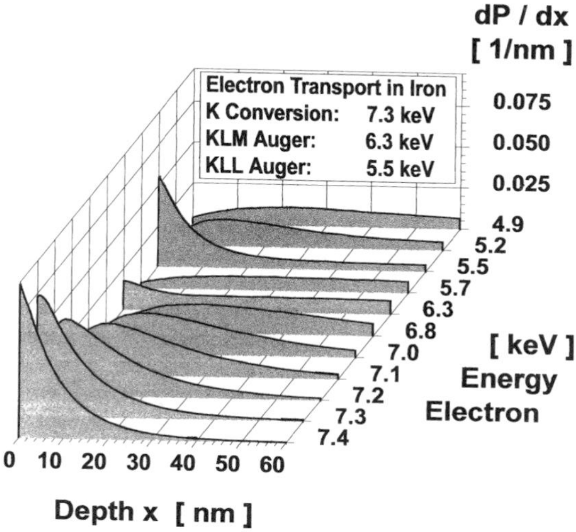

Figure 9. Electron transport in Fe. The calculated curves represent the dependence on depth for several electron energies over the range 4.9–7.4 keV. An acceptance angle of 30–70° and an energy resolution of 1% have been assumed.

30°–70° with respect to the surface normal of the sample, and energy resolution 1%) and the Mössbauer isotope ^{57}Fe. The plotted functions give the weights of the depths for several electron energy settings. Since the whole set of Mössbauer spectra from a DCEMS experiment is fitted simultaneously in the routine FITCEMS, the depth resolution of the method is not given simply by the width of the plotted "weight functions" but by their varying overlap at different depths, dependent still on the statistical quality of the data. The resolution is thus increased significantly and is limited only by the number and the statistical quality of the recorded Mössbauer spectra. It is not possible in general, therefore, to judge the depth resolution of such an experiment since the resolution also depends strongly on the system under investigation. But as a rule of thumb the depth resolution of the method is ca. 2 nm at the outer surface and ca. 30 nm at a depth of 100 nm. A routine for the complete deconvolution of the data by the transport tensor still needs to be developed.

The new experimental arrangement is shown in Fig. 10. On the left-hand side the sample geometry shows the absorption of γ-quanta from the source and the emission of electrons into the acceptance angle of the spectrometer. The electron trajectories are bent in the toroidal magnetic field of the spectrometer, generated by an electric current through an iron-free coil. Electrons with the "correct" energy are focused onto the circular entrance aperture of the single-

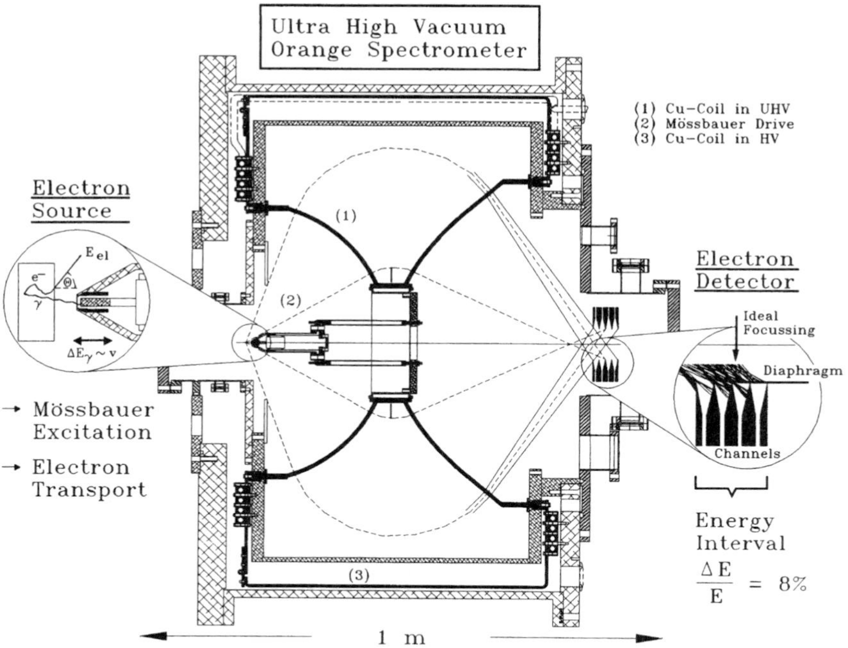

Figure 10. Schematic of the new UHV Orange spectrometer. The Mössbauer source is shown in inset (left). Absorption of a γ-quantum and the emission of an electron are indicated. A schematic view of the multiple-channel detector is shown in inset (right). The detector in the inner UHV chamber is pumped differentially.

channeltron detector. In the parallel mode an arrangement of 40 ceramic channeltrons (individual dimensions $3 \times 10 \times 20$ mm^3) provides five energy channels each split into eight angular channels. In order to establish the necessary UHV conditions with the avoidance of stray magnetic fields, a differentially evacuated double-chamber system made of a special aluminum alloy has been constructed. The earth's magnetic field in the laboratory is compensated by three orthogonal pairs of Helmholtz coils. Electron energies from 10 to 45,000 eV can be resolved by the spectrometer.

In order to visualize the experimental and theoretical requirements in DCEMS analysis, Fig. 11 shows the mesh plot of a function of two variables, i.e., the resonant count rate as a function of both Mössbauer velocity and electron energy. The mesh surface shows the linear interpolation of a set of Mössbauer spectra that were recorded at 17 different electron energies with the UHV Orange spectrometer set to an energy resolution of 2%. The constant

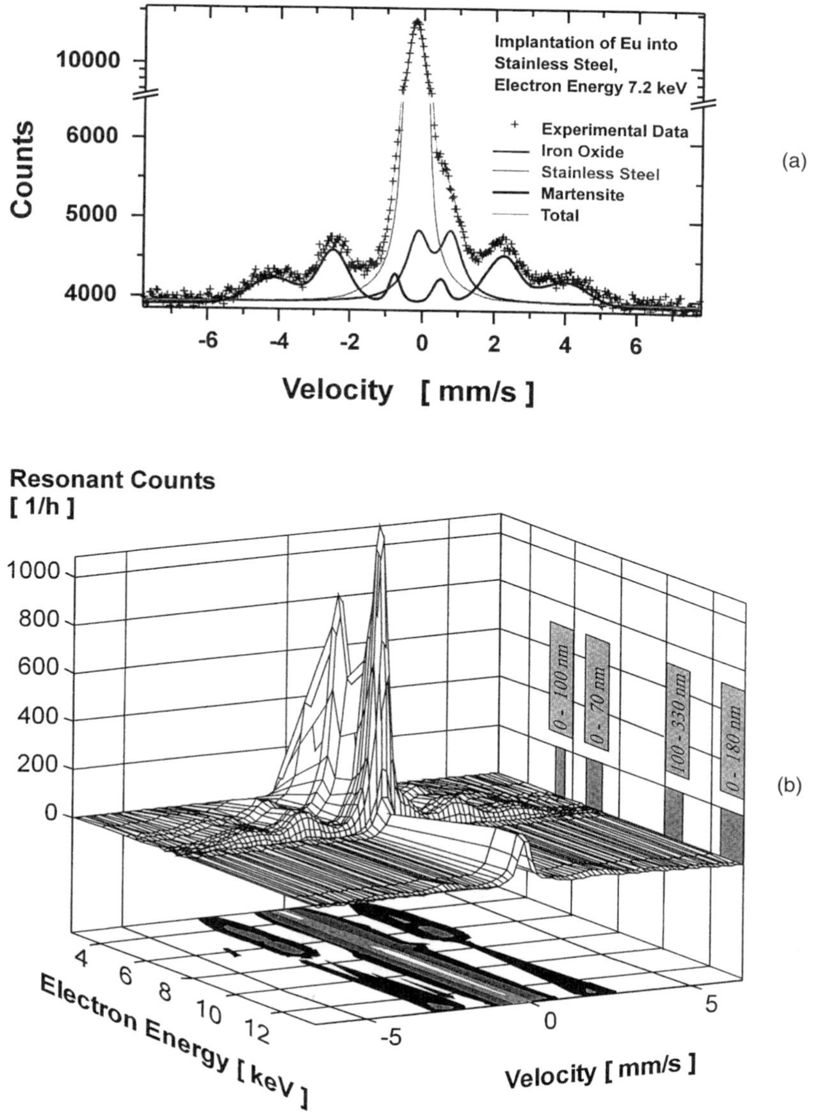

Figure 11. (a) Structure of a Mössbauer spectrum that was measured at an electron energy of 7.2 keV. The asymmetric shape of the martensitic contribution can be seen in the contour plot in (a). A resolvable oxide contribution appears only at electron energies of 7–7.5 keV. (b) Plot of the count rate as a function of the Mössbauer velocity and electron energy. The mesh surface consists of 17 individual Mössbauer spectra that were measured over an electron energy range of 4–14 keV, with a relative resolution of $\Delta E_d/E_{el} = 2\%$. The nonresonant background was prepared by implanting Eu into stainless steel. The contour plot allows the change in the spectral composition with the electron energy to be visualized.

background at each electron energy setting has been subtracted. The sample was prepared by implanting $3 \times 10^{16}cm^{-2}$ Eu into stainless steel at 400 keV. In addition to the single line of the host matrix at velocity –0.20 mm/s relative to Fe, two components appear after implantation (see Fig. 11b): a doublet attributed to an iron oxide and a magnetically split component with a broad hyperfine field distribution attributed to the martensitic transformed phase. The spectral contributions of these three components apparently change with the electron energy and lead to a "phase depth profile" by a least-squares analysis that models a mesh surface as close as possible to the measured one.

It turns out in practice that the magnetically split contribution (→ **martensite**) cannot be described by a single component; the contour plot shown beneath the mesh plot in Fig. 11a gives an indication of the splitting. Contour lines of the function in Fig. 11a were calculated and normalized to the integral over velocity at each electron energy setting. It can be seen that the outer parts of the martensitic component that do not overlap with the stainless steel and oxide contributions are not symmetric about the center. The simultaneous fit of the spectra improves significantly when two different average magnetic splittings are introduced. A further significant improvement is achieved when both contributions are allowed to have different isomer shifts. The result from such a model is that the average magnetic hyperfine field as well as the electron density at the iron probe nuclei change significantly with sample depth.

5. EXAMPLES OF THE APPLICATION OF ION IMPLANTATION

Ion implantation has been applied very widely, especially as a method of studying special problems connected with the modification of surface properties. Nearly all materials groups have been treated with ion beams to modify their material properties. Metals, mainly steel alloys and titanium alloys, have been implanted with N, C, O and metal ions to improve their wear and corrosion behavior [112]. Metals with important prospects in the future have been the objects of only a few investigations [113, 114, 115]. Ceramics, mainly single crystals, have been investigated in order to understand radiation damage [116], whereas composite materials, such as metal/ceramic composites, allow one to learn about diffusion processes [117]. Polymers have been irradiated with ions so as to alter their electrical, optical and mechanical properties, because their importance is increasing due to their light weight and low price. Polymer/metal composite materials have been investigated in order to improve the adhesion between metals and polymer substrates for applications in the electronic industry and as protective coatings.

In this section a number of selected examples coming from different areas of possible application are discussed. The reasons why these examples have been chosen is that all the investigations have been carried out in collaboration with those research institutes and industrial companies that have a particular application or end product already in mind that could be achieved by ion implantation modification. The examples illustrate how certain classes of surface and interface problems arising from industrial and/or technical requirements can be solved by ion implantation methods.

1. The objective (the problem to be solved) was to reduce the wear in medical endoprothesis implants fabricated from a titanium alloy.
2. The objective was to improve the properties of traditional Cr coatings.
3. The objective was to treat polymers so as to produce optical waveguide layers, as well as to explore a promising method for making optical devices needed for optical fiber technology.

5.1. Improved Surface Properties in Medical Endoprosthesis

5.1.1. Introduction

In the field of medicine there is a need for materials for making artificial prostheses that are biocompatible. They should be able to withstand corrosion attack in the human body and the mechanical load in a joint prosthesis. The load-bearing components in an artificial joint prosthesis are made from either a cobalt-base alloy or the titanium alloy Ti6Al4V [118, 119]. The matching components in an artificial joint are the acetabular cup in a hip and the tibia plateau in a knee, which are made from ultrahigh-molecular-weight polyethylene (UHMWPE) [120]. For patients who show allergic reaction to the constituents Ni, Cr and Co in the cobalt-base alloy the titanium alloy Ti6Al4V is more suitable. When unmodified Ti-alloy and UHMWPE are combined in a sliding couple, e.g., in an artificial knee, the tribological properties (that is, the wear resistance of the surfaces) are not adequate [121]. Wear can cause abrasion of TiO_2 particles leading to black staining of the surrounding tissue and failure of the joint [121]. It is necessary to be able to improve the surface properties in such a way that the wear behavior is improved without losing the corrosion resistance. Of course, any additional materials used for the surface treatment must also be biocompatible.

Ion implantation, coating by plasma methods, and anodic oxidation [122] have all been used to improve the wear behavior of Ti6Al4V surfaces. Coating with DLC layers [123] and with TiN by PVD have been successful in decreasing the wear of the spherical alloy components (the "balls") in hip joints and have been extensively investigated [124]. For knee joints with their rather complicated geometry, ion implantation seems to be the preferred method. The effect of N implantation has been demonstrated in several tribological tests using pin-on-disk and annulus-on-disk geometries [125–133] and hip joint simulators [129, 133, 134]. C and O implantation have also been shown to be beneficial [126, 130, 131]. The influence of temperature during ion implantation on the properties of the modified layer has been investigated, with the finding that implantation of N at low temperatures (below 250°C) leads to better wear behavior than at 500°C [133]. Implantation of noble metals is another method of reducing wear provided the implanted layer is adjacent to the oxide layer on the surface [135]. Artificial joints made from Ti6Al4V with surface properties improved by ion implantation of N are already commercially available.

Implantation of N, C, and O, which can form hard bulk phases with Ti and V, has been studied using various ion energies and doses in order to obtain information about how the depth of the implanted profile below the surface influences wear behavior. Additionally, the combinations N + C and C + O have been implanted. Yttrium, which is immiscible with the elements of the Ti6Al4V alloy, has also been implanted to investigate the effects of hardening by lattice distortion. The contribution of intermetallic phases has been studied by the implantation of noble metals and of elements such as hafnium. Hardness measurements were performed under load using a Vickers indentor [135]. Wear tests with pin-on-disk test and annulus-on-disk geometries have been carried out in Ringer solution [137, 138]. In the case of the pin-on-disk test the disk was made from Ti6Al4V and the pin from UHMWPE, while in the annulus-on-disk test the annulus was made from Ti6Al4V and the disk from UHMWPE. A detailed description of the test facility is given in [138]. The electrochemical behavior of the implanted samples was investigated by anodic polarization measurements, described in more detail in [139]. The implanted profiles were analyzed using RBS, HEBS (high-energy back scattering spectroscopy), ERDA and XPS. To understand the induced microstructural changes (phases and precipitates) XRD was used.

5.1.2. Results

The microhardness of a surface is an indication of the resistance to abrasive wear; that of the surface of unimplanted Ti6Al4V has a value of 3000 N/mm^2 [136]. The hardness values obtained under load with a Vickers indentor are summarized in Fig. 12, where the hardness is plotted as function of the applied indentation load. The sample implanted with N showed a more pronounced increase in hardness than those implanted with C. Implantation of N at two energies led to an even greater increase than that at only one energy and also to a thicker layer of increased hardness.

Results showing the improved tribological behavior, as obtained in the annulus-on-disk tests, are documented in Fig. 13. The heights of the two bars represent, respectively, the number of scratches on the annuli and the wear

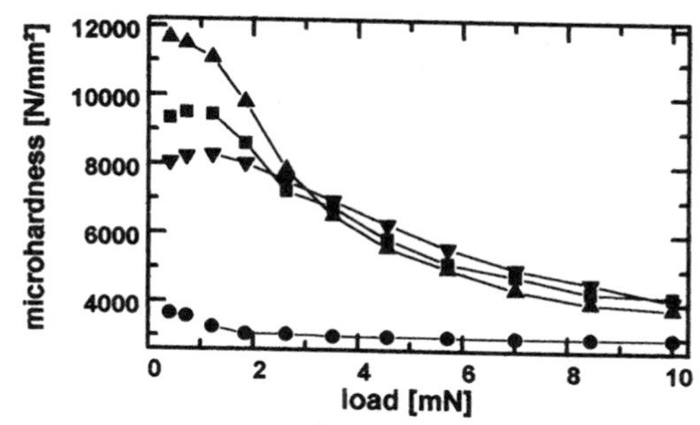

Figure 12. Hardness of implanted Ti6A14V samples measured as a function of the applied load.

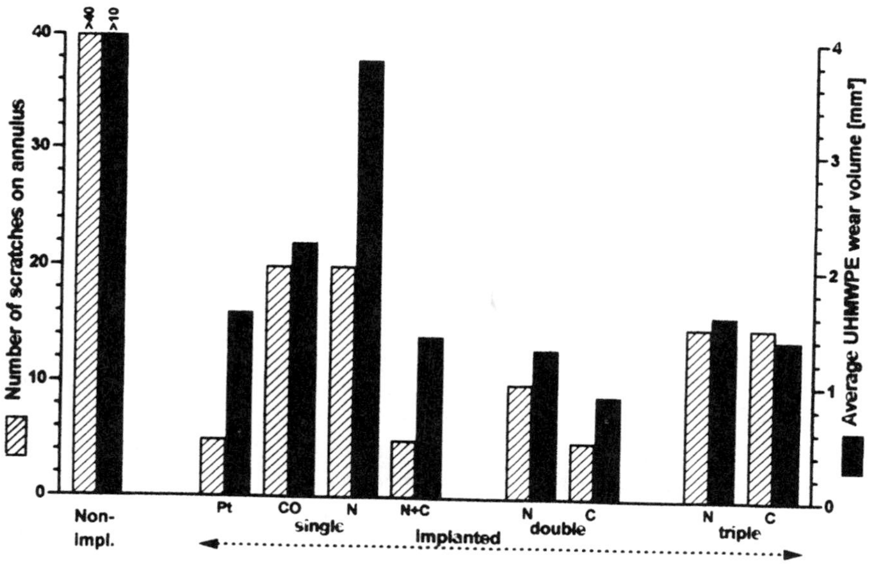

Figure 13. Results of the annulus-on-disk test for various implanted elements and for implantation parameters (ion dose and energy).

volumes of the UHMWPE counterbodies. The micrograph in Fig. 14 shows some examples of scratches on an annulus, their number being a measure of wear. In Fig. 15 counterface made from UHMWPE can be seen showing little wear (right disk) and severe wear (left disk). In addition to the obvious reduction in the wear volume in ion-implanted Ti6A14V annuli, the simulta-

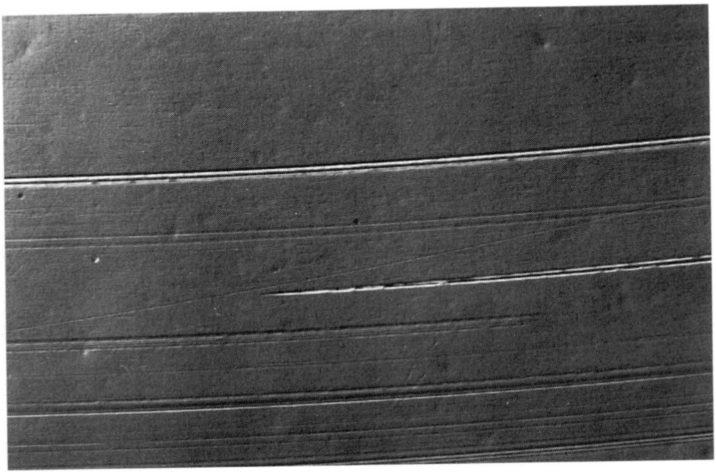

Figure 14. Surface of a Ti6A14V sample after a wear test. Scratches indicate wear at the surface.

neously measured friction torque gave lower values for the ion-implanted samples.

After the initial wear test on each annulus had been in progress for some time, a polyethylene film adhering to the alloy surface was formed, which could be removed with care, whereupon the wear grooves became visible. Gouging of the surface is caused by loose particles of Ti-oxide which become detached from the alloy. It can be seen that ion implantation leads to a significant reduction of the number of scratches. The wear volumes of the UHMWPE disk were also reduced for ion-implanted annuli.

The results of the corrosion tests, based on anodic polarization measurements, are summarized in Fig. 16. Unimplanted Ti6Al4V, which is known to have good resistance to corrosion, remains passive in Ringer solution (0.155 chloride solution, consisting of 8.6 g/L NaCl, 0.30 g/L KCl, 0.33 g/L $CaCl_2 2H_2O$, pH=7.4) up to the highest applied potential of 3000 mV. This is due to anodic oxidation followed by formation of a oxide layer on the surface. Implantation of N and C caused a decrease in the breakdown potential. Further increase in the applied potential led to a breakdown of the passive layer and the occurrence of pinholes. A relationship between the breakdown potential and both ion energy and dose was observed for all implanted elements. N implantation with ion doses exceeding 6×10^{17} cm^{-2} caused a decrease in the breakdown potential. For C the critical dose is 3×10^{17} cm^{-2}, and again a higher ion dose led to a lower breakdown potential. Implantation of C to the same ion dose (6×10^{17} cm^{-2}) as N led to a lower breakdown potential, but if the two ele-

Figure 15. UHMWPE disks after wear testing. Heavy wear is indicated by a deep track, presence of debris and the appearance of black particles (left).

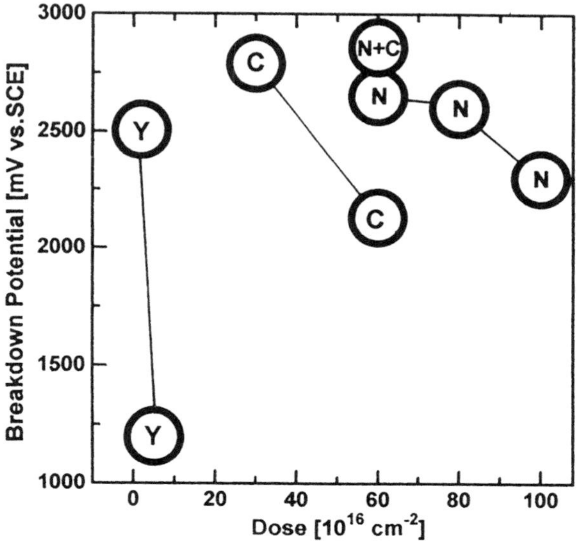

Figure 16. Results of anodic polarization measurements. The effects of increasing implantation dose are indicated by the lines joining the circled implanted elements.

ments were implanted together to the same total dose, the breakdown potential was in fact much higher (2860 mV). Following implantation of C and O together no breakdown of the passive layer occurred even at the highest applied potential of 3000 mV, nor were any pinholes observed.

An example of the distribution of implanted atoms is given in Fig. 17 for the profile of CO^+ implantation, as measured by HEBS. The maxima of implanted C (29 at.%) and O (23 at.%) are reached at a depth of ca. 170 nm with a FWHM of the layer of ca. 200 nm. Due to the large cross sections the yields of C and O are strongly enhanced with respect to that of the substrate. The formation of an oxide layer in air after implantation leads to an O concentration at the surface higher than in the implanted layer. It can also be seen that the implantation profiles are nonzero at the surface, implying that there are implanted atoms in the outer surface layers, an observation that turns out to be important for the wear process.

In earlier work [140] XPS measurements had been performed. It was shown that the distribution of the phases formed as a result of implantation followed the profile of the implanted atoms. This was identified by assigning the shift in the energy of the peak position of the metal peak (Ti) to the energy of the related metal nitride. The XRD measurements shown in Fig. 18 were performed at an incident angle of 0.5°, corresponding to an information depth of about 170 nm, i.e., coinciding with the maximum of the atom distribution. In addition to

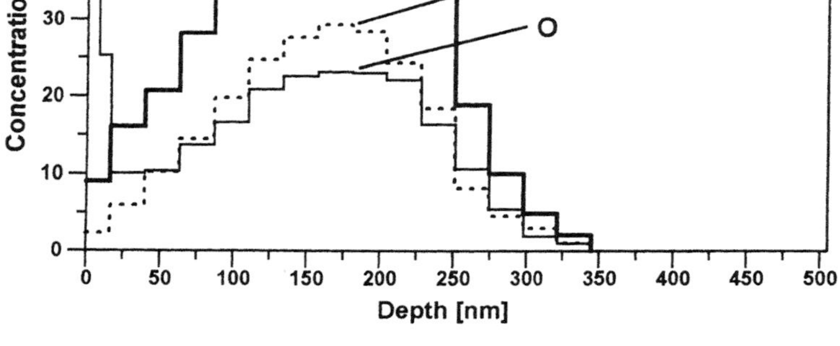

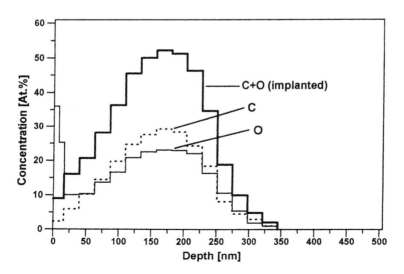

Figure 17. Depth distributions of implanted C and O, obtained by HEBS analysis.

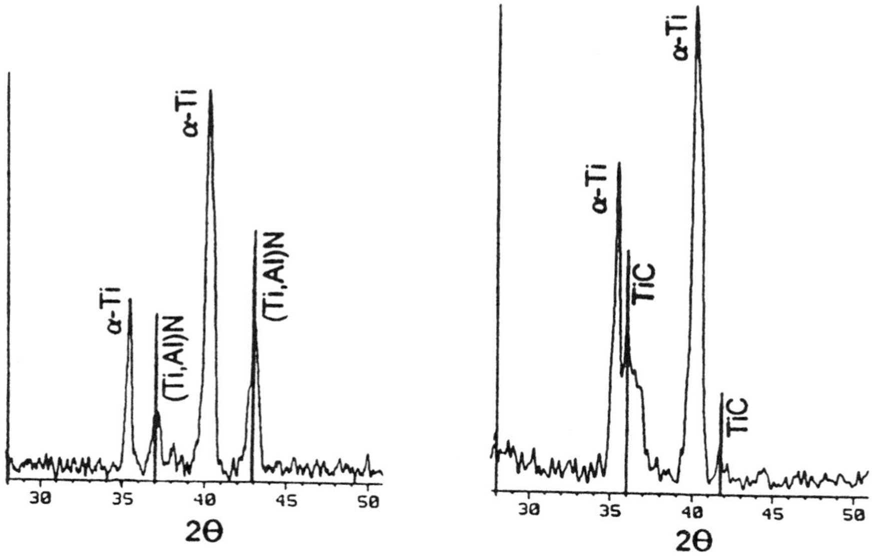

Figure 18. XRD analyses of implanted Ti6AL4V. The intensity axis is in arbitrary units.

the α-Ti phase the formation of (Ti,Al)N and TiC phases can be seen after implantation of N and C, respectively. The implantation of N+C and C+O led to the formation of TiN and TiO phases, respectively, which in both cases are those phases with the lowest formation enthalpy [141, 142]. The β-Ti phase disappeared, which is understandable, since N, C and O are known to stabilize α-Ti. No formation of the ternary phases Ti_3AlC or Ti_3AlN could be observed.

When N and C were implanted at lower energies, continuous layers of TiN and TiC phases were formed [142, 143]. In the case of N implantation at three different energies a continuous layer was again formed and resulted in discoloration of the surface. During implantation of N or C at doses exceeding 2×10^{17} cm^{-2}, finely dispersed TiN and TiC precipitates were formed [143]. After CO^+ and CO_2^+ implantation the precipitates consisted mainly of TiO [135]. The grain size of the titanium compound precipitates was 5–30 nm for low doses [143] and increased with ion dose [144–146]). C or N implantation at low ion energies ($E < 70$ keV) and higher ion doses ($D > 7 \times 10^{17}$ cm^{-2}) led to the formation of continuous TiC or TiN layers, respectively [147, 148]. Implantation of N ions at an angle of 45° led to an enrichment of the implanted atoms in a subsurface layer due to their lower penetration depth, resulting in the formation of a continuous layer and discoloration of the surface [139].

5.1.3. Discussion and conclusions

The best tribological results were obtained after implantation of N at two different energies and of C. This could be attributed to the formation of small precipitates of TiN, (Ti,Al)N or TiC. Implantation at low energies and at doses exceeding ca. 10^{18} cm^{-2} resulted in poor tribological behavior, presumably due to the formation of continuous TiN or TiC layers. Finely dispersed precipitates seem to stabilize the surface layer, preventing the creation of black TiO_2 particles and thus improving the wear characteristics in the sliding against UHMWPE.

The corrosion behavior after implantation showed a different trend. After implantation of C+O and of metal ions such as Pd or Hf [135], no passivation or breakdown and no pinhole formation could be deduced from the current versus potential (mV) curves. After Y implantation a dramatic decrease in the breakdown potential occurred [139].

The Ti6Al4V samples implanted with N and C separately showed a lower breakdown potential than the unimplanted ones, but a larger effect was observed if C and N were implanted together.

The formation of different phases has a considerable influence on the corrosion behavior. Implantation with Pd leads to the formation of intermetallic phases (Ti_2Pd), and with Hf to (Ti,Hf) mixed phases, both of which improve the corrosion behavior. Implantation of Y, on the other hand, causes lattice distortions because of the immiscibility of Y and Ti, leading to a decrease in the breakdown potential and to the formation of many pinholes.

In the case of the implantation of N and C, the microstructural investigations revealed the formation of precipitates that had a beneficial influence on the macroscopic wear behavior, contrary to those implantation conditions in which a continuous layer was formed. If the layer enriched in precipitates formed by ion implantation extends as far as the surface, the number of scratches is reduced and wear behavior is improved. The most pronounced reduction in the number of scratches and in the wear volume of the UHMWPE counterpart was achieved for N or C implantation at energies that allowed the precipitates to be formed immediately beneath the outer oxide layer. If the formation of a continuous layer is to be avoided, the total ion dose should not exceed 10^{18} cm^{-2}.

Implantation with N alone at two energies, not exceeding a total ion dose of 6×10^{17} cm^{-2}, has been found to be effective in the improvement of the surface quality of knee prostheses. Nevertheless combined N and C implantation ought to be investigated more carefully, because of the resulting superior corrosion properties.

The combination of macroscopic wear tests, corrosion tests, and study of the microscopic changes induced by ion implantation suggests a possible quality control procedure based on measurement of the implantation profiles and on corrosion tests. Both can be performed quickly using small test samples mounted alongside the knee joints in the implantation chamber.

Improvements in tribological behavior combined with the possibility of quality control tests should lead to a better acceptance of knee joints made from Ti6Al4V, not just for allergy patients, but generally.

5.2. Modification of Chromium Layers by Nitrogen Ion Implantation

5.2.1. Introduction

Chromium is a technologically important plating material because the films produced are highly reflective, relatively hard, corrosion resistant and have good adhesive properties. However, the use of conventional Cr layers deposited from a Sargent bath [148] is limited to low-temperature applications (generally below 50°C) because of softening at elevated temperatures. Several years ago a new method for the electrodeposition of Cr films was described [149], called the amorphous bright-chromium deposition method (ABCD). The resulting ABCD films are more resistant to corrosion in hydrochloric acid, have smoother surfaces, contain fewer cracks and hole defects, and their hardness increases after annealing at temperatures up to 600°C, thus opening up a wide field of applications for them. N implantation, known as a method of increasing the hardness of steel containing high Cr contents, was used to study the effects of increased hardness after implantation into conventional and ABCD Cr layers.

5.2.2. Experimental procedures

Conventional Cr layers were prepared on low-carbon steel from a solution containing 250 g/L of chromic acid and 2.5 g/L of sulfuric acid at 50°C and a current density of 4×10^3 A/m^2. The samples were cleaned ultrasonically with ethanol and annealed at 600°C in vacuum (1.3×10^{-3} Pa) for 30 min. ABCD Cr layers were prepared from a solution containing -CHO and -COOH groups [150]. Both types of sample were implanted with N ions at various energies, doses and temperatures. AES analysis was performed using a Perkin-Elmer PHI 590 system with a 5 keV primary-electron beam energy and a target current of 1 μA. Depth profiling was carried out using a 4 keV Ar^+ beam with a 4×4-mm^2 sputtered area. In addition, XPS and ISS measurements were made in order to investigate the chemical nature of the implanted N [151]. Surface hardness was measured using a Leitz microhardness tester with a Knoop diamond and an applied load of 10 g.

5.2.3. Results

The hardness of conventional Cr films (unannealed) implanted with N at 90 keV, as a function of the total ion dose and the load, is shown in Fig. 19. For

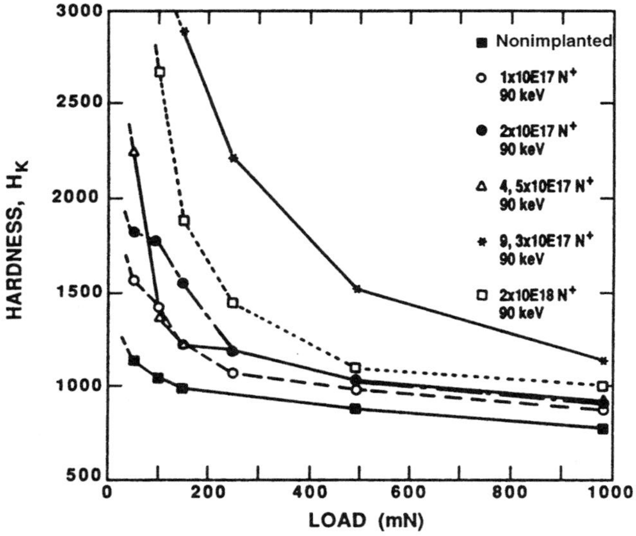

Figure 19. Hardness of N-implanted conventional Cr films as function of load and for various ion doses.

implanted and unimplanted films, the hardness increased as the load decreased, indicating that the near-surface region had a greater hardness than the bulk. The implanted samples gave higher hardness levels both at the surface (low loads) and with depth (increasing load) than the unimplanted, indicating increased hardness due to ion implantation. At high loads the situation is complicated slightly by the fact that the measured hardness values include contributions from the implanted layer and the bulk. This occurs because at high applied loads the indentor penetrates the implanted layer. In further work higher ion energies have been used to obtain thicker implanted layers [152].

Figure 20 shows an AES depth profile taken from the N-implanted conventional Cr layer showing film composition as a function of depth. The predominant constituents of the film are O, Cr and N. The outermost layers of the film consist of oxides. The implantation has produced a broad N profile down to 250 nm with a maximum concentration of 33 at. % at a depth of 70 nm. If a higher ion dose were to be implanted, it would result not in a higher concentration but in a broader profile. The composition of the implanted layer does not quite reach the stoichiometry of CrN, which would correspond to 50 at.% N if all Cr were converted into CrN. It seems that in this region a significant amount of Cr is bound to O.

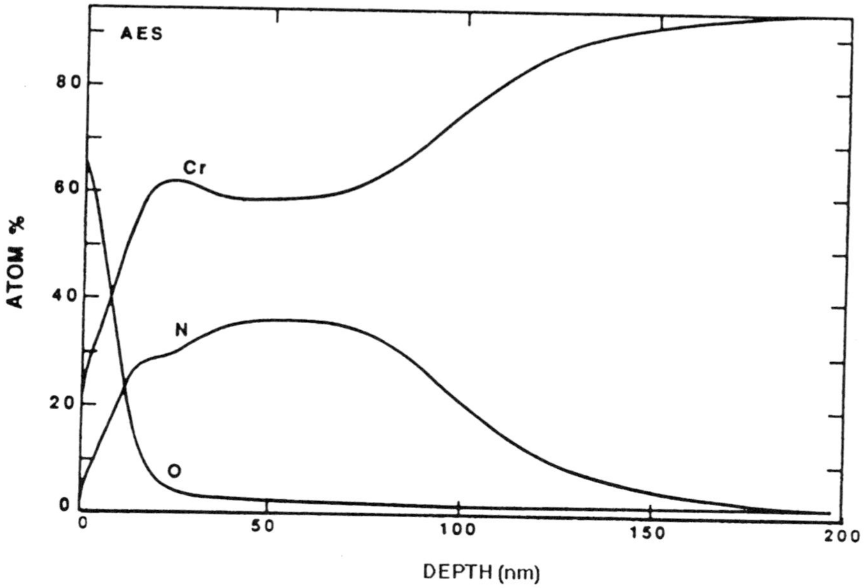

Figure 20. AES depth profiles obtained from an ion-implanted conventional Cr layer. Composition is shown as a function of depth. Zero depth corresponds to the original surface. Only the most prominent changes in Cr, N and O concentrations are shown. [149].

Figure 21 shows an AES depth profile for an ABCD layer implanted with N to a dose of 5×10^{17} ions cm^{-2}. The dominant species are Cr, C and N. It can be seen that the N content exceeds 40 at.%. Additionally a relatively high carbon content can be seen to be associated with the implantation profile. The oxide layer at the surface was removed by sputtering. Most of the N is found in the outermost layer of 160 nm thickness, and its distribution is quite broad with a half-width of ca. 100 nm. Further investigations have shown that the processes of CrN and CrC formation are competitive, but both compounds contribute to the improved hardness [154]. Studies of implantation at elevated temperatures have also been made in order to distinguish the influences of hardening by annealing and by N implantation, respectively, on the tribological properties of the ABCD layers. Thus, it has been found that there is additional hardening of ABCD after annealing at temperatures up to 600°C, but that the contribution of such hardening decreased with further increase in annealing temperature.

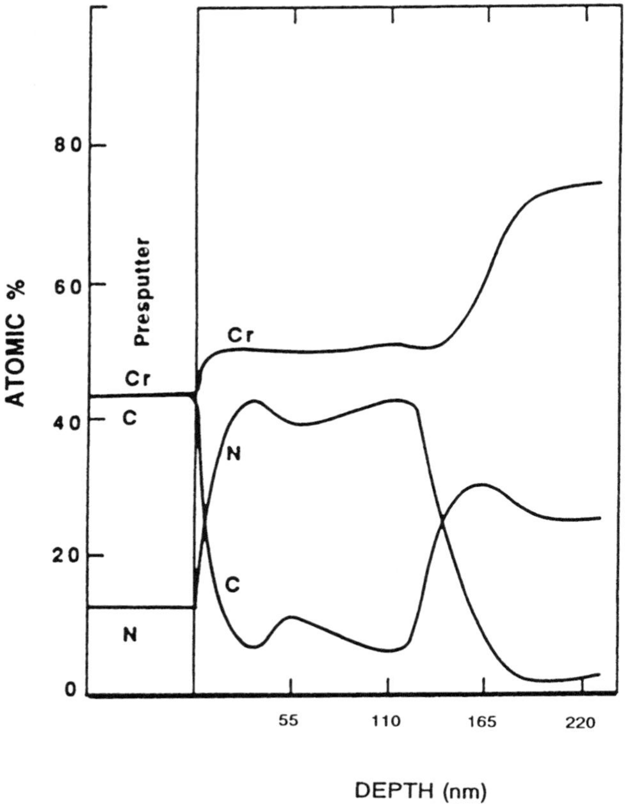

Figure 21. AES depth profiles obtained from an ABCD Cr layer annealed at 500°C and implanted with a total dose of 8×10^{17} N^+ cm^{-2}. Only the most prominent changes in Cr, N and C concentrations are shown [149].

5.2.4. Conclusions

Studies of the combination of coating and ion implantation have been performed with two kinds of Cr layers, both implanted with N ions. It has been shown that N implantation of both conventional and ABCD films results in an increase in the near-surface hardness of the Cr layers. The extent of this increase was greater for those films of both types that had been annealed at 400°C. AES depth profile analysis showed that the concentration of N in the implanted layer should not exceed 40 at.%. Implantation beyond that amount led to a broadening of the nitrogen profile. Analysis of film composition of the implanted Cr layers showed that the films consisted of Cr, C and N in the case of ABCD layers and of Cr, N and O in the case of conventional Cr layers. During implantation N displaces C in the implanted regions of the ABCD films to form CrN, but it was not possible to replace all the C by implanted N. It can be assumed that CrN and CrC each contribute to the increased hardness, but when both are present there is only a marginal increase in hardness.

This example emphasizes that it is important for the understanding of changes in tribological performance after ion implantation to analyze the compositions of the altered layers by methods such as AES or XPS.

5.3. Waveguide Structures by Ion Irradiation of Polymeric Materials

5.3.1. Introduction

The demand for passive optical devices is growing rapidly since fiber optical networks are becoming increasingly prevalent in telecommunication and other technical systems. It follows that waveguides, couplers, multiplexers and demultiplexers fabricated as integrated modules on a single substrate must become readily available. Polymeric materials have been considered as suitable because they exist in a wide range of molecular architectures and meet the requirements of optical transparency, low loss, ease of fabrication, long-term stability and low cost.

Ionizing radiation such as electrons and UV light, as well as ions, have been used since 1959 to modify the properties of polymers for various applications [153–155]. Due to energy deposition, irradiation with ions or UV light causes chemical changes in the polymeric material. Outgassing of damage debris and densification of the remaining material result in an increase in the index of refraction [156]. Ions stop in the substrate and lose their energy according to the well-known energy loss curve, whereas UV light loses its energy exponentially with depth. For ion-irradiated substrates a well-defined layer with increased index of refraction results [156, 157]. On the other hand, UV irradiation produces a broad modified layer, starting at the surface with a modified refractive index which then changes continuously with depth [158, 159].

In previously published investigations the critical parameters for specific ions were found to be the ion energy, which allows the depth of a waveguide layer to be estimated, and the ion dose, which determines whether the layer can act as a single or a multimode waveguide [156]. The required dimensions of a device for technical application are 20 mm long and 30 mm wide, with waveguide structures 7 μm wide. Furthermore the waveguide structures should have a symmetric profile in order to guarantee low-loss coupling between the fiber and the device.

To fulfil these requirements two problems had to be solved: masking the structures 7 μm wide and creating symmetric waveguide layers. For technical applications of a device the coupling between the fiber and the waveguide de-

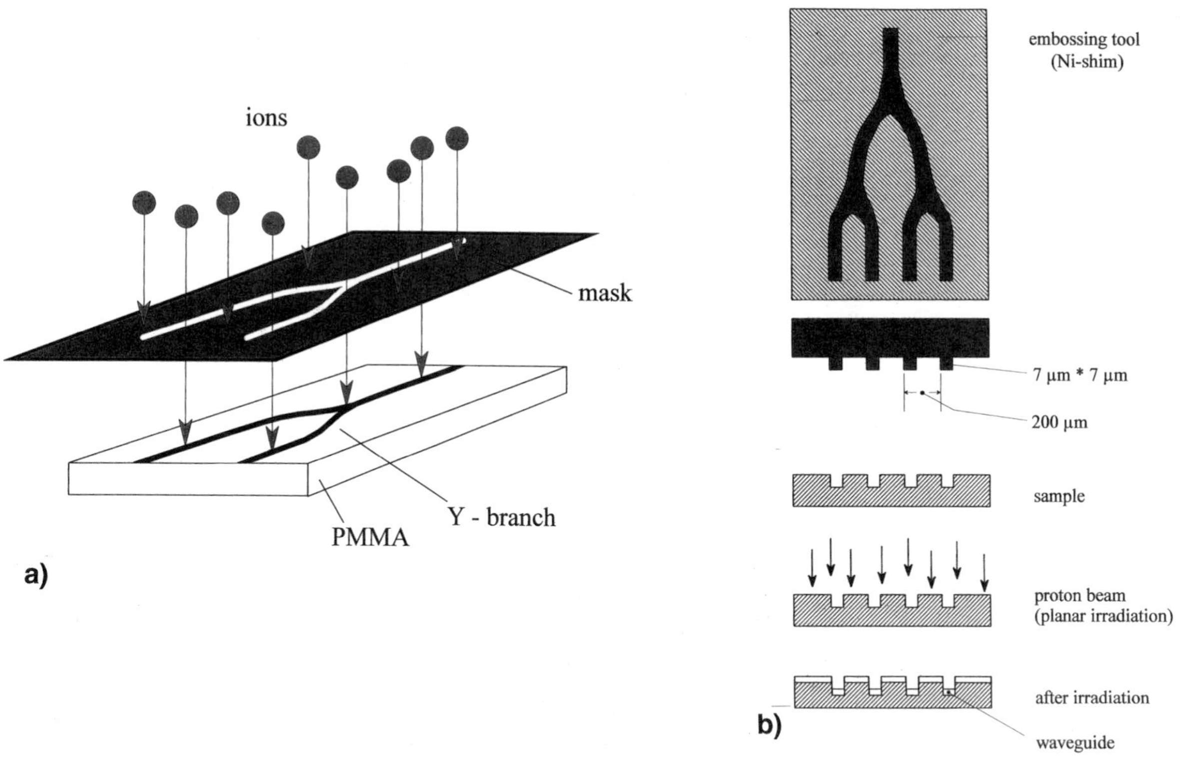

Figure 22. Ion implantation method for production of waveguide structures: (a) implantation through a mask; (b) implantation of preembossed samples.

vice—the fiber-chip coupling—must be simple and capable of automatic fabrication in order to guarantee a low-cost process.

5.3.2. Generation of μm structures

Two possible methods of developing structures with the required dimensions are outlined in Fig. 22: (a) irradiation through a freestanding mask, (b) embossing of the structure directly into the polymeric substrate.

Both methods will be described briefly and the problems discussed. Manufacture of a freestanding mask is described in more detail in [160], so only a short summary of the important features is given here.

Based on a 4-in. silicon wafer a mask is fabricated with a window 2 μm wide etched into the center; the etching is performed by conventional electron beam lithography. The principal difficulties associated with such a mask are the integrity of a structure containing a long slit and the demand for structures with closed loops. To stabilize the structure small 3 μm wide bridges of silicon 15 μm apart are therefore left after etching, and they prevent the closed-loop structures from falling apart. The intensity loss of the guided light measured in such a structure was found to be below the maximum acceptable attenuation value. Such a freestanding mask allows either an individual irradiation of each polymer substrate to provide an individual structure, or a whole set of structures for further study. Figure 23 shows an irradiated splitter device, which was gener-

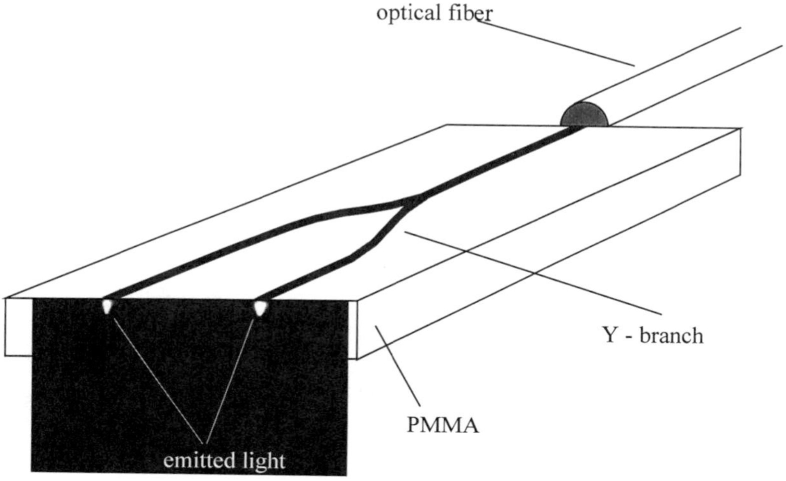

Figure 23. A 1:2 ray splitter planar device. In-plane schematic of the splitter structure (upper); cross-sectional view of the light rays emerging from the waveguide structure (lower).

ated using the freestanding mask. It can be seen that the light is indeed guided through the splitter structure, the white dots at the endplane indicating the emerging light in the two branches.

Rapid change of substrates with a specially developed substrate holder would be necessary for a high production rate. One disadvantage which appeared during the experiments was the sensitivity of the mask to temperature

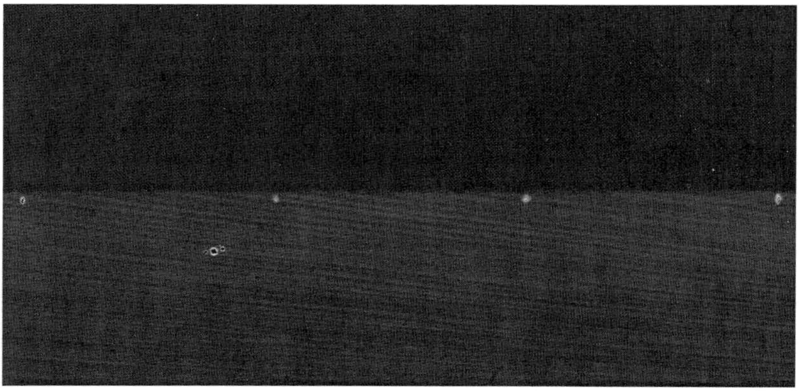

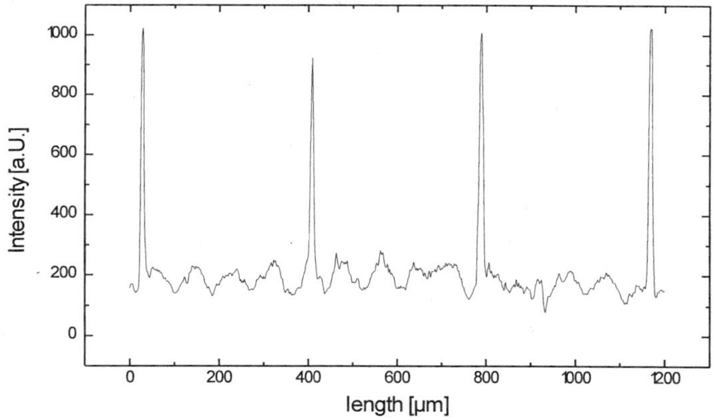

Figure 24. A 1:4 ray splitter planar device. Cross-sectional view of the emerging light (top), and the intensities in the four branches (lower).

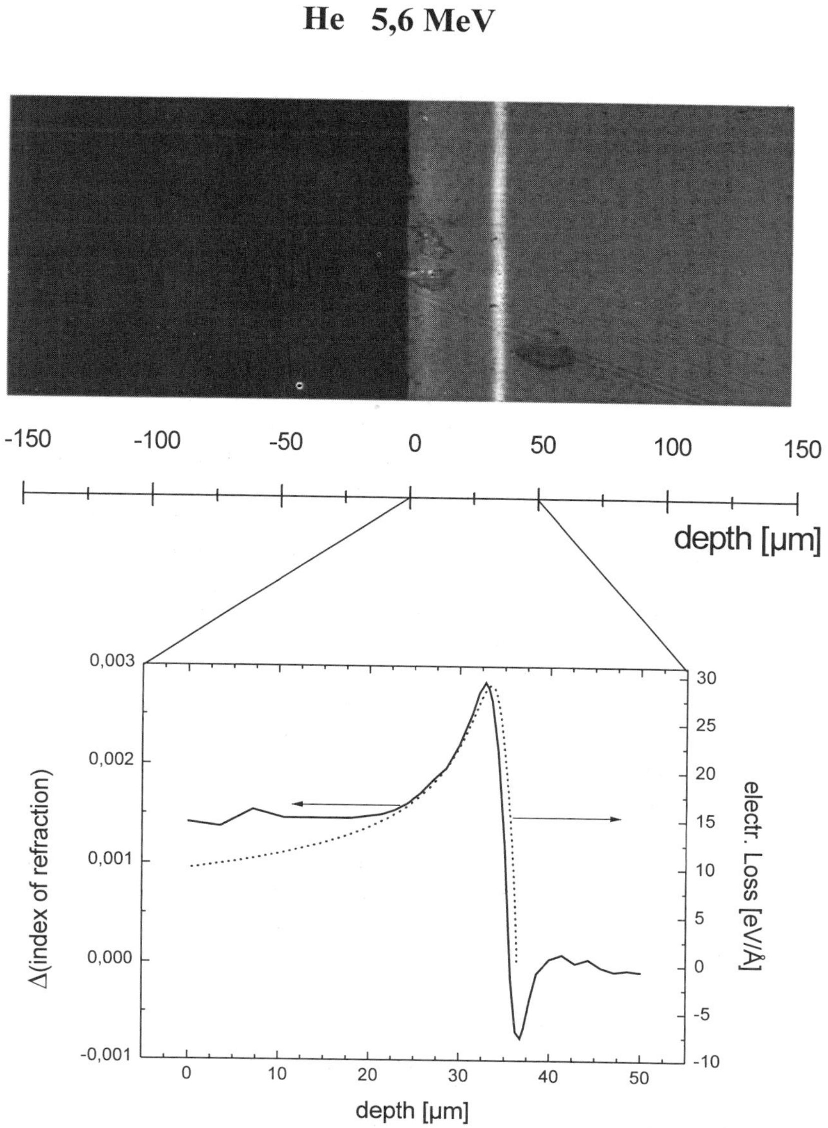

Figure 25. Buried planar waveguide structure produced by irradiation with He^{+} of 5.4 MeV. (upper) Cross-sectional view of the device with the white line indicating the emerging light at 35 µm depth. (lower) Energy loss and index of refraction as functions of depth. The difference between the measured maximum of the intensity of the emitted light and the maximum of the energy loss results from the compaction of 5 µm due to ion irradiation.

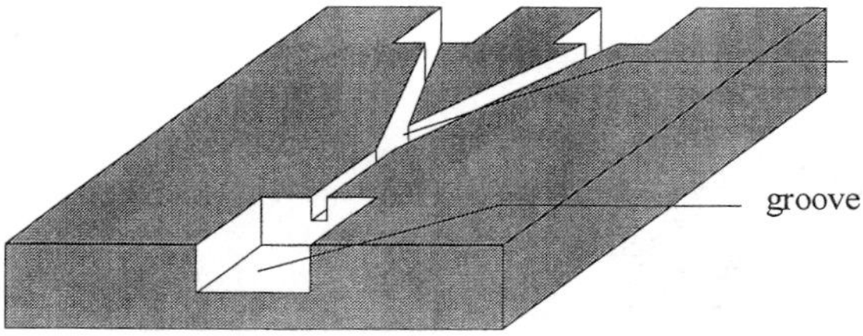

Figure 26. Preembossed device consisting of a 1:2 ray splitter device with rectangular grooves for coupling the device to an optical fiber.

increase as a result of the ion irradiation. This led to the necessity of decreasing the ion current, which consequently increased the irradiation time; on a production line the cost per device would thereby be increased.

The second possibility of producing waveguide structures is by embossing the polymeric substrate with a preshaped Ni embossing tool, followed by planar irradiation of the whole device. The Ni tool is produced using the LIGA technique [160]. Figure 24 shows a 1:4 divider. The lower part of the picture shows the endplane of the substrate with the emerging light. The upper part shows the device itself. Irradiation was performed with 5.6 MeV Ar^+ and a total dose of 10^{12} ions/cm^2. The depth of the embossed structure was about 7 μm relative to the surface of the substrate. As a result of the irradiation a waveguide layer is created at the surface of the substrate and at the surface of the embossed structure. It can be seen that the light is in fact guided and that no undesirable coupling between the surface waveguide layer and the deep waveguide structure occurred.

With this procedure a single embossing tool can generate many structured devices which can then be irradiated simultaneously in large arrays, allowing the possibility of low production costs per device. The principal disadvantage is the high production cost of the shaped Ni embosser by the LIGA technique [160] and the additional process step of embossing the substrate.

5.3.3. Buried waveguide layers

In order to produce buried waveguides, ions of high energy have to be used. Figure 25 shows an example of a PMMA substrate irradiated with 5.4 MeV He^+. Due to the Bragg peak at ca. 40 μm the waveguide layer was produced at a depth of ca. 35 μm in the substrate material. The upper part of the picture shows the device (left) and the endplane with the emerging light (right), and the lower part shows the energy loss and the change of the index of refraction as a function of depth. It can be seen that the change of index follows the electronic energy loss with a maximum at ca. 40 μm. The waveguide is at a depth of about

35 μm due to compaction of the first 5 μm of the surface as a result of densification. Good agreement between theoretical calculation of the energy loss using the TRIM program and the actual depth of the waveguide layer can be seen.

Comparison of the intensity distributions between a surface waveguide and a buried waveguide shows that the intensity distribution in the *y*-direction (normal to the surface) is symmetric only in the case of a buried waveguide [159] and shows a slight asymmetric shape for a surface waveguide.

5.3.4. Coupling between device and fiber: fiber-chip coupling

For technical application of the device in a network it is necessary to solve the problem of coupling the fiber to the device. An optical fiber consists of a core of diameter ca. 7 μm and a shielding sheath of thickness ca. 150 μm. The problem to be solved is to arrange the coupling between the core of the fiber and the waveguide layer so as to avoid coupling losses. Furthermore the coupling mechanism must be fast and easy to handle, which means that individual coupling of each fiber with the help of a microscope is not feasible. Figure 26 shows one possibility of achieving this coupling. Rectangular grooves with the dimensions of the fiber are built into the structure of the Ni embossing tool, leading to the embossed structure shown on the polymer substrate. The whole substrate is then irradiated with ions. It is then not necessary to take into account compaction of both types of groove, since the degree of compaction is identical in both structures.

5.3.5. Conclusions and further developments

It has been shown that by irradiation of a PMMA polymer substrate with ions waveguide structures can be produced. Detailed results of the ion-induced chemical changes that lead to increased index of refraction have been reported elsewhere [156, 158]. A possible solution of the problem of exact coupling between the optical fiber and the device has been suggested.

A disadvantage of the polymer PMMA is the low temperature of continuous operation. New polymers have therefore been investigated and the polymer PMMI (poly(methacrylate methyl imide)), with a temperature of about 170°C, seems suitable. First irradiation experiments have been performed, and it has been found possible to increase its index of refraction by ion irradiation. The results are published elsewhere [159, 161, 162].

ACKNOWLEDGMENTS

The author would like to thank Prof. Kankeleit and his research group at the University of Darmstadt, especially Gertrud Walter and Dr. Branko Stahl for their contribution concerning the topic of Mössbauer spectroscopy, Dr. Harold Schmidt and Andreas Schminke TU Darmstadt and Dr. Ulrich Fink, Nesculap A6 as well as Dr. Hermann Ferber, Daimler-Benz, for helpful discussions.

REFERENCES

1. P. Mazzoldi and A. Miotello, *Mater. Sci. Eng. A115*, 2 (1989).
2. R. Kelly, *Mater. Sci. Eng. A115*, 11 (1989).
3. W. Ensinger, *Surf. Coat. Technol. 83, 84,* 363 (1996).
4. D. E. Alexander and G. S. Was, *Surf. Coat. Techol. 51*, 333 (1992).
5. C. M Gilmore and J. A. Spreague, *Surf. Coat. Technol. 51*, 324 (1992).
6. F. Paszti, *Mater. Sci. Eng. A115*, 57 (1989).
7. D. Fink and M. Müller, *Surf. Coat. Technol. 51*, 352 (1992).
8. S. N. Bunker and A. J. Armini, *Nucl. Inst. Meth B 39*, 7 (1989).
9. G. Terwagne and S. Lukas, *Surf. Coat. Technol. 51*, 373 (1992).
10. A. H. Ommen, *Nucl. Inst. Meth. B 39*, 194 (1989).
11. U. Bussmann, F. H. J. Meerbach and E. H. Te Kaat, *Nucl. Inst. Meth. B 39*, 234 (1989).
12. T. Kimura, S. Yugo, Zhou Song Bao and Y. Adachi, *Nucl. Inst. Meth. B 39*, 238 (1989).
13. A. E. White, K. T. Short, R. C. Dynes, R. Hull and J. M. Vandenberg, *Nucl. Inst. Meth. B 39*, 253 (1989).
14. S. Mantl, *Nucl. Inst. Meth. B 84*, 127 (1994).
15. W. Theodossiu, H. Baumann, A. Markwitz and K. Bethge, *Fresenius J. Anal. Chem. 353*, 483 (1995).
16. R. W. Michelmann, H. Baumann, A. Markwitz, J. D. Meyer and K. Bethge, *Nucl. Inst. Meth. B 108*, 62(1996).
17. J. Z. Wan, D. A. Thompson and J. G. Simmons, *Nucl. Inst. Meth. B 106*, 17 (1995).
18. A. Knecht, M. Kuttler, H. Scheffler, T. Wolf, B. Binberg and H. Kräutle, *Nucl. Inst. Meth. B 80/81*, 37(1993).
19. A. Romano-Rodriguez, A. El Hassain, J. Samitier, A. Perez-Rodriguez, S. Martinez, J. R. Morante, J. Esteve and J. Montserat, *Nucl. Inst. Meth. B 80/81*, 7 (1993).
20. B. Rauschenbach, *J. Mater. Sci. 21*, 395 (1986).
21. R. Leutenecker, G. Wagner, T. Louis, U. Gonser, L. Guzman and A. Molinari, *Mater. Sci. Eng. A115*, 229 (1989).
22. M. Iwaki, Y. Okabe and K. Yate, *Nucl. Inst. Meth. B 45*, 212 (1990).
23. G. Terwagne, M. Piette, F. Bodart and W. Möller, *Mater. Sci. Eng. A115*, 25 (1989).

24. A. Johansen, L. Sarholt-Kristensen, E. Johnson, S. Steenstrup and V. S. Chernysh, *Mater Sci. Eng. A115*, 49 (1989).
25. Y. Okabe, M. Iwaki and K. Takahashi, *Mater. Sci. Eng. A115*, 79 (1989).
26. P. Ballhause, G. K. Wolf and Chr. Weist, *Mater Sci. Eng. A115*, 273(1989).
27. D. Rück, D. Boos and I. Brown, *Nucl. Inst. Meth. B* (1993).
28. S. Saritas, R. P. M. Procter and W. A. Grant, *Mater. Sci. Eng. A115*, 377 (1989).
29. S. Saritas, R. P. M. Procter, W. A. Grant, *Mater. Sci. Eng. A115*, 307 (1989).
30. M. A. El Khakani, G. Marest, N. Moncoffre and J. Tousset, *Surf. Coat. Technol. 51*, 82 (1992).
31. M. Hans, H. Martin, H. Ollendorf and G. K. Wolf, *Surf. Coat. Technol. 51*, 93 (1992).
32. R. Wie, P. L. Wilbur, O. Ozturk and D. L. Williamson, *Surf. Coat. Technol. 51*, 133 (1992).
33. J. Sasaki, K. Hayashi, K. Sugiyama, O. Ichiko and H. Kitamura, *Surf. Coat. Technol. 65*, 160(1994).
34. T. Miyano and H. Kitamura, *Surf. Coat. Technol. 65*, 179 (1994).
35. H. J. Mikkelsen and C. A. Straede, *Surf. Coat. Technol. 51*, 152 (1992).
36. L. Guzmann, L. Ciaghi, F. Giacomozzi, E. Voltohini, A. Peacock, G. Dearnaley and P. Gardner, *Mater. Sci. Eng. A116*, 83 (1989).
37. J. K. Hirvonen, *Mater. Sci. Eng. A116*, 167 (1989).
38. B. Torp, B. R. Nielsen, A. Dodd, J. Kinder, C. M. Rangel, M. F. Da Silva and B. Courage, *Nucl. Inst. Meth. B 80/81*, 246 (1993).
39. W. Ensinger and G. K. Wolf, *Surf. Coat. Technol. 51*, 41 (1992).
40. M. J. Bennett, A. T. Tuson, D. P. Moon, J. M. Titchmarsh, P. Gould and H. M. Flower, Surf. Coat. Technol. *51*, 65 (1992); K. Takahashi and M. Iwaki, *Surf. Coat. Technol. 65*, 57 (1994).
41. W. Ensinger and G. K. Wolf, *Mater. Sci. Eng. A116*, 1 (1989).
42. B. Enders, S. Krauβ and G. K. Wolf, *Surf. Coat. Technol. 65*, 203 (1994).
43. C. J. McHargue, M. E. O'Hern, C. W. White and M. B. Lewis, *Mater. Sci. Eng. A116*, 361 (1989).
44. M. Ohkubo, T. Hioki, N. Suzuki, T. Ishiguro and J. Kawamoto, *Nucl. Inst. Meth. B 39*, 675 (1989).
45. J. E. E. Baglin, *Nucl. Inst. Meth. B 65*, 21 (1992).
46. A. A. Galuska, *Nucl. Inst. Meth. B 59/60*, 487 (1991).
47. L. Wang, N. Angert, D. Rück, C. Trautmann, J. Vetter, Z. Quan and H. Hasntsche, *Nucl. Inst. Meth. B 83*, 503 (1993).
48. D. M. Rück, S. Brunner, W. X. F. Frank, J. Kulish and H. Franke, *Surf. Coat. Technol. 51*, 318 (1992).
49. D. M. Rück, S. Brunner, K. Tinschert and W. X. F. Frank, *Nucl. Inst. Meth. B 106*, 447 (1995).

50. C. Buchal, *Nucl. Inst. Meth. B 59/60*, 1355 (1991).
51. G. Hubler, *Mater. Sci. Eng. A115*, 181 (1989).
52. J. K. Hirvonen, *Mater. Sci. Rep. 6*, 215 (1991).
53. F. E. Smidt, *Int. Mater. Rev. 35*, 61 (1990).
54. G. K. Wolf and W. Ensinger, *Nucl. Inst. Meth. B 59/60*, 173 (1991).
55. R. S. Averback and D. N. Seidman, *Mater. Sci. Forum 15–18*, 63 (1987); L. E. Rehn, *Nucl. Inst. Meth. B 59/60*, 465 (1991).
56. W. Ensinger, *Surf. Coat. Technol. 65*, 90 (1994).
57. A. Kluge, B. H.-S. Javadi, R. Öchsner, H. Ryssel and H. Ruoff, *Surf. Coat. Technol. 51*, 30 (1992).
58. R. Roy, *Surf. Coat. Technol. 51*, 203 (1992).
59. W. Ensinger, A. Schröer and G. K. Wolf, *Surf. Coat. Technol. 51*, 217 (1992).
60. Z. Chen and S. Patu, *Surf. Coat. Technol. 51*, 222 (1992).
61. B. Enders, S. Krauβ and G.K. Wolf, *Surf. Coat. Technol. 65*, 203 (1994).
62. T. Miyano and H. Kitamura, *Surf. Coat. Technol. 65*, 179 (1994).
63. J. R. Conrad, *Mater. Sci. Eng. A116*, 197 (1989).
64. G. A. Collins, R. Hutchings and J. Tendys, *Mater. Sci. Eng. A139*, 171 (1991).
65. G. A. Collins, R. Hutching, K. T. Short, J. Tendys, X. Li and M. Samandi, *Surf. Coat. Technol. 74–75*, 417 (1995).
66. R. Günzel, S. Mändl, J. Brutscher and W. Möller, Second Int. Workshop on Plasma Based Ion Implantation, Sydney, 1995; B. J. Brutscher, R. Günzel and W. Möller, *Plasma Sources Sci. Technol. 5*, 54 (1996).
67. J. Conrad, *Nucl. Inst. Meth. B 106*, 522 (1996).
68. H. S. Wang, K. Q. Cheng, J. Q. Li, M. Geng, X. C. Zheng, D. Z. Xing and Z. K. Shang, *Surf. Coat. Technol. 66*, 525 (1994).
69. D. L. Rey and R. B. Alexander, Los Alamos Report LA-UR 94-312 (1994).
70. W. Ensinger, J. Hartmann, J. Klein, P. Usedom, B. Strizker and B. Rauschenbach, *Mat. Res. Symp. Soc. Proc. 386*, 521 (1996).
71. W. Ensinger and M. Kiuchi, *Surf. Coat. Technol. 84*, 425 (1996).
72. B. Torp, P. Abrahamsen and S. Eriksen, *Mater. Sci. Eng. A116*, 193 (1989).
73. P. J. Wilbur, R. Wie and W. S. Sampath, *Mater. Sci. Eng. A116*, 215 (1989).
74. N. Sakudo, K. Tokiguchi, T. Seki, H. Koike and M. Iwaki, *Mater. Sci. Eng. A116*, 221 (1989).
75. F. J. Körber, W. D. Münz, H. Ranke, St. Reineck, H. J. Füsser and H. Öchsner, *Mater. Sci. Eng. A116*, 205 (1989).
76. B. Torp, P. Abrahamsen, S. Eriksen, P. L. Hoeg and B. R. Nielsen, *Surf. Coat. Technol. 66*, 361(1994).
77. G. Lembert and H. Ferber, Private communication Daimler-Benz and Tribotec Electronic, 1996.

78. B. Torp, B. R. Nielsen, N. J. Mikkelsen, C. A. Straede, *Surf. Coat. Technol. 84,* (557 (1996).
79. J. Biersack, *Nucl. Inst. Meth. 182/183*, 199 (1981).
80. J. Biersack, *Z. Phys. A305*, 95 (1982).
81. J. F. Ziegler, *Handbook of Stopping Cross-Sections for Energetic Ions in All Elements*, Pergamon Press, New York, 1980.
82. J. F. Ziegler, J. P. Biersack and U. Littmark, *The Stopping and Ranges of Ions in Matter*, Vol. 1, Pergamon Press, 1985.
83. A. J. Armini and S. N. Bunker, *Mater. Sci. Eng. A114*, 67 (1989).
84. G. Müller, Diploma Thesis, Technische Hochschule Darmstadt, 1991.
85. S. Kalbitzer, H. Oetzmann, H. Grahman and A. Feuerstein, *Z. Phys. A 278*, 223 (1976).
86. I. G. Brown, M. R. Dickinson, J. E. Galvin, X. Godechot and R. A. MacGill, *Nucl. Inst. Meth. B 55,* 506 (1991).
87. D. Aitken, in *The Physics and Technology of Ion Sources* (I. Brown, ed.), 1989, p. 187.
88. T. Tami, M. Diamond, B. Doherty and P. Splinter, *Nucl. Inst. Meth. B 55*, 408 (1991).
89. R. Keller P. Spädtke and F. Nöhmayer, Proc. Int. Ion Eng. Congr., Kyoto, 1983, p. 25; R. Keller, P. Spädtke and H. Emig, *Vacuum 36*, 833 (1986).
90. B. Torp, B. R. Nielsen, D. M. Rück, H. Emig, P. Spädtke and B. H. Wolf, *Rev. Sci. Instrum. 61*, 595 (1990).
91. B. R. Nielsen, P. Abrahamsen and S. Eriksen, *Mater. Sci. Eng. A116*, 193 (1989).
92. J. R. Conrad, *Mater. Sci. Eng. A116*, 197(1989).
93. R. W. Thomas, B. Seiler, H. Bender, J. Brutscher, R. Günzel, J. Halder, H. Klein, J. Müller and M. Sarstedt, *Nucl. Inst. Meth. B 99*, 569 (1995).
94. W. Ensinger, E-MRS Proceedings (1996) to be published.
95. J. R. Tesmer and M. Nastasi, *Handbook of Modern Ion Beam Materials Analysis*, MRS (1995); *Mater. Sci. Technol.* (R. W. Cahn, P. Haasen and E. J. Kramer, eds.), Characterization of Materials Part II, VCH Weinheim, New York, Basel, 1994.
96. L. R. Doolittle, *Nucl. Inst. Meth. B 9*, 344 (1989).
97. A. Markwitz, H. Baumann, E. F. Krimmel, M. Rose, K. Bethge, S. Logothetis and P. Misaelides, *Vacuum 44*, 367 (1993).
98. D. Kuhn, H. Baumann and F. Rauch, *Nucl. Inst. Meth. B 45*, 252 (1990); D. Endisch, H. Sturm and F. Rauch, *Nucl. Inst. Meth. B 84*, 380 (1994).
99. Gar B. Hoflund, Depth profiling, in *CRC Handbook of Surface Imaging and Visualisation* (A. T. Hubbard, ed.), CRC Press, Boca Raton, FL 1995, chap. 6, p. 63.
100. E. Kankeleit, *Z. Phys. 164*, 442 (1961).
101. T. Bonchev, A. Jordanov and A. Minkova, *Nucl. Inst. Meth. 70*, 36 (1969).

102. K. Nomura, Y. Ujihira and A. Vertes, *J. Rad. Nucl. Chem. 202*, 103(1996).
103. B. Stahl, R. Gellert, O. Geiss, R. Teucher, M. Müller, G. Walter, R. Heitzmann, G. Klingelhöfer and E. Kankeleit, GSI Scientific Report, 1994, p. 180.
104. B. Stahl, Ph.D. thesis, Technische Hochschule Darmstadt, 1995, to be published.
105. R. Gellert, O. Geiss, G. Klingelhöfer, H. Ladstätter, B. Stahl, G. Walter and E. Kankeleit, *Nucl. Inst. Meth. B76*, 29 (1993).
106. D. Liljequist, *J. Appl. Phys. 57*, 657 (1985).
107. J. M. Fernandez-Varea, D. Liljequist, S. Csillag, R. Räty and F. Salvat, *Nucl. Inst. Meth. B 108*, 35 (1996).
108. R. Heitzmann, B. Stahl, G. Walter, D. M. Rück, R. Gellert, O. Geiss, G. Klingelhöfer and E. Kankeleit, GSI Scientific Report, 1993, p. 192.
109. G. Walter, R. Heitzmann, D. M. Rück, B. Stahl, R. Gellert, O. Geiss, G. Klingelhöfer and E. Kankeleit, *Nucl. Inst. Meth. B 113*, 167 (1996).
110. E. Johnson, A. Johansen. L. Sarholt-Kristensen, H. Roy-Poulsen and A. Christiansen, *Nucl. Inst. Meth. B 7/8*, 212 (1985).
111. E. Moll and E. Kankeleit, *Nukleonik 7*, 180 (1965).
112. O. Koefoed-Hansen, J. Lindhard and O. B. Nielsen, *Kgl. Danske Vidensk. Selskab. Mat.-Fys. Medd. XXV*, 16 (1950).
113. Y. Okabe, T. Fujihana, M. Iwaki and B. V. Crist, *Surf. Coat. Technol. 65*, 384(1994).
114. L. E. Toth, *Transition Metal Carbides and Nitrides*, Academic Press, New York, 1971.
115. T. Fujihana, Y. Okabe and M. Iwaki, *Surf. Coat. Technol. 66*, 419 (1994).
116. G. Marletta, F. Iacona and R. Kelly, *Nucl. Inst. Meth. B 65*, 97 (1992).
117. K. Neubeck, C.-E. Lefaucheur, H. Hahn, A. G. Balogh, H. Baumann, K. Bethge and D. Rück, *Nucl. Inst. Meth. B 106*, 589 (1995).
118. R. Van Noort, *J. Mater. Sci. 22*, 3801 (1987).
119. M. Semlitsch, *Clin. Mater. 2*, 1 (1987).
120. H.-G. Willert, G. H. Buchhorn and P. Eyerer, "Ultra-High Molecular Weight Polyethylene as Biomaterials in Orthopedic Surgery" Hogrebe & Huber, Toronto 1991.
121. S. Nasser, P. A. Campbell, D. Kilgus, N. Kossovsky and H. C. Amstutz, *Clin. Orthop. 261*, 171 (1990).
122. C. D. Peterson, B. M. Hillberry and D. A. Heck, *J. Biomed. Mater. Res. 22*, 887 (1988).
123. G. Dearnaley, *Clin Mater. 12*, 237 (1993).
124. A. Cigada, M. Cabrini and P. Pedeferri, 17th Ann. Meeting Soc. for Biomater., Scottsdale, Az., 1991, p. 198.
125. R. A. Poggie, A. K. Mishra and J. A. Davidson, *J. Mater. Sci. Mater. Med. 5*, 387 (1094).

126. P. Sioshansi, R. W. Oliver and F. D. Mathews, *J. Vac. Sci. Technol. A3*, 2670 (1985).
127. F. D. Matthews, K. W. Greer and D. L. Armstrong, in J. M. Williams, M. F. Nichols and W. Zingg (eds.), Mat. Res. Soc. Symp. Proc. Vol. 55, MRS, Pittsburgh, 1986, p. 243.
128. Z. Jianqiang, Z. Xiaozhong, G. Zintang and L. Hedge, in J. M. Williams, M. F. Nichols and W. Zingg (eds.), Mat. Res. Soc. Symp. Proc., Vol. 55, MRS, Pittsburgh, 1986, p. 229.
129. J. Rieu, A. Pichat, L. M. Rabbe, A. Rambert, C. Chabrol and M. Robelet, *Mater. Sci. Technol. 8*, 589 (1992).
130. F. Alonso, A. Arizaga, S. Quainton, J. J. Ugarte, J. L. Vivente and J. L. Onate, *Surf. Coat. Technol. 74–75*, 986 (1995).
131. H. Schmidt, H. E. Exner, D. M. Rück, N. Angert and U. Fink, in *Advances in Materials Science and Implant Orthopaedic Surgery* (R. Kossowsky, ed), Kluwer, Dordrecht, 1995, p. 207.
132. R. Martinella, S. Giovanardi and G. Palombarini, *Wear 133*, 267 (1989).
133. M. Corchia, R. Giorgi, S. Turtu, F. Caccavale, S. LoRusso, P. Mazzolid, G. Abatangelo, R. Cortivo, S. Giovanardi, R. Martinella, A. McCabe, D. Chivers and P. Tweedale, Abstr. 11th Europe. Conf. on Biomaterials, Pisa, 1994, p. 368.
134. J. Lausmaa, T. Röstlund and H. McKellop, *Surf. Eng. 7*, 311 (1991).
135. H. Schmidt, H. E. Exner, D. M. Rück, N. Angert and U. Fink, in *Advances in Materials Science and Implant Orthopaedic Surgery* (R. Kossowsky, ed.), Kluwer, Dordrecht, 1995, pp. 207–221.
136. W. Weiler, *Brit. J. Non-Destr. Test. 31*, 252 (1989).
137. J. Hinterberger, M. Ungethüm and W. Plitz, in *Mechanical Properties of Biomaterials* (G. W. Hastings and D. F. Williams, eds.), Wiley, Chichester, 1980, p. 73.
138. A. Walter and W. Plitz, in *Biomechanics: Current Interdisciplinary Research* (S. M. Peren and E. Schneider, eds.), Martinus Nijoff, Dordrecht, 1985, p. 129.
139. Harald Schmidt, Dissertation Technische Hochschule Darmstadt, 1995 p. 43.
140. D. M. Rück, H. Schmidt, N. Angert, O. Yoda, Y. Aoki and H. Naramoto, in *Ion Beam Modification of Materials* (D. F. Williams, R. G. Eliman and M. C. Ridgway, eds.),Elsevier, Amsterdam, 1996 pp. 1016–1019.
141. R. C. Weast, *Handbook of Chemistry and Physics*, CRC, Boca Raton, 1989.
142. E. A. Brandes and G. B. Brook, *Smithells Metals Reference Book*, Butterworths, London, 1992.
143. R. Martinella, S. Giovanardi, G. Chevallard, M. Villani, A. Molinari and C. Tosello, *Mater. Sci. Eng. 69*, 247 (1985).
144. R. N. Bolster, I. L. Singer and R. G. Vardiman, *Surf. Coat. Technol. 33*, 469 (1987).

145. X. Qiu, R. A. Dodd, J. R. Conrad, A. Chen and F. J. Worzala, *Nucl. Instr. Meth. B 59/60*, 951 (1991).
146. G. Vardiman and R. A. Kant, *J. Appl. Phys. 53*, 690 (1982).
147. R. Hutchings, *Mater. Sci. Eng. 69*, 129 (1985).
148. S. Hoshino, H. A. Laitinen and G. B. Hoflund, *J. Electrochem. Soc.*, Vol. No. 682 (1986).
149. H. Ferber, G. B. Hoflund, C. K. Mount and S. Hoshino, *Surf. Interface Anal. 16*, (1990).
150. H. Ferber, C. K. Mount, G. B. Hoflund and S. Hoshino, *Thin Solid Films 203*, 121 (1991).
151. S. Hofmeister, G. B. Hoflund, C. K. Mount, H. Ferber and S. Hoshino, *Surf. Coat. Technol. 51*, 313 (1992).
152. H. Ferber, G. B. Hoflund, C. K. Mount and S. Hoshino, *Nucl. Inst. Meth. B 59/60*, 957 (1991).
153. E. H. Lee, M. B. Lewis, P. J. Blau and L. K. Mansur, *J. Mater. Res. 6*, 610 (1991).
154. J. O. Choi, *J. Vac. Sci. Technol. B 6*, 2286 (1988).
155. J. P. Harmon, *Polym. Prepr. 35*, 926(1994).
156. D. M. Rück, S. Brunner, W. F. X. Frank, J. Kulisch and H. Franke, *Surf. Coat. Technol. 51*, 318(1992).
157. F. W. X. Frank, J. Kulisch, H. Franke, D. M. Rück, S. Brunner and R. A. Lessard, SPIE Vol. 1559, *Photopolymer Device Physics, Chemistry, and Applications II*, 1991, p. 344.
158. D. M. Rück, S. Brunner, K. Tinschert and W. F. X. Frank, *Nucl. Inst. Meth. B 106*, 447 (1995).
159. F. W. X. Frank, A. Schösser and A. Stelmazek, SPIE Vol. 2851 (1996).
160. W. Ehrfeld and H. Lehr, *Radiat. Phys. Chem. 45*, 349 (1995).
161. S. Brunner and D. Rück, Scientific Report GSI 1995.
162. D. Rück, J. Schulz and N. Deusch, *Nucl. Inst. Meth. B* 26 (1997) (to be published)

10

Introduction to Scanned Probe Microscopy

S. MYHRA Griffith University, Nathan, Queensland, Australia

1. INTRODUCTION

This chapter represents a major break with surface analytical tradition and constitutes an attempt to recognize the winds of change with respect to the tools of the trade in surface science and technology. A book of this kind written only a few years ago could not have anticipated the sudden and dramatic impact of a new family of techniques, scanned probe microscopy (SPM). At the time of writing there are more than 3000 SPM instruments in operation around the world; this number comes from a very low pre-1990 base and is currently increasing at the rate of some 20% per year. The phenomenal growth in installed instrumental capacity, albeit from a low base, is being reflected in the scientific and technical output indices. This is due partly to the unique capabilities of SPM techniques and partly to the relatively low cost of entry. A top-of-the-range instrument operating in air will cost a small fraction (≈10%) of a modern multitechnique surface analytical facility. An SPM operating in an UHV envelope, with ancillary facilities, will be more expensive, but will still be an attractive proposition from the cost-effectiveness point of view, in comparison with a traditional basic surface analytical instrument.

There are many examples of groups, well-established in the "traditional" surface analytical or electron-optical fields, who have taken SPM on board. Conversely, it is true that some of the very best work using SPMs has been done by a new generation of researchers who went directly to the new technology. Consequently there is a danger that members of the two communities will be forced into playing adversarial roles–one seeing itself as the guardians of accepted wisdom, values and culture, and resisting acceptance of the very real merits of the new; the other seeing itself as the new vanguard with a brief to sweep away a moribund and ossified establishment. As in most cases, when there is a clash between generations and paradigms, there is a bit of truth in both descriptions, and a great deal of harm will come from accepting the extremes of either stance. One must also be aware of the pressures from a "drier" and more competitive funding environment, which values relevance, entrepreneurship and mission orientation above all else.

The inclusion of this chapter on SPM, in what otherwise would have been a book on surface analysis dealing with the "traditional" techniques, will hopefully serve the additional purpose of demonstrating the merits of the two communities in being companions rather than competitors. The chapter will attempt to demonstrate the complementarities between the "new" and the "old" techniques (see Table 1) as well as showing the common elements in the method-

Table 1. Complementary Strengths of SPM and Traditional Techniques

Requirement	Merits of SPM	Other techniques
Surface nanostructure	STM/contact AFM real space	LEED (reciprocal space) FIM and ISS (real space)
Morphology	STM/AFM × 1,000,000 good *z*-resolution	AES/SAM/SEM × 10,000 poor *z*-resolution
Grain structure	STM/AFM	AES/SAM/imaging SIMS
Phase structure	—	AES/SAM/imaging XPS/SIMS
Electronic structure	STS (local)	UPS/IPES (nonlocal)
Composition	—	XPS/AES/. . .
Force spectroscopy	AFM/F-d	—
Magnetic structure	MFM ("local")	—
Surface conductivity	SCM (local)	Differential charging (nonlocal)
"Buried" information	—	Profiling XPS/AES/SIMS/. . .
In situ capabilities	UHV/ambient/liquid	UHV

ologies. In the final analysis, the principal rationale for this book is that the "answer" will be the essential index of output and success. The methodologies and the instrumentation are means to this end; this very elementary observation should, if nothing else, be a sufficient motivation for the two communities to act in unison.

1.1. Essential Elements of SPM

At the most schematic level an SPM system (Fig. 1) consists of a surface having structural and physicochemical characteristics, an interaction possessing characteristic strength and range, and a tip which can be located and controlled in the spatial and temporal domains, with its own characteristics. On the basis of this description one can make the elementary observation that, given *complete* knowledge about any two of the three elements, it is possible in principle to obtain *complete* knowledge about the third. This observation demonstrates the intrinsic versatility and inherent richness of the SPM system. In practice, and especially at the present point on the learning/maturing curve, exploitation of the system is predicated on definitions and acceptance of *sufficiency* of knowledge. The consequences of the limitations regarding sufficiency will be explored. Also, note that the terminology of microscopy implies that in the

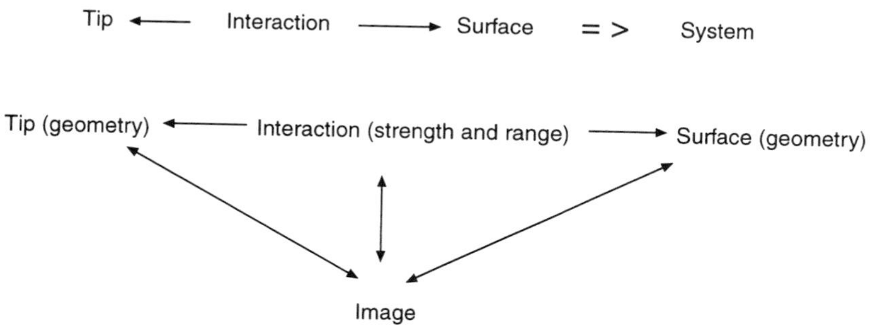

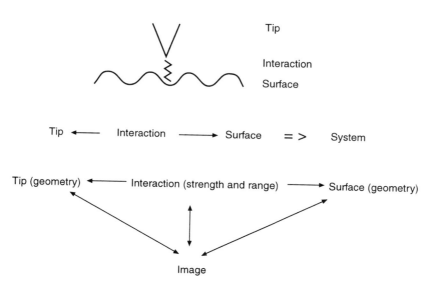

Figure 1. Schematic representation of the essential elements and the image formation process of the SPM system.

overwhelming majority of instances the surface is taken to be the unknown. Thus most of the discussion of the SPM system will concentrate on the conventional microscopy aspects/applications, but some attention will be given to other configurations. The schematic description in Fig. 1 also makes it abundantly clear that a resultant image must represent a complicated convolution of the preceding three elements; more will be said about this.

1.2. Brief History of SPM

The scanning tunneling microscope (STM) was invented by Binnig et al. in 1982 [1]. The two main protagonists, G. Binnig and H. Röhrer, were subsequently awarded the Nobel Prize for physics. Thus began the age of the SPM; however, the event had been anticipated by the developers of the Topographiner some 10 years earlier [2]. Much of the early development and associated excitement generated by the unequivocal demonstration of spatial resolution on the scale of the single atom and of local spectroscopies have been described in the literature [3–5].

The demonstration of the 7×7 reconstruction of the Si (111) surface was arguably the result which put the STM on the scientific map; an example of the resultant images is shown in Fig. 2. Unequivocal evidence for single-atom real-space resolution, or even more convincingly for single *missing* atoms and atomically resolved *I-V* spectroscopy, were also significant events [6]. (Strictly speaking, STM/STS do not map atoms but electron-tunneling probability density functions, but it is useful to speak of "single-atom" resolution.)

The second member of the SPM family, the atomic force microscope (AFM), was demonstrated in 1986 [7]. Its principal claims to fame are those of being more versatile (i.e., being essentially independent of the electrical conductivity of the surface), more forgiving of the operating environment (i.e., providing good service in air) and ultimately more user-friendly, in comparison with the STM. Subsequently, other interactions (e.g., magnetostatic, electrostatic, thermal radiation, etc.) have generated additional members, ably described by Wickramasinghe [8]. As well, most of the original techniques have had offsprings by way of variations on the underlying theme (e.g., contact, noncontact, lateral force and force-constant spectroscopic modes for the AFM).

1.3. SPM Family Tree

In hindsight it may be that the invention of the STM was not, in itself, the most significant aspect of the work of the group at IBM. Rather it was the thinking which led to the STM, and its technical implementation, which paved the way for a plethora of scanned probe techniques. In Fig. 3 the "family tree," in its present and somewhat abbreviated form, has been sketched. For purposes of

Figure 2. STM image of the reconstructed Si(111) 7×7 surface demonstrating single-atom resolution in *real* space. (Courtesy Omicron GMBH.)

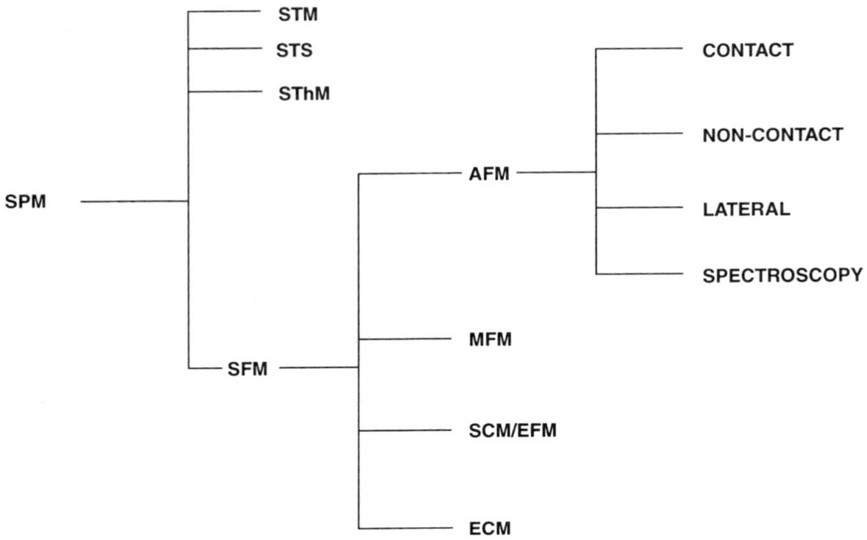

Figure 3. The SPM family tree.

classification, and in the interest of being internally consistent from a definitional point of view, one should note the following: SPM is the name of the family; within this family scanning force microscopy (SFM) constitutes a group for which the interaction is a force acting between a *probe*, consisting of a *lever* and a *tip*, and the surface. The interaction in the case of the STM and its sibling, scanning tunneling spectroscopy (STS), is electron tunneling, while in the case of scanning thermal microscopy (SThM), the "interaction" is the thermal radiative transport of energy between surface and probe.

The SFM branch has emerged as the largest and most popular, in terms of usage and installed instrumental capacity, among the practitioners of SPM. It includes some variants, which are currently less commonly used, such as electrical conductance microscopy (ECM), scanning capacitance (or capacitive, or electrostatic force) microscopy (SCM/EFM), and magnetic force microscopy (MFM). Finally, there is the AFM subgroup, which includes the most widely used technique(s). Thus it merits further subdivision into its various operational modes; contact (when the net force is *repulsive*), noncontact (when the net interaction is *attractive*), lateral force (when the system is controlled in the constant repulsive force mode, but the mapping parameter is responsive to the *lateral*, "frictional," force component), and spectroscopic (when the surface map reflects changes in the *force versus distance* relationship).

In view of the current relative popularities of the various SPM techniques the discussion will be confined principally to STM/STS and to AFM and its variants.

Some of the attributes and features of the various microscopies/spectroscopies are summarized in Table 2. The most topical figures of merit have to do with spatial resolution; at a later stage resolution in the temporal domain will also be considered. Spatial resolution figures are very much subject to judgment and practical considerations, as they often are for other microscopies. The rel-

Table 2. Current SPM Techniques: Features and Attributes

Technique	Interaction	*z*-resolution (nm) *xy*-resolution (nm) Int. vol. (m^3)	Probe/tip	Information attributes
STM/STS	electron tunneling	0.005 0.05 $<10^{-30}$	sharp tip	topography electron spectroscopy "work" function[a]
AFM contact	interatomic forces (repulsive)	0.02 0.15 $<10^{-28}$	cantilever + tip	topography molecular structure
AFM noncontact	(attractive)	0.2 1 $<10^{-26}$	cantilever + tip	topography
AFM contact	(lateral)	n.a. 0.3 $<10^{-28}$	cantilever + tip	tribology friction
AFM spectrosc.		0.2 1 $<10^{-26}$	cantilever + tip	F-d spectroscopy
MFM noncontact	magnetic force	1 5-50 $<10^{-25}$	cantilever + magnetized tip	magnetic topography domains, walls magnetic spectroscopy
EFM/SCM noncontact	electrostatic or capacitive force	5 100 $<10^{-23}$	cantilever + insulating tip	topography patch charge capacitance
ECM contact	electrical conductivity	0.05 50 $<10^{-25}$	cantilever + conducting tip	topography electrical conductivity breakdown voltage I-V spectroscopy
SThM noncontact	heat loss	1 100 $<10^{-21}$	thermocouple tip	topography thermal conductivity emissivity

[a]The work function refers to the effective barrier height; as a parameter it has significance in Eq. (4) where it affects the decay of the exponential tails of the wavefunctions into the barrier region.

evant judgments concern the cutoff limits for (de)localized interactions, effects of tip shape and properties, mechanical and electronic response limits of the instrument, the extent to which the specimen may be considered "difficult," etc. In line with tradition in surface analysis an *interaction volume* will be defined as being the region in space from which averaged information is obtained. In other branches of surface science this figure of merit is determined by the surface specificity and the spot size of the exciting probe (or the lateral dimension of the spot being imaged by a dispersive element). For convenience a similar, but amended, definition will be retained, when transferring to the case of SPM. Thus, a "resolution volume" may be thought of as being related to the "volume" of interaction which connects tip and surface, i.e., effectively the product of z-resolution by the area defined by the lateral resolution. The SPM "interaction volume," on the other hand, will be approximately the product of lateral resolution by the *practical* range of the interaction.

2. PHYSICAL PRINCIPLES

The following will constitute a brief, and necessarily incomplete, account of the essential underlying physical principles which account for the information content of the SPM system. The discussion will be limited to STM/STS and to AFM and its variants.

2.1. STM/STS

Electron tunneling is the elementary process which accounts for the operation of STM/STS. It will be described in accord with the approach adopted by Tersoff and Hamann [9]. Additional material can be found in the review literature [6,10,11]. The energy level scheme of a tunnel barrier is shown in Fig. 4, as are the shapes of the wavefunctions outside and inside the barrier; the exponential dependence inside the barrier is particularly relevant to the problem at hand. The tunnel current is

$$I = \frac{2\pi e}{h} \sum_{1,2} f(E_2)[1 - f(E_1 + eV_T)] \, | M_{1,2} |^2 \, \delta(E_2 - E_1) \tag{1}$$

which describes elastic (hence $\delta(E_2 - E_1)$) tunneling of an electron from an occupied state in the tip (f_2) to an unoccupied state in the surface ($1 - f_1$); the two states are displaced by eV_T, where V_T is the tunnel voltage. The summation is over energetically allowed states. The matrix element $M_{1,2}$ is given, in accord with the transfer-Hamiltonian formalism of Bardeen, by

$$M_{1,2} = \frac{h^2}{2m} \int dA \, (\psi_1^* \nabla \psi_2 - \psi_2 \nabla \psi_1^*) \tag{2}$$

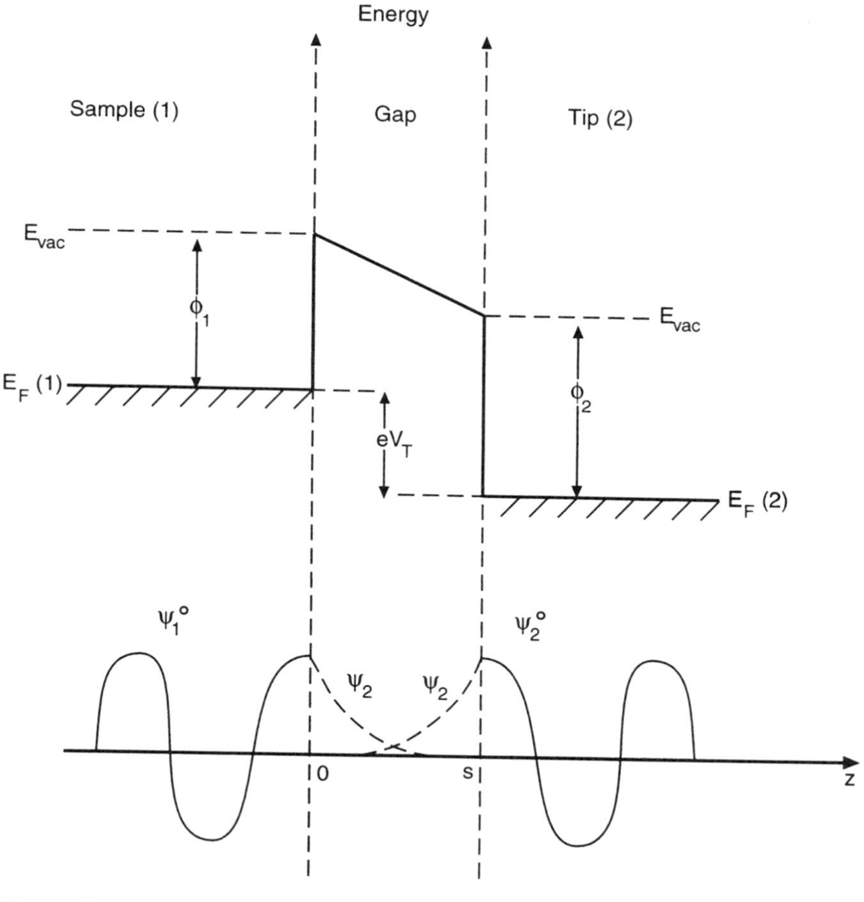

Figure 4. The energy level scheme of an STM tunnel barrier. Subscripts 2 and 1 refer to tip and sample, respectively. The work function is denoted by ϕ, the tunnel voltage by V_T. The wavefunctions are shown schematically; note the exponentially decaying amplitudes inside the barrier.

The integral is evaluated over an arbitrary surface wholly within the barrier. Essential insight into the underlying physics of STM/STS may be gained by considering a simple one-dimensional case of two identical free-electron metals with identical work functions, ϕ. Applying boundary conditions, the wavefunctions in the gap may be written as (Fig. 4)

$$\psi_1 = \psi_1^{\circ} e^{-kz} \qquad \text{and} \qquad \psi_2 = \psi_2^{\circ} e^{-k(s-z)} \tag{3}$$

where the width of the tunnel barrier is s; k is a real number (typically 10 nm^{-1})

$$k = \left(\frac{2m\phi}{h^2} \right)^{1/2} \tag{4}$$

The matrix element M is a weakly varying function, if $eV_T \ll \phi$ and $eV_T \ll E_F$. Thus one may take

$$I \propto \sum_{1,2} |\psi_1^o|^2 |\psi_2^o|^2 e^{-2ks} \tag{5}$$

This result provides a good qualitative description of the tunneling process, irrespective of tip shape, if the wavefunction of the tip can be approximated by an s-wave located in the tip. In the latter case, and if there are only small overlaps of the respective wavefunctions ψ_2 and ψ_1, and if the tails of the two functions are similar, then

$$I \propto \sum_{1} |\psi_1|^2 \delta(E_1 - E_F) \tag{6}$$

Also, the local density of states at the position of the tip is

$$\rho(E_F, \mathbf{r}) = \sum_{1} |\psi_1(\mathbf{r})|^2 \delta(E_1 - E_F) \tag{7}$$

This qualitative description allows one to make a number of observations about the STM/STS system.

a. The exponential dependence in Eq. (5) of the tunnel current I on the width of the tunnel gap, s, accounts for the extreme spatial resolution in the z-direction; a change in s of 0.1 nm will change I by a factor of e^2.
b. Equations (6) and (7) show that an STM image, formed by mapping constant tunnel current, represents a map of constant density of states. Thus, unless there is a spatial one-to-one correlation between the positions of atoms, i.e., nuclei, and maxima/minima in the density-of-states map, then it is the electronic structure that is being revealed.
c. The lateral resolution, in the x-y plane, is due, in part, to the exponential dependence of I on the barrier width; thus the spatial dependence of the density-of-states function in the x-y plane can be resolved. Of equal significance is the fact that the elastic tunnel process picks out particular states in the sample and tip, as a consequence of energy conservation. Thus the tunnel voltage, V_T sets an energy window a few kT wide. Consequently one may choose V_T for maximum resolution, i.e., an energy for which the corresponding electron states have the greatest spatial variation. These observations account for the relative ease with which "atomic" resolution can be obtained for covalent materials (e.g., Si); this is a consequence of the strong

spatial dependence of the density of states across the unit cell at the extrema of bands. Conversely, similar resolution is much more difficult to obtain for metals where the spatial variations are more gentle.

d. The ability to "tune" the energy window of the tunneling process by changing eV_T is the basis for STS. Successive maps of the surface at different values of V_T will reveal the spatial locations of the corresponding equi-energy contours of the density of states. Local spectroscopy can be carried out by fixing the lateral and vertical positions of the apex of the tip with respect to the surface and recording an I_T-V_T curve. The principle of the method is illustrated in Fig. 5, adapted from Ref. 12. The I_T-V_T data will not be interpretable as a simple plot of the density-of-states function, however, since there will be an exponential dependence of the tunnel probability with V_T from the matrix element M. This and other effects can be eliminated, in part, by measuring $(dI/dV)/(I/V)$, which is a useful dimensionless quantity related to the local band structure. As well as being local, STS has the additional merits of extreme surface specificity, literally the first monolayer, and of being able to probe both valence- and conduction-band states by the simple expediency of reversing the polarity of V_T (and thus reversing the tunnel process). Investigations of localized surface states are particularly useful and rewarding with STS. In the context of the traditional surface spectroscopies, one should note that STS can replace the combination UPS/IPES. The additional merits and convenience of STS vis-à-vis UPS/IPES make this an attractive proposition. Particularly nice examples can be found in [6] for the surface chemistry of Si and in [13] for a metal.

e. From Eq. (5) it can be seen that the tunnel current is exponentially dependent on k as well as on barrier width. The decay constant, k, is a function of the effective barrier height and is therefore related to the work function of the surface being probed. Thus it is possible to extract the spatial variation of the work function from the STM map.

f. It will also be apparent that STM/STS can be applied only to materials which are tolerably good conductors. The criterion for goodness is that the effective sample resistance, R_s, must be such that $I_T R_S \ll V_T$. Hence, the technique will work well for clean surfaces of metals, alloys, semimetals, and doped semiconductors. The presence of thin ($\leq$1 nm) insulating surface barriers (e.g., oxide layers) can be accommodated by allowing them to act as tunnel barriers. A consequence of these observations, and of the extreme surface specificity of STM/STS, is that UHV environment is necessary for most applications. One non-UHV area of application in which the STM has much to offer is that of a fluid environment. In particular, electrochemical STM, where the tip can be a local electrode as well as a local probe, has revealed a great deal of detailed information [14].

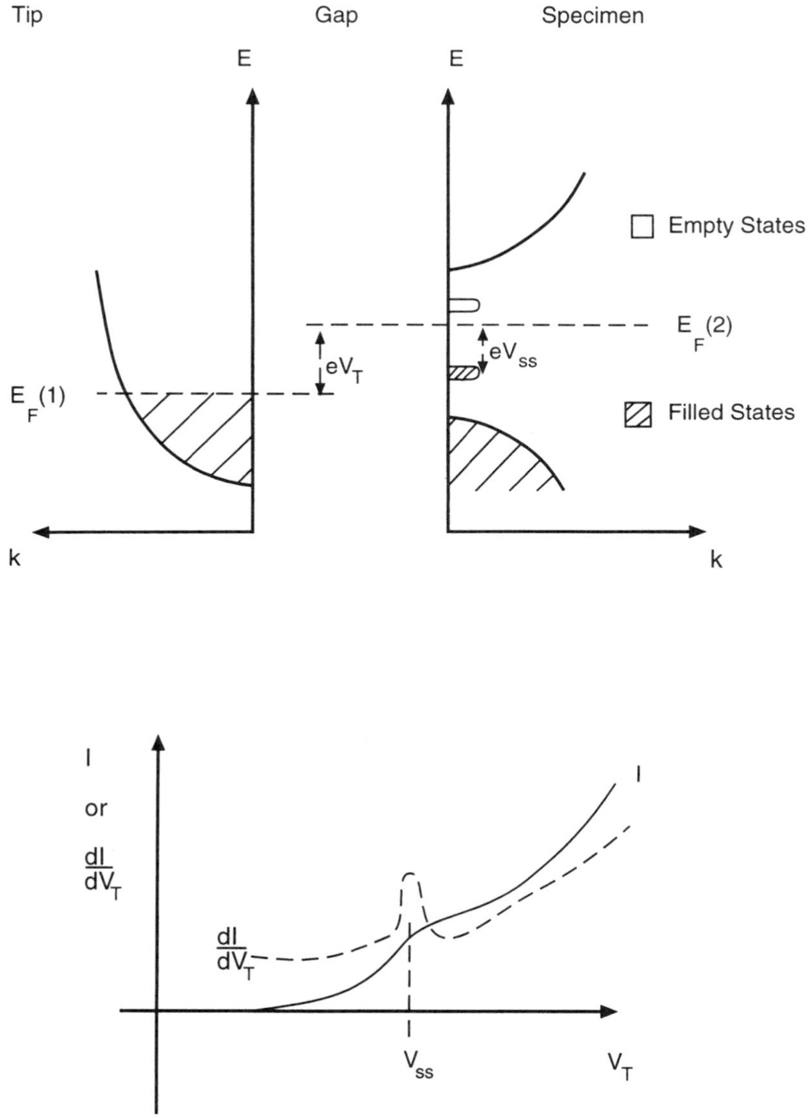

Figure 5. The underlying principles of STS. The diagram shows how surface states in the sample surface are swept past E_F in the tip, thus causing a change in tunnel current when $V_T = V_{ss}$. The effect can be enhanced by taking the derivative, dI/dV_T. The diagram shows that the tunnel current, in general, increases exponentially with V_T. Thus, spectroscopic features may be enhanced by normalization, i.e., plotting $(dI/dV_T)\,(V_T/I)$.

2.2. SFM

A scanning force microscope probes a surface by sensing force, or its gradient, between the surface and a tip; therein is the underlying principle of operation. The attention in this section will be principally with the AFM and its variants.

A description of the SFM system cannot be carried out with the same degree of generality, neatness and unity as for STM/STS. Neither can the image formation process be understood at the same level of detail and physical insight. Continuum theories will provide only a very rough phenomenological description of the system, which may be adequate for most purposes, but the full power of the SFM will only be realized once detailed microscopic theories can be deployed on a routine basis.

It is illustrative to consider some of the general characteristics of interatomic interactions, as shown schematically in Fig. 6. The short-range repulsive interactions combine with the longer-range attractive ones to produce a potential well with a location of lowest energy, the binding energy E_b, and zero net force (Fig. 6a, b). In Fig. 6c is plotted the generalized shape of the force gradient; the magnitude of the force gradient is relevant to the design and characteristics of the force-sensing element of an SPM instrument. The curves demonstrate two operational modes. One is the repulsive force regime, the contact mode, where the force gradient is high, and the repulsion is due to the overlaps of the electron clouds of atoms in close proximity. The simplest description of the interaction in the contact mode is in terms of the Lennard-Jones molecular interaction potential function; it turns out, however, that a proper description of the contact mode requires a full-molecular-dynamics simulation [15]. The other is the attractive force regime, the noncontact mode, where the force gradient is low and the attraction is due to dispersion forces of the van der Waals type and other interactions; relevant geometries and useful expressions for noncontact interactions are summarized in Table 3 (adapted from Burnham et al. [16]). Modeling of forces acting on extended bodies in close proximity is a nontrivial exercise, and is beyond the scope of the present chapter. Most of the current attempts to reconcile theory with practice are based on a seminal monograph by Israelachvili [17]. However, the close proximity of the contact mode, and the shorter range of the forces in this regime, will result in greater spatial resolution than in the case of noncontact modes with greater separation distances and longer-range forces. Also, the lower force gradients in the noncontact modes will place higher technical requirements on the implementation of sensing, control and stability of the feedback loop.

2.3. Force-Distance Spectroscopy

The AFM analog of I-V spectroscopy in STM is force versus distance (F-d) spectroscopy; the force sensing lever is the dispersive element of the system–

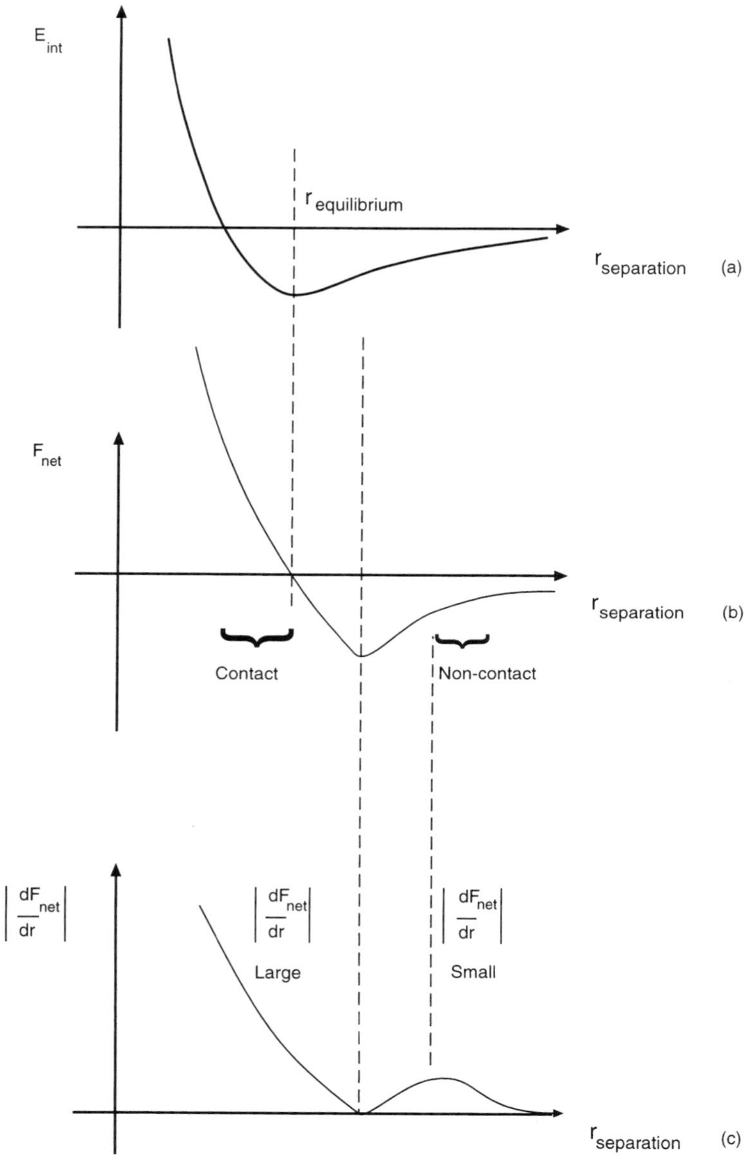

Figure 6. Schematic representation of tip-to-surface interactions in SFM. (a) potential energy as a function of separation distance; (b,c) net force and force gradient, respectively. The repulsive and attractive force regimes have been defined.

Table 3. Force Relationships for Tip-Surface Noncontact Interactions ($D << R$)

Interaction	Geometry	Expression[a]
Capacitance	sphere-flat	$-\varepsilon\varepsilon_0\pi RV^2/D$
	cone-flat	$-\varepsilon\varepsilon_0\pi V^2 \tan^2\theta \ln(1/D)$
Charge-fixed dipole	charge-flat	$\pm(\pi\rho\mu/2\pi\varepsilon\varepsilon_0)\ln((D+t)/D)$
Charge-"free" dipole	charge-flat	$-(\pi\rho q^2\mu^2/3(4\pi\varepsilon\varepsilon_0)^2kT)(1/2D^2 - 1/(D+t)^2)$
Charge-induced dipole	charge-flat	$-(\pi\rho q^2 a/(4\pi\varepsilon\varepsilon_0)^2)(1/2D^2 - 1/(D+t)^2)$
Capillary	sphere-flat	$-4\pi R\gamma_{LV}\cos\theta + 4\pi R\gamma_{SL}$
van der Waals	sphere-flat	$-HR/6D^2$
	cone-flat	$-H\tan^2\theta/6D$
Fixed dipoles	sphere-flat	$\pm\beta((D+2t)\ln(D+2t) + D\ln D - 2(D+t)\ln(D+t))$
	cone-flat	$\pm\eta((D+2t)^2\ln(D+2t) + D^2\ln D - 2(D+t)^2\ln(D+t))$
Patch charges	sphere-flat	$-\delta/(D+A)^2 + \xi/(2D+A+B)^2$

[a] R = radius of curvature of the tip; V = potential; D = tip-to-sample separation; a = length of dipole; q = point charge; θ = half-angle of cone, or contact angle for meniscus; ε_0, ε = permittivity of free space and relative permittivity, respectively; ρ = charge density; μ = dipole moment; t = layer thickness; β = polarizability; $\gamma_{LV/SL}$ = surface energy (tension); LV, SL = liquid-to-vapor and solid-to-vapor, respectively; H = Hamaker constant; ξ, β, η, δ = constants (see Ref. 15); A = location of charge on tip; B = radius of curvature of tip (patch charge model).

in this sense it is appropriate to use the term "spectroscopy." A force-versus-distance curve can be recorded by holding, through sample-and-hold techniques, the tip at a particular *x-y* location far away from the surface. The tip is then driven in toward the surface at a rate which is slow in comparison with the mechanical response of the system but fast in comparison with thermal drift. The net force is sensed during the approach, contact and retraction parts of the cycle. An idealized curve is shown in Fig. 7. Segment A-B reflects the attractive force regime; any structure in this part of the curve can be ascribed to the functional dependence(s) of the attractive force(s) on separation distance. These dependences are weak in comparison with those in other parts of the cycle. The relatively sudden drop at B toward C will generally occur due to meniscus forces (operation in air) and/or adhesion forces. At, or near, point C, tip and surface will be in "contact," and the slope of the segment C-D will be determined by the short-range repulsive force. If the surface is compliant, due to finite hardness, or if there are other kinds of relaxation at the point of contact (e.g., due to adsorbed molecular species), then there may be structure in the F-d curve at the point C, and/or the slope C-D may be dependent on the mater-

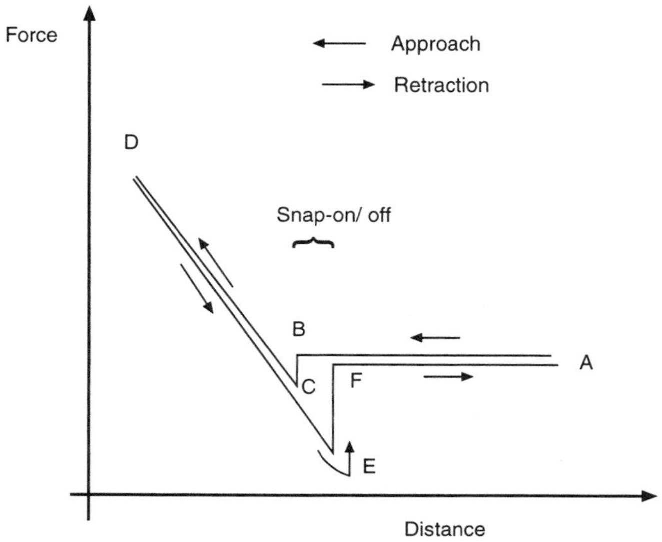

Figure 7. Schematic representation of SFM spectroscopic data, an F-d curve. Note the nearly horizontal sections which represent the noncontact regime; the sloping straight-line segments which represent hard-sphere repulsive contact; and the hysteretic effects due to snap-on and snap-off forces.

ial [18,19]. The principal difference in the curve during retraction will be due to hysteretic effects around the points of making and breaking contact (B-C versus E-F). The making and breaking of a meniscus microbridge clearly will result in hysteresis, as will surface adhesion. As well, molecular rearrangements at the interface may be hysteretic [20]. Some of these effects have been discussed in Refs. 16 and 17. Typical dynamic ranges of F and d will be up to 10^2–10^3 nN and 10^2–10^3 nm, respectively. A consequence of this discussion is the need to match the stiffness of the lever to the compliance of the system being investigated in order to extract maximum information.

3. TECHNICAL IMPLEMENTATION OF SPM INSTRUMENTATION

3.1. Generic Features and Elements

A block diagram of an SPM system is shown in Fig. 8. The "tip" and "sample" elements of the total system are at the heart of SPM and will be dealt with under separate headings. Until about 1987, SPM instruments were constructed in-house. Some in-house design and construction is still taking place, but the overwhelming

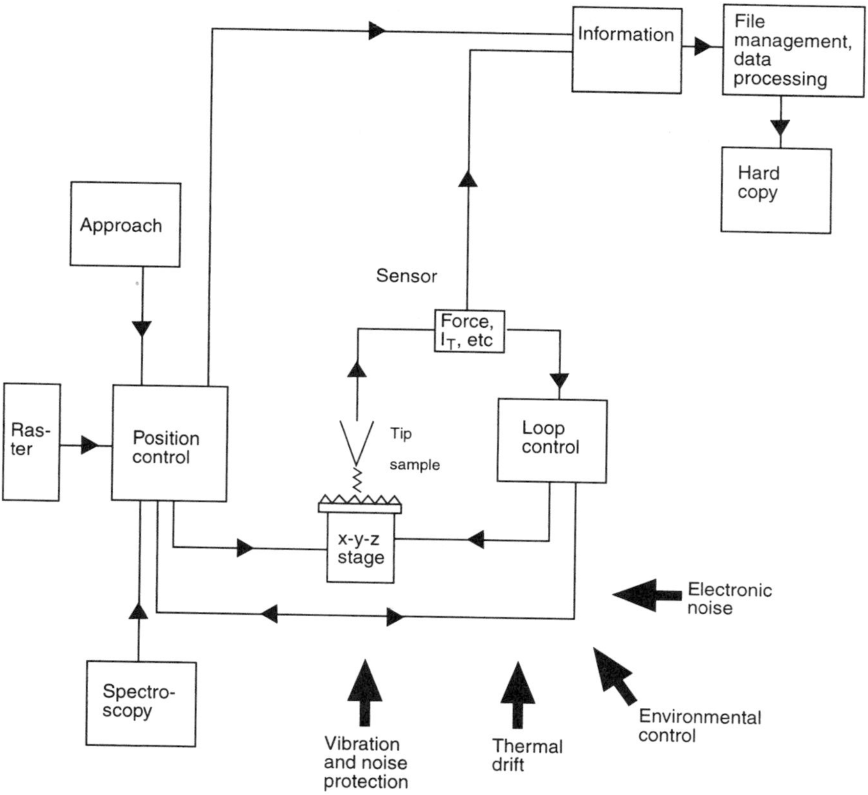

Figure 8. Block diagram functional layout of an SPM instrument.

majority of instruments are now obtained from commercial suppliers. Commercial instruments are the culmination of hundreds of person-years of development and refinement and are becoming increasingly flexible, reliable, sophisticated and user-friendly. Most users will not understand, or need to understand, the details of the engineering of an SPM system in order to exploit most of its functions. However, there are aspects that do need to be understood, such as the image formation process, effects of environmental variables, criteria for optimum selection of probe specifications, selection of optimum scan conditions, etc. SPM is a technique at the early part of the learning and developmental curve. An agreed and documented set of guidelines for best practice is yet to emerge. Thus a practitioner needs to be sufficiently well informed about the technicalities of SPM in order to derive the full benefits from existing instruments, and to recognize the significance of ongoing developments in the field.

3.2. Spatial Positioning and Control

Spatial positioning and control of the SPM probe with respect to the surface need to be considered in three contexts:

1. Coarse lateral movement is required in order to locate promising regions of the surface within the dynamic range of the field of view; this may be done manually, or with a stepper motor under software control, or with a mechanical lead-screw/lever arrangement, or with a so-called inchworm linear drive, or with a combination of these.
2. Coarse and fine movements of tip with respect to the surface in the z-direction are required in order to bring the probe into the dynamic range of the gap control loop (the tip approach sequence). The coarse movement may be implemented, as above, with an inchworm and/or lead-screw drive. The trend for modern instruments is to have the approach sequence fully automated and under computer control.
3. The x-y rastering of the surface relative to the tip needs to be driven by actuators which displace either the sample stage (usually the case for SFM instruments) or the probe mount (usually the case for STM instruments). The raster drive and the fine z-positioning always use one or other types of piezoelectric device.

The piezoelectric building blocks–beam, tube and bimorph–are illustrated in Fig. 9. The displacement, Δx, as a function of applied voltage, V, of the respective configuration will be

$$\Delta x = \frac{d_{31} V l}{h} \quad \text{(beam and tube)} \tag{8}$$

$$\Delta x = d_{31}\left(\frac{3V}{8}\right)\left(\frac{l}{h}\right)^2 \quad \text{(bimorph)} \tag{9}$$

where d_{31} is the piezoelectric coefficient, and the other symbols are defined in Fig. 9. From the two expressions it can be seen that, for given dimensions and voltage, there will be a greater dimensional change for the bimorph than for the other two elements. Thus the bimorph will be favored when a large dynamic range is required. Figure 9 also shows, schematically, the two most commonly used x-y-z translation devices—the tripod and the tube scanner. The inchworm, not shown, is a piezoelectric linear drive comprising two clamps and an expandable body, each of which can be controlled by a sequence of voltage signals. All devices for motion and positional control must meet requirements for resolution and for dynamic range; these are conflicting demands. Typical values are summarized in Table 4. Other figures of merit relate to mechanical eigenmodes of the devices, coefficient of thermal expansion, creep, hysteresis, linearity, orthogonality and aging. Different manufacturers deal with these figures of

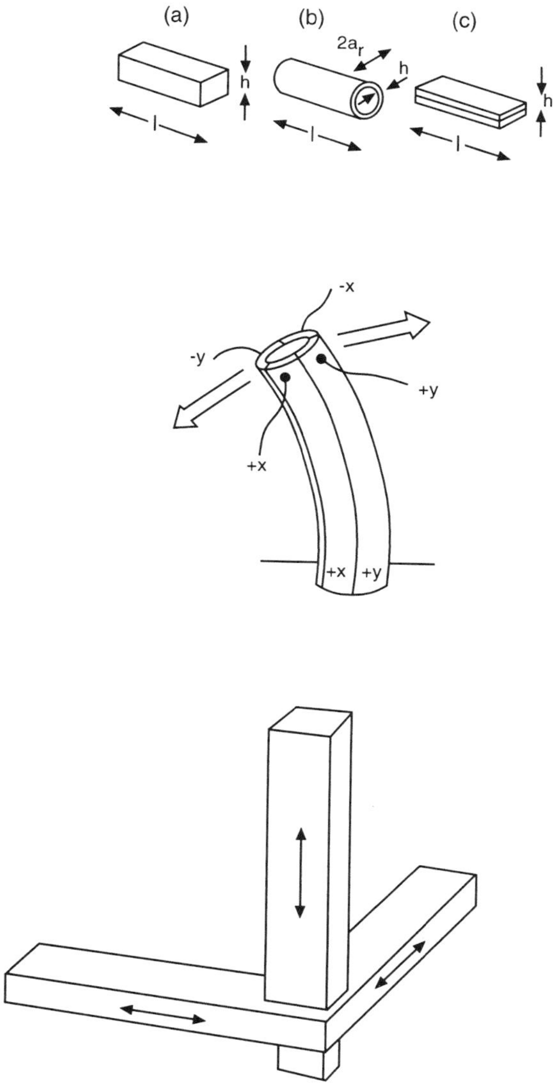

Figure 9. Configurations of piezoelectric elements (a rectangular beam, thin-wall tube, and sandwich bimorph structure) for positional control. (a) and (b) are polarized uniformly, while (c) has opposing polarizations on the two sides of the interface. The electrostatic field is perpendicular to the long dimension (a) and (c), or radial (b). The two common implementations of *x-y-z* positioning are shown; a tube scanner (middle) and a tripod (lower).

Table 4. SPM Position Assemblies

Function	Device type	Dynamic range	Resolution
Coarse *x-y*	lead screw/lever or inchworm	mms	10–20 μm
Coarse *z*-approach	lead-screw/lever or inchworm or shuffler[a]	mms	1-5 μm 50–500 nm
x-y raster	piezotripod	0.1–1 μm	0.03 nm
	piezotube	1–25 μm	0.1 nm
	piezobimorph	10–100 μm	0.5 nm
z-gap control	piezotripod	0.1–0.5 μm	0.01 nm
	piezotube	1–10 μm	0.01 nm

[a]The shuffler is a linear drive which relies on friction during slow movement and then inertial mass to remain stationary, when the track is contracted; also called "inertial drive."

merit in different ways. Experience with manufacturers and agents is that they are glad to provide full documentation on most of these and to argue the relative merits. However, prudence and best practice suggest that, whenever possible, an instrument should be evaluated and calibrated at regular intervals. Most manufacturers supply standard specimens for such use; alternatively, the user may prepare in-house specimens (e.g., ordered arrays of polystyrene spheres; see Refs. 52 and 54).

3.3. Gap Control Loop

The electronic control unit (ECU) closes the loop consisting of surface, interaction and tip. The essential objective is to ensure the best possible tracking by the tip of a contour (e.g., constant tunnel current, attractive force, etc.) in space. To this end one needs to be mindful of the phase shifts and gains of the mechanical components in the loop and of the functional dependence of the interaction on gap dimension. The ECU must thus have facilities for filtering, proportional gain, integral and differential time constants and phase shifting. As well, the ECU must have facilities for generating and mixing in other signals for spectroscopy, etc. These parameters and functions must be under the control of the experimenter so that optimum scan conditions can be set up. In first-generation instruments the control loop consisted of hard-wired analog circuitry, but the current generation of instruments tends to use software-supported digital techniques in combination with A/D and D/A interfacing.

At some point in the control loop there will be a signal, proportional to the excursions of the tip in the *z*-direction, which is required to maintain the position of the tip on the contour. This signal, as a function of the *x*-*y* position, is the primary piece of information that will generate an SPM map. Choosing optimum ECU conditions for SPM scanning remains very much an art form, although there are general guidelines to be found in most instrument manuals. However, as in many other cases, there is no substitute for practice. The design and maintenance of ECU electronics is not for the faint-hearted and is best left to those with the relevant expertise and experience.

3.4. Raster Implementation and Control

An SPM raster is generated in the same general way as for a TV or SEM picture. Two ramp or sawtooth functions of appropriate amplitudes and frequencies drive the tip (or sample) stage along two orthogonal axes. The amplitudes are determined by the requirements for fields of view and pixel resolution. In order to obtain good atomic resolution a typical field of view is chosen which might contain 50–500 unit cells, equivalent to a square with sides of 2–5 nm, and the field might be defined by 50–200 data points per line. The choice of frequencies of the raster waveforms is conditioned by two conflicting demands. High-speed scanning is desirable in order to reduce distortion from thermal drift. Conversely, the upper limit on scan speed is set principally by the lowest eigenmode of the instrument (see below), which must be well above the highest Fourier component in the output signal. As a rule of thumb one may set the scan speed (in nm/s, say) so that its numerical value is a few times greater than the numerical value of the scan width (in nm); thus the fast scan frequency will be 1–10 Hz. The frequency of the slow scan signal will then be scaled down by the number of lines in the image. For a medium-sized field of view ($5 \times 5\ \mu m^2$) some 1–3 min may be required to acquire an image. If high scan speeds are required over a small field of view and a very flat surface, then an STM or AFM may be operated in the constant-height mode, in which the feedback loop is effectively turned off; in this way there will be better rejection of noise from the instrumental eigenmodes, and obviously less distortion from thermal drift. However, the inertia and eigenmodes of the probe may then become limiting factors [21].

3.5. Noise and Drift Management

Ambient noise, acoustic and mechanical, represented major obstacles to obtaining atomic resolution during the pioneering days of STM. At the low-frequency end it was found that building modes (10–100 Hz) represented a serious problem; much clever thinking went into the design of dissipative systems for rejection of low-frequency vibrations [22]. Acoustic noise is much more easily attenuated by the simple expediency of arranging an impedance mismatch with a rigid envelope of the SPM environment. For instance, it will be apparent that an UHV vessel is nearly impenetrable to acoustic noise. All current commercial instruments claim, mostly with justification, to be well protected against environmental noise. Experience suggests, however, that most SPM instruments will benefit from being mounted on a good optical table.

Having dispensed with low-frequency noise/vibrations, it is necessary to consider the effects of internal mechanical eigenmodes [22]. These may be driven high, above the frequency range of normal operation, by increasing the mechanical rigidity of the components of the loop connecting tip and sample, and by decreasing the linear dimensions of the components. A well-designed instrument will have eigenmodes above 10 kHz.

Thermal drift was also a serious problem for first-generation SPM instruments. The absolute rates of thermal drift may be decreased by improving the thermal environment (e.g., smaller temperature differentials, lower rates of temperature change and greater thermal inertia), and the effects can be mitigated by either higher scan speeds, better thermal compensation within the instrument, or better software routines for removing image distortion due to thermal drift.

Finally, it should be mentioned that electronic noise and drift impose limits on resolution, linearity, orthogonality, etc. Again these problems are dealt with in various ways by the different manufacturers, the details being outside the scope of this chapter. It would be fair to say that the most recent designs have largely overcome technical problems of this kind.

3.6. Environmental Control

The extreme surface specificity of SPM techniques requires that the ambient environment should not affect the integrity of the tip-surface system. In the case of the revealing of surface structure by STM/STS with atomic resolution, UHV is the only ambient in which a clean and controlled environment can be ensured. In an earlier section, however, it was pointed out that STM/AFM analysis at a solid/liquid interface is a topical and rewarding area of investigation; most commercial SPM instruments can be supplied with an optional liquid/electrochemical cell for STM/AFM operation in a fluid environment. Indeed one of the real strengths of SPM is its ability to carry out analysis in situ. SFM is in general more forgiving than STM in terms of environmental requirements. Most AFM experiments are carried out under laboratory ambient conditions; some of the consequences of adsorbed moisture and other surface contaminants will be discussed. In some cases it may be necessary to eliminate some or all of the ambient contaminants. It is relatively straightforward to convert an air-ambient

SFM into an instrument operating within a controlled atmosphere envelope (e.g., enclosed in a dry/glove box).

3.7. Data Management

Raw SPM data come under two broad headings: topographic and spectroscopic. In the topographic mode the data consist of the signals from the ECU control loop which maintain constant "strength" of interaction (e.g., tunnel current, force, force gradient, etc.) as a function of position in the *x-y* plane. In the spectroscopic mode, on the other hand, the data refer to changes in a dependent variable (e.g., tunnel current, force, etc.) as a function of an independent variable (*z*-position, tunnel voltage, etc.) at a fixed location in the *x-y* plane. Thus a set of data will consist of information concerning *x-y* locations, as determined from the signals which drive the *x-y* positional elements, the signals which determine the corresponding *z*-position, and/or the signals which map out a spectroscopic curve.

A set of data can be processed in a number of ways by sequential or parallel operations, generally under computer control. Much of the data manipulation comes under the large heading of image analysis. Rather than attempting an extensive discourse on the technicalities of the latter topic, some examples of typical data manipulation are illustrated in Figs. 10 and 11 for images resulting from contact AFM scanning at modest spatial resolution of aperiodic surface structures and at high resolution for a surface with translational symmetry, respectively. The presentation of data in the spectroscopic mode is illustrated in Fig. 7.

4. SPECIFICS FOR SOME SPM TECHNIQUES

4.1. STM/STS Specifics

It has been explained that the fundamental event underlying the image formation process is that of localized electron tunneling between tip and surface. The extreme localization and consequent atomic spatial resolution suggest that STM tips need to be atomically sharp. At first glance it would seem to be extremely difficult to prepare a tip with that attribute. However, an additional benefit of the strong *z*-dependence of the tunnel current is that *any* atom (actually an electron orbital) which protrudes even 0.05 nm beyond any other atom will de facto become the single-atom tip. One may not know precisely where that atom is located, but that is of no concern as long as it is located within the desired field of view. Any electrically conducting object which is terminated by a protruding atom will in principle be capable of resolving atoms on a flat surface. If there are topographic corrugations in the surface (e.g., steps, pores, cracks, pits, pre-

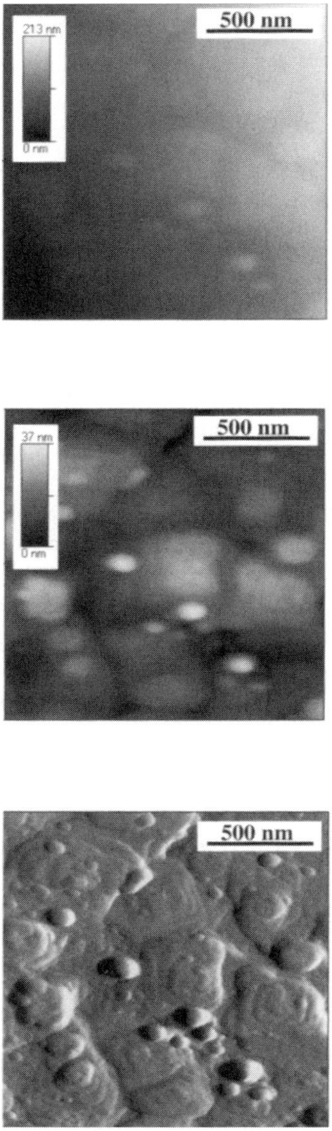

a)

b)

c)

Figure 10. Illustration of the main steps involved in image processing for a surface without long-range order. The data refer to a thin-film specimen of $YBa_2Cu_3O_7$ on an MgO substrate following exposure to an aqueous ambient [23]. Raw data are shown in (a), while successive image enhancements are demonstrated in (b) and (c).

Figure 11. Processed image for a surface exhibiting long-range order. The data refer to a Langmuir-Blodgett film. The inset shows the outcome of two-dimensional fast Fourier analysis (2DFFT).

cipitates, clusters, etc.), then the gross shape of the tip becomes important. The image then becomes a convolution of the geometries of tip and surface; more will be said about this in the context of SPM probes. In Fig. 12 are demonstrated two consequences of tip shape–"forbidden" space and tip "hop." Tip hop, transient or permanent, may also occur due to transfer of material from the surface to the tip, or vice versa; the result may be to translate the location of the tunnel gap by as much as several tens of nanometers [24,25].

Judgments about the quality and merits of STM tips need to be made on the basis of a number of requirements:

a. *Mechanical stability*: The dimensions and rigidity of an STM tip must be such that the frequency of the principal eigenmode for the tip equals, or exceeds, that of the instrument.
b. *Chemical stability*: The tip should be sufficiently inert to resist contamination or chemical alteration (e.g., oxidation) for the duration of an experiment.
c. *Geometrical shape*: As well as being terminated by a single protruding atom the radius of curvature should be smaller than the radii of curvature of surface features. If there are deep corrugations and steep slopes in the surface, then a high aspect ratio (height/width of base) for the conical tip will be required.
d. *Other factors*: The tip must be made from a readily machinable material, and it must be electrically conducting.

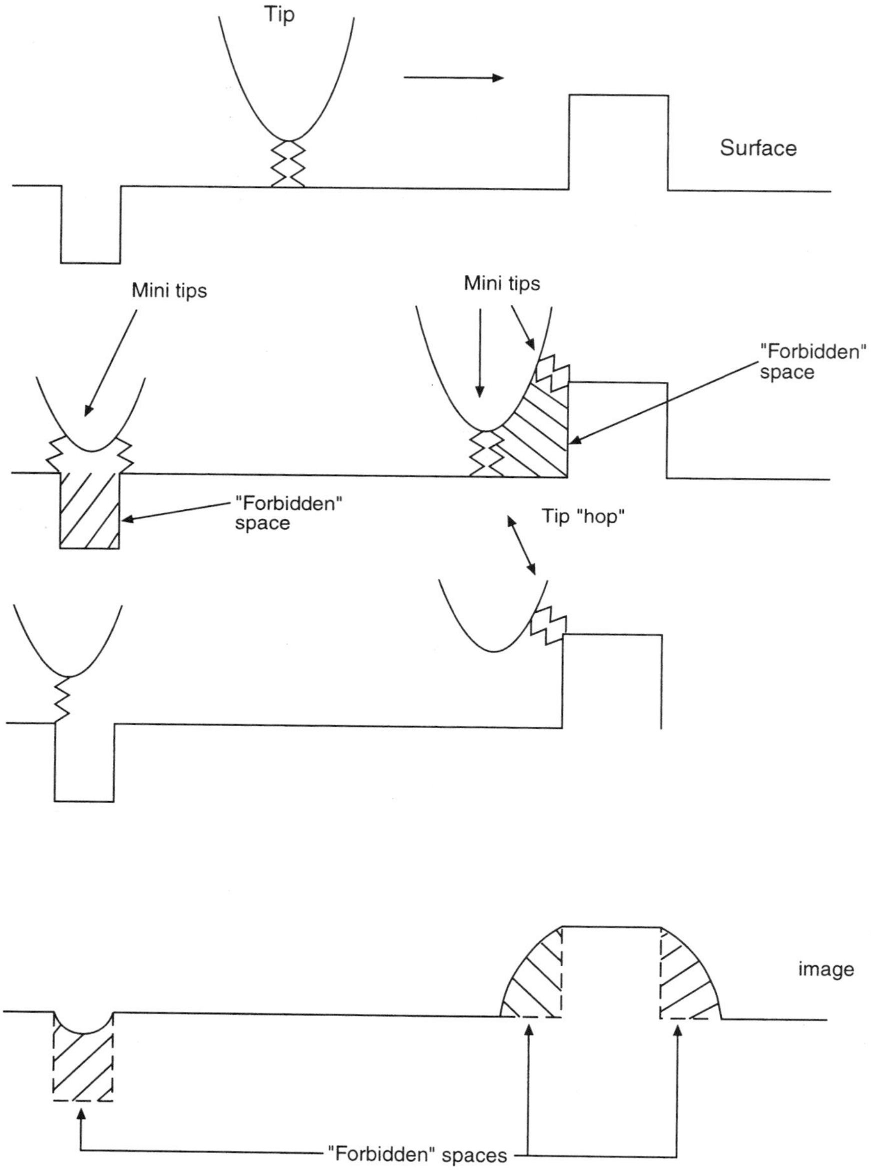

Figure 12. Schematic diagrams showing the geometrical convolution of a tip of finite width with a corrugated surface. Note the existence of "forbidden" space, inaccessible to the tip, and the consequent effect on the image. One reason for "tip hop" is also shown.

In practice, only two materials are in common use. A Pt-Ir (typically 80:20) alloy wire tends to be used for STM scanning in air, while W wire is used as tip material for UHV operation. Electrochemical polishing is the preferred method in both cases for producing good tips with small radii of curvature (≈10 nm) and high aspect ratios (better than 3:1). There are recipes to be found in the literature [26,27]. When W tips are used, it is essential that the tips be cleaned in UHV by field desorption/emission in order to remove the oxide layer and any other contamination.

4.2. SFM Specifics

The implementation of SFM in the contact mode is shown schematically in Fig. 13; for completeness the position of the piezoelectric element which drives the probe in noncontact operation is also shown. The interaction is between tip and surface but sensed by a cantilever; the stronger the repulsive force, the more the cantilever will be deflected, given a particular spring constant for the lever. The most common method for tracking the force, and thus controlling the feed-back loop, is known as the "optical lever." A laser beam is bounced off the back

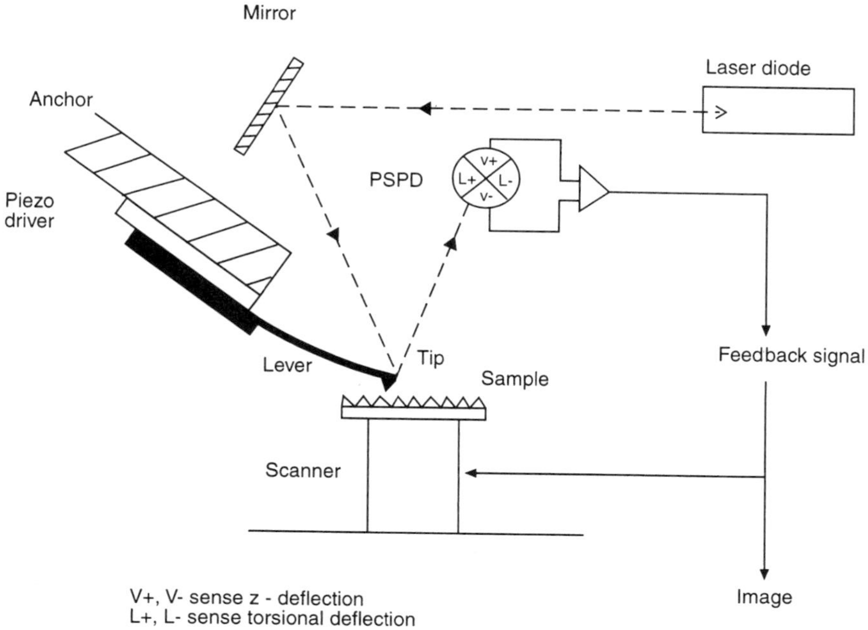

Figure 13. A generic representation of the implementation of contact-mode SFM with the optical beam deflection control method.

of the lever, and the reflected beam is detected by position-sensitive photodiodes (PSPD). Because of the geometrical constraints imposed by alignment of the optical lever, it is clear that the specimen stage must be rastered while the sensor/detector stage remains stationary. A constant force offset can be chosen, and excursions in the difference signal from the set point are then used to generate the topographical map as well as to control the *z*-position of the surface with respect to the tip. The technicalities of this method and others have been described in the literature (e.g., [28]).

Lateral-force AFM is a contact-mode technique. The system is controlled by maintaining constant deflection of the cantilever in the *z*-direction, but the image is generated from monitoring torsional deflections of the lever, in order to derive the horizontal drag on the tip. Torsional deflections lead to off-axis excursions of the reflected laser beam which may be tracked by incorporating two additional detectors in the PSPD assembly, at right angles to the two that sense the normal contact mode. Because the additional hardware requirements are trivial and the additions to the software are also minimal, a contact-mode SFM instrument will include the lateral-force mode as a matter of course.

Non-contact-mode SFM is rather more difficult to implement due to the weaker long-range forces and the much smaller force gradients. Thus ac resonance detection techniques need to be used. A driving element, usually piezoelectric, is introduced into the lever assembly at the fixed end (Fig. 13), which stimulates the lever to vibrate at, or near, its resonant frequency. Small changes in the force gradient will then have the effect of changing the effective mass of the lever, which in turn will lead to a phase shift and to an amplitude change at the free end of the lever. Using one of these variables the loop can be locked to either constant amplitude or phase, so as to track constant contours of the *z*-component of the force gradient across a surface; the control signal is picked off the PSPD assembly. This control mode has been described in the literature as well (e.g., [28,29]). It is illustrative to consider the resonance curves for a driven and damped oscillator when it is 'free'-running (i.e., well away from the surface) and when it is constrained by a force gradient (Fig. 14). The equation of motion for a driven and damped oscillator is

$$\frac{d^2z}{dt^2} + \frac{\gamma\ dz}{dt} + \omega_0^2 z = (m^*)^{-1} F_0 \exp(i\omega t) \tag{10}$$

where $\gamma = b/m$, m^* is the effective mass, $\omega_0 = [k/m^*]^{1/2}$, F_0 is the amplitude with which the anchored end of the lever is driven, and ω is the driving frequency. k is the spring constant and b the drag coefficient. In general, the amplitude and phase angle of the free end of the lever are [29]

$$A(\omega) = \frac{F_0}{m^*}[(\omega_0^2 - \omega^2)^2 + (\gamma\omega)^2]^{-1/2} \tag{11}$$

$$\tan(\delta) = \frac{\gamma\omega}{\omega_0^2 - \omega^2} \tag{12}$$

The resonance frequency under load, ω_{res}, is

$$\omega_{res} = \left[\omega_0^2 - \frac{\gamma^2}{2} \right]^{1/2} \tag{13}$$

and the amplitude at resonance is

$$A_{res} = \frac{F_0 Q}{k(1 - 1/4Q^2)^{1/2}} \tag{14}$$

where $Q = \omega_0/\gamma$ is the quality factor. If the lever is moving in an attractive force field, the resonance frequency will decrease as the lever approaches the surface. We can then define $k' = k - F = k - \delta F_{attr}/\delta z$; the upshot is that the resonance frequency will be dependent on the average force gradient. It can be shown [28] that when the system is operated on the steepest part of the resonance curve, where the sensitivity to change in force gradient is greatest, then

$$\left[\frac{\delta A}{\delta F'} \right]_{\omega = 0.816\omega_0} = \left[\frac{2QA}{3^{4/3} k} \right]_{\omega = \omega_0} \tag{15}$$

where $\omega = 0.816\omega_o$ corresponds to the steepest part of the resonance curve. Also, it turns out that the dependence of the amplitude at this point with respect to change in frequency is

$$\left[\frac{\delta A}{\delta \omega} \right]_{\omega = 0.816\omega_0} = \left[\frac{4QA}{3^{4/3} \omega_0} \right]_{\omega = \omega_0} \tag{16}$$

There are three general figures of merit for an SFM lever; these are the spring constant, k, the "free" resonance frequency, ω_o, and the quality factor, Q. In the case of a contact-mode lever the torsional spring constant (for operation in the lateral-force mode) is also a relevant parameter. The resonance frequency of contact-mode levers is only significant in comparison with that of the eigen-mode of the instrument, and the Q factor is not relevant. Approximate expressions for the two main parameters in the case of a thin beam-shaped lever are

$$k = \frac{Ewt^3}{4L^3} \tag{17}$$

$$\omega_0 = \left[\frac{Ewt^3}{4L^3(m_t + 0.24wtL\rho)} \right]^{1/2} \tag{18}$$

where E is the modulus of elasticity, w is the width, t is the thickness, L is the length, m_t is the mass of the tip and ρ is the mass density. The actual value of

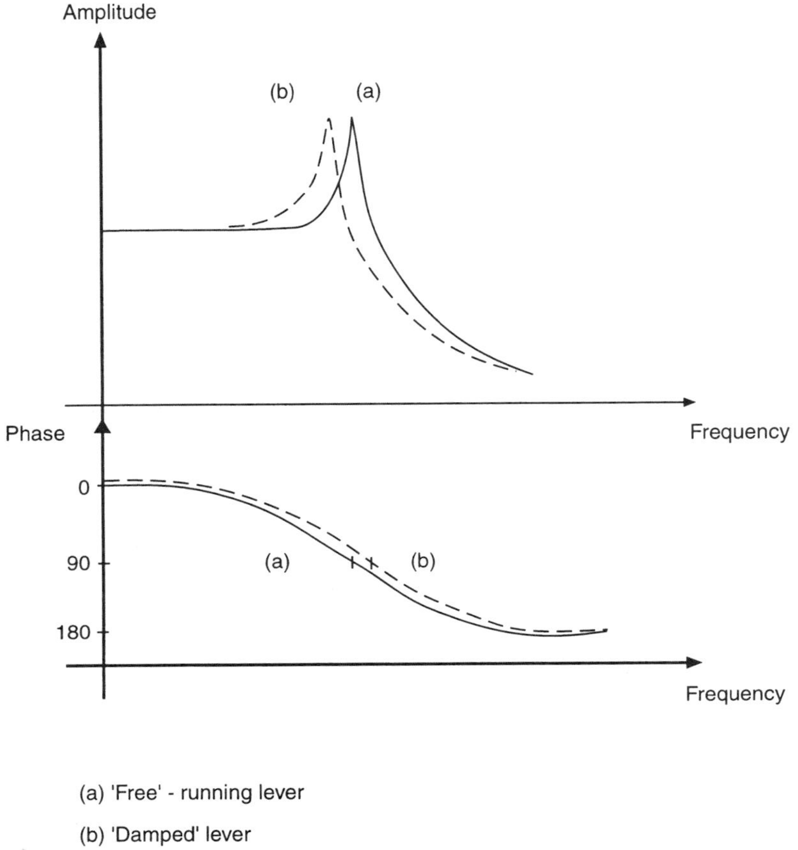

Figure 14. Description of the noncontact ac resonance loop control method. The frequency responses, both amplitude and phase, of a damped and a driven oscillator are shown. Note that both curves are shifted (a to b) by changes in the force gradient.

E for the microfabricated lever will depend on stoichiometry (e.g., Si_3N_4), dopant concentration (e.g., Si) and crystallographic direction (e.g., Si). The intrinsic *Q* value is generally very high, but will be limited to about 10^2 by an ambient atmosphere. Approximate results for the more commonly used V-shaped levers can be arrived at by treating them as two parallel beams. More accurate expressions based on finite element analysis can be found in the literature [30 and references therein].

4.3. SFM Probes: General Considerations

Probes are now available commercially from several suppliers, generally through the instrument makers. These probes, levers with integral tips, are micromachined in bulk quantities using planar technology first developed for production of microelectronic devices. The user can buy single probes, ready mounted, or wafers containing several hundred probes, which must then be separated and mounted. The former is expensive, but convenient, while the latter is less expensive in the long run but less convenient. The design objectives and fabrication methods are described in the literature [31,32]. It is useful to consider the design criteria in full, since they will affect how one chooses particular probes for various applications. An SEM image of a typical lever with integral tip is shown in Fig. 15. The image formation process is intimately connected with, and ultimately limited by, the parameters and properties of the probe.

4.4. SFM Probes; Design Criteria

1. *Force constant*: This must be matched to the operational mode and to the rigidity of the surface being scanned; "soft" surfaces (e.g., organics, polymers, biomembranes, etc.) will be deformed and/or damaged unless a soft lever is used. In the contact mode the usual, and most useful, spring constant is in

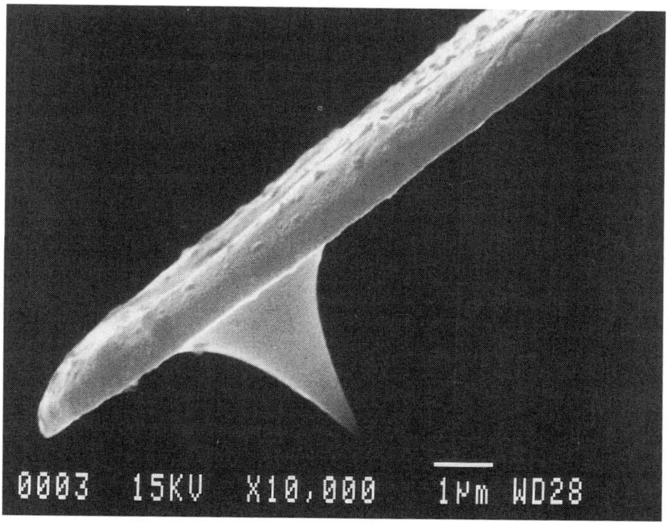

Figure 15. A SEM image of an SFM probe showing the force-sensing lever and the sharp tip.

the range 10^{-2}–10 N/m, while noncontact-mode operation tends to require spring constants in the range 10–100 N/m. Note that the corresponding force loadings can be estimated by the following conversion:

k – [Force constant (N/m)] ⇒ Force – [$k \times 10^{-9}$ (N) assuming 1-nm deflection]

The implication is that a lever with a force constant of, say, 1 N/m will exert a force at the point of contact of 1 nN, if there is a 1-nm deflection.

2. *Resonance frequency*: For contact-mode levers the resonance frequency needs to be no lower than that of the instrumental eigenmode, and preferably around 10 khz; a high resonance frequency will also allow a higher scan rate. For noncontact operation the resonance frequency needs to be at or above 100 khz in order to give a high Q and thus provide better response for the control loop.

3. *Lateral stiffness*: In the normal topographic mode lateral movement of the tip will cause image artefacts due to the effects of lateral forces (often described as "stick-slip"). These effects are particularly important in the contact mode where variable friction may stimulate torsional modes. Thus, contact-mode scanning is usually carried out with a V-shaped lever which has high internal rigidity. If acquisition of lateral-force data is the object of scanning, however, then laterally less rigid levers may be used.

4. *Lever length*: The sensitivity of the optical beam deflection method is increased in direct proportion to the length of the lever, since the angle of deflection (in radians), for small angles, is given by $\Delta z/L$, where Δz is the deflection of the free end of the lever and L is the length. This criterion affects the overall dimension of the instrument. For instance, a lever length of 100 μm will result in z-resolution of better than 0.1 nm for a 1-cm total length of optical path. Typical lengths range from less than 100 μm up to 300 μm, with the other dimensions being 10–30 μm wide and around 1 μm thick.

5. *Reflective coating*: The top surface of the lever, near its free end, must present an efficient and flat mirror surface so as to reflect the incident laser beam without loss or divergence. Also, the thermal stability of the deflection assembly is improved by minimizing energy deposition from the laser beam.

6. *Aspect ratio of tip (height/width of base)*: As in the case of STM, an SFM tip (usually of square pyramidal or conical shape) will be excluded from the "forbidden" space on a corrugated surface due to the geometrical convolution of tip with surface. This topic will be considered further. The total height of the tip is relevant if the sample surface is curved or highly corrugated and should be comparable to the dynamic z-range (typically greater than 5 μm) of the scanning stage for maximum flexibility. Aspect ratios of commercially available probes range from 0.7 for routine pyramidal tips to 3.0 for high-resolution tips of conical shape; the manufacturers will generally quote the

aspect ratio as, say, 3:1. High-aspect-ratio (10) special-purpose tips can be made by growing carbon whiskers in an SEM at the apex of a standard tip [33].

7. *Radius of curvature of tip*: It will be shown that if the radius of curvature of the tip is larger than the radii of curvature of surface features, then the resultant image will reflect tip shape rather than reveal actual surface topography. The radius of curvature ranges from 50 nm for routine pyramidally shaped tips to less than 10 nm for high-resolution tips (e.g., carbon whiskers).

Probes are generally made from Si or Si_3N_4. Some of the relevant materials properties are listed in Table 5.

Other special-purpose SFM probes exist. In the case of MFM the tip of a normal noncontact probe is coated with a magnetic film, which is then magnetized. An MFM tip will also interact as an AFM tip, but the magnetic interaction is of longer range than those ranges which account for the AFM image; hence magnetic structure can be obtained by increasing the average tip-to-surface separation distance in comparison with that normally used for noncontact AFM. This and other factors are considered in the literature [34,35]. A tip used for SCM, on the other hand, needs to be coated with an electrically conducting film; since this is a contact-mode technique, a normal contact probe will constitute the starting point. Yet other probes are optimized for operation in an electrochemical environment by being given a passivating surface coating in order to enhance their chemical durability.

4.5. Probe Calibration and Image Artefacts

Being in the early stages of the learning curve, SFM practitioners are still developing and exploring routines for "best practice." Thus, even though generically derived assumptions about the characteristics and properties of probes have sufficed for many past and present investigations, it is becoming clear that each probe may need to be calibrated before, during and after an experiment in order to demonstrate "quality assurance." Failure to do so, and ignorance of the limitations inherent in the image formation process(es), could invalidate what might

Table 5. Materials Properties for SFM Probes

Material[a]	Mass density (kg/m^3)	Elastic modulus (10^{11} N/m^2)
Si	2330	1.79
Si_3N_4	3100	1.5
SiO_2	2200	0.6

[a]The parameters will depend on doping levels, stoichiometry and crystallographic orientation.

otherwise be a carefully designed and executed study. The major aspects of SFM probes which require calibration are discussed in the remainder of this section.

4.6. Determination of Normal Spring Constant

SFM probes obtained from commercial suppliers are characterized by generic specifications. In the case of the stated spring constant, k_N (it is useful to distinguish the spring constant for normal deflection from that of lateral deflection, k_L, when the lever undergoes torsional deformation, as in the LFM mode), for a particular kind of lever, there may be variations by ±20% between individual levers from a given batch, and even greater differences from one batch to another. As long as the actual value of the constant force of interaction does not affect the interpretation of the results this is not a serious problem. Likewise, there are many cases where only *relative* differences in the outcomes of F-d measurements are important [35–39]. However, if *absolute* measurements of forces, or their gradients, are required, then k_N for the particular lever being used needs to be calibrated *absolutely*. Several methods have been proposed; they are based on measurements of static deflection under known loads [40], on determination of resonance frequency as a function of load [41], on considerations of the resonance envelope resulting from thermal stimulation of the major eigenmode of the lever [42], or on other methods of static loading [42,44,45]. Most of these methods have one or more shortcomings in terms of being onerous, potentially destructive, or unreliable (Table 6).

A reliable and user-friendly method for evaluating k_N for V-shaped levers deserves particular attention. Equation (17) constitutes the lowest-order approximation for purposes of modeling k_N. The expression has been used in several studies with various degrees of success (e.g., [40]). Two correction factors have been derived [46], which offer additional accuracy without unnecessarily adding complexity to the treatment. Application of the first factor to the lowest-order approximation, Eq. (17), results in

$$k_N = \left[\frac{Et^3 w}{2L^3}\right] \cos\theta \left[1 + \frac{4w^3}{b^3}(3\cos\theta - 2)\right]^{-1} \tag{19}$$

where θ is the angle between the two arms of the lever and b is the width of the V at its base. In order to account for any displacement of the point of loading from the end of the lever an additional correction may be applied:

$$k_{N,\text{eff.}} = k_N \left[\frac{L}{L - \delta L}\right]^3 \tag{20}$$

where $k_{N,\text{eff.}}$ is the effective spring constant of the cantilever, and δL is the offset distance of the tip with respect to the end of the lever. The overall precision

Table 6. Methods for Determination of SFM Spring Constant

Method	Ref.	Accuracy	Demerits
Dynamic response methods			
Resonance frequency with added mass	41	≈10%	positioning and calibration of load difficult; potentially destructive
Thermal fluctuations	43	10–20%	temperature control essential; only suitable for soft levers; requires analysis of resonance curve
Simple scaling from resonance frequency	42	5–10%	depends on dimensional accuracy and determination of effective mass
Theoretical methods			
Finite difference calculation	45	10%+	depends on dimensional accuracy and Young's modulus
Parallel beam approximation	40, 46	10%+	depends on dimensional accuracy and Young's modulus
Static response methods			
Static deflection with added mass	40	15%	positioning and determination of load difficult; potentially destructive
Response to pendulum force	40	30–40%	complex and time-consuming procedure
Static deflection with external standard	47	15–40%	requires accurate external standard

of the value of k_N inferred from the foregoing expressions is crucially dependent on uncertainties in the values of the parameters, in particular t and δL; the expressions themselves provide a result with a precision of better than 5%.

Several related methods have been developed which involve deflection of an unknown lever by a precalibrated standard lever. In essence, the unknown lever is deflected a measurable distance by a known force derived from moving the precalibrated standard lever a known distance. The following expression may be used to calculate k_N for the unknown lever:

$$k_N = k_0\left[\frac{C}{D} - 1\right] \tag{21}$$

where k_0 is the spring constant of the standard and D is the slope of the F-d curve, when the unknown is being deflected by the standard lever. The conversion factor C for the optical detection system (relating the differential signal from the PSPD element, in nA, to the deflection of the lever, in nm) can be determined from the slope of an F-d curve when the lever is pressed against a "hard" surface (so that the deflection of the lever must be equal to the relative displacement of the tip with respect to the surface).

An SFM lever may be used as the standard [47]. The accuracy of this procedure is then enhanced by using the imaging capability of the SFM to choose the point of loading as close as possible to the tip of the standard, thus eliminating uncertainty concerning the precise location of the load. In effect, a reverse image [48] of the tip is obtained. The instrumental arrangement is shown schematically in Fig. 16. The spring constant of the primary standard lever can be determined from Eqs. (19) and (20) using dimensions determined by SEM analysis. Secondary standards required to cover the entire dynamic range can

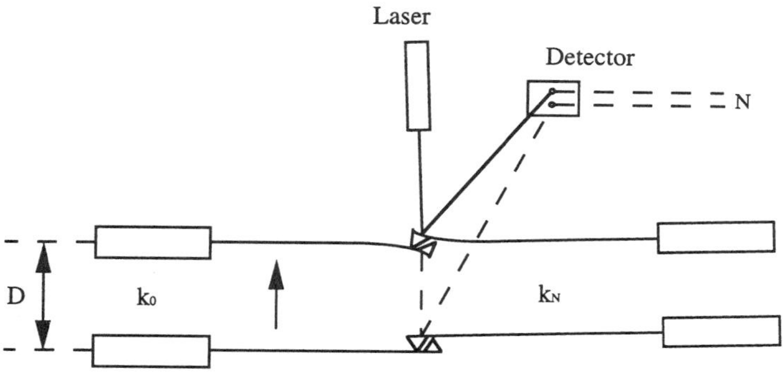

Figure 16. Experimental arrangement showing the "standard" lever, k_0, and the "unknown" lever, k_N, in their initial positions and after the sample stage has been moved a vertical distance, D.

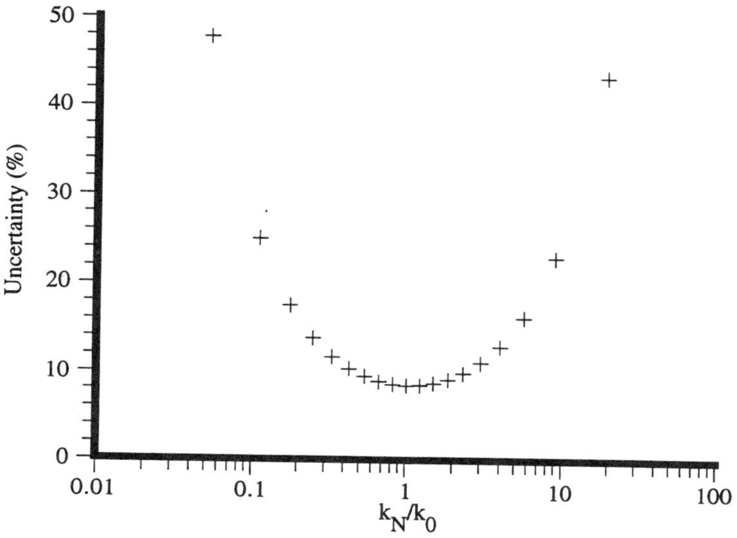

Figure 17. The dynamic range of the method for a single primary standard will be limited, as shown.

be generated by a sequential application of the procedure. The precision deteriorates when k_N is outside the range $0.3k_0$ to $3k_0$, as shown in Fig. 17. The accuracy of the method, given a reliable standard, is 10–20%, with the uncertainty being associated with the factor C/D. Alternatively, a set of primary standards may be prepared once and for all using planar micromachining techniques similar to those currently used to mass-produce SFM probes.

4.7. Determination of Lateral Spring Constant

The LFM system consists of (see Fig. 18) a surface with topographical features and physicochemical properties; a force-sensing/imposing lever which may undergo deformation by normal deflection, in response to topographical excursions in the z-direction, or torsional deformation around an axis aligned approximately with the y-direction, assuming that the specimen is being translated in the x-direction; and a tip in contact with the surface and subject to normal, F_N, and lateral, F_L, forces. The phenomenological aspects of the system are characterized by $F_L = \mu(F_N + SI_A)$, where F_L is the lateral force, F_N is the normal force, μ is the coefficient of friction, S is the area of contact and I_A is the adhesive interaction per unit area. The normal force is $F_N = k_N \Delta z$, where k_N is in the range 10^{-2} to 10^2 N/m, and Δz is the deflection of the free end of the lever when loaded. Since the z-resolution of a typical AFM is better than 0.1 nm, the minimum detectable

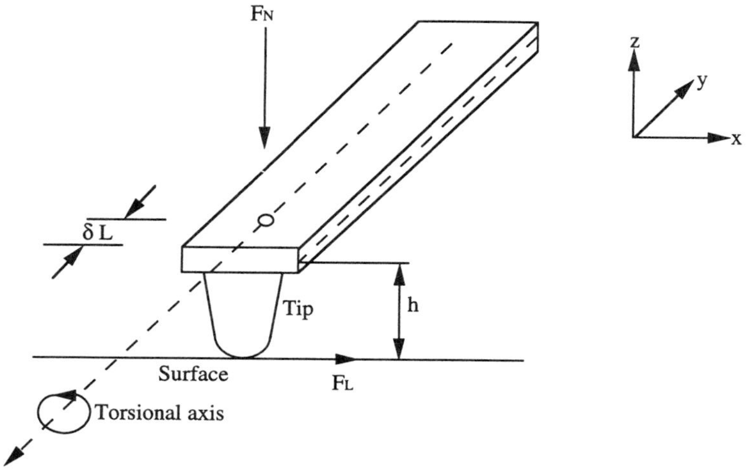

Figure 18. The principal elements of the LFM system.

normal force loading is less than 10^{-3} nN; assuming a maximum compliance of the lever of 10^3 nm, the maximum normal force transmitted through the tip will be 10^5 nN. The corresponding angles of deflection, for a lever of length 100 μm, will be 10^{-6} rad (resolution limit) and 10^{-2} rad (maximum compliance). The lateral force is $F_L = k_L\, \Delta x$ where k_L is related to the torsional spring constant and the height of the tip. The deflection of the tip at the point of contact from its vertical position is Δx for the chosen geometry. The angle of rotation of the lever at the position of the tip will be $\Delta x/h$. Given a limit on resolution of 10^{-6} rad and a tip height of 5 μm, the minimum detectable Δx is approximately 5×10^{-3} nm. Disregarding the effect of molecular interactions at the interface between tip and surface, the coefficient of friction is then

$$\mu = \frac{F_L}{F_N} = \frac{k_L}{k_N}\frac{\Delta x}{\Delta z} \tag{22}$$

A full description of the system requires that the shape of the apex of the tip be known, since that affects both the area over which the normal force is applied and the adhesive force, as well as the scan speed [49].

A recent paper [45] describes modeling of the torsional, longitudinal and normal spring constants of V-shaped levers. The lateral spring constant at the tip is

$$k_L = \frac{Et^3}{3(1+\upsilon)h'^2}[(\tan\theta)^{-1}\log(w'(\delta L\sin\theta)^{-1}) + \frac{L}{w'}\cos\theta - \frac{3}{8}\sin 2\theta]^{-1} \tag{23}$$

where υ is Poisson's ratio, $w' = w\cos\theta$ with w now being the perpendicular width of the beam, $h' = h + t/2$ is the effective height of the tip and L is the

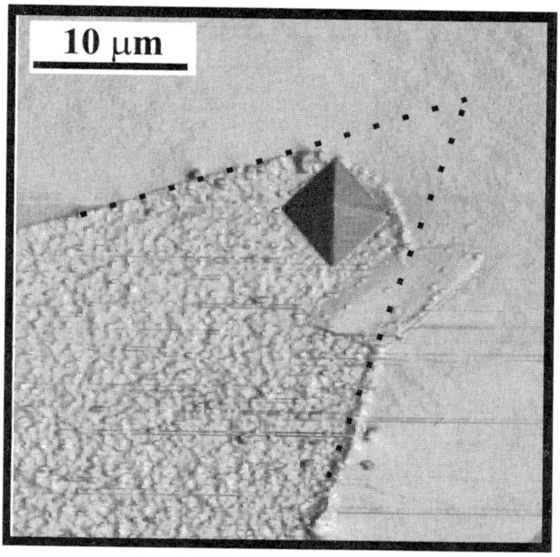

Figure 19. Reverse image of tip and lever illustrating determination of offset distance of tip from the (extrapolated) end of lever. Tip height can be determined from a line profile over the tip.

inside length. As expected, the expression for k_L is highly sensitive to h' and δL. The two parameters can be determined by reverse imaging, as demonstrated in Fig. 19. The factor Et^3 is the other main source of uncertainty; assuming that k_N has been measured by one of the methods in the literature, then Et^3 may be determined independently, thus improving the precision in the value of k_L. With typical values for a Si_3N_4 contact lever, k_L is approximately 300 N/m [50]. Standard probes can allow determination of "coefficients of friction" down to 10^{-3}, while advanced designs could extend the range to 10^{-5}.

4.8. Resonance Frequency

In the contact mode the resonance frequency is not a significant parameter as long as it is higher than the instrumental eigenmode and does not limit the scan speed. In the noncontact mode, however, it determines the resolution of the system, as described before. When ac detection is used, the feedback loop is adjusted until it is locked at a point at or near to the resonance frequency; thus this parameter is adjustable for a particular lever.

4.9. Aspect Ratio

Commercially available levers are supplied with typical average specifications; the aspect ratio is one of these (e.g., 0.7 for routine pyramidal tips or 3.0 for conical etched tips). It is a crucial parameter in terms of interpretation of an image (see previous discussion regarding "forbidden" space). In situations where there are steep edges, deep trenches and other surface features with large z-excursions and small x-y dimensions, convolution with tip shape will produce image artefacts due to limitations of aspect ratio. These effects are particularly prominent in studies of surface precipitates, biosystems on substrates, microelectronic devices, etc. Consequently there is a substantial literature on the topic, most of it concerning contact-mode imaging [51–53, and references therein]. All these studies have been concerned with a surface of known topography, so that the actual tip shape could be inferred from the image. More recently they have been extended to include the effects of tip shape on the image in the lateral and noncontact AFM modes [54,55]. Most promising for the routine determination of aspect ratio is the use of "reverse" imaging techniques, which are based on the observation that if a tip of finite aspect ratio is scanned over a vertical spike, then the resultant image is a true representation of the tip shape [56,57]. A representative reverse image of a pyramidal tip is shown in Fig. 20. This may prove to be the most convenient and direct method for obtaining the aspect ratio for a particular SFM tip.

4.10. Radius of Curvature of Tip

The aspect ratio of an SPM tip will generate image artefacts on a scale from about 100 nm upward; i.e., when vertical excursions of surface features exceed the radius of curvature, then the extent of horizontal broadening and the extent of "forbidden" space is determined primarily by the aspect ratio. Conversely, when surface features have lower vertical excursions, and/or when the radii of curvature of the features are less than that of the tip, then the radius of curvature will be the primary limitation on resolution and the main source of image artefacts. Such ef-

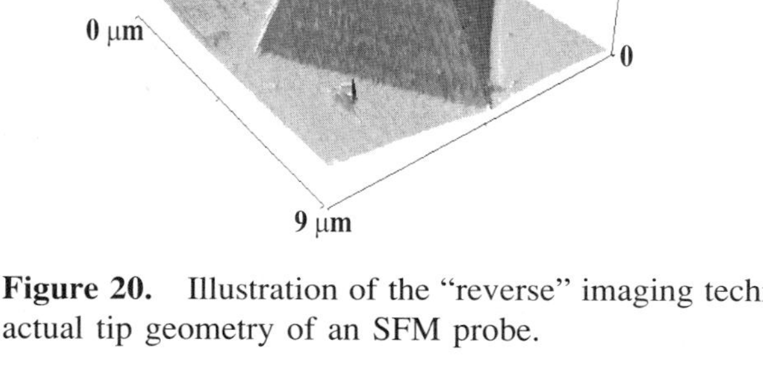

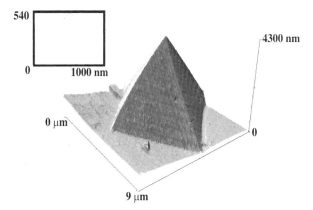

Figure 20. Illustration of the "reverse" imaging technique for determining the actual tip geometry of an SFM probe.

fects have been identified and discussed in many studies, particularly for biological applications of STM and contact-mode AFM. In the biological field it is not possible to prepare the specimen so as to produce flat surfaces without inherently throwing away the objectives of the study. In investigations of biomolecules and biomembranes [58–60], viruses [61–64], and biocrystals [65] the effects of the finite radius of curvature of the tip have been observed and recorded; images and line profiles relevant to TMV (tobacco mosaic virus) are shown in Fig. 21. A simple but useful calculation can be used for contact-mode AFM to relate the apparent width of a spherical or cylindrical object on a flat surface to the actual width:

(a)

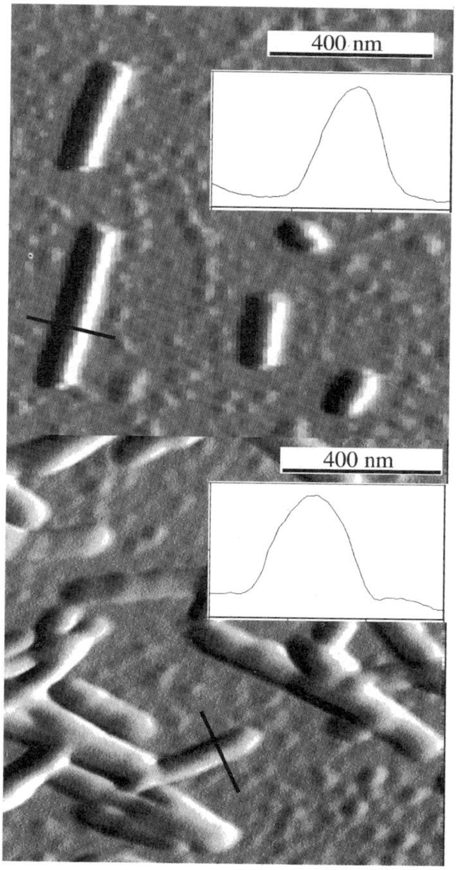

(b)

Figure 21. Images and line profiles of the TMV scanned in the contact (a) and noncontact (b) modes, demonstrating that an AFM tip with an estimated radius of 40 nm gives rise to broadening by factors from 4 to 8, while the height, and thus diameter, of the TMV is measured accurately as 18.5 nm.

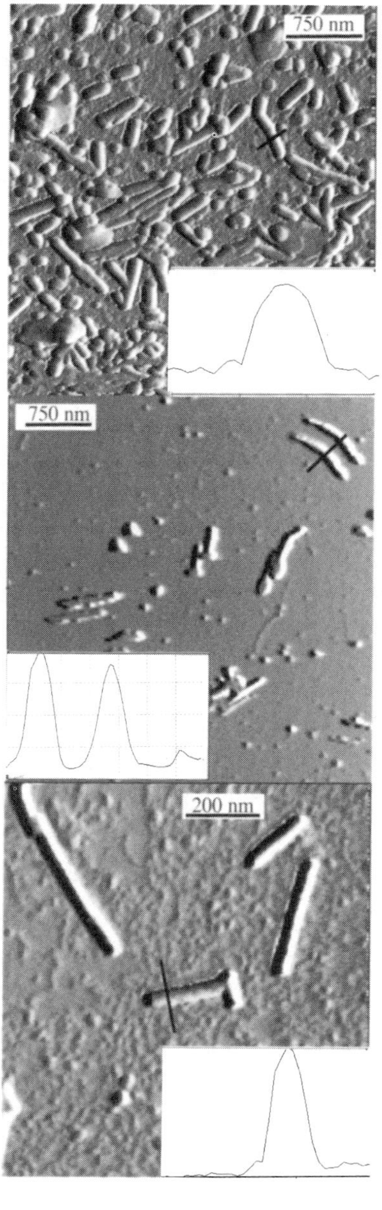

(a)

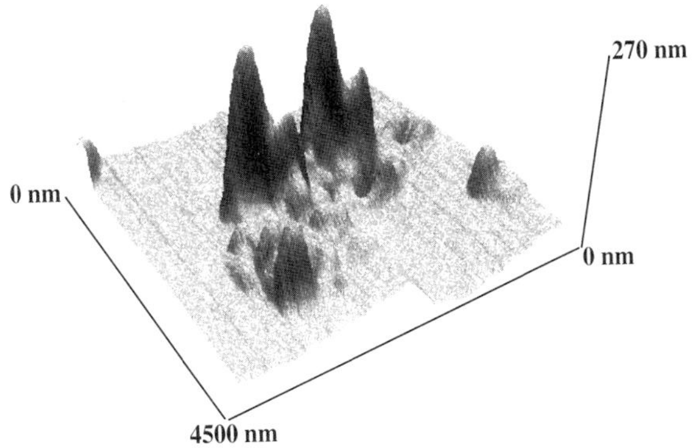

(b)

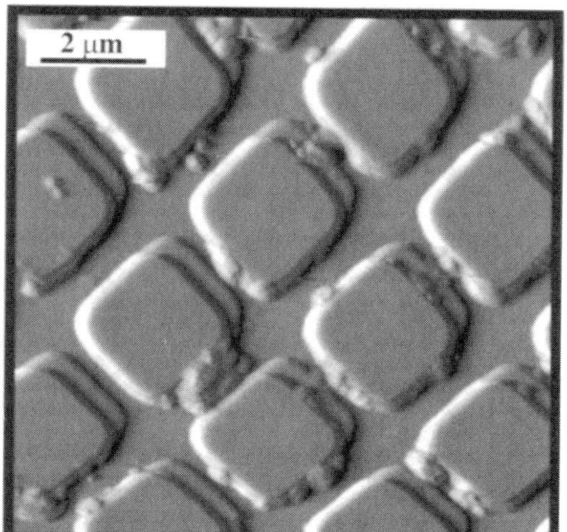

Figure 22. (a) Images and line profiles demonstrating the effect of a blunt tip (upper), a double tip (middle), and a transient sharper tip (lower); the data refer to contact-mode imaging of TMV. (b) Reverse image of a double tip (upper) and the resultant spatially correlated artefacts for a semiconductor grid (lower). (Reprinted from Honors thesis, G. S. Watson, Griffith University, 1996.)

$$W_{\text{apparent}} = 2[R_{\text{object}} + R_{\text{tip}})^2 - (R_{\text{tip}} - R_{\text{object}})^2]^{1/2} \tag{24}$$

Using typical values for the radius of curvature of a particular tip it is straightforward to "deconvolute" an image of a known structure by simple geometrical methods (e.g., [52,55]). When the structure is unknown, deconvolution does require that assumptions are made as to the shape of the tip. It is preferable, however, that the actual shape of a particular tip be known. Examples of the gross effects on an image of TMV of a double tip and other tip shapes are shown in Fig. 22a. An equally striking example is shown in Fig. 22b; this shows a reverse image of a double tip and an image of a semiconductor grid which demonstrates the resultant artefacts.

Calibration of probes for purposes of determining the radius of curvature of a particular tip requires a known surface with features–holes or protrusions–possessing radii of curvature comparable to that expected for the tip. Various kinds of particles are most commonly used (e.g., [66–68]), but holey polymer films have also proved useful [53]; recently it has been found in the laboratory at Griffith University that holey carbon films, often used in TEM applications, will serve the same purpose (Fig. 23). Methods for determining aspect ratio and tip radius are summarized in Table 7.

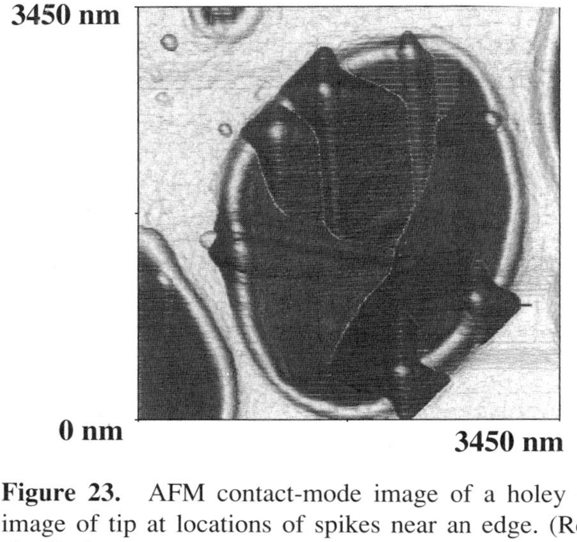

Figure 23. AFM contact-mode image of a holey carbon film. Note reverse image of tip at locations of spikes near an edge. (Reprinted from Honors thesis, G. S. Watson, Griffith University, 1996.)

Table 7. Methods and Procedures for Determining Tip Parameters

Methods/sample	Information content	Ref.	Comments
Cylindrical biological nanostructure, e.g., TMV, DNA	Tip radius	55, 59, 69	Well-defined standard; limited vertical information; expertise required for sample extraction and isolation. Sharper tips may modify or damage the sample. The tip may become contaminated from contact with the sample.
Nanospheres; e.g., colloidal gold and polystyrene beads	Tip radius Some aspect ratio	52, 70, 71	Noncompressible; well-defined standard; limited vertical information. Convolution with large spheres yields less information regarding tip shape at the apex
Holes or pillars Polymer films, calibration arrays, semiconductor grids	Tip radius Aspect ratio Some tip height	53, 72, 73	Provides values for aspect ratio and tip height. Some information lost for some tips due to geometrical constraints (e.g., pyramidal shape).
Delta function spikes, reverse imaging; e.g., copper spikes	Tip radius Aspect ratio Some tip height	56, 57, 74	Some samples require complex fabrication procedures. Information regarding tip shape is localized around a single point, thus more information content.
Other structures; e.g., atomic steps, fabricated arrays	Some tip radius Some aspect ratio Some tip height	75–77	Information is limited to the height of the sample. Limitations imposed by sample shape. Some cases offer limited information.

4.11. Determination of Tip Height and Tilt

Tip height is a relevant parameter when it is exceeded by the range in z-height within a field of view; the lever will then come into contact with the specimen when the tip attempts to maintain contact at the topographic minima. Thus the point of contact will be with the lever rather than with the tip. The resultant anomalies (e.g., sudden lateral shifts, loss of resolution and much higher applied force) are usually readily apparent. One method is to scan across a hole having a diameter greater than the largest cross-sectional dimension of the tip at its base, when the depth of the image will be limited to that of the tip height. The same effect can be achieved by reverse-imaging with a spike of height greater than that of the tip. In either case the relevant height can be measured from the line profile.

It is common practice to mount the lever at an angle to the horizontal to facilitate access of the tip to the chosen field of view (the consequence is that the tip will be tilted with respect to the perpendicular). High ridges or significant tilt of the sample surface will therefore be less likely to interfere with the lever during scanning. The disadvantage is that features affected by tip shape will acquire asymmetry due to the tilt of the tip. The effect is readily apparent in the reverse image of a pyramidal tip in Fig. 24; the line profile demonstrates that the angle of tilt is measurable (equal to 6° in the case illustrated).

5. PROBLEM SOLVING WITH SPM

Subsequent chapters deal specifically with areas of applications and methods for problem solving with SPM in the physical sciences and technologies. Accordingly only some generalities are discussed here.

It should be apparent from the foregoing discussion that the very real strengths of STM/STS can be deployed to best effect in an UHV environment on conducting surfaces when these are atomically flat. Moreover, it is fair to say that STM/STS requires considerable experience, skill and dedication from the experimenter. At the present time neither of these techniques is "routine" and cannot be considered for rapid response general problem solving. Most specimens will arrive in conditions and configurations which are not compatible with STM/STS requirements. However, as a basic research tool and as an instrument for gaining strategic insights into a number of technologically important systems the two techniques have much to offer.

On the other hand, SFM and contact-mode AFM/F-d spectroscopy, in particular, have already become important problem-solving tools. The richness and flavor of this topic will be apparent from a partial list of present areas of applications:

Cell biology and other biosystems [78]
Mineral and ceramic dissolution [23,79]
Semiconductor device characterization [80]
Surface characterization of polymers [81]
Magnetic storage devices [82]

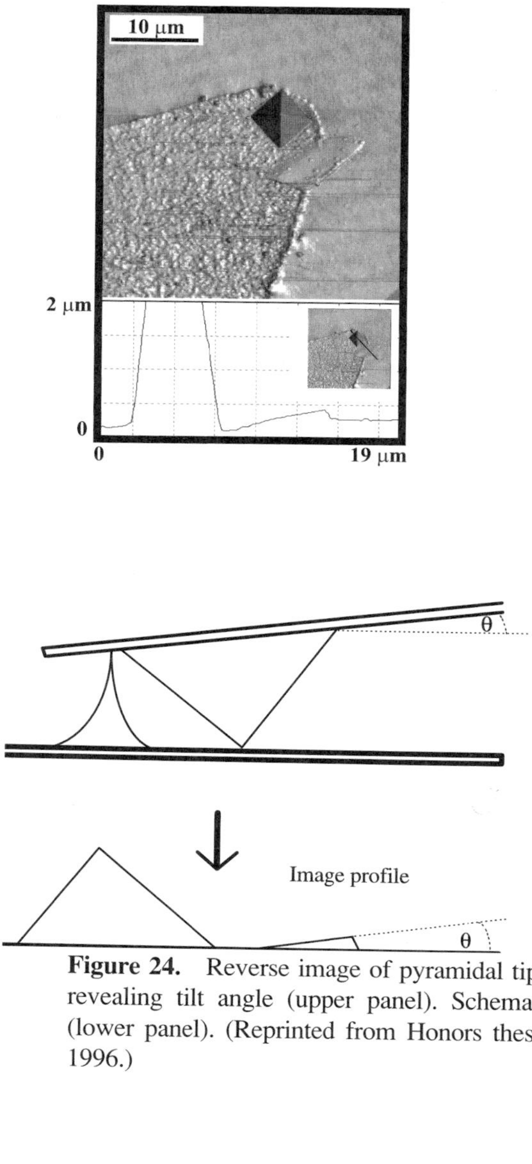

Figure 24. Reverse image of pyramidal tip at the end of a lever and line profile revealing tilt angle (upper panel). Schematic showing the origin of the effect (lower panel). (Reprinted from Honors thesis, G. S. Watson, Griffith University, 1996.)

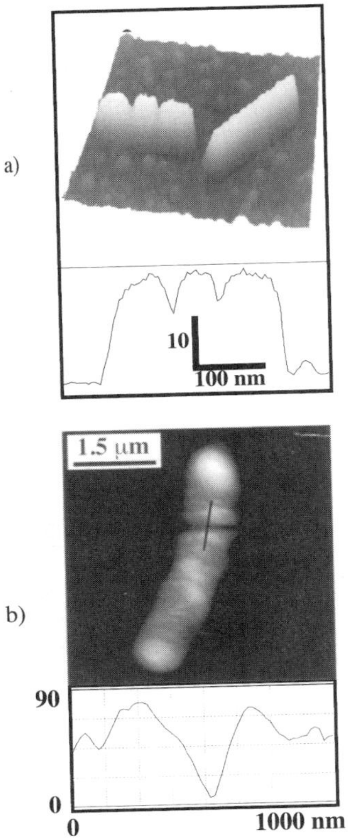

Figure 25. Controlled dissection on the nanoscale of TMV (a) and a bacterium (*Thermus aquaticus*) (b) is demonstrated by pseudo-3D images and line profiles.

5.1. Manipulation on the Nanoscale with SPM

An introduction to SPM would be deficient without brief mention of yet another of the unique capabilities of SPM, namely the ability to manipulate systems on the nanoscale. As a tool for nanomachining, as well as nanocharacterization, SPM will play a central role in the emerging nanotechnologies. It is premature, however, to lay down protocols for problem solving in this area, but it is mandatory to be aware of the rapid strides that are currently being made [83,84]. Some recent examples include manipulation of molecules on zeolites [85], deposition of nm-sized gold particles [86], manipulation of 30-nm GaAs particles [87] and nanoscale writing on chalcogenide films [88]. In the laboratory at Griffith University controlled nanodissection has been demonstrated for TMV virus and bacteria (*Thermus aquaticus*) (Fig. 25a, b).

ACKNOWLEDGMENTS

Much of the insight and knowledge on which this chapter is based has come from work in several laboratories, and from interactions with many industrious and knowledgable colleagues. I owe a great deal to co-workers in the SPM laboratory at Griffith University, namely S. A. Holt, G. S. Watson and C. T. Gibson. As well, many of the activities in this unit would not have been possible without funding from the Australian Research Council and from Griffith University. Also, I must acknowledge fruitful collaboration with the SPM group at AEA Technology (Harwell), J. A. A. Crossley and C. J. Sofield. Early in the history of SPM I was fortunate to visit the IBM laboratory in Zürich; this was the defining event for my interest in SPM. Finally I owe a great deal to Helen Bergen, who graciously admitted into our lives the word processor, the harshest and most demanding of mistresses.

REFERENCES

1. G. Binnig and H. Röhrer, *Helv. Phys. Acta 55*, 726 (1982).
2. R. Young, J. Ward and F. Scire, *Rev. Sci. Instrum. 43*, 999 (1972).
3. See the two special issues, July and September, of *IBM J. Res. Develop. 30* (1986).
4. G. Binnig and H. Röhrer, *Sci. Am. 253*, 50 (1985).
5. G. Binnig and H. Röhrer, *Surf. Sci. 126*, 236 (1983).
6. P. Avouris, *J. Phys. Chem. 94*, 2246 (1990).
7. G. Binnig, C. F. Quate and C. Gerber, *Phys. Rev. Lett. 56*, 930 (1986).
8. H. K. Wickramasinghe, *J. Vac. Sci. Technol. A 8*, 363 (1990).
9. J. Tersoff and D. R. Hamann, *Phys. Rev. B 31*, 805 (1985).
10. L. E. C. van Leemput and H. van Kempen, *Rep. Prog. Phys. 55*, 1165 (1992).
11. J. Shen, R. G. Pritchard and R. E. Thurstans, *Contemp. Phys. 32*, 11 (1991).
12. W. J. Kaiser and R. C. Jaklevic, *Surf. Sci. 181*, 55 (1987).
13. W. J. Kaiser and R. C. Jaklevic, *IBM J. Res. Develop. 30*, 411 (1986).
14. T. R. I. Cataldi, I. G. Blackham, G. A. D. Briggs, J. B. Pethica and H. A. O. Hill, *J. Electroanal. Chem. 290*, 1 (1990).
15. D. Sarid, *Scanning Force Microscopy*, Oxford University Press, New York, 1991.
16. N. A. Burnham, R. J. Colton and H. M. Pollock, *Nanotechnology 4*, 64 (1993).
17. J. N. Israelachvili, *Intermolecular and Surface Forces*, 2nd ed., Academic Press, San Diego, 1992.
18. A. L. Weisenhorn, M. Khorsandi, S. Kasas, V. Gotzos and H.-J. Butt, *Nanotechnology 4*, 106 (1993).

19. M. Heuberger. G. Dietler and L. Schlapbach, *Nanotechnology 5*, 12 (1994).
20. C. D. Frisbie, L. F. Rozsnyai, A. Noy, M. S. Wrighton and C. M. Lieber, *Science 265*, 2071 (1994).
21. H.-J. Butt, P. Siedle, K. Seifert, K. Fendler, T. Seeger, E. Bamberg, A. L. Weisenhorn, K. Goldie and A. Engel, *J. Microsc. 169*, 75 (1993).
22. D. W. Pohl, *IBM J. Res. Develop. 30*, 417 (1986).
23. G. Watson, S. A. Holt, R.-P. Zhao, A. Katsaros, N. Savvides and S. Myhra, *Physica C 243*, 123 (1995).
24. M. E. Taylor, B. Golen, A. W. McKinnon, G. C. Rosolen, S. M. Gray and M. E. Welland, *Appl. Surf. Sci. 67*, 228 (1993).
25. A. Crossley, S. Myhra and C. J. Sofield, *Surf. Sci. 318*, 39 (1994).
26. B. Hacker, A. Hillebrand, T. Hartmann and E. Guckenberger, *Ultramicroscopy 42–44*, 1514 (1992).
27. P. K. Hansma and J. Tersoff, *J. Appl. Phys. 61*, R1 (1987).
28. D. Sarid and V. Elings, *J. Vac. Sci. Technol. B 9*, 431 (1991).
29. G. Y. Chen, R. J. Warmack, T. Thundat, D. P. Allison and A. Huang, *Rev. Sci. Instrum. 65*, 2532 (1994).
30. J. M. Neumeister and W. A. Ducker, *Rev. Sci. Instrum. 65*, 2527 (1994).
31. M. Nonnenmacher, J. Greschner, O. Wolter and R. Kassing, *J. Vac. Sci. Technol. B 9*, 1358 (1991).
32. T. R. Albrecht, S. Akamine, T. E. Carver and C. F. Quate, *J. Vac. Sci. Technol. A 8*, 3386 (1990).
33. F. Zenhausern, M. Adrian, B. ten Heggeler-Bordier, F. Ardizzoni and P. Descouts, *J. Appl. Phys. 73*, 7232 (1993).
34. A. Wadas and H.-J. Güntherodt, *Phys. Lett. 146*, 277 (1990).
35. A. Wadas and H.-J. Hug, *J. Appl. Phys. 72*, 203 (1992).
36. A. L. Weisenhorn, P. Maivald, H.-J. Butt and P. K. Hansma, *Phys. Rev B 45*, 11226 (1992).
37. U. Dammer, O. Popescu, P. Wagner, D. Anselmetti, H.-J. Güntherodt and G. N. Misevic, *Science 267*, 1173 (1995).
38. R. M. Overney, H. Takano and M. Fujihira, *Europhys. Lett. 26*, 443 (1994).
39. M. Radmacher, J. P. Cleveland, M. Fritz, H. G. Hansma and P. K. Hansma, *Biophysical J. 66*, 2159 (1994).
40. T. J. Senden and W. A. Ducker, *Langmuir 10*, 1003 (1994); H.-J. Butt, P. Siedle, K. Seifert, K. Fendler, T. Seeger, E. Bamberg, A. L. Weisenhorn, K. Goldie and A. Engel, *J. Microsc. 169*, 75 (1992).
41. J. P. Cleveland, S. Manne, D. Bocek and P. K. Hansma, *Rev. Sci. Instrum. 64*, 403 (1993).
42. J. E. Sader, I. Larson, P. Mulvaney and L. R. White, *Rev. Sci. Instrum. 66*, 3789 (1995).
43. J. L. Hutter and J. Bechhoefer, *Rev. Sci. Instrum. 64*, 1868 (1993).

44. Y. Q. Li, N. J. Tao, J. Pan, A. A. Garcia and S. M. Lindsay, *Langmuir 9*, 637 (1993).
45. J. M. Neumeister and W.A. Ducker, *Rev. Sci. Instrum. 65*, 2527 (1994).
46. J. E. Sader, *Rev. Sci. Instrum. 66*, 4583 (1995).
47. C. T. Gibson, G. S. Watson and S. Myhra, *Nanotechnology, 7,* 259 (1996).
48. L. Hellemans, K. Waeyaert and F. Hennau, *J. Vac. Sci. Technol. B 9*, 1309 (1991).
49. Y. Liu, T. Wu and D. F. Evans, *Langmuir 10*, 2241 (1994).
50. C. T. Gibson, G. S. Watson and S. Myhra, *Scanning,* in press (1997).
51. U. D. Schwarz, H. Haefke, P. Reimann and H.-J. Güntherodt, *J. Microsc. 171*, 183 (1994).
52. C. Odin, J. P. Aimé, Z. El Kaakour and T. Bouhacina, *Surf. Sci. 317*, 321 (1994).
53. T. O. Glasbey, G. N. Batts, M. C. Davies, D. E. Jackson, C. V. Nicholas, M. D. Purbrick, C. J. Roberts, S. J. B. Tendler and P.M. Williams, *Surf. Sci. 318*, L1219 (1994).
54. M. van Cleef, S. A. Holt, G. S. Watson and S. Myhra, *J. Microsc. 181*, 2 (1996).
55. G. R. Bushell, G. S. Watson, S. A. Holt and S. Myhra, *J. Microsc. 180*, 174 (1995).
56. F. Atamny and A. Baiker, *Surf. Sci. 323*, L314 (1995).
57. L. Montelius, J. O. Tegenfeldt and P. van Heeren, *J. Vac. Sci. Technol. B 12*, 2222 (1994).
58. E. Delain, A. Fourcade, J.-C. Poulin, A. Barbin, C. Couloud, E. Le Cam and E. Paris, *Microsc. Microanal. Microstruct. 3*, 457 (1992).
59. T. Thundat, X.-Y. Zheng, S. L. Sharp, D. P. Allison, R. J. Warmack, D. C. Joy and T. L. Ferrell, *Scanning Microsc. 6*, 903 (1992).
60. W. Wiegräbe, M. Nonnenmacher, R. Guckenberger and O. Wolter, *J. Microsc. 163*, 79 (1991).
61. F. Zenhausern, M. Adrian, R. Emch, M. Taborelli, M. Jobin and P. Descouts, *Ultramicroscopy 42–44*, 1168 (1992).
62. J. G. Manfovani, D. P. Allison, R. J. Warmack, T. L. Ferrell, J. R. Ford, R. E. Manos, J. R. Thompson, B. B. Reddick and K. B. Jacobson, *J. Microsc. 158*, 109 (1990).
63. K. Imai, K. Yoshimura, M. Tomitori, O. Nishikawa, R. Kokawa, M. Yamamoto, M. Kobayashi and A. Ikai, *Jpn. J. Appl. Phys. 32*, 2962 (1993).
64. R. Guckenberger, F. Teran Arce, A. Hillebrand and T. Hartmann, *J. Vac. Sci. Technol. B 12*, 1508 (1994).
65. T. Miwa, M. Yamaki, H. Yoshimura, S. Ebina and K. Nagayama, *Langmuir 11*, 1711 (1995).
66. S. Xu and M. F. Arnsdorf, *J. Microsc. 173*, 199 (1994).

67. J. Barbet, A. Garvin, J. Thimonier, J.-P. Chauvin and J. Rocca-Serra, *Ultramicroscopy 50*, 355 (1993).
68. R. Wurster and B. Ocker, *Scanning 15*, 130 (1993).
69. C. Bustamente, J. Vesenka, C. L. Tang, W. Rees, M. Guthold and R. Keller, *Biochem. 31*, 22 (1992).
70. P. Markiewicz and M. C. Goh, *Langmuir 10*, 5 (1994).
71. J. Vesenka, S. Manne, R. Giberson, T. March and E. Henderson, *Biophys. J. 65*, 992 (1993).
72. J. E. Griffith, D. A. Grigg, M. J. Vasile, P. E. Russell and E. A. Fitzgerald, *J. Vac. Sci. Technol. B 9*, 3586 (1991).
73. P. Markiewicz and M. C. Goh, *Rev. Sci. Instrum. 66*, 3186 (1995).
74. F. Jensen, *Rev. Sci. Instrum. 64*, 2595 (1993).
75. S. S. Sheiko, M. Möller, E. M. C. M. Reuvekamp and H. W. Zandbergen, *Ultramicroscopy 53*, 371 (1994).
76. Y. Umehara, Y. Ogiso, K. Chihara, K. Mukasa and P. E. Russell, *Rev. Sci. Instrum. 66*, 269 (1995).
77. C. A. Goss, J. C. Brumfield, E. A. Eugene and R. W. Murray, *Langmuir 9*, 2986 (1993).
78. J. H. Hoh and P. K. Hansma, *Trends Cell Biol. 2*, 208 (1992).
79. A. J. Gratz, S. Manne and P. K. Hansma, *Science 251*, 1343 (1991).
80. I. Smith and R. Howland, *Solid State Technol. Dec.*, 53 (1990).
81. S. N. Magonov, *Appl. Spectrosc. Rev. 28*, 1 (1993).
82. G. Persch, Ch. Born, H. Engelmann, K. Koehler and B. Utesch, *Scanning 15*, 283 (1993).
83. R. Wiesendanger, *Appl. Surf. Sci. 54*, 271 (1992).
84. P. Zeppenfeld, C. P. Lutz and D. M. Eigler, *Ultramicroscopy 42–44*, 128 (1992).
85. A. L. Weisenhorn, J. E. MacDougall, S. A. C. Gould, S. D. Cox, W. S. Wise, J. Massie, P. Maivald, V. B. Elings, G. D. Stucki and P. K. Hansma, *Science 247*, 1330 (1990).
86. H. J. Mamin, S. Chiang, H. Birk, P. H. Gunther and R. Rugar, *J. Vac. Sci. Technol. B 9*, 1398 (1991).
87. T. Junno, K. Deppert, L. Montelius and L. Samuelson, *Appl. Phys. Lett. 66*, 3627 (1995).
88. H. Kado and T. Tohda, *Appl. Phys. Lett. 66*, 2961 (1995).

11
Metallurgy

R. K. WILD The University of Bristol, Bristol, United Kingdom

1. INTRODUCTION

The failure of a relatively small component in a plant can have catastrophic results. In 1969 a blade failed in one of the turbines at Hinkley Point A Power Station during a routine overspeed test [1]. Figure 1 shows the resultant devastation with components having passed through all the outer casings and through the roof of the turbine hall, some parts ending up several hundred meters distant. Fortunately there were no injuries, but the cost of repair and subsequent prevention ran into millions of dollars. It is therefore important to be able to predict how a metal or alloy will behave in service. Despite all the knowledge concerning embrittlement, examples continue to occur where components have failed as the result of segregation of trace elements in alloys. There is also the important problem of intergranular stress corrosion cracking where components under load in an aggressive environment can fail because small cracks are opened up by the combined effects of stress and oxidation. Knowledge of the grain boundary chemistry can assist in reducing or eliminating both these and other problems. There are many ways in which a metal or alloy can be characterized and the structure and com-

Figure 1. Damage to Hinkley turbine caused by temper embrittlement failure of a rotor blade.

position of the grain boundaries determined. Some techniques analyze the internal surface that has first been exposed, others analyze the internal surface in thin sections of the material, while others section slowly through the material until a boundary is encountered. This chapter outlines some techniques and methods for determining the grain boundary chemistry and, hence, properties of metals and alloys.

1.1. Strength of Materials

The strength of a metal or alloy will be determined by the strength of the weakest bonds between atoms. In some cases, such as single-crystal whiskers with few defects, the theoretical strength can be approached, but in most practical alloys the ultimate strength is several orders of magnitude less than this and is usually determined by the strength of the interface between grain boundaries. The grain boundary strength can be altered dramatically by the segregation of trace elements in the bulk to the boundary, by the enrichment or depletion of bulk elements in the vicinity of the grain boundary, or by the formation of particles, such as carbides, at the grain boundary positions. Thermal treatments and aging in service will modify the grain boundary composition; hence the mechanical properties will vary with time, and therefore the grain boundary composition requires monitoring at various stages throughout the lifetime of the component. In a nuclear power plant the radiation flux will modify the diffusion rates of the various alloying components, further influencing the mechanical properties as a function of time in service.

1.2. Failure Mechanisms

Most metals and alloys are made up of small crystals or grains oriented in a random manner to one another. These grains can vary in size from less than a micron to several hundred microns. The grain size is frequently determined by the heat treatment and mechanical work that the alloy has received. Aging at high temperature allows some grains to grow at the expense of others, resulting in large average grain sizes, while mechanical work breaks up and reorients grains, producing much smaller grains. Within the grains may be found small particles, such as carbides or sulfides, which can have the effect of increasing the hardness and reducing the ductility of the alloy by pinning dislocations and preventing plastic flow. However, the interface between the grains has the greatest effect on the mechanical properties. Grain boundaries may have particles formed on them or elements segregated to the interface. Frequently more than one element will be found segregated to the grain boundary surface. Some segregating elements may increase the strength of the boundary, usually small atoms such as boron, while others, such as phosphorus, tin, antimony or arsenic, reduce the strength. How a metal or alloy will fail is determined by many factors, but if the grain boundaries have high cohesive strength the failure will be ductile (when the metal tears apart leaving a surface looking like the surface of toffee that has been broken on a warm day), or by cleavage (when failure is across sheets of atoms, leaving relatively flat surfaces similar to toffee that has been broken after removal from the refrigerator). If the grain boundary is the weak point, the material will fail by a fracture path following the boundaries between individual grains. Grain boundaries will be randomly oriented, and some will be oriented favorably for failure while others will not. A grain boundary oriented parallel to the direction of fracture has a high probability of failure, whereas a grain boundary oriented normally to the direction of fracture is unlikely to fail. In practice a fracture surface will often contain examples of all types of failure with varying fractions of intergranular, cleavage and ductile fracture.

1.3. Segregation

1.3.1. Thermal

Elements present in metals and alloys in trace quantities can, under certain heat treatments, segregate to the grain boundaries, and there are many examples in the literature. A recent review by Lejcek and Hofmann [2] details many of the segregating species and systems and outlines the theories describing grain boundary segregation. Many materials will fracture in an intergranular manner by impact at low temperature, and for these materials Auger electron spectroscopy (AES) is an ideal analytical tool. The systems studied most extensively by this means are iron and ferritic steels where segregants such as P, S, Sn, Sb and As cause severe embrittlement [3–7]. However, other systems have also been studied. Powell and Woodruff [8] have used AES to study the segregation of bismuth in copper and have shown that bismuth segregated as a single-atom layer at the grain boundary. Chuang et al. [9] have studied antimony in copper in a similar manner. Smiti et al. [10] have shown that carbon could increase grain boundary cohesion in tungsten while oxygen decreased it. Other systems can be induced to fracture in an intergranular manner by first charging the metal or alloy with hydrogen. In some cases the hydrogen-charged alloy will fracture intergranularly by impact [11], but in other cases it is necessary to use a slow tensile pull [12–14]. Austenitic stainless steels [15] and nickel-base superalloys [16] have been studied extensively by this technique. In addition, Ni_3Al containing boron [17], Co–50%Ni containing antimony [18], and nickel-containing indium [19] will all fracture in an intergranular manner following hydrogen charging.

Not all metals and alloys can be induced to fracture along grain boundaries. Some of the pure stainless steels will not fail along such boundaries even after hydrogen charging followed by slow tensile fracture. In these cases other techniques must be utilized. Transmission electron microscopy (TEM) will reveal grain boundaries in suitably thinned specimens. Recently, field emission gun scanning (FEGS) TEM, combined with energy dispersive x-ray (EDX) [20,21] analysis or parallel electron energy loss spectroscopy (PEELS) [22] has proved to be very successful in determining the grain boundary composition in these difficult alloys. Indeed the technique is now used to complement the data from AES in specimens that can be induced to fracture in an intergranular manner. It does, however, suffer from the drawback that it is laborious to produce thin specimens and the thinned region may contain only a few grain boundaries.

Other techniques can be used to detect and determine levels of segregants at grain boundaries in bulk specimens, such as autoradiography [40], radioactive tracer for tellurium in silver [23], field ion microscopy (FIM) for detection of niobium in cobalt [24], and FIM in combination with a position-sensitive detector [25].

Grain boundary segregation in metallurgy has been comprehensively described by Seah [26–29]. McLean [30] analyzed the kinetic behavior of segregation under equilibrium conditions from diffusion theory. He visualized a grain boundary as a bubble raft and considered a lattice with N undistorted sites with P solute atoms among them and n distorted sites with p solute atoms. If the energy of the solute atom in the lattice is E and on segregation is e, the free energy of the solute atoms is.

$$G = pe + PE - kT[\ln n!N! - \ln(n-p)!p!(N-P)!P!] \quad (1)$$

By considering the minimum in G, Eq. (1) may be developed to the familiar McLean equation:

$$\frac{X_b}{1-X_b} = \left[\frac{X_c}{1-X_c}\right]\exp\left[\frac{E_1}{RT}\right] \quad (2)$$

where X_b is the adsorption level as a molar fraction of a monolayer, X_c is the solute molar fraction and E_l is the molar heat of adsorption of the segregant at the grain boundary.

Gibbs [31] related adsorption to the grain boundary energy (γ_b) by the relation

$$\frac{d\gamma_b}{dX_C} = -\left(\frac{RT}{X_c}\right)\Gamma_b \quad (3)$$

where Γ_b is the excess solute at the grain boundary in mole m^{-2}.

Carr et al. [32] have produced a time-temperature diagram (Fig. 2) for the embrittlement of the steel SAE 3140. The diagram indicates that heat-treating to temperatures between 500 and 550°C for times between 1 and 100 h produces

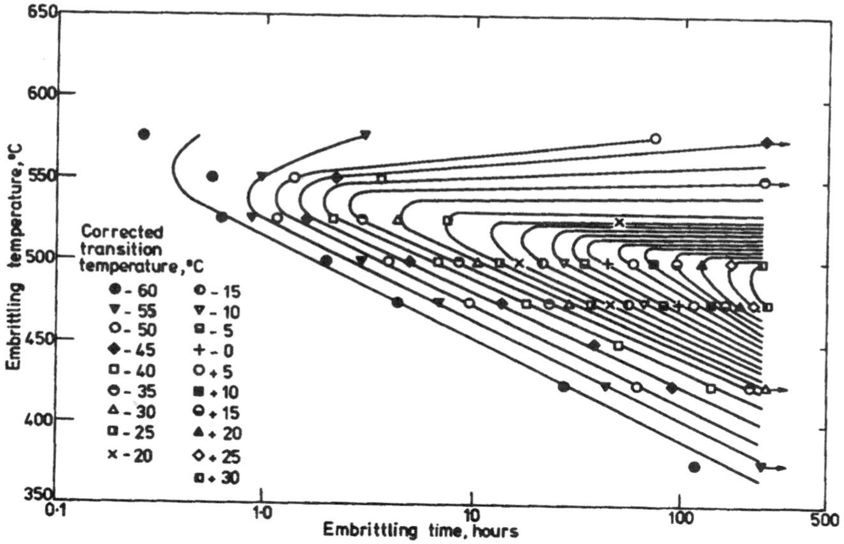

Figure 2. Time-temperature diagram for the embrittlement of steel SAE 3140 [32].

the greatest embrittlement. Seah [33] has used the McLean equation, together with the segregation levels of phosphorus and the equation for the volume diffusivity of phosphorus in iron [34],

$$D = 158 \ \exp\left(\frac{-52{,}300}{RT}\right) \tag{4}$$

to determine the time-temperature diagram for the segregation of phosphorus for the same SAE 3140 steel used by Carr et al. [32], shown in Fig. 3. This plot shows that the grain boundary segregation of phosphorus follows the same time-temperature embrittlement that Carr et al. determined. It can be demonstrated that the addition of other elements such as molybdenum and chromium reduces the diffusivity of phosphorus in iron while manganese increases it, so that Eq. (4) cannot be applied to all steels without using a modified D.

When an element segregates to a grain boundary, it occupies a vacancy site between two adjacent grains. If this element causes the material to be weakened or embrittled there will be considerable enrichment of it compared with its level within the bulk. Seah and Hondros [26] have correlated grain boundary enrichment as a function of atomic solubility for a number of alloy systems (Fig. 4). It can be seen that enrichment ratios up to 10^4 are possible. The extent of the segregation away from the grain boundary has been determined experimentally by Palmberg and Marcus [35], who exposed grain boundaries in materials that had high levels of grain boundary segregation. The segregant was then removed in a controlled manner by bombarding the grain boundary with inert gas ions while the level of segregant was determined. The concentration as a function of time for S, P and Sn on iron and steel grain boundaries is plotted in Fig. 5. The solid line in the figure is the theoretical prediction for a layer of segregant 0.45 nm thick. The good agreement between this line and the experimental points indicates that segregation is confined to a single-atom layer at the grain boundary.

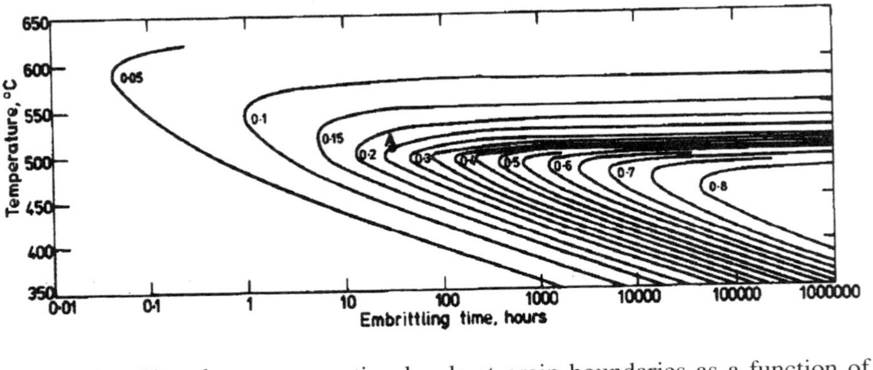

Figure 3. Phosphorus segregation levels at grain boundaries as a function of time and temperature for steel SAE 3140 [33].

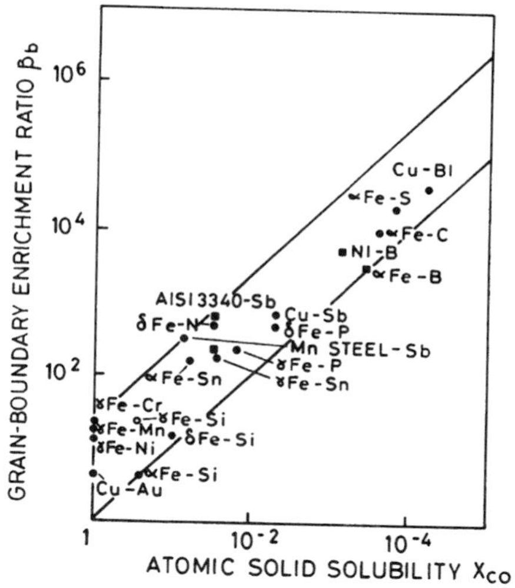

Figure 4. Grain boundary enrichment as a function of atomic solubility for various alloys [26].

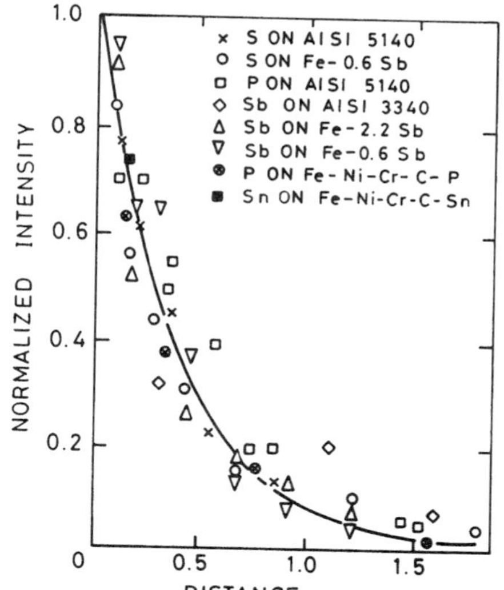

Figure 5. Depth profiles for segregated layers at grain boundaries showing single-atom-layer distribution [35].

Atomic size must be included when considerations are given as to whether an element will embrittle a metal or alloy. In general, large atoms decrease the grain boundary strength while small atoms increase it [29]. Thus, in the case of iron and steel, atoms such as S, P, Sn, Sb and As, which are all larger than Fe, tend to make a steel more brittle (that is, lower its ductile-to-brittle transition temperature), while small atoms such as B, Be and C, which are smaller than Fe, tend to make the steel less brittle.

1.3.2. Irradiation assisted

Irradiation can dramatically alter the level of segregants at grain boundaries. In a nickel-base alloy such as PE16, in the absence of grain boundary carbide formation, the boundaries can be depleted in chromium and iron and enriched in nickel. A mechanism for the observed behavior of the major alloying elements is the inverse Kirkendall effect. Here, fast neutron irradiation of the material produces numerous point defects (interstitials and vacancies). These move through the lattice in a random manner but on encountering a sink, such as a grain boundary, will be annihilated, which produces a net flux of point defects toward grain boundaries. A flow of vacancies toward grain boundaries implies a flow of atoms in the opposite direction; however, the rate of flow depends on the jump frequency, J, of the particular atoms [36]. It has been determined that J (chromium) > J (iron) > J (nickel) [37], so that chromium and iron will flow away from the boundaries at a greater rate than nickel, and the boundaries thus become enriched with nickel by default.

In any attempt to characterize the internal interfaces in a metal or alloy, several steps must be considered. In the following sections some of these steps will be discussed, starting with the simple and nondestructive and finishing with the destructive and complicated.

2. ANALYTICAL METHODS FOR DETERMINING GRAIN BOUNDARY SEGREGATION

2.1. Introduction

In this section a suggested route will be described to determine the type and amount of grain boundary segregation. A flowchart (Fig. 6) will be used to illustrate the sequence of steps that should be taken with any metallurgical investigation that involves surfaces. This essentially starts with techniques that do not require the metal or alloy to be damaged by fracture or thinning but can probe a metallographically polished specimen. Following that surface analytical techniques are considered. The latter require materials that fracture in an intergranular manner in order to be able to expose the grain boundary surface. Finally transmission electron microscopic methods of determining grain boundary

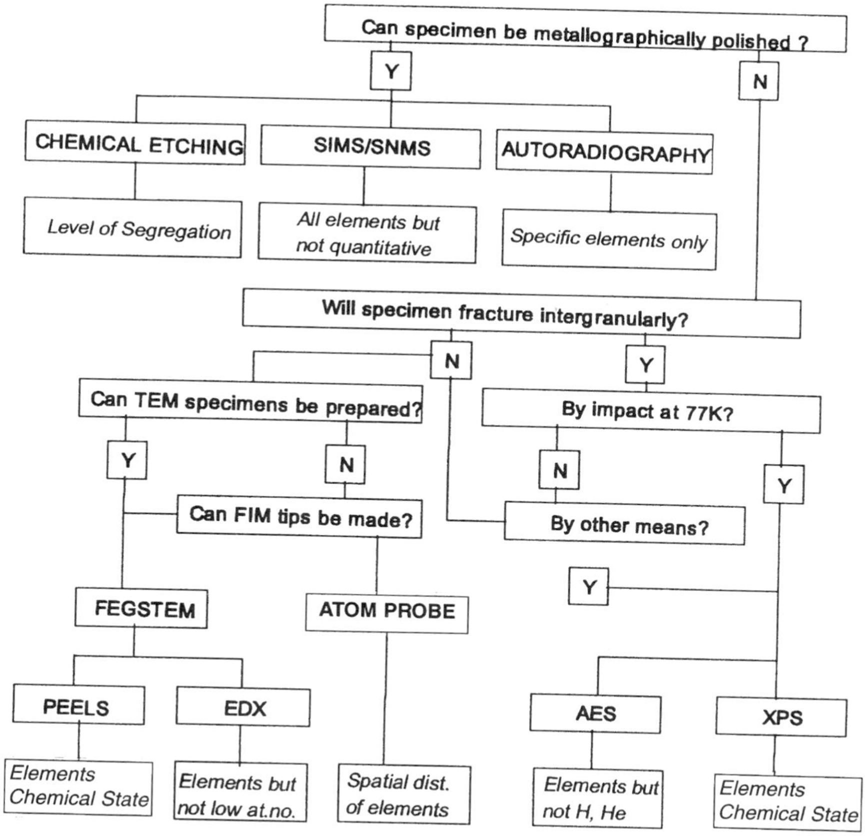

Figure 6. Flowchart for the analysis of metals and alloys.

structure, and more sophisticated techniques such as time-of-flight field ion microscopy, are described.

2.2. Metallographically Polished Specimens

In this section it is assumed that sufficient material is available for a specimen to be mounted in a conducting medium and polished using various grit sizes ending finally with a μm or sub-μm diamond paste.

2.2.1. Chemical etching

As a first approach to the understanding of the properties of a metal or alloy it is useful to prepare a metallographic specimen polished to a mirror finish. This may then be etched to allow both the identification of grain size and to give an indication of segregation levels. The method for detecting segregation in steels has been described by Dryer et al. [38]. The etchant is a saturated picric acid solution containing 1 g of sodium didecylbenzene sulfurate/100 mL. Etching times of approximately 1 h are normally sufficient to expose grain boundaries without significant pitting. Once the ideal etching time for the specimens has been determined, those with different segregation levels at grain boundaries should be etched for equal times. The depth of the grain boundary grooving, which can be measured optically, will then be proportional to the level of segregant. Figure 7 shows a 3.5NiCrMoV steel which has been etched in this man-

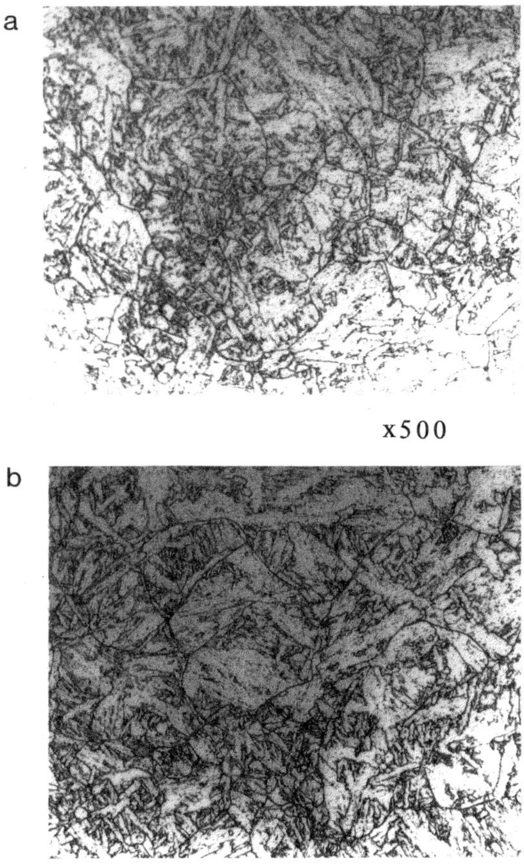

Figure 7. Chemical etching of two specimens of 3.5NiCrMoV steel in (a) the as-received condition and (b) following step cooling, showing increased depth of grooving in the step-cooled specimen.

ner for 1 h in the as-received condition and also after step cooling to increase grain boundary segregation [39]. The increased grain boundary grooving can be seen clearly in the step-cooled specimen.

2.2.2. SIMS

A metallographically polished specimen may be ion-etched and the ejected secondary ions detected by using a mass analyzer. Spatial resolution can be achieved either by rastering a fine focus ion beam over the surface while detecting the secondary ions or by illuminating the area of interest with a broad ion beam and then detecting the secondary ions by ion optical focusing to equivalent points in the image plane. An example of the latter method is given in Fig. 8 in which a carbon map is shown from a region 120 μm in diameter on the surface of a sample of metallographically polished specimen of steel. The instrument used there was a CAMECA 3f which employs an oxygen bleed to increase the secondary-ion yield. This method is very useful for detecting and identifying segregating species. SIMS can detect very small concentrations of specific elements at the ppb level in certain circumstances. However, it has the drawback that the range of sensitivities for different elements is very wide

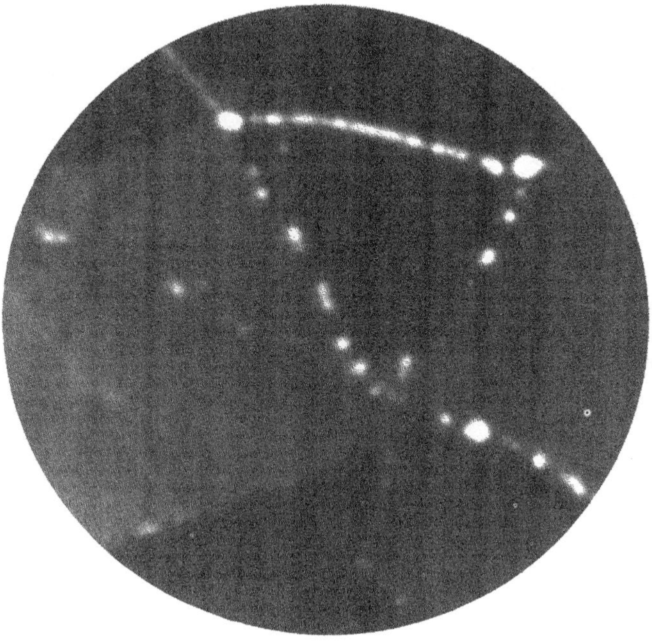

Figure 8. SIMS carbon map over a region of a polished steel specimen.

ranging over four orders of magnitude, since matrix effects are significant, with the result that the technique cannot easily be made quantitative.

2.2.3. Autoradiography

Autoradiography is a technique that can be used to identify certain elements in metals and alloys. It relies on the formation of a radioactive isotope of the element where identification is required following exposure to an intense radioactive source. The decay of the isotope is then used to form an image on a photographic plate placed in contact with the surface. A metallographic sample of the material is prepared, exposed to the radioactive source for a given time, and the photographic plate placed in contact with the polished surface. Autoradiography is particularly useful for detecting the presence of boron in steel, as demonstrated in a study on casts of stainless steel [40]. Figure 9 shows autoradiographs from two casts of stainless steel; cast A contains only 3 ppm of boron, distributed mostly within the bulk, while cast B has a total boron content of 90 ppm, and there is evidence that the boron has segregated to the grain boundaries. Cast A fractures readily in an intergranular manner while cast B is ductile and extremely difficult to fracture intergranularly. Boron is an element

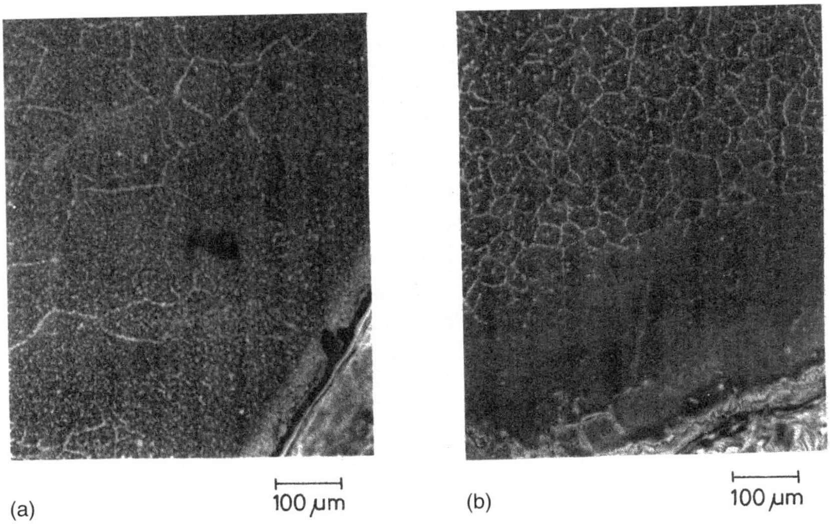

Figure 9. Autoradiographs for two casts of stainless steel containing (a) 3 ppm boron and (b) 90 ppm boron, showing enhanced boron segregation to grain boundaries in (b).

with an atomic diameter that is small compared with the matrix atoms and would be expected to increase the grain boundary energy on segregation.

2.3. Intergranular Fracture

Ferritic steels have a body-centered cubic (bcc) structure and in their most basic form contain carbon additions only, but chromium may be added in quantities up to 12 wt.% to improve corrosion characteristics, while other additions of up to a few wt.% may include Mo, V and Ni. The steels have a ductile-to-brittle transition temperature that is normally below room temperature but above 80 K. The ductile-to-brittle transition temperature tends to rise from this minimum in pure materials to room temperature and above as the grain boundary segregation levels increase. Frequently these steels contain carbides, usually chromium and molybdenum carbides, on the grain boundaries and failure may be through or around them. Copper will also fail in an intergranular manner if impurities such as bismuth [8] and antimony [9] have segregated to the grain boundaries. Molybdenum and tungsten normally contain impurities such as carbon, oxygen and sulfur at grain boundaries and will also fracture in an intergranular manner [10]. These and other similar metals and alloys are therefore suitable for analysis by AES following impact fracture at liquid nitrogen temperature.

2.3.1. Impact at low temperature

A specimen of the material to be fractured is prepared and cleaned by ultrasonical washing in isopropanol. The specimens may be machined to standard forms, which are usually rods of diameter 3–5 mm, length 25–30 mm, and with a notch 1 mm deep at the midpoint. Small specimens down to 1 mm × 1 mm × 10 mm can be fractured by mounting in prepared sleeves. The specimens are introduced into the UHV system through an air lock and positioned for fracture. Alternatively, several may be loaded directly into the fracture stage, which is then evacuated. An example of a low-temperature impact stage is shown schematically in Fig. 10 [56] It is important that the pressure in the system be below 10^{-8} Pa before fracture is attempted in order that the freshly exposed surfaces do not become contaminated. Seah [28] has demonstrated the importance of fracturing in UHV by depositing a monolayer of tin onto a clean surface of iron in UHV, which was then characterized using AES; the spectrum contained intense peaks from tin in the region of 430 eV. Following exposure to air for a short period, the tin peaks were barely detectable. It is good practice to load specimens into the fracture stage at the end of a day and leave the system to pump overnight before starting to refrigerate the next morning. Cooling is normally continued for approximately 1 hour before fracture is attempted. In most systems, fracture is achieved by transmitting a sharp impact to the specimen via

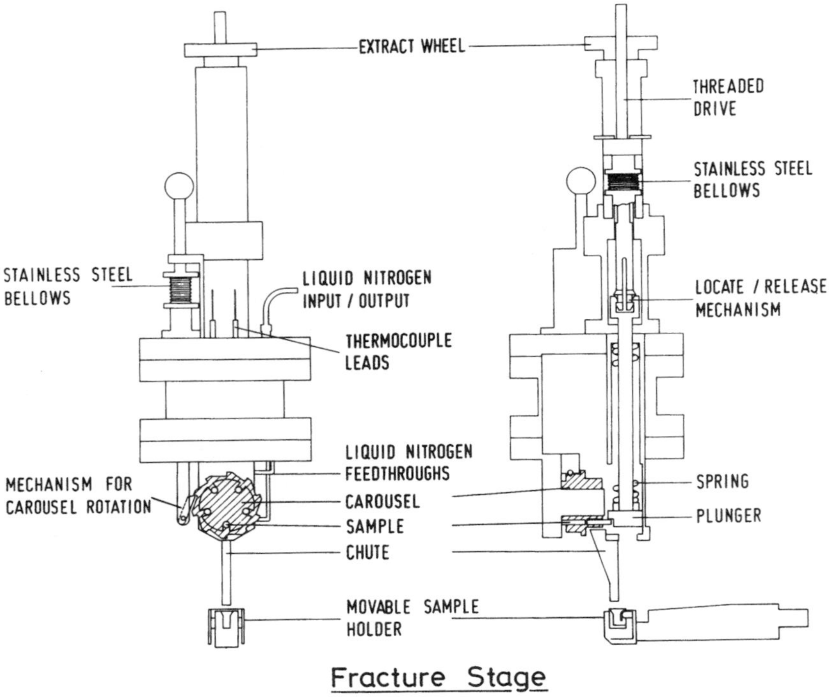

Figure 10. A low-temperature impact fracture stage [56].

a bellows arrangement. A typical fracture surface from an Fe/3% Ni steel which contained 530 ppm of phosphorus and 110 ppm of tin and had been heat-treated at 580°C for 48 h is shown in Fig. 11. The fracture path is almost 100% intergranular with very few ductile or cleavage faces. This is an extreme example; most materials fracture with many more cleavage and ductile facets. Cleavage can be identified from the appearance of a very flat surface containing many "river" lines running across it. Ductile failure has the appearance of warm toffee after it has been pulled apart, with cavities and peaks.

2.3.1.1. AES

Analysis of grain boundaries exposed in UHV is normally carried out using AES. The Auger electrons detected are in the energy range 0–2000 eV, and since electrons of such energies have mean free paths in most materials of only a few nm, the resulting spectrum arises from the top few atom layers of the surface. AES can detect all elements in the periodic table except hydrogen and helium. Chemical-state effects can be detected as shifts in peak position and/or

Figure 11. Impact fracture surface of an Fe/3%Ni steel with phosphorus and tin segregated to the grain boundaries.

changes in peak shape. Most elements give Auger electron yields that are similar for the most intense transitions, and as a result the technique can provide a quantitative analysis of the surface. However, care must be exercised when attempting to quantify the levels of a segregant at a grain boundary. The standard method for determining the surface composition assumes a surface homogeneous in depth, whereas grain boundary segregants almost always consist of a single-atom layer sitting on top of a matrix. A typical Auger spectrum recorded from the grain boundary surface of the material pictured in Fig. 11 is shown in Fig. 12. The spectrum is displayed as the differential of the number of electrons multiplied by the electron energy (that is, $N(E) \times E$) versus the energy of the secondary electrons, E, and shows peaks resulting from Auger transitions within the surface atoms. In this example peaks can be observed at approximately 600, 650 and 700 eV from Fe, at 800 and 850 eV from Ni, at 120 eV from P, and at 430 and 435 eV from Sn.

The extent of segregation at a grain boundary will be determined by the heat treatment that the material has received and by the relative orientations of the adjacent grains. Phosphorus segregation has been found to increase with increasing tilt angle of grain boundary misorientation [41] while others have found that high index planes have a high level, and low index planes a low level, of segregation

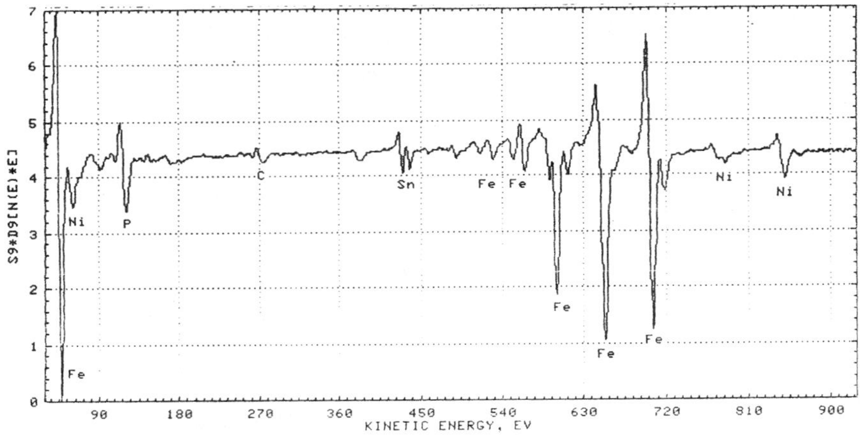

Figure 12. Auger spectrum from the grain boundary surface of the Fe/3%Ni steel with phosphorus and tin grain boundary segregation.

[42]. In practice when a material is fractured intergranularly there will be a distribution of levels of segregation from grain boundary to grain boundary. It is therefore essential that a sufficient number of grains be analyzed on each fracture surface in order to obtain statistics of significance. A recommended number of grains would be between 16 and 25. The time required for each analysis determines in practice the upper limit to the number of grains that can be analyzed, while in many cases it is often difficult to find 16 grains on a fracture face that contains only a small percentage of intergranular fracture.

Quantification of the composition of the grain boundary from the Auger spectrum is not always trivial. For a homogeneous material the composition is

$$X_i\% = \frac{\frac{X_i}{S_i} \times 100}{\sum_j X_j / S_j} \tag{5}$$

where X_i and X_j are the *i*th and *j*th element peak heights, and S_i and S_j are the sensitivity factors for the *i*th and *j*th elements.

In most cases of segregation to a grain boundary the segregating species is present as a single-atom layer on the matrix surface. It is also assumed that half the segregating species remains on each exposed surface following fracture. To calculate the amount of segregant (A) on the surface of matrix (B), it is necessary to know the escape depths of electrons in the segregating species (λ_A) and in the matrix (λ_B) as well as the energy of the incident electron beam, and to determine the amount of electron backscattering (Γ). This is treated in some detail by Seah [43], who has derived an equation which allows the amount of segregant to be calculated from

$$X_i = Q_{AB} \frac{I_A / I_A^\infty}{I_B / I_B^\infty} \tag{6}$$

where

$$Q_{AB} = \left[\frac{\lambda_A(E_A)\cos\Theta}{a_A} \right] \left[\frac{1+\Gamma_A(E_A)}{1+\Gamma_B(E_A)} \right] \tag{7}$$

where a_A is the size of atom A, and Θ is the angle of the incident electron beam relative to the surface normal. For specific segregants, such as phosphorus, on standard matrices, such as iron, using known incident beam energies, it is possible to derive Q_{AB} once and for all, and the value so found may then be applied to the standard equation to give the segregation in monolayers.

Elements may not always segregate to grain boundaries in a uniform manner. Certain elements may induce a grain boundary to form cavities or the element may segregate preferentially to cavity surfaces. Other elements diffusing to the grain boundary may combine to form particles on the boundary. It is therefore important that the boundary be analyzed at high spatial resolution. The advent of field emission guns in both scanning and transmission electron microscopes has increased dramatically the spatial resolution for analysis. The escape depth for Auger electrons is usually only a few atom layers, which ensures that the lateral spatial resolution is determined essentially by the size of the incident probe and not, as is the case with EDX, the size of the incident electron scattering volume. Thus, analysis with a lateral resolution of a few tens of nanometers is possible. The distribution of a segregating species across a fracture surface can be obtained by first recording a secondary-electron image of the surface, then setting the analyzer to detect a major peak from the element of interest together with two backgrounds, one on either side of the peak, while the electron beam is rastered over the field of view. The peak minus background divided by the background is then displayed as a digital map in which the brightness at any point is proportional to the elemental concentration. Dividing the peak height by the background reduces the effect of surface topography on electron yields and hence on elemental distribution. At the top of Fig. 13 is shown a secondary-electron image of the Fe/Ni alloy from Fig. 10 at high magnification, revealing the presence of small cavities varying in size from a fraction of a μm to a few microns. Two elemental maps from the same area for phosphorus and tin are also reproduced. It can be seen that the phosphorus has segregated to the boundary but that the tin has segregated to the cavity surface. Other elemental maps from this surface revealed the presence of boron nitride particles,

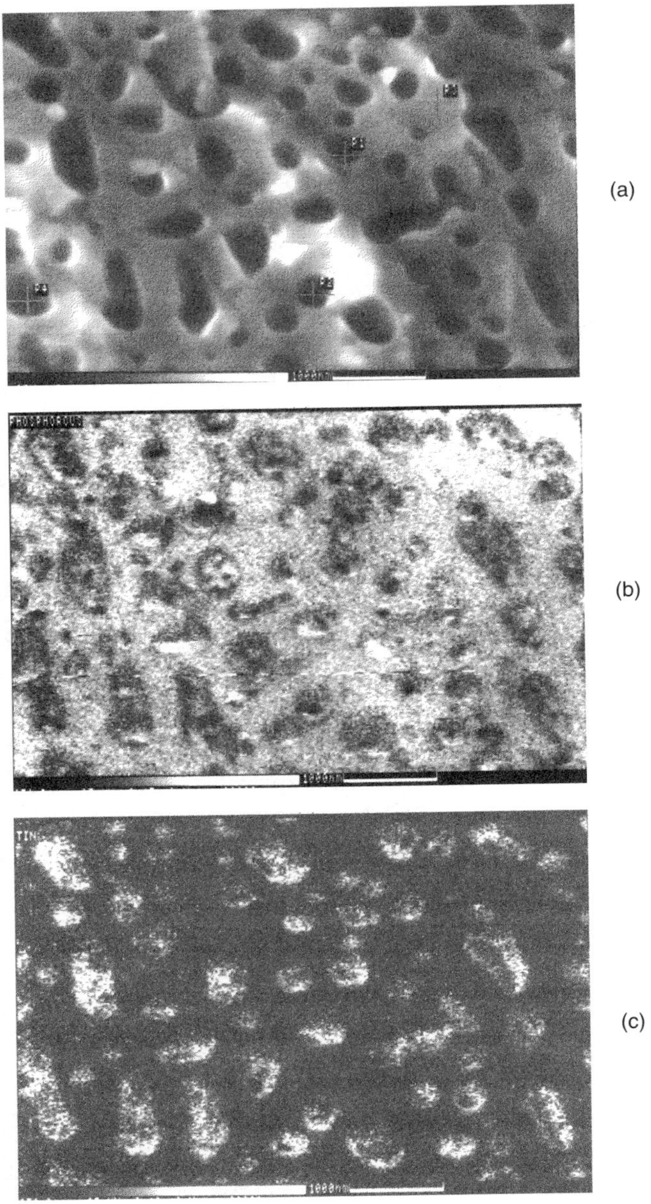

Figure 13. Grain boundary fracture surface of the Fe/3%Ni steel analyzed in a field emission gun scanning Auger spectrometer showing (a) the secondary electron image, (b) the phosphorus elemental map and (c) the tin elemental map.

sulfide particles and the presence of nitrogen on the grain boundary but not on the cavity surface.

2.3.1.2. XPS

AES has good spatial resolution and can determine quantitatively the presence of most elements, but it is not usually able to give much information on the chemical state of atoms at the surface. X-ray photoelectron spectroscopy (XPS), on the other hand, can give chemical-state information but until recently did not have the spatial resolution necessary to distinguish grain boundaries. In the last few years XPS instruments have been developed that can image with a spatial resolution of less than 10 μm and analyze from areas less than 30 μm in diameter. They therefore have the capability of analyzing grain boundary surfaces, many of which are greater than 30 μm. Recently this capability has been demonstrated by fracturing and analyzing the grain boundaries of a ferritic steel [44]. Figure 14 is an example of XPS analysis of the iron/tin specimen described above. It shows the secondary electron image of the surface together with a tin $3d_{5/2}$ map of the surface. Although the spatial resolution is still less than desired, it is possible to identify and analyze individual grains. The XPS spectrum from the tin $3d$ region is shown in Fig. 15 and indicates that in this instance the tin was present in its elemental form and not chemically combined with any other element.

2.3.2. Hydrogen charging

Not all metals and alloys can be persuaded to fracture in an intergranular manner by impact at low temperature. However, a whole group of metals and alloys will fracture in that mode if they are first embrittled by charging with hydrogen. Some will then fracture intergranularly by impact, but others require to be fractured by a slow tensile pull. This group includes the austenitic stainless steels [4], nickel [12], the nickel-base superalloys [14] and nickel aluminum alloys [45].

2.3.2.1. Charging methods

There are several methods for charging a material with hydrogen, but in order to ensure that hydrogen charging does not alter the internal properties of the metal or alloy being charged it is necessary to hydrogen-charge at relatively low temperatures, say less than 250°C. This rules out many of the charging methods that rely on heating a material to high temperature in a gaseous hydrogen/inert gas environment. Probably the most popular and well-tested method is cathodic charging in sulfuric acid [46]. In this method (Fig. 16) the specimen is made the anode in a solution of sulfuric acid of strength less than 1 N containing a small quantity (50 mg/L) of sodium arsenide through which a current,

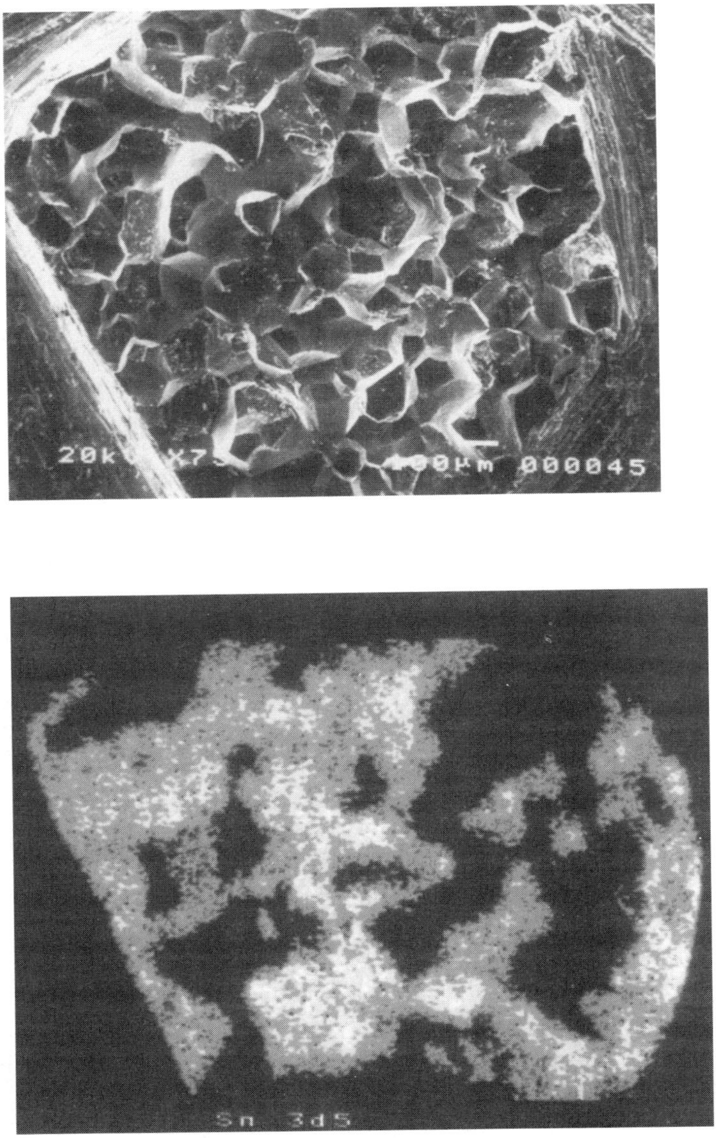

(a)

(b)

Figure 14. The secondary-electron image of the fracture surface of the Fe/3%Ni steel with phosphorus and tin segregated to the grain boundaries (a) together with (b) the imaging XPS tin $3d_{5/2}$ map [44].

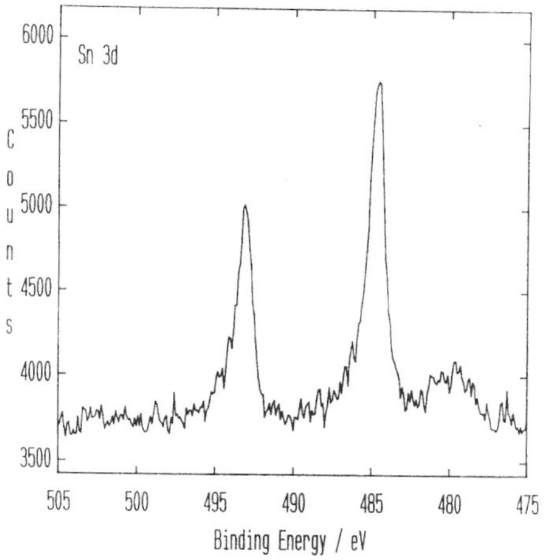

Figure 15. The XPS spectrum for the tin 3*d* region recorded from the grain boundaries of the Fe/3%Ni steel [44].

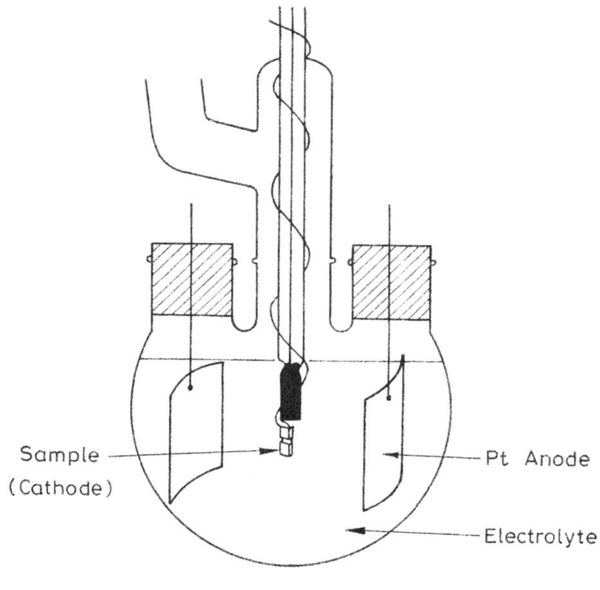

Figure 16. Schematic arrangement for hydrogen charging of metals and alloys.

normally about 50 mA cm^{-2}, is passed to cause the specimen to be surrounded by hydrogen bubbles. The rate of charging can be increased by increasing the temperature so that the solution is normally heated to 50–80°C. When the specimen is removed from the charging solution, hydrogen starts to diffuse out of the specimen. Out-diffusion can be reduced to a minimum by rapid cooling to room temperature or below and/or by plating the specimum with copper, zinc or any other metal that has low diffusion rates for hydrogen. An alternative method is to hydrogen-charge in a solution of molten salts. The electrolyte is then an eutectic mixture of sulfates of sodium and potassium to which water is added continuously. The temperature is maintained at 200°C, which ensures that a specimen 1 mm thick will be charged in less than 2 h [47]. Following hydrogen charging the specimen is cleaned, washed in isopropanol and loaded into the fracture stage in the UHV system.

2.3.2.2. *Impact and slow tensile fracture*

Many metals and alloys will fracture intergranularly by impact following hydrogen charging. However, the method of fracture to be employed is often determined by the grain boundary chemistry, and the exact fracture path may vary between impact and slow tensile methods. Austenitic stainless steels with carbides on the grain boundaries will often fracture by impact, but ultraclean stainless steels may require a very slow tensile pull. Nickel with indium segregated to the grain boundaries will fracture in an intergranular manner by impact [19], but Ni_3Al with boron segregated to the grain boundaries requires a slow tensile pull [17]. In general if a tensile fracture stage is available in the analytical system, the best intergranular fracture results can be achieved.

Figure 17 illustrates a tensile fracture stage developed in the author's laboratory for use on surface analytical systems. It is computer controlled, accepts specimens in standard-shaped holders and allows both fracture faces to be analyzed following fracture [48]. The specimen is loaded into two standard holders back to back, and this assembly is in turn loaded into the fracture stage. The slack in the system is taken up by the computer which moves the jaws apart until a small load is detected on the load cell. The required extension rate is then specified by the operator and the jaws are moved apart at this rate while the force on the load cell is recorded. The computer senses when the specimen has fractured and moves the jaws apart to allow each half of the specimen to be removed and presented for analysis. An advantage of this type of system is that fracture is more controllable than in the impact method, and, furthermore, a load/extension curve is produced which gives additional information regarding the mechanical properties of the material.

Figure 18 shows an example of impact and tensile fracture on the same specimen. A specimen of austenitic stainless steel with carbides on the grain bound-

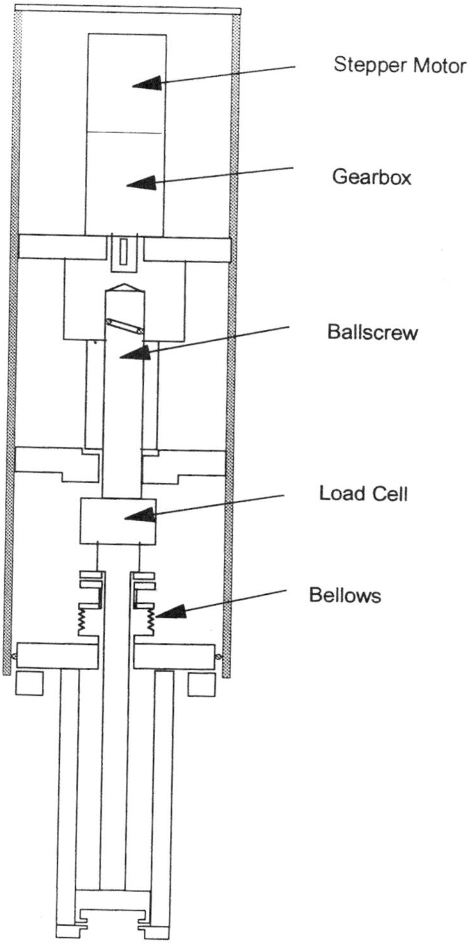

Figure 17. Tensile fracture stage for fracture of hydrogen-charged specimens.

ary was fractured using both methods. Figure 18a shows the fracture surface following impact fracture at liquid nitrogen temperature. There is clearly a high percentage of intergranular fracture but the grain boundary surfaces appear to be fairly rough. The specimen that was fractured in tension (Fig. 18b) shows intergranular fracture only around the edge of the specimen, representing the distance over which the hydrogen had diffused during hydrogen charging. However, the grain boundaries in this specimen are very smooth in contrast to the impact-fractured specimen.

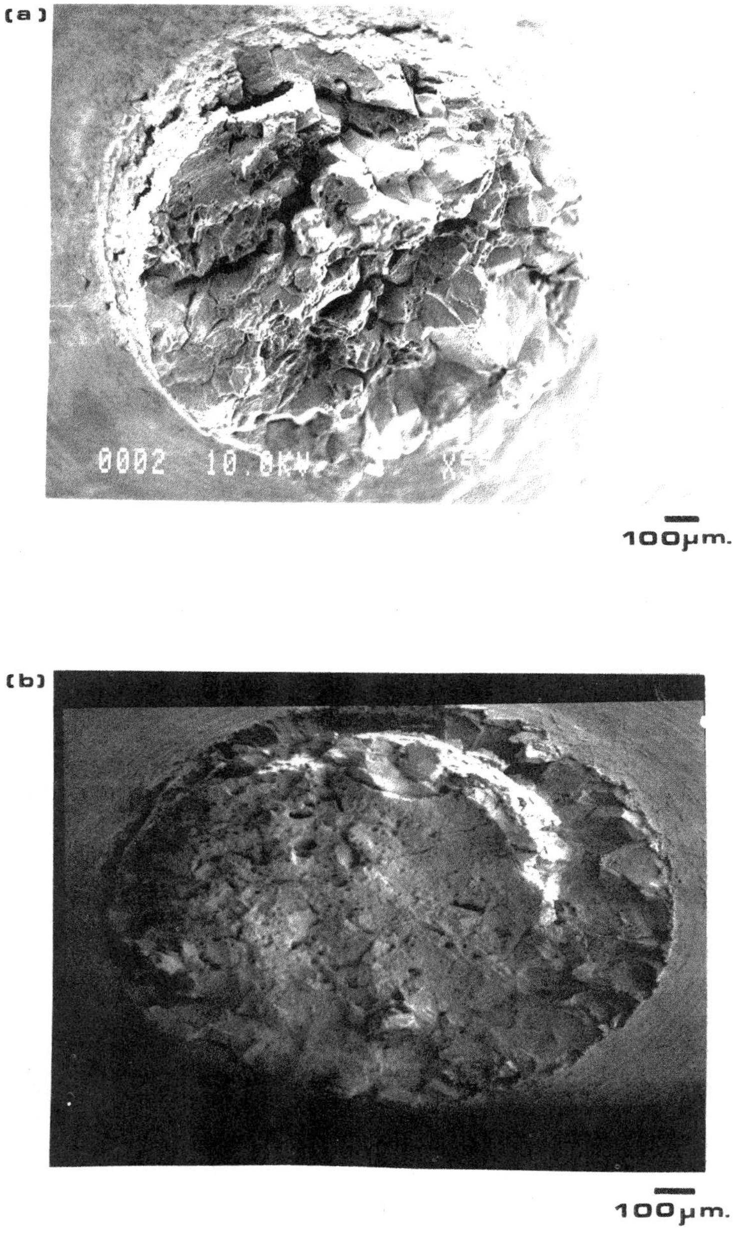

Figure 18. Impact (a) and tensile (b) fracture of austenitic stainless steel with carbides present at the grain boundaries.

Auger analysis has been used to study the effects of irradiation on the segregation of elements to grain boundaries in nickel superalloys. The rod which holds the fuel assembly in advanced gas-cooled reactors (AGRs) is manufactured from the nickel-base alloy Nimonic PE16, and it is important that it have good mechanical stability throughout its life. AES was used to study the effects of prolonged irradiation on the grain boundary composition [49]. The results are summarized in Fig. 19. Three conditions were examined: as-manufactured, following irradiation at a few hundred degrees centigrade, and following an annealing treatment at 873 K subsequent to the irradiation. The compositions as determined by AES of the grain boundary and matrix regions in the as-manufactured condition were similar and corresponded to the quoted bulk compositions. Following irradiation the bulk and trace elemental compositions at the grain boundaries had changed dramatically. Nickel had been enriched while iron and chromium had been depleted and there was significant segregation of silicon and phosphorus. Following the anneal for 1 h at 873 K the bulk and trace element compositions had returned toward their as-manufactured levels.

As described in Section 2.3.2, irradiation can therefore alter dramatically the level of segregants at grain boundaries. The mechanism responsible is the inverse Kirkendall effect, and in a nickel-base alloy such as PE16 the boundaries could be depleted in chromium and iron and enriched in nickel [36,37]. Once the effect of radiation had been removed by annealing at 873 K for 1 h the alloy was able to return to its original state.

2.4. Transmission Electron Microscopy

Not all metals and alloys can be induced to fracture intergranularly by the methods outlined. For example, very pure austenitic 316 stainless steel is very difficult to fracture with any significant percentage of intergranular failure. It is also desirable to compare the Auger and XPS results with independent methods to determine if the fracture path is indeed along the grain boundary and to confirm the levels of segregants. Transmission electron microscopy (TEM) combined with analytical methods offers this possibility, although it can be time consuming.

2.4.1. Production of a thin foil

Sample preparation is often the most time-consuming aspect of transmission electron microscopy. It is relatively straightforward if the sample is a homogeneous metal or alloy and if a thin foil can be produced from any part of it. Problems arise if areas near to the surface or to a boundary between two dissimilar materials are required to be analyzed. Normally a thin section of material 3 mm in diameter is produced mechanically, which is then polished to the minimum practical thickness before being finally thinned to electron transparency. The final stage is carried out using either electrolytic or ion-thinning methods. When

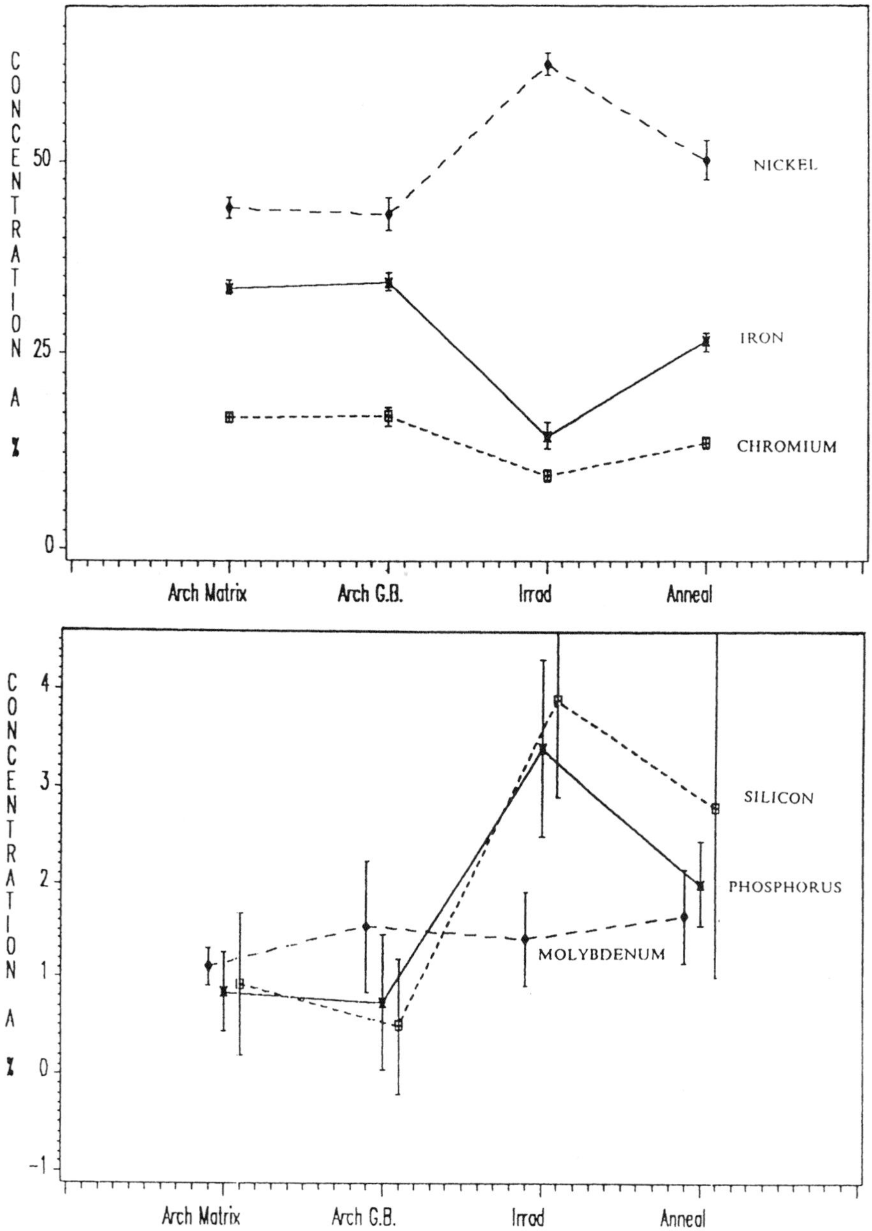

Figure 19. Effect of irradiation on grain boundary composition in the nickel-base superalloy Nimonic PE16 [49].

thinning electrolytically, it is so arranged that the center is thinned at a rate faster than that of the outer edges, and thinning is halted when a hole is observed near the center of the foil. Ion etching is carried out by bombarding the foil with a beam of inert gas ions, usually argon. The ion beam can be stationary and the foil moved, or the ion beam can be rastered to thin uniformly a large area. If the thinnest section is required from an area near the foil surface then it is necessary to build a layer of supporting material onto that surface. Thus a specimen of stainless steel with an oxide on its surface might have a layer of nickel electrodeposited on it. The whole sandwich would then be used to produce a thin foil with the thinned area containing the interface between the stainless steel and the nickel. There are specific instances in which special thinning procedures need to be adopted. To examine radioactive materials it is desirable to keep the volume of active material in the microscope to a minimum in order to reduce operator dose. This is accomplished by making a 3-mm disk in which the outer 1-mm perimeter consists of nonirradiated material, while the irradiated material is restricted to a disk 1 mm in diameter, which is fitted into a hole of corresponding diameter in the nonirradiated material. Since the central region is the thinned area, the dose is reduced by considerably more than a factor of 10.

2.4.2. Field emission gun STEM

The advent of the field emission gun scanning transmission electron microscope (FEGSTEM) has given the electron microscopist a very bright fine-focus electron source. With this tool analysis can be obtained from a thin foil using a beam only one or two nanometers in diameter and without significant spread on passing through the foil (Fig. 20). If an electron-transparent specimen is prepared in which a grain boundary can be oriented so that it is parallel to the incident electron beam, it is possible to use either electron energy loss spectroscopy (EELS) or energy dispersive x-ray analysis (EDX) to determine the composition of a cylinder approximately 2 nm in diameter by the foil thickness. By scanning the beam across the grain boundary, one can obtain a composition profile.

2.4.2.1. Parallel electron energy loss spectroscopy

As the incident electron beam passes through the specimen some of the electrons lose energy by interaction with the atoms. Plasmon losses extend over a region from 0 to 50 eV and are the result of electrostatic interactions between the electrons. Losses greater than 50 eV result from inelastic collisions with inner atomic shells of atoms while the general background results from valence shell excitations. Electrons of a given energy emerging from the thin foil are focused with a magnetic prism to an image point on the detector. Electrons with displaced energy ΔE are brought to a focus position displaced by Δx from the image point. The electrons focused on the detection slit are detected by a scin-

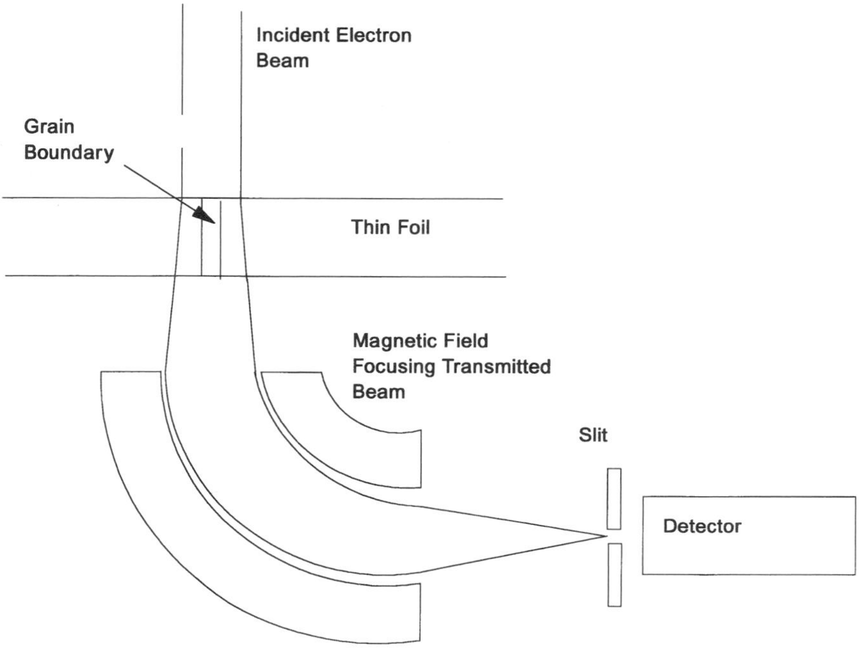

Figure 20. Schematic showing method for EDX and PEELS analysis of grain boundaries in a FEGSTEM.

tillator-multiplier or an array of solid-state diodes and are collected in series or in parallel. Parallel EELS based on solid-state detectors is faster by some two orders of magnitude than serial detection, which increases detection limits, reduces damage to the foil and improves the collection of extended energy loss fine structure. With these systems, information concerning the chemical state of atoms at the grain boundary can be obtained.

2.4.2.2. Energy dispersive x-ray analysis

When the incident electron beam passes through the thin foil it ionizes atoms within the foil. The ionized atoms then decay with the emission of x-rays, and the energy of an emitted x-ray is characteristic of the atom from which it came. Detection of these x-rays with a solid-state detector allows the composition of the volume of the thin foil through which the beam has passed to be determined.

Figure 21 shows the concentrations of chromium, iron, nickel and phosphorus as the incident beam is rastered across a grain boundary in an Inconel 690 alloy. Inconel 690 contains approximately 60 wt.% nickel, 30 wt.% chromium

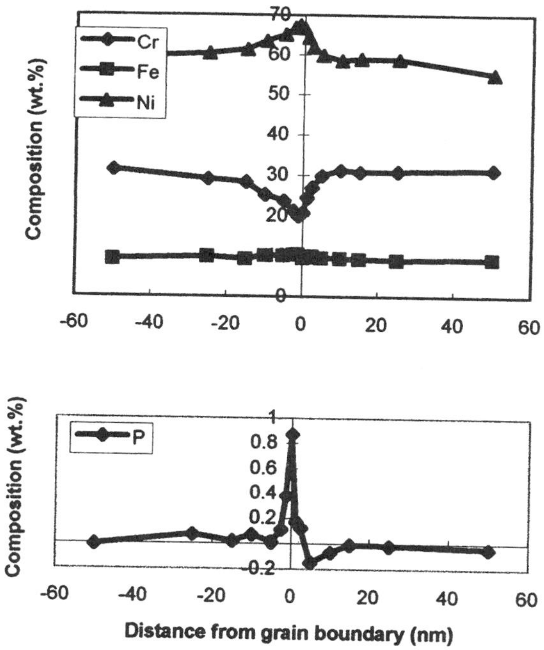

Figure 21. Concentration of Cr, Fe, Ni and P across a grain boundary in nickel-base alloy Inconel 690.

and 10 wt.% iron. The profile shows that at the grain boundary the nickel concentration can be enhanced to 70 wt.% while that of the chromium can be reduced to 20 wt.%. In this case the depleted region extends approximately 10–20 nm on either side of the boundary. Depletion of chromium is thought to be a major factor in stress corrosion cracking. Chromium, when present in steels at concentrations greater than 12 wt.%, will form a protective surface oxide. Microcracks can form at or near a grain boundary at the surface. Depletion in the region of grain boundaries allows the surface to be attacked, and, with no protective oxide able to form, corrosion extends down the grain boundary surface, the crack continuing to open up aided by the stress and the growing oxide. Phosphorus is also shown to be enriched at the grain boundary, but here the profile is much narrower, extending only 1 or 2 nm to either side of the grain boundary. In this case the phosphorus has segregated to the grain boundary as a single-atom layer [50]. The incident beam extends over a diameter of 2 nm and therefore includes matrix material as well as the grain boundary, which is approximately 0.5 nm wide. The FEGSTEM EDX will therefore underestimate the level of phosphorus segregated to the boundary. Phosphorus is a well-known embrittling agent and plays a significant role in stress corrosion cracking. The action of phosphorus at the grain boundary aids the opening of the crack both by applied stress and by oxide formation.

2.4.2.3. Comparison of AES and FEGSTEM

AES and FEGSTEM sample in different ways. AES analyzes directly one side of a grain boundary, sampling mainly the top atom layer but with contributions included from a few layers below the surface. FEGSTEM combined with EDX analyzes a volume that is 2 nm in diameter and thus includes material on either side of the 0.3-nm-wide grain boundary. Each method attempts to determine the actual level of the segregant at the grain boundary, but because of the many assumptions made there may be errors. It is useful to compare the results from the two methods. Fisher [51] has analyzed several

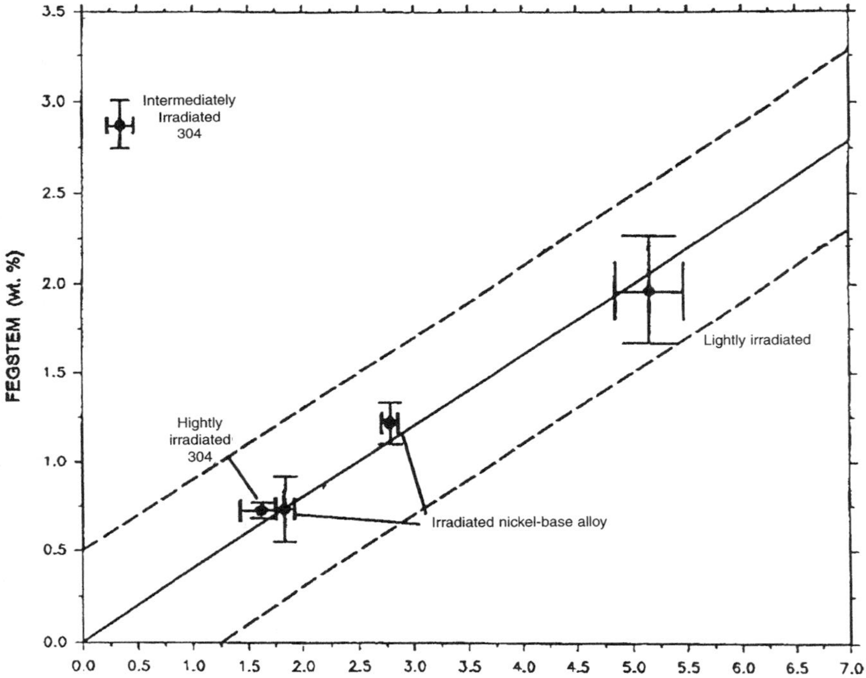

Figure 22. Comparison of the grain boundary composition determined by AES and FEGSTEM EDX showing approximately a 1:2 ratio phosphorus. (Courtesy Fisher [51].)

systems with scanning AES and EDX FEGSTEM, plotting the concentration of one against the other. The results of this are shown in Fig. 22. Converting the signal from AES and EDX each to weight percent of that element in the matrix leads to a relationship in which AES will give approximately twice the concentration of phosphorus as EDX.

2.4.3. Time-of-flight atom probe

The detailed internal composition of a metal can be determined with atomic resolution by using an atom probe combined with a time-of-flight mass spectrometer. Atoms can be stripped from the probe tip layer by layer by pulsing the field in the vicinity of the tip. The ions then travel through the flight tube, with the lighter ions arriving first and the heavy ions last. The ion mass is determined from the time of flight from desorption to detection. The spatial distribution, laterally across the tip, is determined by using a multichannel detector to sense the position of the ion on striking the detector, which position is directly related to the original position of the atom on the tip. Depth information is obtained by stripping individual layers of atoms in a known and controlled manner. The information concerning time and position of arrival is stored by computer, which is subsequently used to map the distribution of elements in three-dimensional space. Clearly this technique is time consuming and the number of specimens that can be analyzed is limited. It can, however, give some unique information. In one example a tip was prepared from a nickel-base superalloy (Astroloy) with composition approximately 52% Ni, 16% Cr, 16% Co, 8% Al and 4% Ti and which contained a grain boundary. The alloy contained 0.11% of boron as a trace impurity. The boundary position was first located with a transmission electron microscope. The tip was then analyzed as described above, and it was possible to identify the positions of the various elements around the grain boundary. Figure 23 shows the distribution for Al+Ti, Cr, Mo and B + C. It can be seen that the boundary was enriched in Cr, Mo and B + C but depleted in Al and Ti. It was also possible to determine the width of segregation at the grain boundary, and in this case the width of the segregated boron layer was 0.5 nm [52,53].

3. CRACKS IN METALS AND ALLOYS

Frequently a material fails in service, and it is necessary to analyze the failed component to determine the causes of failure. While the metallographic and analytical methods described here should be undertaken, the failed surface itself can often yield valuable information. Failure frequently occurs by the gradual opening of a crack for some time before catastrophic failure actually occurs. Often a crack may be detected in the component during routine nondestructive

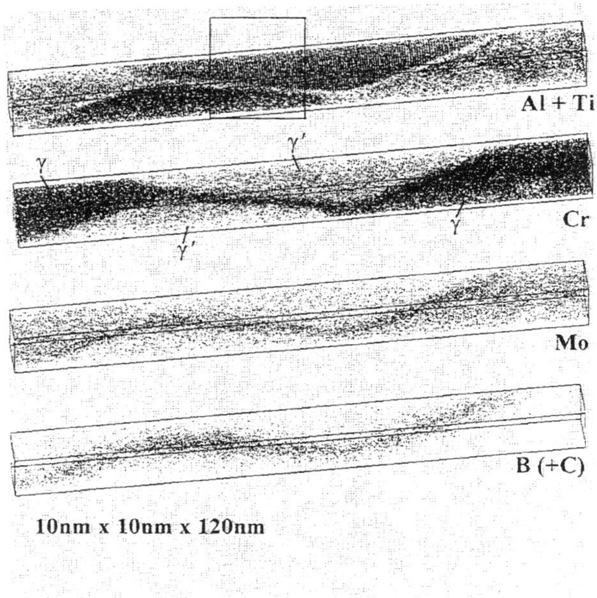

Figure 23. Position-sensitive atom probe images across a grain boundary in the nickel-base alloy Astroloy, for Al+Ti, Cr, Mo and B+C.

testing and before failure has occurred. In these cases it is important to know when the crack started, and, in the case of the crack detected before failure, if it is growing and at what rate.

The oxide thickness can be used to determine the age of different parts of the crack and hence the rate of growth. A freshly exposed metal surface will, in general, oxidize in such a way that the thickness of the oxide increases as the square root of the time of exposure at temperature [54–55], namely

$$x = A + kt^{1/2} \tag{8}$$

where A is a constant and k is the parabolic rate constant.

Frequently the oxide may be many microns thick as in the example of a ferritic steel operating at several hundred degrees Celsius, in which a crack had opened over a period of months or years before final failure. In such cases the oxide thickness may be readily determined from optical examination of a metallurgically prepared specimen. The thickness may then be used with the time-temperature history to build up a picture of the crack history.

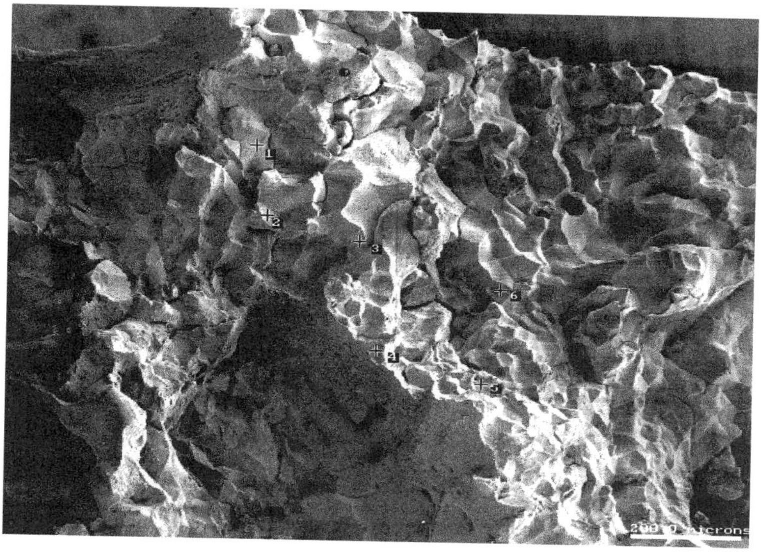

(a)

(b)

Figure 24. Secondary-electron image of a crack tip fractured open in the UHV of the spectrometer (a) together with the oxygen elemental map (b).

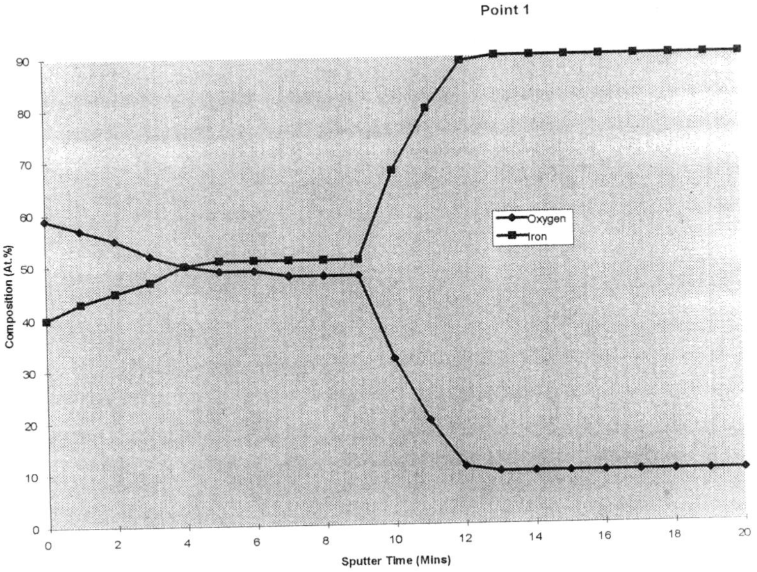

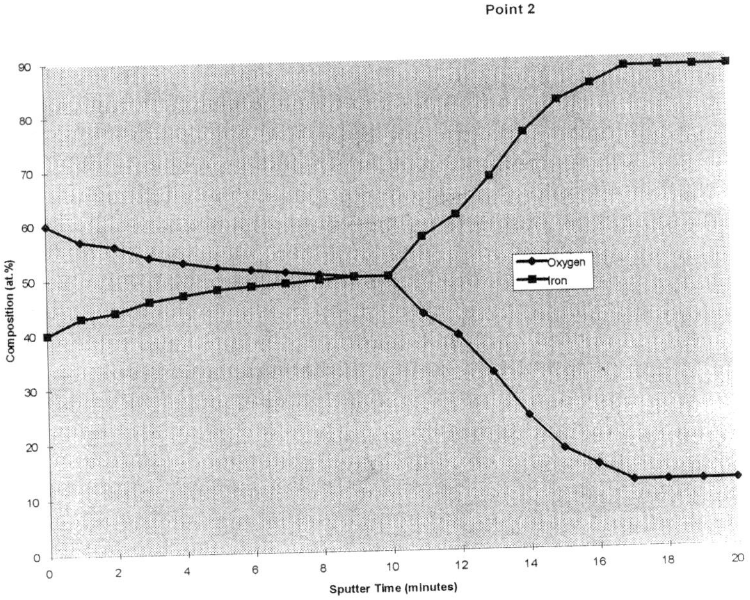

Figure 25. Depth profiles through the oxide at two points on the crack shown in Fig. 24.

However, if the crack has grown in a material which oxidizes only slowly, then the oxide thickness may be too thin to determine by metallographic means. In this case surface analytical techniques may be used to determine the oxide thickness and, in the case of the crack which has yet to cause failure, may also yield some information regarding the causes of crack initiation and growth. Ideally the crack should be opened inside the spectrometer so that any contaminant or segregation can be identified without being first contaminated by air. Sometimes the component may have been cleaned with solvents prior to receipt, and it may be necessary to cut it in order to produce a specimen suitable for fracture in the spectrometer. Both these operations may themselves cause contamination, adding to the difficulties of identifying the cause of the crack. However, it is still preferable to open the crack under UHV conditions. Figure 24 shows a secondary-electron image of a crack tip in an Fe/Cr/Mo/V steel together with an Auger elemental map for oxygen which allows the crack tip to be identified. The oxide thickness can be determined by depth-profiling at regular intervals from the crack tip. These profiles are obtained by determining the peak heights of selected elements, in this case oxygen and iron, at each point on the oxide surface, removing a fixed amount of oxide by bombarding with energetic argon ions, and repeating the process. The operation is controlled by computer, allowing an elemental depth distribution for oxygen and iron at each point to be built up gradually. The time taken to reach the oxide/metal interface is then converted into oxide thickness using calibration data obtained from standard oxides. The depth profiles at two points on the oxide are reproduced in Fig. 25. It can be seen that the oxide/metal interface is very broad, indicating that the grain boundary surface is not completely flat, and that the interface width varies from point to point. The oxide/metal interface was reached at point 1 in about 10 min, indicating an oxide thickness of 400 nm, whereas the interface was reached at point 2 in approximately 13 min, suggesting an oxide thickness of over half a μm. It can also be seen that the Fe/O ratio varies from 40:60 at the outer surface to 50:50 within the oxide, which may indicate that the oxide composition is changing, but which may also be due to ion beam reduction. From a series of such profiles the crack history can be built up, from which it is possible to state whether the crack is continuing to grow or whether it has stopped.

REFERENCES

1. K. N. Akhurst, *CEGB Res. 21*, 5 (1988).
2. P. Lejcek and S. Hofmann, *Crit. Rev. Solid State Mate. Sci. 20*, 1 (1995).
3. H. Erhart and H. J. Grabke, *Met. Sci. 15*, 401 (1981).
4. C. L. Briant, *Acta Metall. 33*, 1241 (1985).
5. M. P. Seah and C. Lea, *Phil. Mag. 31*, 627 (1975).
6. M. Guttmann, *Surf. Sci. 53*, 213 (1975).

7. H. J. Grabke, in *Segregation at Interfaces in Chemistry and Physics of Fracture* (R. M. Latanision and R. H. James, eds.), Martinus Nijhoff, Dordrecht, The Netherlands, 1987, p. 388.
8. B. D. Powell and D. P. Woodruff, *Phil. Mag. A34*, 169 (1976).
9. T. H. Chuang, W. Gust, L. A. Heldt, M. B. Hintz, S. Hofmann, R. Lucic and B. Predel, *Scripta Met. 16*, 1437 (1982).
10. E. Smiti, P. Joufrey and A. Kobylanski, *Scripta Met. 18*, 673 (1984).
11. C. L. Briant, *Met. Trans. A 18A*, 691 (1987).
12. G. P. Airey, *Corrosion-NACE 41*, 2 (1985).
13. A. Elkholy, J. Galland, P. Azou, and P. Bastien, *C. R. Acad. Sci. Paris 284C*, 363 (1977).
14. G. C. Allen and R. K. Wild, *Phil. Mag. A54*, L37 (1986).
15. C. L. Briant, *Surf. Interface Anal. 13,* 209 (1988).
16. D. J. Nettleship and R. K. Wild, *Surf. Interface Anal. 16*, 552 (1990).
17. C. T. Liu and E. P. George, in *Structure and Property Relationships for Interfaces* (J. L. Walter, A. H. King and K. Tangri, eds.), ASM, Metals Park, OH, 1991, p. 281.
18. Y. Ishida, S. Yokoyama and T. Nishizawa, *Acta Met. 33*, 255 (1985).
19. T. Muschik, W. Gust, S. Hofmann and B. Predel, *Acta Met. 37*, 2917 (1989).
20. D. I. R. Norris, C. Baker, and J. M. Titchmarsh, Proc. Radiation Induced Sensitisation of Stainless Steels, Berkeley Nuclear Labs., Berkeley, UK, 1986, p. 86.
21. D. I. R. Norris, C. Baker, C. Taylor and J. M. Titchmarsh, *Effects of Radiation on Materials*, 15th Int. Symp., ASTM STP 1125, (R. E. Stoller, A. S. Kumar and D. S. Gelles, eds.), ASTM, Philadelphia, 1992.
22. O. L. Krivanek, *Ultramicroscopy 28*, 118 (1989).
23. C. Herzig, J. Geise and Y. M. Mishin, *Acta Met. Mater 41*, 1683 (1993).
24. R. Herschitz and D. N. Seidman, *Scripta Met. 16*, 849 (1982).
25. A. Cerezo, T. J. Godfrey and G. D. W. Smith, *Rev. Sci. Instrum. 59*, 862 (1988).
26. M. P. Seah and E. D. Hondros, *Proc. Roy. Soc. Lond. A 335*, 191 (1973).
27. M. P. Seah, *Surf. Sci. 53*, 168 (1975).
28. M. P. Seah, *Proc. Roy. Soc. Lond. A 349*, 535 (1976).
29. E. D. Hondros and M. P. Seah, *Int. Met. Rev. 22*, 867 (1977).
30. D. McLean, *Grain Boundaries in Metals*, Oxford University Press, London, 1957.
31. J. W. Gibbs, *Collected Works*, Vol. 1, Yale University Press, 1948, p. 219.
32. F. L. Carr, M Goldman, L. D. Jaffe and D. C. Buffum, *Trans. AIME 197*, 998 (1953).
33. M. P. Seah, *Acta Met. 25*, 345 (1977).
34. P. L. Gruzin and V. V. Mural, *Fiz. Metal. Metalloved. 17*, 62 (1964).
35. P. W. Palmberg and H. L. Marcus, *Trans. Am. Soc. Metals 62*, 1016 (1969).

36. J. M. Perks, A. D. Marwick and C. A. English, Proc. Radiation Induced Sensitisation of Stainless Steels, Berkeley Nuclear Laboratories, Berkeley, UK, 1986, pp. 15–34.
37. A. D. Marwick, R. C. Piller and M. E. Horton, 1983, AERE-R10895, Atomic Energy Research Establishment, Harwell, UK.
38. G. A. Dryer, D. E. Austin and W. D. Smith, *Metal Progr. 86*, 116 (1964).
39. T. Ogura, A. Makino and T. Masumoto, *Met. Trans. A 15A*, 1563 (1984).
40. R. K. Wild, *Mater. Sci. Eng. 42*, 265 (1980).
41. K. Tatsumi, N. Okumura and S. Funaki, in *Grain Boundary Structure and Related Phenomena* (Y Ishida, ed.), Proc. 4th JIM Int. Symp., *Trans. Jap. Inst. Met., Suppl., Suppl. 27*, 427 (1986).
42. S. Suzuki, K. Abiko and H. Kimura, *Scripta Metall. 15*, 1139 (1981).
43. D. Briggs and M. P. Seah, *Practical Surface Analysis*, Vol. I, J. Wiley & Sons, Chichester, U.K. (1990).
44. K. Hallam and R. K. Wild, *Surf. Interbase Anal. 23*, 133 (1995).
45. A. Choudhury, C. L. White and C. R. Brooks, *Acta Met. Mater. 40*, 57 (1992).
46. E. L. Hall and C. L. Briant, *Met. Trans. 16A*, 1225 (1985).
47. M. Guttmann, Ph. Dumoulin, N. Tan-Tai and P. Fontaine, *Corrosion—NACE 37*, 416 (1981).
48. G. C. Allen and R. K. Wild, Inst. Phys. Conf. Ser. No. 90, *Electron Microscopy and Analysis*, 1987, p. 13.
49. D. J. Nettleship and R. K. Wild, Effects of Radiation on Materials, 15th Int. Symp., ASTM STP 1125 (R. E. Stoller, A. S. Kumar and D. S. Gelles, eds.), ASTM, Philadelphia, (1992), p. 645.
50. D. I. R. Norris, C. Baker, C. Taylor and J. M. Titchmarsh, Effects of Radiation on Materials, 15th Int. Symp., ASTM STP 1125 (R. E. Stoller, A. S. Kumar and D. S. Gelles, eds.), ASTM, Philadelphia, (1992), p. 603.
51. J. Walmsley, P. Spellwood, S. Fisher and A. Jenssen, Proc. 7th Int. Symp. on Environmental Degradation of Materials in Nuclear Power Systems-Water Reactions, Breckenridge, Colorado, (1995), p. 985.
52. D. Blavette, B. Deconihout, A. Bostel, J. M. Sarrau, M. Bouet and A. Menand, *Rev. Sci. Instrum. 64*, 2911 (1993).
53. L. Letellier, M. Guttmann and D. Blavette, *Phil. Mag. Lett. 70*, 189 (1994).
54. C. Wagner, *J. Electrochem. Soc. 99*, 369 (1952).
55. P. Kofstad, *High Temperature Oxidation of Metals*, Wiley, New York, 1966.
56. J. P. Coad, J. C. Rivière, M. Guttmann and P. R. Krahe, *Acta. Metall 25*, 161 (1977).

12

Microelectronics and Semiconductors*

E. PAPARAZZO Istituto di Struttura della Materia CNR, Rome, Italy

1. INTRODUCTION

Few materials in both pure and technologically oriented research have benefited as much from the diagnostic capabilities of surface-specific techniques as have semiconductors and microelectronic devices. Important contributions have been made to the understanding of a wide range of topics extending from the electronic structure of basic materials, such as silicon and III-V semiconductors, to the quality control and failure analysis of microelectronic devices. With the progressive miniaturization of device circuitry has gone the requirement for surface analysis of ever-smaller areas; this requirement has been met over the years by a parallel progressive reduction of the electron beam spot size in AES, but in the near future it is possible that rapid developments in spatially resolved XPS may encroach onto what has up to now been the preserve of AES. Use of XPS will bring the added bonuses of the provision of chemical information and of greatly reduced damage to the specimen surface during analysis.

*This work is dedicated to my sister Piera.

It would be impossible to cover in a limited space all the situations in semiconductor technology in which surface analytical techniques have been applied, so instead the following four typical problem areas have been selected to demonstrate the application. They are typical in that the nature of the information sought in each case is similar to that sought in many other problem areas in the field.

1. Josephson junctions are used as voltage-to-frequency converters; it is essential that no chemical degradation occurs over their operating life, otherwise their characteristics will change, with resultant operational unreliability or failure. The junctions are layer structures, with several abrupt interfaces, and information is required about the effects of degradation (from, e.g., humid air) on both external and internal surfaces. Analysis with lateral spatial resolution is not therefore important, but the ability to probe the interfaces is, and of course both elemental and chemical information are required.
2. In order to protect III-V semiconductors against weathering their surfaces are covered with a passive layer; this layer must be not only chemically passive, for protection, but also electronically passive so that the electronic characteristics of the semiconductor are not altered. Quaternary III-V semiconductors (i.e., containing four constituent elements) are important technologically because the energy of their band gap can be varied by choosing the appropriate chemical composition. Again both elemental and chemical information is required, but because the passive layer is thin and may not be of uniform thickness it is necessary to have spatially resolved information as well.
3. Pure silicon carrying a uniform and homogeneous thin film of SiO_2 is one of the building blocks in many components used in the electronics industry. In its application not only must the thickness, uniformity and homogeneity of the SiO_2 film be known and controllable, but the nature of the Si/SiO_2 interface region, in terms of structure, chemistry and width, as well. There is also the problem of possible chemical modifications occurring as a result of aging, e.g., reaction with ambient humidity over the lifetime of a device. In this case spatially resolved information is not required, but the ability to detect very subtle chemical differences as a function of depth is.
4. Films of InP on SiO_2 are used as substrates for the fabrication of many semiconductor devices. Although there are in general no particular problems associated with the in-service performance of the substrates, it is instructive to examine their surface chemistries, since during fabrication of devices both InP and SiO_2 and their interfaces, become exposed. The various phases that appear are to be found distributed among a series of well-defined surface structures of various shapes and dimensions. Here it is clear that the material is laterally inhomogeneous, and if the local composition and chemistry are to be studied, then spatially resolved information, probably on the scale of a few μm, is necessary.

In describing the ways in which surface analysis has tackled these problems, the emphasis will be in each case on the path chosen, starting with a statement of the particular analytical problem and progressing through the choice of analytical technique(s) to the chosen methodology and ending with presentation of results and their interpretation. The relevance of the results to the materials problem will then be discussed. The surface-specific techniques variously employed are (a) x-ray photoelectron spectroscopy (XPS); (b) synchrotron radiation photoelectron spectroscopy (SRPS); (c) x-ray-induced Auger electron spectroscopy (XAES); (d) (electron-induced) Auger electron spectroscopy (AES); (e) scanning Auger microscopy (SAM); and (f) electron energy loss spectroscopy (ELS).

2. TECHNIQUES

Before describing the surface analysis of the materials listed above, the advantages and limitations of the surface-specific techniques to be used should be discussed. The basic principles, instrumentation and main applications of photoemission [1–3] and Auger [4,5] spectroscopies, as well as of SAM [6,7] and ELS [8–11] have already been described in several reviews, while Seah and Briggs [12] have surveyed in detail the principal features of many surface-specific techniques. Of major interest here are those characteristics of each technique that, on the one hand, may be employed strategically to solve a given analytical problem, but that on the other may affect the reliability of the results. In Table 1 the relative merits of the techniques are rated with respect to those properties likely to be important in the surface analysis of semiconductors and microelectronic devices.

2.1. Surface Specificity

Although all the electron spectroscopies have surface specificities of a few nm, SRPS and ELS are those methods possessing the greatest versatility. By varying the energy of the photon (electron) source, one may (1) achieve the highest possible surface specificity when the kinetic energy (KE) of the photoelectron (inelastically scattered electron) corresponds to the shortest electron attenuation length (AL) or electron inelastic mean free path (λ) [13]; (2) follow, in a continuous fashion, the evolution of spectral features associated with the species of interest as a function of depth over a range from a fraction of a nm to several nm. The latter feature makes both techniques well suited to nondestructive depth profiling, i.e., without resorting to Ar^+ bombardment. Although the range of KEs associated with ELS (ca. 50 to 2000 eV) is currently about twice that available with SRPS [14], photoemission experiments in the KE range 400 to 2000 eV are becoming feasible. Angularly dependent XPS and AES can also provide nondestructive depth profiling, although, in comparison with SRPS and ELS, the variation in surface specificity is relatively limited,

Table 1. Capabilities and Limitations of the Surface-Sensitive Methods Used in this Chapter[a]

Technique	Surface specificity	Elemental sensitivity	Chemical sensitivity	Destructiveness[b]	Quantitation	Lateral resolution	Artefacts
XPS	**	**	**	***	***	—	charging
XAES	**	**	***	***	—	—	charging
SRPS	***	**	***	***	*	—	charging
AES	**	**	**	*	**	—	charging electron backscattering
SAM	**	**	***	**	**	***	charging electron backscattering topo. effects
Scanning ELS	***	*	*	**	—	***	spectral interference

[a](*) = bad; (**) = fair; (***) = good.
[b]As far as "destructiveness" is concerned, the symbols have the following meanings; (*) = high; (**) = medium; (***) = low.

and the results are generally less straightforward to interpret [15,16]. In AES and XPS experiments performed at fixed takeoff angles, the surface specificity can be varied by selecting, for any one element, peaks at different KEs.

2.2. Elemental Sensitivity

By elemental sensitivity is meant the detectability of a given element by positive identification of a measurable signal intensity at the appropriate location in a spectrum. The photoemission and Auger techniques listed in Table 1 can detect all elements of the periodic table except H and He, with a detection limit of about 0.5 at.%. (The presence of hydrogen in OH^- groups can often be inferred qualitatively and quantitatively from O 1*s* photoelectron spectra.) High-flux photon sources at the synchrotron radiation facilities at Berkeley and Trieste have increased further the detection limits by ca. two orders of magnitude, using SRPS. In contrast, elemental sensitivities in ELS are hard to define for multielemental samples, because the energy shifts between the electron loss signals from different chemical phases are not large enough to be able to resolve the individual components, since the latter have inherently broad linewidths. The figure of 0.5 at.% can therefore be taken as a fair average for the sensitivity of surface techniques, although SIMS is able to detect some elements at levels below 1 ppm.

2.3. Chemical Sensitivity

The chemical sensivity, i.e., the capability of identifying the oxidation state and chemical environment of a given element by the KE or the binding energy (BE) of a particular spectral feature, is particularly good in SRPS and XPS (especially monochromated XPS), and possible in SAM and XAES. In rating the techniques under this heading two aspects have been taken into account: the magnitudes of the chemical shifts and the linewidths of the spectral signals. SRPS and XPS have good chemical sensitivity by virtue of chemical shifts sufficiently large to be able to distinguish between different oxidation states for most of the elements that enter the composition of semiconductors. Moreover, the advent of highly monochromatized photon sources combined with appropriate electron analyzers has made it possible to achieve average resolving powers of ca. 10^4 in the energy range 20–1000 eV in third-generation SRPS facilities [17]. AES often has a higher chemical sensitivity than that of XPS and SRPS. For instance, the chemical shift in the Si LVV Auger peak between elemental silicon and silica is more than three times that in the Si 2*p* photoemission peak and is accompanied by a spectacular change in the spectral line shape [18]. However, Auger signals are in general much broader than photoemission peaks due to the convolution of the natural linewidths of the three electronic levels participating in the Auger process (as opposed to the single level giving rise to photoemission), and the resulting signal can be even broader when electrons from the valence band are involved.

A further complication in AES arises from electron backscattering effects (see subsequent section on artefacts) which generally make it difficult for the analyst to ascertain the "true" Auger peak line shape and intensity [19–21]. As a consequence, XPS and SRPS have always been promoted as the electron spectroscopies of choice where chemical sensitivity is concerned. When quantitative resolution (using peak-fitting procedures) is required between individual components contributing to the complex spectral bands of multiphase systems (e.g., coexistence of elemental silicon, silicon suboxides and silica), then the photoelectron spectroscopies must be used. AES has indeed been largely neglected as a chemically sensitive tool [22] in the analysis of semiconductor surfaces, and the diagnostic potential of Auger spectra has been confined mainly to XAES experiments. There are advantages in the latter technique in that electron backscattering effects are absent and that the analyst can combine in situ Auger and XPS spectra to produce Auger parameters [23,24]. The neglect of AES as a chemical indicator originates in the way that the technique was used in the early days [22]. The pioneering work of that time demonstrated [25–27] that Auger spectra could be obtained using LEED optics as a detection system. Although of only low-energy resolution, that experimental arrangement gave the analyst the opportunity to use in situ AES to check the cleanliness of semiconductor surfaces, and sometimes even to quantify the level of surface contamination. High energy resolution was not a requisite for such qualitative application, nor was it for more complex experiments in which Auger spectra were used to monitor quantitatively the evolution of the surface chemistry of a sample as a function of exposure to gases. Since that time, with a few exceptions [28,29], AES has continued to be performed at low-energy resolution, even recently when instruments equipped with dedicated electron analyzers have become available. In fact, provided the resolving power of the analyzer is adequate at all KEs over the range of interest, AES shows, as expected, exactly the same chemical sensitivity as XAES [29]. The same considerations apply to SAM.

2.4. Destructiveness

The potential destructiveness of a technique is important in the surface analysis of semiconductor materials. As a result of exposing the sample to the analyzing probe, destructive effects can take the form either of alteration of the original chemistry of the surface, through chemical reduction and through desorption processes [30], or of the buildup of contamination at the surface, or both together. In general, of the techniques used here, AES is the most destructive. For instance, it has been found that an electron beam can cause not only a buildup of C and O on clean silicon surfaces [31], leading to oxidation in the case of O accumulation, but the formation of Si–Si bonding states as a result of reduction of SiO_2 [32]. Compared to AES, XPS and XAES can be regarded as techniques of low destructiveness; alterations of the sample surface induced by the x-ray beam have been observed, but are restricted to high-valence-state oxide systems ($Cr^{(VI)}O_3$ [33], $Ce^{(IV)}O_2$ [34], etc.), with the cation being reduced to a lower, usually the most stable, oxidation state. Destructive effects can be significant in SRPS using high-brightness photon sources, whereas relatively benign conditions can be chosen for both SAM and ELS analyses by using primary-electron currents of the order of a few nA.

2.5. Quantification

XPS is to be preferred when quantification is needed of the surface chemical composition of semiconductors. Several quantitative XPS methods have been reported in the literature, using either first principles [35,36] or relative elemental sensitivity factor approaches [37]. In both cases the relative accuracy is limited to ± 5–10%, based on the analysis of stoichiometric compounds. AES and SAM can provide approximate quantification, the major problem being the variability of electron backscattering effects in different matrices [38–41] and the difficulties associated with the measurement of the true Auger peak intensity, in both the direct and differential spectra [5]. ELS and XAES provide only rough indications of the elemental concentrations. Because of superior energy resolution, SRPS (and monochromated XPS) are best exploited for the resolution of different oxidation state components contributing to complex bands of a given elemental photoemission transition.

2.6. Spatial Resolution

Good spatial resolution in the surface analysis of semiconductors and microelectronic devices is particularly important. It has always been necessary to investigate the microchemistries of devices with techniques of spatial resolution comparable to the dimensions of the circuitry, and with miniaturization currently at the sub-μm level, there is pressure on technique development to be able to analyze routinely with spatial resolutions of tenths of a μm. SAM [6,7], since the mid-1970s, and, more recently, scanning ELS [8–11], are the two techniques that meet these requirements. Focusing of the primary electron beam onto chosen microspots on the sample (point analysis), or rastering it along straight lines (line scanning), or over areas (imaging), of the sample surface, when combined with an Auger peak intensity in SAM or a characteristic loss feature in ELS, as elemental indicators, can reveal the local lateral elemental distribution with a spatial resolution of some tens of nm in favourable circumstances. Since the late 1980s, there have been increasing efforts to improve the spatial resolution in both XPS [42] and SRPS [43], as described in Chapter 4. Although these efforts have produced spatial resolutions in the several-μm range ("small-area" XPS) and in the 0.1-μm range (synchrotron radiation spectromicroscopy), respectively, possible applications in the field of semiconductors will not be dealt with here.

2.7. Surface Charging and Other Considerations

Except for ELS [106], all the techniques listed in Table 1 suffer from electron charging as the main artefact when analyzing thick insulating layers on surfaces. The main effect of charging is to generate a shift in KE of the Auger and photoemission peaks as a result of the potential arising from either positive or negative excess surface charge created by the imbalance in electron current arriving at, or leaving, the surface [12]. Although not a problem with doped semiconductors (nor, obviously, with metals), charging may affect the analysis of microelectronic devices with thick insulating regions, particularly when using AES at high primary-electron currents, when both positive and/or negative charging can occur [44]. AES and SAM are also affected by electron backscattering effects, which may hamper the quantitative analysis from Auger spectra and the chemical contrast in SAM linescans and images [39–41].

Charging in XPS and XAES is not as serious a problem as in AES, since primary electrons are not involved and a steady state of the noncompensated charge is usually reached. However, if a surface is very inhomogeneous, differential charging may occur [45]. Charging effects can usually be taken into account by referencing the apparent BEs (KEs) to a suitable internal standard. Such problems may also intrude into SRPS when performed with high-brightness photon sources. Unless differential charging occurs, measurement of photoemission and Auger features in the same spectrum, for a given element A, eliminates charging problems since the so-called modified Auger parameter [23,24] α^* (= $BE_A + KE_A$) can be used as a chemical-state diagnostic. Such analysis can sometimes be hampered by spectral interferences for Auger features occurring at KEs close to those of photoemission. The problem is worse when unmonochromated x-ray sources are employed, when Auger transitions gain additional intensity from excitation by the Bremsstrahlung continuum. The interference can, however, usually be removed by changing the primary photon energy, as the author has reported in a surface analysis study of Al_2O_3-SiO_2 systems [46].

In any analysis involving focused electron beams variations in signal intensity from point to point on a surface will arise not only from variations in elemental concentrations but also from topographical effects, if the surface is at all rough. This is a particular problem in SAM, in which the "elemental maps" can be severely misinterpreted if topographical effects are not taken into account. As described in Chapter 5, the problem can be circumvented by applying a correction involving measurement of the background signal B at an energy as close as possible to the elemental signal P, and then displaying a map either of $P - B$ or, in more sophisticated computer programs, of $(P - B)/(P + B)$.

The principal problem with ELS is that of spectral interference. Since the loss features that are recorded in the technique are usually broad, resolution of energy shifts of the same order of magnitude as the widths is difficult, and thus separation of different chemical states is not always possible.

In addition to the above considerations, which are inherent to the techniques themselves, the artefacts induced by Ar^+ etching [47] should be also borne in mind. Although the latter procedure is generally unavoidable when surface chemical analysis has to be extended into the sample, its effects on the alteration of the true chemical composition must be realized and taken into account. The reader is referred to the several reviews dealing with this topic [47] (as well as to Chapters 6–9), while some Ar^+-induced artefacts will be discussed in the description of the surface analysis of Si/SiO_2 interfaces.

3. JOSEPHSON JUNCTIONS

Josephson junctions consist of a base superconducting electrode, a few μm thick, over which a thin (2–3 nm) layer of oxides of the base metal is grown thermally, to form the tunneling insulating barrier, on top of which a superconducting material (metal or alloy) is evaporated so as to obtain the top electrode (counter-electrode) of a few μm in thickness. The samples analyzed here were fabricated at the Istituto Elettrotecnico Nazionale "Galileo Ferraris," Turin, Italy, and were of two different types: Nb/Pb and $Nb/(Pb_{80}\text{-}In_{20})$ junctions [48].

3.1. Problem Specification

When operated for extended periods, the two junctions showed different behavior [48]. Thus, the Nb/Pb type exhibited both an initial stabilization period (five to six months) of the main electrical parameters and then after about three years a progressive deterioration of the Pb film as a consequence of adsorbed ambient humidity. In contrast, the Nb/(Pb-In) junction possessed superior stability of the top electrode against degradation induced by humidity, and its electrical performance was of a higher standard.

3.2. Experimental Approach: Choice of Techniques and Specimen Configuration

It seems likely that the differing macroscopic performances of the two devices are related to the replacement of Pb metal with the Pb-In alloy as the top electrode material. The alloy not only possesses a higher chemical stability, but forms oxides at the interface that are better suited to the desired electrical characteristics, which remain unchanged over longer periods of time.

In deciding which of the techniques listed in Table 1 is (are) the best candidate(s) for providing direct experimental evidence of the differences in behavior, it should be remembered that there is an abrupt change, both chemical and physical, on going from one electrode to the other, in each junction. Because of the very small overall thickness of the devices, fabrication of cross sections would involve very low angle grinding, difficult to perform and handle. Such cross sections would be needed for SAM and ELS. It should also be noted that although SAM and ELS might provide direct access to the chemistry of the single electrodes they would be unable to reveal chemical changes at the interface between the top electrode and the insulating barrier since the interface cross-sectional width is much smaller than their spatial resolutions. Again, the versatility in surface specificity offered by SRPS and ELS could provide only a limited depth resolution compared to the overall thickness of the devices, and chemical quantitative analysis would be poor.

On the other hand, XPS and AES, both combined with Ar^+ etching, appear to be sensible choices in depth-profiling the chemistry of the devices, from the top electrode down to the oxide layer interface. Their combination offers appropriate elemental sensitivity and chemical sensitivity as well as reasonable quantification. Because no charging problems are expected to occur with these conducting samples (even when analyzing the very thin insulating barrier), the main artefacts to guard against are those associated with electron backscattering in AES spectra and with Ar^+ etching. The latter can be evaluated by comparison measurements carried out on standards.

3.3. Results

3.3.1. AES analysis

AES depth profiles of the barrier region of an Nb/Pb junction after 3 and 13 months are given in Figs. 1 and 2. For the younger sample the highest oxygen concentration occurs where the Pb content is still high and the amount of Nb negligible, suggesting a layer of a lead oxide. Below this depth, Nb increases, Pb becomes negligible quite abruptly, and the amount of oxygen decreases in a relatively smooth fashion. The effect of aging on the chemistry at the interface reveals itself in the two intensity maxima in the depth profile of oxygen, one after ca. 2 min sputtering and another, more prominently, after ca. 4 min, and in the intermixing of Pb and Nb. This interdiffusion between the two metals is clearly accompanied by the formation of two discrete oxide layers, marked by the oxygen maxima. Thus, aging induces both broadening and inhomogeneity in the insulating barrier of the Pb/Nb junction, and it is possible that these changes might affect the stability of the electrical parameters of the devices.

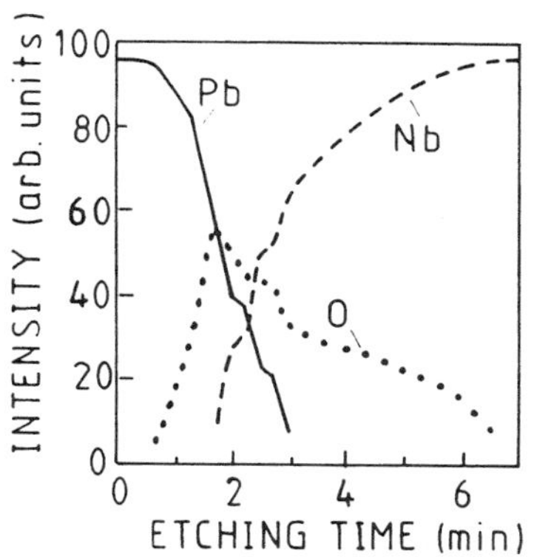

Figure 1. AES depth-profile (Ar^+, 60 μA cm^{-2}, 1.5 keV) of the insulant barrier of the Nb/Pb junction analyzed three months after fabrication. (Reproduced from Ref. 49 by permission of Elsevier Science, Lausanne.)

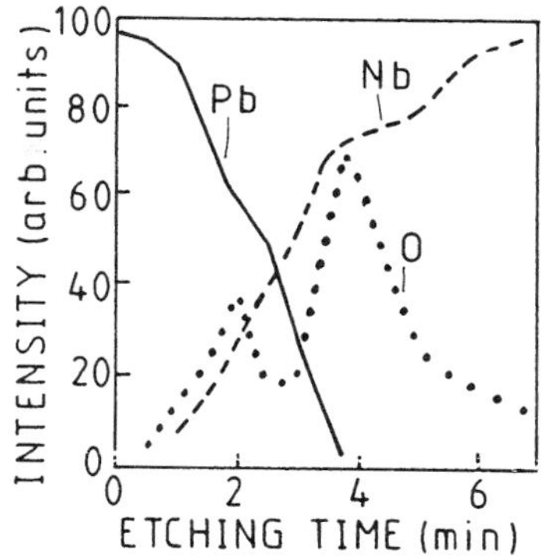

Figure 2. AES depth-profile (Ar^+, 60 μA cm^{-2}, 1.5 keV) of the insulant barrier of the Nb/Pb junction analyzed 13 months after fabrication. (Reproduced from Ref. 49 by permission of Elsevier Science, Lausanne.)

Comparison of the depth profiles in Figs. 1 and 2 might also shed some light on backscattering effects [19] in AES. Thus, the abrupt increase in oxygen in Fig. 1 after ca. 2 min sputtering might have been due partly to the presence of Pb, a strong backscatterer, which might have enhanced artificially the intensity of the O KVV transition. However, such an effect, if present, could not have accounted alone for the oxygen maximum after ca. 4 min sputtering in Fig. 2. This maximum is even higher and occurs where the Pb content is virtually negligible. Nor are backscattering effects significant enough to alter the distributions of the elements present in the Nb/(Pb-In) junction, to judge from the extended sputter profile shown in Fig. 3. For instance, oxygen dominates at the surface of the alloy electrode, due to the presence of oxides, but after a few min of etching, the oxygen content drops to a negligible level, while the intensities of both Pb, a strong backscatterer [19], and In, a medium backscatterer [38], remain high. When the interface is reached (70-min etching), the oxygen content increases dramatically, although the Pb signal drops to nearly zero intensity. In this junction, the insulating barrier (note change in time scale) is much more uniform than Fig. 2, as no sharp oxygen maxima are evident, and it appears to consist mainly of indium oxide. Preferential sputtering effects would undoubtedly have been present during AES depth profiling, yet observation of maxima and minima for oxygen suggests that such effects did not prevent information from being gained on the composition of the insulating barrier.

3.3.2. XPS analysis

XPS was used to study the chemistry of the elements in the two junctions, as well as to evaluate possible Ar^+ artefacts (e.g., chemical reduction). Figures 4 and 5 show the Pb $4f_{7/2}$ spectra at six different depths for the two junctions, respectively, while Fig. 6 shows the corresponding In $3d_{5/2}$ spectra for the Nb(Pb-In) junction. PbO and In_2O_3 were also analyzed for comparison purposes, both as-received and after being subjected to Ar^+ bombardment at energies and doses comparable to those used in the depth profiles. Pb $4f$ spectra from PbO are shown in Fig. 7; the effect of prolonged Ar^+ bombardment is partial reduction to metallic Pb. On the other hand, Ar^+ etching produced no appreciable changes in the In $3d$ spectra, and in particular no metallic phase appeared. The peak at

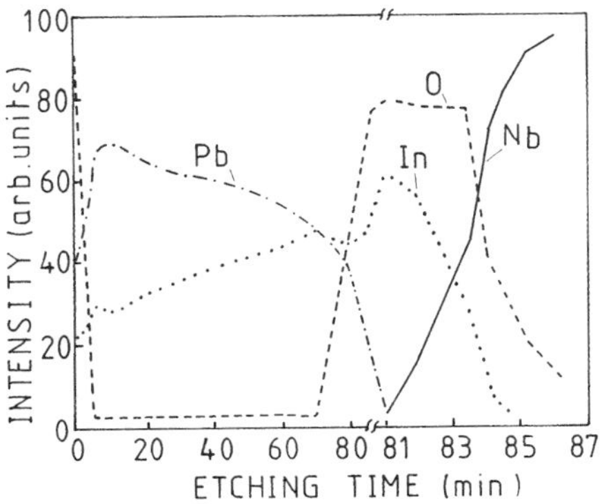

Figure 3. Extended AES depth-profile (Ar^+, 24 μA cm^{-2}, 1.5 keV) of the Nb/(Pb-In) junction. Note the change in scaling of the etching time at 80 min. (Reproduced from Ref. 49 by permission of Elsevier Science, Lausanne.)

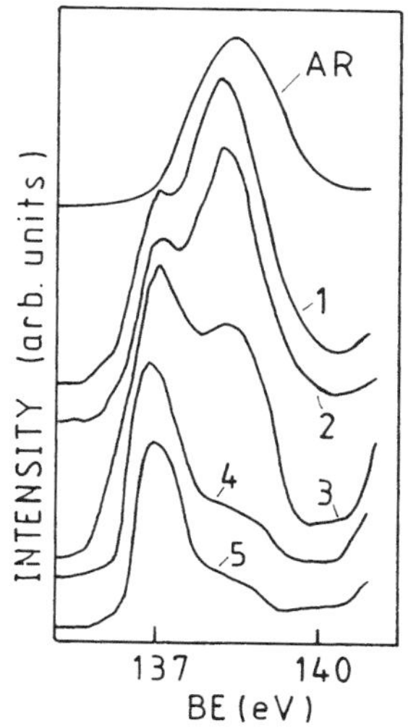

Figure 4. XPS Pb $4f_{7/2}$ spectra for the Nb/Pb junction (13 months old) for the following etching times (Ar^+, 80 μA cm^{-2}, 8 keV): AR = as-received; 1 = 1.5 min; 2 = 3 min; 3 = 6 min; 4 = 10 min; 5 = 14 min. (Reproduced from Ref. 49 by permission of Elsevier Science, Lausanne.)

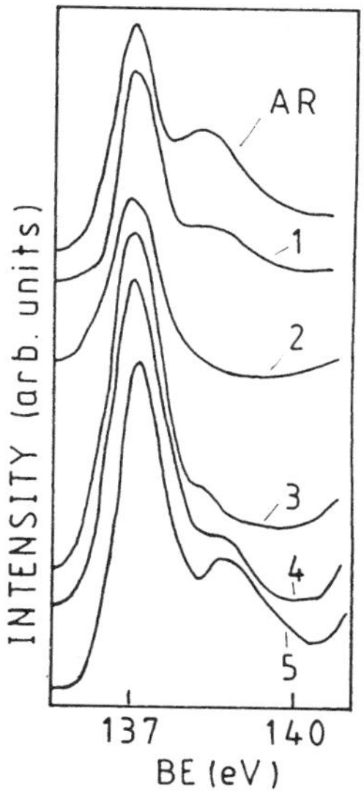

Figure 5. The analog of Fig. 4 for the Nb/(Pb-In) junction. (Reproduced from Ref. 49 by permission of Elsevier Science, Lausanne.)

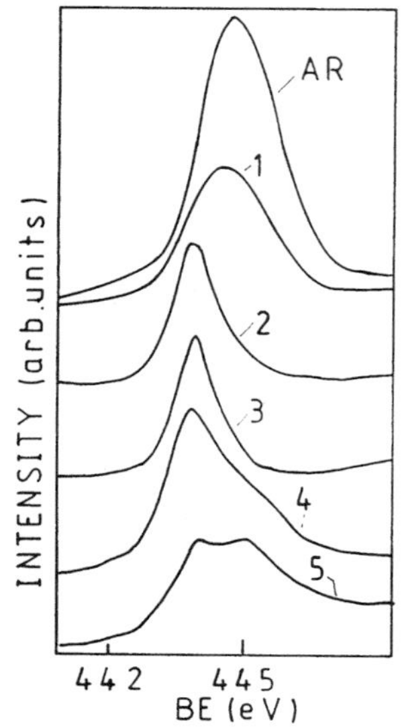

Figure 6. The analog of Fig. 5 for the In $3d_{5/2}$ spectra. (Reproduced from Ref. 49 by permission of Elsevier Science, Lausanne.)

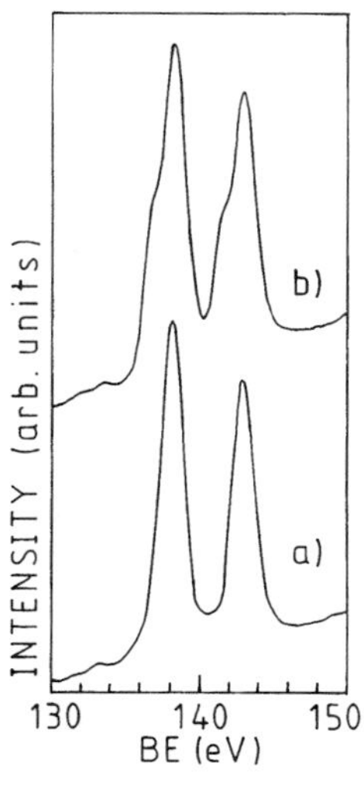

Figure 7. XPS Pb 4*f* spectra obtained from PbO analyzed: (a) as-received; (b) after prolonged Ar^+ etching.

ca. 137 eV in the Pb $4f_{7/2}$ region is assigned to metallic Pb, and that at 138.2±0.2 eV to PbO, for both junctions. For the Nb(Pb-In) junction, the component at 443.8 eV in the In $3d_{5/2}$ spectrum is assigned to metallic In, and that at ca. 445eV to In_2O_3. The O 1*s* spectra consisted of a broad band containing two components; one at about 530.5 eV arising from the oxides, and another at ca. 532.5 eV arising from either hydroxyl species in the as-received condition or to occluded oxygen contamination (occurring during metal film depositions) that appeared during sputtering.

The Nb 3*d* spectra (not shown) contained two maxima at ca. 203 and 206 eV, an energy separation characteristic of the 5/2-3/2 spin-orbit splitting. The 3*d* line shape was always complex because of the presence of various Nb oxides, probably NbO and Nb_2O according to the results reported by Grundner and Halbritter on the oxidation of Nb metal [50].

The evolution with depth of the Pb $4f_{7/2}$ line shape was quite different for the two junctions, as evident from Figs. 4 and 5. That from the Nb/Pb device was dominated by the component from the surface oxide down to the depth reached after the third sputtering sequence, i.e., at the insulant barrier. PbO was still detected subsequently, but with an ever-decreasing content relative to that of Pb metal. In contrast, Pb metal was easily detectable at the surface of the Nb/(Pb-In) device, and its level increased at the expense of PbO up to the second sputtering sequence, when the Pb $4f_{7/2}$ spectrum became characteristic of metallic Pb. At the insulator barrier, however, PbO species appeared again (third to fifth sequence).

3.4. Discussion

Although the chemical states of the Pb species in the two junctions would have been altered during profiling by Ar^+-induced reduction, i.e., $Pb^{2+} \rightarrow Pb^0$ (as found by other workers [51] and as confirmed by the comparison spectra in Fig. 7), the Pb spectra were nevertheless reasonably reliable indicators of varying chemical structure within the two devices. For example, PbO was found at the insulating barrier in Nb/(Pb-In) (Fig. 5), although it had disappeared after the second etching sequence, whereas it was never completely absent in the profile from the other junction (Fig. 6). In other words, although Ar^+ sputtering depleted the *actual* PbO content in the two depth profiles, such depletion did not in fact invalidate the qualitative chemical conclusions from the XPS analysis. Since it was found that Ar^+ sputtering had no measurable effect on In_2O_3, Fig. 6 can be taken as a reasonable indication of the In chemical species that were present in the Nb/(Pb-In) junction. In the light of the results of the analyses and taking into account both e_-^- and Ar^+-induced artefacts, the following interpretation can be made.

The insulating barrier consists not only of the oxides grown on the base Nb electrode but also of oxides the counterelectrode material. The method of oxidation used in the fabrication of both devices (thermal atmospheric growth) results in the conversion of Nb to Nb_2O_5 [48–50], which then reacts with the top electrode according to either $Nb_2O_5 + Pb \rightarrow Nb_xO_y + PbO$ or Nb_2O_5 + (Pb-In) $\rightarrow Nb_wO_z + PbO + In_2O_3$ (see, respectively, Figs. 4 and 5). It can be concluded that the characteristic electrical parameters of the two junctions are very dependent on the barrier chemical composition. In particular, those of the Nb/Pb device are not as "good" as those of the alloy junction, owing to the presence of PbO, and deteriorate with time owing to the progressive separation of two different oxide layers, as indicated by the depth profile of Fig. 2. In addition, the degradation with time of the top electrode of the Nb/Pb device is explained by the oxidation of Pb, not only at the surface, but right through to the insulant barrier (Fig. 4). Addition of In not only protects the counterelectrode from chemical degradation but also enhances the electrical performance of the Nb/(Pb-In) device. The latter effect can be explained by the fact that In forms a more homogeneous and chemically stable oxide layer at the barrier (Fig. 3). The beneficial protective effect of In can be inferred from the presence of Pb metal at the surface and from the absence of PbO in the internal regions (Fig. 5). As discussed in Chapter 19, these results are similar to those obtained in a comparison of surface compositional measurements on pure Pb and on a Pb-Sn alloy. In that case, too, it was demonstrated clearly that tin played a sacrificial rôle through preferential formation of SnO_2, and that the process limited significantly the extensive oxidation of lead.

4. OXIDATION OF $In_XGa_{1-X}As_YP_{1-Y}$ SEMICONDUCTORS BY NO_2

Quaternary materials $In_xGa_{1-x}As_yP_{1-y}$ ($0 \leq x \leq 1$; $0 \leq y \leq 1$) were studied with the following compositions: A, $In_{0.72}Ga_{0.28}As_{0.60}P_{0.40}$, and B, $In_{0.57}Ga_{0.43}As_{0.90}P_{0.10}$. They were prepared at the AT&T Bell Laboratories (Murray Hill, NJ,) using chemical vapor deposition onto InP (100) substrates. Although their band gaps are different, E_g(A) = 0.80 eV and E_g(B) = 0.95 eV, their lattice constants are virtually identical to those of InP. For comparison purposes GaAs and InP were also studied.

4.1. Problem Specification

As stated in the introduction, chemical and electronic passivation of III-V semiconductor surfaces is an important technological process and has given rise to much pure and applied research [52,53]. O_2 itself is a very inefficient passivating agent for GaAs; an exposure of up to 10^6 L is needed to form a substantial fraction of a monolayer of surface oxygen, and the O_2-grown oxide has poor electronic properties [54,55]. Recent studies by workers at the AT&T Bell Laboratories [56,57] and at the University of California at Santa Barbara [58] have shown that NO_2 passivates GaAs at low gas exposures and that the oxide layer is electronically passive. InP has many properties similar to GaAs, including a low oxidation efficiency with O_2 [59]. It is therefore interesting to study the oxidation of InGaAsP materials, InP, and GaAs, by NO_2, with respect to both the surface species formed and the layer thickness, and to compare the results with the surface chemistry of the same materials exposed to the atmosphere.

4.2. Experimental Approach: Choice of Technique and Specimen Configuration

As remarked in the introduction, there are requirements for both elemental and chemical information about the passive layer, but because of the possibility of lateral inhomogeneities, spatially resolved information may be needed as well. The depth distribution of passivation also needs analysis. In Table 1, it can be seen that analysis by XPS, XAES or SRPS would normally be the choice for both elemental and chemical information, but with those techniques no lateral spatial resolution is available. The chosen techniques were therefore SAM and scanning ELS, performed *in situ* in the same experimental chamber, and low-energy resolution AES carried out in a separate apparatus. The choice is appropriate for the following reasons [60]. First, SAM and AES possess sufficient elemental sensitivity to be able to measure oxygen uptake, with AES providing quantitative information. Second, the chemical sensitivity of SAM can be exploited to establish the surface chemical changes caused by NO_2 exposure, in other words which elements present in the surface become oxidized. Third, the surface specificity can be varied by recording Auger and electron energy loss signals extending over a KE range from ca. 100 to 2000 eV, for all samples. Finally, SAM and scanning ELS also provide information on the microchemical homogeneity of the surface.

4.3. Results

4.3.1. AES analysis

Oxidation of the binary and quaternary materials was followed by measuring AES O/(Ga+In+P+As) signal intensity ratios and converting them into oxygen monolayer coverages using elemental sensitivity factors [61]. One monolayer coverage is reached when the number of oxygen atoms equals that of surface atomic sites. Figure 8 shows the oxygen coverage on sample B as a function of exposure to NO_2 and O_2, the latter for comparison purposes. Results from oxidation of the other materials were very similar [60]. In all cases at least

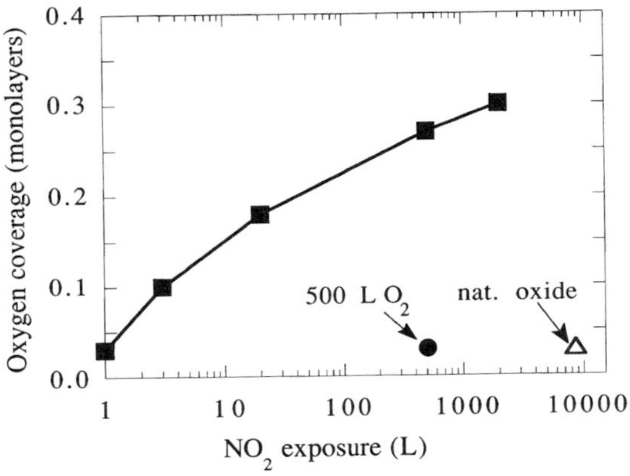

Figure 8. Oxygen coverage obtained from AES analysis for sample B as a function of NO_2 exposure. Also reported are comparison measurements carried out for the air-exposed and 500 L O_2-exposed surface. (Reproduced from Ref. 60 by permission of Elsevier Science, Amsterdam.)

two separate exposure runs were performed to test reproducibility, and the variations were within ±10% for all exposures of both NO_2 and O_2, except for the first 3 L NO_2 exposure in which the variations in oxygen coverage were as much as a factor of 2. The latter variation may have been due to the nonreproducible adsorption of residual gas inside the experimental chamber. There was no significant change in the stoichiometry of the cleaned surfaces throughout the measurements [60,61]. The initial growth of oxide had a slope of 0.03 monolayer of oxygen per 1 L exposure for the first exposure of 3 L. The growth rate leveled off to a saturation coverage of approximately 0.3 monolayer of oxygen.

The oxidation efficiency of exposure to NO_2 can be compared with the native oxide grown on the sample. Oxygen content of the native oxide before sputtering and the oxygen coverage after 500 L exposure are both shown in Fig. 8. Both coverages are lower than those following NO_2 exposure by factors of about 10. The depth profile of oxygen into the NO_2-grown oxide on sample B is shown in Fig. 9, where the oxygen coverage following exposure to 2000 L of NO_2 is plotted against the estimated depth of material removed. The sputter yield was ca. 1.06 atom per ion, according to reports [62,63]. Based on that assumption, it can be calculated, to within a factor of 2, that the thickness of the oxide layer produced by prolonged NO_2 exposure is 1 nm. In all cases, the oxidation and sputtering treatments did not affect significantly the substrate stoichiometry.

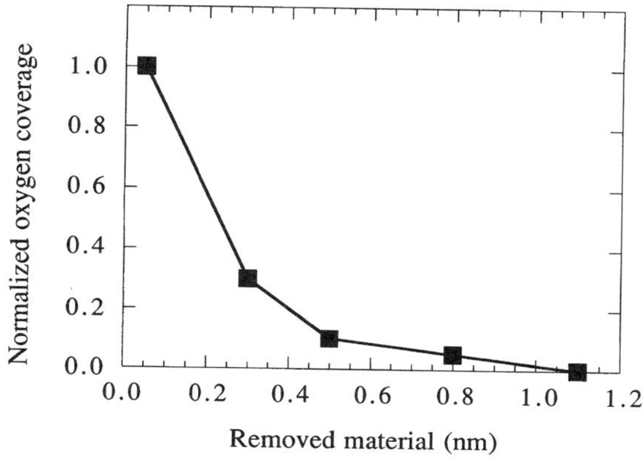

Figure 9. Ar^+ sputter profile of the 2000-L NO_2 treated quaternary B. (Reproduced from Ref. 60 by permission of Elsevier Science, Amsterdam.)

4.3.2. SAM and scanning ELS analysis

The next step involved identification by SAM and ELS of the oxidic species formed on the surfaces following NO_2 exposure. Spectra were recorded corresponding to the conditions: (1) as-received, i.e., with the surface still covered with the native oxide layer; (2) after Ar^+ etching, which removed all the native oxides and other contaminants; (3) and after 2000 L NO_2 exposure. The Auger transitions extended in KE from 120 (P LVV) to about 1850 eV (P KLL), corresponding to inelastic mean free paths in the approximate range 0.4 to 4 nm [64], as seen in Fig. 10. Choice of appropriate transitions thus allowed the surface chemistry over this depth range to be studied. The low-KE Ga (≈35 eV) and As (≈55 eV) signals provided little diagnostic information, since both are inherently broad, and NO_2-induced changes were barely discernible. The $L_3M_{4,5}M_{4,5}$ transitions were therefore used for both As and Ga.

Several Auger point spectra were recorded for each material, showing that all were laterally homogenous except for the occasional contaminant particle. In addition, time-dependent spectra confirmed that the chemistry of the surfaces was not altered by electron irradiation. In general the spectral components were poorly resolved in the expanded spectra, even when the analyzer was operated at a nominal resolution of less than 1 eV. Curve fitting would in principle have assisted in resolving the spectra, but the complex shapes involved in the multi-component spectra would have made interpretation problematic.

Figure 11 shows the P LVV spectra recorded from InP (the corresponding spectra from the two quaternary materials were very similar). In the as-received condition (a), two broad components can just be distinguished. The first, cen-

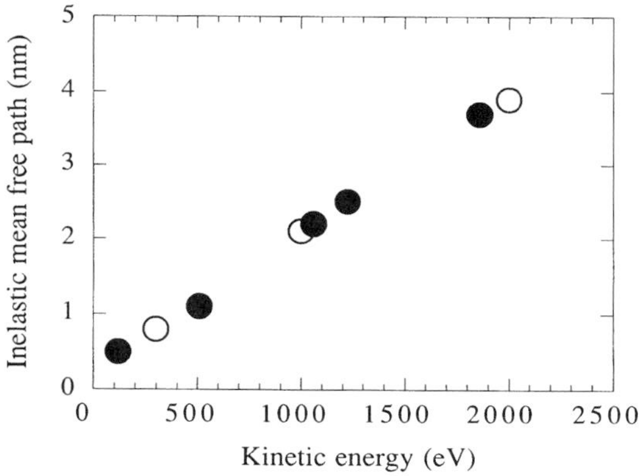

Figure 10. Electron inelastic mean free paths calculated for InP and GaAs from the model of Tanuma et al. [64]. (Reproduced from Ref. 60 by permission of Elsevier Science, Amsterdam.)

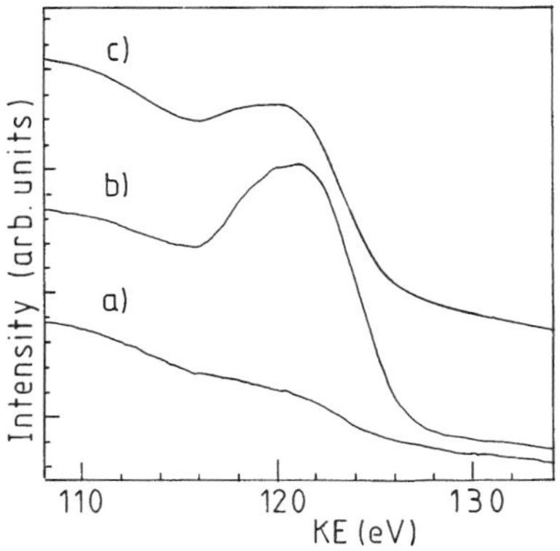

Figure 11. P $L_{2,3}VV$ spectra measured with SAM from InP analyzed in the following conditions: (a) as-received; (b) atomically clean; (c) after 2000-L NO_2 exposure.

tered at about 121 eV, is characteristic of unoxidized InP, whereas the other, at about 110 eV, is ascribed to phosphorus oxide. Trace (b) for the clean surface, was characterized by the unoxidized InP component which dominated the spectrum, but a broad feature was still apparent near 110 eV. Despite its virtual coincidence with the oxidized component in trace (a), the latter feature is in fact due to a convolution of *s-p* and *p-p* orbital states characteristic of atomically clean InP surfaces [65]. Exposure to 2000 L of NO_2 produced, as seen in curve (c), an apparent increase in the oxidized component at the expense of the un-

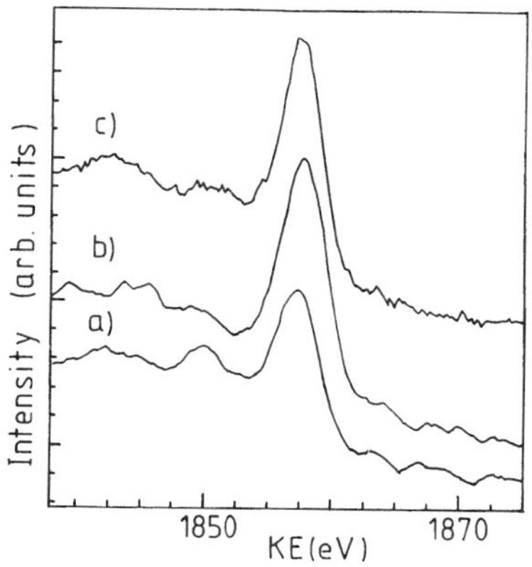

Figure 12. P KLL spectra measured with SAM from InP analyzed in the following conditions: (a) as-received; (b) atomically clean; (c) after 2000-L NO_2 exposure.

oxidized. The P KLL spectra corresponding to the same conditions are shown in Fig. 12. They were recorded with relatively poor energy resolution (3 eV), because at high KEs the transmission of the analyzer reduces dramatically when operated in the FAT (Fixed Analyzer Transmission) mode [66]. All three spectra were dominated by a component at about 1858 eV characteristic of unoxidized InP, by analogy with reports in the literature on the P KLL spectrum of GaP [67]. Another, minor, broad component was visible at ca. 1850 eV, from both the as-received (a) and 2000 L NO_2-exposed (c) surfaces, but not from the clean surface, and can be ascribed to oxidized phosphorus species.

The As $L_3M_{4,5}M_{4,5}$ spectra recorded from sample A are given in Fig. 13. Exposure to NO_2 (spectrum (c)) did not modify significantly, either in shape or location, the spectrum (a) relative to the clean surface, and only a slight chemical shift (ca. –0.5 eV) was observed; this can be accounted for by the change in local chemical environment as nearby species became oxidized [60]. In particular, there was no component at about 6 eV to the lower KE side of the maximum, which would have indicated the presence of arsenic oxides [68]. On the other hand, that component, marked by an asterisk, was observed clearly in the spectrum recorded from the as-received surface. The corresponding Ga LMM spectra for sample A are shown in Fig. 14. NO_2-exposure, spectrum (c), caused formation of Ga oxides [69], evidenced by the presence of a component at about 3.5 eV to the low-KE side of the maximum. The extent of Ga oxidation was greater after NO_2-exposure than in the as-received condition (a). Evidence of NO_2-induced oxidation was also observed in the In MNN spectra from sample B shown in Fig. 15. The spectral changes consisted of an apparent shift of

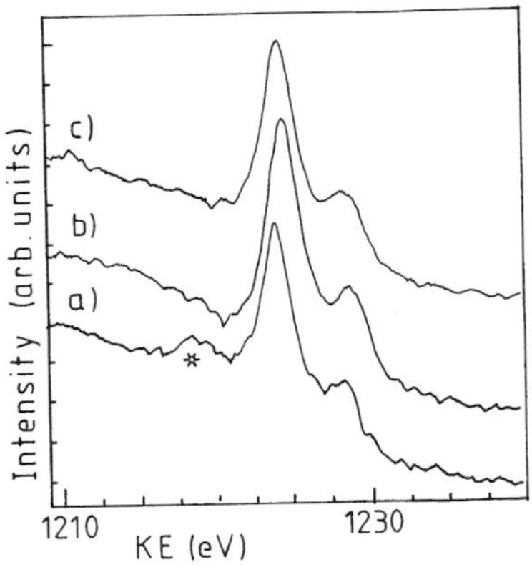

Figure 13. As $L_3M_{4,5}M_{4,5}$ spectra measured with SAM from sample A analyzed in the following conditions: (a) as-received (the asterisk marks the component from oxidized arsenic); (b) atomically clean; (c) after 2000 L NO_2 exposure.

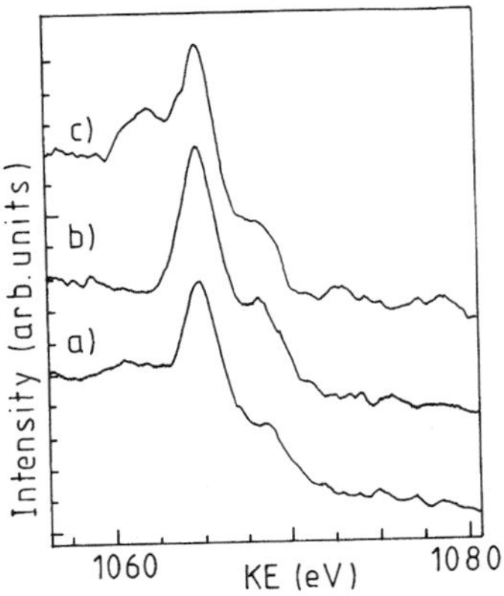

Figure 14. Ga $L_3M_{4,5}M_{4,5}$ spectra measured with SAM from sample A analyzed in the following conditions: (a) as-received; (b) atomically clean; (c) after 2000 L NO_2 exposure.

the band maxima to lower KE, by about 0.5 eV, and an overall broadening of the bands. These changes were due to an overlap of the features characteristic of unreacted InP with the envelope associated with oxidized indium species [70] at lower KE. The In MNN spectrum (a) recorded from the air-exposed sample was indicative of a much greater extent of oxidation.

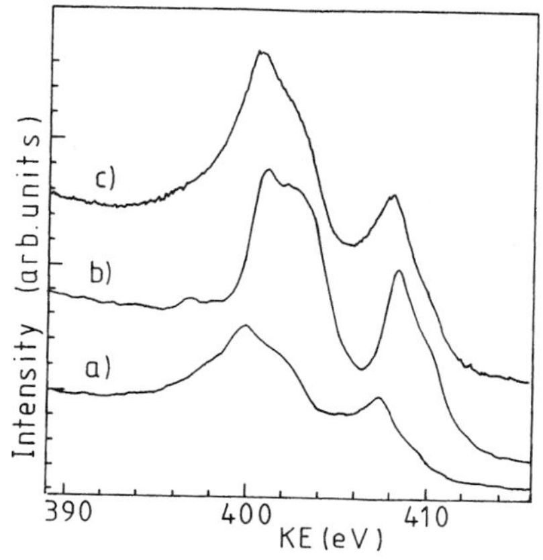

Figure 15. In MNN spectra measured with SAM from sample B analyzed in the following conditions: (a) as-received; (b) atomically clean; (c) after 2000 L NO_2 exposure.

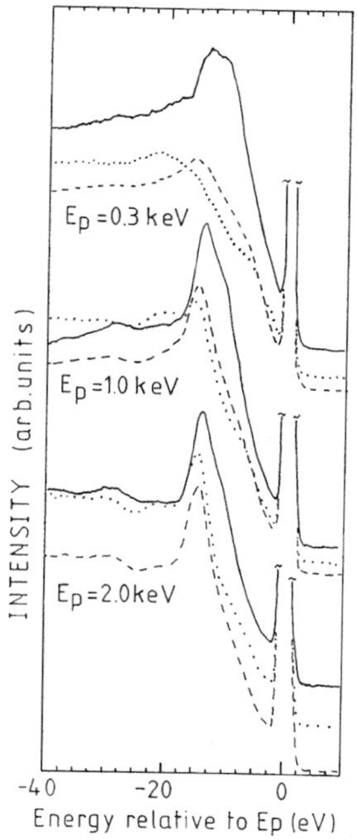

Figure 16. ELS spectra at three different primary energies from sample A analyzed in the following conditions: . . ., as-received; ——, atomically clean; - - -, after 2000 L NO_2 exposure.

NO_2-induced spectral changes were also observed in ELS. Three primary energies (E_p) were chosen: 300, 1000 and 2000 eV, corresponding to λ values of about 1, 2 and 4 nm, respectively (see Fig. 10). Figure 16 shows the results from sample A, which were similar to those from the other materials. The main features in these spectra are the bulk (BP) and surface (SP) plasmons at −14 and −11 eV, respectively. For the oxidized surfaces only, there are features arising from the O_{2p}→CB transition (i.e., one-electron excitation from the oxygen $2p$ orbital to the bottom of the conduction band (CB)) at −6 eV [71], and a series of broad, nonreproducible features associated with carbon contamination. In general, NO_2 oxidation caused the disappearance of the SP, and a shift of the BP to greater loss by about 1 eV. In addition, the feature associated with the O_{2p}→CB excitation, particularly evident at E_p = 300 eV, appeared.

4.4. Discussion

The combination of spectroscopies with different surface specificities provides a clear picture of the oxidation of these materials by NO_2. AES analysis confirms that NO_2 is a very efficient oxidant (Fig. 8) and the thickness of the oxide layer can be estimated (Fig. 9). The only destructive effects associated with the high-current electron beam were limited to the uncertainty of oxide growth in the very early stages of NO_2 exposure of the atomically clean surfaces. SAM measurements have answered the fundamental question as to whether NO_2 oxidation involves all the chemical elements of the quaternaries or only some of them. Thus it seems that As remained unchanged, even after exposure to 2000 L of NO_2, while all the other elements, and in particular Ga, developed oxidic phases. This information is relevant to technological applications of these materials since Ga oxides are electronically passive [58]. The results can be compared with those of Huang et al. [58]. In their elegant study, they also found that NO_2 is an efficient oxidant of GaAs, but their use of low-energy-resolution AES did not enable them to identify which element became oxidized; evidence for the oxidation of Ga was provided by a separate high-resolution electron energy loss (HREELS) analysis [58].

Further information has come from Auger and electron energy loss features spread over an extended KE range which have allowed the depth distribution of oxidized species in the materials to be studied. According to the sputter profile shown in Fig. 9, most of the oxidic species reside within the first 1–2 nm of the surface. This is also the λ value associated with the O_{2p}→CB feature seen at E_p = 300 eV in Fig. 16 after exposure to 2000 L of NO_2. Because of the poor chemical sensitivity of ELS it is difficult to decide how far the latter feature may be considered a quantitative indicator of NO_2-induced surface chemical change. However, its intensity decreased somewhat at E_p = 1000 eV ($\lambda \approx$ 2.5 nm) and became virtually zero at E_p = 2000 eV ($\lambda \approx$ 3.8 nm). SAM has greater chemical sensitivity, and more definite chemical changes were observed in the Auger spectra associated with λ ranging from ca. 0.5 nm (P LVV, Fig. 11) to ca. 2.5 nm (Ga LMM and As LMM, Figs. 13 and 14). On the other hand, evidence of oxidation was much less distinct in the P KLL spectra ($\lambda \approx 3.5$ nm, Fig. 12).

Combined use of SAM and scanning ELS techniques was found to be useful when microchemical analyses were required. All the spectra shown were taken from those areas of the semiconductor surfaces that were free of any obvious foreign material, as monitored by SEM imaging and Auger point spectra. The importance of this is shown in Fig. 17, in which two contaminant grains are seen lying on clean GaAs. Auger and ELS point spectra recorded within the two grains showed that they consisted entirely of carbonaceous species, while outside the grain the spectra were typical of atomically clean GaAs. The microchemical inhomogeneity present in the area of Fig. 17 is confirmed by recording the scanning ELS image of GaAs, based on the bulk plasmon feature, and the Auger carbon image, then overlaying the images as shown in Fig. 18.

5. THE Si/SiO$_2$ INTERFACE

Two samples were studied: (1) a thin film (ca. 150 nm in thickness) of SiO_2 grown on a Si(100) wafer, obtained from a commercial source, and (2) a thin film (about 70 nm thick) of silica on silicon grown thermally at 1000 K for several hours in oxygen in the laboratory. The latter will be referred to as "aged

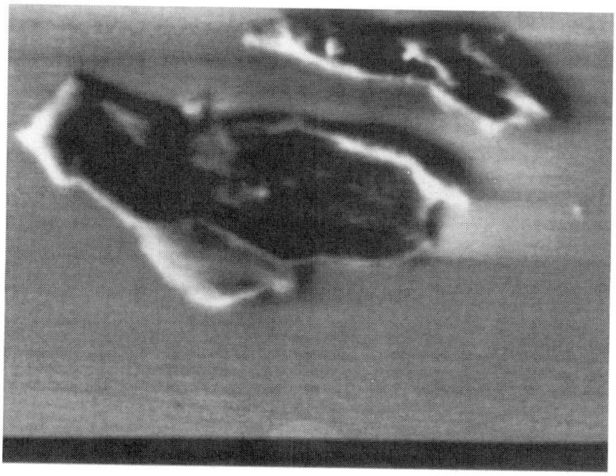

10 μm

Figure 17. SEM image of two contaminant grains lying on the GaAs surface.

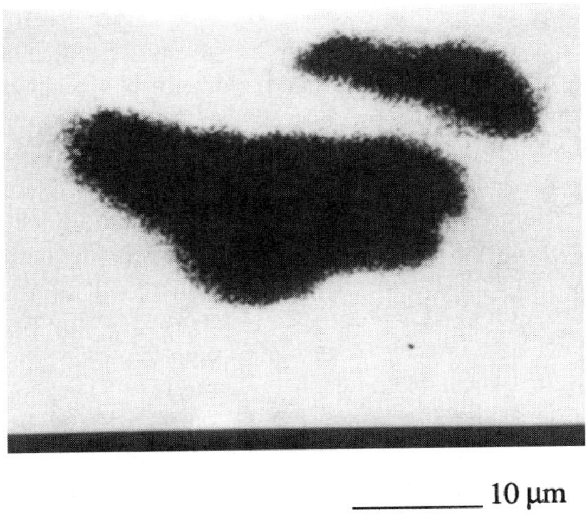

Figure 18. Overlay of the carbon Auger image (black field) and the scanning ELS image (recorded at $E_p = 2$ keV using the bulk plasmon representative of clean GaAs) (white field), from the same area of Fig. 17.

silica" in the following since it was analyzed some three years after preparation. Comparative measurements to study possible Ar^+ induced effects were performed on a high-purity silica powder.

5.1. Problem Specification

The SiO_2/Si system is one of fundamental importance in the microelectronics and semiconductor industry [72–74]. It is now well established that the transition layer between Si and SiO_2 consists of SiO_x ($x < 2$) suboxides [73,75] which enable chemical continuity to be maintained across the interface as a result of Si–Si and Si–O bonding states. The transition layer is normally studied by a combination of surface-specific techniques with Ar^+ etching, which may cause surface damage resulting in chemical reduction, since SiO_x suboxides have been reported in many studies of Ar^+ etched silica surfaces [76–84]. Ar^+-induced SiO_x species are obviously undesirable since they may obscure the true profile of the "genuine" suboxides across the interface. It is therefore important for the analyst to establish any chemical modifications induced by Ar^+ etching in pure SiO_2 as well as in the Si/SiO_2 transition layer.

Another problem is that of aging. As well as an accumulation of carbon contamination, hydrated species have been observed at the surfaces of silica samples prepared by thermal oxidation and then left a few years in contact with ambient humidity [85,86]. The chemical change could be serious in SiO_2-

containing microelectronic devices, since Si–OH bonding states (silanols) might propagate through acid-base reactions [87] along the surface as a function of environmental service conditions. Although the hydration process will not be followed in detail, the qualitative and quantitative chemical characterizations of silanol species are of interest, as also are the vertical and lateral distributions of the species within the silica surface.

5.2. Experimental Approach: Choice of Technique and Specimen Configuration

When studying the modifications induced by Ar^+ on silica surfaces, techniques must be chosen that do not themselves cause any appreciable surface damage. In general, therefore, electron probe techniques such as AES cannot be used, although it is possible that SAM might be acceptable on account of the low electron currents used. However, the chemical information from SAM is limited. The chemical limitation is even worse in ELS, where the distinction between SiO_x and undamaged silica might be unobservable. On the other hand, combined use of SRPS and XPS would have the following advantages: (1) the chemical shift between Si suboxides and silica can easily be observed [81–84]; (2) the depletion of surface oxygen following Ar^+ impact can be accurately quantified [81,83,84]; (3) both O KVV and Bremsstrahlung-induced Si KLL Auger spectra can be recorded using XPS, allowing Auger parameters for both silicon and oxygen to be derived [88]; (4) XPS and XAES combined with variation in takeoff angle allow nondestructive depth profiles to be obtained; (5) in SRPS the photon energy can be chosen so that Si 2*p* photoelectrons have a KE (i.e., ca. 100 eV) corresponding to the minimum in the "universal curve" of ALs [83,84].

For the study of silanol species on aged silica, a combination of photoemission and SAM has advantages [85,86]. In SRPS the photon energy can be tuned in such a way that the Si photoelectron spectra appear at different KEs, corresponding to different ALs, allowing the vertical distribution of Si–OH bonding states within "anhydrous" Si–O bonding states to be measured [85]. The Si 2*p* chemical shift between the two bonding states is readily measurable [85]. XPS quantitative analysis can also be used for measuring the extent of surface hydration, while SAM analysis can reveal lateral inhomogeneities in the distribution of silanol species at the surface [86].

5.3. Results

5.3.1. Effects of Ar^+ bombardment

XPS and XAES spectra were recorded under two conditions: (a) at constant takeoff angle ϕ of 90° (normal emission), using the Bremsstrahlung component of unmonochromatized Al K_α radiation to excite the Auger Si KLL transition;

Table 2. Binding Energies, XAES Kinetic Energies, and Auger Parameters Obtained at Normal Electron Emission for Silica Subjected to Ar^+ etching

Ar^+ energy (keV)	Si 2*p* (eV)	Si 2*s* (eV)	Si $KL_{2,3}L_{2,3}$ (eV)	α^*_{Si}[a] (eV)	O 1*s* (eV)	O KVV (eV)	α^*O[a] (eV)
As-received	103.8	154.8	1607.6	1711.4	533.1	506.5	1039.6
1	103.8	154.8	1607.8	1711.6	533.0	506.7	1039.7
2	103.8	154.8	1607.7	1711.5	533.1	506.7	1039.8
3	103.8	154.8	1607.5	1711.3	533.0	506.8	1039.8
4	104.0	155.0	1607.6	1711.6	533.1	506.7	1039.8
5	103.9	154.8	1607.5	1711.4	533.0	506.7	1039.7
	102.2	153.0	1610.5	1712.7			
6	103.9	154.8	1607.5	1711.4	533.0	506.7	1039.7
	102.2	153.2	1610.5	1712.8			
7	103.8	154.9	1607.7	1711.5	533.1	506.7	1039.8
	102.2	153	1610.6	1712.8			
8	103.9	154.8	1607.5	1711.4	533.1	506.7	1039.8
	102.1	153.2	1610.7	1712.8			

Source: From Ref. 83 by permission of the American Institute of Physics.
[a]The Auger parameters are defined as α^*Si = KE(Si $KL_{2,3}L_{2,3}$) + BE(Si 2*p*) and α^*O = KE(O KVV) + BE(O 1*s*).

and (b) at values of ϕ chosen in the range 20–90° in order to vary the surface sensitivity, using monochromatized Al K_α x-rays to achieve better spectral resolution. Table 2 lists the XPS and XAES results recorded from silica under condition (a) in the as-received condition as well as after bombardment by Ar^+ at energies up to 8 keV. For each bombardment energy a fresh SiO_2 surface was used in order to avoid cumulative damage effects.

As the Ar^+ energy was increased, both photoemission and Auger Si spectra revealed additional components due to Ar^+-induced phases, i.e., suboxides created as a result of the loss of surface oxygen and of subsequent Si–Si recombination of broken Si–O bonds [81–83]. The precise energetic positions of the damage-induced peaks could be identified with confidence only for Ar^+ energies ≥5 keV; they lie at ca. 102.2 eV in the Si 2*p* spectra, at ca. 153 eV in the Si 2*s* spectra, and at ca. 1610.5 eV in the Si KLL spectra. Neither the O 1*s* nor O KVV spectra were affected by Ar^+, in each case maintaining values characteristic of undamaged SiO_2. Similarly an additional α^*Si value of ca. 1712.8 eV emerged for Ar^+ energies ≥5 keV, while α^*O was virtually constant at ca. 1039.8 eV for all Ar^+ treatments. The new value of α^*Si was intermediate between those of silica and silicon. The different behavior of the silicon and oxygen spectra is also shown in Fig. 19 where the FWHM spectral values as a

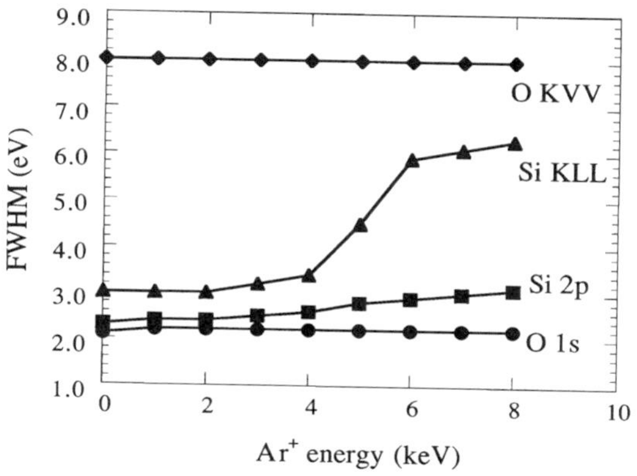

Figure 19. Full-width at half-maximum (FWHM) values of XPS and XAES spectra from silica as a function of Ar^+ energy. (Reproduced from Ref. 83 by permission of the American Institute of Physics.)

function of Ar^+ energy are plotted. As the Ar^+ energy was increased, both the photoemission and Auger Si spectra broadened significantly, particularly for values ≥5 keV, whereas the oxygen spectra were insensitive to all Ar^+ treatment. This constancy of the FWHM for both O 1*s* and O KVV signals implies that differential charging effects can be excluded as being responsible for the broadening of the silicon signals [44,45,81,89,90].

The formation of Ar^+-induced suboxides at the surface was also followed quantitatively using XPS [81,83]. The oxygen-to-silicon atomic ratio was measured using a "first-principles approach," with an estimated accuracy of ±10% [36,46,84,89,90]. Figure 20 shows the n_O/n_{Si} atomic ratio as a function of etching energy. The ratio on the unetched surface was 2.08, which is equal to the theoretical value within experimental uncertainty. There was little change at Ar^+ energies up to 2 keV, but at higher Ar^+ energies the ratio decreased progressively, and after the 8 keV treatment the ratio was as low as 1.28.

In order to enhance surface specificity, SRPS Si 2*p* spectra from 2 keV Ar^+-treated silica using a photon energy of 210 eV were recorded, in addition to monochromatized Al K_α XPS Si 2*s*, Si 2*p* and O 1*s* spectra at a takeoff angle $\phi = 20°$. The corresponding average escape depths, either ALs ($\phi = 90°$) or reduced escape depths, $D = AL \sin \phi$ ($\phi = 20°$), are plotted in Fig. 21. Ar^+ at 2 keV was chosen because in the measurements at normal emission no apparent chemical damage was observed, either qualitatively or quantitatively. However, there was clear evidence for the occurrence of SiO_x suboxides in Ar^+-treated silica even at 2 keV when the measurements were performed with enhanced surface specificity. The higher the specificity, the greater was the extent of chemical

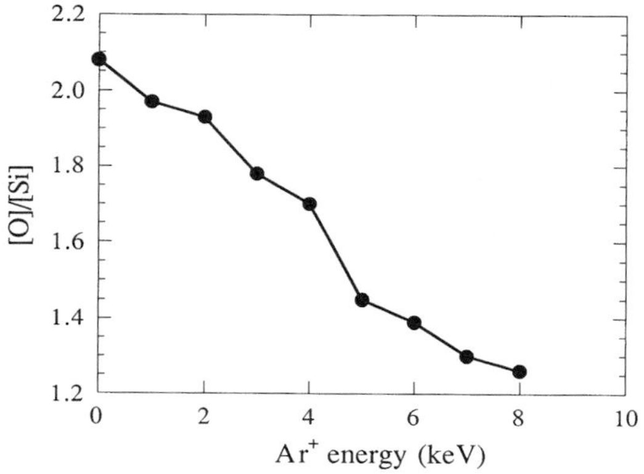

Figure 20. XPS oxygen-to-silicon atomic ratio obtained from silica as a function of Ar^+ energy. (Reproduced from Ref. 83 by permission of the American Institute of Physics.)

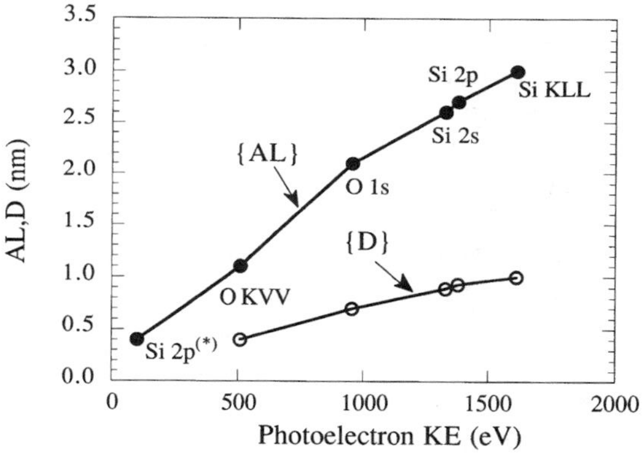

Figure 21. Electron attenuation lengths (AL) and electron escape depths (D, takeoff angle = 20°) of the XPS transitions and XAES transitions recorded from silica. The asterisk marks the AL for Si 2*p* electrons emitted in SRPS experiments. (Reproduced from Ref. 83 by permission of the American Institute of Physics.)

change observed [82,83], as shown clearly in Fig. 22. The Si 2*p* FWHM in the SRPS spectra at $h\nu = 210$ eV ($AL \approx 0.4$ nm) grew from 2.0 to 3.1 eV on going from the as-received to the 2 keV Ar^+-etched condition (Fig. 22a), whereas the Si 2*s* FWHM grew from only 2.38 to 2.72 eV in the XPS spectra recorded at $\phi = 20°$ ($D \approx 1$ nm) after the same treatment (Fig. 22b).

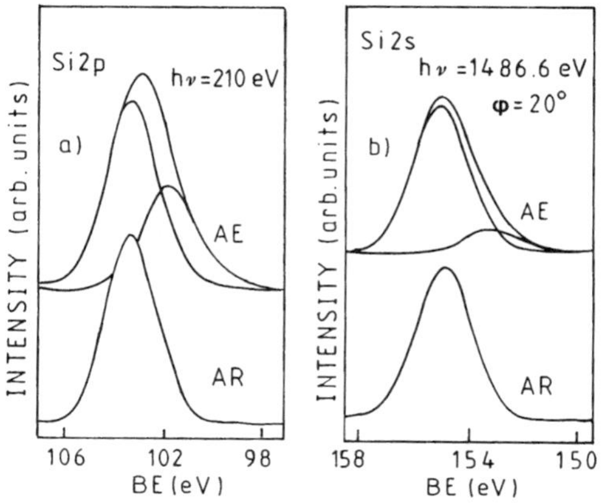

Figure 22. (a) SRPS Si 2*p* spectra of silica analyzed as-received (AR) and after etching (AE). (b) XPS Si 2*p* spectra (takeoff angle = 20°) for the same two conditions. (Reproduced from Ref. 82 by permission of Elsevier Science, Amsterdam.)

These width variations can be explained on the basis of the varying compositions that can be derived using peak-fitting procedures. Thus, an SiO_x/SiO_2 ratio of 0.56 was derived from the SRPS spectrum, compared with a value of only 0.10 from the Si 2*s* XPS spectrum. As before, the XPS O1*s* spectra recorded at enhanced surface specificity showed no changes as a result of the etching since no FWHM differences were observed [82].

From the results so far it can be concluded that the thickness of the damage layer produced by 2 keV Ar^+ is measurable more or less accurately according to the surface specificity of the operational conditions. In addition, it has been found there were no significant changes in the line shapes of any of the spectra following 2 keV Ar^+ bombardment, when recorded at $\phi = 90°$. It therefore seems correct to combine XPS (and XAES) analysis at normal emission with 2 keV Ar^+ in order to depth-profile an SiO_2/Si sample.

5.3.2. Interfacial suboxides

Having established the conditions for Ar^+ depth-profiling of Si/SiO_2 samples, attention was then turned to examining the possible chemical changes originating from the presence of a "genuine" SiO_x species in the transition layer of the interface. Table 3 lists the results obtained at selected stages of a depth profile. Included also are measurements of the energy separation, denoted Δsat, between the Si 2*p* maximum and the first loss feature on the low-KE side [91]. Figure 23 shows the evolution with etching time of the FWHMs of

Table 3. Binding Energies and Auger (XAES) Data Recorded from the Si/SiO_2 Interface as a Function of 2 keV Ar^+ Irradiation Time

Etch time (min)	Si 2*p* (eV)	Δsat[a] (eV)	O 1*s* (eV)	O KVV (eV)	Si $KL_{2,3}L_{2,3}$ (eV)	α*O (eV)	α*Si (eV)	[O]/[Si][b] (eV)
0	103.9	24	533.1	506.5	1607.5	1039.6	1711.5	2.11
1	103.8	24	533.0	506.5	1607.6	1039.6	1711.5	2.04
6	103.8	24	533.1	506.6	1607.6	1039.7	1711.4	1.99
21	103.8	24	533.0	506.7	1607.6	1039.7	1711.4	1.98
140	103.8	24	533.1	506.7	1607.6	1039.8	1711.4	2.04
150	103.8	16	533.0	506.9	1608.5	1039.6	1711.9	1.10
	99.3				1616.4		1715.7	
170	99.3	17	—	—	1616.5	—	1715.8	0

[a]Δsat denotes the energy separation between the Si 2*p* maximum and the first loss feature to the low-E side.
[b]Oxygen-to-silicon atomic ratios derived from XPS intensities.
Source: From Ref. 84 by permission of Elsevier Science, Amsterdam.

the Si 2*p* and Si KLL peaks from oxidized silicon species. The results can be summarized as follows: (1) before the interface (t = 0 to 140 min), no appreciable differences were evident in any of the spectra except for a slight increase in FWHM of both peaks; (2) at the interface (t = 150 min), not only did new spectral features associated with elemental silicon emerge, but much more importantly the FWHMs of the oxidized silicon species became considerably greater than any Ar^+-induced broadening previously observed. In other words, the presence of genuine suboxide species in the transition layer could be observed in the spectra. This observation is made clearer in Fig. 24, where are reported, respectively, Si 2*p* and Si KLL spectra recorded at various stages of the depth profile.

Surface-hydrated species

The idea that Si–OH bonding states might be present on aged silica surfaces was first suggested by unusual SRPS Si 2*p* spectra, which contained two distinct components: the first, dominant, component had a BE characteristic of silica, at 103.9 eV, whereas the second lay at 105.2 eV [85]. Before attempting an interpretation, it is important to rule out differential charging effects as being responsible for the high-BE component. In fact, surface charging was not observed either in SRPS measurements of SiO_2/Si samples with silica films even thicker (ca. 150 nm) than that of aged silica (ca. 70 nm), or in analysis by SAM, i.e., a technique which is much more likely than photoemission to induce

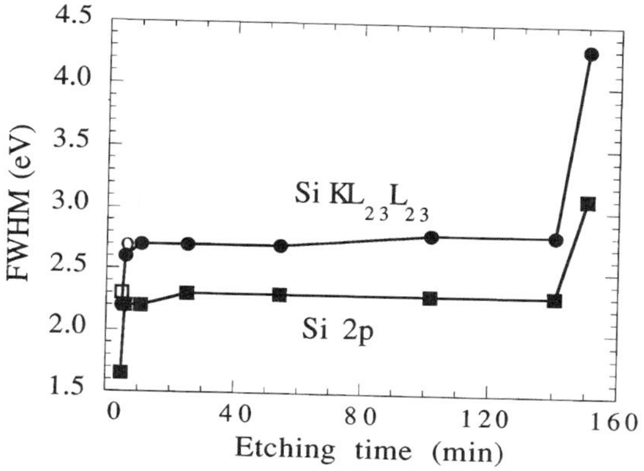

Figure 23. FWHM of XPS Si 2*p* and XAES Si KLL spectra of the oxidized component of silicon at the Si/SiO_2 interface as a function of the 2 keV Ar^+ etching time. Open symbols indicate the values from silica powder analyzed as-received. (Reproduced from Ref. 84 by permission of Elsevier Science, Amsterdam.)

charging effects [4,6,86]. As far as can be ascertained, Si 2*p* photoemission data from silicon hydroxides have not been reported in the literature, and the assignment of the "new" component is therefore based on the following arguments. Both the size and the sign of the chemical shift between the high-BE component and that of "anhydrous" silica compare well with the reported chemical shifts in photoemission M 2*p* spectra between the hydroxides and oxides of such metals as M = Cr, Fe and Ni [92,93]. All these metals possess an electronegativity close to that of silicon [94]. As a consequence it is concluded that the surface of aged silica contains anhydrous $[SiO_4]^{4-}$ groups and partially hydrated $[SiO_n(OH)_x]^{n-}$ groups, where the conditions $0 < x < 4$, $n = 4 - x$ satisfy the electrical balance.

Interestingly, the relative intensity of the high-BE component varied as a function of photon energy, as illustrated in Fig. 25. Since a change in photon energy corresponds to a change in surface specificity, the evolution of the Si 2*p* line shape in Fig. 25 suggests that there was a vertical inhomogeneity in the hydration content of the surface layers of the material. This inhomogeneity could be examined by measuring the [Si–OH]/[Si–O] ratio (the notations of the two species have been modified for the sake of simplicity) as a function of photoelectron KE from 25 to 135 eV. The sampling depths corresponding to that range can be expressed in terms of either the average escape depth ($D = AL \sin \phi$), from the compilation of Seah and Dench [13], or the inelastic mean free path, λ, from the model of Tanuma et al. [64]. The results are summarized in Fig. 26. The [Si–OH]/[Si–O] ratio in-

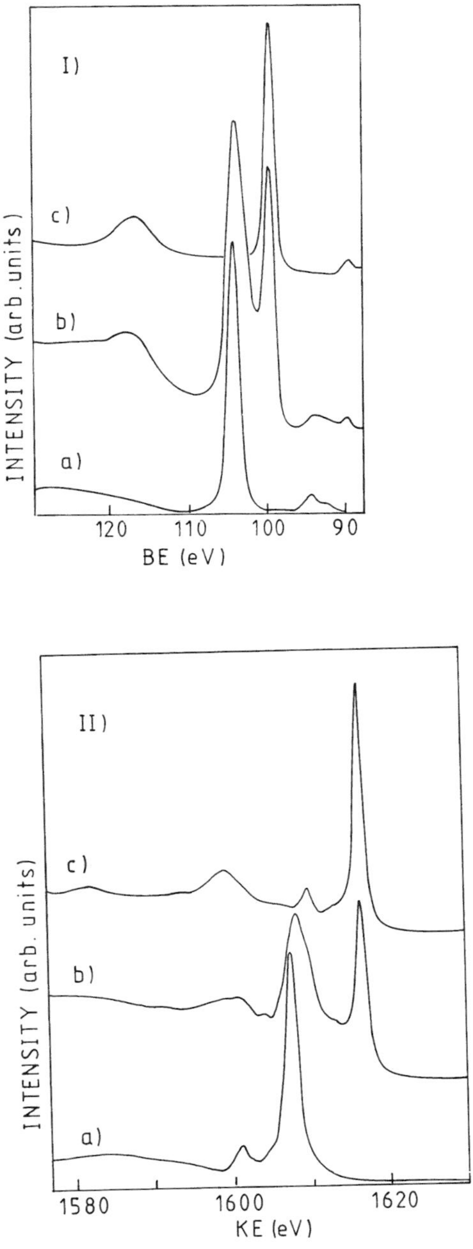

Figure 24. (Panel I) XPS Si 2*p* spectra recorded from the Si/SiO_2 interface after selected 2 keV Ar^+ etching times, *t*: (a) $t = 0$ min; (b) $t = 140$ min; (c) $t = 170$ min. (Panel II): Analog of panel I for XAES Si KLL spectra. (Reproduced from Ref. 84 by permission of Elsevier Science, Amsterdam.)

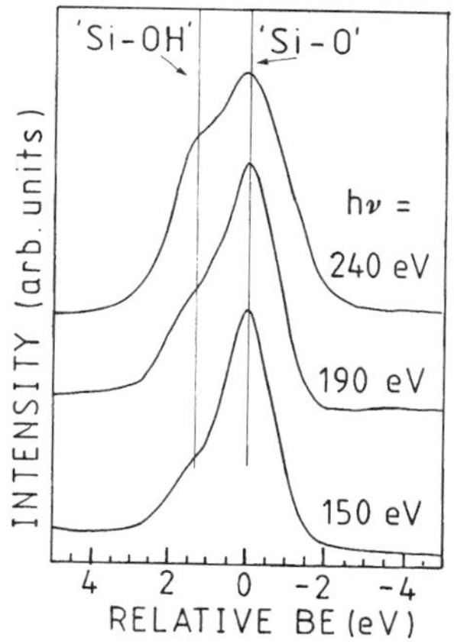

Figure 25. SRPS Si 2*p* spectra from aged silica recorded using different photon energies. (Reproduced from Ref. 85 by permission of the American Institute of Physics.)

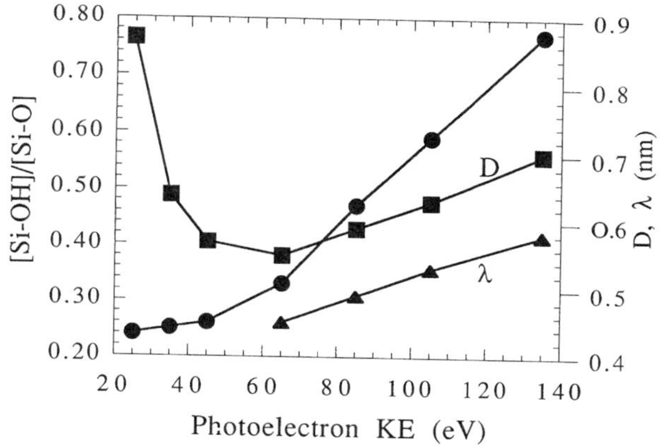

Figure 26. SRPS [Si–OH]/[Si–O] ratios as a function of photoelectron kinetic energy. Also reported are the inelastic mean free paths, λ, and the electron escape depths, *D*, as a function of the same KE range. (Reproduced from Ref. 85 by permission of the American Institute of Physics.)

creased monotonically by a factor of ca. 3 between the KE extremes, and in particular grew from 0.25 to 0.58 in the KE range from 35 to 105 eV, where the average escape depth showed very little change. Note, however, that the accuracy of *D* is not well established in the above KE range, particularly for KE < 50 eV, since the compilation of Seah and Dench is based mainly on best-fit derivations from photoemission spectra recorded at KE > 150 eV [13]. Also, the calculation of λ, which provides a rough guide to actual *AL* values [64], is valid only for KE > 50 eV. However, the ratio does increase markedly between KE = 65 and 135 eV, while in this interval *D* goes only from ca. 0.55 to 0.7 nm.

Based on the above, it is suggested that the silanol species were concentrated mainly in a region ΔT ($T = 3D$, the so-called information depth [95]) about 0.4–0.5 nm wide and located 1.6 to 2.1 nm in from the surface. XPS spectra from aged silica showed no appreciable differences, either in photoemission or in XAES, compared to those from anhydrous silica, except for a slight excess of oxygen, based on quantitative analysis [85]. The n_O/n_{Si} atomic ratio was 2.37, i.e., higher than the theoretical value for stoichiometric silica by approximately twice the experimental uncertainty. Neither was any appreciable difference found when the photoemission and XAES measurements were conducted at a takeoff angle of 20°, which implies $D \approx 0.9$ nm and $D \approx 1.1$. nm for the O 1*s* and Si 2*p* spectra, respectively. It is therefore concluded that the material had a chemical composition which was that of "regular" silica beyond a depth of $T \geq 3$ nm.

Although lateral inhomogeneities in the hydration content were not of prime concern, attempts were made by SAM to study them [86]. Considerable problems were found, in that in neither the Si LVV or KLL Auger spectra nor in the O KVV Auger spectra were there any differences between the aged and the anhydrous silica in respect of the locations of spectral features. Point analyses from some 200 separate regions, each of dimension 5×5 μm^2, on the surface of aged silica could detect no variations. However, when ratios of *signal intensity* were measured, evidence for both excess oxygen and lateral inhomogeneities was found when the Si LVV intensity was used. The Si KLL intensity, at much higher KE and therefore corresponding to much greater sampling depth, could not be used, as shown in Fig. 27. It can be seen that on aged silica the ratio $I_{O\ KVV}/I_{Si\ LVV}$ varied from 0.50 to 0.74 across the surface, whereas the ratio $I_{Si\ KLL}/I_{O\ KVV}$ remained constant at 0.50.

It is possible, of course, that the observed variations in the $I_{O\ KVV}/I_{SI\ LVV}$ ratio might have arisen from carbonaceous impurities. To check this, a contaminant particle was selected and analyzed by SAM. Figure 28a is the SEM image of the particle; Figs. 28b and c are the SAM images in oxygen and carbon, respectively, with the bright areas corresponding to regions of high concentration. The Si SAM map was similar to that for oxygen. It can be seen that the regions of high O and Si lie entirely outside the particle, indicating that high oxygen levels are *not* associated with contamination. The high $I_{O\ KVV}/I_{Si\ LVV}$ ratios must therefore be interpreted as originating from Si–OH species.

Lateral inhomogeneities in hydration content could therefore be studied not by the use of normal SAM mapping but by carrying out a series of point analyses along lines across the surface and measuring $I_{O\ KVV}/I_{Si\ LVV}$ at each point.

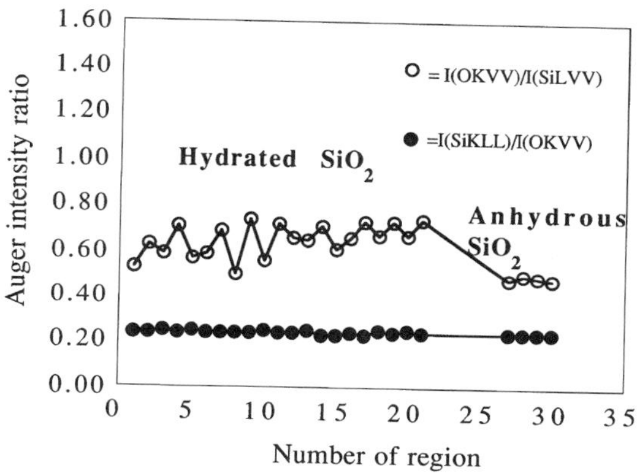

Figure 27. Distribution of signal intensity ratios obtained with SAM from Auger point spectra from aged hydrated silica and anhydrous silica. The full lines are used as a guide for the eye. (Reproduced from Ref. 86 by permission of Elsevier Science, Amsterdam.)

5.4. Discussion

5.4.1. Effects of Ar^+ bombardment and suboxides

It is instructive to discuss the Ar^+ bombardment results in relation to the the chemical sensitivity, surface specificity and lateral resolution of the individual techniques. Thus, the chemical damage caused by Ar^+ etching, as well as the surface hydration, was observed only when the surface specificity of the technique matched the thickness of the Ar^+ damaged layer [79,80,88] and the depth at which Si–OH species resided, respectively. It was important to choose the analytical operating conditions to optimize the combination of chemical sensitivity and surface specificity For example, when using XPS and XAES, an Ar^+ energy of no more than 2 keV should be used to depth-profile the SiO_2/Si system. With that combination the chemical changes in oxidized silicon occurring at the transition layer (Table 3, Figs. 23 and 24) could be studied with only minimal intrusion of damage. For 2 keV Ar^+ these artefacts were limited to a modest and symmetric broadening of both Si 2*p* and Si KLL peaks (Fig. 22), which might have been related to rearrangements of the Si–O–Si bond angle [79,80], or to roughening phenomena rather than to the creation of SiO_x species [84]. Some idea of the influence of roughening can be gained by noting that the Si KLL and Si 2*p* FWHMs for unetched powdered (therefore rough) SiO_2 showed, within 0.1–0.2 eV, the same values as observed for the etched SiO_2 film (Fig. 23) and that they remained virtually unchanged when the powdered material was subjected to 2 keV Ar^+ bombardment.

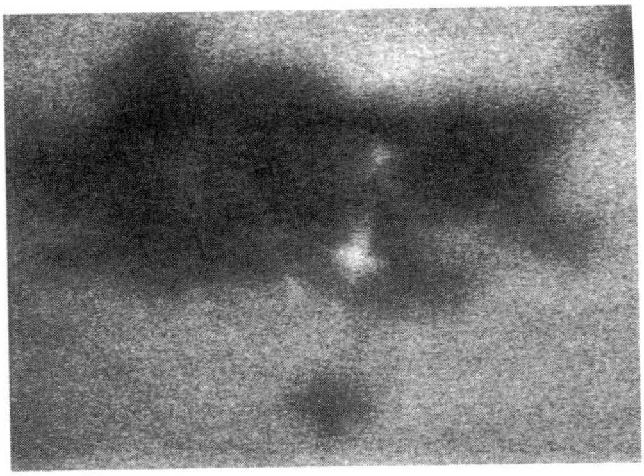

(a)

2 μm

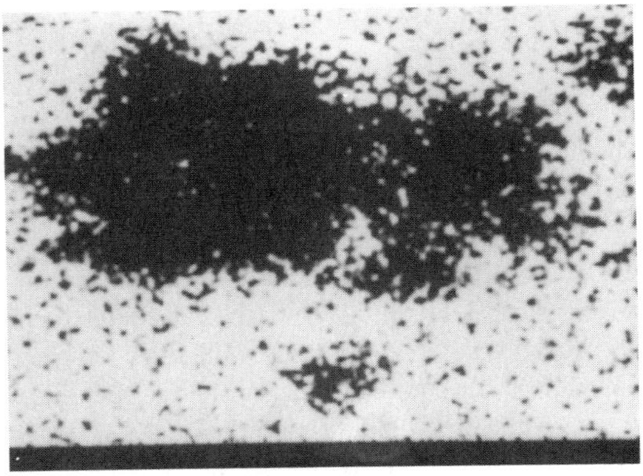

(b)

2 μm

Figure 28. (a) SEM image showing contaminant grains at the surface of aged silica. (b) SAM oxygen image. (c) SAM carbon image. Both SAM images have been recorded after correction by the [peak-background]/background procedure, and were obtained from the area shown in (a). The light field is associated with high abundance, and the dark field with low abundance. (Reproduced from Ref. 86 by permission of Elsevier Science, Amsterdam.)

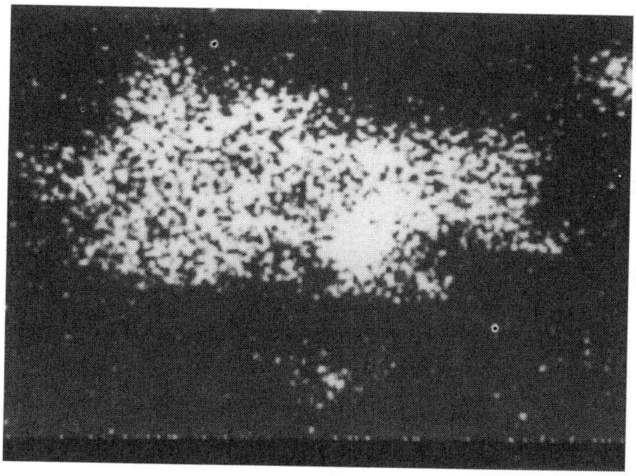

(c)

2 μm

Figure 28. *(continued)*

Again, from Figs. 23 and 24 it can be seen that the Si KLL peak from the silica film showed a chemically induced broadening that was more than three times that due to Ar^+ bombardment while the chemically induced broadening in the Si 2*p* peak was about twice that due to the Ar^+. By "chemically induced" is meant arising from the presence of genuine SiO_x suboxides. The greater spectral change in the Si KLL region can be explained by greater chemical sensitivity since the Si KLL chemical shift between SiO_2 and SiO_x phases is nearly twice that in Si 2*p* (Table 2). Ar^+-induced SiO_x species for Ar^+ energies ≥5 keV can be observed in the Si KLL and 2*p* spectra, because then the depth of damage in silica is about 2.5–3 nm, i.e., similar to the ALs associated with Al K_α-excited Si 2*p* and Si KLL electrons (Fig. 21).

When Ar^+-induced damage is confined to the outermost layers (as with 2 keV Ar^+), it can be detected easily using photoemission techniques, provided the surface specificity is matched, as in XPS at grazing takeoff angle ($AL \approx 1$ nm) or in SRPS Si 2*p* spectra ($AL \approx 0.5$ nm) (Figs. 21 and 22). Under these conditions the intensity of the Ar^+-induced SiO_x component was more than half that of the undamaged component and resulted in a FWHM broadening as large as 1.1 eV. This implies that Si 2*p* spectra recorded in SRPS at low KE are not suitable for studying the depth profile of the SiO_2/Si material because the 2 keV Ar^+-induced broadening in the spectra was comparable to that caused by the coexistence of chemically inequivalent oxidized species at the transition layer. Rivière et al. reached similar conclusions about the inadequacy of the surface-specific Si LVV signal as a chemical indicator of chemical damage at silicon oxynitride surfaces [88]. However, the SRPS Si 2*p* spectra also show that photoemission techniques have a chemical sensitivity sufficient for the detection of SiO_x suboxides formed on etched silica surfaces and comparable to that achieved by Si LVV spectra, provided the experiments are conducted at comparable surface specificities [82].

5.4.2. Surface-hydrated species

Although the extreme surface specificity of the SRPS spectra is a disadvantage in depth profiles of the kind just described, it has proved invaluable in detecting silanol species at the surface of aged silica, as in Fig. 25. Such species had not been detected using XPS techniques, although their occurrence has been documented in infrared spectroscopy [96], photon-stimulated desorption [97] and SSIMS [98] studies. Not only do the SRPS spectra provide evidence for the existence of Si–OH bonding states, but they also give information about the vertical distribution of those states in the external layers of aged silica, with a depth resolution of tenths of a nm (Fig. 26). As far as the chemical sensitivity is concerned, the Si 2*p* spectra can discriminate between Si–O and Si–OH bonding states more easily than can the Auger Si LVV spectra. In fact, the Si LVV spectrum characteristic of silica has an inherently broad line shape (core-valence-valence type) whose principal component is nearly 12 eV wide at half-maximum compared with the atomic-like Si 2*p* line shape (FWHM = 1.7 eV) (Fig. 25). As a result, the Si–OH component is barely resolvable in the Si LVV region. On the other hand, the extreme surface specificity of the Si LVV spectra [88] did reveal the presence of silanol groups via the $I_{\mathrm{O\,KVV}}/I_{\mathrm{Si\,LVV}}$ ratio (Fig. 27), whereas the lower surface specificity associated with the XPS spectra gave but little quantitative evidence of excess surface oxygen. In principle, the higher $I_{\mathrm{O\,KVV}}/I_{\mathrm{Si\,LVV}}$ values might also derive from pickup of ambient oxygen and/or ambient humidity at the surface of SiO_2. Even assuming that dissociation occurs in adsorbed O_2 (H_2O), only a weak interaction, of the van der Waals kind, could be expected to occur between O_{ads} and the silicon ions, the latter being saturated already by four Si–O bonds in the silica network. Conversely, hydration typified by a reaction of the kind $SiO_2 + xH_2O = SiO_{2-x}(OH)_{2x}$, $0 < x < 2$, requires that at least one Si–O bond is replaced by an Si–OH bond. The difference in electronegativity between O and Si induces a local increase in the silicon positive charge, this difference being concentrated only at the Si site of the bond, as compared with two Si sites in anhydrous silica. The increase in the silicon positive charge was verified by the presence of a high-BE component in the Si 2*p* spectra of Fig. 25. The lateral resolution of SAM gave information, via the $I_{\mathrm{O\,KVV}}/I_{\mathrm{Si\,LVV}}$ ratio, on the microchemical distribution of silanol species over the first layers of aged silica (Fig. 28), while the Auger maps of Si and C in Fig. 28 showed, with sub-μm lateral resolution, that the excess oxygen found in the material belonged exclusively to Si–OH species, the level of oxidized carbonaceous contamination being negligible.

6. InP/SiO_2 SYSTEM

The sample consisted of a flat InP matrix in which were grooved parallel trenches and some triangles, as in Fig. 29. The material inside both trenches and triangles was SiO_2.

6.1. Problem Specification

Unlike the materials described in the preceding sections, the notion of a laterally homogeneous sample is not valid here since the chemical composition will depend on the particular surface region considered. Within this lateral inhomogeneity, it is instructive to analyze the local chemistry with respect both to purity and to defectiveness (purposely created by mechanical scraping of some regions of the trenches and triangles), as well as to reveal possible chemical changes at interfaces between InP and SiO_2.

6.2. Experimental Approach: Choice of Technique and Specimen Configuration

Given the need for spatial resolution the techniques of choice are SAM and scanning ELS. SAM can identify the surface chemical composition from Auger point spectra recorded either from the InP matrix or from inside the trenches

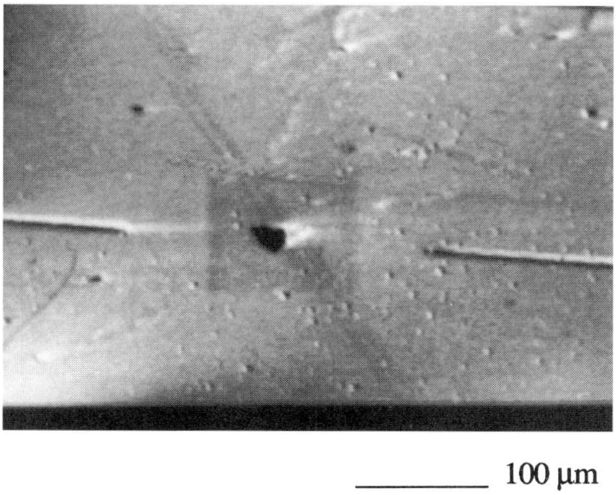

Figure 29. SEM image of the InP/SiO_2 sample showing a triangle and segments of parallel trenches.

and triangles. Chemical inhomogeneities can also be revealed at interfaces that include both InP and SiO_2, e.g., across the boundary of a trench or a triangle. Rastering the electron beam either across areas or along straight lines allows chemical variations to be studied from SAM images or from SAM linescans, respectively. Although ELS does not have the same chemical sensitivity as SAM, this example will demonstrate that scanning ELS can be a useful complement to the more established SAM. Thus, the variable surface specificity offered by ELS, i.e., by varying the primary energy, can be exploited to compare the chemistry of the external and internal layers. The other techniques listed in Table 1 were not considered suitable, because they would have provided only "mean" surface chemical compositions, averaged over all the local chemistries within the area irradiated by either the electron (AES) or the photon (XPS, XAES, SRPS) probe.

6.3. Results

Except where noted all spectra were recorded from flat regions of the surface, free of contaminant grains or structural defects. Figures 30 and 31 show survey Auger spectra recorded from the InP matrix and from inside a trench, respectively, in both the as-deposited condition (a) and after severe (3 keV) Ar^+ etching (b). In the as-deposited condition the InP was partly oxidized, as indicated by the oxygen peak and as confirmed in the high-resolution spectra of P and In (not shown), which were similar to those of Figs. 11a, and 15a, respectively.

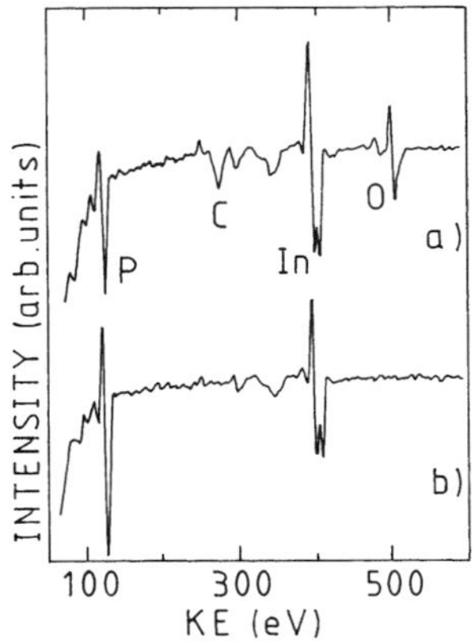

Figure 30. SAM Auger point spectra recorded from a flat region of the InP/SiO_2 sample analyzed: (a) as-received; (b) after 3 keV Ar^+ etching.

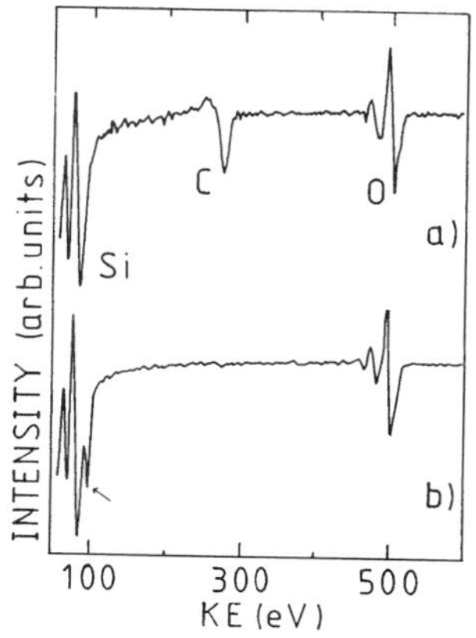

Figure 31. SAM Auger point spectra recorded from the inside of a trench on the InP/SiO_2 sample analyzed: (a) as-received; (b) after 3 keV Ar^+ etching. The arrow in curve (b) marks the presence of Ar^+-induced Si–Si bonding states.

Carbon was the only ambient contamination at the surface of both regions. Although the Ar^+ etching treatment removed all the oxidation and contamination from the surface, it also caused chemical damage in the silica. This is clear from the appearance of a peak at ca. 92 eV in the very surface specific Si LVV spectrum (marked by an arrow in Fig. 31b); there was no change in the Si KLL spectrum (not shown). Despite the occurrence of the damage component, the Si LVV spectrum retained its principal component at 75 eV characteristic of undamaged silica, and the latter was used for both SAM imaging and linescanning. Electron irradiation produced no additional chemical damage at the surface, as checked from time-dependent Auger spectra.

SAM point spectra (Fig. 32) were recorded from several different regions within the area illustrated in Fig. 33, which is an SEM image of a triangle whose silica coating had been partly removed by deliberate mechanical scraping. The spectrum 32b, recorded from the exposed region b, is characteristic of the InP substrate. Spectrum (d) was recorded from point d straddling the boundary of the trench, and contains contributions from both InP and SiO_2. Within the latter area, high-resolution O KVV point spectra recorded (a) from the silica region, (b) at the silica/InP interface, and (c) from the InP region, are shown in Fig. 34. Figure 35 shows SAM images for In (a) and O (b) recorded over an area containing a trench segment, with the corresponding SAM linescans along a line intersecting the trench in Fig. 36.

The effects of surface contamination are shown in Fig. 37, in which the O SAM image recorded from a triangle in the as-deposited condition is shown.

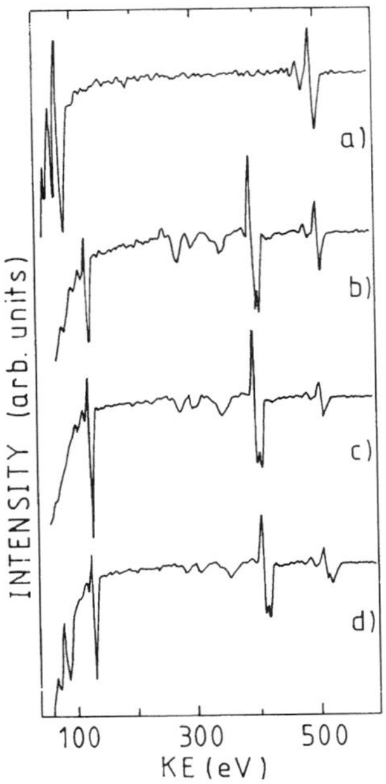

Figure 32. SAM Auger point spectra recorded from four different locations inside and near a triangle on the InP/SiO_2 sample.

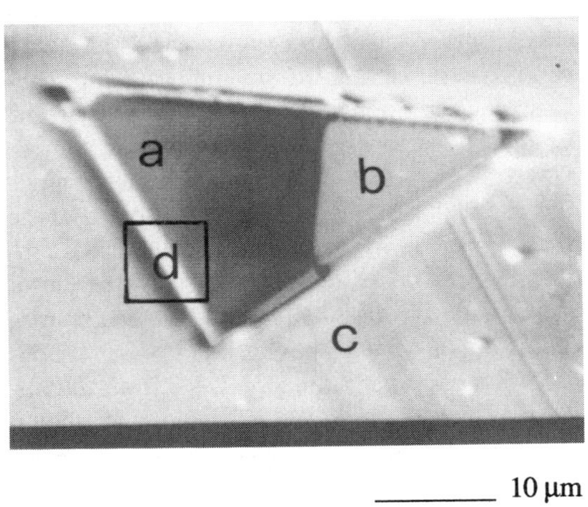

Figure 33. SEM image of the area from which the Auger spectra of Fig. 32 were recorded in regions (a) to (d).

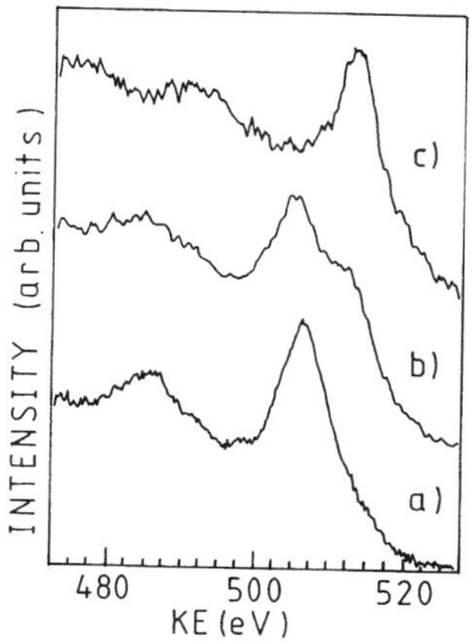

Figure 34. SAM O KVV spectra recorded from the InP/SiO_2 sample in the following regions across the interface denoted (d) in Fig. 33: (a), SiO_2 region; (b) InP/SiO_2 interface; (c) InP region.

Much better chemical contrast is demonstrated in Fig. 38, which contains the scanning ELS map (E_p = 2 keV) based on the maximum in the InP loss feature (Fig. 39). Other comparisons between scanning ELS maps, referenced to the same loss feature, and SAM elemental maps are shown in Fig. 40. There the compositional variations on and near a contaminant grain are imaged; SAM point spectra showed the grain to consist of carbonaceous phases, whose dimensions were of the same order as the trench width. Again, the chemical contrast achieved in the scanning ELS images is sharper than in the SAM images, and this is also evident in Fig. 41, which shows the P image (a), the C Auger image (b) and the scanning ELS image (c) recorded from a total area of 5 × 5 μm^2 containing the "curlshaped" feature marked by an arrow in Fig. 40c.

6.4. Discussion

For this type of material SAM has been able to provide detailed information on the lateral distribution of both composition and chemical state, on a microscopic scale. Thus the two materials present, InP and SiO_2, have been identified whether present in the host matrix or in trenches and triangles, as can be seen in Figs. 30 to 32 and 34 to 37. In particular, high energy-resolution spectra for In and P showed that both elements were present as a mixture of oxidized species in the as-deposited condition, but as clean InP after Ar^+ etching. O KVV spectra

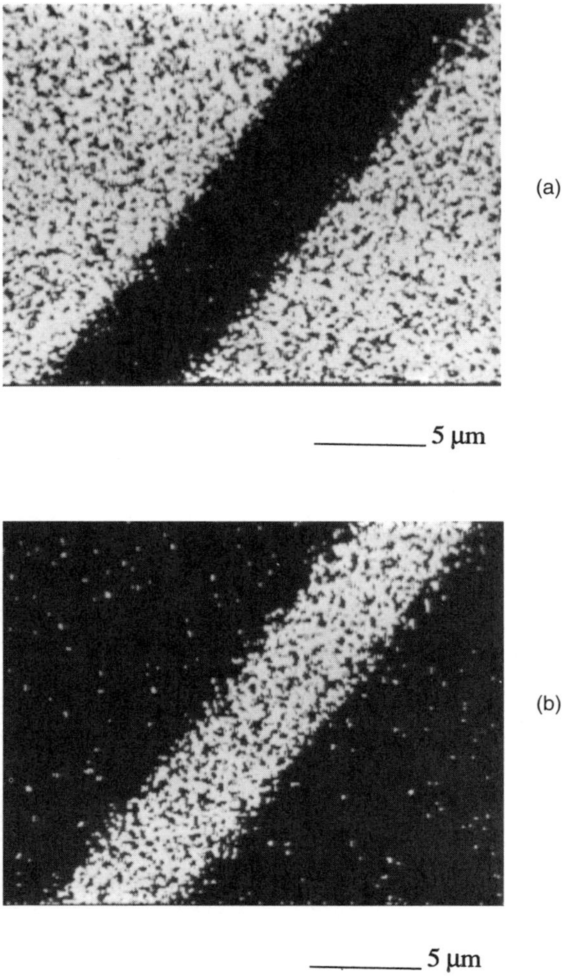

Figure 35. SAM images obtained from the InP/SiO_2 sample for (a) In and (b) O. Both images have been recorded after correction by the [peak-background]/background procedure. The light field is associated with high abundance, and the dark field with low abundance.

recorded at various points across a trench (Fig. 34) in the as-deposited condition were sufficiently different that they could in fact be used as indicators [46,99] for the presence or absence of InP and SiO_2. The author has reported elsewhere [46,99] that the O KVV transition can show chemical shifts as high as 5 eV, confirming that it can be used, as here, for the distinction of different species.

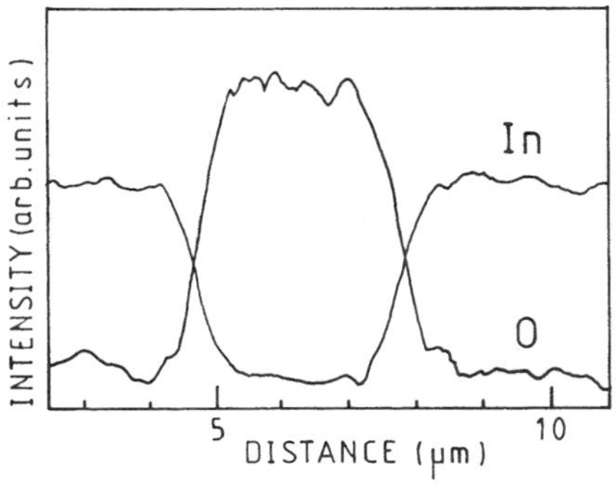

Figure 36. Linescans ([peak-background]/background) recorded along a line across a trench on the InP/SiO_2 sample.

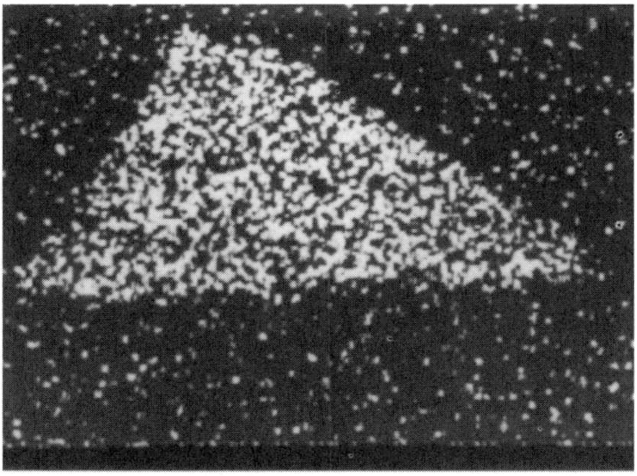

10 μm

Figure 37. SAM image for oxygen ([peak-background]/background) recorded from an area containing a silica triangle on the InP/SiO_2 sample, analyzed in the as-deposited condition.

The microscopic analytical capabilities of SAM are demonstrated in Figs. 35 to 37, in which a spatial resolution of ca. 1 μm was achieved. In Fig. 37 the image is weaker because of a contaminant layer (i.e., as-deposited condition). The degree of attenuation depends on the KE of the elemental feature used (in that case the O KVV), and a stronger image would have been obtained at higher KE.

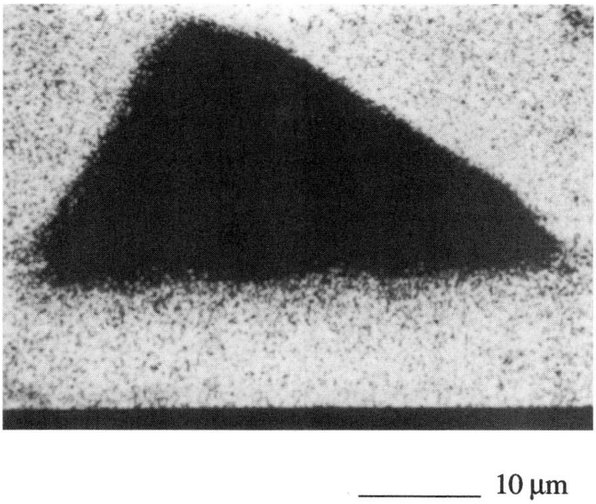

Figure 38. Scanning ELS map ([peak-background]/background), derived from the bulk plasmon of InP, using $E_p = 2$ keV, recorded from the same area of Fig. 37.

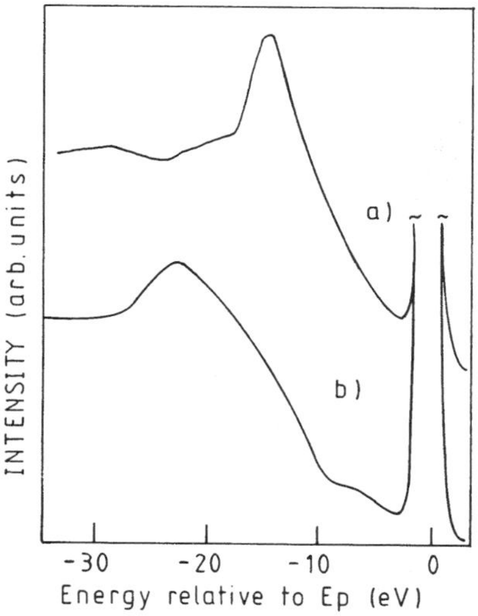

Figure 39. Energy loss spectra ($E_p = 2$ keV) measured from the InP/SiO_2 sample for the regions: (a) InP and (b) SiO_2.

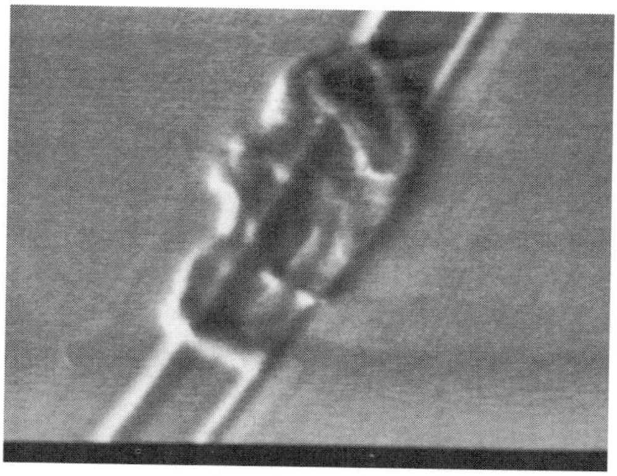

(a)

——— 1 μm

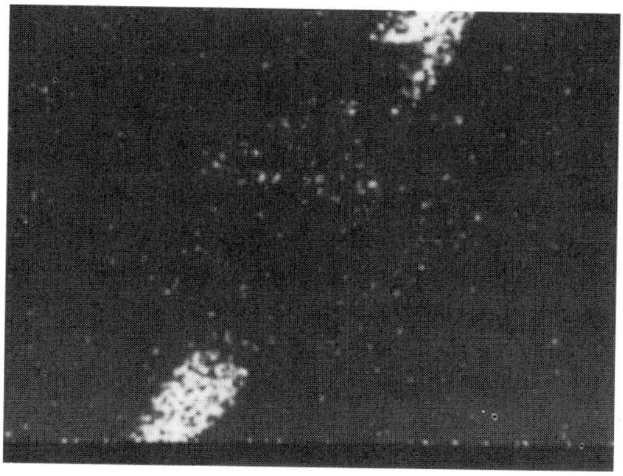

(b)

——— 10 μm

Figure 40. (a) SEM image showing a contaminant grain lying across a trench on the InP/SiO_2 sample. From this area the following SAM and scanning ELS images ([peak-background]/background in all cases) were recorded: (b) SAM image for O; (c) scanning ELS image (E_p = 2 keV; see Fig. 39, curve a) for InP (the arrow marks a curl-shaped feature discussed in the text); (d) SAM image for P; (e) SAM image for C. See this page and two following pages.

(Continued)

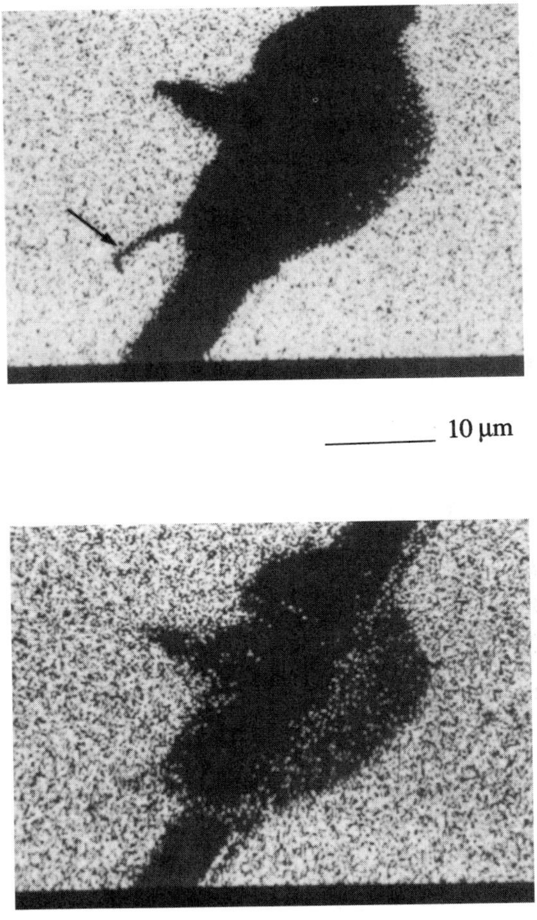

(c)

(d)

Figure 40. *(continued)*

The latter observation emphasizes one of the advantages of scanning ELS. In Fig. 38 from the same area as in Fig. 37 an image is shown based on the strong InP loss feature at an energy loss of ca. −14 eV but at a primary energy of 2 keV. The corresponding inelastic mean free path is about 3 nm, as against about 1 nm for the O KVV transition [64]. Another advantage of ELS is that, according to Bevolo [8,9] and El Gomati and Matthew [10,11], there is a weaker dependence of contrast on electron backscattering than in SAM. In addition, topography, e.g., sharp edges and defects, can affect the interpretation of SAM images drastically [8–11,100–103]. This might explain the better defini-

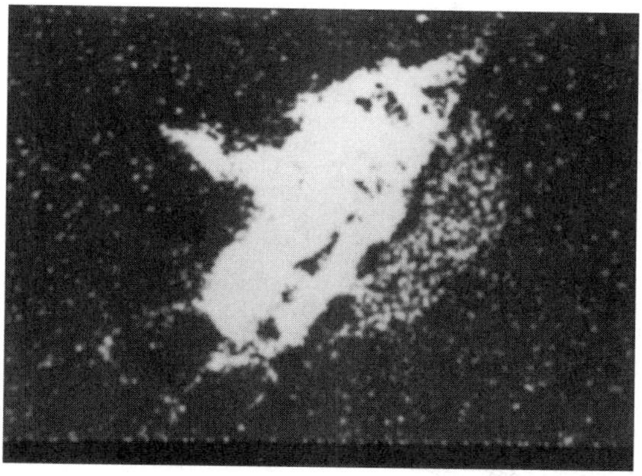

(e)

10 μm

Figure 40. *(continued)*

tion in the scanning ELS images in Figs. 40c and 41c, compared to the corresponding SAM images. The scanning ELS image of the curl-shaped feature, in Fig. 41c, is particularly well defined, with a spatial resolution much better than 1 μm, whereas the SAM images of P and C in Figs. 41a and b, respectively, are much fainter.

In this study the principal limitations of scanning ELS were (1) that only the major InP loss feature shown in Fig. 39 could be used for imaging, since the less intense losses from silica and from carbon contamination were ineffective for this purpose, and (2) that instrumental operation limited the primary energy in ELS to 2 keV, which allowed magnification to only a factor of 2×10^4. Indium SAM images could be obtained at primary energies up to 10 keV, with the possibility of producing Auger maps at magnifications up to 10^5 [104]. The potentials of SAM and scanning ELS as tools for analyzing semiconductor surfaces have been discussed by the author in recent work [105–107].

ACKNOWLEDGMENTS

Ing. Dr. V. Lacquaniti (Istituto "Galileo Ferraris," Turin, Italy) and Dr. M. Cocito (CSELT, SpA, Turin) are thanked for their collaboration and for allowing me to report on the analysis of the two Josephson junctions, which were fabricated as well as characterized as to the electrical parameters at the Istituto "Galileo Ferraris," and studied by AES at CSELT, respectively. The XPS analysis of these materials was performed in collaboration with Dr. Claudio Battistoni (Chairman of the "Comitato Nazionale per le Ricerche Technologiche e l'Innovazione" of the CNR), whom I should also like to thank for

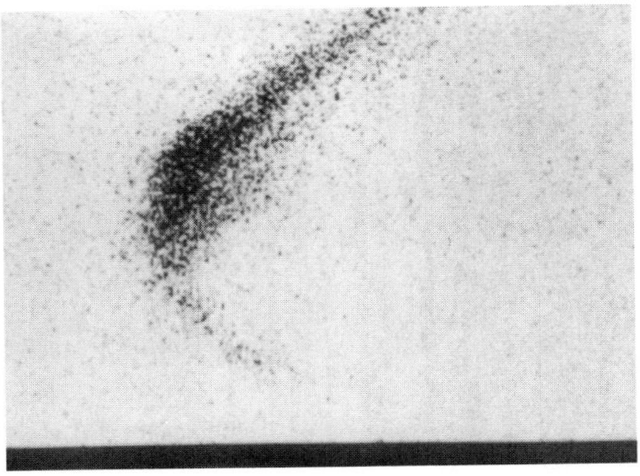

(a)

1 μm

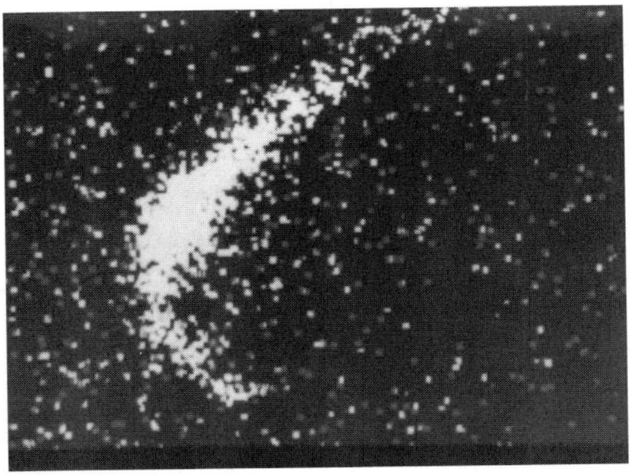

(b)

1 μm

Figure 41. SAM and scanning ELS images recorded from the neighborhood of the curl-shaped feature (marked in Fig. 40c): (a) SAM image for P; (b) SAM image for C; (c) scanning ELS image for InP.

allowing me to report these results. Dr. C. C. Bahr, Jr. (AT&T Bell Laboratories, Murray Hill, NJ) is thanked for allowing me to report on the oxidation of the quaternary semiconductors by NO_2, a project of his own that he pursued in my institute in collaboration with me and with some colleagues of mine, as well as for the kind loan of the InP/SiO_2 sample. I wish also to thank Dr. M. Fanfoni (De-

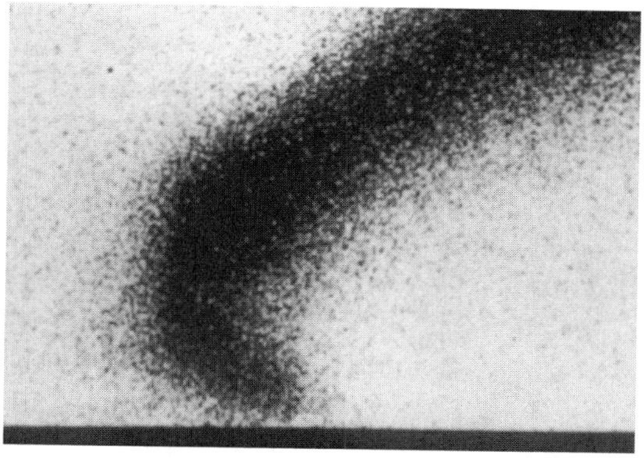

(c)

1 μm

Figure 40. *(continued)*

partment of Physics, University "Tor Vergata," Rome), E. Severini (CSM, SpA, Rome) and S. Priori for their collaboration in the surface studies of the Si/SiO_2 interface and for their friendship. I must also thank L. Moretto for maintaining the Auger microprobe in perfect order. Last, but not least, my deep gratitude goes to M. Brolatti for his skill and immense patience in drawing the figures.

REFERENCES

1. C. S. Fadley, in *Electron Spectroscopy: Theory, Techniques and Applications* (C. R. Brundle and A. D. Baker, eds.), Vol. 2, Academic Press, London, 1978, pp. 5 ff.
2. D. Briggs and J. C. Rivière, in *Practical Surface Analysis: Auger and X-ray photoelectron Spectroscopy*, (D. Briggs and M. P. Seah, eds.), Vol. 1, 2nd ed., Wiley, Chichester, 1990, pp. 85 ff.
3. G. Margaritondo, in *Photoemission and Absorption Spectroscopy of Solids and Interfaces with Synchrotron Radiation* (M. Campagna and R. Rosei, eds.), North-Holland, Amsterdam, 1990, pp. 327 ff.
4. G. E. McGuire and P. H. Holloway, in *Electron Spectroscopy: Theory, Techniques and Applications* (C. R. Brundle and A. D. Baker, eds.), Vol. 4, Academic Press, London, 1981, pp. 1 ff.
5. J. C. Fuggle, in *Electron Spectroscopy: Theory, Techniques and Applications* (C. R. Brundle and A. D. Baker, eds.), Vol. 4, Academic Press, London, 1981, pp. 85 ff.
6. I. F. Ferguson, *Auger Microprobe Analysis*, Adam Hilger, Bristol, 1989.
7. E. Paparazzo, *Microsc. Anal. 11*, 17 (1994).

8. A. J. Bevolo, *Scanning Electron Microscopy*, AMF O'Hare, Chicago, 1985, p. 1449.
9. A. J. Bevolo, in *Analytical Electron Microscopy* (D. C. Joy, ed.) San Francisco Press, San Francisco, 1987, p. 340.
10. M. M. El Gomati and J. A. D. Matthew, *J. Microsc. 147*, 137 (1986).
11. M. M. El Gomati and J. A. D. Matthew, *Appl. Surf. Sci. 32*, 320 (1988).
12. M. P. Seah and D. Briggs, in *Practical Surface Analysis: Auger and X-ray Photoelectron Spectroscopy* (D. Briggs and M. P. Seah, eds), Vol. 1, 2nd ed., Wiley, Chichester, 1990, p. 1 ff.
13. M. P. Seah and W. A. Dench, *Surf. Interface Anal. 1*, 2 (1979). Throughout this chapter we will refer to either *AL* or λ corresponding to electron ejection from the surface induced by photon irradiation or by electron impact, respectively. The reasons for this distinction are discussed in C. J. Powell and M. P. Seah, *J. Vac. Sci. Technol. A8*, 2213 (1990) and A. Jablonski, *Surf. Interface Anal. 15*, 559 (1990).
14. G. N. Greaves and I. H. Munro (eds.), *Synchrotron Radiation Sources and Their Applications*, Redwood Burn, Trowbridge, 1989.
15. C. S. Fadley, R. J. Baird, W. Sieckaus, T. Novakov and S. Å Bergström, *J. Electron Spectrosc. Relat. Phenom. 4*, 93 (1974).
16. R. J. Baird, C. S. Fadley, S. K. Kawamoto and M. Metha, *Anal. Chem. 48*, 843 (1976).
17. C. Quaresima, C. Ottaviani, M. Matteucci, C. Crotti, A. Antonini, M. Capozi, S. Rinaldi, M. Luce, P. Perfetti, K. C. Prince, C. Astaldi, M. Zacchigna, L. Romanzin, and A. Savoia, *Nucl. Sci. Instrum. A364*, 378 (1995).
18. C. R. Helms, Y. E. Strausser and W. E. Spicer, *Appl. Phys. Lett. 33*, 767 (1978).
19. S. Ichimura, R. Shimizu and J. P. Langeron, *Surf. Sci. 124*, L49 (1983).
20. H. H. Madden and J. E. Houston, *J. Vac. Sci. Technol. 14*, 412 (1977).
21. D. E. Ramaker, J. S. Murday and N. H. Turner, *J. Electron Spectrosc. Relat. Phenom. 17*, 45 (1979).
22. Ref. 5, p. 86.
23. C. D. Wagner, *Anal. Chem. 44*, 967 (1972).
24. J. C. Rivière and J. A. A. Crossley, *Surf. Interface Anal. 8*, 173 (1986).
25. R. E. Weber and W. T. Peria, *J. Appl. Phys. 38*, 4355 (1967).
26. P. W. Palmberg and N. T. Rhodin, *J. Appl. Phys. 39*, 2425 (1968).
27. H. E. Bishop and J. C. Rivière, *J. Appl. Phys. 40*, 1740 (1969).
28. G. C. Allen, P. M. Tucker and R. K. Wild, *J. Chem. Soc. Farday Trans. II 74*, 1126 (1978).
29. E. Paparazzo, *Appl. Surf. Sci. 74*, 61 (1994); E. Paparazzo, L. Moretto, C. D'Amato and A. Palmieri, *Surf. Interface Anal. 23*, 69 (1995).
30. M. L. Knotek and P. J. Feibelman, *Phys. Rev. Lett. 40*, 964 (1978).

31. D. E. Ramaker, F. L. Hutson, N. H. Turner and W. N. Mei, *Phys. Rev. B33*, 2574 (1986).
32. G. E. McGuire, *Surf. Sci. 76*, 130 (1976).
33. A. G. Schrott, G. S. Frankel, A. J. Davenport, H. S. Isaacs, C. V. Jahnes and M. A. Russak, *Surf. Sci. 250*, 139 (1991).
34. E. Paparazzo, *Surf. Sci. 234*, L253 (1990).
35. C. J. Powell and P. E. Larson, *Appl. Surf. Sci. 1*, 186 (1878).
36. C. Battistoni, G. Mattogno and E. Paparazzo, *Surf. Interface Anal. 7*, 117 (1985).
37. C. D. Wagner, L. E. Davis, M. V. Zeller, J. A. Taylor, R. H. Raymond and L. H. Gale, *Surf. Interface Anal. 3*, 211 (1981).
38. T. E. Gallon, *J. Phys. D (Appl. Phys.) 5*, 822 (1972).
39. M. M. El Gomati, M. Prutton, B. Lamb and C. G. Tupper, *Surf. Interface Anal. 11*, 251 (1988).
40. M. Crone, I. R. Barkshire and M. Prutton, *Surf. Interface Anal. 21*, 805 (1994).
41. J. C. Greenwood, M. Prutton and R. H. Roberts, *Phys. Rev. B49*, 12485 (1994).
42. M. P. Seah and G. C. Smith, *Surf. Interface Anal. 11*, 69 (1988).
43. C. Kunz, in *Photoemission and Absorption Spectroscopy of Solids and Interfaces with Synchrotron Radiation* (M. Campagna and R. Rosei, eds.), North-Holland, Amsterdam, 1990, p. 93.
44. C. G. Pantano, D. B. Dove and G. Y. Onoda, *J. Vac. Sci. Technol. 13*, 414 (1976).
45. T. L. Barr, *Appl. Surf. Sci. 15*, 1 (1983); *J. Vac. Sci. Technol. A7*, 1677 (1989).
46. E. Paparazzo, *J. Electron Spectrosc. Relat. Phenom. 43*, 97 (1987).
47. S. Hoffman, in *Practical Surface Analysis: Auger and X-ray Photoelectron Spectroscopy* (D. Briggs and M. P. Seah, eds.), Vol. 1, 2nd ed., Wiley, Chichester, 1990, pp. 143 ff, and references therein.
48. V. Lacquaniti, G. Marullo and R. Vaglio, *IEEE Trans. Magn. 15*, 593 (1979).
49. V. Lacquaniti, C. Battistoni, E. Paparazzo, M. Cocito and S. Palumbo, *Thin Solid Films 94*, 331 (1982).
50. M. Grundner and J. Halbritter, *J. Appl. Phys. 51*, 397 (1980).
51. K. S. Kim, W. E. Baitinger, J. W. Amy and N. Winograd, *J. Electron Spectrosc. Relat. Phenom. 5*, 351 (1974).
52. P. J. Grunthaner, R. P. Vasquez and F. J. Grunthaner, *J. Vac. Sci. Technol. 17*, 1045 (1980).
53. J. A. Stroscio and R. M. Feenstra, *J. Vac. Sci. Technol. A6*, 577 (1988).
54. C. H. Alt, *Appl. Phys. Lett. 54*, 1445 (1989).

55. P. W. Yu and D. C. Walters, *Appl. Phys. Lett. 41*, 863 (1982).
56. A. vom Felde, K. Kern, G. S. Higashi, Y. J. Chabal, S. B. Christman, C. C. Bahr and M. J. Cardillo, *Phys. Rev. B42*, 5240 (1990).
57. A. vom Felde, G. Meigs, C. C. Bahr and M. J. Cardillo, *Surf. Sci. 268*, 249 (1991).
58. C. Huang, A. Ludviksson and R. M. Martin, *Surf. Sci. 265*, 314 (1992).
59. Dao-xuan Dai and Fu-rong Zhu, *Phys. Rev. B43*, 4803 (1991).
60. C. C. Bahr, E. Paparazzo, L. Moretto, F. Lama and N. Zema, *Surf. Sci. 316*, 247 (1994).
61. L. E. Davis, N. C. MacDonald, P. W. Palmberg, G. E. Riach and R. E. Weber, *Handbook of Auger electron spectroscopy*, Physical Electronics Industries, Eden Prairie, MN, 1976.
62. Wei-Xi Chen, L. M. Walpita, C. C. Sun and W. S. C. Chang, *J. Vac. Sci. Technol. B4*, 701 (1986).
63. M. Kawabe, N. Kanzaki, K. Masuda and S. Namba, *Appl. Opt. 17*, 2557 (1978).
64. S. Tanuma, C. J. Powell and D. R. Penn, *J. Vac. Sci. Technol. A8*, 2213 (1990).
65. R. Cosso, L. Duò, M. Sancrotti, S. D'Addato, A. Ruocco, S. Nannarone and P. Weightman, *Appl. Surf. Sci. 251/252*, 267 (1991).
66. J. C. Rivière, in *Practical Surface Analysis: Auger and X-ray Photoelectron Spectroscopy*, (D. Briggs and M. P. Seah, eds.), Vol. 1, 2nd ed., Wiley, Chichester, 1990, pp. 19 ff.
67. C. D. Wagner and J. A. Taylor, *J. Electron Spectrosc. Relat. Phenom. 20*, 83 (1980).
68. J. A. Taylor, *J. Vac. Sci. Technol. 20*, 751 (1982).
69. H. Iwakuro, C. Tatsuyama and S. Ichimura, *Jpn. J. Appl. Phys. 21*, 94 (1982).
70. P. A. Bertrand, *J. Vac. Sci. Technol. 18*, 28 (1981).
71. C. R. Bonapace, D. W. Tu, K. Li and A. Kahn, *J. Vac. Sci. Technol. B3*, 1099 (1985).
72. G. Lucovsky, S. T. Pantelides and L. Galeener (eds.), *The Physics of MOS Insulators*, Pergamon, New York, 1978.
73. F. J. Grunthaner and P. J. Grunthaner, *Mater. Sci. Rep. 1*, 65 (1986).
74. P. Blank (ed.), *The Si-SiO_2* System, Elsevier, Amsterdam, 1988.
75. G. Hollinger and F. J. Himpsel, *Appl. Phys. Lett. 44*, 93 (1984).
76. J. S. Johannessen, W. E. Spicer and Y. E. Strausser, *J. Vac. Sci. Technol. 13*, 849 (1976).
77. S. S. Chao, J. E. Tyler, Y. Takagi, P. G. Pai, G. Lukovsky, S. Y. Lin, C. K. Wong and M. T. Mantini, *J. Vac. Sci. Technol. A4*, 1574 (1986).
78. S. S. Chao, J. E. Tyler, D. V. Tsu, G. Lukovsky and M. J. Mantini, *J. Vac. Sci. Technol. A5*, 1283 (1987).
79. S. Hoffmann and J. H. Thomas III, *J. Vac. Sci. Technol. B1*, 43 (1983).

80. J. H. Thomas III and S. Hoffmann, *J. Vac. Sci. Technol. A3*, 1921 (1985).
81. E. Paparazzo, *J. Phys. D (Appl. Phys.) 20*, 1091 (1987).
82. E. Paparazzo, *J. Electron Spectrosc. Relat. Phenom., 50*, 47 (1990).
83. E. Paparazzo, M. Fanfoni, C. Quaresima and P. Perfetti, *J. Vac. Sci. Technol. A8*, 2231 (1990).
84. E. Paparazzo, M. Fanfoni and E. Severini, *Appl. Surf. Sci. 56–58*, 866 (1992).
85. E. Paparazzo, M. Fanfoni, E. Severini and S. Priori, *J. Vac. Sci. Technol. A10*, 2892 (1992).
86. E. Paparazzo, *Appl. Surf. Sci. 72*, 313 (1993).
87. C. J. Brinker and G. W. Scherer, *Sol-Gel Science: The Physics and Chemistry of Sol-Gel Processing*, Academic Press, Boston, 1990, p. 99.
88. J. C. Rivière, J. A. A. Crossley and B. A. Sexton, *J. Appl. Phys. 64*, 409 (1988).
89. E. Paparazzo, *Appl. Surf. Sci. 25*, 1 (1986).
90. E. Paparazzo, *Surf. Interface Anal. 12*, 115 (1988).
91. T. L. Barr, in *Practical Surface Analysis: Auger and X-ray Photoelectron Spectroscopy* (D. Briggs and M. P. Seah, eds), Vol. 1, 2nd ed., Wiley, Chichester, 1990), p. 357 ff.
92. A. Kawashima, K. Asami and K. Hashimoto, *Corros. Sci. 24*, 807 (1984).
93. N. S. McIntyre and M. G. Cook, *Anal. Chem. 47*, 2208 (1975).
94. L. Pauling, *The Nature of the Chemical Bond*, Cornell University, Ithaca, NY, 1986.
95. C. J. Powell, *J. Electron Spectrosc. Relat. Phenom. 47*, 197 (1988).
96. D. Della Sala and G. Fortunato, *J. Electrochem. Soc. 137*, 2550 (1990).
97. B. F. Phillips, D. C. Burkman, W. R. Schmidt and C. A. Peterson, *J. Vac. Sci. Technol. A1*, 646 (1983).
98. N. Niwano, Y. Takakuwa, H. Katakura and N. Miyamoto, *J. Vac. Sci. Technol. A9*, 212 (1990).
99. E. Paparazzo, *Vuoto 17*, 286 (1987).
100. M. Prutton, C. G. H. Walker, J. C. Greenwood, P. G. Kenny, J. C. Dee, I. R. Barkshire, R. H. Roberts and M. M. El Gomati, *Surf. Interface Anal. 17*, 71 (1991).
101. M. Prutton, I. R. Barkshire, M. M. El Gomati, J. C. Greenwood, P. G. Kenny and R. H. Roberts, *Surf. Interface Anal. 18*, 295 (1992).
102. I. R. Barkshire, M. Prutton, J. C. Greenwood and M. M. El Gomati, *Surf. Interface Anal. 20*, 984 (1993).
103. M. Prutton, L. A. Larson and H. Poppa, *J. Appl. Phys. 54*, 374 (1983).
104. E. Paparazzo, L. Moretto, J. P. Northover, C. D'Amato and A. Palmieri, *J. Vac. Sci. Technol. A13*, 1229 (1995).
105. E. Paparazzo, *J. Electron Spectrosc. Relat. Phenom. 76*, 659 (1995).
106. E. Paparazzo, *J. Vac. Sci. Technol. A 14*, 1376 (1996).
107. E. Paparazzo, *Surf. Interface Anal. 25,* 235 (1997).

13

Minerals, Ceramics and Glasses

ROGER ST. C. SMART University of South Australia, The Levels, South Australia

1. INTRODUCTION

The surfaces of minerals and ceramics have many common features. This statement is justified by comparing information arising from their surface analysis (e.g., [1–3]). In fact, tailored ceramics often result from judicious design involving combinations of mineral structures (e.g., [4,5]). Similarities in phase structure and composition, surface structure and surface sites, microstructure and surface reactivity have been demonstrated in numerous studies. Surface reactions involving oxidation, leaching, dissolution, weathering, precipitation and phase transformation are now well documented. Surface modification of minerals as a result of adsorption, reaction, processing (e.g., plasma spraying) and surface coatings (e.g., sol-gel deposition) have been found to be equally applicable to ceramic materials. Hence the methodologies for determining the surface properties that control the mechanisms of reaction and transformation of surfaces of minerals and ceramics will in general require similar surface analytical techniques.

Glass surfaces can also be characterized in terms of composition, adsorption, reaction and surface modification, all of which require the same analytical techniques as are applied to minerals and ceramics. Structural information is less easily obtained by diffraction methods and related imaging techniques, but scanned probe microscopies are useful for definition of surface structure and surface sites. In the case of glasses, there will be less concern with grain boundaries and intergranular films and greater interest in *in situ* reaction, reprecipitation and phase transformation. Changes in composition and chemistry with depth, relating particularly to segregation and reaction, are important parameters for characterizing the reactivity of glass surfaces and for defining mechanisms of adsorption, reaction and surface modification.

These three broad groups of materials therefore require essentially the same approach to their characterization.

In this chapter the approach to problem solving by surface analytical methods will focus first on the nature of the information required and on the surface analytical techniques able to provide that information. In structural terms, phase identification (e.g., lattices, grain size, and size distribution), and information about surface structure and surface sites (from atomic to mm scales), grain boundaries, intergranular films and defects (point, linear, planar) may be required in order to understand the mechanisms controling the behavior of the material. In chemical terms, surface sites and their reactivity, elemental segregation, composition of the surface layers, depth profiles, grain boundaries and intergranular films may all be important. The ability to study adsorption (including surface distribution, molecular form, bonding and layer formation) and surface reaction (e.g., oxidation, dissolution, precipitation, phase transformation), and the mechanisms of these reactions, is central to the design and modification of mineral, ceramic and glass surfaces. Surface modification for controlled interfacial behavior in diverse applications now requires the use of advanced surface analytical techniques not only during the research and development phases but also for quality control and problem solving during fabrication and manufacture.

The next section will therefore be concerned with the choice of technique(s) likely to deliver the required information for a particular material. Such a choice is usually made on the basis of a strategy. A strategy of analysis is often followed which begins with the most readily available and user-friendly instrumentation providing general and/or averaged information over the surface. More detailed and/or specific information then requires higher resolution (in the spatial, temporal or spectral domains), which implies the deployment of more specialized instrumentation. In Section 3 an analytical strategy identifying different techniques appropriate to the provision of different kinds and levels of information will be discussed.

Finally, in the last sections, relevant case studies relating to mineral, ceramic and glass surfaces will be presented. The intention behind these sections is to use specific examples from the literature to demonstrate the utility of the different surface analytical techniques and the type of information that is obtainable in each case.

2. INFORMATION REQUIRED: ANALYTICAL TECHNIQUES

Table 1 summarizes the types of information normally obtained from surface analysis of minerals, ceramics and glasses. Accompanying each category of information is a suggested list of analytical techniques most applicable to the extraction of that information. It can be seen that, for any particular category, more than one technique can be used. In practice, it is almost always useful to study surfaces with at least four analytical techniques. The combination of analytical SEM (i.e., with EDS or WDS), AFM (or STM), XPS and FTIR (or Raman) is gaining acceptance as a basis set for characterization of minerals, ceramics and glasses. This combination, and other analytical techniques providing more detailed information, will be discussed in the next section.

In the problem-solving mode it may be sufficient to use only one or a subset of techniques to obtain the required information. For instance, accelerated corrosion of ceramic or glass surfaces may be the result of an activating agent introduced by contact with a solution or a gaseous ambient. It is often possible to determine the primary cause of the accelerated corrosion using XPS analysis alone [6]. In other cases, it is the molecular structure of the adsorbed layer and its alteration on reaction that is of direct interest; FTIR and Raman spectroscopies may then be required in order to produce the necessary additional information (e.g., [7,8]).

Table 1 constitutes a starting point for the choice of those analytical techniques that will provide the information required for research, development or problem solving. Rigid separation between structural/chemical and compositional information is not intended to be inferred from the list. There are many techniques that provide information in both categories (e.g., SAM, ISS; the latter is also called LEIS). The separate sections on minerals, ceramics and glasses will illustrate the application of the techniques and their complementary information content.

3. ANALYSIS STRATEGY

In the top-down approach taken in this book, it is useful to consider the most efficient use of the surface and analytical techniques for obtaining the information necessary to address the problem at hand. With the focus being on the material rather than on the technique(s), experience suggests that information on structure and composition is most likely to be obtained by proceeding through the sequence in Table 2.

Table 1. Types of Information Required and Relevant Techniques for Mineral, Ceramics and Glass Surfaces

Information	Minerals	Ceramics	Glasses	Techniques
1. Phases				
structure	√	√	X	XRD: STM: AFM: ISS: TEM(ED)
composition (uniformity)	√	√	√	EDS: EPMA: XRF: XPS: SIMS: EXAFS: STS: ISS: EELS
distribution (size)	√	√	X	OM: SEM(BSE,EDS): AFM:STM: TEM
2. Surface structure				
macroscopic (1–500 μm)	√	√	√	OM: SEM: profilometry
microscopic (10 nm–1 μm)	√	√	√	FESEM: AFM: TEM
nanoscopic (<10 nm)	√	√	√	STM: AFM: TEM
3. Surface sites				
structure (defects)	√	√	√	LEED: XRD(AR): EXAFS:XSW: STM: AFM:ISS
chemistry (reactivity, defects)	√	√	√	XPS: SAM: FTIR: Raman: NMR(SS): SIMS: STS: ISS: EELS
4. Grain boundaries				
structure	√	√	X	TEM: STM: AFM
segregation	√	√	X	PEELS(TEM): STS: XPS: SAM: SIMS
5. Intergranular films				
structure	√	√	X	TEM: STM: AFM
composition	√	√	X	PEELS(TEM): EDS: STS: XPS: SAM: SIMS

6. Depth profiles				
structure	√	√	X	XRD(AR): XSW: RBS: EXAFS: LEED: RHEED: TEM (X-section)
chemistry (composition)	√	√	√	XPS(AR): SAM: SIMS: PIXE
7. Adsorption				
molecular form (bonding)	√	√	√	FTIR: Raman: XPS: NMR(SS): SIMS(TOF): EXAFS(AR)
coverage (layers)	√	√	√	FTIR: Raman: XPS (AR): STM: AFM
distribution	√	√	√	SAM: SIMS(TOF): STM: AFM
8. Surface reactions, e.g.				
oxidation (hydrolysis)	√	√	√	SEM: FTIR: Raman: XPS: SAM: SIMS: STM(STS): AFM: ISS
dissolution (leaching)	√	√	√	XPS: SAM: SIMS: STM: AFM: PIXE: RBS: FTIR
phase transformation	√	√	√	XRD(AR): LEED: EXAFS: AFM: STM
9. Surface modification, e.g.				
calcination	√	√	X	XRD: XPS: SEM(BSE): TEM(ED): SIMS: FTIR: AFM
plasma	√	√	√	XPS (Auger): FTIR: XRD: TEM
surface layers (sizing)	√	√	√	FTIR: XPS: SIMS: SEM
flotation separation	√	X	X	XPS: FTIR: Raman: SAM: STM: AFM: SIMS(TOF): SEM(BSE)

Physical imaging of structure can be carried out with the low-cost, readily accessible techniques of optical microscopy and scanning electron microscopy. The usefulness of information obtainable from optical microscopy in reflection mode is underestimated in most approaches to the examination of these materials. In many cases phases can be identified directly from optical microscopy in combination with XRD data. The distributions of phases and their grain size (when greater than 0.5 μm) are also readily obtained. In glasses, devitrified (crystalline) regions can be identified together with structural defects and surface topography. Thin sections investigated in transmission mode with polarized light reveal strain fields, macroinclusions and triple points, and provide evidence for nonuniformity of composition. The optical microscope is an excellent starting point for the examination of the surfaces of minerals, ceramics and glasses, and gives the analyst a much clearer idea of the overall structure of the material at low resolution, for comparison with information from higher-spatial-resolution techniques further down the sequence. The merits of optical microscopy are strengthened by the fact that surface roughness and uniformity on the 1–500 μm scale are parameters of considerable importance in surface reaction, modification, bonding, adhesion and incorporation into other matrices.

The next higher level of spatial resolution (Table 2) is offered by routine SEM. Imaging in secondary-electron (SE), backscattered electron (BSE) and topographic modes extends that resolution to 0.1 μm. The complementary information available on phase identification and distribution from comparisons of SE and BSE imaging will be illustrated in the following sections. Similar images can be obtained by SAM in combination with identification of species and thus composition. The new generation of field emission SEM instruments has taken the resolution down to 2–5 nm in favorable circumstances. The high brightness of these electron sources has an additional advantage in that it allows examination of insulating surfaces without conductive coatings (e.g., C and Au). The accelerating voltage of the electron beam can be reduced, with some loss of resolution, in order to allow images to be obtained from materials as insulating as mullite, alumina and quartz without serious surface charging.

In order to obtain resolution at the unit cell level, one or more of the techniques of HRTEM (requiring thin sections), STM (requiring conducting samples) or AFM (able to deal with nonconductors) can be used. Lattice imaging by TEM can reveal defect structures in microdomains as linear or planar defects, and can provide direct information about phase relationships, grain boundary mismatch and intergranular films at the nm level. STM and AFM can identify individual reactive sites, impurity atoms and defect structures. The full range of imaging techniques is therefore capable of resolving structure on a dimensional scale from atomic to mm. The choice of information can often be

Table 2. Analysis Strategy for Application of Techniques to Give Information in a Top-Down Approach (i.e., macroscopic to microscopic to atomic scales)

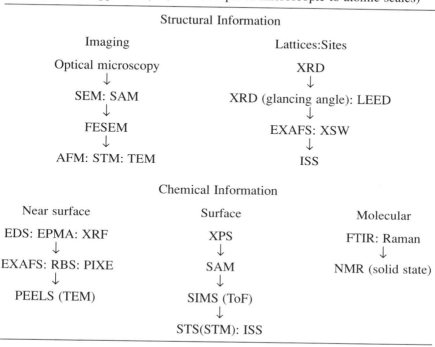

Structural Information

Imaging	Lattices:Sites
Optical microscopy	XRD
↓	↓
SEM: SAM	XRD (glancing angle): LEED
↓	↓
FESEM	EXAFS: XSW
↓	↓
AFM: STM: TEM	ISS

Chemical Information

Near surface	Surface	Molecular
EDS: EPMA: XRF	XPS	FTIR: Raman
↓	↓	↓
EXAFS: RBS: PIXE	SAM	NMR (solid state)
↓	↓	
PEELS (TEM)	SIMS (ToF)	
	↓	
	STS(STM): ISS	

made only after an initial survey of the material at the lower spatial resolutions of optical microscopy and SEM in combination with general information on surface and near-surface compositions from XPS and EDS/WDS.

The second group of techniques relating to structure, but based on diffraction or scattering, can also be used in the top-down approach. A routine examination of the mineral or ceramic sample with XRD is almost always worthwhile. The presence of unexpected phases, possible variations of unit-cell size with impurity incorporation, and identification of potential ideal site structuring in different surface facets can be revealed from a few hours work. If more specific information is required, e.g., the structure of a surface itself, then glancing-angle X-ray diffraction from large cleaved or polished surfaces and/or LEED can reveal any surface restructuring resulting from reaction, relaxation, rumpling and reconstruction. The relatively recently developed, but very promising, synchrotron-based technique of X-ray standing-wave (XSW) studies can probe surface structure at the next level of detail. The XSW method is particularly sensitive to the z-position of the atom. Standing waves generated from different diffraction planes allow the positions of surface atoms to be determined to high accuracy. The method can be applied not only to the structure of the native surface, but also to that of an adsorbate on a surface [10,11] and to the distribution of ions in the aqueous electrical double layer near the solid surface [12]. ISS(LEIS) can, at the atomic level, determine structure and composition of the top atomic layer of a surface [13,14]. The relative positions of the surface atoms can be determined layer by layer in the first few nm of the surface using this extremely surface-specific technique.

Turning now to chemical and compositional information an approach similar to that described above can be taken in order to determine chemical states (e.g., oxidation state, bonding, compound or species identification) and composition (e.g., quantification, for determination of stoichiometry, and for impurity identification).

It is necessary to distinguish between near-surface (i.e., 10 nm–1 μm) and surface regions (i.e., <5 nm) [15] (Table 2). The most readily available technique for establishing composition in the near-surface region is SEM with EDS (EDAX) or WDS capabilities. This technique parallels X-ray fluorescence (XRF) in its ability to determine approximate composition to depths of a few μm. The electron probe microanalysis (EPMA) technique, i.e., SEM with WDS, can provide more accurate quantification in the near-surface region. Recent advances in ultrathin window and windowless detector technology have extended the analytical range to include the low-*Z* elements C, O, N, and F, thus enhancing the method for the purpose of obtaining initial information. It is always useful to carry out EDS/WDS analysis for comparison with surface analysis so that the surface/bulk ratio of individual elemental species can be correlated. This ratio is often important in studies of segregation, leaching, dissolution and reprecipitation, and of surface reactions in general.

In studies of minerals involved in separation by flotation it is of paramount importance to distinguish changes in surface composition from bulk changes before (i.e., conditioned feed) and after (i.e., concentrate and tails) flotation, while in ceramics and glasses surface/bulk compositional ratios are often altered by processing at high temperature or by subsequent weathering reactions. Extended X-ray absorption fine structure (EXAFS) is a synchrotron-based technique capable of revealing the chemical environment of specific atoms in minerals, ceramics and glasses [10,16,17]. Adsorption processes and the formation of new chemical species and phases can be followed directly with EXAFS [16]. Near-edge X-ray absorption fine structure (NEXAFS) is another technique for obtaining information about the crystal and electronic structures, oxidation states and composition in the near-surface region. If higher spatial resolution of the near-surface structure is required, then the combination of TEM with parallel electron energy loss spectroscopy (PEELS) can be particularly valuable. The related, but not TEM-based, technique of HREELS has poor spatial resolution, but is useful for providing information on the molecular vibrational structure of adsorbed species, complementary to that obtainable from FTIR and Raman spectroscopies.

The most commonly used technique for obtaining chemical and compositional information from surfaces of minerals, ceramics and glasses is XPS. Assignment of chemical states from shifts in BE and quantitative analysis (to a precision of ca ±10% for most elements) can be obtained by routine analyses from regions of diameter 0.1–10 mm. Imaging XPS instruments have become available (see Chapter 4) and are capable of mapping the lateral distributions of specific chemical species with spatial resolutions at the 5–50 μm level, depending on the type of instrument and the system under investigation. There is real promise of extension of the lateral spatial resolution to 0.1 μm in the most recent generation of instruments. XPS can be applied to both conductive and nonconductive materials, making it ideally suitable for minerals, ceramics and glasses. If greater lateral spatial resolution is required, then compositions may be mapped in the 50–100 nm range by SAM with LaB_6 electron sources; extension to a few nm is possible with the more recent field emission sources. Correlation of point analyses and maps of elemental distribution with physical images in either the SE or the BSE modes can help to elucidate the behavior of minerals, ceramics and some glasses. A major disadvantage of SAM is that it is not in general suitable for insulating surfaces because of charging effects, but many minerals, ceramics and reacted glasses do have sufficient surface conductivity to allow analysis by that technique.

The next category of detailed information on surface chemistry and composition is that available from ToF-SIMS. With that technique the positive and negative secondary-ion mass spectra correspond to the fragmentation patterns from adsorbed molecules and reaction products, with contributions from sputtered substrate atoms. Although the spectra can be very complex, their analysis has allowed the mechanisms of surface bonding, reaction between adsorbed molecules, and substrate reactions to be studied [18]. The scanning imaging mode, using very low ion beam current densities and liquid metal ion guns (with beam diameters down to 0.1 μm), can provide direct information on the distribution and surface coverage of adsorbates and reaction products [19–21] (see also Chapter 6).

Detail at the atomic level of spatial resolution concerning chemical states and composition is available from STM operated in its various modes—scanning Kelvin probe and scanning tunneling spectroscopy (STS) (Table 2). Single-atom resolution and site identification with these techniques are applicable only to reasonably conductive samples. The alternative is LEIS from which a combination of local atomic structure and quantification of surface composition, especially for light elements, can be obtained [14].

For information on molecular structure, in particular as it relates to adsorbed molecules and their reactions on mineral, ceramic and glass surfaces, the most commonly used techniques are FTIR and Raman spectroscopies [22]. FTIR is used increasingly in the diffuse reflectance (DRIFT) mode for samples in air, and in the attenuated total reflectance (ATR) mode, particularly as circle cell attachments, for *in situ* adsorption, and for reactions in solution [23,24]. These techniques can reveal the functional groups of adsorbed molecules bonded to the surface and can track changes in molecular structure during kinetically controlled reactions at the surface. Quantitative estimates of adsorption, displacement by competitive adsorbates and the roles of acid and base surface sites on the substrate can also be followed. Another useful technique is solid-state NMR, from which detailed information has been derived about the behavior at specific sites (e.g., ^{29}Si, ^{27}Al) of hydroxyl groups and organic adsorbates [25,26].

In summary, it will be apparent that the normal starting point for an analytical strategy would be the obtaining of information with the readily accessible techniques, i.e., optical microscopy, SEM/EDS, XRD, XPS, and FTIR or Raman spectroscopy. Based on the outcome of the initial survey, the researcher/analysts should then be able to make a more informed choice of the detail and nature of information required subsequently, in terms of spatial resolution, atomic imaging, molecular structure, chemical states and lateral distribution. The requirements will also dictate whether the analysis should go in the direction of either greater spatial or better spectral resolutions.

In the next three sections, the strategy sketched out will be illustrated with examples demonstrating the merits of different techniques and methodologies.

4. MINERALS

The approach taken in this section will be to choose examples of the information generated by the techniques listed in the table. Rather than following in strict order the categories of information in Table 1, some typical problems will be addressed, and the contributions made by different techniques to the understanding required to resolve the questions will be discussed. In general, the approach will follow the scheme in Table 1, but it will become apparent that any technique may be able to contribute to more than one category. Reference to more comprehensive reviews, book chapters and books will be provided so that the reader may have access to additional details on experimental and instrumental conditions.

4.1. Phase Structures

The use of optical microscopy as a first step in understanding the structure of minerals and their surfaces has been discussed in the previous section. The type of information available from this technique, in combination with image analysis enhancement, and sample preparation and instrumental conditions, has been reviewed in Refs. [9 and 27]. In summary, it is possible to obtain the following information:

Particle size range (from image analysis) and shape
Phase identification (combination of transmitted, reflected and polarized light)
Phase distribution (volume fractions and spatial distribution)
Mineral association/liberation (down to an average dimension of ca. 0.5 μm)
Surface texture, morphology and porosity
In mineral separation (e.g., flotation), identification of losses due to liberation, size and surface chemistry [27]

Quantitative analysis of multiphase specimens can determine the proportion of different phases present in an ore, a mineral product or, indeed, a fabricated ceramic. Routines for Rietvelt analysis of X-ray (and neutron) diffraction patterns are now available in software packages for PC platforms [28,29]. A least-squares fit of a diffraction pattern, calculated from the crystal structure parameters of the known phases, is made to the observed pattern of the multiphase mixture. The intensities of all peaks from one phase are proportional to a sample scale factor Z and to the mass of that phase. The Rietvelt program [29] determines the scale factors and masses of the crystalline phases present. For instance, the wt% of each of six minerals (rutile, galena, pyrite, sphalerite, chalcopyrite and quartz), as inferred from neutron diffraction analysis, has been matched to the respective known percentages in a prepared mixture with an error of less than 5% in the final measure of fit [30]. The method is particularly useful for the determination of abundances of phases with the same nominal chemical composition but having different crystal structures. As an example, the relative proportions of the cubic, tetragonal and monoclinic phases of zirconia in partially stabilized zirconia ceramics can be determined by this method [30]. The masses of different phases of crystalline silica can also be measured and, in some cases, the proportion of amorphous silica inferred from the difference between the calculated and weighed amounts [30]. The method is unreliable when preferred orientation, due to directional grain growth or disposition, or when high concentrations of amorphous material are present. The results from Rietvelt analysis are sometimes difficult to interpret when major changes in the intensities of particular diffraction peaks are the result of incorporation of impurities, or when there are variations in stoichiometry between grains of the same phase within a single sample or between different samples. Nevertheless, Rietvelt analysis is now a well-established tool for determining crystalline phases in minerals and ceramics. The proportion of different phases likely to be present in surface analysis can be determined directly so that surface/bulk ratios can be measured.

Identification of different phases and their distribution can be made in mineral mixtures, ores, calcined products and ceramics using SEM/EDS. A comparison of the types of information available from SE, BSE and topographic images is presented in Fig. 1. The SE image reveals pores, cracks, grain edges and other topographical features. The typical BSE detector is an annulus split into two semicircles. In the BSE image, using the sum of the two signals, most of the contrast is due to compositional variations arising from the atomic number

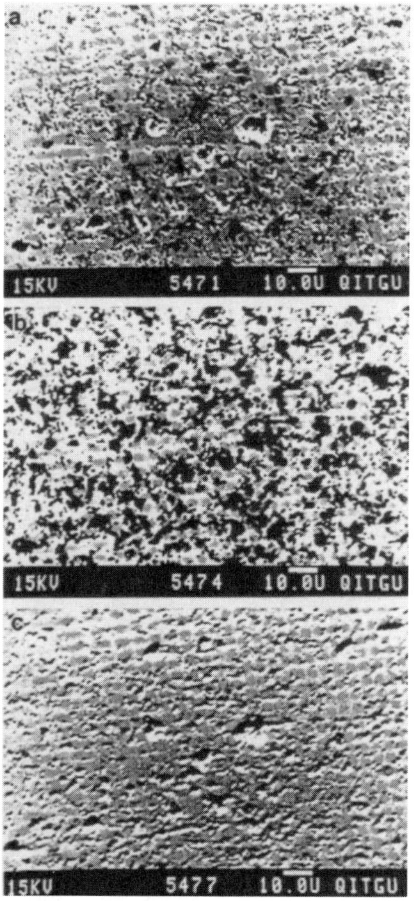

(a)

(b)

(c)

Figure 1. Comparison of (a) SE, (b) BSE and (c) topographic images from SEM micrographs of the same area of a polished ceramic surface [15] showing grain pullout and porosity. The scale bar is 10 μm. Note that the BSE image, in addition to porosity, indicates regions of low (dark) and high (white) average atomic number.

dependence of the yield of backscattered electrons. In many cases it is possible to distinguish different phases by the difference in contrast and thus to map the distribution of each phase. Also it can be seen that the BSE image does not reveal the structure of pores, cracks and other topographical features. The third image is obtained by taking the difference between the two BSE signals thereby suppressing the compositional dependence. This mode is sensitive predominantly to topography. The image shows the broad features of surface roughness at the cost of increased noise level [31].

A "locked," or composite mineral, particle comprising two or more phases can be analyzed by use of the sequence SE imaging, BSE imaging, and EDS analysis. Figure 2 shows, as an example, a muscovite (mica) and pyrite locked particle in the BSE imaging mode in which the higher average atomic number of the pyrite phase (light contrast) can be clearly distinguished from the lower average atomic number muscovite (dark contrast). EDS point analysis confirms this differentiation by recording signals from Fe, S, and O in the partially oxidized pyrite (light region) and from K, Si and O in the region with dark contrast. A version of this analysis is now used extensively in minerals processing

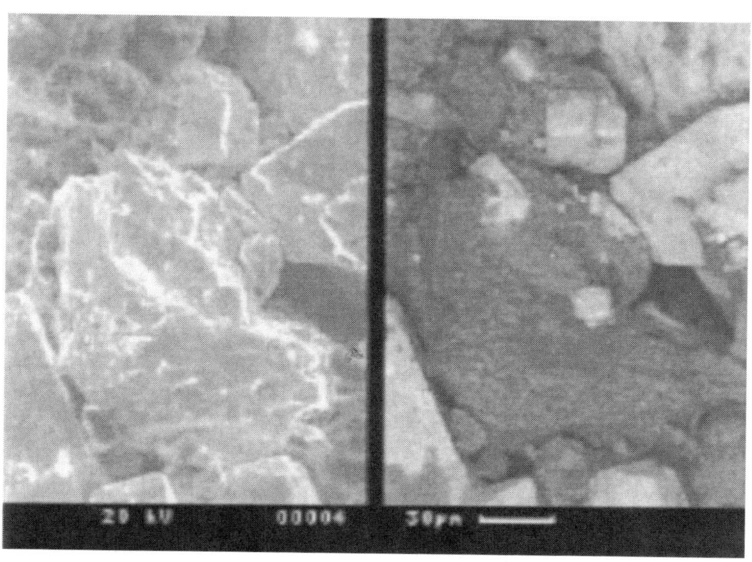

Figure 2. Comparison of SE (left) and BSE images (right) from the same area of pyrite and muscovite (mica) grains showing composites of the two minerals. The central muscovite particles have locked regions of pyrite (light contrast in BSE image) not liberated during grinding of the ore.

under the acronym QEM*SEM [32]. The technique measures BSE images and EDS spectra stereoscopically and refers the analytical information to a data base of compositions in order to identify the mineral phases. It can provide quantitative estimates of the percentage of each phase in different particle size fractions, and the proportion of each phase in locked composite particles. It is used widely for the assessment of ores, and for evaluation of mineral separation processes which include flotation, electrostatic and magnetic separation, and separation by gravity.

TEM of sections (usually <300 nm thick) can be used to study the internal microstructure, crystal structure and surface features of mineral crystals. Electron diffraction from areas of diameter down to 300 nm, as selected by an aperture (SAD), can be observed by imaging at the back focal plane of the objective lens, which gives specific identification of the mineral structure from the sharp intensity maxima, corresponding to Bragg diffraction by the periodic lattice. Amorphous films or regions give rise to broad rings in the diffraction pattern with radii which correspond to the most probable interatomic spacings in multiphase samples, or in samples with intergranular films, triple-point regions, microdomains and nonuniform grains. Information can be obtained about the extent of crytallinity and about the orientation and crystal structure of individual grains. Weak diffraction spots arising from surface structures can, in some cases, be used to form dark-field images which enhance surface detail. For instance, phase contrast imaging of MgO crystals after exposure to water vapor can reveal small (1–2 nm) rectangular depressions and protrusions which had been formed across the original smooth {100} surfaces [31]. Lattice imaging of very thin sections with the electron beam orientated down a crystal symmetry axis can be related directly to the crystal structure, with resolution corresponding to the spacing of the planes giving rise to the Bragg reflections. An example of an image showing point defects and dislocations is shown in Fig. 3. Profile imaging with the TEM [33] can reveal detail on the atomic scale of the surface structures of facets at the thin edges of crystals. Matching the experimental images to those computed from dynamical electron scattering models has established this technique as a complement on the atomic scale to SPM [34].

The composition of individual phases in the near-surface region is normally obtained by EDS or by the more accurate WDS while surface composition can be determined by XPS or SAM. Examples of the latter will be discussed later in this section. Other techniques, gaining in importance, which can be applied to compositional phase analysis are EXAFS and NEXAFS. Examples of application of these techniques can be found in the review by Greaves [10]. For instance, the ratio of monovalent to divalent copper in tetrahedrites (e.g., $(Cu,Fe)_{12}(Sb,As)_4S_{13}$) can be determined from NEXAFS peaks recorded at different X-ray absorption thresholds (which are dependent on oxidation state) [35]. Chemical state imaging of regions having the same element in different oxidation states by col-

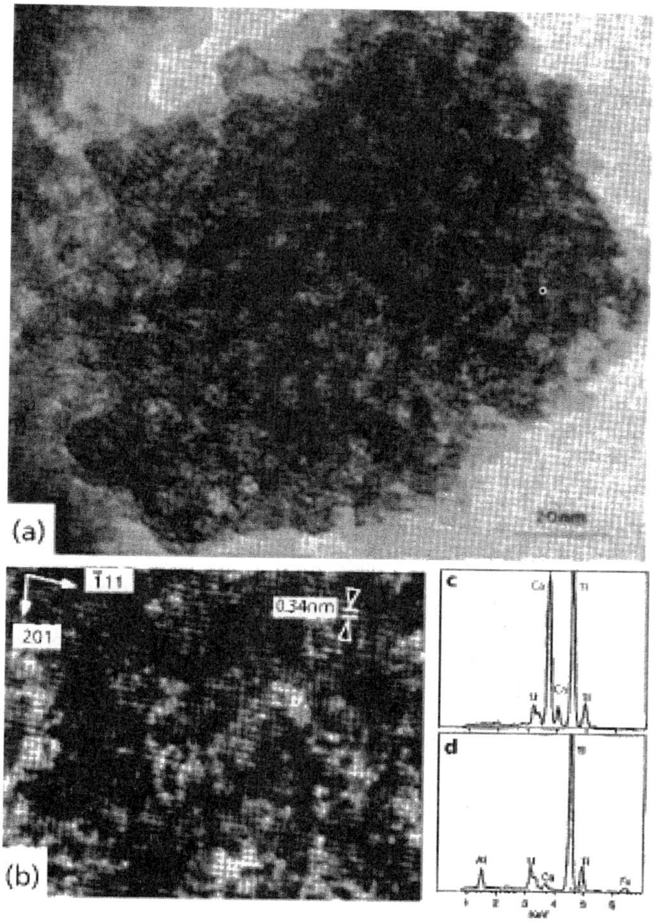

Figure 3. (a) HREM of brookite crystal precipitate on perovskite substrate. Note the poorly developed facets. (b) Enlargement of a portion of (a) illustrating lattice strain and incomplete crystallization. (c) EDS from perovskite substrate. (d) EDS from brookite showing some Al, U, Ca and Fe in solid solution.

lecting microfocus XRF data above and below the Fe K-edge can reveal haematite and magnetite inclusions in sintered iron ore [36]. Impurity sites (e.g., incorporating Ti and Mn in staurolite, $(Fe, Mg)_4Al_{17.3}(Si_{7.6}Al_{0.4})O_{48}H_3$ have been identified by refinement of the near-edge structure, and by comparison with model compounds. It was shown that the Ti species was associated with octahedral Al-sites and the Mn with tetrahedral Fe^{2+} sites; this was in direct contrast to the conclusions drawn from XRD refinement of the structure [10]. The implications of studies such as these which determine the expected site occupancy in the surfaces by impurity atoms is of considerable importance for studies using SPM.

4.2. Surface Structures

The use of SPM in studies of mineral surfaces has been reviewed by Hochella [37]. At the atomic level of detail, a combination of STM and STS has been used to reveal the positions of surface atoms and electronic structure of rutile (TiO_2) single-crystal surfaces [38,39]. The sites of surface titanium cations, but not those of the anions, were revealed clearly. A series of STM studies of sulfide mineral surfaces has also been completed by Hochella et al. [40–46]. Figure 4 illustrates the extreme sensitivity of this technique in the case of the (001) surface of galena (PbS); the atomically resolved image was recorded in the constant-height mode. In Fig. 4a, a surface unit cell is outlined with an edge dimension of approximately 0.6 nm. With a sample bias of –0.6 V (Fig. 4a) the high-intensity maxima in the image arise from tunneling contributions from the 3*p* electron states of individual sulfur atoms, whereas with a positive sample bias of +0.2 V (Fig. 4b) the main maxima still have sulfur character, but the subsidiary maxima now show the locations of electron states associated with lead cations [40].

An image from the (100) surface of pyrite (FeS_2) has been obtained with unit-cell resolution, with the main maxima being due to the Fe 3*p* states [42]. Steps on the surface with dimensions of half a unit cell were imaged. Imaging at atomic resolution of haematite (Fe_2O_3) [43] has revealed similar information about monatomic steps and the arrangement of oxygen atoms. Atomic positions in the STM map could be matched to theoretical models of a stepped (001) haematite surface, showing rows of oxygen atoms being offset as a step is crossed. The atomic structure and morphology of the more complex, nonconductive albite {010} surface has been studied with a combination of AFM and electron diffraction [47].

These examples serve to illustrate the power of SPM, in combination with spectroscopic information (e.g., STS), in defining the structure of surfaces, defects at lattice points, corners, steps, edges, dislocations, etc.

SPM can also be used to study directly the reactivity of mineral surfaces. The oxidation of sulfide mineral surfaces, weathering (e.g., leaching and dissolution) of nonsulfide minerals, sorption of cations and anions, and reactions taking place during surface modification can be studied on the atomic scale. Eggleston and Hochella [40,41,44] have proposed models for oxidation of the galena surface on the basis of information from images in which individual atoms were observed to react on contact with air or water. Figure 5 shows the

(a)

(b)

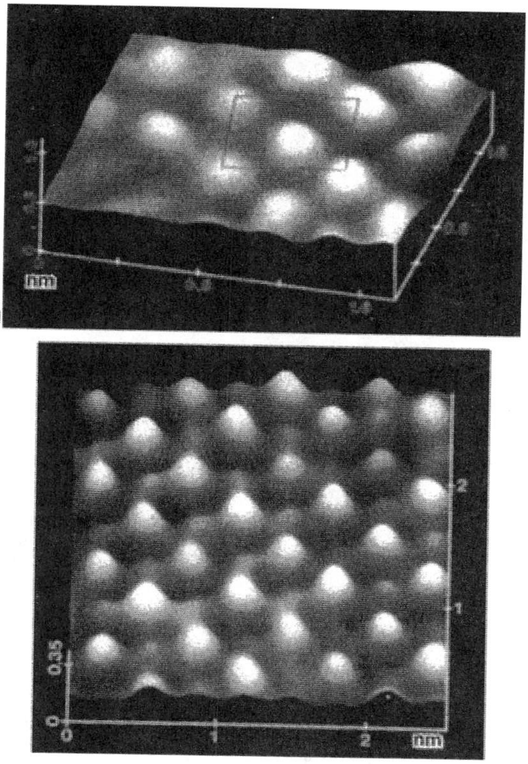

Figure 4. (a) STM image of the (001) surface of galena taken in the constant-height mode –600 mV sample bias and 2.9 nA tunneling current. The peaks in the image are the tunneling contribution for 3*p* electrons from individual S atoms on the surface. The box outlines the surface unit cell, ca. 0.6 nm on an edge. (b) STM image of the same surface as in (a) but taken at +200 mV sample bias and 1.8 nA tunneling current. The major peaks still have S character, and the smaller peaks in between have Pb character. All scales are in nm for both images [40].

(001) surface of galena after exposure to water for 1 min. On oxidation, the tunneling maximum from a surface sulfur atom disappears due to alteration of the local chemical structure. In the image (Fig. 5), the apparent vacancies are the sites of oxidized sulfur atoms. In their work, and in another study described below, it is clear that, when a surface S atom is oxidized, atoms in adjacent sites are activated, thus promoting further reaction rather than initiating new reaction at S sites further away. There is an observed trend for [110] directionality at the boundaries of the oxidized patches. Eggleston and Hochella [44] suggest that

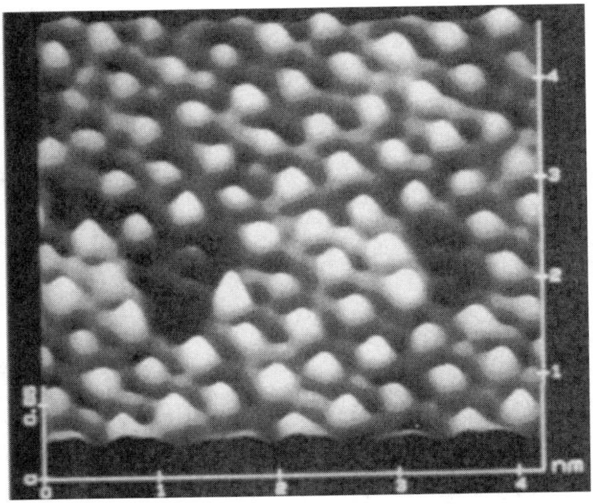

Figure 5. STM image of the (001) surface of galena taken after exposure to water for 1 min. The conditions of image collection were the same as in Fig. 4b. The peaks represent surface S, and the lower maxima in between are Pb atoms. The apparent vacancies are oxidized sulfur sites; the tunneling current to these sites is very low [37].

this is due to the association of only one nearest-neighbor oxidized S atom with an adjacent unoxidized S across the [110] front, whereas across a [100] front there are two nearest-neighbor oxidized S atoms adjacent to an unoxidized S. This suggests that the [100] fronts move faster and ultimately become lost.

Sorption of gold (Au(III)) from solution has been followed directly on galena surfaces over time scales ranging from seconds to several days [41,44]. The metallic nature of the small islands (of dimensions <10 nm) was confirmed with STS information from gold foil standards and in an XPS study by Bancroft and Hyland [48]. Island growth on unoxidized galena surfaces again shows [110] directionality and is faster by a factor of 3 than growth on preoxidized PbS surfaces. In similar studies combining STM, XPS, FTIR and SAM, the sorption of uranium on pyrite and galena surfaces [48,49] has been examined. Reduction of U(VI), coincident with oxidation of the sulfide surface to polysulfides (for PbS), or to Fe(III) oxides/hydroxides (for FeS_2), at inhomogeneously distributed reaction sites can be followed and is thought to be associated with surface roughness and defect sites.

Precipitation reactions involving the formation, growth, and coalescence of surface nucleation sites, are important in studies of the processes of dissolution, solution speciation and saturation followed by precipitation. AFM has been used to follow the appearance and development of the nucleated growth of calcite on a calcite crystal substrate in solution [50]. Growth was observed to take place along steps 1–3 monolayers in height. The effect of inhibitors such as phosphate on the growth mechanism suggested preferential adsorption at steps, edges and corners which blocked sites for incipient nucleation and crystalline growth. The transition from surface nucleation to spiral growth is apparently then delayed until after nearly 2 h in solution.

At an intermediate level of spatial resolution, studies of oxidation and reaction processes using SPM can also be of fundamental importance for understanding the mechanism(s) and progress of alteration of a surface. Continuing with the model (100) PbS surfaces, a systematic combination of STM and XPS investigations of oxidation in air of a 70×70 nm^2 area illustrates the usefulness of information at this level of detail. Figure 6 shows the progressive growth of oxidation products; the original surface features with lateral dimensions of less than 0.6 nm eventually evolve into overlapped regions greater than 10 nm in diameter [51]. The final stage is that of a nearly continuous oxidized surface with "holes" of dimensions of a few nm through which continued reaction can apparently proceed [52]. XPS information, obtained in parallel, could be correlated with the initial appearance of hydroxide (and possibly peroxide (O^-)) species followed by growth of the predominant product lead hydroxycarbonate from reaction with CO_2. It is interesting that, even after reaction with air for a duration of 120 min,

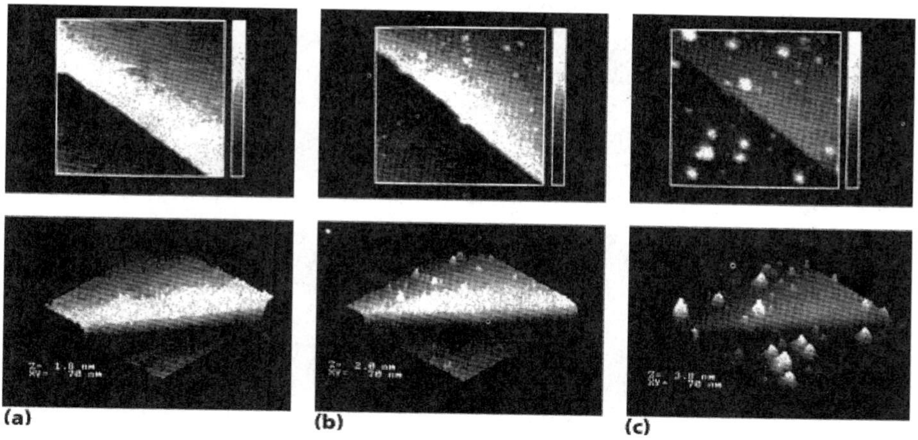

Figure 6. STM images from a 70×70 nm^2 area of (a) freshly cleaved galena surface; (b) the same surface after 70 min standing in air; (c) after 270 min in air. The upper row are gray scale images; the lower row are three-dimensional (rotated) images with the vertical scales of 1.8, 2.0 and 3.8 nm, respectively [51].

the S 2*p* spectra still showed only sulfide character; i.e., sulfur-oxygen species were not formed despite the interaction with the lead ions. This may have been due to attenuation of the signal from any sulfur-oxygen species by the overlying lead hydroxy carbonate or, as suggested by Buckley and Woods [53], to diffusion of lead ions into the metal-deficient surface. In this work, there was also an apparent absence of preference for initiation of reaction at low-coordination sites (i.e., corners and steps), but instead an apparently random initiation at sites on the (100) surfaces. In a subsequent investigation [54], comparison of natural galena (being the subject of earlier studies) and high-purity synthetic galena revealed that the difference was likely to have been due to the presence of impurities in the surface of natural galena. Growth of oxidation products on the synthetic sample occurred preferentially at step edges and dislocation sites instead of on (100) faces as on the natural galena surface. The rate of surface oxidation of the synthetic galena, as inferred from both STM images and XPS O/Pb ratios, was lower than that for natural galena by factors greater than 10. The XPS spectra showed in both cases that lead hydroxide was the main oxidation product. Exposure beyond 6 h in air generated, for the natural galena but not for the synthetic galena, small signals from sulfate and carbonate species. The effects of impurities, which dominate the oxidation process of the galena surface, have important consequences for the mineral processing industry since the impurity concentrations in natural galena from different ores, and even from different locations within the same ore body, can be very variable thereby causing wide differences in rates of oxidation and in subsequent processing behavior.

It is also possible to use SPM techniques for *in situ* studies of the mechanisms of oxidation and reaction in solution. A different mechanism for oxidation of galena occurs in aqueous solution when compared with that in air [55]. STM imaging over 500×500 nm^2 fields of view is shown in Fig. 7. The development with time of pits of sub-nm dimensions is illustrated. The *z*-dimensions of these pits correspond to half- and full-unit-cell parameters of galena [55]. AFM imaging under the same conditions revealed a closely similar development of pits. The main surface process occurring is that of congruent dissolution; this has been confirmed by XPS, which shows unaltered Pb 4*f* and S 2*p* spectra during the initial stages. The rate of formation of dissolution pits is strongly dependent on the concentration of impurities (as above), on pH and on the gas used to purge the aqueous solution. High-purity synthetic galena reacts much more slowly. Analysis of the STM images allows quantification of the rate of development of the dissolution pits in terms of area (*x, y* dimensions) and depth (*z*-dimensions). Figures 8 and 9 illustrate the dependence of the reaction rate on pH and on the reactive gas.

In the same sequence of studies, STM was used to observe the deposition of colloidal (ca. 50×20 nm^2) lead hydroxide structures (confirmed by XPS analysis) on a galena surface from a 10^{-3}M solution of Pb ions at pH 7. Figure 10

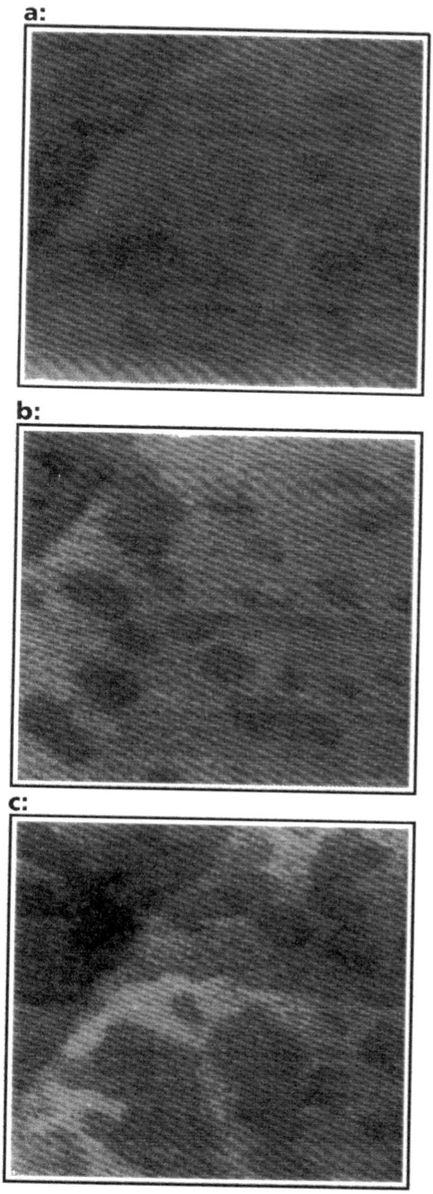

Figure 7. Top view STM images of galena as a function of time in air-purged water at pH6 (*xy*-scale = 500 nm): (a) 40 min (*z*-scale = 3.8 nm); (b) 60 min (*z*-scale = 2.6 nm); (c) 120 min (*z*-scale = 2.8 nm) [55].

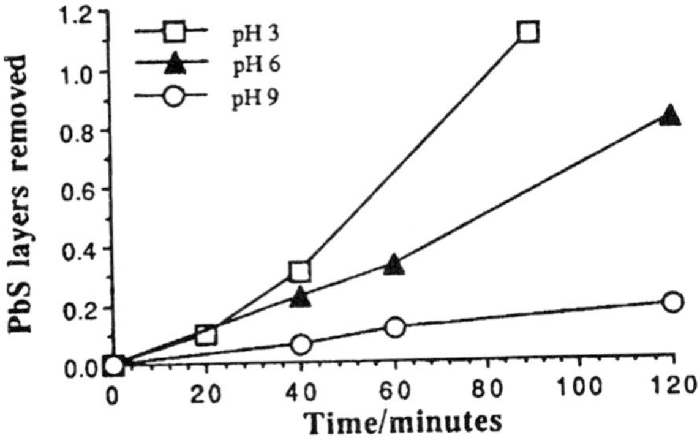

Figure 8. Number of equivalent PbS monolayers removed from the galena surface as a function of time in air-purged water at different pH values [55].

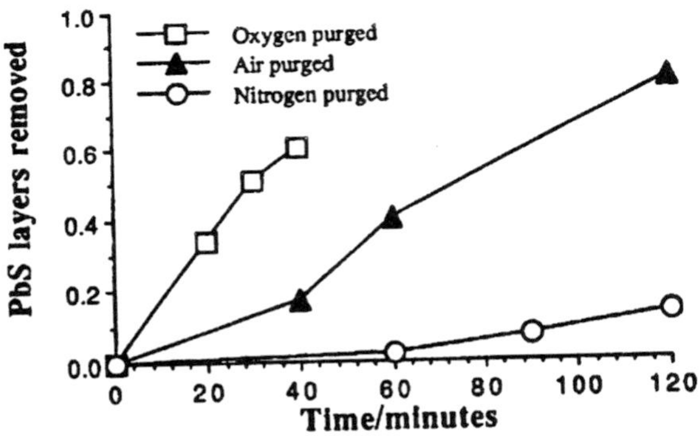

Figure 9. Number of equivalent PbS monolayers removed from the galena surface as a function of time in water at pH 7 with different purging gases [55].

illustrates the [110] directionality of the precipitation. The presence of oxidized lead regions on the surface is likely to have implications for flotation separation of galena in minerals processing. The reactions of surfaces with collector molecules used for separation of minerals can be followed by *in situ* imaging. Adsorption of molecules, such as ethyl xanthate, is used to render the sulfide surfaces hydrophobic, thus promoting bubble-particle attachment and consequent concentration of the desired sulfide in the flotation froth which is collected at the top of the flotation cell. *In situ* STM images of galena surfaces in the pres-

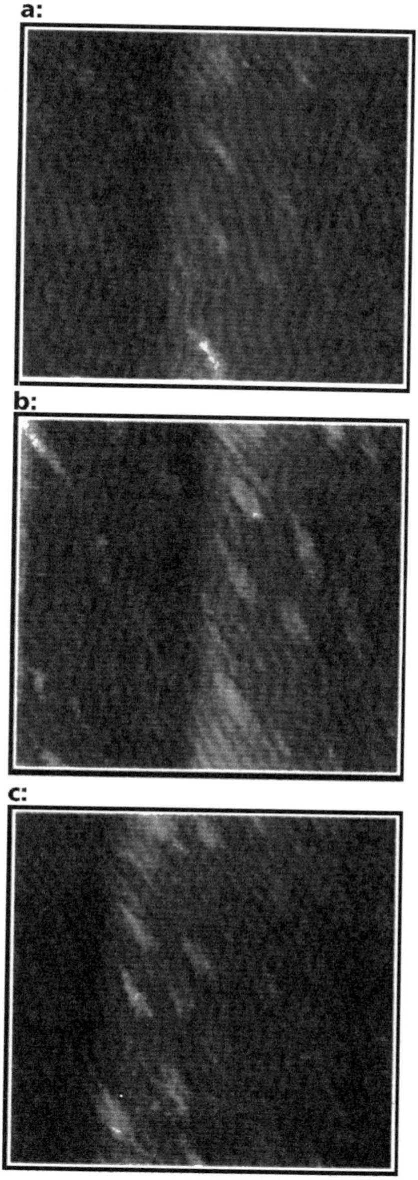

Figure 10. Top view STM images of galena treated with 10^{-3} M lead (II) ions at pH 7, as a function of time (*xy*-scale = 500 nm): (a) 10 min (*z*-scale = 11.2 nm); (b) 30 min (*z*-scale = 13.7 nm); (c) 50 min (*z*-scale = 15.4 nm) [55].

ence of ethyl xanthate showed attached colloidal projections, which were confirmed by XPS and FTIR to be lead ethyl xanthate. The XPS results suggested that the formation of that species displaced some hydroxide from the oxidized galena surfaces at pH 10. The rate of formation and extent of surface coverage could be measured directly and were controlled by ethyl xanthate solution concentration and by the extent of preoxidation of the galena surface [56].

These examples may be related specifically to the information categories in Table 1 under the headings of phase structures, surface structure (from atomic to 500 nm scale resolution), surface sites (e.g., structure, chemistry, reactivity and defects) adsorption (e.g., distribution, coverage, monolayer versus multilayer, colloidal, molecular form and bonding) and surface reactions (e.g., oxidation, dissolution, precipitation and phase transformation).

4.3. Surface Sites

Structural investigation of surface layers, which aims to define the positions of atoms in the surface phase, can be carried out by LEED and, more recently, by glancing-angle X-ray diffraction and EXAFS. LEED methodologies have been dealt with in the review literature [57]. Information on glancing-angle XRD and EXAFS methodologies is scarcer, but the techniques have received some attention [10,58]. Of relevance to this section, it is noteworthy that an overlayer of α-Fe_2O_3 of thickness 9.0 ± 1.5 nm on a γ-Fe_2O_3 substrate has been identified [59]. XRD and EXAFS techniques have a major advantage in that experiments can be performed in air or in liquids (provided that the liquid is less dense than the solid). They can therefore be used to study reactions *in situ*.

A good example of the information on surface sites which is available from the combined use of XPS and SAM is the series of studies of iron sulfide mineral surfaces carried out by Pratt et al. [60,61]. They have published the first spectroscopic evidence for the existence of Fe(III) on pyrrhotite (Fe_7S_8) surfaces. Careful calculation and curve fitting of the multiplet and satellite structures associated with each of the Fe(II) and Fe(III) atoms bonded to sulfur indicated the presence of 32% Fe(III) surface sites, in close agreement with stoichiometry (which suggests 29% Fe(III) in the particular pyrrhotite studied). The fitted data are illustrated in Fig. 11 for the Fe $2p_{3/2}$ and S $2p$ spectra. The sulfur atoms are present primarily as monosulfide (S^{2-}) with minor contributions from disulfide (S_2^{2-}) and polysulfide (S_n^{2-}). After 6.5 h oxidation in air, 58% of the iron is in the Fe(III) state and bonded to oxygen, while most of the Fe(II) remains bonded to sulfur. The sulfur spectra show a range of oxidation states from sulfide to sulfate. ARXPS and AES depth profiles revealed three compositional zones, with an outer zone composed of iron oxyhydroxides, an underlying zone of iron-deficient, sulfur-rich, reacted pyrrhotite, and then a continuous compositional grading to bulk pyrrhotite. The outer zone was 0.5–1.0 nm in thickness, and was separated by a sharp interface from the reacted

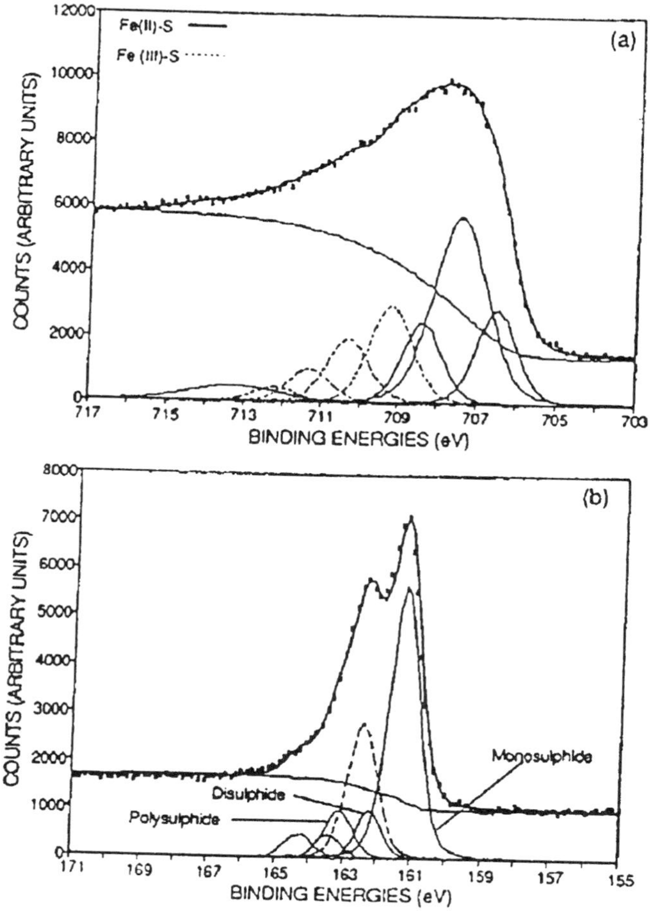

Figure 11. XPS spectra of the pyrrhotite sample fractured under high vacuum, 10^{-3} Pa: (a) Fe $2p_{3/2}$ spectrum; the series of peaks with solid lines represent Fe(II) (multiplets and satellite) bonded to S; (b) S $2p$ spectrum of pyrrhotite surface with S $2p_{3/2}$ peaks plotted as solid lines and S $2p_{1/2}$ peaks plotted as dotted lines [60].

layer, which had a thickness of ca. 3 nm. The mechanism of Fe diffusion reaction with oxide, hydroxide and water, and the production of disulfide species in the S-rich zone beneath the iron oxyhydroxide layer, could all be demonstrated using the combination of the two techniques.

It is also noteworthy [62] that the adsorption of water alone on the pyrrhotite or pyrite surfaces does not result in oxidation. The O 1*s* spectra are then composed of contributions from hydroxide and water species, but there is no evidence for the formation of iron oxyhydroxide or oxidized sulfur products. Careful analysis of the multiplet and satellite contributions to the Fe $2p_{3/2}$ structure reveals that the interaction of water with the two minerals occurs as a result of fundamentally different processes. Pyrrhotite reaction involves the transfer of electronic charge through iron vacancies, whereas on pyrite water interacts with the Fe3*p* (e_g) molecular orbital, suggesting that hydrogen bonding with the disulfide sites may be important.

Similarly detailed information on the chemistry of atoms at surface sites and on the relationships of such sites to surface reactivity may be obtained for oxide, silicate, aluminosilicate and other minerals using XPS in association with EXAFS and adsorption studies. The reviews by Bancroft and Hyland [48], and Brown et al. [16], provide many examples of the types of information obtainable for these surfaces.

Although not an experimental surface analytical technique, recent advances in the application of molecular modeling for the determination of location, electronic structure and chemistry of atoms at surface sites on mineral and ceramic surfaces should be noted in this section. Modeling has now reached a stage of maturity from which, in some cases, it is possible to predict experimental results ahead of actual measurement, i.e., theory is leading experiment. An early example concerns the prediction that sites on nondefective (100) MgO surfaces would not cause dissociation of small molecules such as CO and H_2O [63]; this was later verified experimentally using atomically flat MgO smoke particles [64]. Dissociation and reaction at, and activation of, adjacent sites were predicted and found to occur at defects, and at low-coordination sites at corners, steps and edges [65]. Tossell [66] provides a useful review of the various quantum mechanical methods for calculation of the free energy of mineral surfaces. Examples dealing with surface reconstruction of the [111] face of sphalerite (ZnS), the [001] face of Al_2O_3, and the electronic structure and spectra of the reduced oxide TiO are presented in the review. Major reviews of oxide surface models have also been provided by Henrich [67] and Kunz [68]. The time may be approaching when it will be possible, if not routine, to match experimental observations of site structure and chemistry with theoretical models of the electronic structure of the sites and the surface. Such correlations will clearly lead to an increasing understanding of reaction mechanisms at mineral and ceramic surfaces.

4.4. Grain Boundaries and Intergranular Films

The topic of segregation to grain boundaries and surfaces is of particular importance to the behavior of ceramic materials during the fabrication process. Solid-state reactions, sintering, grain growth and resulting mechanical (e.g., fracture strength, toughness, high-temperature creep and wear resistance) and electrical properties (e.g., conductivity and dielectric loss), are all affected directly by structural and compositional factors. Some examples of relevant studies of grain boundaries and intergranular films in ceramics will be given in Section 5.4. In principle, information on segregation, structure and composition in grain boundaries and intergranular films is of similar importance in the study and behavior of minerals, particularly where multiphase composites or ores are involved.

The methods for studying grain boundaries have been reviewed by Burggraaf and Winnubst [69]. The most direct method of observation is HRTEM in combination with PEELS or EDS analyses. Those techniques allow the grain boundary to be imaged so as to identify any lattice mismatch of adjacent grains and the presence of segregated elements or impurities. Grain boundaries and intergranular films can also be studied by exposure of fracture faces, if the predominant mechanism of fracture is intergranular rather than transgranular. In this case, a variety of surface analytical techniques, including XPS, SAM, SSIMS and ISS, may be used to determine composition, distribution and chemical states at the fracture interface. Fracture in the vacuum environment of the spectrometer, or carried out externally under liquid nitrogen with the use of a sample transfer vessel, may be required in order to avoid oxidation or other reactions of the fracture face with the ambient.

It is important to realize that segregation factors (i.e., surface/bulk elemental ratios) can range from 2 to more than 10^3, and that equilibrium segregation at grain boundaries and surfaces is normally limited to regions of depth some tens of nm, but in cases of nonequilibrium segregation much thicker layers can be found. Studies have revealed that the driving force of segregation can arise from strain relaxation, when there is a large misfit of ionic radii of the elements within the mineral phases, or from space-charge effects with impurities of differing valency, or from a combination of both mechanisms. Hence, in naturally formed minerals, the structure and composition of fracture faces may be significantly different from the bulk mineral, thereby affecting surface chemistry and reactivity.

Grain boundaries can also be important in the formation and stabilization of microinclusions. A specific example can be found in the work of Pring and Williams [70] in which TEM studies revealed microinclusions as small as a few nm of galena (PbS) in pyrite (FeS_2). Structures of this type have potential implications for selective separation of the minerals, and may affect the ultimate grade of mineral concentrates.

Properties of intergranular films with widths of a few nm, and with composition and structure distinctly different from those of the mineral phases, have been studied by XPS, SIMS, SAM and related techniques. A specific example concerns the presence of relatively thick (i.e., >25 nm) surface coatings of graphitic carbon on iron sulfide fracture faces after grinding a copper/lead-zinc sulfide ore

[71]. A proportion of the iron sulfide particles with this type of surface layer floats naturally, which is an undesirable effect during the concentration of the copper, lead and zinc sulfides by mineral flotation separation. The carbonaceous layers are formed during ore genesis and may vary considerably from one deposit to another. Additional studies have shown that the form of the carbon can be either graphitic (i.e., conducting), saturated hydrocarbon (i.e., insulating) or humic/fulvic (i.e., oxidized hydrocarbons and therefore insulating) residues depending on the degree of reduction experienced by the ore in its genesis [72].

4.5. Depth Profiles

Depth profiling using ARXPS (a nondestructive method), or by ion beam sputtering in combination with XPS, SAM, or SIMS, is often used in order to study the variation of composition with depth, as in the examples above dealing with pyrrhotite and pyrite carbonaceous layers. Additional examples of the multitechnique approach to the study of the composition of surface layers can be found in a series of papers describing the analyses of weathered or dissolved feldspar mineral surfaces [73–76]. A combination of TEM, XPS, SIMS and SEM was used. Structural and compositional profiles through the surface layers provided strong evidence for the formation of leached layers from which Ca and Al had been removed, leaving a residual porous structure enriched in Si. The thicknesses of the altered layers were dependent on the pH of the reactant solution, and on the composition of the plagioclase feldspar. The dependence on pH of the depth profiles obtained by SIMS for labradorite in pH 3.5 and 5.7 is illustrated in Fig. 12. The altered layers increased in thickness from a few tens to over 100 nm in the order albite < oligoclase < labradorite < bitownite. SEM and TEM micrographs of labradorite specimens showed lamellae of calcium-rich and sodium-rich phases of widths ca. 70 nm. The calcium-rich phase was weathered preferentially to a greater average depth (135 nm) than that of the sodium-rich phase, thus producing a corrugated surface. The XPS analysis provided support for a mechanism of hydrolysis of the Si–O–Al bonds, which allowed ion exchange during the initial leaching process. This process has a well-established analog in the weathering of glass surfaces [4]. The surface reaction process can be followed in the spatial, structural and chemical domains using the multitechnique approach with angle-resolved analysis and depth profiling by ion etching.

4.6. Adsorption

Information concerning adsorption according to Table 1 may be obtained as follows. Spatial distribution and surface coverage of adsorbed molecules can be measured from the atomic to sub-μm level of resolution using STM (with STS) or, for

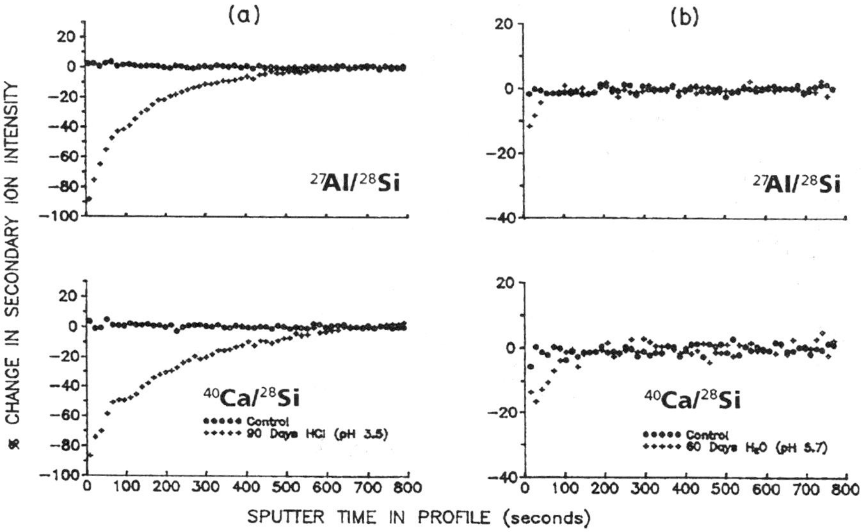

Figure 12. SIMS depth profiles for (a) dissolution of labradorite for 90 d in an aqueous solution of HCl (pH 3.5), and (b) dissolution of labradorite for 60 d in doubly distilled and deionized water (pH 5.7) [53].

nonconductive surfaces, with AFM as shown in Section 4.2. For imaging at lower spatial resolution, ToF-SIMS, particularly when used in the SSIMS mode, has proved to be a valuable tool. For instance, the spatial distribution of the molecular ion of diisobutyl dithiophosphinate on galena surfaces has been observed. Figure 13 shows a comparison of a positive-ion SIMS image of Pb abundance with the negative-ion image of the abundance of the molecular ion of the adsorbed dithiophosphinate. The observed uneven spatial distribution suggests face-specific adsorption [19], and is in accord with similar images from previous work where ion fragments were imaged under nonstatic conditions with a quadrupole SIMS spectrometer [20]. The 200 nm spot size of the Ga^{69} ion gun operated in a pulsed mode was used in combination with charge neutralization (i.e., the electron flood gun was pulsed out of phase with the ion gun). Areas of dimension from 60 to 250 μm were examined. ToF-SIMS has also been used to determine directly surface compositions of mineral particles from concentrator plant operation [21]. Mineral particles ranging in dimension from 20 to 100 μm were probed in order to obtain surface composition. The initial ion pulse is claimed to provide submonolayer depth resolution. Spot analyses (from areas of 250 nm diameter) determined the presence of flotation-activating species (i.e., Cu and Pb) and oxidation products on sphalerite, pyrrhotite, pyrite and quartz. Spatial distributions

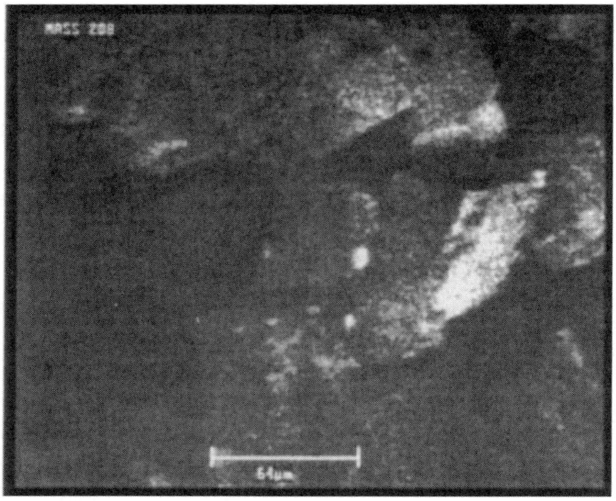

(a)

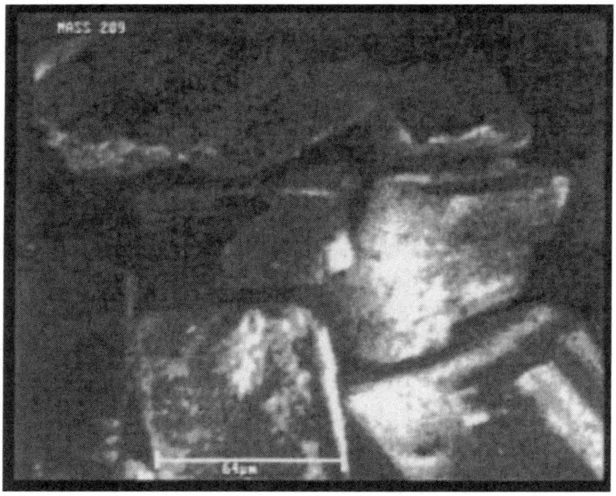

(b)

Figure 13. (a) Positive-ion SIMS image of Pb at 208 amu from floated galena crystals [19]. (b) Negative-ion SIMS image, from the same field of view, of the molecular ion of diisobutyl dithiophosphinate (209 amu) adsorbed onto the surface of galena.

across the surfaces were mapped. The distributions of adsorbed collector molecules, namely amyl xanthate and diisoamyl dithiophosphate were also mapped on surfaces of laboratory-prepared galena. Uneven distributions were again observed of both the adsorbed cations and the adsorbed xanthate or dithiophosphate. Specificity, in terms of crystal facet, of the adsorption of xanthate and dithiophosphate was confirmed in images of the molecular ion and ion fragments. Correlations of the adsorbed molecules with surface concentrations of hydroxyl groups were carried out in both sets of experiments. There is still controversy as to the association of these two species, but it is likely that more detailed information available from ToF-SIMS will ultimately allow resolution of the debate.

Information about surface coverage (e.g., monolayer, multilayer, or colloidal), molecular form and bonding can be obtained from a combination of XPS and FTIR analyses, and electrochemical experiments. Some of the methods used to model the surface of PbS with xanthate adsorption have been summarized [77]. Careful analysis of the results from a number of studies has shown that there is good agreement between XPS and FTIR, with the latter used in the ATR mode, when surface adsorption is measured at potentials equilibrated in air and under controlled potential conditions. The techniques have established that there is a submonolayer coverage of adsorbed lead ethyl xanthate at potentials below those at which the species is expected to form in bulk solution. It was also established that, even at relatively high concentrations of xanthate in solution, there was still significant coverage of the surface by adsorbed oxygen species. Orientational effects, evident from the XPS spectra, as well as small differences in the positions of IR absorption bands corresponding to adsorbed product and crystalline lead ethyl xanthate, suggested that the adsorbed molecules had perpendicular orientation with respect to the surface. In this case, multilayers of adsorbed molecules were apparently not formed. Examples of multilayer formation will be discussed, however, in Section 6.4 on glass surfaces.

4.7. Surface Reactions

Some of the entries under this heading in Table 1 will be illustrated with particular examples. The variety of surface reactions of minerals, and the techniques used for their study, is far too broad for all aspects to be included in this chapter. Several monographs [1–3,15, 67] and a review [4] provide many relevant case studies.

Surface oxidation of sulfide minerals has been reviewed recently by Smart et al. [78]. Studies of the physical and chemical forms of oxidation products by SAM, XPS, STM, AFM, SEM and ToF-SIMS have revealed several different processes of oxidation. The seminal work of Buckley et al. (e.g., [79,80]) was the first to identify the process of formation of oxyhydroxide products on underlying metal-deficient, sulfur-rich layers of similar forms to those described in Section 4.3. Other oxidation products have been observed directly, such as polysulfides, elemental sulfur, oxidized fine sulfide particles, colloidal hydroxide particles and flocculated aggregates, as well as continuous surface layers of oxyhydroxide species (of variable depth), and sulfide and carbonate species. Different spatial distributions (i.e., isolated, patchwise or face-specific products) have already been described in previous examples in this chapter. The review also identifies the actions of adsorbed molecules in several different modes, such as: adsorption at specific surface sites; adsorption or precipitation of colloidal species from solution; detachment by competitive adsorption of small particles from surfaces of larger particles; detachment of small oxyhydroxide particles; removal of adsorbed amorphous oxidized surface layers; inhibition of surface oxidation; disaggregation of larger particles; and patchwise or face-specific coverage. Examples of multitechnique studies of each of these processes will be presented.

Several reviews have described leaching, dissolution and precipitation reactions for oxides [81,82], silicates, aluminosilicates [16,17,83] and titanates [4,84]. These reviews point in particular to the necessity for studying both solution and surface phases in order to understand fully the mechanisms governing each reaction. For instance, while solution analysis may suggest that simple ion exchange is occurring at the surface, surface analysis has shown that bond-breaking reactions in many cases precede ion exchange and leaching, with the products of the reaction being retained *in situ* at the surface. The kinetics of leaching may therefore be governed by that of the hydrolysis reaction (often base-catalyzed), and not by ion exchange (i.e., due to diffusion limitations). *In situ* phase transformation and precipitate formation are also observed to take place without significant alterations to the solution phase, as in the transformation of $CaTiO_3$ to TiO_2 as brookite and anatase. Precipitation on the same site of layers with different compositions has been observed in SAM depth profiles, which show that sequential saturation of the solution will occur with nucleated growth at crystallites established previously [84].

An example of surface phase transformation can be found in the work of Jones et al. [85]. Pyrrhotite ($Fe_{1-x}S$) surfaces in acid solution restructured to a crystalline, defective, tetragonal Fe_2S_3 surface phase due to loss of Fe to solution. The metal-deficient, S-rich product was identified by XRD in combination with XPS. Linear chains of sulfur with a nearest-neighbor distance similar to that of elemental S were observed, but with an S $2p$ binding energy 0.2–0.6 eV less than that of S_8. As a result of the oxidation process, hydrophilic iron oxides were found at the surface.

4.8. Surface Modification

Two examples of surface modification of minerals will be discussed in order to illustrate the merits of a combined-technique approach. The breadth of the subject constrains the discussion to that dealing with strategies for gathering information and for choosing the most relevant surface analytical techniques.

Surface modification of the kaolinite mineral arising from low-temperature water vapor plasma reactions followed by adsorption can produce tailored interfaces ranging from fully hydrophilic to fully hydrophobic [86]. The surfaces of the basal planes of kaolinite, containing siloxane (Si–O–Si–) and aluminium hydroxide (Al–OH) structures, are relatively inert, with the edge sites of the platelet being most reactive for the unmodified kaolinite. Reaction in a water vapor plasma (for a duration of 3 h) renders the surfaces reactive due to an increase in the number of sites now apparently located on the basal planes, as well as at edges of the kaolinite. The reaction could be monitored by FTIR analysis based on the absorbance of a new peak at 1400 cm^{-1}. The reacted surface had a contact angle of 0°, and the kaolinite then underwent total dispersion in aqueous solution. The activated surface could be rendered irreversibly hydrophobic by reaction with conventional adsorbates, such as monochlorosilanes. The uptake of the latter could be monitored by measuring the two resolvable contributions to the Si 2*p* XPS spectrum. SEM and TEM images failed to reveal any observable changes to the morphology of the kaolinite platelets as a result of plasma reaction or adsorption. Recent molecular modeling studies [87] have suggested that the plasma reaction leads to the attachment of –OH species at the Si sites in the siloxane basal plane rather than the breaking of Si–O–Si bonds. The energetics of the attachment route are more favorable than those of the bond-breaking reaction, in contradiction to an initial interpretation of the plasma reaction mechanism. These studies illustrate clearly the power of molecular modeling for arriving at alternative interpretations of experimental results. AFM information at the nm level of spatial resolution would also assist considerably the description of the mechanism along the lines of the strategy in Table 1.

Ultrafine particles from bauxite ores consist of gibbsite, boehmite, kaolinite, haematite, goethite, anatase and quartz, all with phase dimensions less than 2 μm. TEM, SEM and XPS analyses have shown that iron is present in three forms in the particulate material: as separate haematite and goethite particles; as small (1–3 nm) particles of iron oxide adhering to the surfaces of gibbsite, boehmite and kaolinite; and as lattice-substituted iron in the latter three minerals [88]. The material has been shown [89] to have catalytic properties of superior activity and selectivity for Fischer-Tropsch synthesis of short-chain alkenes from steam-reformed natural gas, with surface composition and structure as critical parameters for the performance of the catalyst. A combination of XPS, FTIR, SIMS, SEM, TEM and XRD was used to characterize the parameters during the preparation of the catalyst. In particular, the surface concentration of Fe could be followed during the process of decomposition at 520°C in air and subsequent calcination at high temperatures. The initial surface was investigated by XPS, which gave surface/bulk ratios for iron in the range 1.14–1.22. SIMS depth profiles showed a surface enhancement of iron in the first 20 nm, relative to bulk abundances, by factors greater than 5. After heating at 1100°C, XPS analysis showed that there was less iron in the surface (surface/bulk ratio 0.45–0.5), while TEM/EDS showed an apparently uniform distribution of Fe, Al, Ti and Si within and between the particles. Aggregates of platelike and quasi-cubiform crystallites in the relatively open arrays produced substantial porosity. Both Fe and Si appeared to have become incorporated into the bulk structure of the calcined material. θ- and α-alumina were shown by XRD to be present in the final bulk phases. The calcination treatment appeared to lead to incorporation of the small (2–3 nm) iron-containing particles into the surface layers, and thus to produce the reactive sites for catalytic synthesis. Full characterization of the starting materials, and of the final catalyst, clearly required the use of a set of techniques, among which surface analytical techniques were of paramount importance.

5. CERAMICS

5.1. Phase Structures

Multiphase titanate ceramics designed to immobilize high-level nuclear waste have been characterized by SEM, TEM, electron diffraction, XRD, XPS, SIMS, SAM and dissolution studies [4,90–92]. SEM characterization at lower magnification using SE and BSE imaging modes in order to reveal phase distribution has been illustrated in Fig. 1. Phase distribution at higher spatial resolution can be revealed by TEM analysis; a schematic representation of five phases is shown in Fig. 14. Microencapsulated spherical grains can also be imaged, as shown in Fig. 15. Identification of the phase structure of individual grains by EDS has been described [90–92]. In the same publications, other examples involving twinned structures and phase transformations, and the use of special techniques such as superlattice reflections in SAD patterns and lattice imaging, are described. Resolution of phase structures within individual grains at the level of the unit cell, are similar to those described in Section 4.2 using STM and AFM techniques.

The application of nuclear methods to studies of ceramic surfaces and interfaces is a well-developed methodology, although not as yet applied extensively to mineral surfaces. Relevant applications of RBS, PIXE, NRA, ERDA and high-resolution α-spectroscopy have been described in the review by Matzke [93]. Analyses by RBS, used in the channeling mode, and NRA have revealed structural defects in U_4O_{9-y}. RBS is sensitive mainly to the metal atom sublattice, whereas NRA can be used selectively to study the nonmetal sublattice. It was shown that the U sublattice is nearly identical to that of UO_2, whereas significant cluster formation occurs in the O sublattice; these observations have implications for surface sites in the material. Investigations of surface reactions (e.g., leaching and dissolution) of titanates and UO_2 by nuclear techniques have

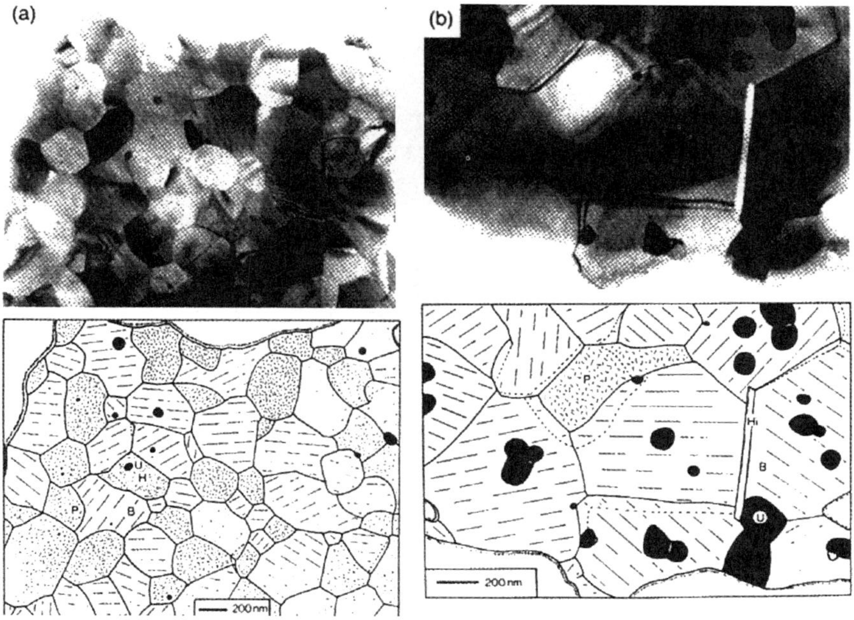

Figure 14. TEM micrograph and schematic representation of (a) uraninite-poor region and (b) uraninite-rich region: (B) betafite, (P) perovskite, (U) uraninite, (H) hollandite type, (Hi) hibonite [90].

also been described. Studies by ERDA, with He^+, showed uptake of hydrogen in the first ca. 80 nm, in accord with earlier SIMS analysis of the hydration of titanates in water at 150°C [94]. The effects of ion beam damage in ceramics, particularly the formation and annealing of defects introduced during ion implantation, can be studied by RBS channeling techniques. High-dose ion implantation may cause ceramic surfaces to become metamict (radiation-damaged) or amorphous, as demonstrated by Matzke and Whitton [95] for Al_2O_3, TiO_2, U_3O_8 and $CaTiO_3$. In the case of $CaTiO_3$, after Pb-implantation, loss of crystallinity was observed to a depth of ca. 0.2 μm. After annealing at 425°C the amorphous/crystalline interface, which was sharp and planar over lateral dimensions of 100 nm, moved some 0.1 μm toward the surface. Channeling analysis [96] was able to identify the movement of two interfaces in the surface region; an amorphous phase crystallizing into γ-Al_2O_3; and the transformation of the substrate from the γ- to the α-phase at a well-defined interface.

Hence, in the study of phase structures related to surface properties, the combination of techniques described in Section 4.1, and the exploitation of the ad-

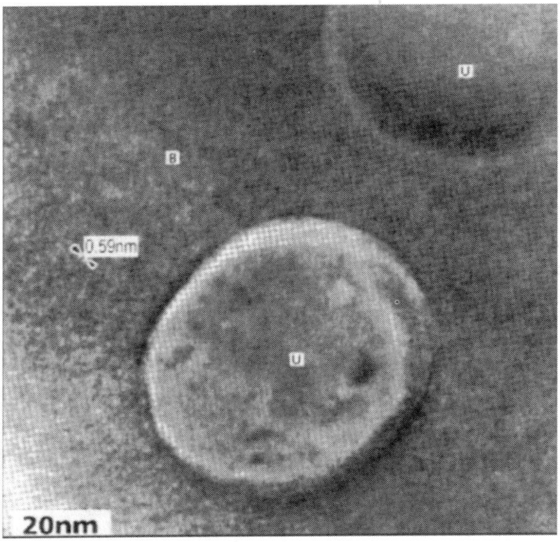

Figure 15. HREM of a [100] betafite zone in which uraninite grains are encapsulated [90].

ditional methodologies described in this section, can give a reasonably complete picture of the structure, composition, and grain size, and the phase distribution and uniformity from the mm to atomic scale of spatial resolution, for both single-phase and multiphase minerals and ceramics.

5.2. Surface Structures

The information that can be derived for ceramic surface structures, and the necessary methodologies, are similar to those described in Section 4.2 for mineral surface structures. Practical applications of AFM observations of ceramic surfaces have recently been described by Yamana et al. [97]. The fast Fourier transform (FFT) method for enhancing atomic resolution has been demonstrated in the case of a fluoride-containing mica surface of a mica glass ceramic as seen in Fig. 16. A zirconia-dispersed mica glass ceramic, formed from zircon ($ZrSiO_4$) added to the mixture of starting materials (resulting in both mica and zirconia grains in the fluorine-containing glass matrix), was also examined [97]. The topography of a surface resulting from cutting, grinding and polishing can reveal patterns characteristic of the fracture or abrasion mechanisms. It was demonstrated that fracture and cutting occur not only at grain boundaries but also within the mica grains. Net-

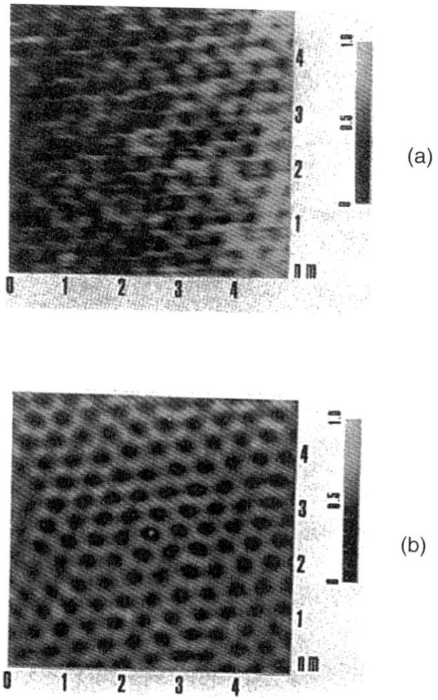

Figure 16. (a) AFM image of fluoro-mica cleavage. Note that noise in recording of the image makes atomic detail difficult to analyze [97]. (b) AFM image after removal of noise by FFT filtering [97].

works of cracks were formed by compressive pressure during grinding, and the loss of zirconia grains appeared to be due to polishing.

5.3. Surface Sites

The examples discussed in Section 4.3 in connection with minerals are also illustrative of the information required about surface sites in ceramic materials. The techniques applied on a routine basis in order to obtain this information are essentially the same.

It is worthwhile, however, emphasizing the extensive work that has been carried out with ISS on ceramic surface sites [13,14]. The review by Brongersma et al. [14] summarizes studies of composition, structure, diffusion, segregation, growth, adsorption, desorption, depth profiling, sputtering, neutralization, annealing, oxidation/reduction and dispersion for materials such as SiO_2, Al_2O_3, MgO, ZrO_2, VO_x, SnO_2, FeO_x, and high-temperature superconductors, and for other ceramics and oxides. In addition, cation and anion sites on MgO (001) surfaces, surface-segregated Ca impurity sites, and surface defects have been investigated. Site labeling in SnO_2 with oxygen isotopes has allowed identification of the positions of O atoms bridging Sn sites selectively removed by heating for 3 min in vacuum at 700 K. Amorphous and polycrystalline ceramic materials have also been studied. In the case of spinels, cations in tetrahedral and octahedral sites can be identified. Surface segregation taking place during oxidation, e.g., Fe in Al_2O_3, and as a result of ion bombardment (e.g., alkali metal diffusion) can cause profound changes to surface sites, which will affect subsequent reactions. The phenomena of growth and wetting during oxidation of metals and semiconductors, or during growth of oxides on oxide supports, have also been considered in the literature dealing with the use of ISS.

As for mineral surfaces, the matching of theoretical models to experimental data by molecular modeling methods is a powerful tool for the prediction of adsorption and surface reactivity on ceramics. The reviews by Henrich [67], Kunz [68], Stoneham and Tasker [98], and Colbourn and Mackrodt [63,99] discuss the applications of theoretical techniques and correlations with experimental results.

5.4. Grain Boundaries and Intergranular Films

Studies of interfaces in ceramic materials have a long and distinguished history. The merit of using TEM to observe boundaries between grains of the same phase is illustrated by Fig. 17 in which the lattice spacings are evident in adjacent Magneli phases (Ti_nO_{2n-1}) when using SAD [90]. The boundaries were found to be coherent. The presence of intergranular films can be recognized, even if of width only 1–3 nm. Other examples include the studies of grain boundaries in ceramics by Ruhle [100]. A series of systematic studies of intergranular films by Clarke [101–103] established the formation of interphase regions differing in structure and composition from the adjacent crystalline grains; the thicknesses ranged from a few nm to some μm in materials as different as alumina and multiphase titanate ceramics. An example of the imaging of films a few nm wide can be seen in Fig. 18 in which the apparently amorphous film, ca. 3 nm wide, between perovskite ($CaTiO_3$) and magnetoplumbite ($(Ca,Si)Ti_{12}O_{19}$), can be seen during fracture (crack identified by an arrow) to remain attached to the perovskite grain [104]. Intergranular films often promote concentration of impurities and some minor constituents by segregation. The subject of equilibrium segregation at such interfaces has been reviewed by Cabane and Cabane [105].

The application of surface analytical techniques to fracture faces, at which grain boundaries and intergranular films are preferentially exposed, also has a well-

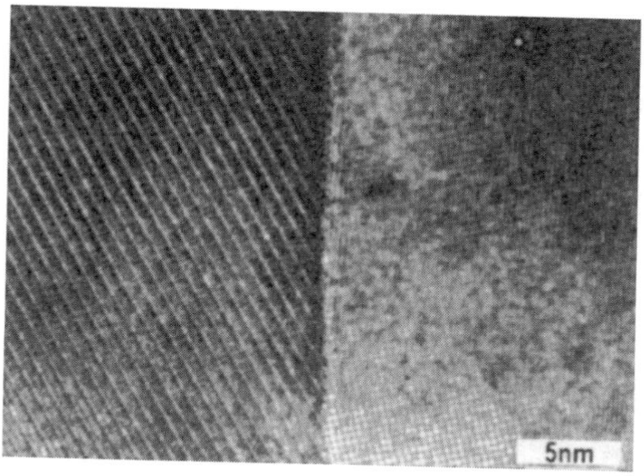

Figure 17. HREM of a grain boundary between Magneli phases [90].

established literature [e.g., 104,106]. Observations by SSIMS of alkali metal segregation into intergranular regions of ceramics and glasses [107] have demonstrated enhancement of Cs and Na, but not of Ca or Ti, as shown in Fig. 19. SAM elemental maps of the distribution of Cs around microvoids in fracture faces, before and after removal of 15 nm by ion etching, show high concentrations of the volatile Cs species (Fig. 20) due to trapping in the voids and segregation into the intergranular film regions [104]. XPS can also be used to determine composition, chemical states and bonding of elements exposed in the fracture faces. If the intergranular films are very thin (i.e., less than 2 nm), XPS is likely to detect contributions from the underlying crystalline grains, but consistency between SSIMS, XPS and SAM analyses has been demonstrated in these studies.

5.5. Depth Profiles

Examples of SSIMS depth profiles obtained from ceramics and glasses have been discussed (e.g., Fig. 19) [104,107]. Examples of XPS depth profiles obtained from similar titanate fracture surfaces are shown in Fig. 21 [104]. An enhancement of Cs, Al and, to a minor extent, Ca and Mo on the fresh fracture face, and to a depth of approximately 5 nm, is inferred from these profiles. Exposure to water removes the surface enhancement.

Depth profiling, with nuclear techniques such as RBS channeling and with ARXPS, on large flat single-crystal surfaces can be carried out without suffering the uncertainties introduced by ion beam sputtering. The former technique is especially applicable to studies of segregation [93], while the latter is useful for characterizing oxidized layers on reactive substrates where the depth of oxidation is less than 10 nm.

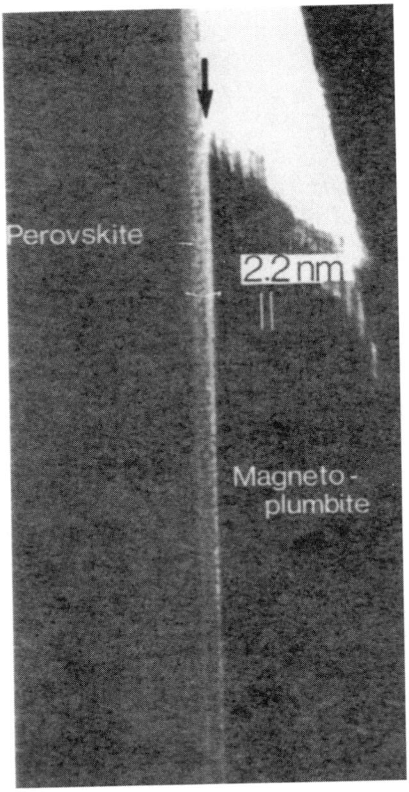

Figure 18. HRTEM micrograph of an intergranular film lying at the interface between perovskite and magnetoplumbite-type crystals. Cleavage has begun, with a crack (arrowed) running between the crystallites and an intergranular film adhering to the perovskite grain [104].

5.6. Adsorption

Studies of the adsorption of long-chain alcohols and surfactant molecules on silica surfaces, using a combination of XPS, FTIR, SEM, SAM and contact-angle measurements, have provided information on adsorbed layers [86,108–110]. Adsorption of primary alcohols on silica powder surfaces can be measured quantitatively using DRIFT in absorbance, rather than in Kubelka Munk, mode. Long-chain alcohols and aldehydes adsorb strongly and are partially oxidized to carbonyl species. In some cases oligomeric products appear to be formed at multiple adsorption sites, thus inhibiting desorption from those sites by hydrolysis. The

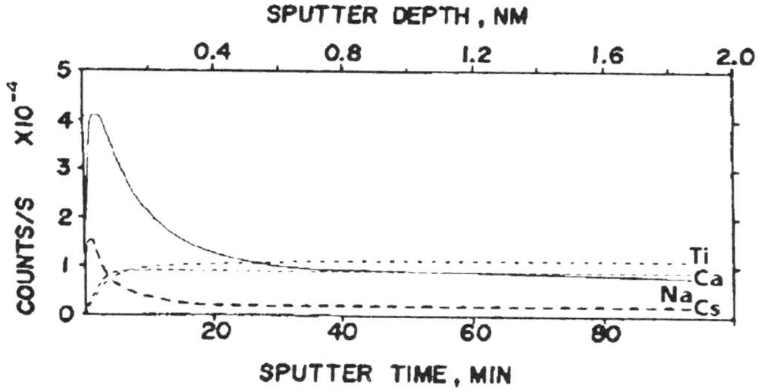

Figure 19. SSIMS depth profile of a fresh fracture face of the ceramic synroc C, demonstrating enhancement of Cs and Na, but not Ca and Ti, in the intergranular region [104,107].

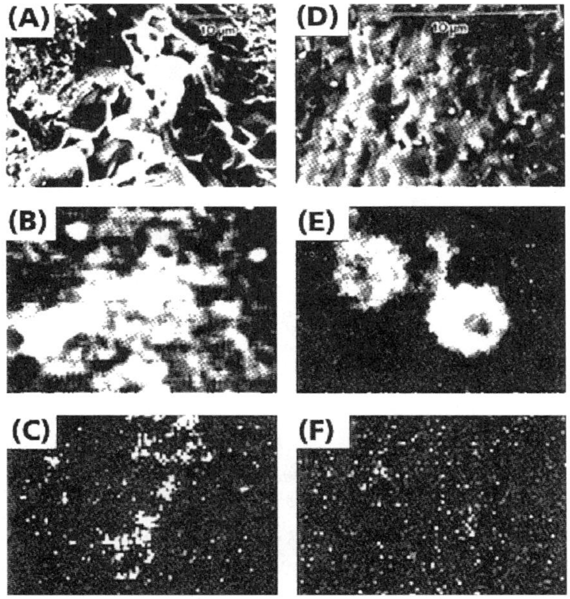

Figure 20. (A) SEM micrograph taken in SE mode of the microvoid in the fracture surface of the ceramic synroc C, and SAM Cs maps of the same area (B) before and (C) after removal of 15 nm by ion etching. (D) Micrograph of the fracture face containing Cs-rich regions and corresponding SAM Cs maps (E) before and (F) after removal of 375 nm by ion beam etching [104].

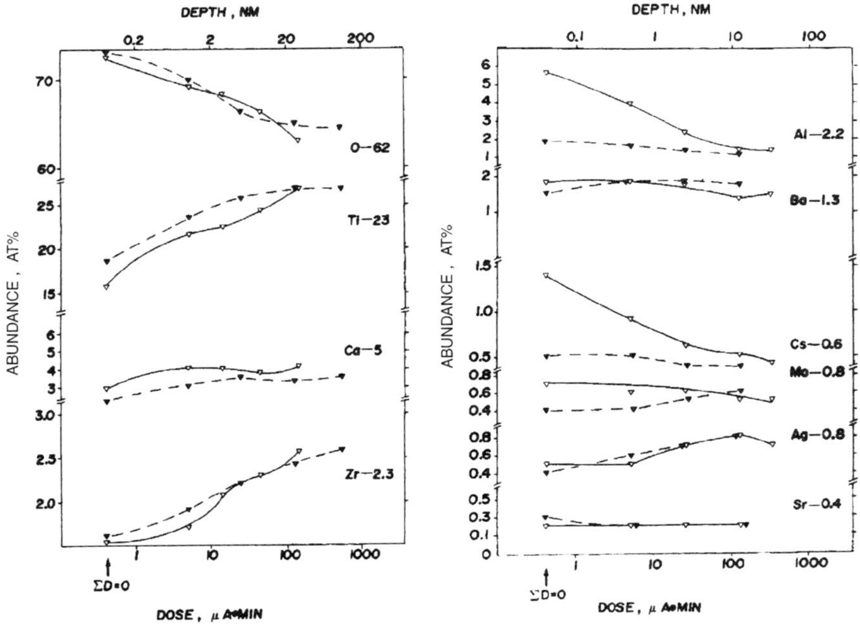

Figure 21. XPS depth profiles of synroc C fracture surface (△) before and (▲) after dipping in doubly distilled water. Enhancement of Cs, Al, and, to a lesser extent, Mo, in the fresh fracture face is apparent. The surface species are removed by immersion in water. Numbers on right indicate bulk atomic abundance [104].

density of adsorbed alcohol molecules can be compared to the number of reactive sites per nm^2. For instance, quantitative measurements on stearyl alcohol ($C_{18}H_{37}OH$) give densities of between 0.9 and 1.4 molecules nm^{-2}, depending on the absorbance band chosen for quantification (e.g., 3740, 2930 or 1460 cm^{-1}). Pyrolysis experiments give an average value of ca. 0.8 molecules nm^{-2}. Previous studies of this type of silica surface suggest that there should be 4.6 reactive silanol groups nm^{-2} accessible for reaction. DRIFT and pyrolysis results therefore suggest that less than 25% of the silanol groups had reacted with adsorbed stearyl alcohol molecules. Estimations of film thickness and surface coverage, derived from overlayer equations applied to XPS signal intensities from both the adsorbed long-chain alcohols and the silica substrates, gave values from 2 to 3 alcohol molecules per nm^2. The measurements required that a methodology be developed for estimation of attenuation length (e.g., for C 1*s* an experimental value of ca. 4 nm can be compared with the empirical value of 4.3 nm). A series of long-chain alcohols of differing lengths were adsorbed on silica plates. XPS at a fixed angle combined with variation of the film thickness, and in the angle-resolved mode at constant film thickness, was used to determine the thickness of the adsorbed layer. The combined results from XPS, FTIR, pyrolysis and contact angles suggest that adsorption is in the form of patchy, disordered and multilayered structures, rather than as an ordered oriented monolayer. For instance, even at an exit angle of 5° with a sampling depth of less than 1 nm (using $C_{22}H_{45}OH$), ARXPS revealed 3 at.% of Si and 11 at.% of O, indicating that exposed areas of silica had not reacted with the alcohol molecules.

A second method for estimating the relative film thickness of adsorbed layers has been developed [111] which does not require determination of the attenuation length, but is based on comparisons of relative atomic concentrations. Cetyltrimethyl ammonium bromide (CTAB) provides a model adsorbate for which the mechanism of adsorption is well-established [86,112]. Below the critical micelle concentration (cmc) XPS showed that bromide ions were not adsorbed with the CTA ions. At the cmc a distinct change was observed in the measured atomic concentrations of C, Si, O, N and Br. ARXPS observations suggested that, above the cmc, the CTAB layer was disordered, but *in situ* measurements of contact angle agreed with those expected of an ordered bilayer structure [112]. The methods developed in this work have also allowed comparisons to be made of the adsorbed layer thicknesses of quaternary ammonium surfactants, long-chain alcohols and conventional silanes, with differing chain lengths.

5.7. Surface Reactions

Reactions involving hydrolysis, leaching, dissolution, precipitation and *in situ* phase transformation have been discussed in reviews by Smart et al. [4], Myhra et al. [84], Blesa et al. [82], Casey [17], Brown et al. [16] and Schoonheydt [83]. The strategies, methodologies and techniques are essentially the same as those described in Section 4.7 and, for this reason, no additional examples will be given in this section.

5.8. Surface Modification

Examples of surface modification of ceramic materials resulting from adsorption, plasma and sol-gel reactions have been reviewed by Smart et al. [86]. There are many other examples [2,3,15,67,113].

A specific example of the use of the modified Auger parameter, α^*, in XPS is that of the study of silicate structures produced in thin oxide layers in metal surfaces arising from low-temperature plasma reactions [86,114]. Oxide layers of predetermined thickness can be produced with an H_2O vapor plasma followed by reaction with silane:H_2O plasma products. Mass spectra [114] of the plasma have demonstrated the presence of reactive SiO entities that produce a sequence of silicate structures ranging from orthosilicate (SiO_4^{2-}), pyrosilicate ($Si_2O_7^{6-}$), and other oligomeric structures to layer silicates and bulk silica depending on the duration of reaction. The chemical states of these products have been derived from the changes in α^* with increasing Si concentration in the surface layers, as illustrated in Fig. 22. The parameter is defined as

$$\alpha^* = \mathrm{KE(Auger)} + \mathrm{BE(XPS)} \tag{1}$$

where KE is the kinetic energy of the most intense Auger transition and BE is the binding energy of (normally) the most intense core photoemission peak. The parameter is often more sensitive to the chemical state than either the BEs of XPS peaks or the KEs of Auger peaks. It is also independent of surface charging and does not depend on the reference energy (i.e., Fermi or vacuum) used in the spectral analysis. It reflects directly changes in the screening of atoms in different chemical environments. In addition to identifying the changes in chemical states with progressive reaction, α^* can also be correlated with a parameter representing the change in concentration of oxygen in the film. There is an increase in the O/Si ratio as the plasma reaction proceeds from $SiO_{1.5}$ to SiO_2.

Ceramic silicate surface layers are strongly resistant to hydrolysis and acid attack. The dissolution rate in pH2 acid at 60°C is reduced by factors of more than 300 in comparison with the uncoated oxidized metal. Earlier studies, correlating XPS, FTIR, XRD and TEM observations, also established that the silicate reaction takes place on bulk oxide surfaces [115]. More direct evidence for the change in chemical form of the silicate species in the ceramic layer is given by α^*, however.

6. GLASSES

Many of the examples concerning mineral and ceramic surfaces can be used equally well to illustrate information, measurements and processes for glass surfaces. The reader will find it useful to consider examples under the same headings in Sections 4 and 5 in conjunction with those presented in this section. The studies discussed here have been selected to illustrate processes that have some generality in glass surface reactions and modifications.

6.1. Surface Composition

The bulk structural constituents of glasses consist of bridging and nonbridging oxygen atoms (e.g., –Si–O–Si– and –Si–O), other network-forming elements replacing Si (e.g., Al and B), network modifying cations (e.g., Fe and Ca) and ion-exchangeable cations (e.g., Na and K). The structure of glasses is relatively complex, and yet flexible, in comparison with crystalline structures; the locations of alkali and alkaline earth ions in holes in the Si–O structure exhibit an

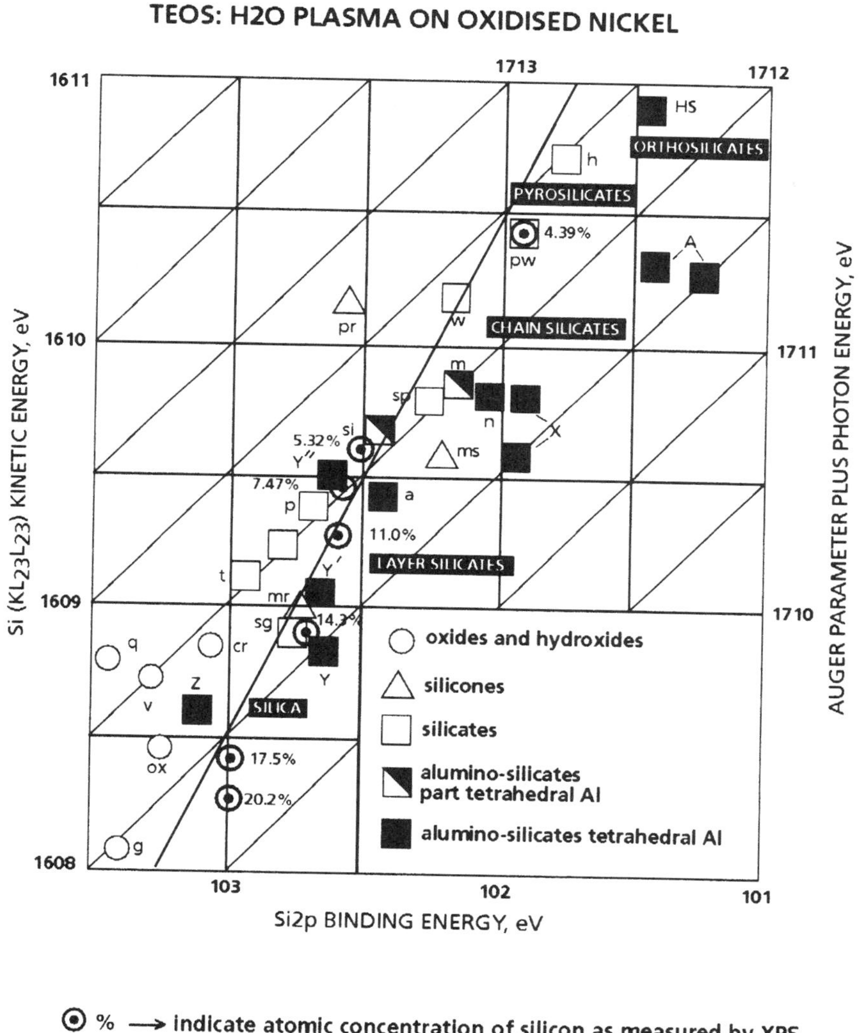

Figure 22. Modified Auger parameter plot for a number of silicates (from Ref. 137) and plasma-reacted (H_2O:silane) nickel surfaces ⊙ % values show the atomic concentrations of Si measured by XPS. Note the transition from pyrosilicate to bulk silica species as the reaction progresses [114].

irregular pattern of repetition and there is therefore no characteristic chemical composition or short-range order. Discussion of the relationship between structure and composition can be found in reviews, such as those by Paul [116] and Kruger [22].

Surface composition and the nature of surface sites are governed by the fact that planar termination of the bulk structure, even for flat defect-free surfaces, does not produce an equilibrium structure or composition similar to that of the bulk material. The surface and subsurface layers exhibit gradual changes down to relatively large depths, which may correspond to thousands of atoms, from the surface into the bulk, thus causing even a rigid glass to allow ingress and hence reaction with small molecules at those depths. Reacted surface layers with thicknesses in the μm range will result from the "microporosity" of the subsurface layer. Another consequence of the structural changes with depth is that small-diameter glass fibers differ in structure and composition from the glass material used in their manufacture; their chemistries may therefore not resemble that of the bulk glass.

The observations above are illustrated by XPS compositional analysis of E-glass surfaces of fibers and plates; the results and a comparison with nominal bulk composition can be found in Table 3 [117]. The surface compositions of the fibers and plates are significantly different from the nominal composition of the glass; there is depletion of B, Mg, Al, Na and, particularly for the plate, Ca

Table 3. Composition of E-glass as Determined from the Constituents Used in the Manufacturing Process and by XPS[a]

	Composition (at.%)		
		XPS	
Element	Manufacture	Fiber	Plate
O	61.7	66.2	72.7
Si	18.8	21.9	17.9
Ca	7.9	7.1	2.6
Al	5.9	4.3	4.7
Na/K	1.1	0.5[b]	0.4[b]
B	4.4	<0.05	1.7
Mg	0.3	<0.05	<0.05

[a]The XPS percentages were adjusted so that the carbon contamination was taken as zero.
[b]Atomic percentages of sodium only; no potassium was found on the glass fiber or plate surface.

species. Fibers showed higher surface concentrations of Si and O in comparison with bulk composition, while the plate showed surface enhancement of O (as hydroxyl groups) without Si enhancement. The surface elements were apparently lost, or depleted by segregation or dissolution, during manufacture of the fiber and plate specimens. The glasses included all three types of species, namely network formers, network modifiers and ion-exchangeable cations. FTIR studies of surface functionality, as related to surface composition and surface sites, have been covered in the literature, much of which has been reviewed by Kruger [22]. Assignments of bridging and nonbridging oxygens to hydroxyl groups associated with different cations can now be made with reasonable confidence using FTIR. Dehydroxylation during heating to higher temperatures has also been followed by FTIR. SIMS [107], NMR [118], ISS [14] and nuclear techniques [93] complement XPS and FTIR studies of the surface compositions of glasses.

In view of the discussion above it is clear that processing and pretreatment, as well as exposure to air, will result in the compositions of surfaces of glass being different from those of the bulk materials.

6.2. Surface Sites

Much of the information from studies of surface composition can be related to the chemical environment of different surface sites. AFM with nm resolution gives information on the distribution of surface sites, and on surface topography and surface reactions, including those leading to the formation of hydrolyzed, leached surface layers [97]. Because glasses are insulating, it is difficult to assign sites unequivocally to specific atoms, or to map the distribution of a particular elemental species on a glass surface. Thus, STS(ETS) and Auger techniques cannot be used. Much of the information on the chemistry and reactivity of specific surface sites has to be inferred from techniques which average over relatively large surface regions, e.g., XPS, FTIR and ToF-SIMS. ISS offers the most atom-specific information about surface composition and individual sites.

6.3. Depth Profiles

Depth profiling glasses with ion beam techniques is also difficult due to surface charging problems. Pulsed ion beams in ToF-SIMS with relatively long interpulse periods have been used successfully, however. ARXPS can be used to study variation in composition and chemical states over the outer 10 nm. Nuclear techniques, such as PIXE, NRA, RBS and ERDA, are particularly useful in studying the composition of deeper lying altered layers to depths of μm. The latter techniques generally have depth resolutions of the order of 10 nm which is sufficient in most cases [93].

6.4. Adsorption

Adsorption on glass surfaces is an important area of investigation, especially when there is concern about the coupling agents designed to reduce the effects of water on the mechanical properties of composites consisting of glass fibers in inorganic or polymer matrices. Section 6.1 noted that glass surfaces may undergo hydrolysis during manufacture and as a result of exposure to air; the result is production of a surface layer with a chemical composition different from that of the rest of the glass [117]. Diffusion and reaction at the glass-matrix interface may lead to delamination and to loss of strength of glass-reinforced composites [119,120] unless a suitable adsorbed layer is incorporated at the interface. Silanes are the most common coupling agents used to provide a water-resistant interfacial bond between glass fibers and organic polymer matrices. Surface modification using siloxane alternatives, instead of silanes, will be discussed in Section 6.6.

For many applications, however, it is sufficient to render the glass surface hydrophobic by adsorption of simpler and cheaper reagents. Long-chain alcohols adsorb strongly on silica surfaces (see Section 5.6); adsorption of these alcohols on E-glass plates, fibers and powders has been studied. Alcohols with chain lengths C_{12}, C_{14}, C_{18} were reacted at 130°C with the glass surfaces [117,121]. Analysis by FTIR of the methylene absorption peak at 2950 cm^{-1} shows that the C_{14} and C_{18} alcohols adsorb strongly on the glass surface with the C_{12} alcohol adsorption resulting in lesser surface coverage. Even after seven extraction cycles in warm cyclohexane it was found that all three alcohols retained an absorbance value of about 0.01, and high contact angles (i.e., 85–97°) were measured. The contact angles corresponded to values expected for methylene groups rather than to values (i.e., 112°) characteristic of highly oriented monolayers which expose predominantly methyl groups. Quantitative DRIFT analysis produced results agreeing with the 0.01 methylene absorbance and indicated a relatively high surface coverage of one to five monolayers, in comparison to known alcohol concentrations in standards. These estimates carried the same uncertainty as for silica surfaces (Section 5.6); the assumption is that the extinction coefficient for the above vibrational mode is the same for both bulk alcohol and adsorbed alcohol molecules. Nevertheless, disordered multilayer adsorption of the alcohols was indicated by the DRIFT results. ARXPS analyses of C 1*s* and Si 2*p* signals suggested that complete coverage of the glass surface was unlikely. ARXPS was carried out on E-glass plates and fibers before and after seven cyclohexane washing cycles. Figure 23 presents the results obtained from clean and from alcohol-reacted plates after the seven cycles. The clean surfaces of the plates showed the "normal" hydrocarbon contamination acquired during contact with air (i.e., ca. 25 at. % C at 45°). The concentrations of hydrocarbons on the alcohol-reacted surfaces were considerably greater. There was a systematic increase in carbon concentration before washing, with increasing alcohol chain length, at all takeoff angles. Washing had the effect (Fig. 23a) of producing similar thicknesses of hydrocarbon for each of the alcohols, suggesting that the stable residual adsorbed layers, or regions, had similar structures with respect to thickness and molecular orientation.

Analyses of the Si 2*p* spectra (Fig. 23b) showed lower silica signals at all takeoff angles on the alcohol-treated samples in comparison with clean samples. However, it can be seen that there was considerable presence of silica at the 10° angle, i.e., 25–55% of the signal obtained from clean surfaces, even though the depth of analysis was of the order of the molecular size of the alcohol molecules.

The combined results from DRIFT, contact angle and ARXPS analyses showed no evidence for a high degree of orientation of the adsorbed alcohols; instead disordered multilayers in separate regions or islands could be inferred. Nevertheless, major changes in surface composition and hydrophobicity were measured despite apparently incomplete monolayer coverage.

A previous study combining XPS, streaming zeta potential and Wilhelmy wetting measurements [122] confirmed that substantial adsorption and alteration of surface properties took place with only fractional surface coverage at very low surfactant concentrations (i.e., 0.1 wt.%). These results again support an interpretation in terms of patchwise adsorption in relatively thick multilayered regions.

6.5. Surface Reactions

Most studies of surface reactions on glasses have been concerned with the three processes of leaching or ion exchange, base-catalyzed hydrolysis of the glass network, and *in situ* phase transformation or reprecipitation. Reviews of the techniques applicable to physical characterization of glass surfaces can be found in the work of Hench and Clark [123,124], Myhra et al. [4], Kruger [22], and references therein.

The reaction processes can be described by a combination of results from leach tests and SIMS. The nominal composition and leach rates (g m^{-2} d^{-1}) of a glass formulation, PNL 76–68 (Pacific North-West Laboratories), used as the standard high-level nuclear waste glass matrix [4], are shown in Table 4. The definition of leach rate used normalizes the loss (in g) to the proportion of that element in the nominal bulk composition. It will be seen that the leach rates for Cs, Na, Mo, Si and B exceeded the overall rate of mass loss from the glass surface, whereas the elements Fe, Zn and Ti did not show significant loss to solution under these conditions. There were also several elements, e.g., Ca, Ba, Cd and Sr, that were neither rapidly leached nor apparently retained in the surface layers. Studies described in the above-mentioned reviews, using XPS, FTIR, SIMS, SEM and dissolution rates, have established clearly that the primary reaction occurring in solution is the bond-breaking attack by OH^- at Si (or Al and

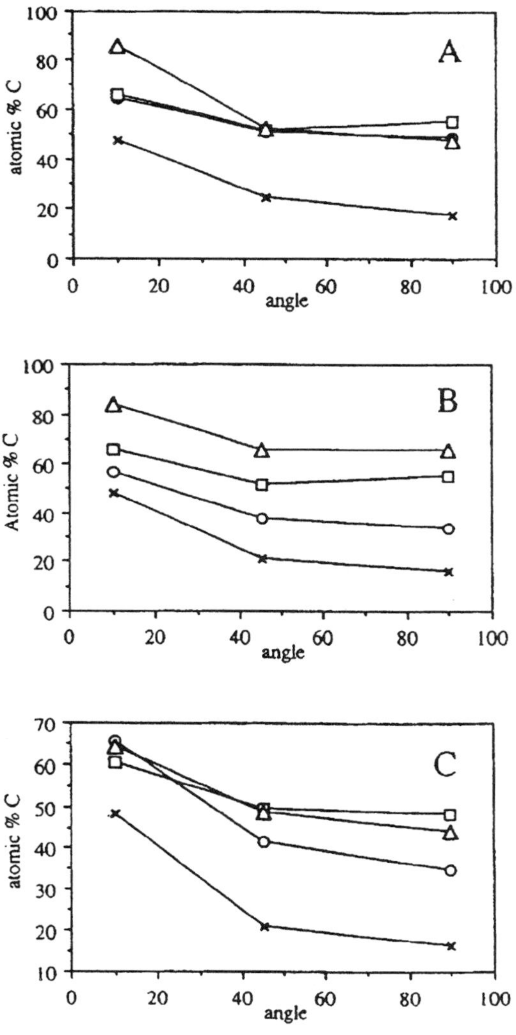

Figure 23(a). Atomic percentages of C, from C 1*s* ARXPS spectra, for (A) clean plate and alcohol-reacted plate (washed seven times); (B) clean fibers and alcohol-reacted fibers (unwashed); (C) clean fibers and alcohol-reacted fibers (washed seven times). (x) Clean plate or fibers; (○) C_{12}, (□) C_{14} and (△) C_{18} alcohol-treated plate or fibers. [117]

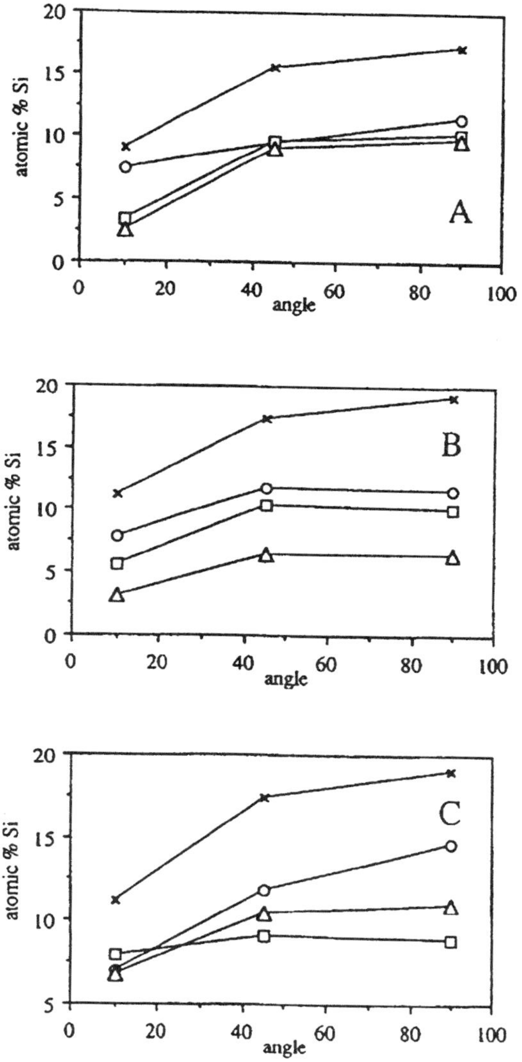

Figure 23(b). Atomic percentages of Si, from Si 2*p* ARXPS spectra, for (A) clean plate and alcohol-reacted plate (washed seven times); (B) clean fibers and alcohol-reacted fibers (unwashed); (C) clean fibers and alcohol-reacted fibers (washed seven times). (x) Clean plate or fibers; (○) C_{12}, (□) C_{14} and (△) C_{18} alcohol-treated plate or fibers [117].

Table 4. Nominal Composition (wt.%) of Elements in PNL 76-68 Glass (at.% values in parentheses) with Leach Rates (g m^{-2} d^{-1}) of Selected Elements in Water at 90°C for 7 d

	Composition	Leach rate
B	2.9 (4.2)	1.74
O	41.1 (59.0)	
Na	9.6 (10.8)	1.94
Si	18.6 (17.0)	1.47
P	0.2 (0.1)	1.06
Ca	1.4 (1.4)	0.43
Ti	1.8 (1.0)	0.002
Cr	0.3 (0.1)	
Fe	7.3 (2.0)	0.008
Ni	0.2 (0.1)	
Zn	4.0 (2.4)	0.003
Rb	0.1 (<0.05)	
Sr	0.3 (0.2)	0.53
Y	0.2 (<0.05)	
Zr	1.4 (0.4)	
Mo	1.6 (0.3)	1.79
Cs	1.0 (0.2)	2.1
Ba	0.4 (0.1)	0.15
La	0.5 (0.1)	
Ce	1.3 (0.2)	
Pr	0.6 (<0.05)	
Nd	4.7 (0.4)	
Sm	0.4 (<0.05)	
Eu	0.1 (<0.05)	
Mass	100.0	1.01

B) atoms of the glass network (i.e., –Si–O–Si–). The resulting –Si–OH and –Si–O^- species then react further with water to produce two –SiOH groups and an accompanying release of an OH^- for further reaction. This bond-opening mechanism allows ion exchange of mobile ions, particularly alkali metal cations. Hence, the initial rates of loss to solution reflect release of the exchangeable cations, while the rate of loss of Si is closer to that of the overall loss rate of mass. The surface layer will become depleted in alkali metal cations and hydrolyzed to a siliciceous hydrogen-bonded structure. Those elements which exhibit very low leach rates are retained in the surface as new and relatively insoluble products.

SIMS profiles from a disk of waste glass, recorded before and after hydrothermal attack, are presented in Figure 24. The loss of Na, and the partial loss of Si and B, is apparent from Fig. 24a. On the other hand, the profiles show that Fe, Ti and Zn were retained in the surface layer with maximum abundances in the near-surface region. SIMS and FTIR show the formation of SiOH groups, and SIMS also reveals hydroxylated ions such as $CaOH^+$ and $SiOH^+$. While XRD did

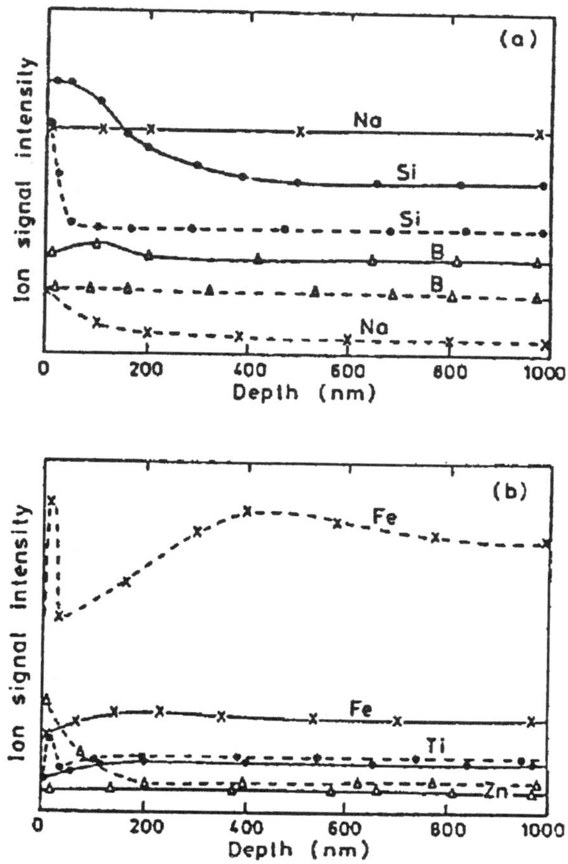

Figure 24. SIMS profiles from the surface of a polished disk of the glass (Table 4) before and after hydrothermal attack in DDDW water for 7 d at 300°C. SIMS conditions: 1 keV Ar^+, 1.25×10^{-5} A cm^{-2}, charge compensated by low-KE electron flood gun; 2×2 mm^2 raster, 20% gated; etch rate ca. 10 nm min^{-1}.

not reveal evidence of crystallinity in the reacted surface layer, TEM imaging in combination with SAD found crystalline products such as β-cristoballite and two pyroxene structures similar to wollastonite and rhodonite. Several studies of this and other glasses [125–127] have found crystalline products containing Fe and Zn [127]. SEM images of a Zn-containing glass attacked hydrothermally at 170°C for 60 h showed small crystallites of μm dimensions, which were shown by TEM and electron diffraction to be zinc hemimorphite and zinc orthosilicate. Not all of the retained elements were in crystalline form, however, since crystalline structures retaining Mg, Ni, Nd and Ce have not been identified despite their accumulation in the near surface, as shown by depth profiles from SIMS and XPS.

6.6. Surface Modification

Surface modification of glasses can be the result of adsorption and surface reactions, and can be described by methods similar to those covered in the previous two sections. The focus will now be on forms of surface modification in which the lateral interactions are at least as important as the bonding to the glass surface. The characterization of these "sheathing" layers again requires a combination of surface analytical techniques. Separation of the mechanisms of lateral bonding from those involving interaction with the glass surface can be difficult even with information from several different techniques. For instance, silanes are believed to adsorb initially through hydrogen-bonding interactions, with subsequent condensation and lateral reactions generating siloxane structures. There are examples in which lateral polymerization occurs, apparently without the formation of bonds to the substance [128], and the siloxane film has been shown to consist of multiple layers [129–132].

An example of this type of surface modification is that of "sizing" of glass fiber surfaces, which involves both adsorption of coupling agents and formation of relatively thick adhering overlayers. XPS can be used to monitor the thickness of the overlayer by providing the ratio of abundances of substrate elements before and after application of the coupling agents or overlayers. For instance, the use of less expensive siloxanes, instead of silanes, as the interacting species, has been studied with E-glass [133,134]. Because the polymerized siloxane is the final product of the lateral reaction of adsorbed silanes, the application of the adsorbate to the surface in this form may be more direct, less expensive and of equal value in surface modification. There is also potentially greater control and reproducibility. Adsorption of six types of siloxane (i.e., hydrido, methacryl, epoxy, unsaturated amino, pendant amino and terminal amino) was studied. The siloxanes bearing amino, hydrido and methacryl functional groups bound strongly to the E-glass and resisted removal more effectively by a variety of organic solvents than the silane coupling agent used extensively in industrial applications (i.e., vinyltrimethoxyethoxy silane). It is noteworthy that the functional groups on the three reagents do not undergo the same hydrolysis reaction that silane coupling agents are believed to require when bonding to glass surfaces. Nevertheless, they are either equally, or more strongly, bound which suggest that other mechanisms may be equally important for the formation of the final polymerized siloxane structures resulting from silane adsorption.

The thicknesses (t) of the surface modifying layers can be inferred relatively easily from the calcium concentration in the underlying E-glass substrate and with the formula [135]

$$t = -\lambda_{\mathrm{Ca}} \ \ln\left[\frac{I_{\mathrm{Ca}}}{I^*_{\mathrm{Ca}}}\right]\frac{2}{\pi} \qquad (2)$$

where I_{Ca} is the measured concentration (proportional to the area under the Ca 2p envelope) after formation of the modifying surface layer, I^*_{Ca} is the corresponding concentration for a clean E-glass surface, and λ_{Ca} is the attenuation length for Ca 2p photoelectrons. The factor $2/\pi$ accounts for the curvature of the fiber, which increases the effective path length of photoelectrons in the surface layer. The attenuation length can be estimated from the empirical determinations by Seah and Dench [136] to be 2.6 nm for Ca 2p electrons (KE is 907 eV). There is an inverse correlation, as expected, between the abundances derived from the C 1s and Ca 2p intensities with increasing surface coverage. Ion etching of the sample increases systematically the Ca concentration as the overlayer is removed, re-exposing the E-glass surface. Layer thicknesses between 1.2 and 4.1 nm were measured for different functionalized siloxanes at different surface coverages.

The coupling agents and overlayers applied to the glass surface must be able to interact with both the glass surface and the polymer matrix into which the fibers are incorporated. It is now well established that interactions require multiple layers of silane coupling reagents in order to generate an interpenetrating network within the polymer matrix, thus optimizing the strength of the resulting composite [130]. Aminohydroxysiloxanes have many hydrolyzable alkoxy groups along the polymer backbone and have been shown recently [134] to produce multiple-site attachment and close-packed structures, analogous to those resulting from adsorbed silane reagents. DRIFT and XPS techniques were used in a study of these reagents. The work demonstrated that, when the mechanisms of surface attachment and lateral polymerization were understood, it was possible to design alternative reagents to perform the same surface modification.

ACKNOWLEDGMENTS

The work from our research group, described in this chapter, has been carried out by a large number of scientists, and I express my gratitude for their many and varied contributions. They include Professors John Ralston, Tim White and

Slavek Sobieraj, Drs. Stephen Grano, William Skinner, Andrea Gerson, Belina Braggs, Janis Matisons, Pawittar Arora and Tom Horr, and Messrs. Angus Netting, and Darren Simpson. My previous association at Griffith University (Brisbane, Australia) with Professor Robert Segall, Drs. Peter Turner, Sverre Myhra, Roger Lewis, David Cousens and Mr. Colin Jones is also gratefully acknowedged. The collaboration with the Åbo Akademi (Finland) which involved Professor Jarl Rosenholm and Dr. Heidi Fagerholm, has contributed to the work on glass surfaces.

The award of a Senior Research Fellowship by the Australian Research Council (1992–1996) has been of central importance for much of the work described in this review.

REFERENCES

1. D. J. Vaughan and R. A. D. Pattrick (eds), *Mineral Surfaces*, Min. Soc. Series No. 5, Chapman and Hall, London, 1995.
2. J. Nowotny (ed.), *Science of Ceramic Interfaces*, Mat. Sci. Monographs No. 75, Elsevier, Amsterdam, 1991.
3. J. Nowotny (ed.), *Science of Ceramic Interfaces II*, Mat. Sci. Monographs No. 81, Elsevier, Amsterdam, 1994.
4. S. Myhra, R. St. C. Smart and P. S. Turner, *Scanning Microscopy 2*, 715 (1988).
5. T. J. White and I. A. Toor, *J. Mater. 48*, 54 (1996).
6. P. S. Arora, A. K. O. Netting and R. St. C. Smart, *Mater. Forum 17*, 293 (1993).
7. C. A. Prestidge, J. Ralston and R. St. C. Smart, *Int. J. Min. Proc. 38*, 205 (1993).
8. J. R. Mycroft, G. M. Bancroft, N. S. McIntyre, J. W. Lorimer and I. R. Hill, *J. Electroanal. Chem. 292*, 139 (1990).
9. C. D. Gribble and A. J. Hall, *Optical Microscopy: Principles and Practice*, UCL, London, 1992, pp. 1–34.
10. G. N. Greaves, in Ref. 1, p. 87.
11. J. C. Woicik, T. Kendelewicz, K. E. Miyano, P. L. Cowan, C. E. Bouldin, B. A. Karlin and W. E. Spicer, *Phys. Rev. Lett 68*, 341 (1992).
12. H. D. Abruna, G. N. Bommarito and D. Acevedo, *Science 280*, 69 (1990).
13. D. J. O'Connor, Low energy ion scattering in *Surface Analysis Methods in Materials Science* (D. J. O'Connor, B. A. Sexton and R. St. C. Smart, eds.), Springer Ser. Surf. Sci. No. 23, Springer-Verlag, Berlin, 1992, p. 245.
14. H. H. Brongersma, P. A. C. Groenen and J.-P. Jacobs in Ref. 3, p. 113.
15. C. Klauber and R. St. C. Smart, in Ref. 13, p. 1.

16. G. E. Brown Jr., G. A. Parks and P. A. O'Day, in Ref. 1, p. 129.
17. W. H. Casey, in Ref. 1, p. 185.
18. J. C. Vickerman, in *Methods of Surface Analysis* (J.M. Walls, ed.), Cambridge University Press, (1989), p. 169.
19. J. S. Brinen and F. Reich, *Surf. Interface Anal. 18*, 448 (1992).
20. J. S. Brinen, S. Greenhouse, D. R. Nagaraj and J. Lee, *Int. J. Min. Proc. 38*, 93 (1993).
21. K. G. Stowe, S. L. Chryssoulis and J. Y. Kim, *Min. Eng. 8*, 421 (1995).
22. A. A. Kruger, in Ref. 67, p. 413.
23. J. M. Cases, M. Kongolo, P. deDonato, L. Michot and R. Erre, *Int. J. Min. Proc. 30*, 35 (1990).
24. J. O. Leppinen, *Int. J. Min. Proc. 30*, 245 (1990); J. W. Stojek and J. Mielcarzski, *Adv. Coll. Interf. Sci. 19*, 309 (1983).
25. P. F. Barron, R. L. Frost, J. O. Skjemstad and A. J. Koppi, *Nature 302*, 49 (1982); R. L. Frost and P. F. Barron, *J. Phys. Chem. 88*, 6206 (1984).
26. J. O. Skjemstad, R. L. Frost and P. F. Barron, *Aust. J. Soil Res. 21*, 539 (1983).
27. J. R. Craig and D. J. Vaughan, *Ore Microscopy and Ore Petrography*, Wiley Interscience, 1981.
28. H. M. Rietvelt, *J. Appl. Cryst. 2*, 65 (1969).
29. R. J. Hill and C. J. Howard, *A Computer Program for Rietvelt Analysis of Fixed Wavelength X-ray and Neutron Powder Diffraction Patterns*, Aust. Atomic Energy. Commission Report AAEC/M112, July, 1986.
30. C. J. Howard, R. J. Hill and M. A. M. Sufi, *Chem. Aust. 367* (1988).
31. P. S. Turner, in Ref. 13, p. 79.
32. D. Sutherland, R. Creelman, P. Gottlieb, R. Jackson, V. Quittner, G. Wilkie, M. Zuiderwyck, N. Allen and T. Maclean, in CHEMECA 87, Proc. Chem. Eng. Conf. (Aust. Inst. Min. Met. Publ.), Vol. 2, (1987) pp. 106.1–106.6; P. Gottlieb, B. J. I. Adair and G. J. Wilkie, Proc. Mill Op. Conf. (Aust. Inst. Min. Met. Publ.), Melbourne, (1994), pp. 5–13.
33. L. D. Marks and D. J. Smith, *Nature, 303*, 316 (1983); D. J. Smith, *Surf. Sci. 178*, 462 (1986).
34. L. D. Marks, *Surf. Sci. 139*, 281 (1984).
35. G. van der Laan, R. A. D. Pattrick, C. M. B. Henderson and D. J. Vaughan, *J. Phys. Chem. Solids 53*, 1185 (1992).
36. A. Iida, T. Noma, S. Hayakawa, M. Takahashi and Y. Gohshi, *Jpn. J. Appl. Phys. 32*, 160 (1993).
37. M. F. Hochella, Jr., in Ref. 1, p. 17.
38. F. Fan and A. J. Bard, *J. Phys. Chem. 94*, 3761 (1990).
39. S. E. Gilbert and J. H. Kennedy, *Langmuir 5*, 1412 (1989).
40. C. M. Eggleston and M. F. Hochella, Jr, *Geochim. Cosmochim. Acta 54*, 1511 (1990).

41. C. M. Eggleston and M. F. Hochella, Jr, *Science 254*, 983 (1991).
42. C. M. Eggleston and M. F. Hochella, Jr., *Am. Mineral. 77*, 221 (1992).
43. C. M. Eggleston and M. F. Hochella, Jr., *Am. Mineral. 77*, 911 (1992).
44. C. M. Eggleston and M. F. Hochella, Jr., *Am. Mineral 78*, 877 (1993).
45. C. M. Eggleston and M. F. Hochella, Jr., in *The Environmental Chemistry of Sulphide Oxidation*, ACS Symposium Series, 550 (C. N. Alpers and D. W. Blowes, eds.), American Chemical Society, Washington DC, 1994, p. 201.
46. M. F. Hochella, Jr, C. M. Eggleston, V. B. Elings, G. A. Parks, G. E. Brown Jr., C. M. Wu and K. Kjoller, *Am. Mineral. 74*, 1235 (1989).
47. M. F. Hochella, Jr, C. M. Eggleston, V. B. Elings and M. S. Thompson, *Am. Mineral. 75*, 723 (1990).
48. G. M. Bancroft and M. M. Hyland, in *Mineral-Water Interface Geochemistry* (M. F. Hochella, Jr. and A. F. White, eds.), Reviews in Mineralogy, Vol. 23, Mineralogical Society of America, Washington, DC, 1990, p. 511.
49. P. Wersin, M. F. Hochella Jr, P. Persson, G. Redden, J. O. Leckie and D. W. Harris, *Geochim. Cosmochim. Acta 58*, 2829 (1994).
50. P. M. Dove, M. F. Hochella, Jr. and R. J. Reeder. in *Water-Rock Interaction* (Y. Kharaka and A. Maest, eds.), Balkema, Rotterdam, 1992, p. 141.
51. K. Laajalehto, R. St. C. Smart, J. Ralston and E. Suoninen, *Appl. Surf. Sci. 64*, 29 (1993).
52. G. F. Cotterill, R. Bartlett, A. E. Hughes and B. A. Sexton, *Surf. Sci. Lett. 232*, L211 (1990).
53. A. N. Buckley and R. Woods, *Appl. Surf. Sci. 17*, 401 (1984).
54. B. S. Kim, R. A. Hayes, C. A. Prestidge, J. Ralston and R. St. C. Smart, *Appl. Surf. Sci. 78*, 385 (1994).
55. B. S. Kim, R. A. Hayes, C. A. Prestidge, J. Ralston and R. St. C. Smart, *Langmuir 11*, 2554 (1995).
56. B. S. Kim, R. A. Hayes, C. A. Prestidge, J. Ralston and R. St. C. Smart, *Colloids Surf. A*, in press.
57. P. J. Jennings, in Ref. 13, p. 275.
58. G. N. Greaves, *Adv. X-ray Anal. 34*, 13 (1991).
59. T. C. Huang, *Adv. X-ray Anal. 33*, 91 (1991).
60. A. R. Pratt, I. J. Muir and H. W. Nesbitt, *Geochim. Cosmochim. Acta 58*, 827 (1994).
61. J. R. Mycroft, H. W. Nesbitt and A. R. Pratt, *Geochim. Cosmochim. Acta 59*, 721 (1995).
62. S. W. Knipe, J. R. Mycroft, A. R. Pratt, H. W. Nesbitt and G. M. Bancroft, *Geochim. Cosmochim. Acta 59*, 1079 (1995).
63. E. A. Colbourn and W. C. Mackrodt. *Surf. Sci. 117*, 571 (1982); *Solid State Ionics 8*, 221 (1983); E. A. Coulbourn, J. Kendrick and W. Mackrodt, *Surf. Sci. 126*, 550 (1983).

64. C. F. Jones, R. A. Reeve, R. Rigg, R. L. Segall, R. St. C. Smart and P. S. Turner, *J. Chem. Soc. Faraday* I, *80*, 2609 (1984).
65. C. F. Jones, R. L. Segall, R. St. C. Smart and P. S. Turner, *Proc. Roy. Soc. A374*, 141 (1981); *J. Mater. Sci. Lett. 3*, 810 (1984); *J. Chem. Soc. Faraday I 74*, 1615 (1978).
66. J. A. Tossell, in Ref. 1, p. 61.
67. V. E. Henrich, in *Surface and Near-Surface Chemistry of Oxide Materials* (J. Nowotny and L.-C. Dufour, eds.), Mat. Sci., Monographs No. 47, Elsevier, Amsterdam, 1988, p. 23.
68. A. B. Kunz, in *External and Internal Surfaces in Metal Oxides* (L.-C. Dufour and J. Nowotny, eds.), Mat. Sci. Forum Vol. 29, Trans. Tech. Publ., Switzerland, 1988, p. 1.
69. A. J. Burggraaf and A. J. A. Winnubst, in Ref. 67, p. 449.
70. A. Pring and T. B. Williams, *Min. Mag. 58*, 453 (1994).
71. S. Grano, J. Ralston and R. St. C. Smart, *Int. J. Min. Proc. 30*, 69 (1990).
72. R. St. C. Smart, *Min. Eng. 4*, 891 (1991).
73. I. J. Muir, G. M. Bancroft and H. W. Nesbitt, *Geochim. Cosmochim. Acta 53*, 1235 (1989).
74. I. J. Muir, G. M. Bancroft, W. Shotyk and H. W. Nesbitt, *Geochim. Cosmochim. Acta 54*, 2247 (1990).
75. W. H. Casey, M. J. Carr and R. A. Graham, *Geochim. Cosmochim. Acta 52*, 1545 (1988).
76. W. P. Inskeep, E. A. Nater, P. R. Bloom, D. S. Vandervoort and M. S. Erich, *Geochim. Cosmochim. Acta 55*, 787 (1991).
77. K. Laajalehto, P. Nowak, A. Pomianowski and E. Suoninen, *Coll. Surf. 57*, 319 (1991).
78. R. St. C. Smart, J. Amarantidis, W. M. Skinner, C. A. Prestidge, L. La-Vanier and S. Grano, *Scan. Microsc.* in press.
79. A. N. Buckley, *Coll. Surf. A93*, 159 (1994).
80. A. N. Buckley, I. C. Hamilton and R. Woods, in *Flotation of Sulfide Minerals* (K. S. E. Forssberg, ed.), Elsevier, Amsterdam, 1985, p. 41.
81. R. L. Segall, R. St. C. Smart and P. S. Turner, in Ref. 67, p. 527.
82. M. A. Blesa, A. E. Regazzoni and A. J. G. Marito, in Ref. 68, p. 31.
83. R. A. Schoonheydt, in Ref. 1, pp. 303–332.
84. S. Myhra, D. K. Pham, R. St. C. Smart and P. S. Turner, in Ref. 2, p. 569.
85. C. F. Jones, S. LeCount, R. St. C. Smart and T. J. White, *Appl. Surf. Sci. 55*, 65 (1992).
86. R. St. C. Smart, P. Arora, B. Braggs, H. M. Fagerholm, T. J. Horr, D. C. Kehoe, J. G. Matisons, J. Ralston and J. B. Rosenholm, in *Interfaces of Ceramic Materials: Impact on Properties and Applications* (K. Uematsu, Y. Moriyoshi, Y. Saito and J. Nowotny, eds.), Key Eng. Mat. Vols. 111–112, Trans. Tech. Publ., Switzerland, 1995, p. 361.

87. A. R. Gerson, A study of the surface modification of kaolinite using molecular modeling, Proc. PacRim2, Aust. Ceramic Soc., Sydney, in press.
88. K. Kuys, J. Ralston, R. St. C. Smart, S. Sobieraj, R. Wood and P. S. Turner, *Min. Eng. 3,* 421 (1990).
89. R. St. C. Smart, B. G. Baker and P. S. Turner, *Aust. J. Chem. 43*, 241 (1990).
90. C. J. Ball, W. J. Buykx, F. J. Dickson, K. Hawkins, D. M. Levins, R. St. C. Smart, K. L. Smith, G. T. Stevens, K. G. Watson, D. Weedon and T. J. White, *J. Am. Ceram. Soc. 72*, 404 (1989).
91. W. J. Buykx, K. Hawkins, H. Mitamura, R. St. C. Smart, G. T. Stevens, K. J. Watson, D. Weedon and T. J. White, *J. Am. Ceram. Soc. 71*, 678 (1988).
92. J. A. Cooper, D. R. Cousens, R. A. Lewis, S. Myhra, R. L. Segall, R. St. C. Smart, P. S. Turner and T. J. White, *J. Am. Ceram. Soc. 68*, 64 (1985).
93. H. J. Matzke, in Ref. 2, p. 457.
94. S. Myhra, D. K. Pham, R. St. C. Smart and P. S. Turner, Scientific Basis of Nuclear Waste Management XIII, Mat. Res. Soc. Symp. Proc., Vol. 176, 1991, p. 249.
95. Hj. Matzke and J. L. Whitton, *Can. J. Phys. 44*, 995 (1966).
96. C. W. White, L. A. Boatner, P. S. Sklad, C. J. McHargue, J. Rankin, G. C. Farlow and M. J. Aziz, *Nucl. Instrum. Meth. Phys. Res. B32*, 11 (1988).
97. K. Yamana, M. Miyamoto, S. Nakamura and K. Kihara, in Ref. 86, p. 57.
98. A. M. Stoneham and P. W. Tasker, in Ref. 67, p. 1.
99. E. A. Colbourn, *Surf. Sci Rep. 15*, 281 (1992).
100. M. Ruhle, *J. Phys., Colloq. (Orsay) 43*, 115 (1982).
101. D. R. Clarke, *J. Am. Ceram. Soc. 63*, 104 (1980).
102. D. R. Clarke, *Ultramicroscopy 8*, 95 (1982).
103. D. R. Clarke, *Ultramicroscopy 4*, 33 (1979).
104. J. Cooper, D. R. Cousens, J. A. Hanna, R. A. Lewis, S. Myhra, R. L. Segall, R. St. C. Smart, P. S. Turner and T. J. White *J. Am. Ceram. Soc. 69*, 347 (1986).
105. J. Cabane and F. Cabane, in *Interface Segregation and Related Processes in Materials* (J. Nowotny, ed.), Trans. Tech. Publ., Switzerland, 1991, p. 1.
106. A. B. Harker, D. R. Clarke and C. M. Jantzen, Mat. Sci. Res., Vol. 14, *Surfaces and Interfaces in Ceramic and Ceramic-Metal Systems* (J. Pask and A. Evans, eds.), Plenum Press, New York, 1981, p. 207.
107. R. St. C. Smart, *Appl. Surf. Sci. 22/23*, 90 (1985).
108. T. J. Horr, J. Ralston and R. St. C. Smart, *Coll. Surf. 64*, 67 (1992).
109. T. J. Horr, J. Ralston and R. St. C. Smart, *Coll. Surf. 92*, 277 (1994).
110. T. J. Horr, J. Ralston and R. St. C. Smart, *Coll. Surf. A97*, 183 (1995).

111. T. J. Horr, P. S. Arora, J. Ralston and R. St. C. Smart, *Coll. Surf. A102*, 181 (1995).
112. B. H. Bijsterbosch, *J. Coll. Interf. Sci. 47*, 186 (1974).
113. K. N. Strafford, R. St. C. Smart, I. Sare and C. Subramanian (eds.), *Surface Engineering: Processes and Application*, Technomic, 1995.
114. P. Arora and R. St. C. Smart, *Surf. Interf. Anal., 24*, 539 (1996).
115. W. R. Pease, R. L. Segall, R. St. C. Smart and P. S. Turner, *J. Chem. Soc. Faraday Trans. I, 76*, 1510 (1980).
116. A. Paul, *J. Mater. Sci. 12*, 2246 (1977).
117. H. M. Fagerholm, J. B. Rosenholm, T. J. Horr and R. St. C. Smart, *Coll. Surf. A 110*, 11 (1996).
118. D. W. Sindorf and G. E. Maciel, *J. Am. Chem. Soc. 105*, 1489 (1983).
119. A. G. Atkins, *J. Mater. Sci. 10*, 819 (1975).
120. J. O. Outwater, *J. Adhesion 2*, 242 (1970).
121. J. M. Olinger and P. R. Griffiths, *Anal. Chem. 60*, 2427 (1988).
122. H. M. Fagerholm, C. Linsdjö, J. B. Rosenholm and K. Rokman, *Coll. Surf. 69*, 79 (1992).
123. L. L. Hench, D. E. Clark and E. Lue Yen-Bower, *Nucl. Chem. Waste Manage. 59*, 1 (1980).
124. D. E. Clark and L. L. Hench, *Nucl. Chem. Waste Manage. 2*, 93–101 (1981).
125. C. Q. Buckwalter and L. R. Pederson, N. W. Labs Report PNL-SA-9940, Hanford, 1983.
126. G. L. McVay and C. Q. Buckwalter, N. W. Labs Report PNL-SA-10474, Hanford, 1984.
127. R. A. Lewis, S. Myhra, R. L. Segall, R. St. C. Smart and P. S. Turner, *J. Non-Cryst. Sol. 53*, 299 (1982).
128. C. P. Tripp and M. L. Hair, *Langmuir 8*, 1120 (1991).
129. E. P. Pleuddemann, *Silane Coupling Agents*, Plenum Press, New York, 1982.
130. J. M. Park and R. V. Subramanian, *J. Adhesion Sci. Technol. 5*, 459 (1991).
131. K. W. Allen, *J. Adhesion Sci. Technol. 6*, 23 (1992).
132. E. K. Drown, H. Al Moussawi and L. T. Drzal, *J. Adhesion Sci. Technol. 5*, 865 (1991).
133. D. R. Bennett, J. G. Matisons, A. K. O. Netting, R. St. C. Smart and A. G. Swincer, *Polymer Int. 27*, 147 (1992).
134. L. Britcher, D. Kehoe, J. Matisons, R. St. C. Smart and A.G. Swincer, *Langmuir 9*, 1609 (1993).
135. C. G. Pantano and T. N. Wittberg, *Surf. Interface Anal. 15*, 498 (1990).
136. M. P. Seah and W. A. Dench, *Surf. Interface Anal. 1*, 2 (1979).
137. C. D. Wagner, D. E. Passoja, H. F. Hillery, T. G. Kinisky, H. A. Six, W. T. Jansen and J. A. Taylor, *J. Vac. Sci. Technol. 21*, 933 (1982).

14

Composites

PETER M. A. SHERWOOD Kansas State University, Manhattan, Kansas

1. INTRODUCTION

Surface analytical techniques have been especially useful for the study of composite materials because the properties of these materials are so strongly influenced by the nature of the internal and external interfaces. Composite materials consist of a low-modulus continuous-phase matrix strengthened by an embedded discontinuous reinforcement phase of high modulus and strength. In such a material the stress is carried in proportion to the moduli of the constituent phases, weighted by their respective volume fractions. The reinforcement phase is generally a fiber, which thus bears the principal load, while the matrix not only binds the composite together, but deforms under load thus distributing the majority of the stress to the fibers. Isolation of the fibers in the reinforcement phase reduces the possibility that the failure of an individual fiber will lead to catastrophic failure. These mechanical properties, combined with those of light weight, corrosion resistance, ability to be fabricated into complex geometries, and other desirable ones have meant that composites have found many important applications.

The fibers used in composite materials may be carbon fibers, fibers of inert materials such as boron-carbide-coated boron fibers, silicon carbide (Nicalon), and glass fibers. The matrix material may be an epoxy resin, a plastic, a ceramic material such as carbon or a glass, or metal. Two interfaces are important, the first being that between the surface of the fiber and the matrix material, and the second between the outer surface of the composite and the ambient. The characteristics of the interface between fiber and matrix can have a major impact on the mechanical properties of the composite, while the surface of the composite may be important when applications of the composite are considered. A serious problem from oxidation can result in the failure of composites used at elevated temperatures (>1000°C), and considerable effort has been expended in attempting to reduce or eliminate such oxidation. Carbon-carbon composites are a good example of an oxidation-sensitive system, since they are generally employed at high temperatures. Oxidation protective films may be applied to the surfaces of high-temperature composites, but the films may crack and allow composite oxidation if the coefficients of thermal expansion are different for the protective coating and the composite. Such is the case for many oxidation protective films on carbon-carbon composites. Thus the interface between the composite and any oxidation protective film is also important.

Fibers used in composites present some special practical considerations for the surface scientist, and this chapter will focus upon such considerations. There will be no attempt to review the large body of work in this area [1–5], but the emphasis will be on the problem-solving methods relevant to the study of composites.

2. PRESENTING FIBERS FOR SURFACE ANALYSIS

The analysis of fibers by surface-specific techniques presents some special problems which will be discussed in Sections 2.1–2.3.

2.1. Presentation of Multiple Fibers for Analysis

Fibers are generally small. Most commercial fibers are greater than 1 μm in diameter, though submicron fibers are of interest, and may present special problems. Typical carbon fiber sizes are 5–10 μm, while other fibers may be as large as 50 μm in diameter. In all cases, however, it is not normally practical to examine single fibers. Surface analytical techniques that lack high spatial resolution, such as standard XPS, generally require a sample of adequate size (e.g., 5 mm^2) to produce an adequate signal. Surface analytical techniques with high spatial resolution such as scanning Auger microscopy (SAM) can examine small samples, but the narrower the probe beam (and hence the greater the current density), the greater the possibility of sample damage. In practice then it is normal to group fibers. The most convenient unit is a “tow” of fibers. Fibers are manufactured in tows in which a large number of fibers are bundled together. Typical carbon fiber tows contain 3000 fibers, though tows containing as many as 12,000 fibers have been used in applications requiring cheaper fibers. Samples may thus be presented as a tow attached to be spectrometer sample holder. A good arrangement is to use a "brush" of fibers, where one end of the fiber tow is attached to the holder; the end to be attached is wrapped in aluminum foil and then clamped in the sample holder (Fig. 1a). The problem with this approach, however, is that the fiber tow often "splays out" at the end (which may also lead to some loss of sample), and it is advisable to wrap the other end of the sample as well with aluminum foil to prevent splaying (Fig. 1b). When this is done, it is important to ensure that the surface analytical probe does not "see" either the aluminum foil or the sample holder.

Sample orientation is also a consideration. The easiest arrangement is that of the sample aligned normal to the direction of the surface analytical probe, allowing attachment to a rotatable shaft, as illustrated in Fig. 1c. Such an arrangement is compatible with most sample insertion systems such as those of the rod in an insertion lock or of the end of an x,y,z manipulator. Where angle-resolved experiments are to be attempted the takeoff angle from the sample surface must be varied. In that case, often used in XPS, the sample should be mounted so that the fibers are aligned normal to the the direction of rotation, as shown in Fig. 1d. The only problem that arises then is that of holding the sample in such a way that no signal is derived from the sample holder; clearly, for angle-resolved experiments the sample length must be greater.

2.2. Problems in the Study of Conducting Fibers

Conducting fibers such as those of carbon have the advantage that surface charge does not accumulate on them in techniques such as XPS. On the other hand, the small size of the fibers coupled with their conducting nature presents a very serious practical problem. If sample mounting is not carried out so that the fibers are very firmly anchored, then individual fibers can be lost into the instrument. Once in the instrument they can cause havoc by short-circuiting electron lenses, multipliers, insulators, feedthroughs, etc., and they are very difficult to remove. The first problem is to establish exactly where the lost fibers are without having to dismantle the instrument. If they can be located and then connected to an external power source, they can be burnt away by passage of a large current. Destruction of the offending fibers in this way is, however, not always possible. Other methods involve breaking the vacuum in the system. Sometimes the fibers can be blown out of a critical location by a sudden increase in pressure from vacuum in that part of the system. When all else fails, as can happen unfortunately, there is then nothing for it but to dismantle the instrument completely and examine each dismantled part for the offending

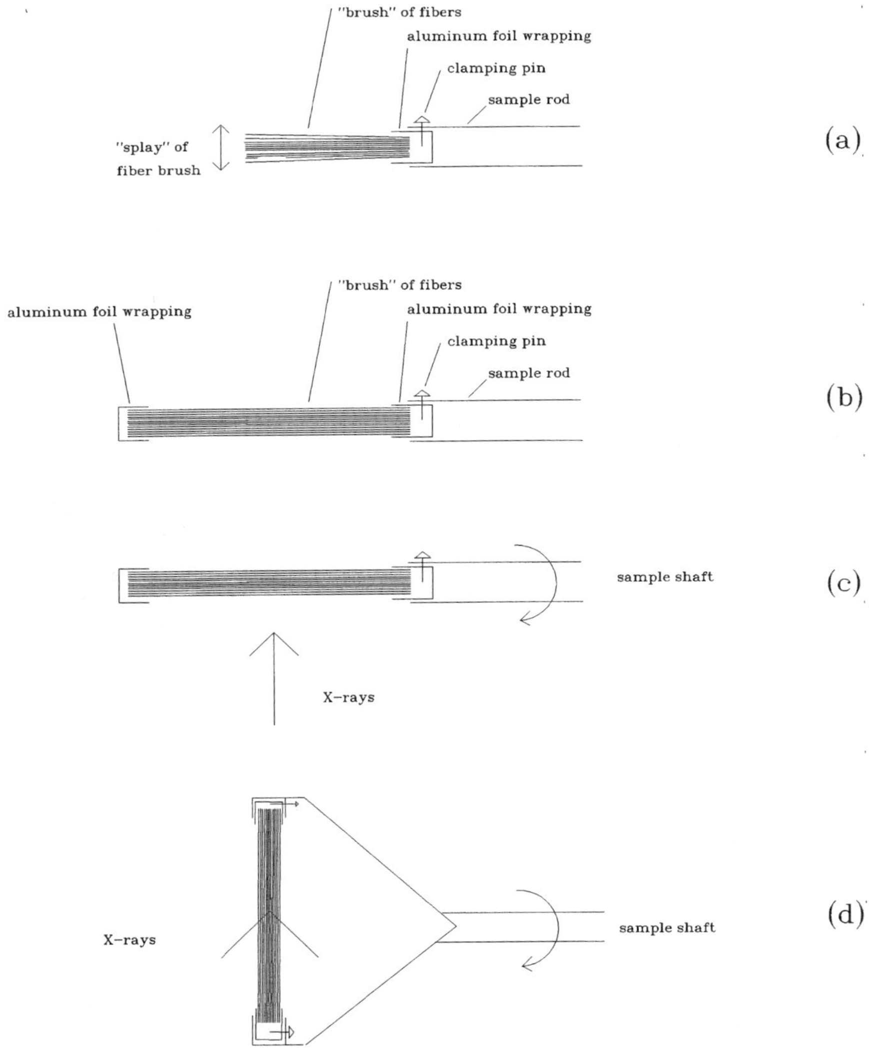

Figure 1. Mounting methods for fibers for surface analysis. (a) "Brush" of fibers clamped at one end after wrapping in aluminum foil. Unfortunately in this method the fibers "splay" out at the end of the brush. (b) "Brush" clamped at one foil-wrapped end, with the free end also wrapped in aluminum foil. (c) A typical arrangement for XPS in which the fibers are arranged with their axes along the sample shaft axis. (d) In this arrangement the fibers are mounted so that their axes are normal to the sample shaft axis.

fiber(s). Such action may sound excessive to some readers, but the author is aware of a number of laboratories in which that approach has had to be taken.

2.3. The Question of Fiber Decomposition

Any surface analytical approach must be evaluated critically with respect to the possibility that surface decomposition might occur during analysis; if it does, then the surface analyst is recording data from a surface whose chemistry has been influenced strongly by the surface analytical conditions. Many fibers used in composites, and the composites themselves, are relatively inert. Unfortunately the surface chemical groups formed at the important interfaces described above may be highly sensitive to surface decomposition. For example, on carbon fibers surface oxidation leads to –C/O functionality which can be lost by the combination of heat, high vacuum, and high-energy radiation. Electron beams, especially those focused for high spatial resolution, can cause substantial damage. Such damage may consist not only of the loss of surface functionality but also of changes in lattice arrangement. It is this problem that seriously limits the number of techniques that can be applied to composites and their component parts. The least-damaging approach is the use of XPS, but even the heat and the Bremsstrahlung radiation from an X-ray gun can cause significant surface decomposition. By far the best solution is to record XPS data using an X-ray monochromator in which the white-hot X-ray target is of the order of 2 m from the sample rather than the 1–2 cm common to most achromatic X-ray sources. Monochromatic X-ray sources also remove Bremsstrahlung radiation which can cause decomposition.

Figure 2 indicates how sensitive carbon fibers are to decomposition. Figure 2a shows the C 1*s* XPS spectrum [2] from a fiber that had been exposed to 10 min of Ar^+ etching, to be compared with the spectrum (Fig. 2b) of the same fiber before etching. The considerable broadening is the result of damage to the lattice caused by the etching process. The relationship between the C 1*s* linewidth and lattice order is discussed below.

3. PRESENTING COMPOSITES FOR SURFACE ANALYSIS

The external surfaces of composites can be examined in the normal way by surface analysis, but usually there is greater interest in the internal properties of a composite. To gain analytical access to the interior, composites are therefore fractured, either outside the surface analytical instrument or within it. Clearly the latter approach would be preferable since any atmospheric changes could then be eliminated. Some surface analytical instruments have facilities for internal fracture, but a note of caution must be sounded, since such fracture will

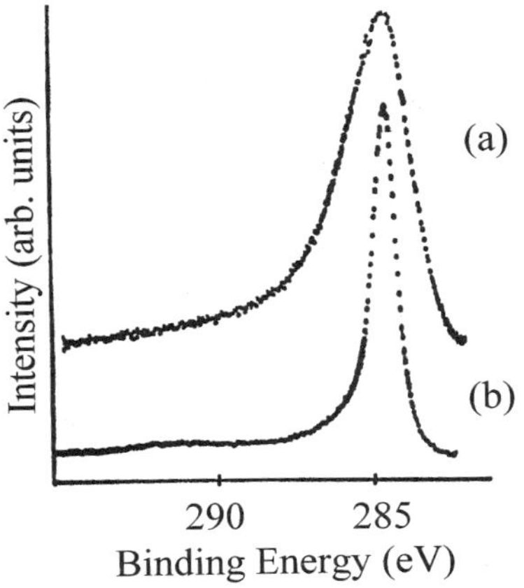

Figure 2. Damage caused to fibers by ion beams. The C 1*s* XPS spectrum of a PAN-based carbon fiber is shown (b) before etching and (a) after a 10-min Ar^+ etch. (From Ref. 2.)

almost certainly lead to the loss of fibers into the instrument with the possible consequences discussed in Section 2.2.

4. SURFACE ANALYTICAL TECHNIQUES FOR COMPOSITES AND FIBERS

There are a range of surface analytical techniques that can be applied to composites. Some of these are inherently surface specific, such as XPS, SIMS, and ISS, while others, such as XRD and IR spectroscopy, may provide surface analytical information in appropriate cases.

4.1. X-ray Diffraction

XRD has been used extensively for the examination of thin films. For the crystalline fibers used in composites, XRD can provide useful information on the extent to which surface treatment of the fiber has affected the bulk of the material. It is possible to use XRD in a thin-film mode, employing very small takeoff angles, to derive some surface information, but generally speaking it must be regarded as a bulk structural technique. Bulk, or near surface, structural information is valuable because there are several surface treatments that can cause changes affecting the interior of the fiber. Examples of the application of XRD to the study of carbon fibers can be found in the author's own work [6–8]. Figure 3 shows that the XRD data showed little change when surface oxidation was taking place (a to d), but when substantial oxidation occurred (e to g), changes had clearly penetrated

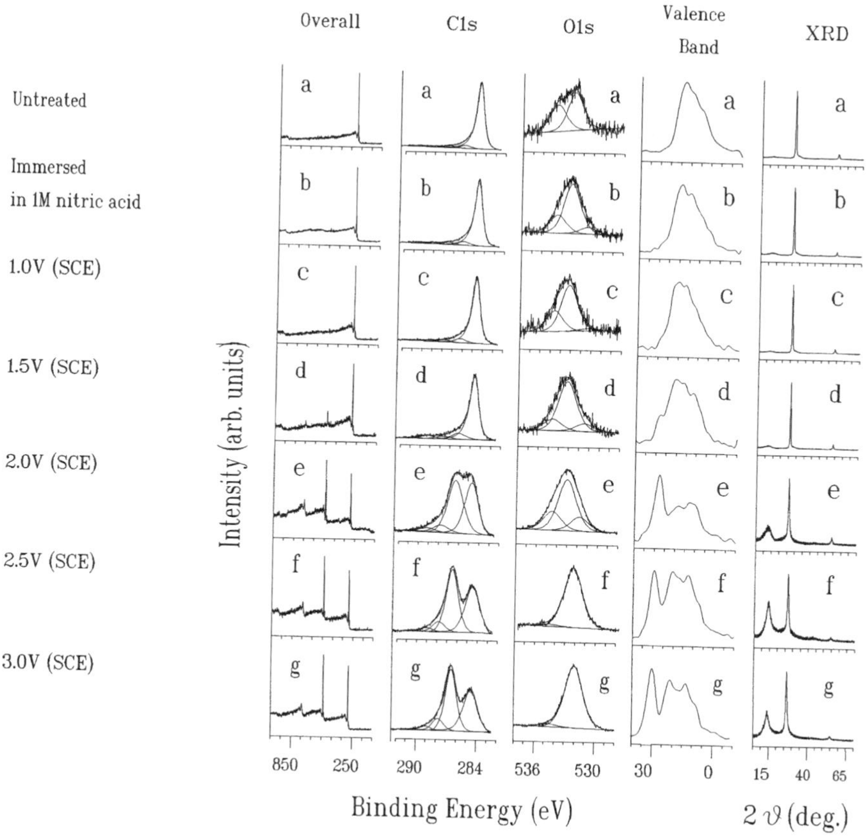

Figure 3. XPS and XRD results for a high-modulus pitch-based carbon fiber before, and after, immersion in, and electrochemical treatment in, 1 M nitric acid. Each treatment consisted of 20 min at a series of increasing polarization potentials. The spectra labeled "overall" are also known as "survey" or "wide scan." (From Ref. 7.)

into the bulk of the fiber; the main peak broadened and a new peak at about 15° (2θ) appeared which might have been associated with lattice disorder.

4.2. FTIR and Raman Spectroscopies

FTIR and Raman spectroscopies can in principle provide information about surface functionality on fibers and composites. In both techniques, however, the sampling depth is considerable, which means that if there is any chemical functionality in the interior of the fiber or composite, then the information is ambiguous. Nevertheless it can sometimes be a useful complement to the information from surface analytical techniques [3,4], as for example when investigating carbon functionality. Raman has been used more often than FTIR spectroscopy, and that includes studies with the Raman microprobe, which allows individual fibers to be examined. Raman spectroscopy can also be used to determine the extent of lattice order in fibers. For example, it has been found that surface-treated carbon fibers have a more disordered nature than untreated fibers [9].

4.3. SEM

SEM has been used extensively for the study of composites, and is the principal method used to determine sample topography. It has been applied to both whole composites and fibers. SEM can provide good evidence for the nature of the overall interaction between fiber and matrix. Thus, good wetting between fiber and matrix—an essential requirement for a composite with desirable qualities—can be studied easily by SEM. Composites prepared with surface-treated fibers have shown generally much better wetting and adhesion than those made from untreated fibers [2]. The examples are so numerous that no attempt will be made here to review them. A good example of the application of SEM to microstructural changes is given by a recent study of fiber-matrix interaction in carbon-carbon composites [10]. Surface treatment of fibers can lead to significant topographical changes on the fiber surface, as for example when different electrochemical or plasma treatments are used [7,11,12].

4.4. STM and AFM

STM and AFM are useful complements to SEM, since they can provide microstructural information at the sub-μm level of lateral resolution where SEM becomes resolution limited. With their use important information has been obtained about fiber microtopography and the effect of surface treatment [13]. The topography of protective surface coatings can also be observed; in Fig. 4 a 5×5 μm^2 area of a silicon carbide coating on a carbon fiber, recorded by AFM, is shown [14].

4.5. Wavelength Dispersive X-ray Emission in an Electron Microprobe

The distribution of elements on a fiber sample can be obtained by using wavelength dispersive X-ray emission (WDS) in an electron microprobe. This approach has the advantage that it is possible to investigate the locations of carbon, oxygen and other light elements which cannot be studied with the normal energy dispersive X-ray emission (EDAX) technique found on most SEM instruments, and has been applied in carbon fiber studies [15].

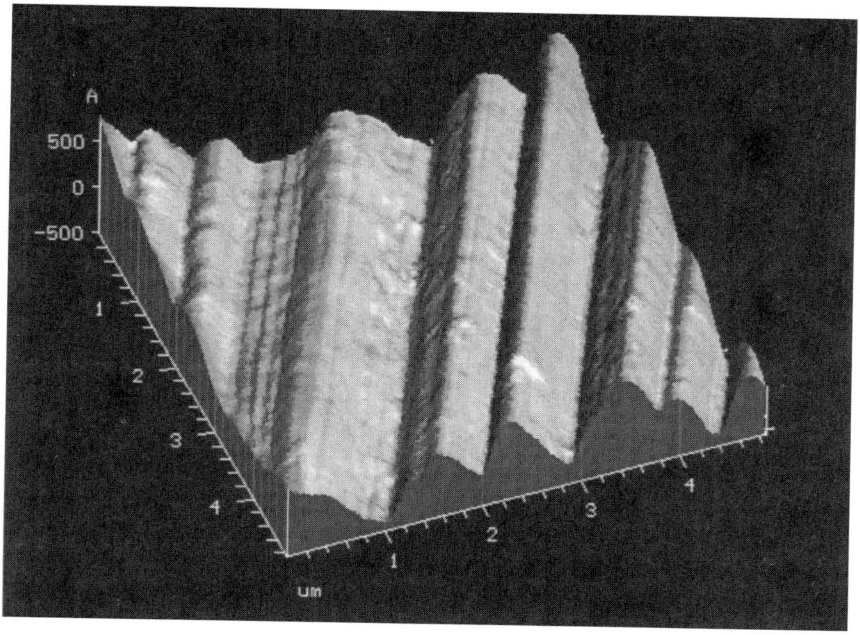

Figure 4. AFM image of a 5×5 μm^2 area of a high-modulus pitch-based carbon fiber coated with an SiC protective film. (From Ref. 14.)

4.6. Surface Energy

The surface free energy can be obtained either by measuring the contact angle for a series of liquids [16], or by inverse gas chromatography (i.e., by placing fibers in a column and passing probe molecules through the column [17]), or by surface tension, or by calorimetry [18]. It can be considered to be composed of a dispersive and a polar component. Liquids chosen for contact-angle studies have known values for these two components. Results show typically an increase in the polar component of the surface free energy on fiber surface treatment, although sometimes the dispersive components increases. The changes in the polar and dipersive components have been related [16–18] to different surface functionalities.

4.7. Titrimetric Methods

Functional groups on carbon fiber surfaces can be analyzed by carrying out titration reactions with appropriate reagents [19], but the method is generally not very sensitive.

4.8. Mass Spectrometry

An important probe of carbon surface functionality is mass spectrometry, in monitoring the nature of the gaseous phase produced from heated carbon fibers by temperature-programmed desorption (TPD). In the author's group TPD has been applied by heating the fibers electrically while monitoring the temperature by optical pyrometry and the gaseous phase by mass spectrometry; XPS data were recorded at the same time [8].

4.9. SIMS

SIMS has considerable potential for the study of composites and is particularly effective for polymers [20,21], which often form the matrix material of composites. The advantages of the technique include the ability to identify hydrogen containing fragments, to distinguish different isotopes and hence to use isotopic labeling, and to provide spatial information by the use of scanning SIMS. As can be seen from Fig. 2, ion beams can cause substantial sample damage, and SIMS has been found to cause such damage when a liquid metal ion gun with a beam diameter of 50 nm was used. Briggs [21] examined a carbon fiber/thermoplastic fracture surface and found an intense negative-ion peak at 26 amu (due to $C_2H_2^-$), characteristic of the carbon fiber surface. A peak was also found at 24 amu (due to C_2^-) which could not be used to discriminate between the fiber and the thermoplastic polymer used in a fiber composite.

4.10. ISS

ISS can also be used for studies of composites and their component fibers, e.g., carbon fiber surfaces [22,23]. ISS is an especially surface-specific technique and is good at detecting and quantifying light elements as in surface impurities (e.g., sodium, potassium and calcium), as well as in the outer monolayers (e.g., carbon, oxygen and nitrogen).

5. X-RAY PHOTOELECTRON SPECTROSCOPIC STUDIES OF COMPOSITES AND FIBERS

5.1. Introduction

Much of the rest of this chapter will be devoted to the use of XPS in the study of composites and fibers. XPS has been able to provide a great deal of valuable chemical information about these materials. There are, however, a number of important practical considerations that need to be considered if the quantity and quality of this information is to be optimized.

Both the core and the valence-band regions in XPS can be used to extract the desired information about the surface region (to depths of no more than 10 nm) of fibers and composites. The core region (electron binding energies >30 eV) is the easier to interpret since each type of atom has core electrons (i.e., electrons that play no significant role in chemical interaction) with a set of binding energies characteristic of that atom; thus atomic identification is immediate. Information about the chemical environment of an atom can be provided by the chemical shift in the binding energy.

The valence-band region (0–30 eV), on the other hand, will contain features arising from all species present on the surface. Since all energy levels contributing to the valence band are involved in chemical interaction, and since that interaction will be specific to the species present at or near the surface, it follows that valence-band spectra in XPS will be highly sensitive to surface chemical states. Core electron chemical shifts are sometimes insufficient to be able to distinguish between subtle chemical differences, and the valence-band region can thus play an important role. In order to understand the valence-band region, reliable spectra from model compounds are needed as well as some means of reliable prediction of the spectra. Spectral prediction is especially important because many compounds have surface composition and structure unrepresentative of the bulk (e.g., most lithium compounds have a surface consisting of lithium carbonate) and misleading information can result from a reliance on model compounds alone.

XPS core electron chemical shifts can be interpreted using a potential model which relies upon point charge values for the atoms present. In its most effective form, the relaxation potential model, the relaxation energy is included. The author has used such calculations to predict shifts associated with carbon surface functionality [24]. Since the C 1*s* photoelectron peak exhibits significant shifts with changes in the carbon oxidation state, changes occurring in the core region spectra can provide considerable chemical information. The shifts can be measured accurately in carbon fiber spectra since there is always an intense C 1*s* feature at the lowest binding energy, associated with the graphitic backbone, and the *separations* between this peak and the other C 1*s* features in the same spectrum can be identified easily. The measurements are not therefore subject to the normal calibration problems associated with XPS.

Valence-band XPS spectra require more sophisticated methods of calculation. The author has found that scattered-wave Xα and ab initio molecular orbital calculations are good approaches, which he has developed so that they may be used to generate a predicted valence-band spectrum simultaneously with the experimental observations [25]. Agreement with experiment has been found to be excellent. In the valence band, the separation of spectral features, and their relative intensities, which can be compared with calculations and model compounds, provide the chemical information. As in the core region such separations can be measured accurately and are not subject to the uncertainties of calibration.

Figure 3 gives an example of how a comparison of core and valence-band spectra in XPS with each other and with changes in X-ray powder diffraction (XRD) data (to be discussed later) can provide useful information. The spectral and diffraction data are from a high-modulus pitch-based fiber (Dupont E-120) first immersed in 1 M nitric acid and then oxidized electrochemically for 20 min at different voltages (versus the saturated calomel electrode (SCE)). SCE is used as a reference electrode in many electrochemical studies; it has a potential of 0.242 V versus the normal hydrogen electrode (NHE) whose internationally accepted reference potential is 0 V. The overall (often called "survey") spectra show an intense peak in the C 1*s* core region, with an increasing contribution from the O 1*s* feature as the extent of oxidation increased. Detailed analysis of the C 1*s* and O 1*s* regions by spectrum fitting, as shown in Fig. 3, revealed a number of overlapping features corresponding to different chemical functionalities. The valence-band spectra (which have had nonlinear backgrounds subtracted) also showed significant differences. The surface chemical information that can be extracted from the data in Fig. 3 will be discussed at greater length later.

5.2. The Question of Surface Charging

Sample charging arises because the positive charges that result from the photoelectron process are not neutralized by electrons flowing from earth. Differential sample charging occurs when different parts of a sample experience different degrees of charging. Any application of XPS must be conducted carefully in order to avoid differential surface charging. When grounded conducting samples such as carbon and carbon fibers are examined, surface charging is not a problem as the surface charge is instantly eliminated by the flow of electrons from ground. On the other hand, when insulating materials such as the polymers and fibers found in some composites, or in coatings of carbon composites and fibers, are studied, then differential surface charging can be a problem. Calibration methods, such as the use of an external calibrant, or of an internal core peak as a calibrant, can be effective [26]. Differential sample charging can cause what is really a single peak to *appear* as though it were a mixture of overlapping peaks [27].

Figure 5 illustrates how the problem can be demonstrated by devising an experiment in which the differential sample charging situation is deliberately induced. Spectrum 1 in Fig. 5 shows the Ag $3d_{5/2}$ region for a silver iodide sample mounted on a fine copper grid. Spectrum 2 results from the generation of a differential sample charge by biasing the grid 4 V negative. It can be seen that part of the peak has shifted the full 4 V, but another part has shifted by a lesser amount,

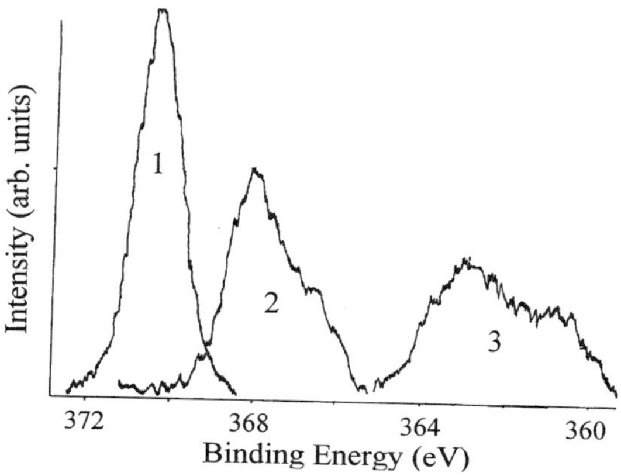

Figure 5. Ag $3d_{5/2}$ XPS spectra from a silver iodide sample placed on a fine copper mesh. Spectra: (1) no bias, (2) −4 V dc bias, and (3) −10 V dc bias. (From Ref. 27.)

giving the impression of an asymmetric peak corresponding to the overlap of at least two component peaks. Spectrum 3 shows the result of exaggerating the differential sample charging by the application of a bias of 10 V negative. It might be thought that the spectrum could now be resolved into three or more peaks. Of course in all cases only one Ag $3d_{3/2}$ peak should be seen.

An example of a situation in which differential sample charging can occur in fibers and composites is that of an oxidation protective film on a fiber or composite. Such films are often thick (a few μm) and may consist of insulating materials. In checking as to whether or not differential charging is present on the surface of a sample, and its effects, it is advisable to apply *both* positive and negative bias voltages. This is illustrated in Fig. 6 in which the C 1*s*, O 1*s*, N 1*s* and Si 2*p* spectra from a carbon fiber coated with a film of silicon nitride about 2 μm thick are shown [28]. Spectra (d) and (a) were recorded before and after biasing, respectively, while in (b) and (c) the specimen was biased +15 and −15 V, respectively. The appearance of single peaks in the C 1*s* and N 1*s* regions for +15 V indicates that differential charging was largely eliminated when the specimen was given a positive bias. On application of a negative bias there seems to have been sufficient stress built up in the silicon nitride coating to cause the coating to crack and expose the underlying fiber. It is suggested that the different behavior with respect to positive and negative biasing arises from the electrical properties of silicon nitride, the electrical current being able to flow from coating to fiber during positive bias, but unable to do so on application of a negative bias (i.e., a type of diode effect). The resultant electric field across the silicon nitride coating may have been high enough to cause

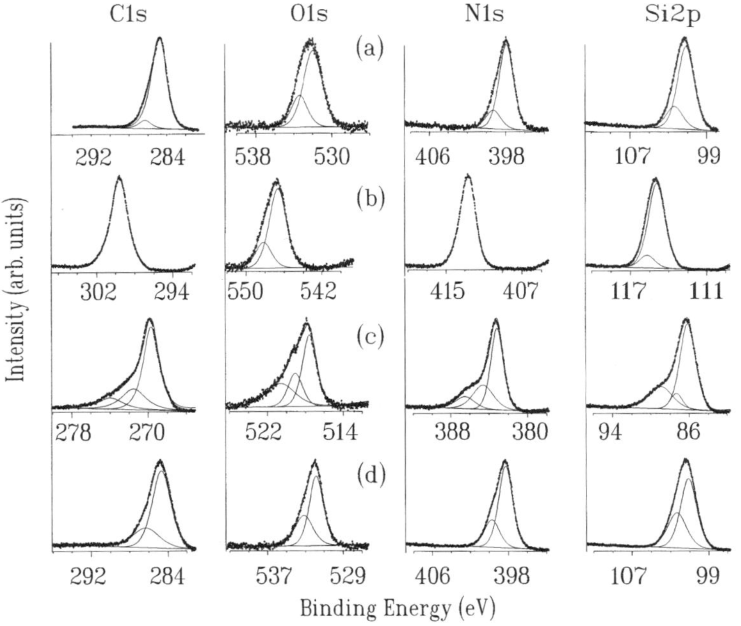

Figure 6. XPS core spectra from Ar^+ etched silicon-nitride-coated carbon fibers. (a) Spectra at the conclusion of the biasing experiment, (b) at a bias of +15 V dc, (c) at a bias of −15 V dc, and (d) the initial spectra. (From Ref. 28.)

cracking. The differences observed in this experiment should alert the experimentalist to the potential problems of differential sample charging.

5.3. Depth Profiling of Carbon Composites and Fibers

Depth profiling by the usual Ar^+ bombardment cannot be used because of the damage caused to the fibers and/or the composite (see discussion above and illustration in Fig. 2). There is information about the variation with depth of the surface functionality available in XPS spectra, however, based on the relationship between photoelectron kinetic energy and the average electron escape depth. Thus the more energetic the photoelectrons, the greater the depth from which spectral features will arise, which means that the valence-band region will always correspond to greater average depths than the core regions. The same relationship can also be exploited by using a range of X-ray sources of different photon energies so that any one characteristic core line then appears at various kinetic energies, corresponding again to a range of escape depths.

Angle-resolved photoemission has also been used to study differences with depth of fiber surface functionality. The method needs flat surfaces to work properly since photoemission from the surface layers is favored preferentially at low takeoff angles, and a rough surface would therefore blur the dependence of spectral information on depth. When fibers are to be studied, the approach is often based on the sample orientation shown in Fig. 1d. The principle is that the fiber looks to the energy analyzer more like a flat surface when turned at right angles to the X-ray beam than it does when normal to the direction of the beam. However, any fiber orientation (either Fig. 1c or d) is only an approximation to a flat surface, and so the results of angle-resolved experiments are less informative than they would be for a truly flat surface. In fact, both orientations (Fig. 1c and d) will provide some information since mutual geometrical "shadowing" by neighboring fibers will promote the angular selectivity.

5.4. Decomposition of Surface Functionality during Spectral Collection

An important point is that surface functionality is usually sensitive to heat and may therefore be lost by decomposition caused by heat from the X-ray gun (see Section 2.3). Such decomposition is illustrated in Fig. 7, which shows how the C 1*s* spectrum from a PAN-based carbon fiber oxidized in 0.22 M nitric acid at 2.0 V (versus SCE) changed after spectral recording for 10 and 90 min. The data were fitted using the same peak positions as those in Table 2 (discussed later) and a nonlinear background (discussed later). The amount of oxidized carbon fell from about 50% to about 40% during the period of data acquisition [11].

Figure 8 shows similar data for a pitch-based carbon fiber oxidized in 1 M nitric acid at 3.0 V (versus SCE) [7]. In this case the C 1*s*, O 1*s* and valence-band regions of the photoelectron spectra are shown. The kinetic energies of the photoelectrons in these regions are, in descending order, valence band > C 1*s* > O 1*s*, indicating that the surface sensitivity should be in the order O 1*s* > C 1*s* > valence band. Thus, if it is assumed that any x-radiation damage would have had the greatest effect on the outer surface, then decomposition should have occurred from the outer surface inward and its effects should have been more noticeable on the O 1*s* and C 1*s* spectra than on the valence-band spectra. Figure 8 shows that that is exactly what was observed. For instance, compare the C 1*s* spectrum after 1.3 h exposure (Fig. 8b) and the valence-band spectrum after 1.5 h exposure (Fig. 8a) with the C 1*s* spectrum after 24 h exposure (Fig. 8d) and the valence-band spectrum after 28 h exposure (Fig. 8c). Whereas the C 1*s* spectrum had altered dramatically after 24 h exposure, there were only minor changes in the valence-band spectrum over approximately the same time.

Detailed discussion of the features in the core and valence-band regions will be left until later. However, some simple observations can be made now. The

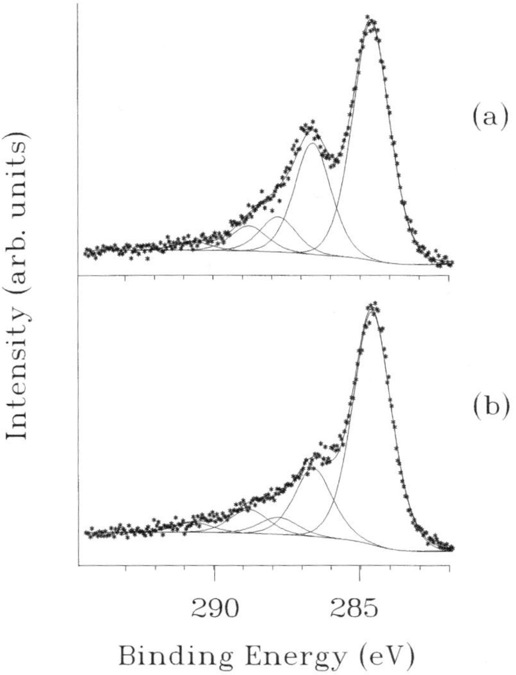

Figure 7. C 1*s* XPS spectra from a PAN-based carbon fiber electrochemically oxidized in 0.22 M nitric acid to 2.0 V (vs. SCE) for 20 min. (a) Spectrum after 10-min X-ray exposure, and (b) spectrum after 90-min exposure. (From Ref. 38.)

peak at highest binding energy in the valence band is due largely to the O 2*s* contribution, while that at the next highest binding energy arises mostly from the C 2*s* band; in the C 1*s* core-level spectrum the component peak at the highest binding energy is due to oxidized carbon. It is clear that the fall in the intensity of oxidized carbon in the C 1*s* region is much greater than in the valence-band region, supporting the suggestion that decomposition occurs from the outer surface inward into the bulk. The data have been fitted to the same peak positions as those in Table 2 (discussed later). In the O 1*s* region two principal peaks can be identified, that at higher binding energy being due to –C bridged functionality and that at lower binding energy to –C=O functionality (from –C=O or $-CO_2X$). In the C 1*s* region the corresponding features in Fig. 8a are that at the highest intensity, –C bridged, and those at the next two highest binding energies, –C=O functionality. The relative changes in the ratios of these peak areas are shown in Table 1. Although the spectra were recorded for different lengths of time, it is clear that the amount of –C bridged appeared

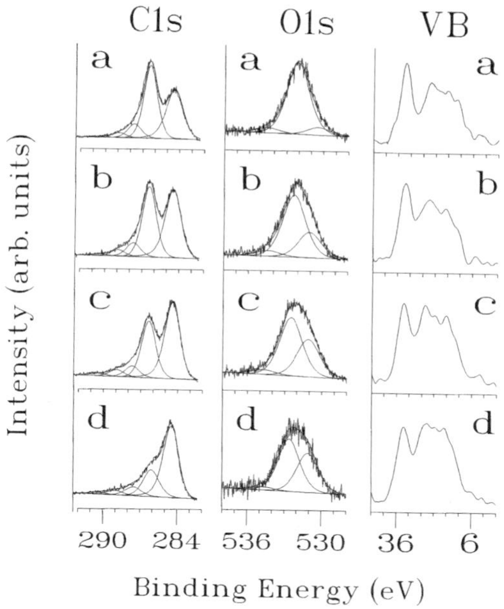

Figure 8. C 1*s*, O 1*s* and valence-band XPS spectra from an E120 pitch-based carbon fiber electrochemically oxidized in 1 M nitric acid at 3.0 V (vs. SCE) for 20 min. The X-ray exposure times for the C 1*s* spectra were (a) 12 min, (b) 1.3 h, (c) 3 h and (d) 24 h; for the O 1*s* spectra, (a) 6 h, (b) 20 h, (c) 44 h, and (d) 67 h; and for the valence-band (VB) spectra, (a) 1.5 h, (b) 7 h, (c) 28 h, and (d) 48 h. (From Ref. 7.)

to fall more rapidly with respect to –C=O, based on the changes in the O 1*s* spectrum, than in the C 1*s* spectrum. This difference presumably reflects both the greater surface specificity of photoemission in the O 1*s* region and the greater tendency of the –C bridged functionality to decomposition under the influence of heat and/or x-radiation.

The data in Figs. 7 and 8 were collected with achromatic x-radiation. The decomposition would have been avoided if monochromatized X-rays had been used.

5.5. XPS Data Analysis and Interpretation of Core Chemical Shifts

Careful data analysis can yield important chemical information that might otherwise be missed [29–31], and is essential for the extraction of such information in a number of systems, especially those of fibers and composites. Their

Table 1. Curve Fitting C 1*s* and O 1*s* Spectra of Oxidized Carbon Fibers Exposed to X-radiation for Different Lengths of Time

Exposure time (h)	O 1*s*/C 1*s* area ratio	Area ("bridged"/–C=O) in C 1*s* spectra	Area ("bridged"/–C=O) in O 1*s* spectra
0.5	0.76	3.6	All bridged
3.0	—	2.7	—
6.0	0.43	—	17.57
20	0.43	—	2.83
24	—	2.00	—

core electron spectra contain a number of overlapping features associated with different surface functionalities. In many papers [32–35], the author and coworkers have shown how such overlapping features can be resolved by combining data analysis methods with the requirement that, where the surface chemistry of a sample is changing in a continuous fashion (e.g., *by electrochemical oxidation), then the fit of individual spectral components to the resultant series of core spectra must be entirely consistent.*

5.5.1. Fitting C 1*s* spectra

For many fibers and composite materials the C 1*s* spectrum is the most important. Care must be taken, however, in the interpretation of the C 1*s* region because fibers and composites are subject to hydrocarbon contamination. In experiments using single-crystal metal samples such contamination can be removed by Ar^+ etching and heating and by other cleaning methods, but such methods are quite inappropriate for fibers and composites because of the resulting sample decomposition. In carbon fibers and composites analysis of the as-received surface can give very important information about surface treatment and interfacial chemistry. Fortunately, carbon fibers have a rather low affinity for hydrocarbon contamination so that the original surface can be analyzed reliably in most cases. Some manufacturers treat carbon fibers with size, which is very difficult to remove [36]. If the size is not removed, then the C 1*s* spectrum will be characteristic of the size and not the fiber.

Depending on their origin, untreated and unsized carbon fibers can show significant differences in their C 1*s* spectra. In general, high-modulus fibers have the narrowest C 1*s* spectra. Typical data using achromatic XPS are shown in Fig. 9 [37], in which the C 1*s* spectra from five different types of fiber are compared with that from pyrolytic graphite. It can be seen that the spectra from the untreated fiber and from the graphite can be curve-fitted by including components due to oxidized carbon. The main difference between the various spectra

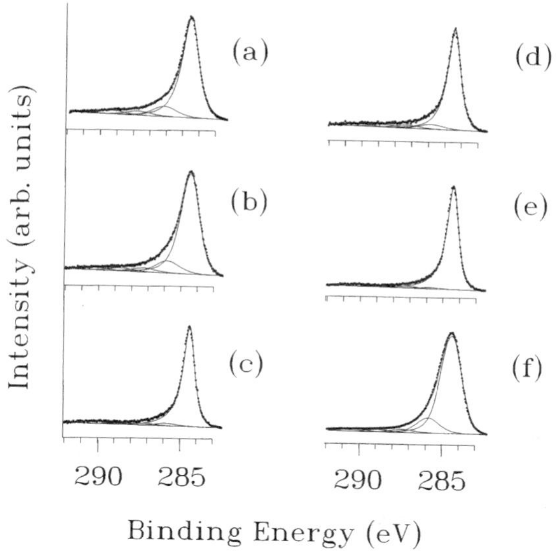

Figure 9. C 1*s* XPS spectra from various forms of untreated and unsized carbon fiber and from carbon. (a) A type II PAN-based fiber stored for 18 months; (b) an AU4 PAN-based fiber; (c) a P55X pitch-based fiber. (d); an E-35 pitch-based fiber; (e) an E-120 pitch-based fiber; (f) pyrolytic graphite. (From Ref. 37.)

in Fig. 9 is that the C 1*s* linewidth varies considerably, though in all cases the principal peak has an exponential line shape. The latter arises from conduction-band interaction, typical of conducting materials. The peaks also have substantial Lorentzian character. When carbon fibers are subjected to considerable oxidation, the exponential line shape is lost since the fiber surface is then no longer conducting. Examples of this condition will be discussed below. The differences in peak width in Fig. 9 can be related [37] to differences in the degree of graphitic character and in the degree of order in the lattice, and may be compared with measurements of bulk order by XRD [37]. Ar^+ etching would be expected to disrupt the graphitic order and cause a considerable broadening of the C 1*s* peak, as seen in Fig. 2.

Figure 3 illustrates how the C 1*s* and O 1*s* regions can be fitted by a number of component peaks, each peak corresponding to a different surface functionality. The curve-resolved spectra in Fig. 3 have been derived by consistent fits to a sequence of spectra from progressively chemically modified carbon fiber surfaces, i.e., surfaces subjected to electrochemical oxidation at increasingly positive potentials. Note how the component peaks change in relative intensity with oxidation—a low-intensity peak at low potential grows into a major peak at high (oxidizing) potentials. In the C 1*s* example the feature at lowest binding energy corresponds to the fiber carbon, that at the higher binding energy to various different oxidized carbon species. It is tempting to correlate the C 1*s* spectra with the same features at the same intensities as those resolved in the O 1*s* region since correspondence might be expected. However, it should be remembered that the depths sampled by photoemission from these two core levels are different and differing information would therefore be given by each if the surface were not homogeneous.

The comparisons made in Fig. 10 help to illustrate how the peak separations in the C 1*s* spectrum can be related to carbon surface functionality. The separations can be predicted by calculation as mentioned earlier, but model compounds can also give helpful information. In Figs. 10b and c are shown the spectra of solid hydroquinone at 77 K (an aromatic molecule with –C–OH functionality) and of solid benzoquinone (an aromatic molecule with –C=O functionality), respectively. It is clear that the separation between the C 1*s* peaks arising from the aromatic ring (at lower binding energy) and from the –C–OH functionality is less than that between the former and that due to

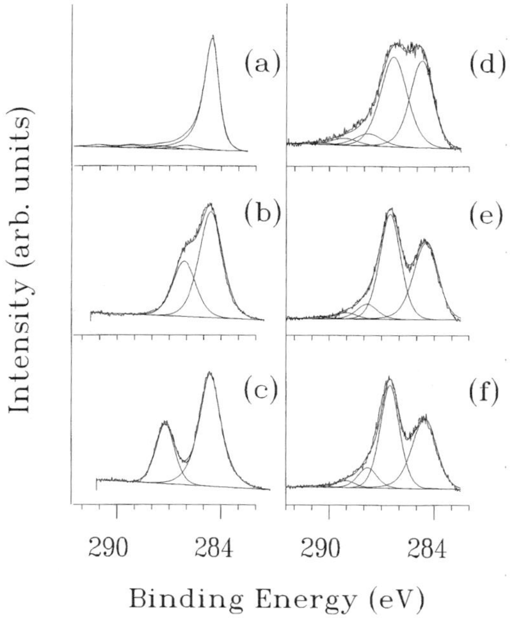

Figure 10. C 1*s* XPS spectrum from an untreated pitch-based carbon fiber (a) compared with spectra from oxidized fibers (d) at 2.0 V, (e) at 2.5 V, and (f) at 3.0 V, and from solid hydroquinione (b) and solid benzoquinone (c). (From Ref. 25.)

–C=O functionality. A high-modulus pitch-based carbon fiber oxidized to 2.0 V (Fig. 10d) shows a C 1*s* peak shift comparable to that seen in hydroquinone, so it is reasonable to assume that the surface functionality on that surface is due to –C–OH groups. The same fiber oxidized to 2.5 V (Fig. 10e) or to 3.0 V (Fig. 10f) produces a shift intermediate between those due to –C–OH in hydroquinone and to –C=O in benzoquinone. It has been suggested [25,38] that such an intermediate shift corresponds to an intermediate functionality involving a "bridged" structure, discussed below. Typical shifts between the C 1*s* peak from the oxidized sample and the "graphitic" C 1*s* peak are 1.5 eV for –OH functionality, 2.0 eV for the bridged structure, 3.2 eV for a C=O-type functionality, and around 4.2 eV for carboxyl ($COOH/CO_2^-$) or ester (COOR)-type functionalities. Features due to $\pi \rightarrow \pi^*$ transitions occur at shifts of around 6.1 eV.

5.5.2. Detailed fitting considerations

Curve fitting of the type discussed has the advantage that an observer can see clearly how the summation of the fitting components compares with the original experimental data. The key point about curve fitting is that there is *never a unique solution to fitting the data*, and it is essential that as much chemical and physical reality as possible be incorporated in the overall procedure. It is also important that as much additional supporting information as possible be included.

An important starting point is the selection of an appropriate fitting function. In XPS the basic peak shape is Lorentzian, but it is modified by instrumental and other factors (such as phonon broadening) whose combined effects are equivalent to a Gaussian contribution. In some cases, especially for electrically conducting materials, the final peak shape is asymmetric due to various loss processes, such as the exponential tail on the C 1*s* peak discussed. Different functions have been used in XPS over the years, and the author has reviewed this matter in more detail elsewhere [29,30]. A function that combines Gaussian with Lorentzian character and has the ability to represent asymmetric peak features is essential if overlapping peaks are to be correctly identified. Such a function was introduced by the author [29,30,39] and is used throughout this chapter.

Most XPS data are recorded using achromatic x-radiation, for which the X-ray linewidths are ca. 0.8 and 1.0 eV for the popular magnesium and aluminum X-ray sources, respectively. If achromatic x-radiation is used, then there will appear associated with each XPS peak a series of weak peaks resulting from the weaker X-ray satellite radiation in the achromatic X-ray source. In this chapter all the spectra generated by achromatic x-radiation have been fitted in such a way that these satellite features are included, using the function previously described [39] with a 50% Gaussian/Lorentzian character.

In core-level XPS data from fibers and composites the background of the spectrum represents a significant unknown. Where the slope of the background is small, the background can be fitted effectively by a line, and such a simple background has been used in Figs. 3, 8–10. In other cases the spectra of a mixture of surface components may contain substantial background contribution due either to the inhomogeneous distribution of the various components in the mixed surface layer region, or to the presence of surface contaminants, or to the particular surface topography. Fortunately there have been considerable advances in the understanding of such backgrounds, in particular the ideas and formulations of Tougaard ([40,41]) and others (see Chapter 5, references therein and [30]). Background analysis is generally performed over a large energy range, but curve fitting is done over a small energy range. The reason for this is that the range of chemical shifts is generally no more than 10 eV and often much less. The spectral range chosen for curve fitting needs to be large enough to include the main features, especially any arising from the $K_{\alpha 3.4}$ satellite peaks from achromatic x-radiation, but also small enough to allow data with good statistics to be collected in a reasonable length of time. The reality is that the energy range chosen for curve fitting, being significantly smaller than that needed for background analysis, makes subtraction of a correct background more difficult. The author has found recently that a nonlinear iterative background [42] gives the best results for spectra from composites and fibers [43]. The background should be included in the curve fitting process as has been carried out in all the examples of curve fitting in this chapter, thus allowing simultaneous display of the fit and of the original experimental data.

Spectra from composites and fibers are often in the form of a series changing progressively with surface treatment. An example is given in Fig. 3, in which the set of spectra corresponds to increasing oxidation of carbon fibers. The challenge in curve fitting is then to achieve consistency throughout the series of changing spectra. Such consistency might be achievable, for example, by fixing the energetic positions of the component peaks but varying their intensities. Although few chemical systems produce photoelectron spectra in which the component peaks show no variation in position, the approximation is often a good one in electrochemical oxidation processes. If a suitable system is chosen, electrochemical treatment, such as that leading to oxidation, for example, produces changes in electrode surface chemistry that are highly controllable and reproducible, so that such treatment is very useful in the study of surface chemistry. In Fig. 11 are shown the results of electrochemically oxidizing carbon fibers at increasing oxidizing potentials. The C 1*s* spectra in Fig. 11 were recorded with achromatic radiation from PAN-based carbon fibers which were oxidized electrochemically in 2.7 M nitric acid for 20 min at various potentials. The spectra have been fitted with five component peaks, four of which correspond to different surface functionalities on the carbon fiber. Clearly any attempts to fit the data at 0.5 (a) and 1.0 V (b) would have been difficult without additional information. Now the spectra at 2.5 (e) and 3.0 V (f) could be re-

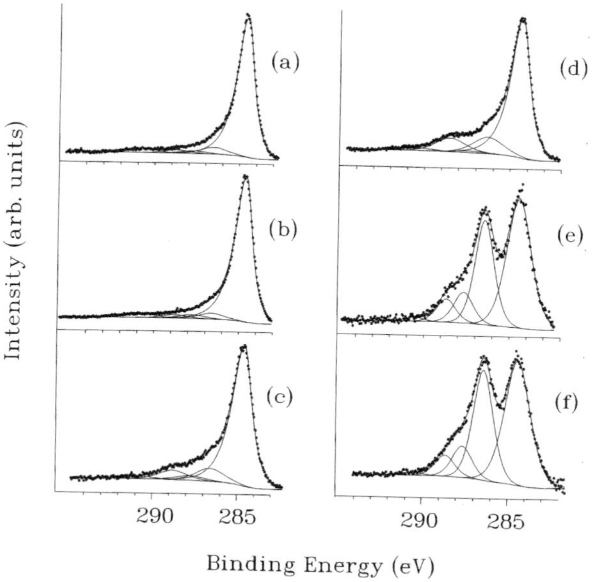

Figure 11. C 1*s* XPS spectrum from a PAN-based carbon fiber after potentiostatic polarization at different potentials (vs. SCE) for 20 min in 2.7 M nitric acid solution. The spectra have been fitted with five peaks. Four peaks (G/L mix = 0.5) are separated from the principal fiber carbon peak (at 284.6 eV with G/L mix = 0.8) by 2.0 eV (carbon attached to "bridged" oxygen), 3.2 eV (–C=O carbon), 4.2 eV (–CO_2H/R) and 6.1 eV (carbonate/$\pi \rightarrow \pi^*$ shakeup processes). The FWHM of the peaks arising from oxidation were at about 2 eV (for polarization potentials 0.5 to 2.0 V) and 1.4 eV (for polarization potentials 2.5 and 3.0 V) (e) and (f). A nonlinear background has been included in the fit. Results of the curve fitting are shown in Table 2. (From Ref. 43.)

solved into at least three features and (using additional information) one or two weaker features could be added as well. The peak positions established in (e) and (f) could then be used to fit the whole series giving the results shown (the peak positions were fixed at those values and all the features arising from oxidized carbon were placed in the same "group" with the same peak widths). Table 2 shows that the steady decrease in the intensity of the peak at 284.6 eV, corresponding to carbon in the fiber, with the simultaneous growth in the intensity of the two most intense features due to oxidized carbon, is what would be expected from the electrochemistry. The shifts are the same as those listed in the preceding subsection. Although there is no unique fit to the data, the chosen fit is consistent with the chemistry of the process. Note how the inelastic

Table 2. Curve Fitting a Series of C 1*s* XPS Spectra: Changes in Surface Chemistry

Shift from principal "graphitic" peak (eV)	Percent of total area due to peak after various electrochemical treatments Electrochemical potential (volts vs. SCE)					
	0.5	1.0	1.5	2.0	2.5	3.0
0	86.8	86.9	78.8	74.8	50.4	49.5
2.0	5.4	4.4	9.3	12.1	31.7	33.7
3.2	2.2	2.9	3.4	1.2	9.5	9.5
4.2	2.5	2.6	6.7	9.9	7.1	6.3
6.1	3.1	3.2	1.8	2.0	1.3	1.0

tail on the "graphitic" carbon peak is lost following oxidation levels at 2.5 and 3.0 V, corresponding to the loss of graphitic (and therefore conducting) character in the surface region as the graphite sheets become highly oxidized.

Figure 12 shows how the fit to the C 1*s* spectrum from an epoxy resin on a carbon fiber, where the background has been fully incorporated into the fit, varies with the choice of background type. Although the fits are of comparable quality, Table 3 shows that the relative amounts of the different components show significant variation depending on the choice of background. Clearly there is no way of assessing which is the most chemically realistic fit in this particular case, but in a review of other cases [43] the author found that the iterative nonlinear background [42] gave the best result.

5.5.3. The use of monochromatic X-radiation

Monochromatic x-radiation has a number of advantages, already discussed. In the case of carbon fibers decomposition is virtually eliminated, and when used with the best instrument resolution, a spectacular improvement in linewidth can be obtained. Figure 13 compares the C 1*s* spectra from an untreated and unsized Toray M40 PAN-based carbon fiber recorded with achromatic (Fig. 13a) and with monochromatic (Fig. 13b) radiation [44]. The spectrum in Fig. 13a has been fitted to the component peaks listed in Table 2 *plus* an additional peak shifted by 1.0 eV from the principal "graphitic" carbon peak at 284.6 eV. This additional peak is in fact a nongraphitic contribution to the principal peak. Of course, it would have been possible to fit the spectrum in Fig. 13a without the extra peak simply by increasing the slope of the exponential tail of the principal "graphitic" peak. It is believed, however, that there is a genuine peak in that position, based on examination of the spectrum obtained with monochromatic radiation. The use of monochromatic radiation reduces the FWHM of the C 1*s*

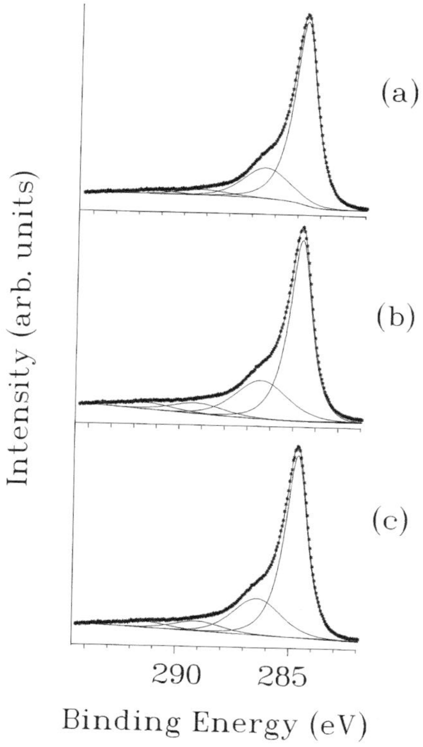

Figure 12. C 1*s* XPS spectrum from a type II HT PAN-based carbon fiber coated with a thin film of epoxy resin after potentiostatic oxidation for 20 min at 2 V (vs. SCE) in 2 M ammonium bicarbonate solution, fitted with various backgrounds. (a) A nonlinear background, (b) a Tougaard background, and (c) a linear sloping background. X-ray satellite features have been included in the curve fitting. Area ratios and the quality of the fit are given in Table 3. (From Ref. 43.)

"graphitic" peak from around 1 to 0.32 eV. This reduction allows further spectral features to be identified. There is a clear shoulder marked by an arrow in the expanded part of the spectrum in Fig. 13b, believed to be due to β-carbons. The latter are carbon atoms bonded to other carbon atoms to which in turn oxygen atoms are attached. The author suggested earlier that such peaks might be present [45], but as is clear from Fig. 13a they could not be identified unambiguously when achromatic radiation was used. Obviously the way in which the β-carbon peak has been included in Fig.13a overestimates its contribution. Since that contribution could be adjusted easily by merely increasing the

Table 3. Curve Fitting the Same (Table 2) C 1*s* Spectrum but with Different Background

	Percent of total area due to different types of peak			
Background used	Carbon	Oxide I	Oxide II	Oxide III
Nonlinear ($\chi^2 = 1140$)	65.6	24.0	6.5	3.9
Tougaard ($\chi^2 = 1117$)	61.8	26.6	7.2	4.4
Linear sloping ($\chi^2 = 925$)	72.4	21.4	4.0	2.2

amount of exponential tail, it is clear that the fit in Fig. 13a is ambiguous, illustrating how difficult it is to fit correctly a β-carbon peak when using achromatic radiation. Figure 13b also indicates that chemically shifted features, especially the peak shifted by 2 eV, are much more readily identifiable than in the data recorded with achromatic radiation. In fact the only evidence in Fig. 13a of oxidation is based on subsequent information from which it is known that continued oxidation causes an increase in the contributions of peaks at the positions used in the fit.

When monochromatic radiation is used to examine the effect of electrochemical oxidation of fibers, the spectra obtained are of better quality than with achromatic radiation, allowing more, and more precise, information to be extracted. For example, in Fig. 14 are shown the C 1*s* spectra recorded from the same fibers as in Fig. 13 following galvanostatic oxidation in 1 M nitric acid in a pilot plant [8] after exposures for 40 s at various currents [44]. Spectra from the oxidized fibers have the same general shape as those in Fig. 12, for which achromatic radiation was used, but a much improved spectral quality. Three features are clearly resolved in addition to the principal peak, with the β-carbon peak being included effectively in the fit. The components were fitted according to the positions given in Table 2, together with a plasmon feature at a shift of about 6.9 eV. A nonlinear background was also included. The valley between the principal peak and that shifted by about 2 eV from it is clearly deeper than in Fig. 12. Oxidation caused an increase in the width of the principal "graphitic" peak (to around 0.9 eV), although the peaks arising from oxidized carbon were narrower than with achromatic radiation. The increase in width of the graphitic peak was expected since oxidation increases the disorder in the lattice, as discussed in Section 5.5.1.

5.5.4. Fitting O 1*s* spectra

The O 1*s* region of the photoelectron spectrum from organic materials is generally rather broad, but can be resolved into three principal features as seen in Figs. 3 and 8. The features correspond to –C=O at around 531.8 eV, to

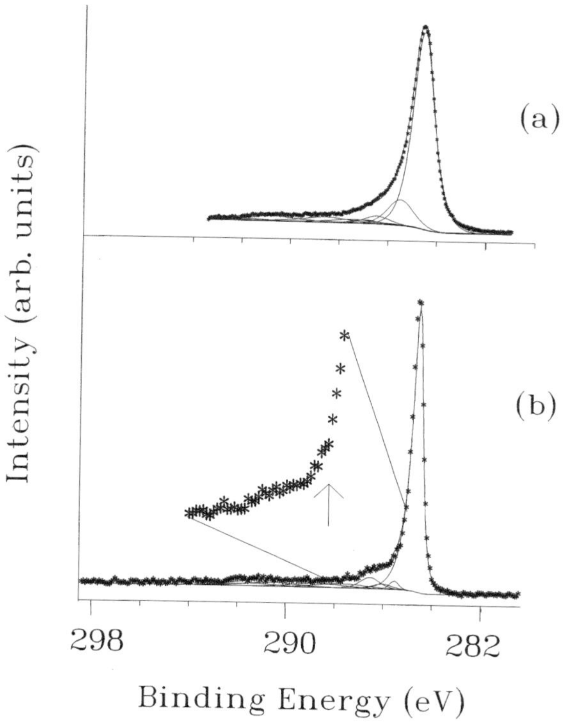

Figure 13. C 1*s* XPS spectra from a PAN-based untreated and unsized carbon fiber. Spectra obtained with (a) achromatic Mg x-radiation, and (b) monochromatic x-radiation using an instrumental resolution of 0.2 eV. (From Ref. 44.)

–C–OH/C-bridged at around 533.3 eV, and to adsorbed water and some chemisorbed oxygen at around 536.1 eV. Analysis of O 1*s* spectra can complement the information provided by analysis of C 1*s* spectra, although it should be remembered that since the O 1*s* photoelectron kinetic energies are lower than those of the C 1*s*, the O 1*s* sampling depth is smaller and therefore the O 1*s* spectra are slightly more surface specific.

5.5.5. Fitting N 1*s* spectra

Although the N 1*s* spectrum from an organic material is often of low intensity, and therefore difficult to resolve accurately, it is capable of giving some distinctive information [46]. On the other hand, inorganic nitrogen, that might for example be deposited on a surface from adsorbed nitrate ions as a result of surface treatment, can be distinguished readily from organic nitrogen by the much

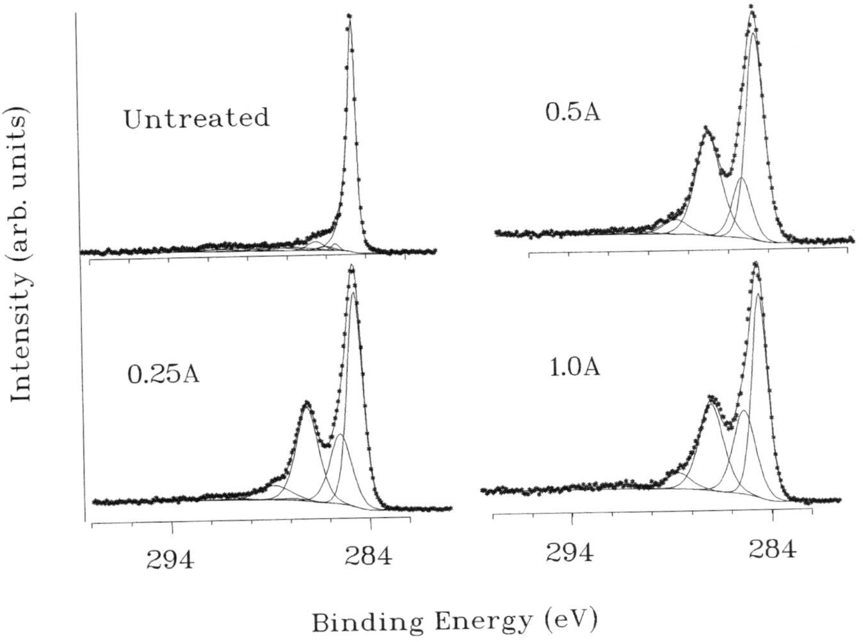

Figure 14. C 1*s* XPS spectra from a PAN-based carbon fiber galvanostatically oxidized in a pilot plant for 40 s in 1 M nitric acid at different currents. (From Ref. 44.)

higher binding energies of inorganically combined nitrogen (e.g., 407.3 eV for nitrate ions).

5.6. Interpreting the Valence-Band Spectrum

Valence-band spectra in XPS can be especially valuable in the study of fibers and composites because the analysis of such spectra may be able to assist in the interpretation of ambiguous spectral features in the core regions. The effects of achromatic x-radiation are also much less obvious in valence-band spectra for the reasons discussed in Section 5.4. With the help of the valence-band spectrum it is possible to distinguish between unfunctionalized hydrocarbons [47], while useful information about polymers can also be extracted [31]. When comparing valence-band data with calculation, background and X-ray satellite removal from the experimental data is usually helpful, and can be performed effectively by data processing methods [29–31].

5.6.1. Using calculations to predict valence-band spectra

The valence band can be modeled by performing Xα, ab initio, or band structure calculations on appropriate structural units. For Xα, or ab initio calculations, the unit is the repeat unit for a polymer or a substituted coronene structure for a carbon fiber. The calculations allow the energy level positions within the band to be calculated and the intensities of the component peaks in the spectrum to be estimated; the latter are found by taking the molecular orbital coefficients (or the partial density of states in the case of band structure calculations) from the calculation and multiplying them by their photoelectric cross sections. The component peaks are then set to have equal widths, comparable to the experimental resolution. The calculations thus do not include any experimental "adjustment" factors and can be used for predictive purposes. Calculations of this type allow the experimental and theoretical spectra to be compared, especially the separations and relative intensities of the peaks. Good agreement between experiment and theory gives reasonable confidence in assigning a particular compound to an experimental spectrum.

Three examples of the above approach are given in Figs. 15–17. In Fig. 15 the experimental valence-band spectrum from a Kynol novoloid fiber [48] is compared with the spectrum calculated from an ab initio calculation using the polymer repeat unit. Figure 16 shows the calculated spectra for two components of the repeat unit in a Nomex aramid fiber, *m*-phenylenediamine and isophthaldehyde; if the compounds are combined they produce the predicted spectrum in Fig. 17b, which compares well with the experimental spectrum [48]. Figure 18 shows the predicted spectra for oxidized carbon fibers using multiple scattered wave Xα calculations on a set of substituted coronene units [25]. The use of the band structure approach in polymers has been demonstrated in a study of polystyrenes [49].

5.6.2. Understanding the valence-band spectra of carbon fibers

The calculated spectra of Fig. 18 indicate that two important pieces of information can be obtained from the valence-band spectra of carbon fibers:

1. the energy region around 20–30 eV corresponds to C 2*s*/O 2*s* mixing, and the separation between the two major peaks, one at the highest binding energy, (O 2*s*), and the other distanced from it by some 8–11 eV to lower binding energy (C 2*s*) is sensitive to surface functionality. Thus the –O– and the –OH functionalities may be differentiated by their different separations, 10.58 and 8.77 eV, from the O 2*s* peak. Such differentiation is not possible from spectra in the C 1*s* and O 1*s* core regions. Note how the spectrum of carboxyl (CO_2^-) has *two* features in the O 2*s* region (arising from the ability of the O 2*s* orbital to form bonding and nonbonding interactions with C 2*s* orbitals).

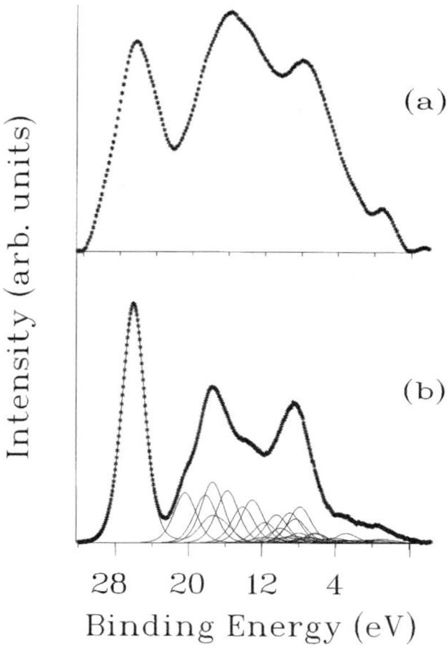

Figure 15. Experimental valence-band spectrum (a) from Kynol novoloid fibers compared with the calculated spectrum (b) using *ab initio* calculations. The experimental spectrum was smoothed and had a nonlinear background subtracted. (From Ref. 48.)

2. The region below a binding energy of 18 eV serves as a "fingerprint" region. Calculations show that it contains a number of "peaks" resulting from the overlap of mixed O 2*p* and C 2*p* components. Note how that region changes substantially with changes in surface functionality. In order to be able to compare the calculated valence-band spectra with the experimental spectra from oxidized carbon fibers, the photoelectron cross sections have been adjusted to account for variable amounts of carbon and oxygen [25].

The C-bridged structure referred to earlier is shown in Fig. 18d with a hydrogen atom placed symmetrically between two oxygen atoms. The C–O bond length used (0.12825 nm) was intermediate between that of –C=O (0.1215 nm) and –C–OH (0.135 nm). The real situation may be more complex, probably with a double potential well, but the "bridged" structure is regarded as a reasonable model for this type of oxidation.

The ability of the valence-band region to provide more chemical information than the core region is demonstrated by the author's work on the electrochemical oxidation of pitch-based fibers in ammonium carbonate solution [50]. Both

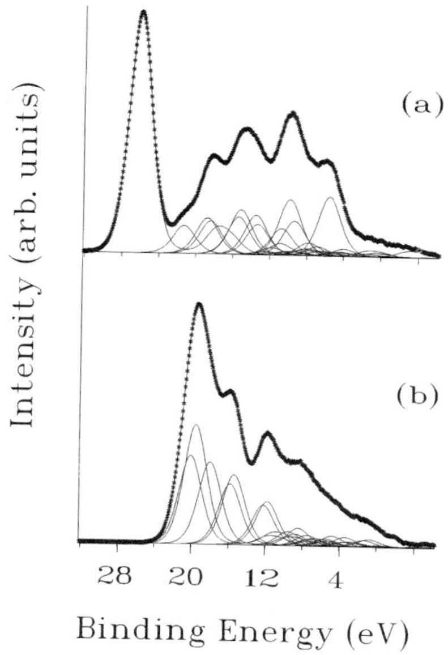

Figure 16. Valence-band spectra of repeat unit components for Nomex aramid fibers, generated by *ab initio* calculations: (a) *m*-phenylenediamine and (b) isophthaldehyde. (From Ref. 48.)

potentiostatic (holding the potential constant during electrochemical treatment) and galvanostatic (holding the current constant during electrochemical treatment) oxidation led to the appearance of two peaks in the O 2*s* region, as in Fig. 18a. The two peaks are believed to correspond to epoxide and/or –C–O–C– (near 30 eV) and –C–OH (near 25 eV).

5.6.3. The use of UV rather than X-radiation

If UV radiation (photon energies typically of 20–40 eV) were to be used rather than X-rays to excite emission from the valence band, the resultant spectra might be more intense, but since the photoelectrons would have much lower kinetic energies the spectra would correspond to emission from the outer surface layers only. The process would therefore be particularly sensitive to the presence of impurities and contamination on the surface. In addition, UV excitation involves transition probabilities and ionization cross-sections different from those appearing in X-ray excitation, which has the consequence that spectral analysis by calculation has to include both ground and virtual states, and is therefore more difficult.

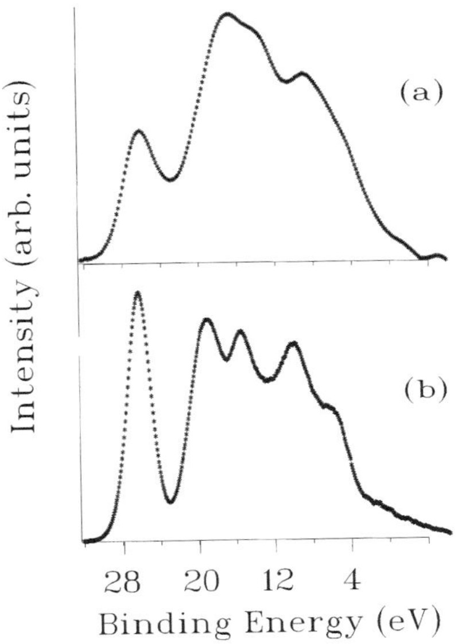

Figure 17. Experimental valence-band spectrum (a) of Nomex aramid fibers compared with a theoretical spectrum (b) generated by combining the two calculated spectra in Fig. 14. (From Ref. 48.)

5.7. Interfacial Studies

Interfaces are very important in the study of composites, especially the buried interfaces mentioned in the introduction. Such an interface is that between a fiber surface and the surrounding matrix in a composite material, and its study is important because it is susceptible to oxidation in composites exposed to high temperatures under atmospheric conditions. The author [50,51] has developed a method that allows a buried interface to be examined without damage by arranging for the interface to be close to the surface and within the sampling depth of valence-band photoemission (see Section 5.6). The method involves preparation of an interface by coating the surface of a fiber with a very thin film of the matrix material, and is carried out by immersing the fiber in a solution of the matrix material and allowing the solvent to evaporate. The spectrum of the interface can then be derived from the following spectra:

1. Spectrum of matrix alone = M.
2. Spectrum of fiber surface alone = F.
3. Spectrum of the fiber coated with a very thin layer of matrix material = S. It is essential that the spectrum of the underlying fiber can be seen in the XPS data, to make sure that the buried interface is within the sampling depth.

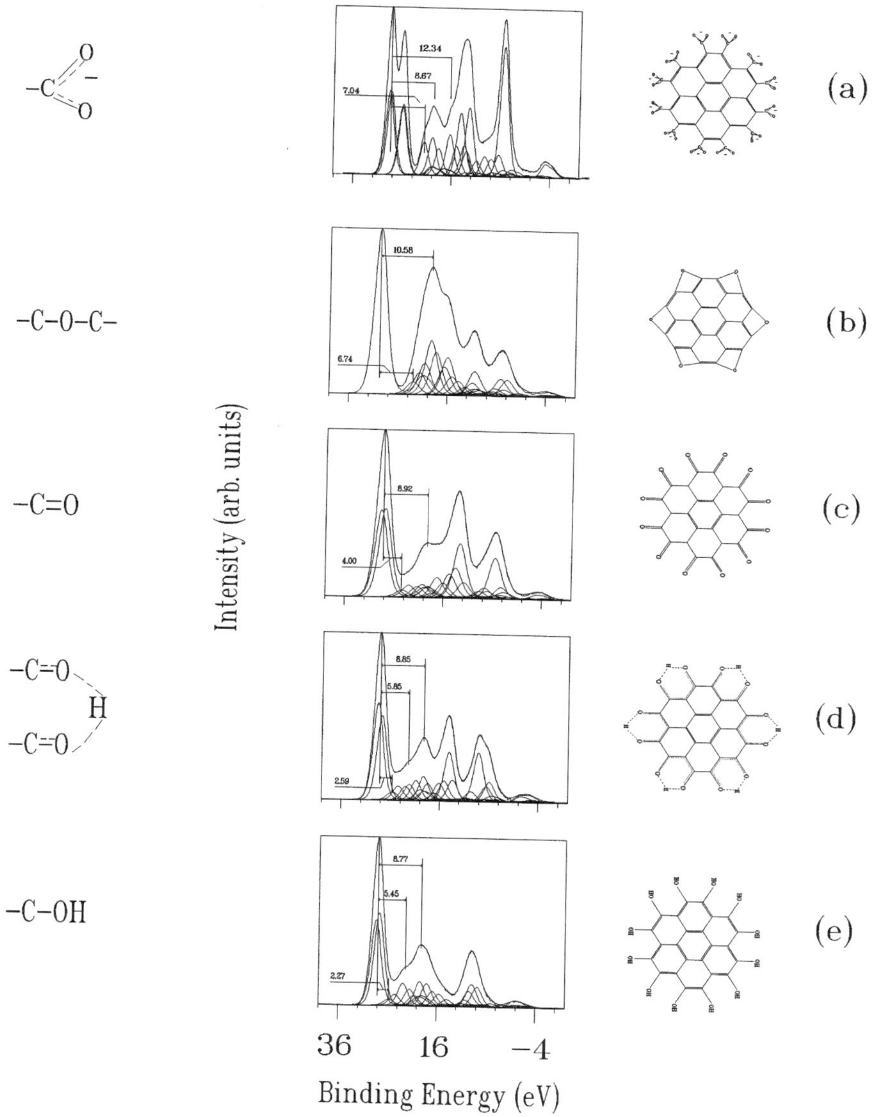

Figure 18. Valence-band spectra for oxidized carbon fibers calculated by multiple scattered wave Xα method. The functional groups shown in the left-hand column and attached to the coronene nucleus in the right-hand column are (a) carboxyl, (b) epoxide, (c) carbonyl, (d) bridged and (e) hydroxide. (From Ref. 36.)

The situation can be illustrated schematically:

Vacuum

Matrix

Buried Interface

Fiber

If there is *no* interaction at the fiber/matrix interface,

$$S = M + F$$

If there is *chemical interaction* at the fiber/matrix interface,

$$S \neq M + F$$

In the latter case the spectrum of the interface region can be found from a difference spectrum [29], that is, by subtracting M and F from S.

An example of this approach is illustrated in Fig. 19. The difference spectrum (d) is the result of subtraction of the spectra (a) of an oxidized carbon fiber and (b) of a phenolic matrix from the total spectrum (c) of the fiber and matrix coupled with a titanium alkoxide coupling agent (TOT=tetrakis(2-ethylhexyl)titanate) [51]. The difference spectrum is characteristic of the –O–TiR$_2$–O– group which might be formed by reaction of the titanium alkoxide with both the oxidized fiber surface and the phenolic resin according to the scheme.

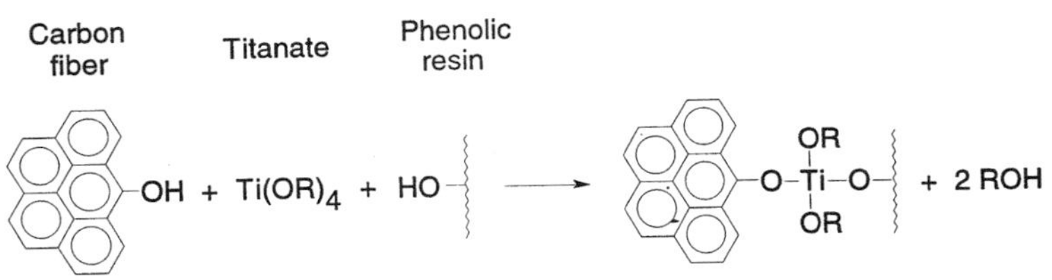

6. CONCLUDING COMMENTS

Surface analysis has an important role to play in the study of composites and fibers. Many surface analytical methods are available, but XPS is the most extensively used because surface chemical information is essential. The two most important practical points are that samples be mounted and handled very carefully in the spectrometer and that the data be analyzed thoroughly. Complete problem solving cannot be achieved unless the surface analyst uses the full extent of information available (e.g., by careful curve resolution and by making use of valence band spectra). This chapter has given many examples of the type

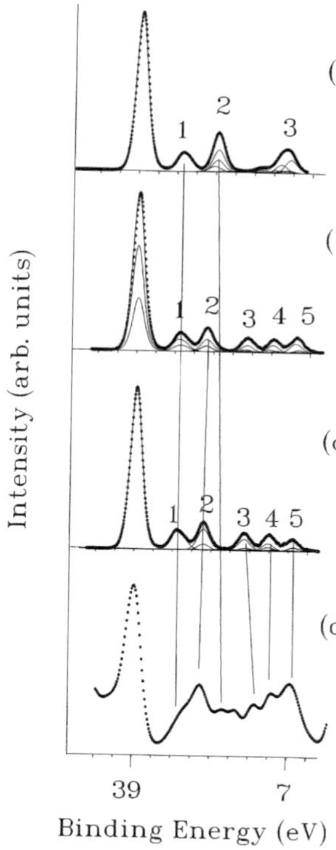

Figure 19. Illustration of the method of extracting the valence-band spectrum of the buried interface between an oxidized carbon fiber and a phenolic resin, with an intermediate titanium alkoxide agent. (a) Valence-band spectrum of oxidized carbon fiber alone, (b) valence-band spectrum of phenolic matrix alone, (c) valence-band spectrum of the composite, and (d) difference spectrum—the result of subtracting spectra (a) and (b) from (c). Spectrum (d) represents the desired valence-band spectrum of the interface. (From Ref. 51.)

of information obtainable and has discussed factors that need to be considered in conducting an analysis in the most effective manner.

ACKNOWLEDGMENTS

The material quoted in this chapter is based upon work supported by the National Science Foundation, the Air Force Office of Scientific Research, and NASA. The monochromatic XPS data were recorded on an instrument funded by the National Science Foundation under grant CHE-9421068. The author acknowledges his co-workers in studies of carbon fibers, Dr. Andrew Proctor, Dr. Carol Jones, Dr. Yaoming Xie, Dr. Cara Weitzsacker, Dr. Tiejun Wang, Dr. Michael Bellamy, Dr. Tom Schuerlein, Dr. Michael Rooke, Oliver Franklin, Nather Havercroft and Hema Viswanathan.

REFERENCES

1. P. M. A. Sherwood, *J. Electron Spectrosc. Relat. Phenom.*, in press.
2. P. M. A. Sherwood *Mater. Res. Soc. Proc. 270 (Novel Forms of Carbon)* 79 (1992).
3. J. Matsui, *Critical Rev. Surf. Chem I*, 71 (1990).
4. M. S. Dresselhaus, G. Dresselhaus, K. Sugihara, I. L. Spain and H. A. Goldberg, *Graphite Fibers and Filaments*, Springer Series in Materials Science, Vol. 5, Springer-Verlag, Berlin, 1988.
5. J. E. Castle and J. F. Watts, The study of interfaces in composite materials by surface analytical techniques, in *Interfaces in Polymer, Ceramic, and Metal Matrix Composites* (H. Ishida, ed.), Elsevier, Amsterdam, 1988.
6. Y. Xie and P. M. A. Sherwood, *Appl. Spectrosc. 44*, 797 (1990).
7. Y. Xie and P. M. A. Sherwood, *Appl. Spectrosc. 44*, 1621 (1990).
8. Y. Xie and P. M. A. Sherwood, *Appl. Spectrosc. 45*, 1158 (1991).
9. G. Katagiri, H. Ishida and A. Ishitani, *Carbon 26*, 565 (1988).
10. C. Ahearn and B. Rand, *Carbon 34*, 239 (1996).
11. C. Kozlowski and P. M. A. Sherwood, *J. Chem. Soc., Faraday Trans. I. 80*, 2099 (1984).
12. C. Kozlowski and P. M. A. Sherwood, *J. Chem. Soc., Faraday Trans. I. 81*, 2745 (1985).
13. W. P. Hoffman, W. C. Hurley, T. W. Owens and H. T. Phan, *J. Mater. Sci. 26*, 4545 (1991).
14. M. A. Rooke and P. M. A. Sherwood, *Carbon 33*, 375 (1995).
15. Y. Xie and P. M. A. Sherwood, *Chem. Mater. 6*, 650 (1994).
16. L. T. Drzal, J. A. Meschen and D. L. Hall, *Carbon 17*, 375 (1979).
17. J. Schultz, L. Lavielle and H. Simon, *Surface and Adhesion Properties of Carbon Fibres*, Proc. Int. Symp. on the Science and New Applications of Carbon Fibers '84, Toyohashi University of Technology, Japan, 1984, p. 125.
18. B. Rand and R. Robinson, *Carbon 15*, 311 (1977).
19. J-B. Donnet, H. Dauksch, J. Escard and C. Winter, *C. R. Acad. Sci. Paris 275*, 1219 (1972).
20. D. Briggs, *Surface Analysis*, in *Encyclopedia of Polymer Science and Engineering, Vol. 16*, Wiley, New York, 1989, pp. 399–442.
21. D. Briggs, *Surf. Interface Anal. 9*, 391 (1986).
22. L. T. Drzal, *Carbon 15*, 129 (1977).
23. D. J. D. Moyer and J. P. Wightman, *Surf. Interface Anal. 14*, 496 (1989).

24. A. Proctor and P. M. A. Sherwood, *J. Electron Spectrosc. Relat. Phenom. 27*, 39 (1982).
25. Y. Xie and P. M. A. Sherwood, *Chem. Mater. 3*, 164 (1991).
26. M. P. Seah, in *Practical Surface Analysis by Auger and X-ray Photoelectron Spectroscopy*, 2nd ed. (D. Briggs and M. P. Seah, eds), Wiley, Chichester, 1990, Appendix 2, pp. 541–554.
27. T. Dickinson, A. F. Povey and P. M. A. Sherwood, *J. Electron Spectrosc. Relat. Phenom. 2*, 441 (1973).
28. M. A. Rooke and P. M. A. Sherwood, *Surf. Interface Anal. 21*, 681 (1994).
29. P. M. A. Sherwood, in *Practical Surface Analysis by Auger and X-ray Protoelectron Spectroscopy*, 2nd Ed. (D. Briggs and M. P. Seah, eds.), Wiley, Chichester, 1983, Appendix 3, pp. 445–475.
30. P. M. A. Sherwood, in *Practical Surface Analysis by Auger and X-ray Photoelectron Spectroscopy*, 2nd ed. (D. Briggs and M. P. Seah, eds.), Wiley, Chichester, 1990, Appendix 3, pp. 555–586.
31. P. M. A. Sherwood, in *Surface Analysis of Advanced Polymers* (L. Sabbatini and P. G. Zambonin, eds.), VCH, Weinheim, 1993, Chap. 7, pp. 257–298.
32. P. M. A. Sherwood, *J. Vac. Sci. Technol. A 9*, 1493 (1991).
33. P. M. A. Sherwood, *J. Vac. Sci. Technol. A 11*, 2280 (1993).
34. S. Thomas and P. M. A. Sherwood, *Anal. Chem. 64*, 2488 (1992).
35. S. Thomas and P. M. A. Sherwood, *Surf. Interface Anal. 20*, 595 (1993).
36. C. L. Weitzsacker, M. Bellamy and P. M. A. Sherwood, *J. Vac. Sci. Technol. A12*, 2393 (1994).
37. Y. Xie and P. M. A. Sherwood, *Chem. Mater. 2*, 293 (1990).
38. C. Kozlowski and P. M. A. Sherwood, *Carbon 25*, 751 (1987).
39. R. O. Ansell, T. Dickinson, A. F. Povey and P. M. A. Sherwood, *J. Electroanal. Chem. 98*, 79 (1979).
40. S. Tougaard and P. Sigmund, *Phys. Rev. B 25*, 4452 (1982).
41. H.S. Hansen and S. Tougaard, *Surf. Interface Anal. 17*, 593 (1991).
42. A. Proctor and P. M. A. Sherwood, *Anal. Chem. 54*, 13 (1982).
43. P. M. A. Sherwood, *J. Vac. Sci. Technol. A14*, in press (1996).
44. H. Viswanathan, M. A. Rooke and P. M. A. Sherwood, *Surf. Interface Anal. 25*, 409 (1997).
45. A. Proctor and P. M. A. Sherwood, *Carbon 21*, 53 (1993).
46. Y. Xie and P. M. A. Sherwood, *Chem. Mater. 1*, 427 (1989).
47. P. M. A. Sherwood, *J. Vac. Sci. Technol. A10*, 2783 (1992).
48. L. E. Hamilton, P. M. A. Sherwood and B. M. Reagan, *Appl. Spectrosc. 47*, 139 (1993).
49. E. Orti, J. L. Bredas, J. J. Pireaux and N. Ishihara, *J. Electron Spectrosc. Relat. Phenom. 52*, 551 (1990).
50. Y. Xie, T. Wang, O. Franklin and P. M. A. Sherwood, *Appl. Spectrosc. 46*, 645 (1992).
51. T. Wang and P. M. A. Sherwood, *Chem. Mater. 7*, 1031 (1995).

15

Corrosion and Surface Analysis: An Integrated Approach Involving Spectroscopic and Electrochemical Methods

N. S. MCINTYRE, R. D. DAVIDSON and I. Z. HYDER University of Western Ontario, London, Ontario, Canada

A. M. BRENNENSTÜHL Ontario Hydro Technologies, Toronto, Ontario, Canada

1. INTRODUCTION

Corrosion may be defined as the chemical reaction of a solid with its environment [1]. Such reactions may or may not be detrimental to the solid. Iron metal, for example, will react with a moist-air atmosphere to produce an orange rust; this reaction will continue indefinitely with eventual destruction of the solid. On the other hand, exposure of chromium metal (or even a chromium-containing alloy) to a similar environment will result in formation of a very thin oxide film (invisible to the eye) that protects the active metal from further deterioration. Both are corrosion reactions, but only one is destructive.

It is clear that both types of corrosion reaction are of immense concern to society: the former results in unimaginable levels of destruction of the world's mechanical structures, while the latter reaction shows that all that stands between oxidative deterioration of a particular solid is often a few atom thicknesses of a particularly stable oxide.

While many of the best-known examples of corrosion involve metallic solids and oxidation-reduction chemistry, the surfaces of all solids, including polymers and ionic crystals, can undergo some degree of chemical reaction with their environment. Whether involving metals or nonmetals, oxidation and reduction processes occur inevitably during corrosion. For this reason, the measurement of electrical charge transfer processes has always been important in modeling corrosion reactions, and knowledge of electrochemistry is important. However, the identification and diagnosis of new corrosion processes are usually uncovered using surface and microscopic techniques. This chapter is intended to assist the novice surface analyst in preparing samples for analysis which bear possible evidence of a corrosion process. In addition, some advice on the most appropriate analysis methods is offered. Finally, the story of a major corrosion investigation is presented in order to show the reader the present relationship between information gained from surface and microanalytical investigations and that obtained from electrochemistry. It can be seen that the role of electrochemistry is increasing and it will soon be true that the surface analyst involved in corrosion must also have a good appreciation of electrochemical measurements.

1.1. Types of Corrosion Process

Three general types of corrosion can be defined: chemically induced uniform corrosion, localized corrosion and stress corrosion cracking. In all cases, the net chemical reaction occurring is considered as the sum of oxidation and reduction half-reactions: an anodic reaction involving oxidation of the metal, and a cathodic reduction of oxygen or water. Such reactions occur at separate locations on the sample surface, sometimes as close as one atomic spacing, sometimes many nm distant. Different sites have tendencies to act either as anodes or cathodes because of their relative reactivities toward oxygen or water. If there is a greater concentration of anodic or cathodic sites, the electron current density will be uneven across the surface and more chemical activity will be observed at sites associated with the lower concentration reaction couple.

Uniform or near-uniform corrosion of a surface occurs in cases where there are few regions of the surface which exhibit unusual reactivity toward the oxidizing medium and the concentrations of anodic and cathodic areas are approximately equal. This is usually true in cases of dry oxidation (tarnishing), active dissolution in acid media, and anodic oxidation [1]. Localized corrosion has a number of forms and causes. Metal alloys consisting of more than one metallurgical phase will often exhibit preferential corrosion attack of one of the structures [2]. Some good examples are the dezincification of brass [2] or the preferential attack of chromium precipitates in the alloy Stellite [3]. Metallic impurities in the form of microscopic inclusions can act as cathodic or anodic sites, to promote localized attack of the surface. In its extreme form, such attack is known as pitting corrosion and can result in highly localized perforations without significant damage to surrounding areas [4]. A region that has undergone cold work will contain a higher concentration of dislocations and corresponding surface kink sites, which will react more rapidly than surrounding regions, probably as a result of impurities attracted to them [5]. Grain boundaries are often preferentially corroded (a process called intergranular corrosion) [6], when they collect or lose elements to create a chemical composition substantially different from that of the bulk of the metal. Certain austenitic stainless steels may be made susceptible to intergranular corrosion by heat treatments which cause depletion of chromium from grain boundaries with a corresponding loss of corrosion resistance in that region. The exposed crystallographic orientation of a phase can also affect the rate of solution attack. Surfaces of metals such as steels and titanium exhibit preferential attack in directions perpendicular to the direction of rolling or working. Finally, enclosed regions on a surface may undergo accelerated corrosion, because the solution concentration within them may differ from that in contact with the rest of the surface, thus causing a form of electrochemical concentration cell to be established. This is known as crevice corrosion and frequently results where the geometry of a structure leaves confined volumes in which free exchange of solution is not possible, e.g., welded joints, particulate surface deposits. Stress corrosion cracking (SCC) is an effect in which surface electrochemistry and tensile stress factors combine to cause intergranular cracking in high-strength aluminum alloys and steel which have been subjected to continuous stress at a level often much lower than the tensile strength of the alloy. A related effect, corrosion fatigue (CF), occurs when cyclic stresses combine with a chemical environment to produce degradation in a wider range of materials than those affected by stress corrosion. Both SCC and CF can lead to catastrophic failures.

1.2. Corrosion and Surfaces

Corrosion begins on surfaces, usually at many microscopic locations. From there it may propagate across the entire surface, down a grain boundary, or remain confined within pits or narrow crevices. Thus identification and analysis of the type and origin of the corrosion problem will involve a range of possible strategies to take into account possible locations. The strategy must include both surface and microscopic tools which can characterize the properties of internal structures as well as those of the outermost surfaces. In practice, most samples resulting from industrial corrosion failure are likely to require more emphasis on studies of internal interfaces, while laboratory-scale controlled experiments will derive greater benefit from analyses of outer surfaces.

Analysis of the products of corrosion has its historic origin in the light-microscopic studies of metallographic cross sections that started in the 1950s. These were supplemented with TEM and SEM investigations in the 1960s of cross sections and plan view samples. Considerably more structural and compositional information became available in the 1970s and 1980s with the introduction of "true" surface techniques such as XPS (or ESCA), AES and SIMS, particularly when used in depth profiling modes. By the 1990s, more comprehensive information about corrosion-related structures, such as the porosity and extent of hydration of corrosion products, was being required. Some promising developments in these directions are suggested later in this chapter.

2. PROTOCOLS FOR CORROSION FILM ANALYSIS

2.1. Preliminary Sample Handling

As suggested, requests for analysis of a corrosion specimen will arise either from an unexpected failure of a system component or from the result of an experimental exposure of a representative sample to controlled corrosion conditions. Different storage conditions are required for each. In the latter case, it is assumed that experimental conditions surrounding the corrosion potential (E_{corr}), solution pH, solution flow and temperature history have been carefully controlled by corrosion scientists or engineers and are on record. It is then particularly important that specimens so treated do not undergo additional reaction while awaiting analysis. Reactions of corrosion test "coupons" in electrochemical cells or in open-circuit pressurized autoclaves could leave the sample surface susceptible to oxidative changes if left exposed to the ambient atmosphere (oxygen and water). The analyst should therefore counsel those involved in removing sample coupons from an experiment to place the coupons directly into glass vials under an inert atmosphere, such as nitrogen or argon. Sample desiccation is unnecessary and may even alter the hydrate content of the surface; for the same reason, vacuum storage is also to be discouraged. The analyst should remind those who are running the corrosion experiment that the procedures by which the sample is removed from its experimental situation may alter the surface composition. Samples which remain in the test solution during cooldown of the autoclave or after the electrode potential has been turned off cannot fail to have their surfaces altered to some extent by such transient conditions.

On the other hand, samples from industrial failure have inevitably come in contact with an oxidizing atmosphere and probably been subjected to other interference; such conditions must be accepted as a fact of life. However, to compensate for this, the analyst must attempt to play the role of detective, piecing together as many as possible of the events surrounding the failure. Some questions that might be asked are concerned with (a) details of the chemical environment surrounding the specimen at the time of the failure, (b) any changes in those conditions which have occurred recently, (c) the method of manufacture and history of the component, (d) the operating history of other identical components in the system (perhaps one of these might be secured to serve as a reference), and, (e) the method of sample storage since the failure incident.

In situations where bacteria are present, microbiologically influenced corrosion (MIC) may have taken place. If it is suspected that microorganisms have had a role in the corrosion process, then additional precautions have to be taken during the handling and preparation of samples. The samples should be kept in a moist condition prior to fixing with a 2% glutaraldehyde solution because it is essential in the diagnosis of MIC that a spatial relationship be established between microorganisms, substratum metal and corrosion. MIC cannot be verified by morphology of the corrosion alone. SEM/EDX can be used to reveal bacteria in corroded areas and to determine surface chemistries resulting from MIC. However, examination of the surfaces with conventional SEM/EDX is possible only if the sample has been subjected to a commonly used biological sample preparation process known as critical point drying. Exposure of a sample which has not been so prepared leads to the evaporation of the liquid in the cells, which causes damage to the microorganisms. Recently, environmental scanning electron microscopes (ESEM), capable of working in pressures up to several kPa, have been employed for examination of MIC samples. In such cases, the need for critical point drying is eliminated.

Sometimes the analyst is asked about protocols for taking additional samples from the failed component (or even the reference). Whenever possible samples should be cut by methods in which neither heat nor fluids contaminate the surface to be analyzed. Cutting methods of choice therefore include hacksaws and heavy-duty metal shears. Where use of these is impractical, torch cuts should be as far as possible from the failed area and water, not oil, should be used as a lubricant with cutting saws. Samples from failures can normally be stored under air in closed containers.

2.2. Contaminants

Many corrosion specimens from either of the origins described above may contain a plethora of chemical elements when subjected to surface or microanalysis. Some of these could be considered to be contaminants because they arose from processes having little or nothing to do with the corrosion. For example, the internal walls of high-pressure autoclaves are ready sources of cross-contaminants from previous experiments, particularly alkaline and heavy elements. Similar elements precipitate in many industrial heat transport circuits from sources as varied as valve packing to water treatment columns. In the case of the controlled experiment, it is best to remove the source of the contamination and then repeat the run. No such option is available with most failure specimens.

2.3. Preliminary Examination

One of the first decisions the analyst must make is the choice of regions on the sample(s) to be investigated. For example, if the corrosion is localized, it will be necessary to choose at least three such localized regions in order to assure oneself that similar processes are taking place in each. In addition, three other areas not containing evidence of localized corrosion should be analyzed. But, how does one make these decisions at the start of an investigation? Normally, one would start with a visual examination of the plan view surface using light microscopy. The stereomicroscope (or macroscope) produces a spatial image of three-dimensional objects at relatively low magnification (5–25 times). It is similar to compound microscopy, but uses two separate beam paths which traverse a compound main objective. This causes each eye to observe the object from a different direction, which increases the depth of field whose lack is normally the major drawback in compound microscopes. Depth of field is improved with the use of small apertures.

Surface analysis by light microscopy has improved greatly in the past few years, partly due to the improved light-gathering powers of newer optics, but more to the routine availability of high-sensitivity solid-state CCD array cameras for image collection, and of high-resolution video printers for instant display. This is particularly important in allowing accurate color rendition of microscopic features which may be partly obscured in a crevice or pit or under a thin overlayer of crystallites. The newer microscope optics are also capable of achieving spatial resolutions close to the theoretical limit for white light (0.25 μm). Illumination is performed using visible light from a broad-spectrum tungsten halide lamp which is passed through fiber optics to a ring illuminator surrounding the objective lens, or is allowed to illuminate the sample from a fixed direction. Both should be tried when examining a new type of sample. Illumination intensity and direction are important for the recognition of features on plan view surfaces. Intensity that is too high will wash out the contrast needed to see or record features of near μm size. Varying the angle of direction will produce the shadowing that is so helpful in visualizing three-dimensional features. However, care must be taken to record all images under identical lighting conditions or comparisons will be meaningless. Several magnifications should be used in areas containing corrosion features, with each one retaining some recognizable features in common with the others.

Viewing the surface with a stereomicroscope can reveal gross topographic features, such as pitting, cracking or other surface patterns which can provide information about the corrosion mechanism, as well as about the areas where further investigations will be concentrated. Figure 1a shows a plan view image of an alloy surface containing a few, but severe, corrosion pits. An additional benefit of having light microscopy as the first stage of the investigation proto-

(a)

(b)

Figure 1. Light optical micrographs of (a) plan view of a corrosion pit in a nickel alloy tube, with concentric ring illumination, at a magnification of 10X, and (b) cross section of a corrosion pit in copper metal with the same illumination and 25X magnification.

col is that it will not compromise any future work by altering the specimen surface. In general, investigations should proceed in an order that is likely to preserve all structures through the greatest number of investigations.

After the plan view imaging is completed, some strategic decisions concerning the next investigative steps can be made. The following selection sequence is suggested for differing sample conditions:

a. *No evidence of any surface film compared to an unreacted or unfailed surface, or at most a light tarnish.* The corrosion film is probably less than 50 nm thick; it thus requires analysis by XPS and AES, followed by depth profiling using either technique (AES is faster). Mapping of surface elemental distribution by AES is possible.
b. *Evidence of only occasional local attack with presence of pits or roughening.* SEM should be carried out on the local areas, accompanied by EDS analysis. Cross sectioning of one or two local areas, followed by SEM and optical microscopy, may help to identify the type of corrosion. Small-area XPS or AES could provide valuable added information on the outer surface composition of the corroded areas.
c. *Fairly widespread light corrosion with areas of intense roughening or pitting.* The protocol in (b) is suggested, with in addition AES depth profiling of the lightly corroded film looking for multiple oxide layers.
d. *Extensive roughening and discoloration of the surface.* Careful cross sectioning (perhaps using tapered sections) is required, followed by thorough microscopic analysis. For this, EDS and SIMS should be used to identify both major and trace elements responsible for the corrosion. If extensive cracking is revealed, an attempt at fracturing the sample along the direction of the crack could be made. Subsequent small-area XPS or AES studies could be carried out in order to confirm the presence of particular elements or of structures along the surface of the crack.
e. *Thick oxide layer with some evidence of spallation.* Sectioning of the film should be carried out, taking care to preserve the integrity of the film during polishing. Loose oxide could be removed with adhesive tape and subjected to powder XRD analysis. The protocol in (d) is then suggested for the cross-sectioned samples.

2.4. Cross Sectioning of Oxide Surface Films

The analysis of corrosion films in cross section requires preparation procedures that take into account the friable and discontinuous nature of corrosion oxides. Some areas of the oxide will be less stable mechanically than others due to growth mechanisms. Therefore, it is important that any mechanical procedure for sectioning the oxide preserves all portions of it. The section could be prepared by polishing in a standard metallographic preparation, with the requirement that the polishing mount material needs to be at least as hard as the softest material in the oxide. Preparations such as Bakelite are appropriate choices. It is also important for the mounting material to be free of S and Cl, because evidence of these elements is frequently looked for in the section. In addition, it is important that the supporting mount material infuse into oxides which are particularly porous so that no air remains entrapped in the oxide and contributes to mechanical instability during polishing. Sections containing oxides are therefore often prepared under a reduced pressure to force the epoxy into interstices in the oxide film. Another procedure is to mount two corrosion surfaces face to face in the polishing mount in order to prevent breakaway of either oxide film. Typical polishing sequences are the same as those for preparation of metallurgical mounts; a series of silicon carbide papers from 280 to 600 grit, followed by 6 to 1 μm diamond, followed by an aqueous slurry of fine alumina, is a common sequence. One advantage of metallurgical preparation routines for oxide analysis is that the relationship between near-surface grain structure and oxide microstructure is observed. Such relationships are particularly important in depicting intergranular corrosion. If chemical etchants need to be used to reveal grain structure, they are usually applied *after* microscopic examination of the oxide structure.

Polished mounts prepared as above can be used readily in studies by light microscopy, SEM, SIMS and EPMA. Bright-field light microscopy is used normally; phase contrast and dark-field techniques do not usually reveal additional details of the oxide structure. A light-microscope cross section of a corrosion pit in a copper surface is shown in Fig. 1b; oxidic and metallic components are both distinguishable from the mount material itself on the basis of contrast. If analysis of this sample by AES or small-area XPS were contemplated, it must be remembered that such mounts are electrically nonconducting and that the resin components in the epoxy have high vapor pressures. Furthermore, the outer surface would have been contaminated by the polishing process. Imaging information from polished surfaces using SIMS can be quite useful, but the mounting material does present considerable charging problems. A partial solution is to coat the mount surface with a conducting metal such as gold; another is to use a low-melting metal (such as Wood's metal) in place of the plastic for mounting. However, this can cause significant cross-contamination of the sample during polishing, as a result of smearing.

An alternative method of sectioning oxide films is to fracture the sample at cryogenic temperatures; the fracture will occur along a direction where structural bonding is weakest. An example of such fracture is shown in Fig. 2; it is that of an oxide of zirconium on a zirconium alloy surface, the oxide having fractured along the direction of oxide columnar growth. An advantage of this method is the appearance of a rich near-surface microstructure which can be readily detected using SEM, particularly if a field emission (FE) SEM is used,

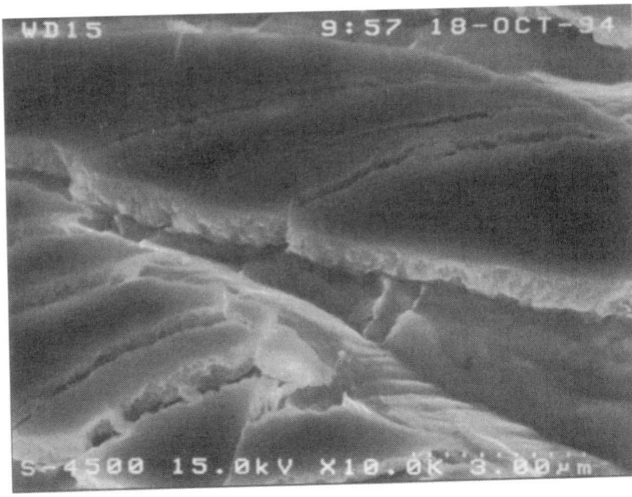

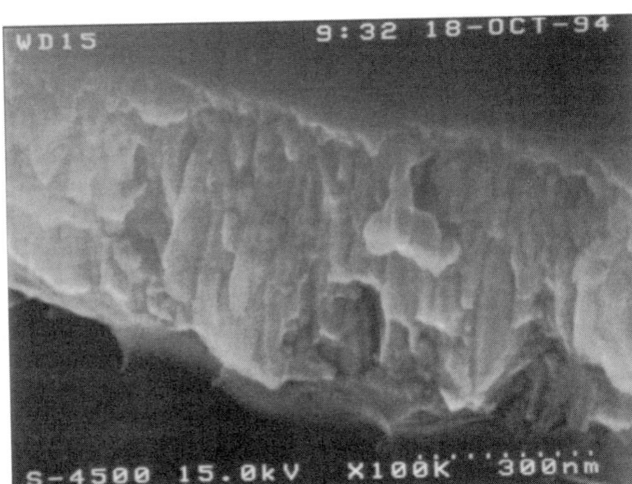

Figure 2. SEM micrographs of a fracture section across a corrosion film of Zr oxide on Zr metal. The fracture was performed at cryogenic temperature and reveals columnar growth of the oxide film above a structureless oxide at the interface.

due to its capability of using low-energy electron beams. Stereomicroscopy on such samples is more difficult because the surface relief will be excessive for its limited depth of field.

Both of the above sectioning techniques have an important place in the measurement of film thickness. For films that are much thicker than 1 μm, polishing seems to be the preferred approach. However, for thinner corrosion films (even as thin as a few hundred nm) fracturing at cryogenic temperatures can often produce thickness values that are more accurate than depth profiling using AES, XPS or SIMS, and is certainly faster.

2.5. Pressure Restrictions on Sample Analysis

Some failure specimens come from highly contaminated environments which may lead to problems during analysis when using surface analytical instruments that require the sample to be placed in high vacuum. Organic fluids such as light oils, glycols, fluoro- and chlorocarbons will certainly cause contamination of UHV systems used for XPS, AES, FESEM and SIMS, and samples should not be introduced into such systems without prior removal of the organic. This is best done by immersion of the sample in a dry methanol ultrasonic bath. Ordinary SEM systems may tolerate occasional brief exposures to such organics.

Some porous oxide specimens may contain extensive hydration which is released only slowly under vacuum; it may be impossible to analyze such specimens in UHV unless they are outgassed in a poorer vacuum for a number of hours. This is also a good reason to keep sample size as small as possible when dealing with failure analyses. Another aspect of sample hydration is that the loss of water under electron bombardment and/or vacuum conditions can often result in destruction of the microstructure itself due to the presence of "soft" structures. An example of this is shown below. One remedy for such hydration effects is critical point drying.

2.6. SEM and EDS Analyses

The SEM is one of the most versatile instruments for investigating corrosion oxide films and their relationship to the metallurgical microstructure of the base metals. With the development of the FESEM, the useful lateral resolution has improved to under 5 nm at accelerating voltages of 5 kV. This makes it possible to view molecular aggregates as small as 100 atoms, a major step forward in understanding of early-stage corrosion growth mechanisms. Moreover, FESEM images can be obtained at electron accelerating voltages as low as 500 V; this means that the surface thickness sampled is much less than that achievable previously. The use of such low-energy electrons will also reduce the extent of charging on electrically insulating corrosion oxide surfaces. A significant change in EDS analysis has also occurred in the past few years with the widespread use of detector systems capable of measuring the major low-*Z* elements such as C, O and N. Prior to this, there had been a tendency to use AES in some applications solely for the purpose of detecting low-*Z* elements that were undetectable by EDS units of the day. Another development has been the improvement in computer routines for quantitative analysis of materials by EDS. The characterization of corrosion oxides has been assisted greatly by the ability to carry out quantitative EDS measurements on oxides, sometimes even allowing the determination of oxide stoichiometries.

Recent advances in SEM techniques have led to the development of a new form of microscopy known as orientation imaging microscopy (OIM). OIM uses low-light television technology for the rapid access of backscattered Kikuchi diffraction images. The technique allows measurement and characterization of surface texture (grain crystallographic orientation) of materials. The orientation of individual grains, and hence the misorientation at grain boundaries, can be determined. Because corrosion processes can be highly dependent on crystal texture, particularly at the grain boundaries, OIM is a powerful surface analytical tool for corrosion studies.

SEM analysis of a plan view corrosion specimen should be relatively straightforward to most analysts having even minor experience with the technique. The two main choices for the analyst who wishes to obtain high-quality images are the working distance and accelerating voltage. For rough surfaces, containing deep crevices and much general relief, a fairly long working distance should be chosen in order to increase the depth of field, and the accelerating voltage should be a compromise between improved spatial resolution (higher voltage) and low penetrating power (low voltage). The best lateral resolution is achieved in the voltage range >5 kV, but depth of film sampled is then at least several hundred nm. In order to study films of thickness <10 nm, accelerating voltages under 1 kV must be used, but inevitably lateral resolution is somewhat degraded. If quantitative EDS spectra are to be recorded, accelerating voltages of at least 15 kV, to excite the K lines of typical first row metal oxides, must be used. Some qualitative work on thin films is possible using L lines and accelerating voltages of 2–3 kV.

It is possible to get some feeling for these criteria by considering the conditions required to obtain the best SEM images and EDS spectra of the pitted surface shown in Fig. 1a. The pit represents something of a challenge because of the relatively long distance between the mouth of the pit and its furthest extension. The SEM images in Fig. 3a were obtained using long and short working distances in order to show the differences in depth of field. It was possible to measure the pit depth (penetration) from the shift in lens potential required to have each level in focus; the lens potential shift is calibrated for a known vertical distance. This is called differential focusing. SEM imaging of the thin

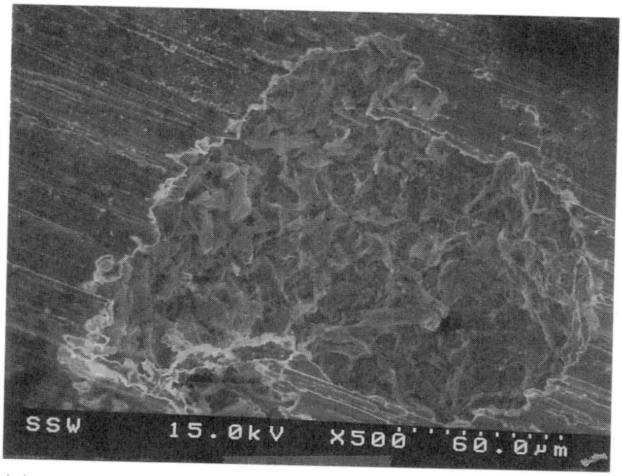

(a)

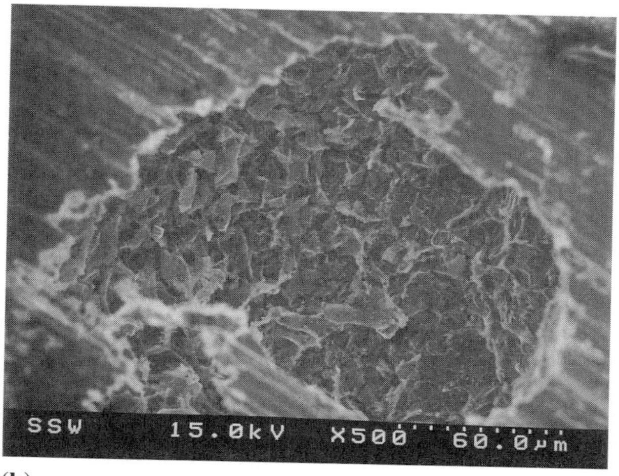

(b)

Figure 3. SEM micrographs of the pit in Fig. 1a using long (30 mm) and short (5 mm) working distances showing the difference in depth of field and resolution, and (b) higher resolution image of the anodic oxide surrounding the pit shown in Fig. 1a.

anodic oxide surrounding the pit (Fig. 3b) is best performed using a shorter working distance and low accelerating voltage. As was the case with the images taken by light microscopy, it was advisable to take a series of images over a range of magnifications. EDS analysis of this film required a sufficiently low accelerating voltage in order to keep most of the interaction volume within the thickness of the film. Figure 4 shows two spectra from the film recorded with (a) high and (b) low accelerating voltages.

SEM imaging of corrosion cross sections has also been extremely useful in identifying possible sites of corrosion initiation at the metal/oxide interface. In practice, structures within an oxide film which are sub-μm in size can be examined visually by SEM. If EDS microanalysis is required, it must again be remembered that the diameter of the interaction volume places a limit to lateral resolution of 0.5–1 μm. Some elemental differentiation at higher resolutions can be carried out using contrast levels caused by backscattered electrons (BSE mode). For films >10 μm in thickness, EDS is the clear choice for understanding the elemental distribution both laterally and as a function of depth. Figure 5 shows such a cross section for an epoxy coating on a galvanized steel; one of the samples had experienced a corrosion failure. Chloride incursion is detected by the mapping and has caused underscale corrosion.

2.7. XPS

XPS performs best in its role in determining the elemental and chemical compositions of the outer surfaces (i.e., 1–5 nm) of both thin and thick corrosion films. Information on outermost surface chemistry is the key to determining which chemical elements are being released to or adsorbed from any solution with which the surface has been in contact. For such surfaces, XPS provides usually reliable quantitative analysis, good differentiation of many oxide and hydroxide species in different oxidation states [7,8], and even some measure of the degree of hydration of the surface. A good example of an XPS analysis of such a corrosion film is shown in Fig. 6; the spectrum is from the surface of a crevice which had appeared during active corrosion of the Alloy 600 in an aqueous heat exchanger (boiler), and the film is sufficiently thin that a contribution from the underlying metals is discernible. Because of the small size of the crack, the analysis was performed with an X-ray spot size of 300 μm. More recent spectrometers are capable of analyzing areas as small as 5 μm in diameter. The spectrum shown in Fig. 6 should allow the analyst to carry out a quantitative analysis of all elements present in the outermost 1–4 nm (except hydrogen). Moreover, it is possible to distinguish many of the chemically shifted species, even from the low-resolution survey scan. The collection of spectra of the individual photoelectron lines may provide greater detail, but often adds surprisingly little to the initial chemical and elemental assessment of

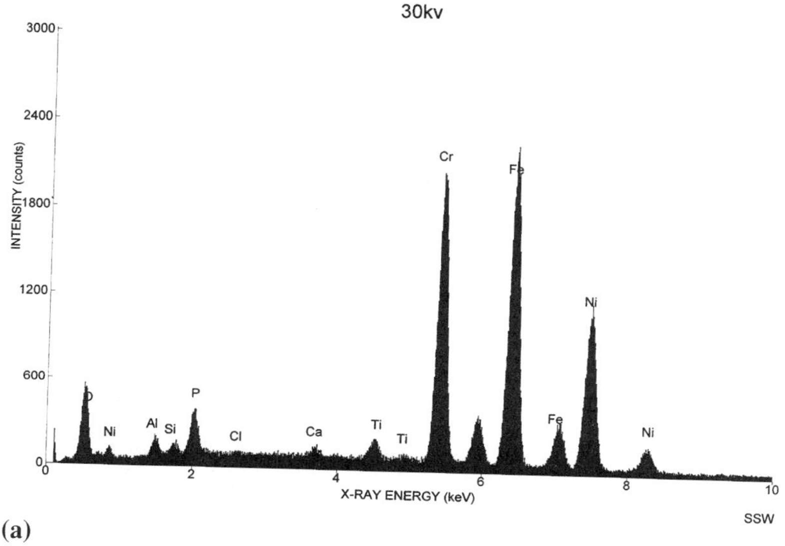

(a)

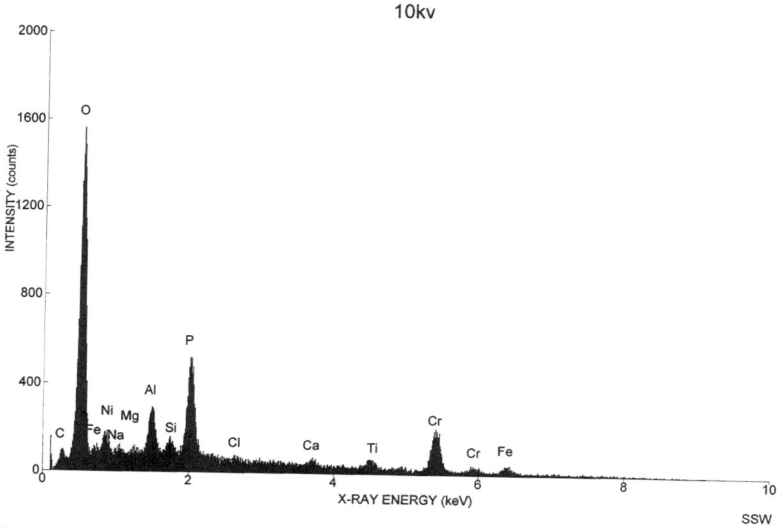

(b)

Figure 4. EDS spectra taken from the anodic area imaged in Fig. 3b using high (a) (30 kV) and low (b) (10 kV) excitation energies. The elements P, Cr, C and O are found in higher concentration in the low energy spectrum as a result of ZAF quantitative analysis.

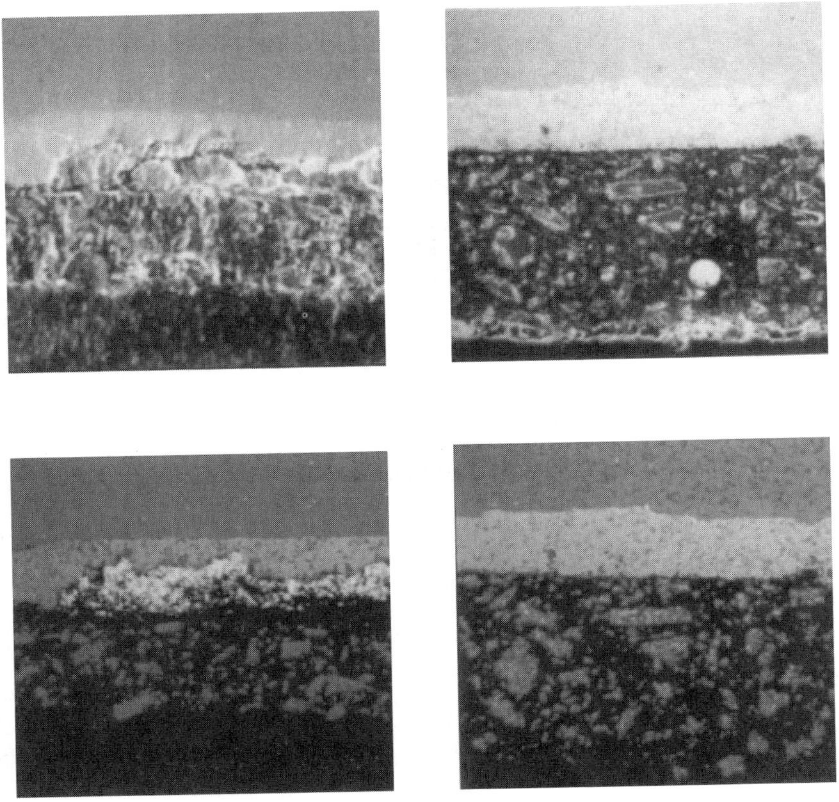
(a) (b)

Figure 5. Energy dispersive X-ray maps of cross sections of (a) failed and (b) integral epoxy-coated galvanized steel.

the surface. It is better to spend valuable counting time on acquiring one low-noise XPS spectrum with all elements present. Suitable counting times for achieving the noise level seen in Fig. 6 may vary from 5 min to 2 h, depending on the efficiency of the particular spectrometer and on the cleanliness of the sample surface. The sensitivity of XPS to minor (or even trace) elemental species has been underestimated. Providing spectral backgrounds are reasonably low, detection limits for crucial corrosion precursors such as Cl (0.01 at.%) and Pb (0.005 at.%) are respectable. By contrast, detection limits for AES are seldom better than 1 at.%.

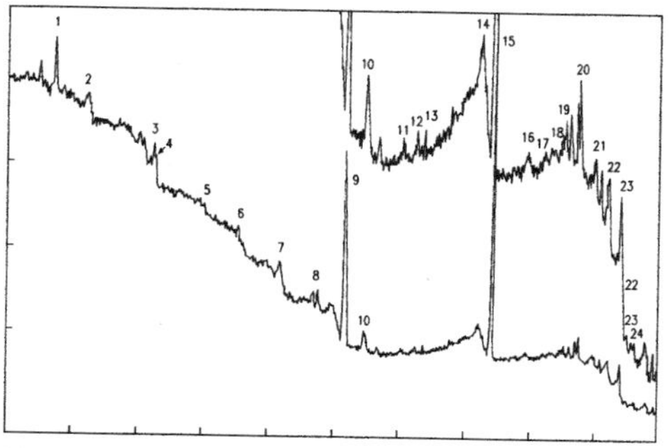

Figure 6. Broad-range XPS scan of the surface from a thin corrosion film on a fracture surface of Alloy 600. The rising background is due to the presence of metals. The numbers refer to peaks identified in Table 1.

For a quantitative analysis of the spectrum in Fig. 6, the following sequence of steps was taken:

a. All photoelectron peaks were analyzed for peak position and intensity. Computer routines to accomplish this are now quite common. For quantitative analysis, peaks were chosen for which the photoelectrons had a restricted range of KEs, corresponding to a BE range of 30 to 530 eV. Such peaks are less attenuated by contamination overlayers than those at higher BEs, because the associated attenuation lengths are greater.
b. Detailed visual inspection of the broad range spectrum on the computer screen then allowed the analyst to see that the Ni 2*p* and Ni 3*p* peaks both had metallic and oxide components. Computer-aided analysis of other lines showed them to be single species, and they are identified in Table 1.

The outcome of the analysis of the integrated peak intensities shown in Fig. 6 is shown in Table 1 after corrections for different photoelectron cross sections and for the different inelastic mean free paths for the photoelectrons. The C 1*s* peak was not included in the quantitative balance since it was felt to represent a contamination overlayer which was not an integral part of the corrosion film. Previous XPS intensity studies of reference oxides showed reasonably good agreement with expected stoichiometries [9]. From the analysis, it can be seen that this film contained more oxygen than was accountable by stoichiometry; the balance is believed to be water of hydration.

Depth profiling can be carried out in conjunction with XPS analysis. The procedure, however, is time consuming and what chemical information was present

Table 1. Spectral Analysis of the XPS Spectrum in Fig. 6 of a Crevice Surface

Identifying number	Binding energy	Relative area	Assignment
1	1021.6	235	Zn $2p_{3/2}$ (oxide)
2	990.1	200	C KLL
3	855.1	62	Ni $2p_{3/2}$ (oxide)
4	853.0	50	Ni $2p_{3/2}$ (metal)
5	776.7	24	Co $2p_{3/2}$ (oxide)
6	712.2	99	Fe $2p_{3/2}$ (oxide)
7	641.3	1530	Ni LMM
8	576.9	92	Cr $2p_{3/2}$ (oxide)
9	531.2	4130	O 1*s*
10	499.5	140	Zn KLL
11	452.2	17	Ti $2p_{3/2}$
12	413.6	12	Pb $4d_{5/2}$
13	399.6	18	N 1*s*
14	307.1	36	Mg KLL
15	285.2	9115	C 1*s*
16	226.7	330	S 2*s* (multiple)
17	199.6	2	Cl 2*p*
18	167.6	135	S 2*p* (sulfate)
19	163.3	70	S 2*p* (sulfide)
20	138.6	967	Pb 4*f* (oxide)
21	102.1	307	Si 2*p*
22	89.2	484	Zn 3*p*
23	67.7	580	Ni 3*p* (metal)
24	55.9	92	Fe 3*p* (oxide)
25	43.0	122	Cr 3*p* (oxide)

Calculated composition of outermost surface (not including carbon): ZnO 31%, SiO_2 20%, NiO 19%, SO_4^{2-} 8%, Cr_2O_3 8%, Fe_2O_3 5%, S^{2-} 4%, PbO 4%.

in the original surface analysis is sometimes compromised by the fact that many oxides can undergo some decomposition in the ion beam. The fact that the area sampled by XPS is often several hundred μm in diameter also complicates the interpretation of the depth profile. As the depth profiling ion beam approaches the metal interface, some regions of the metal will be uncovered before others and the interface will appear to be very broad. Profiling using AES has fewer uncertainties and is much faster. On the other hand, gentle mechanical sectioning of thicker oxide films using a very sharp (and clean) knife edge can often be successful. The resulting chemical information will be free of the anomalous decomposition effects and preferential sputtering produced by ion beams.

One noteworthy characteristic of XPS that is particularly useful to corrosion science is the relatively low degree of damage that is imparted to "soft" surface structures, containing hydrates or chemisorbed species. The flux of low-energy X-ray photons produced by all but the most recent XPS systems is sufficiently low that strongly hydrated structures appear to remain intact even after hours of exposure. Since some passivating corrosion films may contain hydrates, it is important to have a spectroscopic technique such as XPS which may provide future support for the role of water in such films.

The measurement of film thicknesses in the range 1–6 nm is often possible with XPS. This results from the ability to distinguish oxide and metal peaks for an element, often at an intensity ratio of 20:1, coupled with the fact that the inelastic mean free paths for many photoelectron lines are in the range of 2–3 nm. Recently it has been possible to measure the growth of oxide on Al and Mg metals in the presence of water vapor over an exposure range of 12 orders of magnitude [10].

2.8. AES

From an historical perspective, AES was first developed as a tool of the metallurgical sciences. This meant that the technique saw early practical use in solving corrosion problems and was used extensively in conjunction with depth profiling in order to probe structures of thin corrosion layers. Such situations suited the characteristics of AES; the technique with its focused beam of exciting electrons can be used conveniently to sample (sub)-μm surface areas, and the high flux of exciting electrons results in a correspondingly strong Auger spectral signal which can be collected in as little as 30 s. Thus, the technique lends itself to making many microscopic analyses in a short time; this suits one of the requirements, which is to understand the nature of highly localized corrosive attack in some metals, based in turn on an understanding of the statistical probabilities of finding certain corrosion conditions on a given surface. On the other hand, AES has higher background intensity than XPS, and can never hope to produce the sensitivity required to detect low-concentration elements. Furthermore, the straightforward formalisms required for XPS quantification, and the relatively benign excitation conditions are other clear advantages for XPS. Quantification with AES from first principles is complicated by electron backscattering contributions; empirical relative sensitivities (derived from metal spectral intensities) are frequently used, with results that are sometimes unpredictable.

Because of its form of excitation, AES has a close relationship to SEM. In fact, the analysis will usually be preceded by an SEM analysis of the plan view surface in order to locate different grain or topographical structures on the corrosion surface whose surface compositions are required. Modern AES systems include an SEM capability that is moderately good, but do not quite compare with the best resolution of an FESEM. Thus areas within an SEM image are chosen for analysis by AES using a focused electron beam with an energy of ca. 3 kV. The area analyzed can be as small as the beam diameter itself (<100 nm in modern systems), or somewhat larger when defined by a raster pattern. Because in corrosion oxides are usually being studied, there is a high likelihood that beam-induced charging will broaden the spectrum under some conditions, such as too high a beam current and the use of a high angle of beam incidence. Sometimes use of a rastered beam will eliminate or reduce charging. Typical beam currents for corrosion studies range from 1 to 10 nA.

It is normal when commencing Auger analysis of a corrosion film to record several "survey" scans across that KE range (0–1000 eV) that will detect most elements in the periodic table (except hydrogen). Such scans can be obtained within 5 min, and should provide a basis for determining the best places to perform depth profiles. Traditionally, most AES scans are presented in the differential mode. Nowadays, with amplifiers of improved stability, there is no good technical reason for doing that, but it is still common practice to present spectra in that form. An example of the alternative integral form is shown below; conceptually, it is easier to understand, particularly if peaks are overlapping.

In contrast to XPS, a certain amount of depth profiling with a rastered Ar^+ beam is performed in most AES investigations. The reason for this is that most Auger lines fall in a KE range that renders them more susceptible to attenuation from surface contamination than is the case for XPS. Also, the rapid rate of data acquisition in AES makes it easy to review the results of a particular surface treatment, such as ion bombardment.

The results of an AES point analysis of corrosion products on a surface is shown in Fig. 7. A stainless steel 304 surface had been in a high-temperature (290°C) circulating water loop in alkaline conditions for 500 h of corrosion testing. An earlier depth profile carried out over a large (2 mm diameter) area had not revealed any of the Cr surface enrichment which might have been expected of stainless steels under these conditions. However, selected-area analysis of some of the unsputtered oxide was more successful. The many individual crystallites on the surface were determined to be Fe-Ni oxides with no detectable Cr. However, the featureless oxide underlying the crystals was found to contain significant concentrations of Cr. The mechanism of formation of each oxide is different: the base oxide forms by solid-state diffusion, but the crystallites form by dissolution/precipitation, a process that clearly does not involve Cr.

With such a complex surface composition, it is not surprising that a larger-area analysis revealed little. With a multipoint survey the compositional statis-

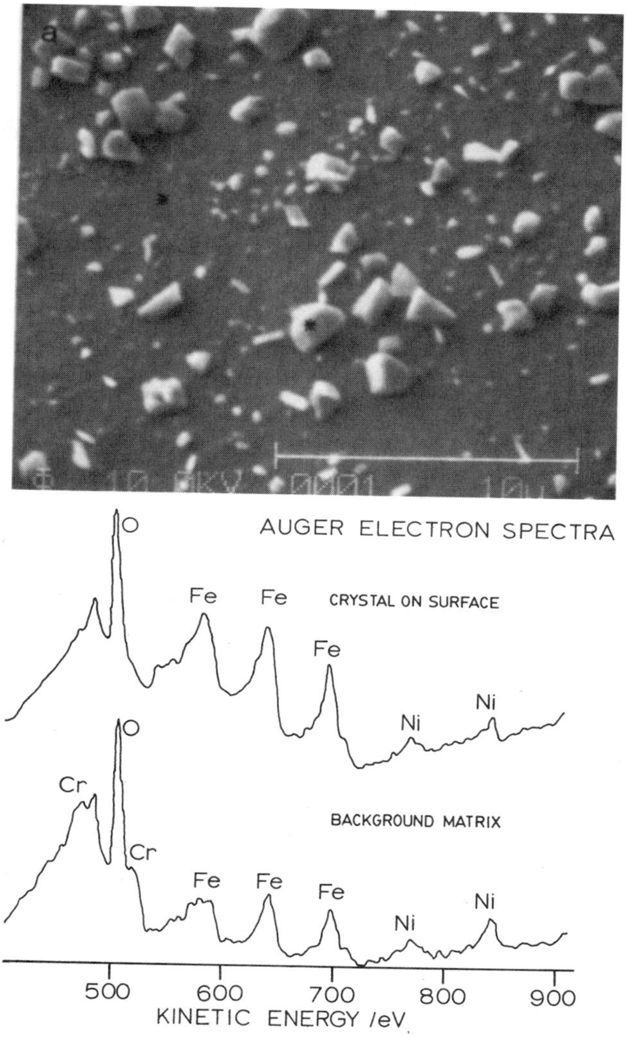

(a)

(b)

Figure 7. Point AES analysis of corrosion deposits on a stainless steel surface. The SEM image and spectra were taken with a 3 kV, 10 nA electron beam. Spectra are shown in (b) for a crystallite on the surface as well as for the background.

tics associated with every major type of feature can be assessed before additional time is invested in profiling. When profiling is carried out on such complex surfaces, one should consider what the process of ion bombardment is likely to do to the surface. Raised areas, such as the summits of the crystallites, will have their outermost oxide layers removed quite efficiently; thus one would expect to detect any difference in composition between, for example, that of a thin coating on the crystallite surface and the crystallite beneath. On the other hand, the compositional profile of the base oxide could be compromised by deposition of material sputtered from more exposed areas. Thus there are limitations to the information which could be extracted from an extended depth profile of such a surface, even when carried out on limited areas.

Another analysis option provided by most AES instruments is the capability of producing maps of the Auger electron intensity distribution for each element, in a manner entirely analogous to EDS mapping. Attempts to map corrosion product surfaces such as that in Fig. 7 will be time consuming and subject to anomalous effects due to the topography and they should be avoided.

On the other hand, some corrosion surfaces, particularly those not involving solution transport processes, are much more amenable to both extended depth profiling and mapping. One example of this is shown in Fig. 8 where comparative small-area AES depth profiles were recorded from two grain faces of a Ni-Cr alloy oxidized at 600°C in low-pressure oxygen. On one grain face (VII), significant forward migration of Cr was detected as the result of oxidation; corresponding depletion of Cr in the unoxidized alloy underlying can also be seen. On the other grain face, no evidence of preferential migration is seen. The reasons for this difference are not completely understood, but probably relate to the differences in surface energy of the two grain faces. Even on this surface, mapping might prove difficult because the real differences show up *within* the oxide, and mapping is most often used on surfaces to provide direction for further studies. In cases of thin surface tarnishes, mapping is a more reasonable option.

AES depth profiling has often been used to measure oxide film thickness. As indicated above, other cross-section methods exist which may be more accurate and faster. In Fig. 8 the two depth profiles have the position of the oxide/metal interface indicated as "I"; the position itself is usually taken to be the depth at which the metal abundance becomes >50%. In Fig. 8 there appears to be little doubt about the position of the interface as so defined. However, in many profiles of significant depth, the interfaces become broadened to the point that the term "interface" loses its meaning. This is particularly true for films arising from aqueous corrosion. It follows that there is little likelihood of observing significant compositional changes in the inner oxide of such films, and, for that reason, other methods such as SIMS must be used if such information is desired.

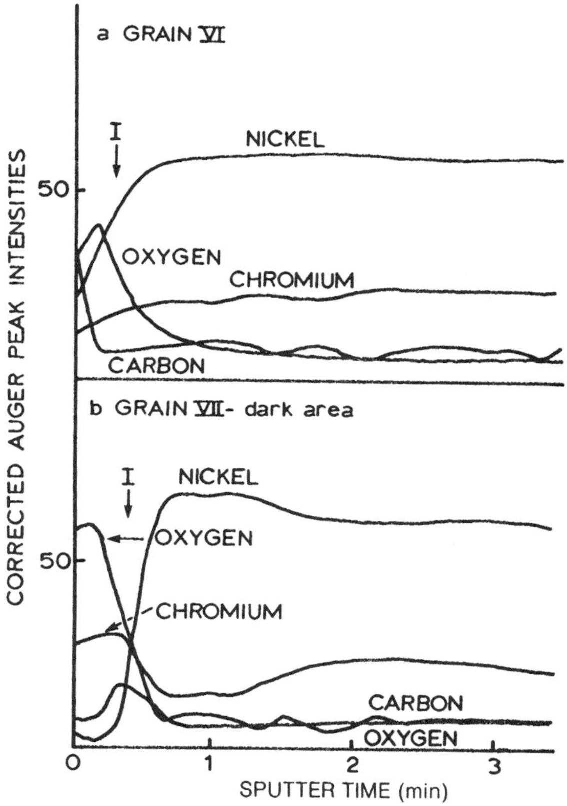

Figure 8. Comparative AES depth profiles of microscopic areas of a 90% Ni, 10% Cr alloy exposed to low-pressure oxidation at 600°C. On grain VI, the Ni/Cr ratio in the oxide is similar to that in the alloy. By contrast, on grain VII, the oxide is strongly enriched, and the underlying alloy depleted, in Cr. The two grain faces have different crystallographic orientations.

2.9. SIMS

SIMS has not been much used as a tool for studying corrosion. The main reason for this seems to be that SIMS spectra and depth profiles are frequently difficult to interpret in terms of clearly defined diffusion or deposition mechanisms. Spectra from oxides contain a complex (and frequently overlapping) series of molecular ion masses; secondary-ion depth profiles are complicated in the oxide and interface region by matrix-effected changes in ion yield which could easily be interpreted by a nonexpert as a change in elemental concentration.

In spite of this, SIMS does have properties which lend themselves well to the characterization of corrosion films. Not the least of these is the capability of generating large amounts of relevant data. Recently, routines for more rapid and effective analysis of these data have come into active use, particularly for instruments which are capable of collecting entire secondary ion spatial distributions during a depth profile. Instruments produced by Cameca, Charles Evans and Associates, and, more recently, Physical Electronics, fall into this category. Second, molecular ions have been found to be very sensitive to changes in phase structure; the intensities of oxide molecular ions containing one or more metals from the corrosion system are found to respond well to changes in structural phase, as well as to the major reduction in oxygen concentration as the profiling beam moves from the oxide phase into the metal substrate. SIMS is one of the few techniques capable of detecting hydrogen (and its isotopes) and measurement of this important corrosion by-product adds a dimension to this technique not originally envisaged. The high dynamic range of SIMS for trace element analysis also provides an unparalleled opportunity to measure the behavior of possibly important minor or trace corrosion species, such as Cl or Pb. Finally, the collection of ion images during profiling helps to identify local corrosion phenomena, just as the point analysis does during AES measurements.

SIMS systems capable of imaging can be used to study corrosion samples in plan view or in cross section. The choice of primary ion should be of one that enhances the detection of the major products; Cs is the most suitable since it enhances the production of electronegative species such as O, H and oxide molecular fragments. The beam currents used should be such that the entire oxide thickness can be profiled in a reasonable time; typical sputter erosion rates in oxides are 3–5 μm/h for a 1 μA beam rastered over a 250 μm^2 area. Ion images collected should reflect the possible oxide combinations, e.g., FeO^- for a simple Fe oxide, or $FeCrO^-$ for an oxide believed to contain both Fe and Cr.

A good example of the use of SIMS for depth profiles is shown in Fig. 9. Alloys 600 and 800 were exposed in an autoclave at 300°C to acidic sulfate excursion conditions, followed by a return to a normal pH of 10. The surface oxides were profiled by SIMS imaging, in which a series of lateral intensity distributions of the molecular ions NiO^- and CrO^- were measured as a function of depth through the oxide and into the underlying metal. After the images were "stacked" in the computer, a small section 0.8 μm in diameter was taken through the stack and the intensities plotted for the two secondary ions for each alloy. It can be seen in Fig. 9 that the intensity of oxide ions such as NiO^- decreased sharply by two to three orders of magnitude at the oxide metal interface. For Alloy 600 the ratio of CrO^- to NiO^- clearly increased as the interface was approached, suggesting an enrichment of Cr in that region (see arrow). For

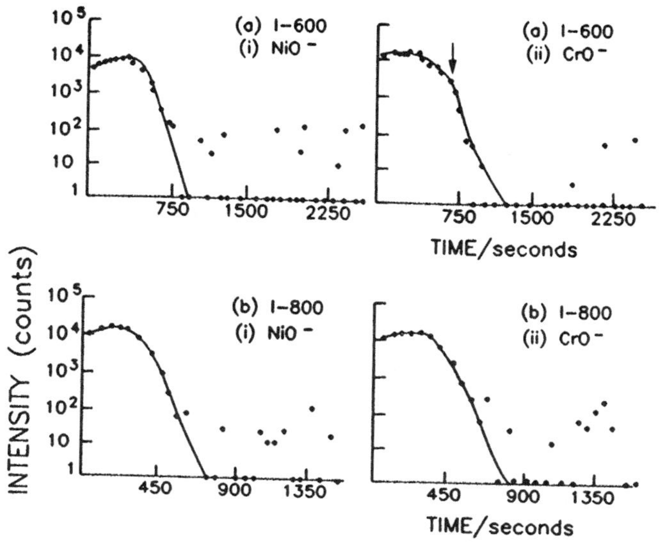

Figure 9. Comparative SIMS molecular ion depth profiles of oxide films on Alloys 600 and 800. The arrow indicates an enrichment of chromium oxide at the interface.

Alloy 800 no such enrichment was observed, probably because the Cr content of that alloy was lower. The technique thus appears to provide profiling information of a type impossible to obtain by AES depth profiling.

2.10. Other Methods

Some surface characterization techniques, such as Raman and infrared reflectance spectroscopies (see Appendix 3), have been used extensively in corrosion experiments in some laboratories, but the techniques are not considered sufficiently "universal" to be discussed here. Moreover, instrument operation and data interpretation are, at the moment, sufficiently complex and specialized that production of a series of protocols to suit most corrosion situations would be difficult. Other techniques such as XRD have been used frequently for routine characterization of thick corrosion layers (often after mechanical separation from the substrate). However, XRD has not been used on films much thinner than 2–3 μm, and, where it is used, a major problem has been the inability to determine the precise location of the various phases whose XRD patterns were reflected from the surface. Grazing-incidence XRD may provide some of this depth resolution in the future. Another technique on the horizon is SPM, which is described within the major corrosion application below.

As part of the same application, the uses of electrochemistry are discussed. Electrochemical investigations can range from generalized studies aimed at understanding the mechanisms of corrosion of a metal in a particular medium to measurements of defect areas of failed specimens from the field. For the latter, defect areas can be masked off from noncorroded areas on the metal surface and numerous electrochemical procedures can then be applied in order to determine the film porosity and the chemical activity of the region. The studies on Monel 400 alloy corrosion described below represent a comprehensive approach to a corrosion problem. While, by itself, the work cannot claim to have "solved" the problem at hand, it has helped to develop confidence in the mechanisms proposed earlier, which were based on much less solid evidence. The experimental approach followed is based on some of the most practical choices facing a scientist or engineer who wishes to employ the techniques of modern surface science and electrochemistry to solve a practical problem.

3. BACKGROUND TO THE PROBLEM: A WORKING HYPOTHESIS

For the past 20 years all heat exchanger tubing in the main boilers of Pickering Nuclear Generating Stations (PNGS) A and B has been constructed of Monel 400 (Monel 400, containing 66 wt.% Ni and 32 wt.% Cu, is the trademark of the International Nickel Company (INCO) and its Unified Numbering System designation is N04400). The alloy was chosen because of its low corrosion rate when in contact with deoxygenated water at 300°C. In the boilers, the water in both primary and secondary coolant streams is adjusted to pH 10 at room temperature; this gives a pH of 7 at 300°C.

In PNGS A, most boilers have operated for about 25 years with very little detectable corrosion. By contrast, within six years of the start-up of the PNGS B reactors, serious corrosion problems leading to perforation were reported in the Monel boiler tubing. Failure analysis by cross-section metallography and SEM/EDS showed localized depletion of the metal of the type which is associated with acid attack of Ni. The alloy itself has been regarded as analogous to Ni metal in its corrosion behavior because of the high Ni content and the fact that Cu, with its higher solubility in water, is unlikely to play a major part in any passive surface film.

Further examination of failed tubes taken from the boilers revealed two important facts: first, the relative Ni-to-Cu concentrations from PNGS B were slightly outside ASME specifications (see Table 2); second, the grain boundaries of "B" tubing had considerably elevated concentrations of B, which had migrated from the bulk phase of the alloy. The latter fact was determined using SIMS imaging of metallographic cross sections of "A" and "B" alloys (see Fig. 10). The high sensitivity of SIMS allows B inclusions to be detected, even

Table 2. Elemental Analysis of Monel 400 Alloy Specimens from A and B Sources

Alloy	Wt.% Ni	Wt.% Cu	Wt.% Mn	Wt.% Fe	Ni:Cu
A	65.9 ± 0.35	31.5 ± 0.22	1.03 ± 0.092	1.60 ± 0.21	2.09
B	64.5 ± 0.12	33.2 ± 0.45	1.08 ± 0.13	1.14 ± 0.17	1.94
Nominal	66.5	31.5[a]	1.0	1.15	2.1

[a]Maximum copper allowed by UNS is 32.2%.

though their total concentrations would probably not exceed 1000 μg/g. The grain boundary segregation in the "B" alloy suggested that grain boundary chemistry may have been altered sufficiently to change the corrosion properties of the Monel surface.

4. EXPERIMENTAL STRATEGY

With the above hypothesis in mind, a series of surface analytical and electrochemical experiments was devised to answer several questions.

1. Can differences in the corrosion properties of the "A" and "B" alloys be detected after only a short exposure to boiler water conditions? Positive findings might allow such exposures to be used as qualification tests.
2. What is the relationship between the hydrothermal films and films formed in room-temperature water? The latter are much more amenable to electrochemical tests which would yield more quantifiable results than surface analysis.
3. Since local acid attack was a suspected mechanism, how do films corroded in acid media compare with those corroded in mildly alkaline conditions?

The strategy required two sets of experiments: those carried out in an autoclave to simulate boiler conditions, and those carried out in room-temperature aqueous media. The autoclave was a replenishing type in which a weak sodium hydroxide deoxygenated solution (pH 10) was fed into an Inconel pressure vessel held at 300°C. Samples of archived "A" and "B" Monel tubing were suspended in the autoclave and after 2 d of exposure were removed and stored under a protective atmosphere. No *in situ* electrochemical control or measurements of the autoclave experiments was attempted, although such measurements are becoming more practical.

In preparing the surfaces of specimens for corrosion experiments, it is always difficult to achieve a balance between "authentic" field conditions and those conditions that are contrived to accentuate certain aspects of the corrosion

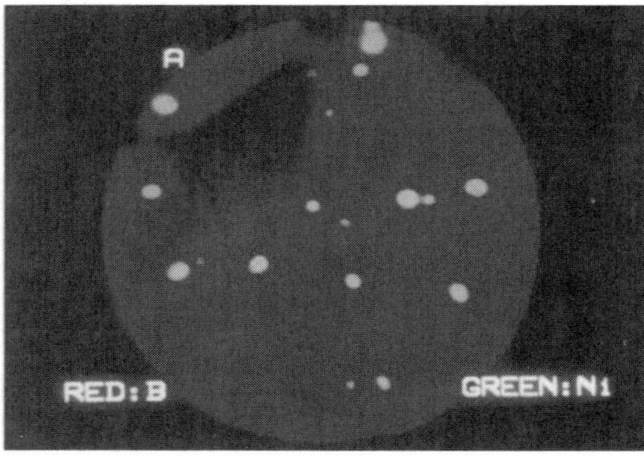

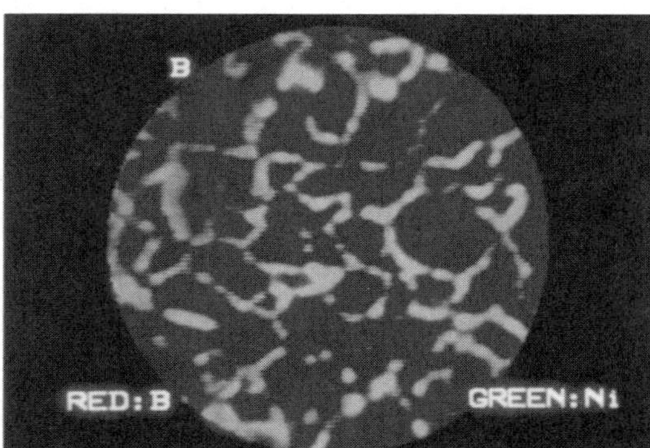

Figure 10. SIMS images of the distribution of B (bright) and Ni (dark) in polished cross sections of Monel 400 from "A" and "B" sources. The imaging field is 150 μm. The grain boundaries of "B" are enriched in B, while in "A," B is largely found in inclusions in the bulk of the alloy grains. Ni concentration is found to vary somewhat within certain grains of the alloy. The faces of some grains are darker than others because of orientation effects on contrast.

process. For example, rough surface topography, frequently found in field specimens, could contribute to highly localized surface chemistries which might be responsible for a major corrosion process; on the other hand, on such complex surfaces it may be difficult or impossible to identify one particular process among many. In practice, it is useful to analyze surfaces prepared in more than one fashion in order to identify the effects of surface preparation [11].

In this study, the field specimens exposed to autoclave treatment had 10 μm deep grooves resulting from the tube-drawing process. The cleaning process consisted of a methanol rinse to wash off drawing lubricants and to remove air-borne organic contaminants.

Aside from the autoclave studies, the other (more contrived) experiments were conducted in electrochemical cells at room temperature. In such cases it is important to eliminate ambiguous surface reactions by keeping the electrode surfaces as reproducibly flat as possible. After exposure to boiler or ambient aqueous conditions, the samples were analyzed by a range of surface analytical techniques. Some of these were more effective than others in elucidating distinctive changes under particular treatment conditions.

5. ELECTROCHEMICAL TECHNIQUES FOR SURFACE CORROSION STUDIES

Electrochemistry can be used to control the nature and progress of an aqueous reaction at a metal electrode and to measure the nature of the surface which has been altered by the corrosion reaction. Both modes are used in this study. In order to understand better the complementarity of modern electrochemistry with surface analytical techniques, it is useful to review some of the processes involved, as well as some of the terminology. The reader is referred elsewhere for a full treatment of the subject [12,13].

5.1. Basic Electrode Kinetics

For a metal at equilibrium in an aqueous solution, the rates of the oxidizing (anodic) and reducing (cathodic) half-reactions are equal. These rates may be expressed as current densities I_a and I_c, whose algebraic sums are equal:

$$I_{net} = I_a + I_c = 0 \qquad (1)$$

The magnitude of I_a or I_c is a measure of the balanced Faradaic activity and is called the exchange current density, I_0:

$$I_a = -I_c = I_0 \qquad (2)$$

The potential of the metal at which this occurs is called the equilibrium potential, E^0. If the system deviates from equilibrium, say through active dissolution, a new potential is established, which is related to the equilibrium potential by a term called the overpotential:

$$\eta = E - E^0 \qquad (3)$$

The relationship between overpotential and current can be expressed as

$$\eta = \beta \log \frac{I}{I_0} \qquad (4)$$

β, the Tafel slope, is equal to $nF/\alpha RT$, where F is Faraday's constant and α is the transfer coefficient. Near $\eta = 0$, the expression can be approximated by the linear form

$$I = I_0 \frac{nF\eta}{RT} \tag{5}$$

The potential-current curve in this region can thus be considered linear, with slope equal to $RT/I_0 nF$. This slope takes on the form of a resistance, R_{ct}, called the charge transfer resistance:

$$\eta = IR_{ct} \tag{6}$$

R_{ct} is a good measure of the rate of transfer of charge to and from an electrode under nonequilibrium conditions.

In a metal corroding in water with no external voltage applied (open circuit), a number of different half-reactions can occur; at equilibrium the potential which represents their average potential is called the corrosion potential E_{corr}, while I_{corr} is the corrosion current. Thus at E_{corr}, the current associated with the dissolution of a metal, e.g.,

$$M \rightarrow M^{2+} + 2e^-$$

would be equal to the current produced by the reduction of hydrogen,

$$2H_3O^+ + 2e^- \rightarrow H_2 + 2H_2O$$

5.2. Electrochemical Techniques

5.2.1. Linear polarization

Linear polarization takes advantage of the linear region of the *E-1* curve, near the corrosion potential, to extract values of R_{ct}. Within 10–20 mV of the corrosion potential, current varies linearly with potential, with the slope equal to R_{ct}. At high values, R_{ct} is related inversely to the corrosion rate.

Linear polarization techniques are often used to conduct an initial electrochemical characterization of a metal or alloy, prior to more complex investigations. E_{corr} is first determined relative to a reference electrode, usually the standard calomel electrode (SCE). A small potential is then applied and swept from about 20 mV below E_{corr} to 20 mV anodic to it. The current density is measured and R_{ct} is calculated.

5.2.2. Anodic polarization

In anodic polarization studies the electrode potential is swept through anodic potentials which deviate significantly from E_{corr}, e.g., up to 2 V, producing entirely new electrochemical reactions. As before, the current density is followed as a function of applied potential. From this type of test, changes in the contributions of different corrosion reactions can be identified. At some potentials, for example, active dissolution can cease because of the growth of an oxide film that passivates the surface. Such active-to-passive transition potentials lead to an understanding of the surface protective processes.

5.2.3. Electrochemical impedance spectroscopy

An electrochemical cell can be modeled by some common circuit elements corresponding to the electrical response of certain structures at or near the interface. In one particularly useful model, shown in Fig. 11a, a resistor is in series

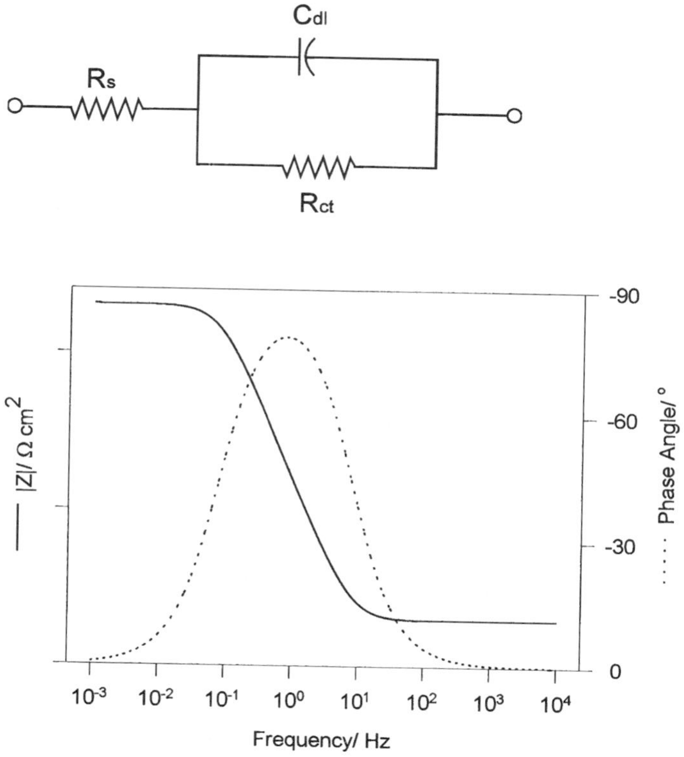

Figure 11. (a) A commonly used circuit model representing resistive and capacitive elements in a corrosion cell, and (b) graphical derivation of values for R_s and R_{ct} from a Bode plot of the modulus and phase angle values as a function of applied sweep frequency in an electrochemical impedance spectroscopy experiment.

with a second resistor and a capacitor, the latter two being in parallel. The first resistor, R_s, represents the solution resistance, the second, R_{ct}, the charge transfer resistance described above. Capacitance effects of the electrical double layer or of any oxide films or coatings are accounted for in C_{dl}.

If instead of a direct potential an alternating (i.e., sinusoidal) signal is applied to the model electrode circuit in Fig. 11a, the complex impedance can be measured. The impedance Z is the quotient of E and I:

$$Z = \frac{E}{I} = \frac{E_m \sin(\omega t)}{I_m \cos(\omega t + \phi)} \tag{7}$$

where E_m and I_m are the voltage and current amplitudes respectively, ω is the angular frequency ($= 2\pi f$), and ϕ is the phase shift between E and I.

The impedance comprises contributions from both purely resistive and capacitive components in the electrical system. The impedance is measured by applying a small amplitude (10–20 mV peak-to-peak) sinusoidal potential and measuring the current response. An impedance spectrum is obtained by measuring the impedance as the applied frequency is swept, typically from 1 mHz to 10 kHz. Use of a frequency response analyzer allows the modulus of the impedance and the phase angle to be obtained as a function of frequency. Solution and charge transfer resistance values can be obtained graphically from the zero-slope regions of the modulus plot at high and low frequencies, respectively (see Fig. 11b). C_{dl} can be obtained from the relationship.

$$\log |Z| = -\log C_{dl} \tag{8}$$

which is true for $\omega = 1$. R_{ct} can be taken as a measure of the protectiveness of the film, while C_{dl} is often found to be a function of the film thickness.

6. RESULTS AND ASSESSMENT

6.1. Initial Characterization

The "A" and "B" Monel alloys, supplied for the study by the electrical utility company, Ontario Hydro, were first characterized metallographically in this laboratory to ensure that each had the same metallurgical properties as specimens studied in the past by Ontario Hydro corrosion engineers.

The specimens, after being polished to a 0.05-μm diamond finish and etched in a nitric acid–ethanol mixture (nitol), gave characteristic structures, shown in Figs. 12a and b. Grain structure was equiaxed in both "A" and "B" alloys, because the tubing had been cold-drawn and recrystallized during manufacture. Average grain size is somewhat larger in the "B" alloy (ca 20 μm diameter). Both alloys contained inclusion "stringers"; in the "A" alloy these were composed of MgS as established by EDS and were readily etched; in "B" the stringers were

(A)

(B)

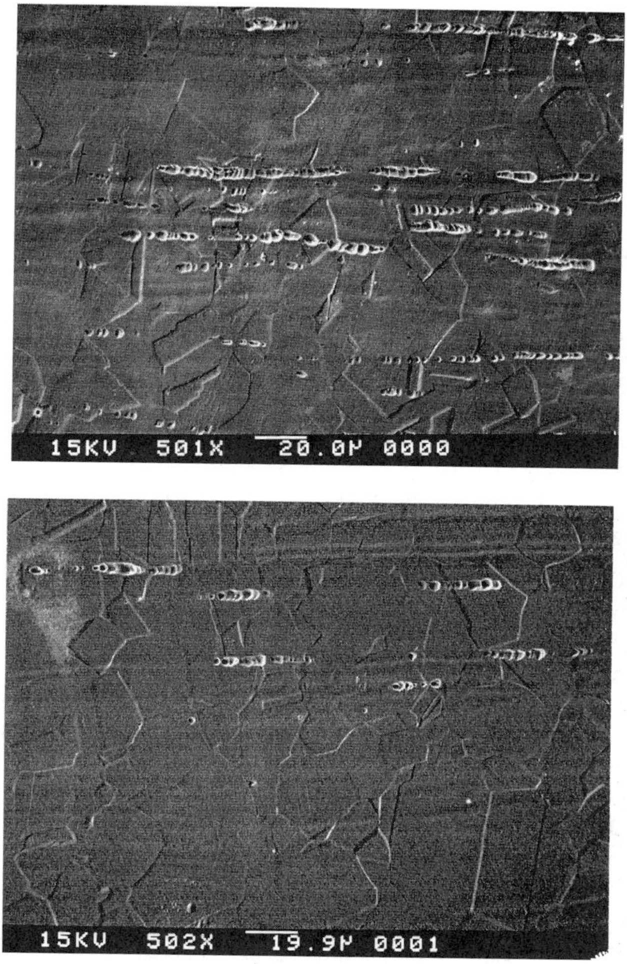

Figure 12. SEM micrographs of polished sections of "A" and "B" after etching in a mixture of nitric acid and ethanol.

found to be calcium sulfide and were not as heavily etched. These results confirmed that the alloys were similar to those previously studied.

6.2. Boiler Simulation Corrosion Experiments

The "A" and "B" samples removed from the autoclave after 2 d of exposure at 300°C were first studied by a standard SEM instrument, i.e., a system with a spatial resolution and surface sensitivity commonly found in most laboratories over the past decade. The images for "A" and "B" surface corrosion films are compared in Figs. 13a and b. The image of the "A" film shows an even distribution of 2–4 μm diameter hexagonal platelets which are oriented generally in a vertical direction. These are virtually identical to corrosion films observed on Ni-rich alloys such as Inconel 600 exposed under similar conditions [14]. Such

(A)

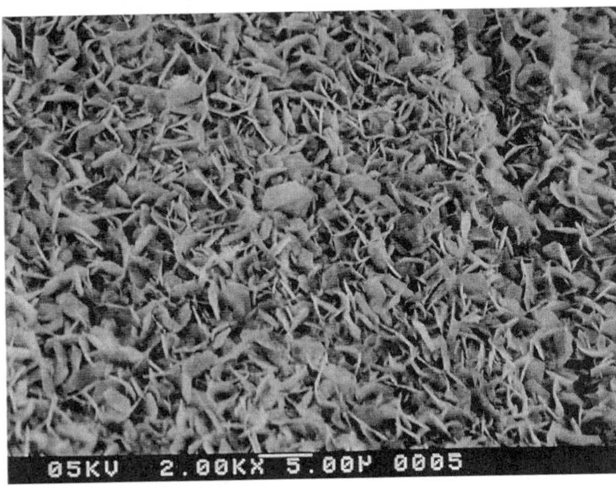

(B)

Figure 13. SEM micrographs of the outer surfaces of "A" and "B" alloys after a 48-h exposure to pH 10 conditions at 300°C.

crystals have been identified previously as $Ni(OH)_2$ in a Brucite structure. The microstructural appearance of the "B" film is quite different; individual crystallites, though still platelike, are randomly oriented, and have a less regular shape. Moreover, there are distinct regions (appearing dark in Fig. 13b) where crystals appear to be less densely formed. The shapes of these regions correspond to the general outlines and dimensions of the grains shown in Fig. 12; one might speculate that corrosion activity specifically associated with grain boundaries is responsible for the change in crystal growth pattern.

The compositions of these films have been probed as a function of depth using a three-dimensional (3D) SIMS imaging procedure [15]. Images of the spatial distributions of secondary ions from the film are recorded sequentially, as a function of depth, and are "stacked" to form a 3D or volumetric image. The latter can be very effective in visualizing both the consistencies and variations expected in a typical corrosion film structure.

SIMS 3D profiles, shown in Figs. 14a and b, compare the compositions and structures found on the corroded surfaces of "A" and "B" alloys. The secondary ions NiO and CuO were chosen to measure the relative concentrations of oxidized Ni and Cu in the film. These are preferable to the analogous negative metal ions since the oxide ions virtually disappear when the depth profile reaches the oxide/metal interface; thus changes in composition near this interface can still be detected clearly when oxide ions are used [14]. It goes without saying that the measured oxide intensities do not necessarily represent "NiO" or "CuO," but only some form of oxide. Under the conditions used in these profiles the relative sensitivity factors (RSFs) for CuO^- and NiO^- are approximately equivalent. However, the sputter rates of the "A" and "B" oxide films were found to be dissimilar. Using profilometry the thickness of the oxide on "B" was found to be twice that of "A," which should be taken into account in viewing the images in Fig. 14.

The 3D images in Fig. 14 show that there are major qualitative differences in the "A" and "B" films. The "A" film contains largely oxidized nickel. Subsequent XPS measurements showed it to be a mixture of $Ni(OH)_2$ and NiO. Whatever oxidized Cu is present is found to be distributed near the oxide/metal interface. By contrast the "B" film has a significant concentration of oxidized Cu, distributed fairly evenly through the film structure. Some measure of the differences in Ni/Cu ratio at individual points in the volume, can be gained from the intensity plots in Fig. 14, where NiO^- and CuO^- intensities are compared. Also, the distribution of oxides on the "B" alloy surface shows them to be less densely packed and more irregular. In addition, the crystallite size is larger.

It appears, then, that the "A" alloy has an oxide that is thinner with smaller crystallites and which is enriched in nickel oxide species at the outermost surface. The thicker "B" oxide has an irregular network of larger crystals with oxidized Cu distributed throughout the film. Furthermore, it appears that oxide

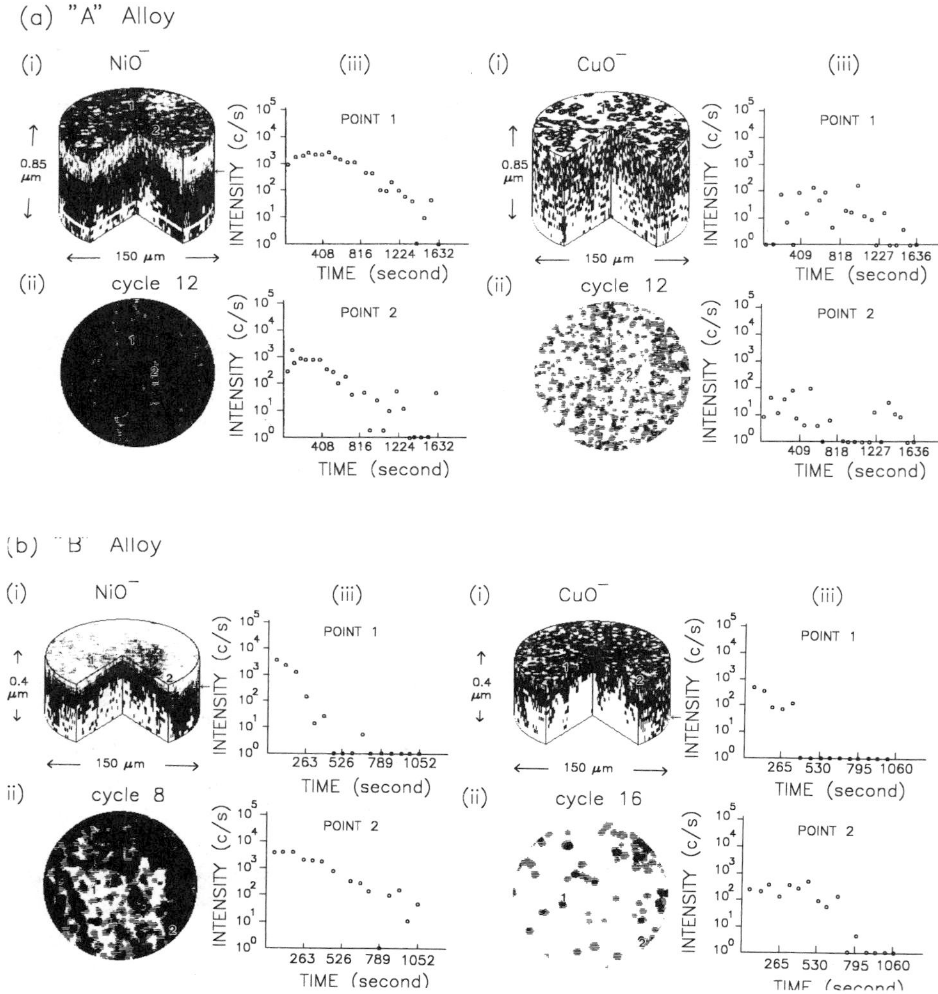

Figure 14. SIMS 3D depth profiles of (a) Monel alloy "A" and (b) Monel alloy "B" after autoclave exposure to a pH 10 solution for 48 h at 300°C. Each profile contains graphic information on NiO and CuO. For each ion, four pieces of information are shown: (i) a 3D profile, (ii) a horizontal slice from the 3D profile at the point on (i) indicated by the arrow, and, (iii) and (iv) intensity depth profiles at two points, 1 and 2, selected on the outer surface. The resultant intensity profiles represent the intensities in a vertical line of pixels intersecting these two points. Each point represents an area 1 × 1 μm².

growth is particularly sparse on areas adjacent to grain boundaries. Thus, of the two oxide films, that on the "A" alloy gives all the appearances of being a more passivating one. Evidence of the more soluble Cu alloy component is not found in that part of the oxide most recently formed, implying that dissolution has virtually terminated by the end of the autoclave experiment.

One may wonder about the consequences of using a depth profiling technique such as SIMS in order to analyze highly irregular crystallite surfaces such as those shown in Fig. 13. Many chemical changes are certainly produced by the interaction of an intense ion beam with very thin crystallite platelets of $Ni(OH)_2$. Microscopy shows that many such crystals are transformed into larger agglomerates or that some facets are preferentially sputtered. Even so, many of the structural differences specific to each corrosion mechanism are preserved, so that a qualitative understanding of the effects of film structure on corrosion passivation is possible.

Depth profile analysis of such complex oxide thin film structures is not as easy using techniques such as XPS or scanning AES. Both techniques normally present data in terms of average elemental concentrations within an analytical volume. Any ion beam interaction with uncovered regions of the metal substrate will result in a redeposition of those substrate elements onto oxide surfaces still remaining. Thus XPS and AES depth profile information obtained near the oxide/metal interface is often compromised by the presence of sputtered products. SIMS data collected using oxide secondary-ion signals are less affected by such artefacts since most of the redeposited metal phase goes undetected [11]. The high dynamic detection sensitivity of SIMS is also important in the identification of minor phases, such as the oxidized copper component in the corrosion film on the "A" alloy.

After the "A" and "B" alloys were removed from the autoclave, they were subjected to electrochemical impedance spectroscopy (EIS) measurements under ambient conditions in pH 10 solutions. It was anticipated that such measurements might help to support the corrosion mechanisms hypothesized from the surface analysis of alloys exposed to actual boiler conditions. Analysis of the ambient EIS data gave R_{ct} film resistances of 1.2×10^6 and 1.1×10^6 $\Omega\text{-cm}^{-2}$ for the "A" and "B" alloys, respectively, at their corrosion potential. Both are very high resistances, which suggest that a passive film can form on either alloy *under these electrochemical conditions*. However, in the light of their significantly different behavior under simulated boiler conditions, it can be concluded that these particular EIS tests may not be pertinent to an understanding of the corrosion problem at hand.

6.3. Contrived Corrosion Experiments on Monel

6.3.1. Electrochemical measurements at pH 10

All further studies were conducted at ambient temperature with a view to finding an electrochemical procedure which would identify failure mechanisms specific to those observed in the field for the "B" alloy. For these studies, the alloy surfaces were prepared in ways that differed from the as-received surfaces used in the above simulation studies. For some microscopic studies, surfaces were polished to a "mirror" finish using 0.05 μm alumina abrasive, which was useful in allowing the identification of incipient corrosion films. For most other studies, the surfaces were polished to a rougher 600 grit finish; this is a finish commonly used by corrosion scientists, because it is felt to produce surface structures more representative of those found in the field.

Anodic polarization studies of both "A" and "B" alloys were carried out at 25°C in pH 10 deaerated NaOH solutions. The alloys were immersed in the solution for 1 h prior to the polarization studies. Potentiodynamic polarizations were performed using a Hokuto Denko HA-501G potentiostat/galvanostat with an arbitrary function generator used to create the change in applied voltage. The current-voltage polarization curves shown in Fig. 15 display the major electrochemical process occurring in this potential range. All potentials given are measured relative to the SCE. The entire potential range was scanned within a

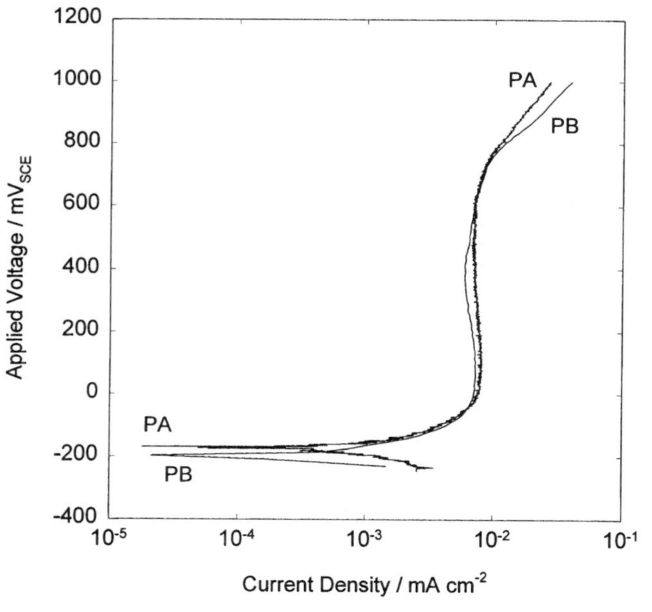

Figure 15. Potentiodynamic polarization curve for a specimen of Monel alloy "A" in pH 10 solution at 25°C. All voltages are given with respect to the standard calomel electrode. The alloy was first exposed to open circuit corrosion for 1 h prior to the voltage sweep.

period of 15 min or less. The sharp inflection point observed near −200 mV for both alloys represents the corrosion potential—the voltage at which some anodic processes become possible and a small current starts to flow. Actual breakdown of the passive film begins above the knee in the curve, i.e., at about +800 mV. This is a reasonably high potential for the occurrence of breakdown, an indication of the relative stability of the oxide film. The polarization curves for "A" and "B" are identical within the reproducibility of the technique. Thus, no significant indication of specific corrosion problems pertinent to the "B" alloy can be found from the potentiodynamic anodic polarization curves.

The above results were obtained for films which had not grown for very long periods on the alloy surfaces. Because development of surface films with different passivating characteristics is clearly a function of the growth kinetics of the film, it was felt that electrochemical studies of more mature films might produce more definitive differences between the alloys. Indeed, some potentiostatic experiments suggested that exposure periods of at least 1 h would be necessary to produce a stable film. Oxide films were then grown on the two alloys at specific applied potentials: −180mV (the corrosion potential) and +300 mV. For both potentiostatic conditions films were allowed to grow for specific periods of time: 1, 3, 7, 12, 24, and 28 h. The structure of the films was then studied by EIS, as well as by a range of other surface analytical techniques.

EIS measurements were performed with an in-house-designed potentiostat controlled by a Solartron 1055 frequency response analyzer. Experiments were carried out on samples which had not been moved from the cell after potentiostatic film growth, and the growth potential was maintained on the sample during EIS. For such studies a 10 mV ac potential was imposed on the dc potential and the frequency of the ac signal was swept from 30 kHz to 0.5 Hz over a period of 150 s. The frequency response analyzer provides "modulus" and "phase shift" parameters for each frequency; these can be analyzed to give specific values of R_{ct} and C_{dl} specific to a particular film. These values are given in Table 3 for the exposure times and potentials studied.

For both alloy films, the R_{ct} values are high, even for short-term exposures. Further, the R_{ct} values initially increase with exposure time, which implies the development with time of a more passive film on both alloys. However, after a particular exposure time (24 h for "A" and 7 h for "B"), R_{ct} no longer increases but begins to decrease, implying some loss of passivity at that point, perhaps due to a change in structure of the film. Values for C_{dl} did not provide any information about film thickness; they all fell in the capacitance range in which most or all of the contribution comes from the electrical double layer at the solution interface rather than from the film itself.

EIS measurements carried out at the more oxidizing potential (+300 mV) were more definitive. For "A," R_{ct} values rose rapidly as a function of time and remained high even after a 12 h exposure, indicating the formation of a passive

Table 3. Charge Transfer Resistance and Double-Layer Capacitance Values Derived for Alloy "A" and "B" Oxides from Electrochemical Impedance Spectroscopy Measurements in pH 10 Solutions at 25°C

Potential (SCE) (mV)	Exposure time (h)	C_{dl} (μF cm^{-2})		R_{ct} (Ω-cm^{-2})	
		Alloy "A"	Alloy "B"	Alloy "A"	Alloy "B"
−180	1	20	17	9.1×10^5	2.1×10^5
−180	3	19	14	1.5×10^6	4.3×10^5
−180	7	20	13	2.4×10^6	8.3×10^5
−180	12	20	14	3.8×10^6	7.1×10^5
−180	24	20	13	1.1×10^6	6.3×10^5
−180	28	19	14	7.7×10^6	5.9×10^5
+300	1	17	15	5.0×10^5	5.0×10^5
+300	3	19	18	1.1×10^6	7.7×10^5
+300	7	20	20	1.0×10^6	7.1×10^5
+300	12	21	19	2.8×10^6	7.7×10^5

film stable with time. For "B," the R_{ct} value rose initially but remained significantly lower than for "A," possibly a sign of greater porosity of the oxide film. From this it may be concluded that, at higher oxidizing potentials, EIS measurements in an ambient-temperature pH 10 experiment may be useful in distinguishing surfaces which have differing corrosion characteristics at boiler temperatures.

6.3.2. Microscopy studies of oxides from pH 10 exposures

Atomic force microscopy (AFM) was the principal technique used to visualize structural changes occurring on the surfaces of "A" and "B" as a result of the above electrochemical experiments. A Topometrix Explorer AFM was employed, with a probe having an integral standard pyramidal tip. A relatively low force constant was used to avoid damage to the "soft" surface film. Figures 16a–g show a series of images obtained for "A" and "B" alloys which had been polished initially to a mirror finish and were treated electrochemically as described above. Surfaces were analyzed in air immediately after their removal from solution.

Figure 16a shows a 1000×1000 nm^2 field of view of a surface of "A" after polishing. Despite the mirror finish, many polishing ridges are apparent at this high magnification. Figures 16b–d show the film structures developing at the corrosion potential on "A" as a function of time. The equivalent images for "B" are shown in Figs. 16e–g. What is most notable is that, even after only a 3 h exposure, a film is clearly detectable on both alloy surfaces, having a thickness sufficient to obscure the original surface scratches. For "A," two features can be distinguished: very small (15 nm) crystallites and larger (80 nm) globular structures. For "B" only globular features are observed, clearly separated by narrow crevices. After 12 h exposures, differences in the appearances of these two films continue to widen. An even distribution of 30–40 nm diameter crystallites has developed on the "A" alloy; in contrast, on "B," the major feature appears fairly similar to the globular structure observed after 3 h. Finally, after a 24 h exposure, the appearance of both films seems changed, and the original surface scratches are clearly evident. On the basis of height measurements made by AFM, as well as from the appearance of the features, there seems to have been a "collapse" in both film structures after 24 h of exposure. However, the film on "A" appears to have retained more of its microcrystallinity than that on "B."

The observed film structures may, in fact, have some correspondence to the measured R_{ct} values at the corrosion potential. Initial increases in these values seem to follow an increase in thickness of the films, as measured by AFM. After longer exposures, R_{ct} then decreases at about the same time as when the film structures are seen by AFM to collapse.

FESEM images taken at a low primary-beam voltage (1 kV) reveal more about the early growth films on "A" and "B." No evidence of the globular structures, seen clearly by AFM after 3 h exposures, can be detected by FESEM. It may be that the combined effects of vacuum and electron beam damage are sufficient to destroy the structure of the fragile film, underlining the importance of the use of nonvacuum techniques in exploring the nature of passive surface films. FESEM images of the surface structure on "A" after a 12 h exposure show some partial correspondence to the film seen by AFM in Fig. 16c. The FESEM image of this structure, shown in Fig. 17, also reveals 30 nm diameter crystallites, but they are not the same as the continuous structures seen in Fig. 16c. Thus, most of the crystallites must have been volatilized during the FESEM analysis.

What is the thickness of these oxide films which produce R_{ct} values as high as 10^6 Ω-cm^{-2}? The best estimate seems possible for the even and integral layer of crystallites on "A," seen in Fig. 16c. Assuming that only a single layer of these is present, the layer thickness would then be ca. 30 nm.

6.3.3. Elemental and chemical surface compositions of oxides formed at pH 10

Determination of the elemental composition of these films can now be considered, as well as consideration about their chemistries. Such analyses inevitably require that the sample be placed in a vacuum and be exposed to electron or photon radiation. On the basis of the above microscopic experiments, it is to be expected that the oxide will undergo some degree of decomposition. If volatile components such as water are the only ones removed in this process, then one

(a)

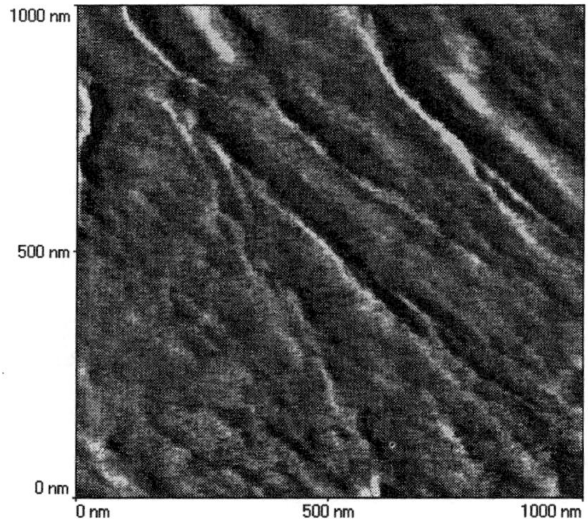

Figure 16. AFM images of "A" and "B" surfaces before and after a series of exposures to pH 10 solutions at 25°C at the corrosion potential: (a) alloy "A" surface before exposure, (b)–(d) "A" following exposure to the above solution for periods of 3, 12, and 24 h, and (e)–(g) "B" following exposure to the solution for periods of 3, 12, and 24 h.

can hope that the relative distributions of heavier elements are not disturbed. XPS is particularly useful in determining the compositions of the outermost surfaces of such films, because it appears to cause less damage to the oxygen component of the film than do electron beam techniques. The XPS system used for this investigation was an SSL SSX-100 Spectrometer equipped with an excitation source of monochromatized Al K_α radiation. No charge compensation was required to produce narrow and consistent photoelectron lines for all major surface species.

XPS-based surface compositions are shown in Table 4 for the films grown on "A" and "B" alloys during the ambient temperature pH 10 electrochemical experiments detailed above and reported in Table 3. Quantification was carried out using the higher KE Cu and Ni 3*p* photoelectron lines, the line energies corresponding to an analysis depth of ca. 3 nm, which suppresses somewhat the attenuating effects of outer surface carbon contamination. Photoelectron intensities were converted to elemental compositions using Scofield cross sections (see Chapters 4 and 5). In presenting XPS quantitative data in simple tabular form, the (usually erroneous) assumption is made that the surface layer analyzed is homogeneous. Each XPS analysis refers to a 600 μm diameter region of the surface.

(b)

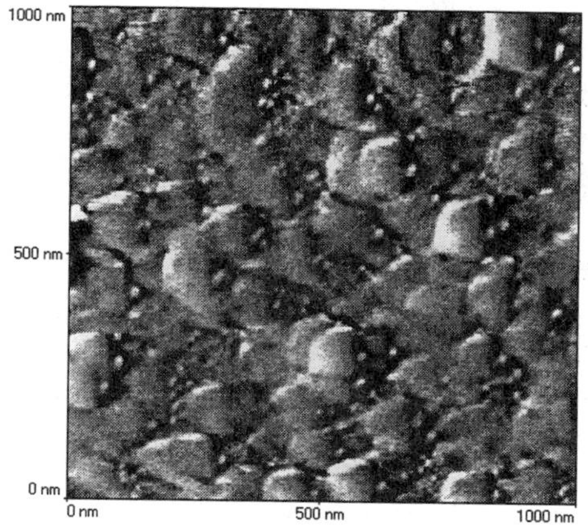
1000 nm
500 nm
0 nm
0 nm
500 nm
1000 nm

(c)

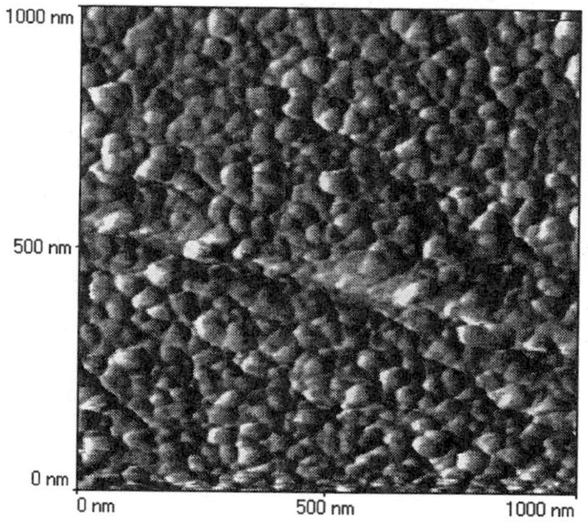
1000 nm
500 nm
0 nm
0 nm
500 nm
1000 nm

(d)

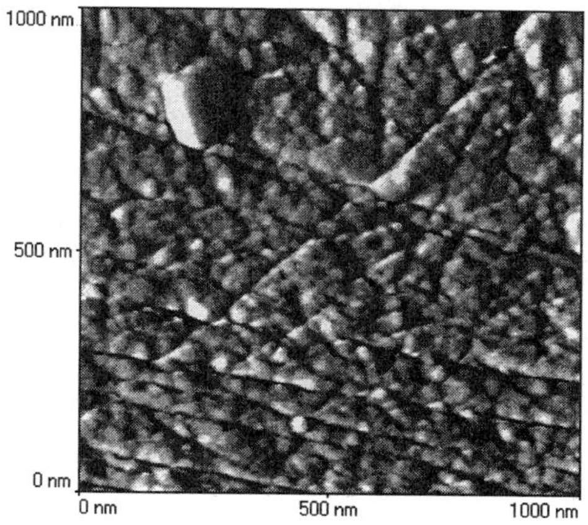
1000 nm
500 nm
0 nm
0 nm
500 nm
1000 nm

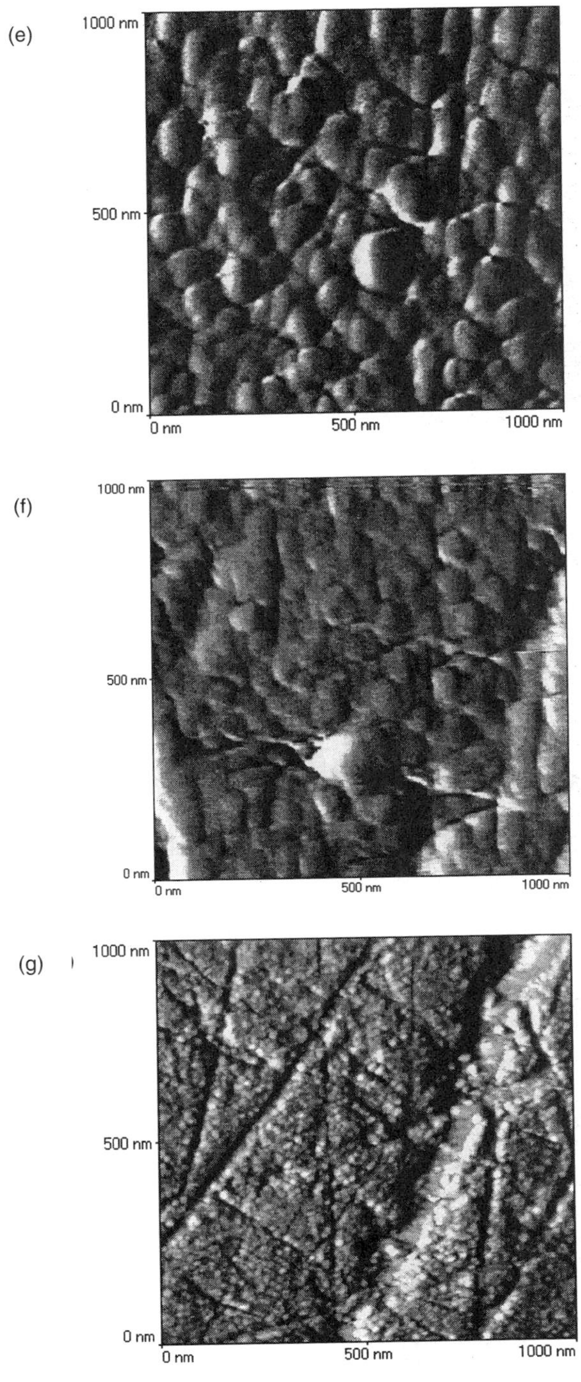

Figure 16. *(continued)*

(a)

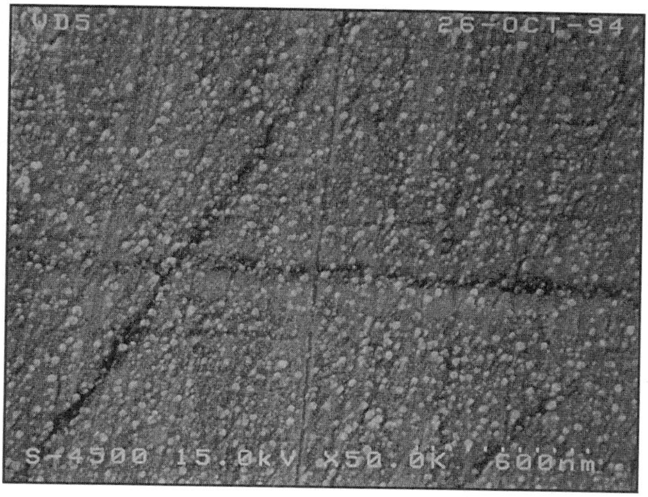

(b)

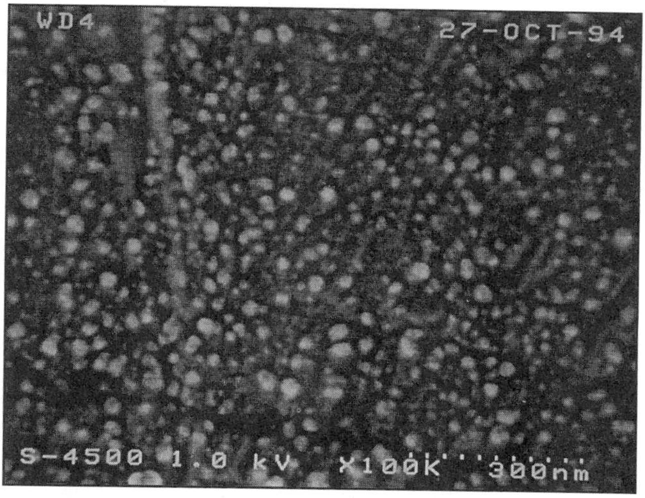

Figure 17. FESEM image of a surface of "A" after a 12 h exposure to pH 10 solution at 25°C at the corrosion potential. The crystallite structure has been altered by vacuum exposure, by comparison with the AFM image in Fig. 16 which was recorded in air.

Table 4. XPS Surface Compositions of Oxide Films from Ambient Temperature pH 10 Exposures at the Corrosion Potential

Exposure time (h)	Alloy "A"				Alloy "B"			
	At.% O	At.% Ni	At.% Cu	Ni:Cu	At.% O	At.% Ni	At.% Cu	Ni:Cu
3	48	14	39	1:3	39	10	51	1:5
3[a]	40	6.6	53	1:8	40	9	51	1:5
7	66	20	15	4:3	52	24	23	1:1
12	44	9	47	1:5	46	12	42	2:7
24	48	13	38	1:3	50	10	40	1:4

[a]Duplicate.

With the exception of the specimens exposed for 7 h, all surfaces have a preponderance of copper. There is no apparent difference in the surface compositions of "A" and "B" specimens, nor is there any evidence of an evolution in surface composition with exposure time. The high Cu content of the films contrasts with the high Ni content of the surfaces of Monel exposed to boiler conditions. At room temperature, the solubility of copper is so low that most of it precipitates back onto the Monel surface; on the other hand, at boiler temperatures most of it remains in solution. High-resolution XPS scans of the Cu $2p_{3/2}$ and Ni $2p_{3/2}$ line shapes are given in Figs. 18a and b for a "B" alloy surface after a 3 h exposure. From the absence of the well-known shakeup structure associated with Cu^{2+} compounds [8] and from the presence of a small satellite peak 14.0 eV above the main line, the Cu can be identified as Cu_2O. The Ni $2p_{3/2}$ spectrum shows a peak representing metallic nickel from the alloy substrate, as well as a peak representing $Ni(OH)_2$. This observation would infer an oxide thickness which is less than the sampled depth; on the basis of AFM results, the contrary is known to be the case. Thus the surface analyzed by XPS has clearly lost some of its thickness due to processes such as vacuum-induced desiccation. Despite this, the oxide compositions in Table 4 and the $Ni(OH)_2$ and Cu_2O species identified in Fig. 18 can be reasonably well reconciled, with some additional oxygen assigned to strongly bound hydrates. The presence of a Cu^{1+} oxide would be expected at the corrosion potential on the basis of the predictions of the Pourbaix diagram for copper. $Ni(OH)_2$ is the expected product from nickel precipitation at this pH.

Attempts to use scanning AES for point analyses of these films were unsuccessful. The oxygen concentration of the corrosion product was much lower than that found by XPS, probably the result of electron-beam-induced decomposition.

SIMS studies of these same films, however, did add useful information. The film thicknesses were too low to be able to use a high-current Cs primary beam for depth profiling. An Ar beam was used instead at low currents, which would not have introduced many chemical perturbations into the film. The results of such profiling are shown in Figs. 19a–c for samples of alloy "A" after exposures in pH 10 solution at the corrosion potential for periods of 3, 12, and 24 h. The intensities of the oxide secondary ions NiO^- and CuO^- are shown as functions of equivalent sputter time. The depths profiled in these instances were so shallow that it was impossible to gauge them by profilometry; thus only the product of sputter time and current density is given on the abscissa. As before, the RSFs for the two ions are considered to be approximately equal.

The SIMS profiles in Fig. 19 show that, after a 3 h exposure, the depth distributions of the Cu and Ni oxides in the film seem to be the same. With increased exposure, however, there is a clear separation of them, with the region of Cu oxide located closer to the outer surface. Also, the relative concentration of Cu oxide in the film seems to increase with length of exposure. This was not

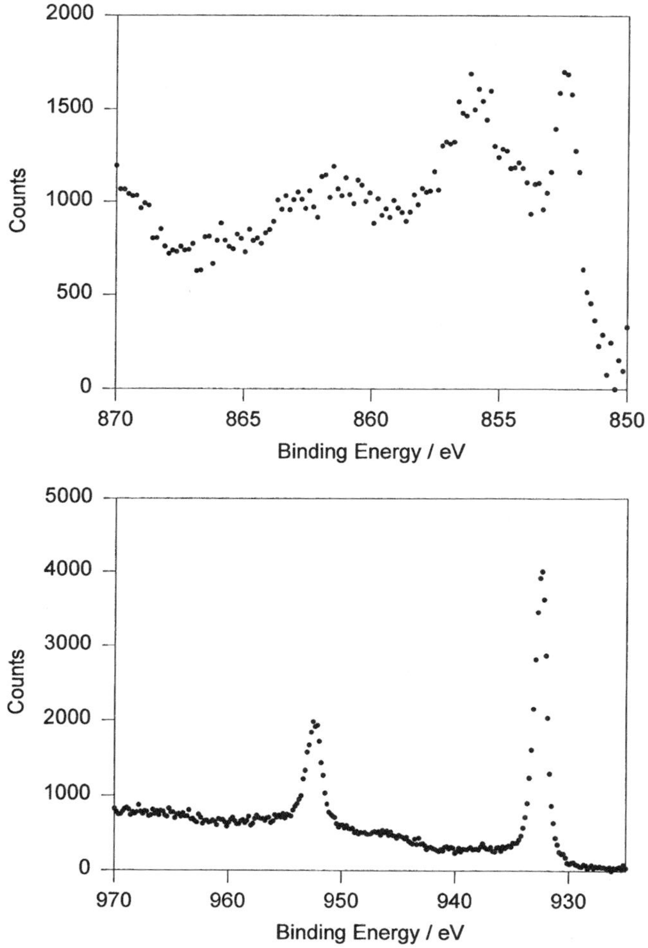

Figure 18. High-resolution XPS spectra of (a) Cu $2p_{3/2}$ and (b) Ni $2p_{3/2}$ lines from a "B" alloy surface after a 3-h exposure at 25°C in pH 10 solution.

evident in the XPS spectra, which measured film compositions only in the outer surface. No significant differences could be found between the SIMS profiles on "A" and "B" films.

6.3.4. Electrochemical and microscopy studies of alloys exposed to pH 1

Studies equivalent to those described above were also carried out on alloy samples exposed to contrived corrosion in pH 1 solutions produced by a mixture of nitric and acetic acids. Potentiodynamic polarization studies revealed much

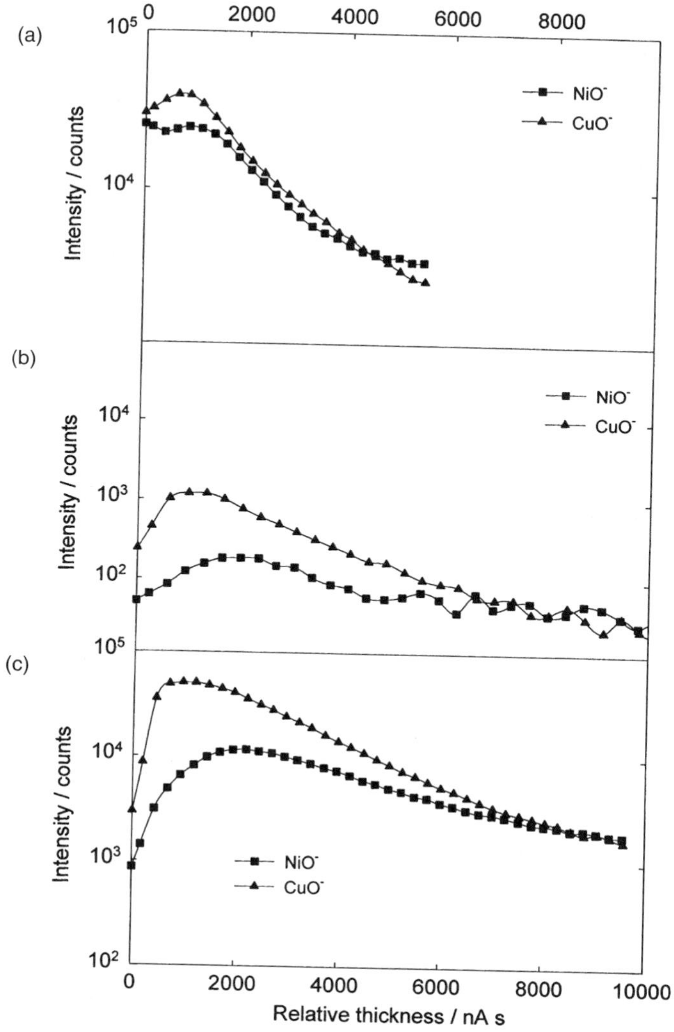

Figure 19. SIMS depth profiles of alloy "A" after exposures to a pH 10 solution at 25°C for (a) 3, (b) 12, and (c) 24 h.

higher passive current densities, compared to those found in pH 10 studies. This is normal since the film at pH 10 is much more passivating than that formed by exposure to highly acidic solutions. Values of R_{ct}, obtained by EIS measurements, were also higher. No significant difference in these values was found between "A" and "B" alloys. Both gave typical R_{ct} values of 1.2×10^{-6} Ω cm^{-2} after 3-h exposures.

In an effort to understand this lack of distinctive behavior of the alloys at acid pH, some additional sample treatment procedures were introduced. Samples of "A" and "B" alloys were heat-treated at 900°C, and quenched in cold water. One heat-treated sample for each alloy was then annealed at 500°C for 100 h. These treatments were intended to determine if the two alloys exhibited different corrosion behaviors as a result of a heat-treatment procedure that would alter the distribution of impurities at the grain boundaries. The original high-temperature heat treatment would be expected to cause sensitizing impurities (such as B) to diffuse away from the boundaries; an anneal at a lower temperature, on the other hand, would result in impurities drifting back to the boundaries if there were no structure in the bulk alloy capable of trapping the impurities. The latter condition is believed to be relevant to alloy "B," where the phase composition (see Table 2) is close enough to the phase boundary to affect the solubilities of many impurities.

All four heat-treated alloys were subjected to the same EIS study in pH 1 solution. The changes in R_{ct} values were followed for a lengthy period. Figure 20 shows the evolution of these values as functions of time. Heat treatment of both alloys results in an improvement in resistance to acid corrosion, presumably through the growth of a more stable oxide layer over the grain boundaries. The oxide is more stable because of the lower chemical activity at the boundary occasioned by a lack of chemically sensitizing elements in the boundaries themselves. However, when "B" is annealed, the R_{ct} values decrease dramatically, while those for "A" remain unchanged. These observations confirm that the bulk structure of

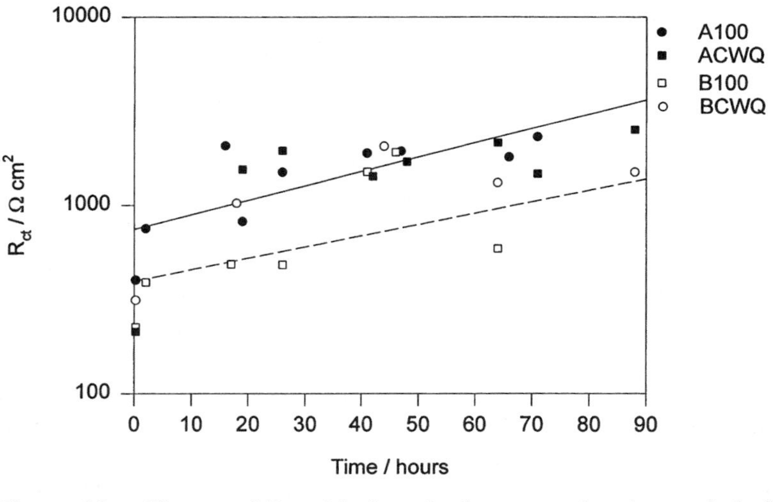

Figure 20. Change of R_{ct} with time for heat-treated and annealed alloys exposed to pH 1 solution at 25°C.

"B" is incapable of retaining impurities away from the grain boundaries if it is maintained at a temperature close to boiler operating conditions.

Although this heat treatment test reveals a dramatic difference in the corrosion behavior of the two alloys, one may still wonder why no significant difference was apparent in the EIS studies of the untreated alloys. This may be due to the significantly higher currents passing through a film formed under acid conditions. Acidic corrosion is usually localized at grain boundaries and a few other microscopic features. A significant difference in grain boundary chemistry is thus necessary to alter the R_{ct} values under acid conditions. Under alkaline conditions, in contrast, the values are affected by the condition of the oxide surface covering the entire grain.

The effect of the acid exposure on the surfaces of the two annealed alloys can be compared in the SEM micrographs in Figs. 21a and b. The extent of localized grain boundary attack on the annealed "B" surface is striking; on "A," however, grain boundary attack is considerably reduced.

XPS analysis of these surfaces provided little additional information. All surfaces had a similar 1:1 Ni:Cu compositional ratio. The real differences would be likely to be found in the grain compositions, such as were observed in Fig. 10. Unfortunately, no similar measurements were made of the heat-treated alloys.

7. CONCLUSIONS

What have we learned about the differences in corrosion behavior of these two Monel specimens? It appears from the contrived experiments that the structural differences in these two alloys actually provoke different corrosion behaviors in both acidic and basic solutions. The acidic case is clearer; localized grain boundary attack is observed under conditions in which impurities in the defective "B" alloy are free to migrate to those boundaries. However, in a basic solution, there is also evidence that the oxide film structures on "A" and "B" are somewhat different and that this difference influences corrosion properties. The difference is apparent in their appearance and in their response to the passage of an electrical current. Grain boundary chemistry is the major difference between the alloys, and it is indeed possible that the boundaries could provoke different galvanic responses on each of the two surfaces.

Under ambient conditions in basic solution, the film that forms comprises mostly of Cu oxides that result from solution precipitation of the more soluble Cu alloy component. Whatever Ni oxides are formed, do so beneath the Cu oxides and may, in fact, result from solid-state oxidation of the undissolved Ni component of the alloy. It is important to note that this film has the capacity to passivate the surface.

Under boiler conditions, in basic solution, it is the Ni solubility which governs the passivity of the surface; whatever Cu is released to solution from the surface remains in solution. The resultant film can be passivating at boiler tem-

(A)

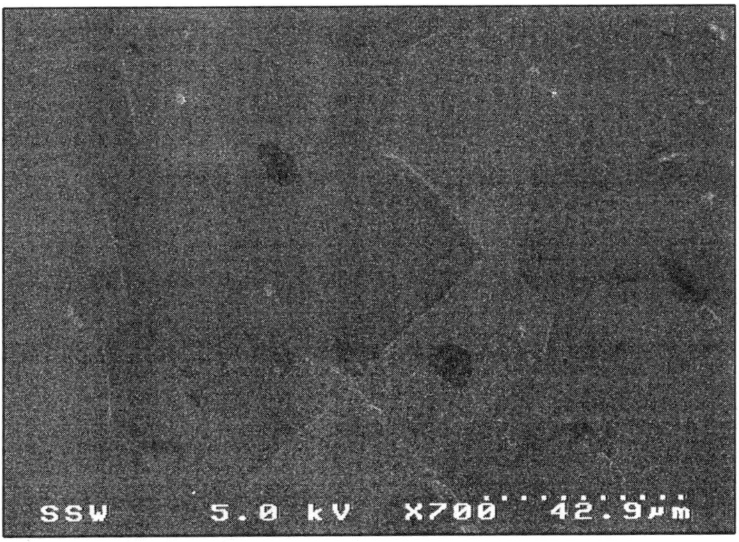

(B)

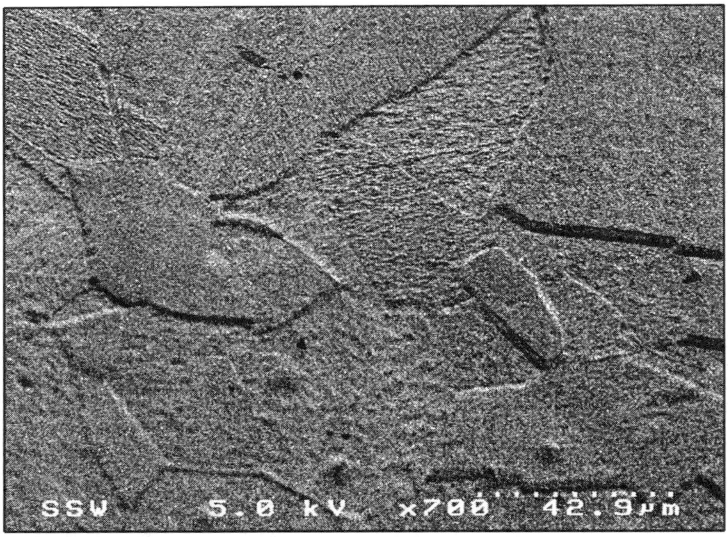

Figure 21. SEM micrographs comparing the corrosion behavior in acid solution of "A" and "B" alloys which had been heat-treated and annealed. Exposure occurred in pH 1 solution at 25°C for 7 h.

perature. There would seem little likelihood that the two passivating films have much in common. However, it appears that each is influenced in some manner by grain boundary impurities. Notwithstanding this common trait, it would have been better to have chosen contrived conditions in which the corrosion mechanisms in force were similar. For example, it might have been preferable to have carried out EIS tests at ca. 200°C. Such tests are necessarily more complicated than those carried out under ambient conditions.

What lessons may be drawn about the appropriate use of techniques in approaching this corrosion problem? First, it must be realized that the measurements made by an electrochemical technique, such as EIS, are much more likely to have direct relevance to a corrosion process than any surface analytical technique. Corrosion itself is an electrochemical process. Thus EIS measurements made in the contrived ambient experiments have more credibility than surface analysis. However, individual surface techniques were invaluable in uncovering the mechanisms associated with the chemical processes. In particular, AFM appears to have an important future in the visualization of films whose fragility is just beginning to be understood. The use of SIMS for chemical characterization of corrosion films also seems to be underappreciated.

Where experiments are conducted which simulate process conditions, the effectiveness of electrochemical testing is more controversial. Postreaction EIS studies of alloys exposed to autoclave corrosion were not able to illuminate differences in the performance of the alloys. By contrast, imaging SIMS provided useful insight into the structural and compositional differences, leading to a characterization that might possibly be used to qualify other alloy specimens. Alternatively, one could consider the surface techniques to be at their best when performing a postfailure analysis.

When one is faced, however, with choosing the most appropriate test of the corrosion worthiness of a future Monel alloy specimen, there is no doubt that the contrived exposure in acid conditions would be the most reliable. All these studies have identified the grain boundary impurities as the root of the corrosion problem in the alloy. The experiments in acid are specific to the boundaries and illustrate the merits of an electrochemical approach.

The complementarity of electrochemical and surface analysis has been described. In principle, both should be accessed regularly in assessing any new corrosion problem; in practice, electrochemists and surface scientists are often ignorant of the power of their respective techniques. It is hoped that new microscopic and spectroscopic methods will provide added inducement by their ability to examine the fragile water-filled films that are believed to be so important to passivation. This could open up a greater dialogue than has been possible in the past.

The example offered here as an illustration of the coordinated approach to the understanding of a corrosion process, is, of course, much more extended than is normally practical for 90% of industrial corrosion problems. In particular, the lengthy electrochemical studies undertaken here would not be feasible under pressing circumstances. Nevertheless, there will be many more corrosion problems in the future that will be tackled with a complementary battery of techniques that includes some type of electrochemical assessment.

ACKNOWLEDGMENTS

The authors gratefully acknowledge the comments of T. L. Walzak on EIS measurements and the technical assistance of M. Lee Wagter.

REFERENCES

1. L. L. Shrier, Basic concepts of corrosion, in *Corrosion-Metal/Environment Reactions* (L.L. Shrier, R. A. Jarman and G.T. Burstein, eds.), Butterworth, Oxford, 1994.
2. R. W. Revie and H. H. Uhlig, *Corros. Sci. 12*, 669 (1972).
3. N. S. McIntyre, D. Zetaruk and E. V. Murphy, *Surf. Interface Anal. 1*, 105 (1979).
4. U. R. Evans, *The Corrosion and Oxidation of Metals*, Arnold, London, 1960, p. 12.
5. H. H. Uhlig, *Corrosion and Corrosion Control*, Wiley, New York, 1971.
6. L. L. Shrier, Localised corrosion, in *Corrosion-Metal/Environment Reactions* (L. L. Shrier, R. A. Jasman and G. T. Burstein, ed), Butterworth, Oxford 1994.
7. S. Hofman, *Proc Electrochem. Soc. 91*, 366 (1991).
8. N. S. McIntyre and T. C. Chan, Electron spectroscopies in corrosion science, in *Practical Surface Analysis VI*, 2nd ed. (D. Briggs and M. P. Seah, eds.), Wiley, Chichester, 1990, pp. 485–529.
9. N. S. McIntyre, R. D. Davidson and J. R. Mycroft, *Surf. Interface Anal. 24*, 591 (1996).
10. C. Chen, S. J. Splinter, T. Do and N. S. McIntyre, *Surf. Sci. Lett., 382*, L652 (1997).
11. N. S. McIntrye, R. D. Davidson, T. L. Walzak, A. M. Brennenstühl, F. Gonzalez and S. Corazza, *Corros. Sci 37*, 1059 (1995).
12. F. Mansfeld and W. J. Lorenz, Electrochemical impedance spectroscopy—applications in corrosion science and technology, in *Techniques for Characterization of Electrodes and Electrochemical Processes* (R. Varma and J. R. Selman, eds.), Wiley, New York, 1991, p. 581.
13. J. R. Scully, D. C. Silverman and M. W. Kendig (eds.), *Electrochemical Impedance: Analysis and Interpretation*, ASTM STP 1188, American Society for Testing and Materials, Philadelphia, PA, 1993.
14. S. F. Lu, G. R. Mount, N. S. McIntyre and A. Fenster, *Surf. Interface Anal. 21*, 177 (1993).
15. N. S. McIntyre, R. D. Davidson, C. G. Weisener, K. F. Taylor, F. Gonzalez, E. M. Rasile, and A. M. Brennenstühl, *Surf. Interface Anal. 18*, 601 (1992).

16

Problem-Solving Methods in Tribology with Surface-Specific Techniques

CHRISTOPHE DONNET Laboratoire de Tribologie et Dynamique des Systèmes, CNRS UMR 5513, Écully, France

1. TRIBOLOGY AND SURFACE-RELATED PHENOMENA

Tribology is involved in both industrial and basic scientific research. In industry, tribological considerations arise in all requirements and applications relating to the control of friction and wear in mechanical components. This domain includes extension of the lifetime of components, increase in their mechanical and energy efficiency, and improvement in their safety in, for example, the transportation, chemical, textile, food, plastic, biomedical and space industries. The widespread application of tribology in industry is evident in a wide variety of situations. Historically, the optimization of contact geometries, and the choice of bulk materials and lubricants, have been the main technological thrusts in the design of specific remedies for industrial applications.

From a basic research point of view, the field of tribology involves the contact of surfaces in relative motion, leading to the introduction of surface science concepts, which are coupled with those from other more traditional scientific fields, such as materials science (including metallurgy) and continuum and fluid mechanics. Even if tribology may be considered a multidisciplinary field, embracing mechanics, physics and chemistry, it can also be described as a scientific discipline in its own right, with its own methodologies. When two solid surfaces are in dynamic contact, the material phases between them may exist in either gas or liquid forms or as a solid that is thick enough for the interface media to be described adequately in terms of traditional concepts and to be amenable to the experimental techniques of bulk materials science. However, numerous situations occur in which the interface responsible for overall tribological properties can be so thin (as little as a monomolecular layer) that extreme surface analysis methods must be used to determine its nature and to quantify its intrinsic properties.

For readers not familiar with this field, the basic aspects concerning the lubrication of two contacting surfaces should be summarized briefly, in order that the surface analysis requirements and methodologies carried out in tribology can be understood [1,2]. When two surfaces undergo sliding or rolling under load, energy dissipation leads to friction in which the production of wear particles and the plastic deformation of the contacting surfaces are the main causes of serious material loss. In the case of sliding friction, the three basic phenomena of concern are adhesion, plowing (due to wear particles or to asperities on the harder counterface) and asperity deformation, as depicted in Fig. 1. Lubrication consists of separation of the moving surfaces with an interposing film of solid, liquid or gas, characterized by low shear resistance and minimization of surface damage. Various lubrication modes can be identified [1], of which the principal ones are as follows.

Hydrodynamic lubrication is based on the formation of a thick lubricant film (typically of thickness from 1 to 100 μm) which inhibits contact between the moving surfaces. This lubrication mode is governed by the bulk physical properties of the lubricant, mainly the viscosity, and by the speed of the relative motion.

Elastohydrodynamic lubrication (EHD) occurs when the extent of surface deformation is comparable with the lubricant film thickness, so that a heavy load causes local elastic deformation of the contacting surfaces, but without any significant asperity interaction. Taking into account the low lubricant thickness (typically from 0.01 to 10 μm) and the high contact pressures (typically in the GPa range), the lubricant properties differ from those of a traditional bulk liquid, since a strong viscosity increase may occur, when the lubricant behaves more like a solid than a liquid. Moderate temperature rises may occur, thus inducing some thermochemical reactions between the surfaces and the lubricant additives, but no tribochemical phenomena are involved, unlike the next lubrication mode.

Boundary lubrication is probably one of the most complex phenomena studied by tribologists, in view of the numerous experimental parameters which influence the tribological behavior. Considerable asperity deformation is usually observed, since the contacting surfaces move very close to each other. The role

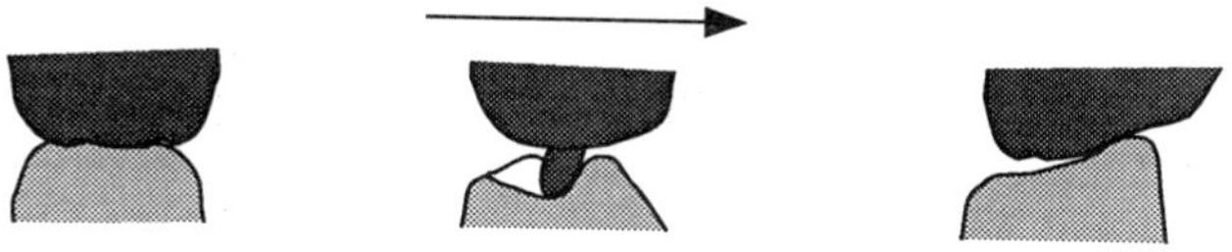

Figure 1. The three basic phenomena in sliding friction: (a) adhesion, (b) plowing and (c) asperity deformation.

of the thin-film lubricant (thicknesses from 1 to 100 nm) adsorbed on the solid surfaces is that of inhibition of asperity welding, thus reducing wear, and of lowering of the friction coefficient, as a result of the formation of a friction-induced tribochemical film with a shear strength lower than that of the bulk solid material. Generally speaking, the boundary lubricant modifies the solid-solid interactions, due to various solid-lubricant reactions (physisorption, chemisorption, chemical reactions) which depend strongly on the environmental conditions and on the lubricant composition. As pointed out in an early publication by Bowden and Tabor [2], a frictional interface under boundary conditions can be considered as a specific chemical reactor where the combined effects of pressure and temperature, and of the fresh surfaces created by the friction process in the presence of gaseous, liquid or solid phases acting as lubricants or contaminants, are unique to the particular system. Lubricants contain additives which act as packages with several functions, including improvement of the pressure-velocity (rheological) properties of the lubricant, control of friction, and inhibition of oxidation of contacting surfaces and of oil at the highest operating temperatures. Understanding of the mechanisms of friction and wear requires a precise identification and quantification of the tribochemical phenomena, which in turn allows the correct selection of materials, surface treatments, surface morphology, lubricants with additives, and operating conditions, for a given engineering device. The complex phenomena occurring in boundary lubrication go some way to explain why no model for tribochemical reaction exists, thus emphasizing the crucial role of surface and interface analysis in the investigation of tribosystems and in the understanding of their behavior. Nevertheless, there are two nonexclusive schools of physicochemical thought. The first states that the tribochemical process is nonequilibrium, and that materials in nonequilibrium condition are generated in the contact [3]. This might explain the amorphous states of most tribochemical films formed in boundary lubrication [4]. The second school asserts that the effects of friction (increase of temperature, formation of fresh metal surfaces, mechanical deformation) only enhance, accelerate and stimulate reactions that would occur in the same but static interaction without friction [5]. With this latter assumption, it is believed that the final products are thermochemically stable and that the film composition can be predicted from the thermochemical considerations. Thus one of the major motivations in boundary lubrication is the identification of the chemical products and an understanding of the chemical reactions, leading to formation of the beneficial tribochemically induced lubricant films [6].

Mixed lubrication occurs when the contact behavior is governed by a mixture of EHD and boundary lubrication. Asperity contact and deformation may occur, even when the surfaces are separated by a lubricant film from 0.01 to 1 μm thick.

Solid-film lubrication, more recently developed, is used when conventional lubricants can no longer operate, as in extreme conditions, such as at very low (cryogenic) or very high temperatures (>300°C) and in vacuum. The principle consists of the interposition of a solid film (of thickness usually less than 10 μm) between the two sliding or rolling surfaces, prior to friction. Shear takes place within and across the film or between the film and the sliding surfaces. The film is applied directly by conventional coating deposition techniques, or it can also be formed by reaction between the triboactivated sliding surface and the environment [7,8].

From this well-established and generic classification, it appears that there is a strong need for surface analysis to investigate tribological contact and to understand interfacial and surface interactions, particularly in the boundary, mixed and solid lubrication modes. Tribologists agree that recording of the friction force and measurement of wear rates are necessary but not sufficient to investigate such complex tribosystems. Since the pioneering work of Buckley in the 1970s [9], the combination of tribology and surface science has allowed progress in the development of low-friction and wear-resistant contacts, using optimized liquid or solid lubricants, in various mechanisms and devices. What is now needed in particular is the identification and classification of the surface analytical methodology in the examination of tribological contacts, and these will be described in the next section.

2. SURFACE ANALYSIS REQUIREMENTS FOR TRIBOLOGY

. . . If an understanding of the nature of surfaces calls for such sophisticated physical, chemical, mathematical, materials and engineering studies in both macro- and molecular terms, how much more challenging is the subject of . . . interacting surfaces in relative motion?

D. Dowson, History of Tribology, Longman 1979, p. 3

2.1. Overview

This section discusses the surface and interface analysis methodology needed for the investigation of a tribological contact. It will not describe the potentialities, performances and limitations of the various surface analysis techniques which have been applied in this field, since that description can be found in Chapters 4 and 5. Nor is it intended to be exhaustive with respect to the numerous studies that have been reported. However, it will emphasize a logical procedure leading to an integrated top-down approach, since surface analysis will be considered as a means to a complete understanding of a tribological contact and to the elucidation of triboinduced surface modifications. Several examples, entitled generic studies, will then be used to illustrate, in Section 3, the different procedures that couple a tribological experiment with relevant surface investigation of the contacting bodies. Each generic study has been selected on the basis of how various analysis techniques have been brought together to extract crucial information from a tribological point of view. A more exhaustive and detailed presentation of surface investigations in tribology is given in the excellent book *Surface Diagnostics in Tribology*, edited by Miyoshi and Chung [10], in which the power of surface analytical techniques in the probing of complex solid surfaces and lubricants, and in the understanding of their interactions in lubricant systems, is emphasized.

Basically, surface analysis is applied in tribology in two ways:

1. *Post mortem* examination of failures in real working systems. This helps to identify the origins of the failures, with the possibility of recommendation for remedial or preventative action.
2. Systematic surface analytical studies on model or ideal systems. These are performed in order to understand the mechanisms and underlying physics and chemistry of various phenomena, such as the reactions promoted by additives, the effects of adsorption on friction and adhesion, and the degradation of a thin lubricant coating in a severe environment.

The first of these requires a classical surface analysis methodology which is not necessarily specific to the field of tribology. The second is more specific and requires a detailed methodology, based on the following three basic criteria that must be considered when a tribologist has to follow a coherent methodology in the investigation of surface-related phenomena in a tribological contact:

1. *Dimensional criterion.* In tribological contact processes both lateral and depth dimensions are important, the former because of the size and shape of wear debris and the latter because of the presence of thin films. It is therefore vital to match the characteristics of a technique to the dimensions of the analysis.
2. *Time-scale criterion.* In most tribological tests, friction is measured continuously, while, on the other hand, surface composition and topography are determined only at the completion of the friction test. Moreover the surface phenomena cannot only vary continuously during the friction experiment, but can also be altered by air exposure between the end of the tribological

test and analysis, due to environmental effects such as oxidation and hydration. Careful thought must thus be given to the chosen surface analysis configuration in order to compensate for these experimental difficulties, as far as possible.

3. *Information criterion.* The surface analytical information necessary for elucidation of the tribological mechanisms in terms of physicochemical, structural, morphological and rheological modifications of the contacting surfaces must be identified precisely.

These three criteria will now be examined in detail. Sections 3.1–10 will describe 10 generic studies. They have been chosen to illustrate the complementarity of surface and interface investigations, in relation to the above three criteria, in the implementation of the top-down approach to the field of analytical tribology.

2.2. Dimensional Criterion

Surface analysis applied to tribology requires the dimensions of the various surface elements and related phenomena to be borne in mind, when two surfaces are probed which have undertaken relative motion under contact pressure. This consideration has to be coupled with others concerning the nature of the probed information, as indicated in Section 2.4; chemical information, such as the nature of the oxidation states of the surface elements, is present in both a thin film of a few molecular monolayers (in boundary lubrication) and in a film of μm thickness (in thin-film lubrication by a solid lubricant coating). Here the depth dimension is of great importance in selecting the appropriate technique and analysis procedure; XPS on its own, with a depth resolution of less than 5 nm, is suitable in the first case, whereas XPS with sequential sputtering of the coating is required to analyze the entire thickness of the solid lubricant film, with the usual caveat that damage induced by ion beam sputtering must be minimized.

Figure 2 (from Ref. 7, p. 37) shows an overview of the size range of surface elements and surface-related phenomena in tribology. A continuous size distribution is observed of surface-related phenomena characteristic of tribological behavior, since the pertinent data cover approximately six orders of magnitude, ranging from 10^{-9} m to the scale of design tolerances (10^{-6}–10^{-3} m), and then to the size of engineering components (from 10^{-3} m). Monolayer films influencing friction and wear behavior in boundary lubrication are in the 10^{-9}–10^{-8} m range. Natural oxide films, multilayer films and typical sliding wear particles range in size from 10^{-9} to 10^{-7} m. As stated in Section 1, elastohydrodynamic films (EHD) range in thickness from about 10^{-8} to 10^{-5} m. This size range includes the typical sizes of rolling wear particles and microjunctions. Further up the scale, hydrodynamic films cover a size range from 10^{-6} to

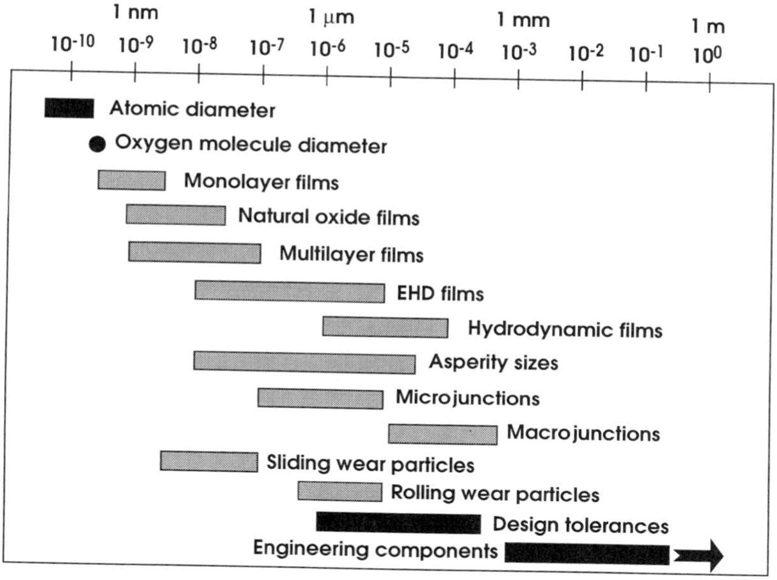

Figure 2. Magnitudes of the relevant surface related phenomena in tribology. (From Ref. 7, p. 37, with permission.)

10^{-4} m, whereas asperity heights cover a wide range from 10^{-8} to 10^{-4} m, depending on the surface geometric finish. Finally macrojunctions observed in catastrophic adhesive wear can reach the 10^{-3} m range, which is of the same order of magnitude as some less restrictive design tolerances.

This dimensional criterion strongly influences the choice of analytical technique for obtaining the required surface information, taking into account also the available spatial resolution. The reader should refer to Chapter 4 for information about the spatial resolution range of the technique(s) identified as being suitable for acquiring the surface information classified in Section 2.4.

2.3. Time-Scale Criterion

This criterion concerns the different chronologies within which the friction experiment and the surface analysis of the contacting surfaces can be arranged. Different configurations, also called modes, can be considered on the basis of Fig. 3. Whatever the configuration, *prefriction analysis* must be performed in order to have complete knowledge of the experimental system prior to the tribotest. Both qualitative and quantitative analyses are required of the surfaces that will be in contact, and of the lubricant phase, as depicted in Section 2.4. This

Ex situ Outside the tribometer	***Post mortem*** After friction	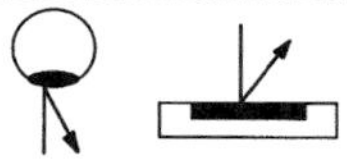	See generic studies : 2-6-7-8-9
In situ Inside the tribometer Outside the contact	***Pre mortem*** During friction	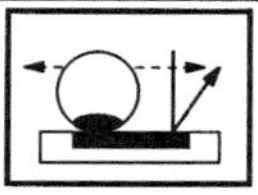	See generic study : 5
	Post mortem After friction		See generic studies : 2-3-4-6-8
In vivo Inside the tribometer Inside the contact	***Pre mortem*** During friction	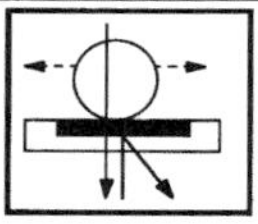	See generic studies : 1-2
	Post mortem After friction	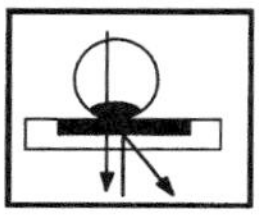	See generic study : 10

Figure 3. Time-scale criteria combining friction experiments and the surface analysis in different modes. Letters in the right-hand column refer to the generic studies in Section 3 which illustrate the use of the corresponding technique.

routine step needs only moderate spatial resolution, since virgin surfaces prior to friction are usually laterally homogeneous. Prefriction analysis is performed either outside the tribometer with traditional surface analysis apparatus, or inside the tribometer if it is equipped with dedicated analytical instrumentation.

The time-scale criterion can be broken down into three main analysis modes for the investigation of a tribocontact. Compared to the analysis carried out prior to friction, the three modes all require spatial resolution consistent with the size of the area affected by friction and wear (which is often less than a few 100 μm^2), assuming that the intrinsic lateral resolution of the technique to be used is adequate. The three modes are the following:

1. *Ex situ analysis*: surface analysis performed outside the tribometer, after the friction test.
2. *In situ analysis*: surface analysis performed inside the tribometer, inside the wear scars but outside the contact. *In situ* analysis can also be performed during (*pre mortem*) or after (*post mortem*) the friction test.
3. *In vivo analysis*: surface analysis performed inside the tribometer, inside the wear scars and inside the contact. *In vivo* analysis can also be performed during (*pre mortem*) or after (*post mortem*) the friction test.

These three modes allow the identification of friction-induced surface modifications, such as the development of transfer films, tribochemical phase formation, or the removal of material and lubricant films by wear. Correlations may be revealed which would help to elucidate the friction and wear mechanisms. There follows a description of these modes, with emphasis on their interests, potentialities, and limitations, as illustrated by the generic studies developed in Section 3, and summarized in Fig. 3.

1. *Ex situ analysis* is performed outside the tribometer and obviously after the friction test, with analytical instruments independent of the tribometer. Direct observation of the surfaces may therefore be coupled with physicochemical and structural investigations performed inside the wear scars of the surfaces, which have been separated from each other at the end of the tribotest. Third-body products, such as wear particles, can be extracted from the contact for independent examination. The main advantage of *ex situ* analysis is the possibility of using the great potentiality of various complementary modern surface analysis instruments, providing sensitive measurements from ever more restricted areas. The first drawback is the inevitability of air exposure between the tribometer and the analytical instrument, thus leading to adsorption and/or reaction of surfaces with air, water vapor or other mineral and organic contaminants, complicating the tribological process under investigation. A second drawback is the usual size limitation of samples for analysis in conventional analytical instruments, which are often unadapted to surface investigation of larger tribotested samples or components. The generic studies described in Sections 3.2 and 3.6 to 3.9 illustrate the surface analysis needs and the use of various complementary *ex situ* methods.

2. *In situ analysis* is performed *inside the tribometer*, which incorporates the surface probe. This mode allows exploration of the friction and adhesion of clean solid surfaces, the effect of adsorbed species or the nature of the lubricant phase, in order to elucidate the triboinduced surface modifications in as direct a manner as possible. Depending on the analytical technique used, the nature of the environment during the tribotest is not necessarily the same as that during analysis; *in situ* Raman spectroscopy can be performed in ambient air, whereas *in situ* XPS and AES require ultrahigh vacuum. This mode can be performed in either of two configurations according to the time-scale criterion.

a. Analysis can be performed *after the friction test* (*in situ post mortem* analysis) by direct observation inside the wear scars. Possible limitations arise from the geometry of the analyzed surfaces, since some techniques (e.g., LEED for crystallinity study) require flat surfaces, thus prohibiting the analysis of transfer films on the spherical balls often used in tribological tests. XPS/AES can be performed on both the plane and the ball, the latter having curvature diameters in the range of several mm or more. Examples are given in the generic studies in Sections 3.2 to 3.4, 3.6 and 3.8.

b. Analysis can also be performed *during the friction test* (*in situ* pre mortem analysis), but is less systematic, since it places greater demands on the coupling between the tribotest and the analysis procedure. Such a configuration requires rotational motion of the contact, with analysis carried out on the rotating disk, inside the wear scars but outside the dynamic contact. Ideally the analysis should be carried out during rotation of the disk counterface, but there are obvious difficulties: (a) the analyzed area changes continuously, with a period related to the rotation speed, and (b) most of the analytical techniques require acquisition times longer than the normal rotation periods. That is probably why such a direct analytical configuration has never been attempted. However, in earlier work it was approximated by a sequential procedure, that is, by stopping the rotation every so often to allow analysis by AES (see the generic study in Section 3.5). In that way a succession of tribotests/AES analyses were performed in order to study continuous friction-induced surface modification.

3. *In vivo analysis* consists in performing spatially resolved surface analysis directly inside the contact. This mode imposes drastic conditions on both the tribometer configuration and the performance of the analytical technique. First, the use of electron or X-ray transparent materials is necessary in order that the probe may reach the interface with only minimal interaction with the bulk contacting bodies. Second, a dedicated tribometer must be compatible with the analytical configuration. As in the case of *in situ* analysis, two time-scale configurations are possible.

a. Analysis can be performed *after the friction stage* (*in vivo post mortem* analysis), thus providing analytical data relating to the contact with minimal perturbation, since the two contacting bodies are not separated; this is illustrated in the generic study in Section 3.10.

b. Analysis can be performed *during the friction stage* (*in vivo pre mortem* analysis): this mode is the dream of all tribologists, since it corresponds to the ultimate goal in recording real-time and spatially resolved information from the friction interface, which can then be correlated with simultaneous friction forces, surface forces or wear rate measurement. Of course, as for the *in situ* pre mortem analysis mode, a balance has to be struck between the dynamic of the friction movement on the one hand, and the spatial resolution/acquisition time of the analysis on the other. More details are given in Section 3.1 where ultrathin thickness measurements of a lubricant film inside a dynamic contact are described. Attempts to record physicochemical and structural data during friction are also mentioned in Section 3.2.

2.4. Information Criterion

There is now an extensive body of accumulated experimental work concerning relevant surface analytical data in tribology [10]. Three categories of information can be listed.

2.4.1. Physicochemical and Structural Information

This heading includes the

Nature of the elements and their chemical states.
Nature of the chemical bonds and orbital hybridization.
Nature of the chemical phases.
Structure, related to the spatial arrangement of atoms, and groups of atoms or molecules. This ranges from long-range order, characteristic of crystalline compounds, and established by conventional diffraction techniques, to friction-induced short-range order (local order) established by more specialized techniques, as detailed hereunder.

The above types of information are of great importance in boundary and solid-film lubrication, since the lubricant phase and the surface composition can change continuously during the friction test, depending on the experimental conditions (contact pressure, velocity, temperature, nature of the surrounding environment). Acquisition of such information is vital for clarification of the relationships existing between the nature of the friction-induced surface and of the interface modifications and the interpretation of the tribological mechanisms. Only specialized techniques can provide the physicochemical and structural information, taking into account the dimensional and time-scale criteria discussed above. Table 1 compares qualitatively the capabilities of the analytical techniques most commonly used in tribology, with references to the generic

Table 1. Capabilities of Some Analytical Techniques in Terms of Physical, Chemical and Structural Data

	Chemical bonding[a]	Oxidation state[a]	Orbital hybridization[a]	Local order[a]	Crystalline structure[a]	See generic studies[b]
XPS	●	●	○	○	○	3.4–3.8
AES	●	●	●	○	○	3.2, 3.4–3.6, 3.8
Raman	●	●	●	○	○	3.10
XAS	●	●	●	●	○	3.2
EELS	●	●	●	●	○	3.2, 3.4, 3.6
IR	●	●	●	○	○	3.9
TEM	○	○	○	●	●	3.2, 3.6, 3.8
XRD	○	○	○	○	●	3.7
CEMS	●	●	●	●	●	3.2

[a]● good; ○ poor
[b]Right-hand column refers to generic studies illustrating in Section 3 the use of the corresponding techniques.

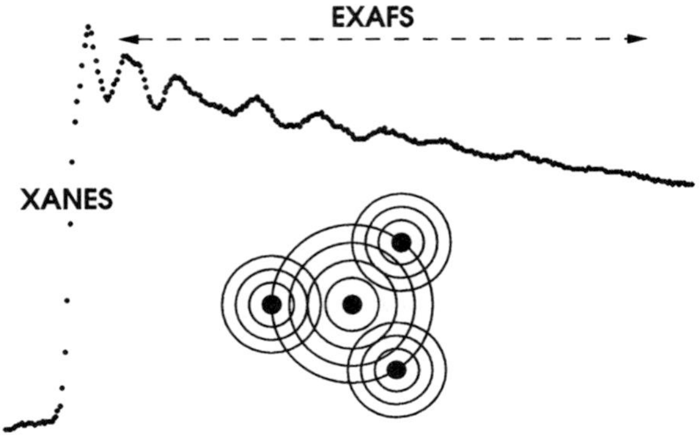

Figure 4. General shape of the ionization K-edge of an element as obtained by X-ray absorption spectroscopy (XAS). XANES gives chemical and electronic information; EXAFS reflects the short-range order in the material. Schematic view of the diffraction of the electron wave.

studies described in Section 3. More detailed information regarding the characteristics and capabilities of each technique can be found in Chapter 4. From the literature [11], it can be seen that spectroscopic techniques such as XPS and AES have been used extensively in the field of tribochemistry, as illustrated in the generic studies 3.2, 3.4–3.6, 3.8 and 3.9. More recently, X-ray absorption spectroscopy (XAS) [12] and electron energy loss spectroscopy (EELS) [13] have revealed unique capabilities in terms of chemical, electronic and crystallographic information on the atomic scale. The chemical bonding, oxidation state and hybridization of selected atoms can be extracted from a careful analysis of the near-edge fine structure of the corresponding absorption (or ionization) edge (Fig. 4). This is called X-ray absorption near-edge spectroscopy (XANES) for X-rays and electron (energy) loss near-edge spectroscopy (ELNES) for electrons.

Local order can be studied by processing the extended fine structure found above an absorption edge (Fig. 4), called extended X-ray absorption fine structure (EXAFS) for X-rays and extended electron loss fine structure (EXELFS) for electrons. The EXAFS arises from the oscillatory variation of X-ray absorption as a function of photon energy beyond the absorption edge. When the X-ray photon energy is tuned to the binding energy of the core level of an atom of a given material, an abrupt increase in absorption (called the absorption edge) occurs. For isolated atoms (gas, for example), the absorption coefficient decreases monotonically as a function of the energy beyond the edge. For atoms in a molecule or in a condensed phase (such as a lubricant), the variation of the absorption coefficient at energies ranging from 40 to 1000 eV above the absorption edge displays a complex fine structure (Fig. 4). A single-scattering short-range order theory is adequate to explain the structure beyond 50 eV from the threshold (EXAFS region). The technique is particularly successful in deriving the precise local order around selected atoms; the result of the EXAFS processing is the so-called radial distribution function (RDF), which is a function of the local environment of the selected atoms or, in other words, the number of neighboring atoms versus the radial distance. In a crystalline structure, two or three coordination spheres are typically observed, whereas in an amorphous structure only the first atomic shell is usually detected in the RDF.

XAS requires synchrotron radiation and a relatively large amount of material but no vacuum condition. On the other hand, EELS can be performed directly using an electron spectrometer fitted to a scanning transmission electron microscope (STEM). Here, the main advantage is the high spatial resolution attainable. (The incident electron beam can be as small as 1 nm in diameter.) EELS can also be coupled with conventional transmission electron microscope (TEM) facilities and particularly high-resolution transmission electron microscopy (HRTEM) and energy dispersive X-ray spectroscopy (EDS).

A general strategy for applying these analytical techniques to a tribological test has been developed [14], as depicted in the generic study in Section 3.2. Bulk lubricants can be analyzed using a liquid cell, at different temperatures, and in various environments, since no vacuum is required. Local analysis of wear debris collected at the end of the test (*ex situ*) is always possible and can be carried out by XAS in the transmission mode, preparation being very easy in that case (Fig. 5a). Alternatively, when high spatial resolution is needed, selected wear fragments can be analyzed in the STEM. For this purpose, they are deposited on a holey carbon film supported by a copper grid.

The analysis of surfaces is also possible by either X-rays or electrons (Fig. 5b). If photoelectrons are used (as in XPS) their mean free paths are very small (approximately 2 nm) and the outer surface only is probed, thus requiring high-vacuum conditions. The techniques are referred to as SEXAFS (surface EXAFS) and SEELS (surface EELS). If high-energy secondary electrons or Auger electrons are detected instead, then the corresponding mean free paths indicate that the depth being analyzed is a few tenths of a nanometer, which is very suitable for tribochemical films. Finally, if X-rays are detected from fluorescence XRF in the material, the information comes from much deeper regions.

The analysis of friction interfaces in motion is feasible (*in vivo* configuration, see Section 2.2) due to the high transparency of the matter to X-rays (Fig. 5c). Martin et al. have developed a tribometer which can be coupled to both X-ray absorption spectroscopy (XAS) and Raman spectroscopy (Fig. 6) [14], illustrating the desirable coupling of the time-scale criterion mentioned above with the information from the analysis. For this purpose, X-ray-transparent sup-

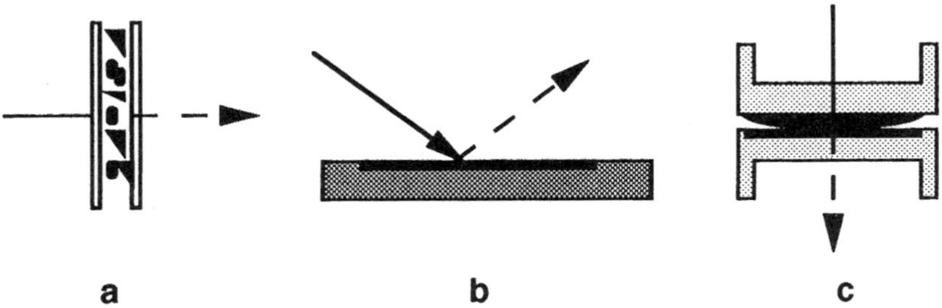

Figure 5. X-ray absorption spectroscopy as applied to tribology: (a) *ex situ* wear particle analysis in the transmission mode, (b) *in situ* surface analysis (total electron yield, X-ray fluorescence or photoelectrons), (c) *in vivo* frictional interface analysis in the transmission mode.

porting materials were machined, which allowed the X-ray beam to probe the friction interface even while sliding, thus providing *in vivo* pre mortem experimental conditions, as depicted in the previous section. The test rig consists of a compact reciprocating pin-on-flat tester with the X-ray beam crossing the contact area. The motion is actuated by a vibrator, with a maximum amplitude of 5.0 mm and a frequency range from 0.2 to 20 Hz. The normal load is applied by a spring system. The beam dimension at the exit of the synchrotron monochromator has a rectangular section that can be collimated by shutters in both directions.

From a triboanalytical point of view, the EXAFS data, in terms of the RDF, allow identification of the location of the shearing process in the friction interface. For example, if the friction process involves only a few easily sheared planes, as in lamellar compounds, the EXAFS parameter of the whole interface material (wear debris plus surface films) will be practically unchanged. This is because the EXAFS signal is averaged over all identical atoms while shear planes involve only a few. Alternatively, if the velocity accommodation occurs on the "atomic scale," implying that all atoms are continuously and perhaps only slightly displaced from their initial crystal sites during friction, then the EXAFS parameter will be drastically modified. Thus EXAFS is able to tell us about the localization of the shearing process and to specify the velocity accommodation mechanisms.

XANES (or ELNES) reflects the electronic structure of selected atoms and can give vital information on electron transfer processes, due to the well-known "chemical shift" of the ionization edge energy. Just as in the XPS technique, tribochemical reactions can be carefully analyzed. Moreover, the hybridization of light elements (carbon, nitrogen, oxygen) can be deduced from the near-edge structure because that structure is a function of the distribution of antibonding

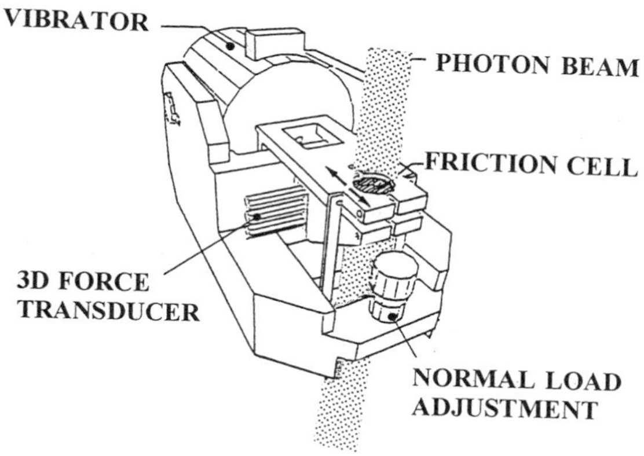

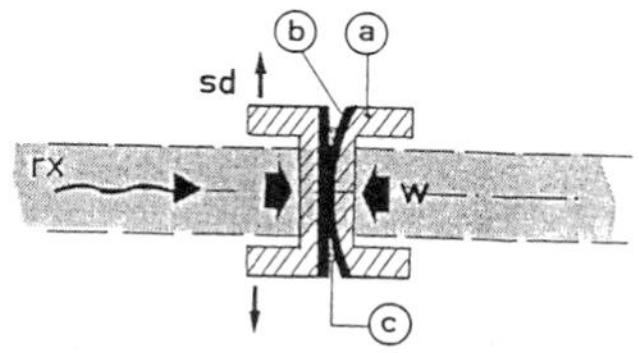

Figure 6. Schematic view of the *in vivo* EXAFS tribometer. (top) general arrangement. (bottom) the friction cell, consisting of (a) low X-ray absorbent carbon supports, (b) thin iron foils ($2 \times 7\ \mu m^2$) and (c) the lubricant or coating suffering friction. *sd* is the sliding direction and *W* the applied normal load. The photon beam, represented by the gray line (rx), is transmitted through the friction cell.

orbitals (or unoccupied levels). As both structural and chemical modification often takes place in many tribochemical processes, this kind of analysis is unique in that all the information is contained in the XAS (or EELS) spectrum and can be processed accurately. The generic study in Section 3.2 describes investigations of tribochemical films in boundary lubrication performed by this methodology.

2.4.2. Surface Morphology

Surface roughness strongly influences the tribological behavior of contacting bodies in engineering devices and occurs over a wide dimensional range. At one extreme, the microscopic surface roughness (up to 10^{-5} m) depends on the process by which the surface was produced. At the other extreme, investigations by the atomic force microscope (AFM) and the scanning tunneling microscope (STM) show surface roughnesses on an atomic scale. Surface topography anisotropy can also appear over a wide size range and can influence the friction and wear process. At the grossest level, some surface preparatory processes can induce anisotropy (e.g., turning), whereas others do not (e.g., lapping). At the finest level, the surface crystallographic structure can also induce anisotropy of the crystalline orientation. Whatever the scale of anisotropy, surface topographic investigations are crucial, since not only can surface morphology influence tribological behavior, but conversely the tribological process can modify surface morphology, due to local stresses and to heating of the contacting asperities during friction.

The techniques used extensively (but not exclusively) for obtaining surface topographical information are stylus profilometry, stereomicroscopy, optical interferometry and the scanning probe microscopies, including STM and AFM. Chung [15] has made a comparison of the principles, strengths and weaknesses of these techniques. As mentioned by Bhushan et al. [16], recent use of the probe microscopies has allowed a systematic investigation of surface and interface phenomena at the atomic and molecular scales in tribology. The STM, which provides images of electrically conducting surfaces with atomic resolution, is used for the imaging of clean surfaces and lubricant molecules. The AFM provides topographical measurements of electrically conducting or insulating surfaces on the nanoscale. Moreover, when coupled with computational techniques for simulating tip-surface interactions and interfacial properties, the scanned probe microscopies provide a better fundamental understanding (in terms of adhesion and friction) of interacting surfaces in relative motion, as well as a guide and a methodology for rational design in the lubrication of microdevices (see the next subsection). Since direct contact exists between the probed surface and the AFM tip, the latter technique also allows the measurement of ultrasmall surface forces. Specific modifications of the AFM have led to the development of the friction (or lateral) force microscope (FFM), which measures forces transverse to the surface, allowing atomic and microscale studies of friction. Reference 16 reviews recent developments in the use of tip-based microscopies in the field of tribology.

2.4.3. Physical, Mechanical and Frictional Surface and Interface Properties

New fundamental insights into boundary and thin-film lubrication have been provided recently thanks to increased interest in the measurement of the properties of molecular thin liquid films sandwiched between two contacting surfaces [16,17]. Since the scope of such studies does not match exactly classical surface analysis, it is not proposed to go far in that direction. Nevertheless the study of confined thin films is a traditional area which appeared first in colloid and adhesion science prior to tribology. In tribology, it has considerable practical importance since the interaction between loaded engineering counterfaces occurs through the close proximity of high spots or asperities, with locally high contact pressures and temperatures. Homola [17] has reviewed the different experimental approaches to the direct measurement of the physical properties of thin liquid films with thicknesses approaching the dimensions of the liquid molecules themselves. Bhushan et al. have given an overview dealing with the development of experimental and computer simulation techniques for the study of these phenomena on the atomic scale, thus providing an emerging global understanding of the molecular mechanisms of tribology in thin films and at surfaces [16]. Experimental methods included the surface force apparatus (SFA) for measuring very weak forces, submicroscopic surface geometries and surface separation on the nanoscale. The SFA, originally developed to study static or equilibrium interfacial forces, has been modified recently to measure the dynamic shear response of liquid confined between sliding surfaces [18–22]. The objective here is to ascertain how the physical properties of liquid localized in small spaces (pores, crevices, thin films) differ from those of the bulk material, and how these properties vary with the size or the thickness of the confined molecules, in order to provide a description of the transitions from continuum to molecular behavior of very thin layers. When film thicknesses are globally greater than 10 molecular diameters, both static and dynamic properties can be described in terms of their bulk properties. However for thinner films, the shear behavior becomes progressively more solidlike and their bulk properties are altered drastically, due to relationships between the molecular ordering (or glassing) and the structural (molecular architecture), physical and mechanical properties of the interface. For example, SFA experiments carried out with complex fluid or polymer films indicate that branched-chain molecules lubricate better than straight-chain molecules, even though the former have much higher bulk viscosities; the symmetrically shaped straight-chain molecules are prone to ordering and freezing, which dramatically increases their shearing resistance, whereas the irregularly shaped, branched molecules remain in the liquid state even under high loads [16]. This explains why straight-chain molecules are particularly appropriate when high friction is required, such as in clutch mechanisms.

From a technological point of view, the interest in this kind of study is in the possibility of controlling friction and wear in real conditions approaching boundary lubrication, by choosing judiciously the lubricant fluid, and by chemically grafting chainlike molecules such as surfactants or polymers to surfaces. In the latter case, the absence of traditional fluid lubricant owing to the self-lubricating grafted surfaces indicates that surface investigations on the nanometric scale in molecular tribology open up the possibilities of new types of lubricant systems for many applications, including microdevices.

3. GENERIC STUDIES

Ten generic studies have been chosen to illustrate how a variety of surface and other analytical techniques can be combined intelligently to investigate tribological systems, taking into account the dimensional and time-scale criteria along with the nature of the required surface information, as discussed in the previous section. The common objective is to arrive at the tribological mechanisms responsible for the observed friction and wear behavior. Table 2 sets out the subjects of the selected studies, whose relevant results are summarized in the following.

3.1. Ultrathin Boundary Lubricant Films [23,24]

This section addresses the question, how does one measure the minimum lubricant film thickness in boundary lubrication?

As mentioned earlier, boundary lubrication is undoubtedly the least well-understood of the liquid lubrication règimes. Several complex and interactive phenomena occur (adsorption, tribochemical reaction) at film thicknesses below the minimum at which quantitative measurements of most relevant surface properties can be made. One of the most influential tribological parameters in the lubrication mode of an engineering device is the thickness of the lubricant film, as explained in Section 1. The historical background contains considerable controversy as to whether boundary films were considered thin (few molecular thicknesses) or thick (above 10 to 100 molecular thicknesses). Since the 1980s, Johnston et al. have developed the technique of ultrathin film interferometry (UFI), consisting of an extension of conventional optical interferometry, allowing lubricant film thickness measurements in the hydrodynamic mode to be made. UFI has been refined recently to the extent that it can be used to measure film thicknesses down to about 1 nm, i.e., of molecular dimensions. This tool is an interesting and powerful means of measuring the thicknesses of boundary lubricating films directly in realistic, rubbing contacts, thus illustrating the *in vivo* pre mortem analysis configuration performed in tribology.

The principle of UFI can be described on the basis of Fig. 7 (from Ref. 23). A steel ball is loaded hydraulically against the underside flat surface of a rotating glass disk to form a circular concentrated contact. The ball is contained in a lubricant bath and is driven by the disk in nominally pure rolling motion. The test rig can be heated to temperatures up to 200°C, which is consistent with temperatures related to very low lubricant thickness due to low viscosities. The contact area is illuminated by white light. Compared to conventional optical interferometry, the originality of this apparatus consists first in a spacing layer of transparent silica about 400 nm thick coated on the underside of the chromium-plated glass disk. Acting like a supplementary oil film, it enables optical inter-

Table 2. Overview of the Generic Studies in Section 3, as Illustrations of the Original Analytical Methodologies Designed to Investigate Surfaces and Interfaces in Tribology

Study	System	Methodology	Information sought	Technique	References
3.1	Ultrathin boundary lubricating films	*In vivo pre mortem*	Film thickness	Thin-film interferometry	23, 24
3.2	Tribochemistry of antiwear additives	*Ex situ*	Composition/bonding	TEM/EELS	4, 14, 25–28
		In situ post mortem	Hybridization/phases	XAS/CEMS	
		In vivo pre mortem	Local structure	AES	
3.3	Tribochemical activity of nascent surfaces	*In situ post mortem*	Chemical activity	Mass spectrometry	29
3.4	Nature of surface vs. tribochemsitry	*In situ post mortem*	Composition/bonding	XPS/AES	
				HREELS/STM	30
3.5	Monolayers—dry friction	*In situ pre mortem*	Composition/bonding	AES	31
3.6	Tribochemistry of SiC vs. O_2 partial pressure	*Ex situ*	Composition/bonding	AES	32
		In situ post mortem	Phase/local structure	TEM/EELS	
				EXELFS	
3.7	Durability vs. microstructure MoS_2 coatings	*Ex situ*	Structure	XRD	33
3.8	MoS_2 sliding in UHV	*Ex situ*	Composition/bonding	XPS/AES	34–36
		In situ post mortem	Nanocrystallinity	HRTEM	
3.9	Microtribology of carbonaceous films	*Ex situ*	Composition/bonding	FTIR/SPM	39
			Phase		
3.10	Tribochemistry of C_{60}	*In vivo post mortem*	Hybridization/phase	Raman	28

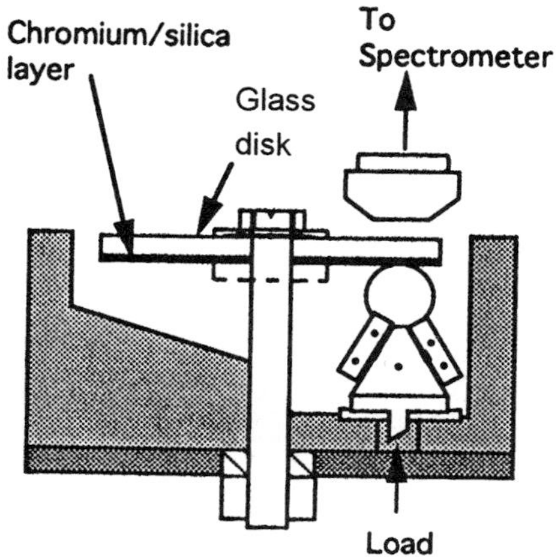

Figure 7. The test rig for ultrathin layers used for *in vivo pre mortem* thickness measurements in boundary-mixed lubrication. (From Ref. 23, with permission.)

ference to be observed even when the actual oil film present is very thin. The second improvement comes from the presence of a spectrometer dispersing the reflected interfered light, thus allowing a great increase in the precision of the thickness measurement. The dispersed spectrum is computer-analyzed to identify with which wavelength of light there has been most constructive interference. In addition, the rig used for traction coefficient measurements is the same as that employed to measure film thicknesses, except that the glass disk is replaced by a smooth, hardened, polished stellite disk, with a surface finish similar to that of the glass disk.

Experimental investigations carried out with various kinds of mineral- and organic-based lubricants show that some of them can form boundary layers up to 20 nm thick, with viscosity different from the bulk fluid. These layers may be formed either by adsorption of monolayers of polymers on surfaces or by the accumulation of more polar, low-molecular-weight species near the solid surface due to van der Waals attraction. Deviation from classical EHD film-forming behavior is introduced by these triboinduced layers, resulting in the observed film thickness variations. The boundary layers can also shift the onset of mixed lubrication, as observed by friction coefficient changes. Thus, the UFI technique leads to considerable improvement in clarifying the chemical and rheological contributions of the different additives and base oils to the surface separation of contacts under realistic conditions, including the highest temperatures of use consistent with the lowest viscosities and therefore thicknesses.

3.2. Tribochemistry of Antiwear Additives in Boundary Lubrication [4,14,25–28]

This section addresses the questions, what kind of physicochemical interactions occur between antiwear additives and the contacting surfaces, and how do they depend on the nature of the environment during friction?

The role of antiwear additives in boundary lubrication has been mentioned briefly in the general presentation (Section 1). The way in which zinc dithiophosphate (ZDDP) acts as an antiwear additive is almost certainly that of the elimination of the abrasive wear contribution of crystallized iron oxide species as a result of their tribochemical reactions with the additive molecules and/or various degradation products. The use of XAS and TEM/EELS together with complementary CEMS and AES represents a powerful analytical combination for the elucidation of the tribochemical reactions responsible for antiwear mechanisms. The first part of this section summarizes analytical results obtained with friction tests carried out in the ambient atmosphere. The second part describes some original results obtained with friction tests and *in situ* post mortem AES analysis in ultrahigh vacuum. In this way the intrinsic frictional properties of the solid reaction film itself, without the effect of the environment, and the ability of friction to modify the nature and the structure of the surface films are investigated.

3.2.1. Ex situ Surface Analytical Investigations [4,14,25,28]

The solid wear particles from ZDDP films produced in lubricated contacts contain phosphorus, sulfur and zinc from the ZDDP molecule and oxygen mainly from the surrounding air environment. They also have a low iron content. *Ex situ* examination by XAS, EELS and CEMS of these particles (configuration shown in Fig. 5a) has been carried out to provide local analysis of the iron atoms, since their localized environment in the surface film is of great interest; it is related directly to the wear of the steel surface and can also play a key role in the adhesion mechanisms of the film. In order to investigate this aspect, a study was made by XAS (EXAFS plus XANES) of a collection of wear debris from two lubricated tests, with and without ZDDP in the lubricant base, respectively. The processed EXAFS data presented in Fig. 8 show the RDFs of iron atoms (noncorrected phase shifts) in four samples; data from the standards, pure crystalline iron and iron oxide have been included for comparison. From this EXAFS study, some important results can be deduced [4]:

Compared to the metallic substrate, iron atoms have become bound to oxygen in the wear debris, irrespective of the lubricant. The main peak in the RDF is attributed to an Fe-O separation of 0.19 nm. In the sample without ZDDP, the second peak corresponds to Fe-Fe separations of 0.29 nm.

In the presence of the additive, and during the steady-state friction regime, iron atoms in the wear debris are found to be isolated from each other, because the second Fe-Fe peak in the RDF has decreased considerably, indicating a high level of crystallographic disorder.

To complete these investigations, XANES experiments undertaken by recording the Fe-K near-edge structure and ELNES experiments in the TEM at the O K-edge indicated, respectively, the presence of a fully oxidized Fe^{3+} state in octahedral symmetry and the presence of some residual iron oxide Fe_3O_4 [4]. Additional CEMS measurements confirmed the presence of octahedrally coordinated ferric cations [28].

Based on these results, the transition from the abrasive wear to the low-wear regime has been explained in terms of velocity accommodation mechanisms in the interface film. The most important phenomenon is a complex polyphasic tribochemical reaction responsible for the formation of a solid transition-metal phosphate glass material as an adherent thin film. This acts as a protective layer against wear due to its superplastic behavior when it is formed in the contact

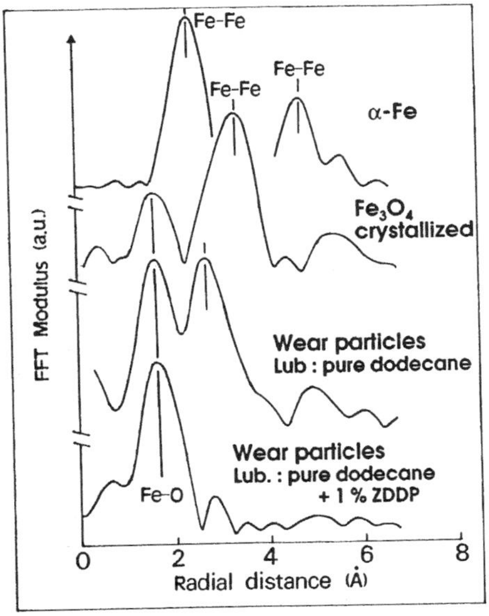

Figure 8. RDFs of iron (obtained at the Fe K-edge) in different local environments, as recorded by XAS from alpha metallic iron, crystalline Fe_3O_4, and wear particles after friction tests, with pure dodecane and with dodecane + 1% of ZDDP antiwear additives as the lubricant phases. (From Ref. 14, with permission.)

area. Crystalline iron oxides have very rigid atomic skeletons with selective plastic deformation processes and are abrasive. By comparison, the triboinduced phosphate glass has chemical bonding angles which are flexible in all directions and so can undergo slight deformation; in addition, ionic iron atoms can diffuse easily in the multiply connected channels. Consequently, the accommodation of velocity during the shearing process occurs by distribution throughout the film thickness due to the accumulation of atomic-scale displacements, somewhat similar to the hydrodynamic regime. The abrasive wear at the early stage of the friction test is thus practically eliminated since the crystalline iron oxides are digested progressively in the phosphate glassy material, which is able to support much greater deformation because of its superplastic behavior.

A fundamental assumption that can be made is that, at the end of the friction test, the state of the wear fragment material is "fixed," the rapid relaxation of shear and pressure being equivalent to thermal treatment, and its analysis can therefore reveal interesting features with regard to what was on the surface before and during friction. The validity of this assumption can be checked by studying the possible differences between the microstructure of wear debris (Fig. 5a) and the surface film (Fig. 5b) formed in a single friction experiment. This has been done by comparing the environment of zinc atoms present in both ZDDP wear fragments and surface films, with the advantage that the same analytical tool can be used. Wear debris are normally analyzed by XAS in the transmission mode, while worn surfaces are analyzed in the reflection mode using the total electron yield, including high-energy Auger electrons (in this case the thickness analyzed is approximately 20 nm). EXAFS was recorded at the Zn K-edge (9659 eV).

The analysis shows that the zinc atoms have a similar environment both in wear fragments and on the top of the surface film, suggesting that the wear mechanism is dominated by nonequilibrium states, the tribochemical reaction products being quenched and the microstructure being fixed at the end of contact, due to the rapid relaxation of shear and pressure [4].

3.2.2. In vivo Pre mortem Surface Analytical Investigations [26]

The localized analysis of zinc atoms in the operating interface (*in vivo* pre mortem analysis) has been obtained successfully from the tribometer whose design is shown in Fig. 6. Although the lubricant quantity that is analyzed in the operating interface is very small (say a few microliters), the zinc signal is strong enough to be processed, despite lack of quality due to poor signal/noise ratio. The frictional motion does not disturb unduly the *in vivo* analysis (the acquisition time was 5 min in the sequential mode, but it could be reduced to a few seconds with a parallel detection system). From these experiments, it can be said that the analysis of interfaces in motion is now feasible, but the tribochemical reaction of zinc atoms could not be observed in this first experiment at the Orsay synchrotron line, since the initial Zn-S bonding of the molecule remained Zn-S after a friction experiment of 7 h duration. No reaction film was formed on the surfaces. Future studies involving real-time EXAFS on the frictional interface in motion will be possible with a parallel detection system which will allow much shorter spectral recording times.

3.2.3. In situ Post mortem Surface Analytical Investigations in Ultrahigh Vacuum [27]

Some fundamental aspects of the tribochemistry of metal DDP have been investigated by performing ultrahigh vacuum tribotests on selected chemisorbed films previously formed on steel bearing surfaces. In order to isolate the effect on friction of the outer surface contamination layers, the authors checked their tribological behavior by coupling friction tests with *in situ post mortem* AES microanalysis before and after ion etching. They found that the native oxide film of nanometer-scale thickness on the steel surface had high wear resistance but very high friction in UHV. However, friction was reduced at the beginning of the test by the carbon-rich contamination layer, which was removed mechanically by a few reciprocating friction passes without reaction with the surfaces. Friction between the pure metals (after ion etching) was lower than unity, but adhesive wear and considerable plastic deformation took place.

The authors then checked as to whether tribochemical reactions really existed, which would provide evidence of any friction-induced chemical transformations of the chemisorbed species. Zinc and molybdenum dithiophosphate additives (ZDDP and MoDDP) were chemisorbed onto steel plates prior to the friction test, by a conventional immersion procedure in an additive-containing oil. Two immersion times (2 and 24 h) and one immersion time (5 h) were chosen, respectively, for the ZDDP and MoDDP additives. The chemistry of the treated steel surfaces was investigated by XPS/AES and the friction test was carried out in UHV just after the analysis. At the end of the test, *in situ* post mortem AES microanalysis was performed both inside and outside the wear scars on both the pin and the flat. Unfortunately XPS microanalysis was not possible with the current equipment, so that complete characterization of chemical bonding could not be achieved.

Figure 9 shows the average friction coefficient versus the number of reciprocating cycles, in UHV, for the AISI 52100 steel without (a) and with a ZDDP film adsorbed (b) for 2 h and (c) for 24 h and with a MoDDP film adsorbed (d) for 5 h. The XPS analysis prior to friction (not shown here) indicated that a very thin sulfate-rich film was formed after a 2 h immersion, whereas a thicker metal (phosphate + sulfide) layer appeared after 24 h, with the ZDDP adsorbate films. A similar layer was formed with the MoDDP film but containing Mo instead of Zn, and no MoS_2 was identified. Friction tests were car-

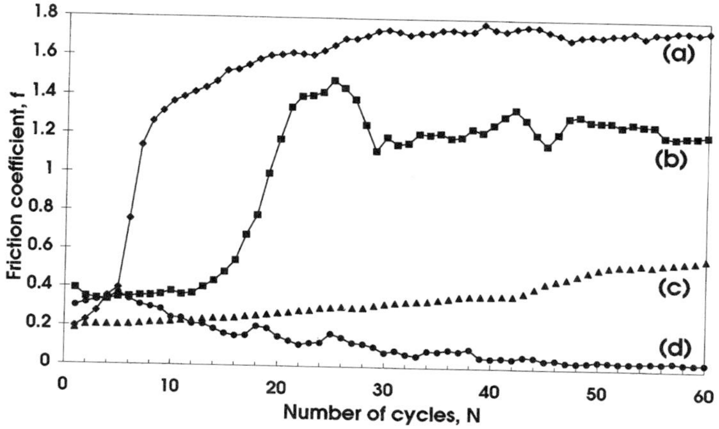

Figure 9. Average coefficient of friction versus the number of reciprocating cycles, of two sliding steel surfaces in UHV (a) without and (b and c) with ZDDP adsorbed films (b) for 2 h, (c) for 24 h, and (d) with an MoDDP adsorbed film for 5 h. (From Ref. 27, with permission.)

ried out with a normal contact pressure of 0.54 GPa, a wear track length of 3 mm and a sliding speed of 0.5 mm/s. The period between two consecutive passes of the pin over a given point of the plane was therefore 12 s. Figure 10 shows the *in situ* post mortem AES microanalysis performed after the end of the friction tests, outside the wear track of the flat and inside the wear track of both the flat and the pin in the three configurations of adsorbed films mentioned above. By considering Figs. 9 and 10 together, the authors observed that a thin chemisorbed film of ZDDP molecules decreased the friction to 0.2. This decrease correlated with a transfer of sulfur from the flat to the pin. Where phosphorus was present on the treated surface, it was eliminated from the contacting areas by the wear process. By comparing results from the two immersion times, it was also concluded that the higher the sulfur content the lower the friction. This demonstrated that the effect of sulfur (in contrast to that of oxygen) was to reduce friction in UHV conditions. Moreover, thicker chemisorbed films of MoDDP molecules induced a noticeable progressive decrease in the friction, from 0.3 during a running-in period to the 10^{-2} range in the steady-state regime. The AES microanalysis, coupled with *ex situ* Raman and TEM/EELS analyses (not reproduced here), indicated that a tribochemical reaction had taken place with a selective transfer mechanism: an antiwear phosphate glass (of the same nature as that described in Section 3.2.1), identified by TEM/EELS, remained on the flat surface, whereas a friction-reducing

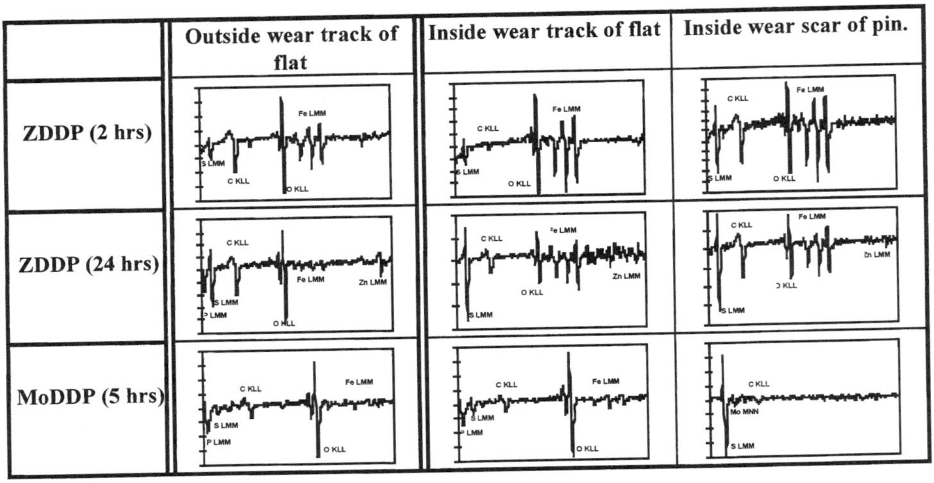

Figure 10. *In situ* post mortem AES microanalysis on chemisorbed metal-DDP films, inside and outside the wear tracks, at the end of the friction tests of Fig. 9. Selective sulfur transfer onto the pin is observed with the ZDDP films, whereas a MoS_2 film is formed and then transferred onto the pin with the MoDDP film (From Ref. 27, with permission.)

MoS_2 film (identified by Raman) was formed during friction and transferred preferentially onto the pin.

Since the effect of the environment had been eliminated, the overall results demonstrated that tribochemical reactions took place that were selective with respect to the pin-on-flat geometrical configuration, during friction in an UHV environment. A judicious combination of analysis by *ex situ* and *in situ* methods thus allowed the complex tribochemical mechanisms responsible for friction and wear behavior in boundary lubrication to be elucidated.

3.3. Tribochemical Activity of Nascent Surfaces [29]

This section considers the question, what is the role of friction-induced fresh surfaces in boundary lubrication?

As mentioned above, tribological behavior in boundary lubrication is strongly affected by surface reactions as well as by chemisorption of additives on contacting surfaces. The molecular structure of the additives is not the only parameter responsible for any particular tribochemical behavior. During the friction process, contamination layers on the outer surface are removed mechanically, leading to the formation of fresh, nascent surfaces characterized by enhanced chemical activity. It is the chemical nature of these nascent surfaces that is one of the most important parameters, since such surfaces are active sites for tribochemical reactions. The author has undertaken *in situ* quantification of the chemical adsorption activity of fresh surfaces created in controlled environments (i.e., nature and pressure of ambient gases), following a standardized scratch test of the studied surfaces. The chemical adsorption activity of a given gas on a given scratched surface is estimated from the time dependence of the pressure decrease during the adsorption process, since the slope of the pressure-time curve is proportional to the sticking coefficient. A systematic investigation of the chemical activity of various metals and alloys with respect to gaseous organic compounds has been performed. Systematic enhanced activities were observed on the nascent surfaces. A comparison of the chemical reactivities of various organic compounds as a function of the nature of the surfaces has been made, as shown in Fig. 11. The gas/surface chemical reactivity depends on the polarity of the organic compound and on the chemical nature of the solid surface, i.e., either nascent or oxidized; nonpolar compounds chemisorb on nascent metallic surfaces more easily than on oxidized metal surfaces, whereas polar compounds chemisorb easily on oxidized metal surfaces. From Pearson's hard and soft acids and bases principle, the nonpolar compounds are classified as soft bases and interact more strongly with soft acids such as clean metal surfaces rather than with hard acids, i.e., oxidized metals. Now metal surfaces are

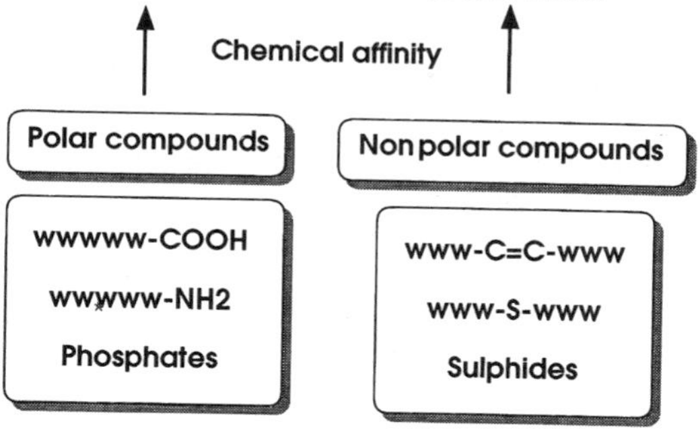

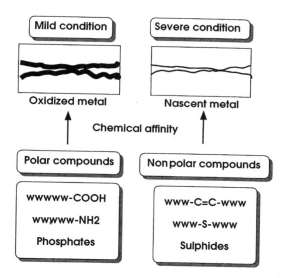

Figure 11. Tribological conditions and surface chemistry in boundary lubrication (From Ref. 29, with permission.)

covered with oxide layers, which under severe conditions can be removed by friction. Thus the chemical nature of metal surfaces is closely dependent on the lubricating conditions, that is, oxidized under mild conditions and surfaces under severe conditions. Consequently, the additives that should be effective under mild conditions will be polar compounds such as fatty acids and phosphates, whereas the additives likely to be effective under severe conditions will be nonpolar compounds, such as organic sulfides.

This experimental contribution confirms previous tribological investigations, which showed that phosphate additives were effective under low load conditions (mild wear), whereas sulfides were effective under high load conditions (severe wear). The results are of particular interest in understanding the beneficial roles of some specific additives, in relation to the nature of the contacting surfaces. A major contribution of this study has therefore been identification of the additives that should be chosen for antiwear under extreme pressure conditions, as functions of wear conditions and of the particular metals and alloys in contact. It contains also some interesting observations on how gas/solid reactions can be mediated by tribological activation (i.e., via a scratch), which enhances the reactivity.

3.4. Influence of the Nature of the Surface on the Tribochemistry of Various Tribomaterials [30]

This section addresses the questions, what kinds of tribochemical reactions occur at the friction interface, and how do they depend on the nature of the contacting bodies and of the lubricant?

Friction measurements in UHV and controlled gaseous environments, coupled with *in situ* post mortem XPS/AES and *ex situ* STM investigations have been performed with various types of contacting material, including metal/metal (Fe/Fe), metal/ceramic (Fe/Al_2O_3), ceramic/ceramic (B_4C/B_4C) and steel/steel, with organic lubricants. The goal of the study was an understanding of the influence of the surface chemistry on the friction of two contacting solids, using the above solid/solid contact model systems. Friction was measured in well-controlled environments, ranging from ultrahigh vacuum to pure gases and even liquids. With a purpose-built pin-on-flat device which allowed contact interchange in UHV, *in situ post mortem* XPS/AES were performed inside the tribometer. The clean and lubricant-covered steel surfaces were also examined by HREELS to study the evolution of molecular bonds in the metal/organic compound interface, after various thermal treatments carried out independently of the friction test, in order to separate thermochemical and tribochemical phenomena.

The friction of Fe/Fe contacts in UHV decreases when the superficial oxide thickness increases within a controlled range, from zero in the clean metallic state to about 4 nm. O/Fe ratios were established by AES linescans on both sides of each wear track that corresponded to a different oxide thickness. The nature of the thin oxide interfaces in metal/metal contacts was found to govern the friction process, as shown already in the two previous generic studies. Attempts to understand the surface chemical and mechanical interactions of Fe with films formed from the reaction of a model perfluorodiethylether (PFDE) were also made. The fluorine films were found to be able to bear loads much better than thin Fe oxide films, as evidenced by lower friction values and by STM profiles of the wear scars.

The friction between Fe/Al_2O_3 in UHV depends both on the presence of an oxidized outer layer on the iron pin surface and of a carbonaceous contamination layer on the alumina plate. This was revealed from AES investigations carried out on the transfer film formed on the iron pin; the film contained relatively pure carbon in some areas and iron oxide plus carbon in others. Thus the amount of carbon transferred from the alumina outer surface to the iron was identified as a parameter having considerable influence on the level of friction.

B_4C is one of the hardest ceramic materials in the temperature range 20 to 1800°C. Its friction behavior has been investigated in UHV and ambient air, for clean and air-exposed surfaces prior to friction. The friction coefficient values ranged from 0.2–0.3 to 0.8–1.8, depending on the environment during friction (low friction in air, high friction in UHV) and on the surface cleanliness (i.e., either air-exposed or cleaned prior to UHV friction). AES linescan analyses of the scars revealed that the compositions of both the pin and the flat wear track were quite different from that of the noncontacted area. For the contact whose surface had been cleaned prior to friction, an increase in the boron signal and a decrease in the carbon signal were observed. For the surface exposed to oxygen prior to friction, there was evidence of a tearing of the oxide layer near the center of the scar.

Finally, the authors studied interfacial aspects of the tribochemical properties of high-temperature lubricant molecules, seeking answers to the following questions of interest in both tribology and surface science: (1) What is the nature of the chemical interactions between thin layers of adsorbed lubricants and both clean and oxidized surfaces? (2) Do enhanced surface chemical reactions occur at liquid/metal interfaces under the influence of friction? (3) If so, are the reaction products due to momentary thermal excursions or to other consequences of friction such as mechanically induced bond breaking to be regarded as tribochemical phenomena? In order to separate unambiguously the thermochemical effects and the tribochemical effects, surface chemical and friction studies were carried out in parallel on steel (M50) surfaces either oxidized in an almost atomically clean condition after preparation in UHV. The thermal surface chemistry was studied at different temperatures (from 90 to 750 K) using HREELS and XPS, while the tribochemical interactions were investigated by *in situ post mortem* XPS subsequent to friction tests in UHV.

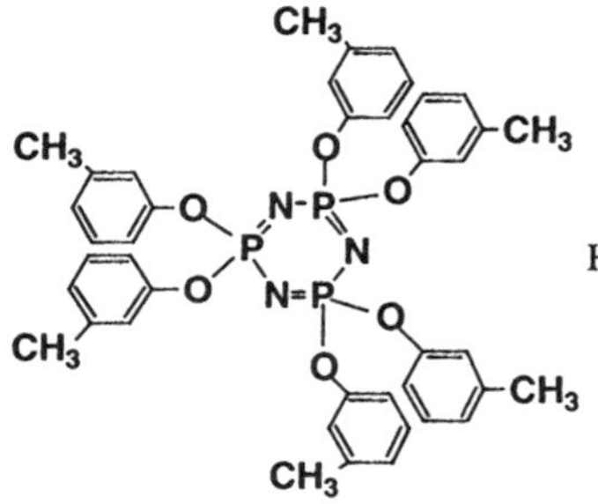

HMPCP

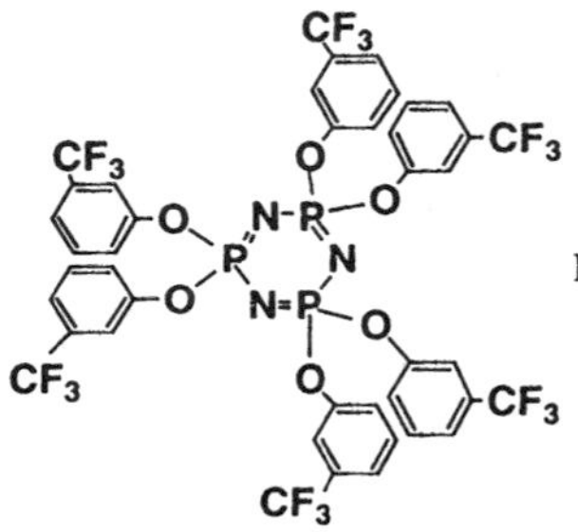

HTFMPCP

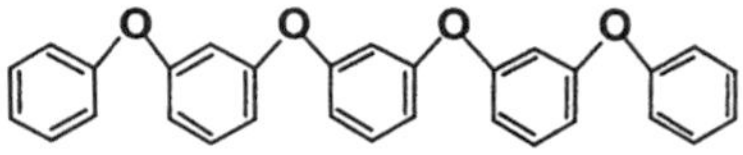

PPE

Figure 12. Molecular structures of the three organic lubricant fluids used for thermochemical and tribochemical investigations, as described in Ref. 30.

Three fluids which are strong candidates for lubrication under high-temperature conditions were used; a classical polyphenyl ether (PPE) and two aryloxicyclotriphosphazenes (HMPCP and HTFMPCP), with the same molecular structure except for the nature ($-CH_3$ or $-CF_3$) of the group on the phenoxy ring (Fig. 12). Despite the similarity of the two phosphazene molecules, substantial differences in the thermochemistry and tribochemistry were found, on the basis of the surface analytical investigations. Thermally induced chemistry was observed on the clean and oxide-covered surfaces for both phosphazenes after thermal treatment above 470 K. The CF_3-containing molecule reacted both at the aromatic ring (as did PPE) and at the CF_3 group. Surprisingly, the CH_3-containing molecule did not show any reactivity at the aromatic ring. Breaking of the aryl-O bond was observed for both phosphazenes. In tribological behavior, the CF_3-containing molecule and the PPE exhibited tribochemical phenomena, unlike the CH_3-containing molecule.

The tribochemical effects were identified by *in situ* post mortem XPS which provided direct analysis of the interface, taking into account the low thickness of the lubricant film (<2 nm). Evidence for the formation of metal fluorides and of metal oxides was found for the CF_3-containing molecule and for the PPE, respectively. On the other hand, no chemical change was identified as being associated with the CH_3-containing molecule. Since the tribochemistry was not believed to be thermally induced, because of the low loads and sliding speeds, the most likely mechanism for the observations was that in which the shear stress due to sliding caused mechanical bond breaking (that is, a mechanochemical effect) for both the CF_3-containing molecule and the PPE. If only one layer of fluid molecules separated the two sliding surfaces, all the shear stress (except that dissipated by plastic deformation of the solid surfaces) must have been concentrated in this single monolayer. Now the PPE and the CF_3-containing molecules were strongly anchored to the surface, so some of the energy dissipation must have induced bond breaking in the fluid molecules, leading to metal-O or metal-F triboinduced bondings. By contrast, the stress applied to the more weakly bonded CH_3-containing molecules might have been relieved by "plowing" through the lubricant film, which would have broken the surface-fluid bonds without inducing any tribochemical changes in the molecular structure.

Studies of this kind emphasize the role of surface chemistry and the reactivity of organic compounds on solid surfaces in separating the thermochemical from the tribochemical effects in boundary lubrication.

3.5. Effect of Adsorbate Monolayers on Dry Friction [31]

This section deals with the question, what is the effect of adsorbed monolayers on friction [31].

As mentioned above in connection with the time-scale criterion, the use of surface analysis techniques separately from a friction rig requires the exposure of the specimen to air between the friction test and the analysis procedure. That is the reason why tribologists have developed *in situ* analytical tribometers which allow direct surface chemical investigations. The triboanalytical coupling may combine the friction device and the analytical technique in the same chamber, ideally in a *pre mortem* configuration. An example of this mode is shown in Fig. 13 (from Ref. 31). Direct *in vivo* AES analysis inside the dynamic contact is not possible in this case, unlike in the configuration described above in Fig. 6. Thus the choice of coupling mode linking the tribological test and the analysis procedure depends strongly on the surface analytical technique chosen: *in vivo* AES is physically impossible, although XAS performed with X-ray transparent substrates is possible. In the study from the above reference, *in situ*

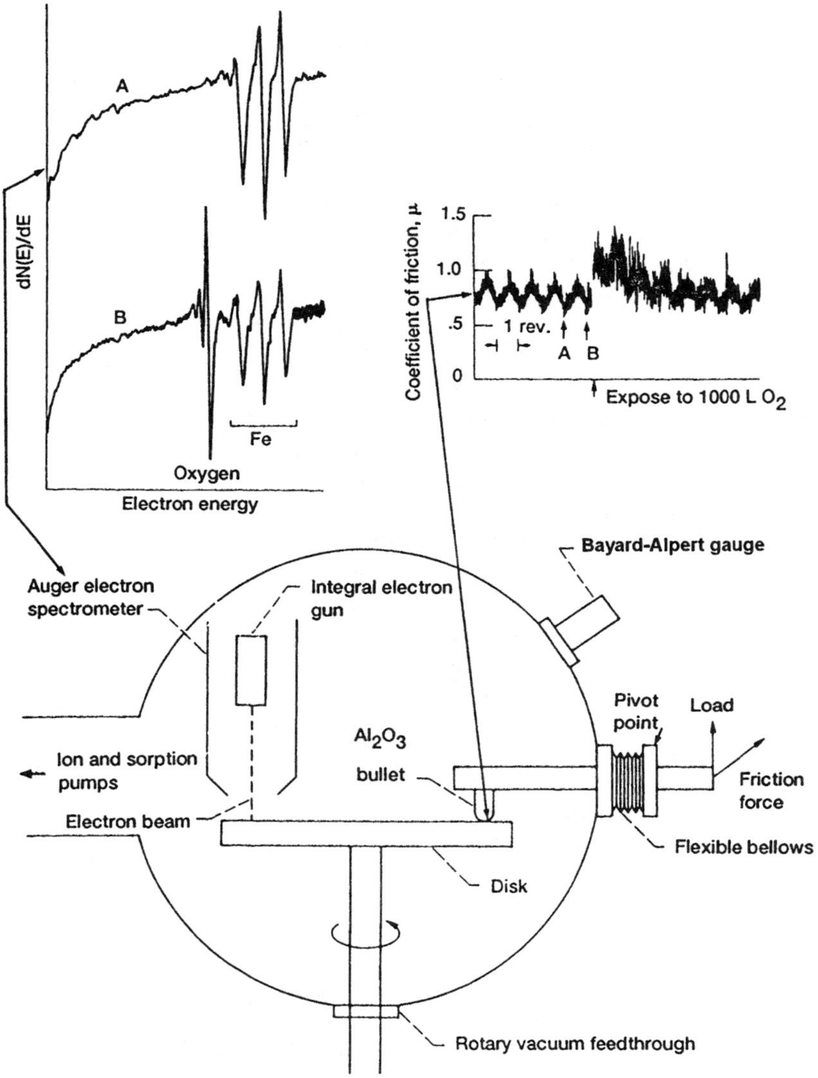

Figure 13. An example of *in situ pre mortem* analysis in a tribological experiment. (From Ref. 31, with permission.)

pre mortem AES was used to demonstrate the effect of monolayer amounts of adsorbates on friction in UHV and in various pure gases, for aluminum oxide sliding against iron, nickel and copper. Because of the acquisition time required for AES, the surface wear scar analysis was performed sequentially by stopping the rotational motion just long enough to record the analytical signal. Clean metal surfaces were exposed to O_2, Cl_2, C_2H_4 and C_2H_3Cl, and the changes in friction due to the adsorbed species were measured. Figure 13 shows that an oxide-free iron surface (as shown by AES spectrum A) has a friction coefficient of about 0.7. Exposure to an oxygen partial pressure of about 10^{-7} hPa allows oxygen to adsorb on the clean iron (AES spectrum B) and results in an increase in friction, demonstrating the significant effect of adsorbate monolayers on friction. It was found also that systems exposed to Cl_2 exhibited low friction, interpreted as being due to van der Waals interactions between the alumina and the metal chloride. The generation of metal oxides by oxygen exposure resulted in an increase in friction, interpreted as being due to strong interfacial bonds formed by the reaction of metal oxide with alumina to form a complex spinel. The only effect of the C_2H_4 was to increase the friction of the Fe system, but C_2H_3Cl exposures decreased friction in both the Ni and Fe systems, indicating the dominance of the chlorine over the ethylene radical on the surface.

3.6. Tribochemistry of SiC/SiC under a Partial Pressure of Oxygen [32]

This section addresses the question, is it possible to lubricate ceramic materials by a reactive gas phase?

Some fundamentals of the surface-chemistry-related tribological properties of alpha silicon carbide (SiC) have been studied using an analytical UHV tribometer equipped with XPS/AES for *in situ post mortem* surface investigations. The experimental configuration was the same as that described in Section 3.2.3. Experimental work was carried out at room temperature in order to isolate the role of tribochemistry from that of any static interaction of the SiC surface with oxygen prior to friction. Spatially resolved *in situ post mortem* AES analyses were coupled with *ex situ* TEM/EELS and HRTEM analysis of the wear fragments. Figure 14 shows the friction and associated AES results, leading to the following observations.

Under high vacuum (10^{-10} hPa), SiC friction was dominated by high adhesion and by fracture, compaction and attrition of the spheroidal grains in the interfacial region. No preferential shear plane appeared to form, but amorphization of SiC took place mainly at the edges of the grains. Consequently the friction was high (0.8).

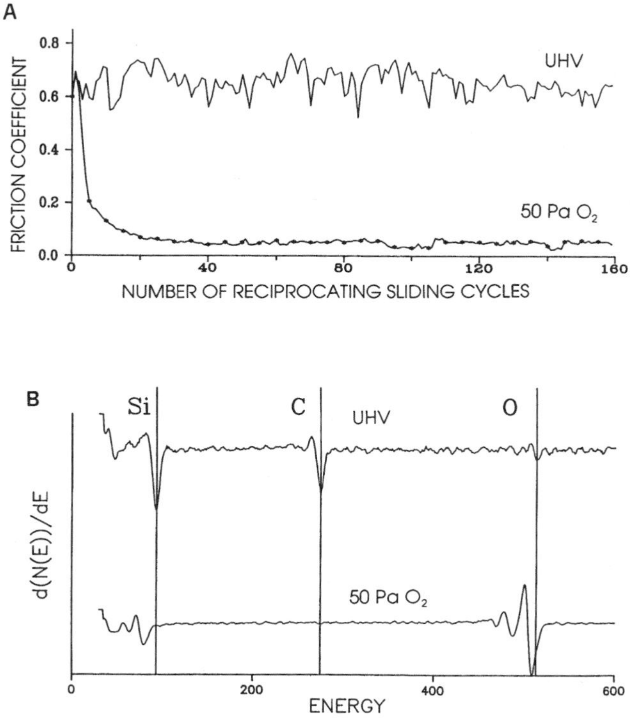

Figure 14. (A) frictional dependence of an SiC/SiC sliding contact on the nature of the surrounding environment (UHV or pure oxygen) during 160 cycles of sliding friction. (B) Corresponding *in situ post mortem* AES analyses inside the flat wear scars. Similar AES analyses were performed outside the wear scars and showed persistence of the Si and C signals, for both environments.

Under 0.5 hPa O_2 partial pressure, SiC quickly oxidized at the interface, producing silicon oxide, as identified by AES in Fig. 14. *Ex situ* TEM/EELS and HRTEM examination (not shown here) identified the presence of rolling pins inside the wear track at the end of the friction test. These wear fragments consisted of bulk amorphous silicon oxide with a pregraphitic carbon outer layer. This pregraphitic component, having planes with low shear strength in the presence of oxygen, seemed to be responsible for the low shear strength of the interface film. Consequently, friction was lowered (<0.1) and corrosive wear was predominant on the SiC pin surface.

Thus friction of alpha silicon carbide appears to be tribochemically controlled and very dependent on the reactive gases present in the surrounding environment. The simple tribochemical reaction involved in this experiment was identified by a judicious combination of *in situ* and *ex situ* surface analysis.

3.7. Relationship of Durability to Microstructure of IBAD MoS_2 Coatings [33]

This section addresses the question, what are the correlations between the microstructure of a solid lubricant and its durability?

MoS_2 is one of the most widely studied solid lubricants, since it is in general use in conditions of high vacuum or extreme temperature, when conventional liquid lubricants fail. One of the determining factors in choosing a solid lubricant coating is the durability; once applied to the contacting surfaces, the coating must either lubricate throughout the system lifetime (especially in space mechanisms) or until possible renewal. The lubricating properties of MoS_2 are related to its layer structure (Fig. 15). Covalent bonds join sulfur and molybdenum atoms in planar arrays of hexagonal S–Mo–S sandwiches, in which weak van der Waals interactions between adjacent and superimposed sulfur planes allow easy shear parallel to the sliding direction. The factors determin-

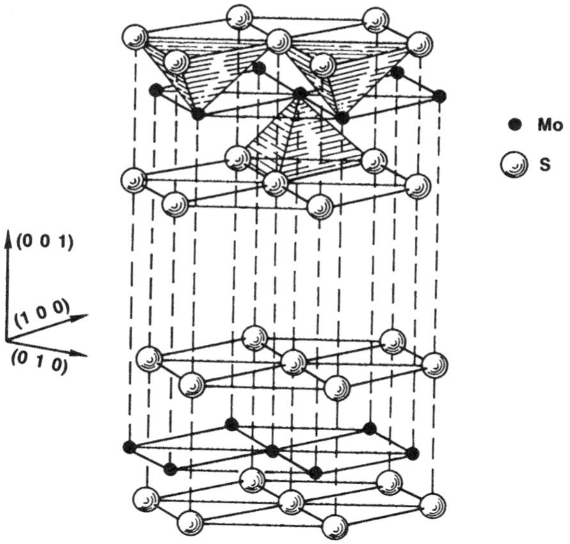

Figure 15. Crystalline structure of the 2H-MoS_2 solid lubricant, showing the layered-sandwich atomic arrangement and the crystallographic axes. The (0001) basal surfaces are perpendicular to the (001) axis.

ing the durability of solid lubricants are complex and include the nature of the substrate, the coating thickness, the adhesion, the coating morphology and the impurities dispersed in the lubricant layer. The purpose of the study referred to above was to address the relationship between the coating structure and the durability for MoS_2.

Ion beam assisted deposition (IBAD) of various MoS_2 coatings was carried out on steel substrates in such a way that careful control of the crystallite orientation could be maintained by variation of the deposition conditions. Friction tests (rotational motion) in dry air were performed with standardized and reproducible contact pressure and speed conditions, and friction values in the 0.02 range within the steady-state regime were measured. *Ex situ* X-ray diffraction (XRD) established the precise crystalline orientation before friction, which provided very good correlation between the crystalline structure and the lifetime as defined by the duration of friction to coating failure. The grain size of the IBAD deposited coatings was below 10 nm, based on the width of the characteristic MoS_2 diffraction peaks. From the XRD experiments, the (002) basal and the (100) edge intensity ratios were determined, taking into account the thickness and substrate variations from one coating to another and using an intense steel substrate peak as a reference. These intensity ratios were then used to define orientation since they were a measure of the relative number of crystals with either a single element or both elements exposed in the outer surface. The MoS_2 crystal structure is ordered hexagonal with each basal plane containing only one elemental species. A high basal intensity ratio meant that the (002) basal planes were preferentially oriented parallel to the surface, and therefore to the sliding direction.

Coatings with high basal intensity ratios generally had high durability (>100,000 cycles), while those with low basal intensity ratios tended to have lower lifetimes. Thus, coatings with no edge-orientated crystallites exhibited the best durability. In addition, no relationship between durability and the coating thickness or the nature of the substrate or the presence of a TiN interlayer was identified. This work illustrates the analytical needs required to study the optimization of the lifetime of a typical solid lubricant as a function of its preparative procedure.

3.8. Frictionless Sliding of Pure MoS_2 in UHV [34–36]

This section addresses the question, what is the origin of the superlow frictional behavior in solid lubrication by a pure MoS_2 coating?

Whereas the previous generic study placed emphasis on the durability of MoS_2 as a lubricant, the present one focuses on its low friction properties. Many tribological and analytical investigations have been devoted to this subject, in particular by Didziulis and Fleischauer [37] and by Roberts [38]. In this example the vital contribution made by surface analysis in providing complementary information is well illustrated. It has helped to arrive at an understanding of why a particular MoS_2 structure should have an unusually low coefficient of friction under UHV conditions. The coefficient was found to be in the 10^{-3} range, in fact the lowest value ever observed in solid lubrication. This frictional behavior was compared to that of a conventional MoS_2 coating, as shown in Fig. 16. The coefficient of friction of the conventional coating was generally about 10 times greater than that of the superlubricant. In both cases, a transfer film buildup was observed on the pin counterface. Analysis of the coating composition prior to friction and of the wear scar and wear particle structure after friction (*in situ post mortem* and *ex situ* examinations) revealed the origin of the differences in frictional behavior. From a theoretical point of view, frictionless sliding between two atomic planes needs three conditions to be satisfied:

1. Weak interaction forces between sliding atoms.
2. Atomically clean surfaces.
3. Incommensurate sliding atomic lattices; that is, the friction force is minimized when a misfit angle shifts two superimposed sliding crystal planes with respect to each other, as visualized in Fig. 17.

Condition 1 is satisfied in the MoS_2 structure, which accounts for the ultralow coefficients of friction reported in the literature (10^{-2} range) and confirmed here with the conventional coating. Condition 2 requires UHV during friction

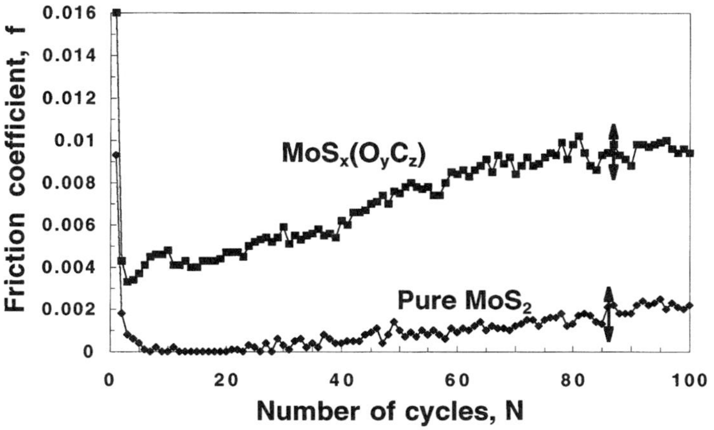

Figure 16. Average coefficient of friction versus the number of reciprocating sliding cycles of pure and contaminated molybdenum disulfide coatings, in UHV (10^{-7} Pa).

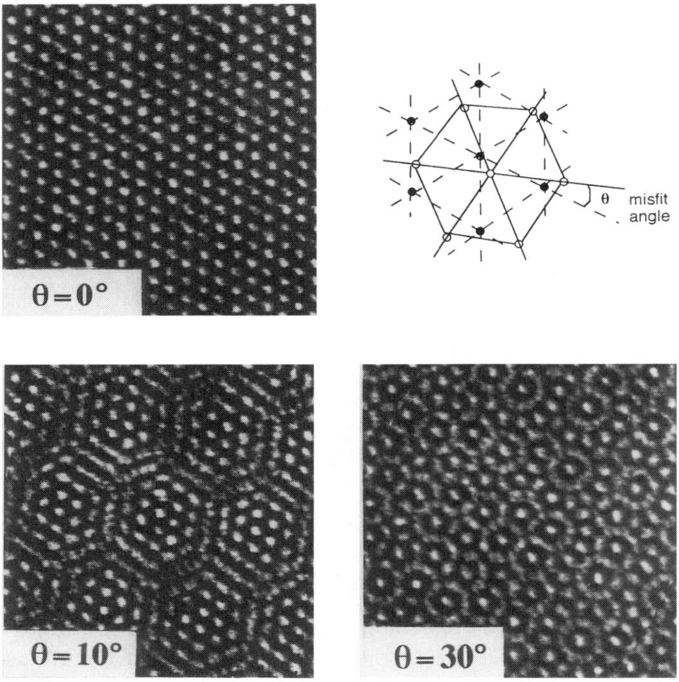

Figure 17. Visualization of two superimposed hexagonal lattices (representing adjacent sulfur planes in MoS_2), with no misfit angle ($\theta = 0$) and with misfit angles of $\theta = 10°$ and $30°$. The frictional force is known to be minimum at $\theta = 30°$, because of the hexagonal symmetry of the crystalline structure.

together with the absence of oxygen contaminant. The oxygen content of both coatings prior to friction was measured by nuclear reaction analysis (NRA) using the resonant nuclear reaction $^{16}O(\alpha,\alpha')^{16}O$ stimulated with α particles at 7.5 MeV. The oxygen concentration versus depth for both coatings was also deduced from the NRA spectra, as shown in Fig. 18. The conventional MoS_2 coating contained about 13 at.% of oxygen, whereas the nearly pure MoS_2 coating contained only 4 at.% of oxygen at its outer surface (probably due to surface oxidation subsequent to coating synthesis). The nuclear method did not of course indicate the site of the oxygen species in the conventional coating. The oxygen might have been present either as substitutes for sulfur atoms, or in small molybdenum oxide precipitates, or in water molecules dispersed in the coating. The first hypothesis is contradicted by the fact that oxygen substitution on sulfur lattice sites expands the *c*-axis, thus lowering the van der Waals

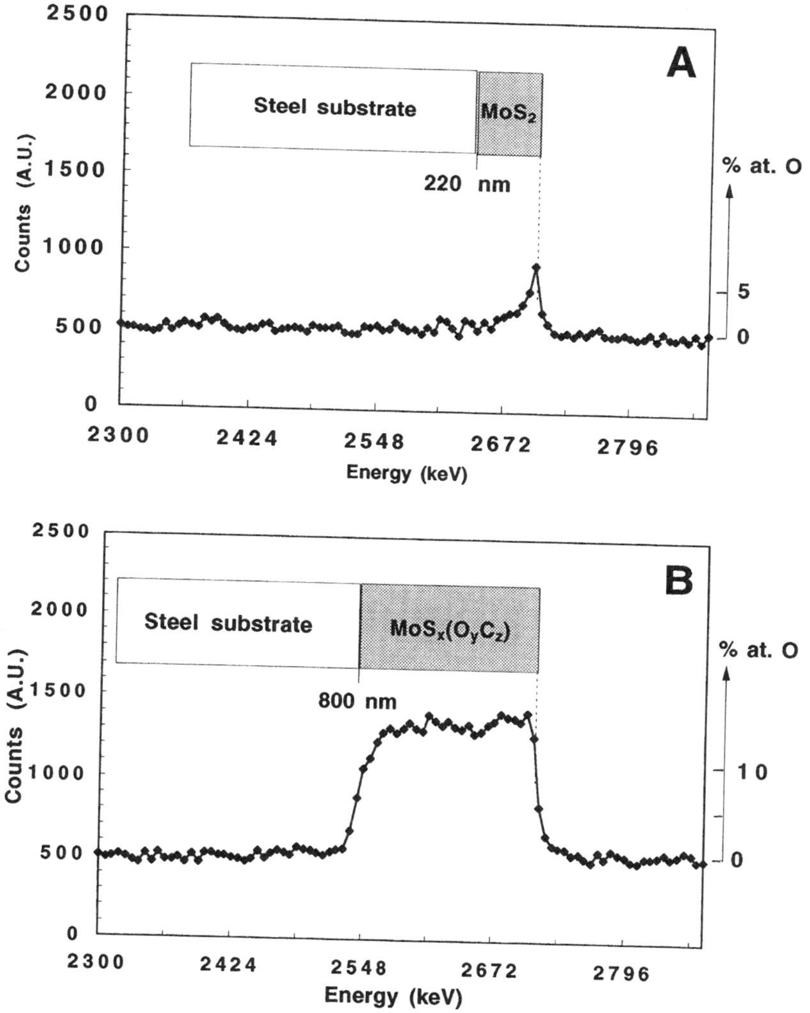

Figure 18. Quantitative oxygen depth profiles in pure (A) and contaminated (B) molybdenum disulfide coatings, as recorded by nuclear reaction analysis.

interaction between chalcogen atoms, leading to a decrease in the frictional force. Since the opposite phenomenon (higher friction) was observed, the second hypothesis is more plausible. EXAFS investigations would be necessary to take further the correlation between frictional behavior and the nature of the coatings.

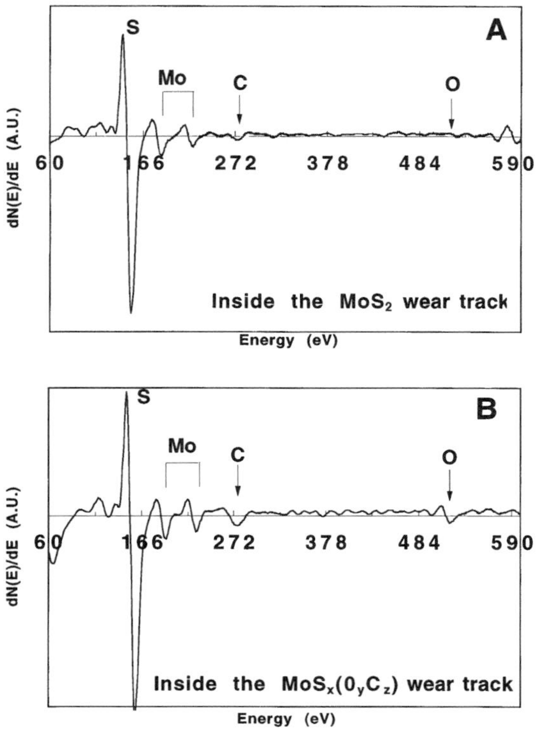

Figure 19. *In situ post mortem* AES microanalysis performed inside the wear scars of the superlubricant pure MoS_2 coating (A) and the contaminated coating (B) exhibiting higher friction.

The role of the oxygen contaminant was also established by *in situ post mortem* AES microanalysis performed inside the wear tracks of both coatings in the same UHV conditions as during friction. Figure 19 compares the two AES spectra, which show clearly the absence of oxygen and carbon on the outer surfaces of the wear scars of the superlubricant MoS_2, in contrast to the commercial coating which contained both carbon and oxygen.

Condition 3 was checked by a combination of *ex situ* AFM and *ex situ* HRTEM carried out respectively inside the wear scars and on the wear particles extracted from the contact area. AFM images (not reproduced here) confirmed the cleanliness of the external surface inside the wear scar, with the (002) easy-shear basal planes oriented parallel to the sliding direction. HRTEM micrographs from the wear fragments (Fig. 20) showed the existence of Moiré patterns characteristic of superimposed crystals with a misfit angle between them, as simu-

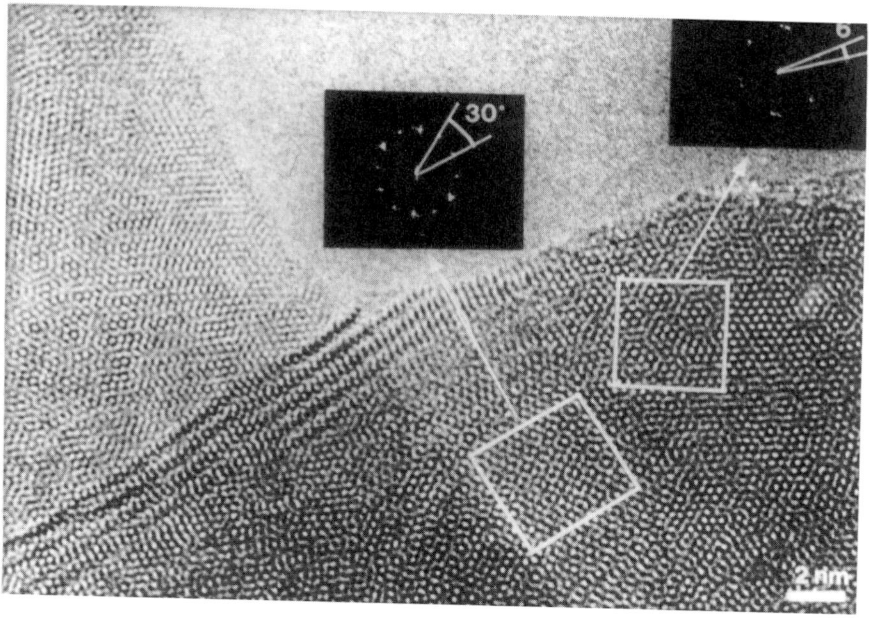

Figure 20. HRTEM micrograph of an MoS_2 wear particle, with atomic resolution. The electron beam was parallel to the *c*-axis. The calculated diffractograms correspond to the image frames showing Moiré patterns. Areas where atomic coincidence has disappeared are characterized by a misfit angle of 30° between two superimposed crystals.

lated in Fig. 17. Such information on the atomic scale thus helped to confirm the atomistic origin of the superlow frictional behavior and showed how it is possible for basic research to advance knowledge about solid lubrication.

3.9. Tribology of Carbonaceous Coatings [39]

This section addresses the question, what is the influence of the composition of carbonaceous films on their lubrication properties?

Amorphous diamond-like coatings (DLC), whose primary use is as hard coatings, possess some of the lowest friction coefficients ever measured in a wide range of environments from high vacuum to ambient air and dry air [40]. For the last 15 years, a great deal of work has been devoted to triboinvestigations of DLC coatings synthesized by physical and chemical vapor deposition processes (PVD, CVD), with a considerable spread in the friction and wear results. This may be explained by the diversity of structures and compositions, according to the preparative procedures and parameters. DLC coatings can be subdivided into two categories. The first is amorphous carbon (a-C, with negligible hydrogen content), usually produced by physical processes from a solid carbon precursor. The second, amorphous carbon containing hydrogen (a-C:H, with a hydrogen content ranging from a few to about 50 at.%), is prepared by chemical processes from a gaseous hydrocarbon precursor. Dopant elements (such as Si, F, metals) can be added to enhance adhesion, to decrease the surface energy, and to modify some mechanical properties, such as hardness and residual stress [41]. In particular, fluorinated Si-containing films are thought to be useful in reducing the atomic-scale wear and friction, as evaluated by scanning probe microscopy (SPM) [39]. From ultralow load friction and wear tests, it was shown that the microtribological characteristics of an a-C:H:Si:F film are improved by the silicon inclusions and the fluorination. Silicon inclusions increase the hardness of the film, and reduce microwear on the atomic scale. Fluorination decreases the surface energy and reduces atomic-scale microwear, as well as the microfriction force.

Of particular interest in the use of DLC is the deposition temperature, which can be lower than 200°C, thus making it possible to deposit coatings on most types of engineering materials, including polymers. Adhesion properties are generally improved by the use of intermediate layers, such as TiC, SiC, Ti or a TiN/TiC composite, the choice depending on the nature of the substrate. The strong atmospheric dependence of the friction and wear of DLC coatings has been the subject of many studies [41].

One of the principal motivations in the search for DLC coatings with low friction and wear under the widest range of environmental conditions is the need for reliable solid lubricants in technological fields that require the very highest reliability, for example magnetic storage devices, and in vacuum and space technology. In particular, space mechanisms are required to function not only in vacuum but also during the assembly, test and storage phases, which take place in ambient atmosphere during the months preceding space-vehicle launching. Although MoS_2 coatings systematically exhibit coefficients of friction higher than 0.1 in ambient air, due to irreversible tribo-oxidation, many studies have indicated the potentialities of DLC in sliding friction with coefficients less than 0.1, and with low wear rates under atmospheric conditions. Nevertheless few published works have appeared in which tribological investigations of a particular DLC coating in both vacuum and air conditions are reported [40]. In addition, no systematic correlation has been established between tribological phenomena and coating properties and characteristics, in terms of hydrogen content, carbon hybridization ratio, dopant element concentration and mechanical properties such as hardness and intrinsic stresses.

In the study considered here [39], tribochemical reactions have been identified by *ex situ* IR microanalysis of the contacting surfaces, transfer film and particles, thus allowing a better understanding of the tribological mechanisms as a function of the nature of the environment (air or vacuum) during friction. Analytical measurements in the reflection mode were performed using a micro-FTIR spectroscope equipped with a wire-grid polarizer. Two kinds of DLC coatings were studied using this methodology: a-C:F coatings produced by RF fluoropolymer sputtering, and a-C:H and a-C:H:Si coatings obtained from the decomposition of gaseous precursors.

The a-C:F coatings exhibited complex frictional behavior in an air environment: a running-in period (the first 100 sliding cycles) with coefficients of friction in the 0.21–0.35 range, a stable period (the next 300 cycles) in which the coefficient stabilized at 0.20, a transient period (the next 600 cycles) in which the value increased from 0.20 to 0.65, and finally a failure period characterized by decomposition of the coating. *Ex situ* IR measurements at the end of each friction regime provided correlation of these frictional fluctuations with molecular transformations of the C:F structure. It was found that the halogenated carbon framework decomposed into amorphous carbon and activated carbon, which then reacted with fresh metal through bond cleavage between halogen and carbon. The carbon-carbon bond cleavage also reduced the molecular size. The low-friction state occurred during the decomposition process, due to the presence of molecules containing fluorinated C=C moieties, such as polydifluorinated acetylene (PDFA), which were oriented in the sliding direction (as seen by the polarized IR probe).

The a-C:H and a-C:H:Si coatings exhibited frictional behavior strongly dependent on the hydrogen content (two different values, not quantified in the study) and the silicon content (10 and 40 at.%). The tribological tests were performed in air and in vacuum. It was found that amorphous carbon coatings containing a large amount of hydrogen had extremely low coefficients of friction in vacuum (about 0.01) but rather greater coefficients in air (about 0.2). A lower hydrogen content did not lead to any change in the friction in air, but to an increase up to 0.4 in vacuum. Other studies have attributed the origin of the observed low friction in UHV to the weak van der Waals interactions between hydrogen atoms [40]. The lowest coefficients of friction in vacuum (0.007) were recorded for an a–C:H:Si coating with a silicon content of about 10 at.%. Nevertheless even this ultralow frictional state was followed by a transient phase characterized by an irreversible increase in the coefficient, up to 0.3–0.4. Micro-IR investigations were performed at the point corresponding to the lowest coefficient of friction. This temporary high-lubrication performance was found to be attributable to the formation and adherence of sp^3 hydrocarbons produced on the ball surface from the rubbed film and oriented along the sliding direction.

Studies of this kind show how powerful the vibrational spectroscopies are in determining both the initial structures and the friction-induced modified structures of carbonaceous films, which are representative of typical thin coatings requiring specialized analytical methods, due to their composition (light elements) and their amorphous structure.

3.10. Tribochemistry of C_{60} Coatings [28]

This section addresses the question, do new carbonaceous compounds exhibit lubrication properties?

Carbon has always been an element of interest in tribological applications [42]. C_{60} (fullerene) molecules have unique properties in terms of stability, high load capacity, low surface energy, weak intermolecular bonding and spherical shape. Moreover, thin films of fullerenes can be obtained easily either from evaporation of a solvent solution or directly by sublimation of the powder in high vacuum. Fullerenes are therefore good candidates for solid lubrication, but little is known of their tribochemistry. Tribochemical modifications and frictional behavior of C_{60} coatings were investigated by an analytical tribometer coupled to a laser Raman optical microprobe (Fig. 6). *In vivo* post mortem Raman spectroscopy was carried out inside a static contact in order to study the structural modifications of C_{60} solid lubricant films in the contact area between two sapphire substrate transparent to the laser probe at the wavelength used.

In the following example, fullerene thin films were deposited by sublimation of C_{60} powder at 450°C. Raman tribometry analysis was performed with a micro-Raman spectrometer on C_{60} deposited on sapphire sliding against C_{60} deposited on steel in air at room temperature. Friction tests were carried out with an average contact pressure of 45 MPa, a radius of contact of 100 μm, and a normal force of 1.6 N. The contact zone was recorded by a video device which allowed the argon-ion laser beam (of power less than 1 μW at 541 nm) to be focused to a spot of 1 μm diameter. Because of the sensitivity characteristics of the detector and of the nature of the coatings, *in vivo pre mortem* analysis was not possible. With an acquisition time of 20 min, the sliding velocity would have been too low. Raman spectra were therefore acquired while the sliding motion was halted temporarily.

Figure 21 shows a series of Raman spectra acquired before friction, and after friction tests of 15 min (coefficient of friction 0.82 after 15 min) and a total of 110 min (coefficient of 0.40 after 110 min). Contact was maintained during analysis. Before the beginning of the friction test, the Raman spectrum (1) revealed the presence of the characteristic fullerene peaks, located at 1425, 1459, 1469 and 1570 cm^{-1}, in agreement with published results [43] measured from C_{60} in powder or solution form. After 15 min (470 cycles) of friction (spectrum (2)), the same Raman peaks could be distinguished but their intensities were much lower. From optical observations, the presence of rolling pins in the interface was noted, evidence for C_{60} film degradation. Spectrum (3) corresponds

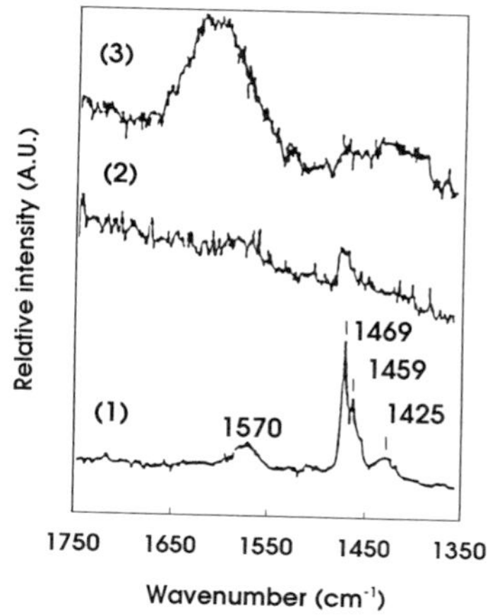

Figure 21. *In vivo post mortem* Raman analysis of a typical C_{60} film (1) before, (2) after 15 m, (3) after 110 m of friction, as detailed in Section 3.10. (From Ref. 28, with permission.)

to the Raman analysis after 110 min (3470 cycles) of friction. The C_{60} characteristic peaks can be seen to have disappeared and to have been replaced by two broad bands centered around 1430 and 1600 cm^{-1}, characteristics of a disordered sp^2 carbon structure. Simultaneously a large number of rolling pins were observed inside the contact.

A correlation between the coefficient of friction and the C_{60} lubricant film tribochemistry can be deduced from these experimental results; rolling pin formation produces a progressive decrease in the coefficient of friction, and the lubricant film modification corresponds to the amorphization of the C_{60} molecules.

4. SYNTHESIS AND CONCLUSION

Analytical tribology consists basically of the judicious coupling of surface analysis with tribological experiments. There are several essential types of information that are needed, such as surface chemistry, structure, morphology and other physical properties, on different dimensional scales in order to elucidate the tribological mechanisms and thus to be able to recommend specific mechanical designs in various technological fields. Analytical tribology can take full advantage not only of the substantial advances in the well-established surface analytical techniques but also of the emergent scanned probe microscopies. However, the said coupling between tribology and surface analysis requires a top-down approach as the current best practice procedure to guide the problem-

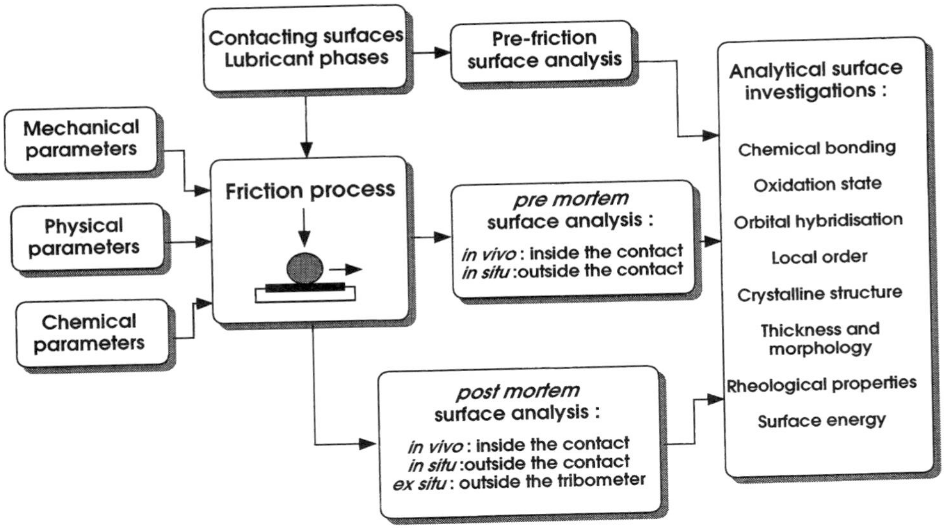

Figure 22. Synopsis of a typical top-down approach for surface analytical investigations in tribology.

solving methodology into the most rational, direct and valid analytical paths. This chapter gives an overview that can be summarized by the global generic scheme, illustrated in Fig. 22. Prefriction analysis is often required in order to have precise knowledge of the contacting surfaces and the lubricant phase prior to any dynamic contact between the tribopartners. The dynamic friction process must be perfectly controlled, in terms of mechanical parameters (contact geometry, contact pressure, vibrations), physical parameters (temperature, radiation) and chemical parameters (nature and pressure of the ambient atmosphere). The lack of standardization in tribological investigations is a real obstacle in reviewing studies in this field. However, the well-standardized surface analytical procedures allow the many generic results obtained by the numerous research laboratories working in this field to be correlated, thus providing significant progress in the control of friction and wear in various technological configurations. Surface analysis of contacting surfaces is performed more commonly after the end of the friction experiment (*post mortem* examination), requiring spatially resolved analysis inside the wear scars. Ideally the pre mortem examination would allow a direct real-time surface investigation of the contact, but it requires specific triboanalytical combinations in terms of the nature of the materials and/or the tribometer configuration. In both cases, the analysis can be performed inside the tribometer either outside (*in situ*) or inside (*in vivo*) the contact. Moreover *ex situ* examinations can be carried out systematically, which allows advantage to be taken of the full capabilities of the various surface analysis techniques, in spite of indirect examination due to air exposure of the tribosurfaces, which separates the friction experiment and the analysis step. Whatever the configuration, the analytical tribologist has to define precisely the nature of the pertinent surface analytical information required to clarify the tribological mechanisms and thus to propose significant improvements for controlling friction and minimizing wear in engineering devices.

ACKNOWLEDGMENTS

The author is grateful to his direct collaborators Professors J. M. Martin and J. M. Georges, Dr. P. Kapsa, M. Belin and T. Le Mogne for helpful discussions and valuable comments on the manuscript.

ACRONYMS

AES	Auger electron spectroscopy
AFM	Atomic force microscopy
CEMS	Conversion electron Mössbauer spectroscopy
CVD	Chemical vapor deposition
EDS	Energy dispersive (X-ray) spectroscopy
EELS	Electron energy loss spectroscopy
EHD	Elastohydrodynamic (lubrication)
ELNES	Electron (energy) loss near-edge spectroscopy
EXAFS	Extended X-ray absorption fine structure
EXELFS	Extended (electron) energy loss fine structure
FFM	Friction force microscopy
HREELS	High-resolution electron energy loss spectroscopy
HRTEM	High-resolution transmission electron microscopy
IBAD	Ion-beam-assisted deposition
IR	Infrared
MoDDP	Molybdenum dithiophosphate
PVD	Physical vapor deposition
RDF	Radial distribution function
SEELS	Surface electron energy loss spectroscopy
SEXAFS	Surface extended X-ray absorption fine structure
SFA	Surface force apparatus
STM	Scanning tunneling microscopy
TEM	Transmission electron microscopy
UFI	Ultrathin film interferometry
UHV	Ultrahigh vacuum
XANES	X-ray absorption near-edge spectroscopy

XAS	X-ray absorption spectroscopy
XPS	X-ray photoelectron spectroscopy
XRD	X-ray diffraction
ZDDP	Zinc dithiophosphate

REFERENCES

1. B. Bhushan and B. K. Gupta (eds.), *Handbook of Tribology*, McGraw-Hill, New York, 1991, p. 229.
2. F. P. Bowden and D. Tabor, *The Friction and Lubrication of Solids*, Part 2, Clarendon Press, Oxford, 1964, p. 210.
3. G. Heinicke, *Tribochemistry*. Carl Hanser Verlag, München, 1984.
4. J. M. Martin, M. Belin, J. L. Mansot, H. Dexpert and P. Lagarde, *ASLE Trans. 49(4)*, 523 (1986).
5. I. L. Singer, *Surf. Coat. Technol. 49*, 474 (1991).
6. S. M. Hsu and R. S. Gates, *Tribochemistry and Lubrication*, Proc. Int. Tribology Conf., Yokohama (Japan) 1995, Satellite Forum on Tribochemistry, p. 43. To be published in *Tribol. Lett.*
7. K. Holmberg and A. Mathews, *Coatings Tribology*, Tribology Series, 28 (D. Dowson, ed.) Elsevier, 1994.
8. I. L. Singer, Solid lubrication process, in *Fundamentals of Friction: Macroscopic and Microscopic Processes* (I. L. Singer and H. M. Pollock, eds), Kluwer, Boston, 1991, p. 237.
9. K. Miyoshi and D. H. Buckley, *ASLE Trans. 22*, 245 (1979).
10. K. Miyoshi and Y. W. Chung (eds.), *Surface Diagnostics in Tribology*, World Scientific, Singapore, 1993.
11. K. Miyoshi, in Ref. 10, p. 93.
12. B. K. Teo and D. C. Joy, *EXAFS Spectroscopy: Techniques and Applications*, Plenum Press, New York, 1979.
13. R. F. Egerton, *EELS in the Electron Microscope*, Plenum Press, New York, 1986.
14. J. M. Martin and M. Belin, *Thin Solid Films 236*, 173 (1993).
15. Y. W. Chung, in Ref. 10, p. 33.
16. B. Bhushan, J. N. Israelachvili and U. Landman, *Nature 374*, 609 (1995).
17. A. M. Homola, in Ref. 10, p. 271.
18. J. N. Israelachvili, P. M. McGuiggan and A. M. Homola, *Science 240*, 189 (1988).
19. J. M. Georges, D. Mazuyer, J. L. Loubet and A. Tonck, in Ref. 8, p. 263.
20. S. J. Hirz, A. M. Homola, G. Hadziioannou and C. W. Franck, *Langmuir 8*, 328 (1992).
21. S. Granick, *Science 253*, 1374 (1991).
22. J. Klein, D, Perahia and S. Warberg, *Nature 352*, 143 (1991).

23. H. A. Spikes, *Boundary Lubrication Films*. Proc. Int. Tribology Conf., Yokohama (Japan) 1995, Satellite Forum on Tribochemistry, p. 49. To be published in *Tribol. Lett.*
24. G. J. Johnston, R. Wayte and H. A. Spikes, *STLE Tribol. Trans. 34*, 187 (1991).
25. M. Belin and J. M. Martin, *STLE Tribol. Trans. 32*, 410 (1989).
26. M. Belin and J. M. Martin, *In situ Structural Changes of Lubricated Surfaces, as Studied* by EXAFS, Proc. Leeds-Lyon Symp. on Tribology–Wear Particles (D. Dowson et al., eds.), Elsevier 1992, p. 413.
27. J. M. Martin, T. Le Mogne, L. Frelin and T. Palermo, *How to Study the Respective Rôle of Adsorption and Friction in the Tribochemistry of Antiwear Additives*, Proc. Int. Tribology Conf., Yokohama (Japan) 1995, Satellite Forum on Tribochemistry, p. 67. To be published in *Tribol. Lett.*
28. J. M. Martin, M. Belin, T. Le Mogne, C. Donnet, J. M. Millet and N. Millard-Pinard, *Towards the in situ Analysis of Tribological Surfaces*, 6th Conf. on Applications of Surface and Interface Analysis, Montreux (Switzerland), 1995. To be published in *Surf. Interface Anal.*
29. S. Mori, *Tribochemical Activity of Nascent Metal Surfaces*, Proc. Int. Tribology Conf., Yokohama (Japan), 1995, Satellite Forum on Tribochemistry, p. 37. To be published in *Tribol. Lett.*
30. B. M. DeKoven, in Ref. 10, p. 299.
31. S. V. Pepper, *J. Appl. Phys. 47*, 2579 (1976).
32. J. M. Martin, T. Le Mogne and M. N. Gardos, *Friction of Alpha Silicon Carbide under Oxygen Partial Pressure: High Resolution Analysis of Interface Films*, Proc. Japan Int. Tribology Conf., Nagoya, Japan, 1990, p. 1407.
33. L. E. Seitzman, R. N. Bolster, I. L. Singer and J. C. Wegand, *Tribol. Trans. 38*, 445 (1995).
34. T. Le Mogne, C. Donnet, J. M. Martin, N. Millard-Pinard, A. Tonck, S. Fayeulle and N. Moncoffre, *J. Vac. Sci. Technol. A1 2(4)*, 1998 (1994).
35. J. M. Martin, C. Donnet, T. Le Mogne and T. Épicier, *Phys. Rev. B48(14)*, 10583 (1993).
36. J. M. Martin, H. Pascal, C. Donnet, T. Le Mogne, J. L. Loubet and T. Épicier, *Surf. Coat. Technol. 68/69*, 427 (1994).
37. S. V. Didziulis and P. D. Fleischauer, in Ref. 10, p. 135.
38. E. W. Roberts, *Trib. Int. 23(2)*, 95 (1990).
39. S. Miyake, in 10, p. 183.
40. C. Donnet, *Condensed Matter News 4(6)*, 9 (1995).
41. A. Grill and V. Patel, *Diamond Rel. Mater. 2*, 597 (1993).
42. H. O. Pierson, *Handbook of Carbon, Graphite, Diamond and Fullerenes*, Noyes Park Ridge, NJ, 1993, p. 356.
43. B. Chase, N. Herron and E. Holler, *J. Phys. Chem. 96*, 4262 (1992).

17
Catalyst Characterization

W. E. S. UNGER and T. GROSS Bundesanstalt für Materialforschung und -prüfung (BAM), Berlin, Germany

1. INTRODUCTION

The analytical characterization of heterogeneous catalysts occupies a central position in catalysis research, whether fundamental, applied or industrial. The various activities are guided by the overall objective of gaining a detailed chemical and structural picture of the catalysts themselves. The information obtained can be used to develop better understanding of the function of a catalyst, leading to the formulation of new catalysts and the improvement of existing ones. In industry the results of characterization are often useful for quality management and for control of process economy. Consequently, there are two approaches to catalyst characterization. One includes all efforts to establish empirical relations between catalyst performance and such parameters as elemental composition and chemical state, particle size and dispersion, etc. The other more fundamental and academic approach, is to determine the catalyst's surface composition and structure on the atomic scale, preferably under reaction conditions but with a simplified catalyst model. The latter studies are often carried out on well-defined surfaces such as those of single crystals. To match the overall aim of this handbook the emphasis here will be on the characterization of real catalysts and on appropriate models.

Nowadays it is becoming common practice to bring multiple analytical techniques to bear on the characterization of catalysts. Very often bulk and surface analysis methods are combined to yield detailed complementary information. A typical scheme of catalyst parameters to be determined and of appropriate analytical methods has been presented by Delmon [1], who reviewed the potential of surface analysis in the characterization of hydrodesulfurization (HDS) catalysts. This scheme, redrawn, is shown in Fig. 1. Unfortunately, due to certain inherent problems, the whole range of well-established surface spectroscopic methods cannot be applied in every case. Another disadvantage is that the application of surface-specific spectroscopies requires the transfer of catalyst samples into vacuum, thus restricting the possibilities for genuine *in situ* surface analysis of working catalysts.

Statistical analysis of papers covering catalyst characterization has revealed that XPS is the most frequently applied surface analysis technique in the field [2]. For that reason this chapter will deal mainly with applications of XPS.

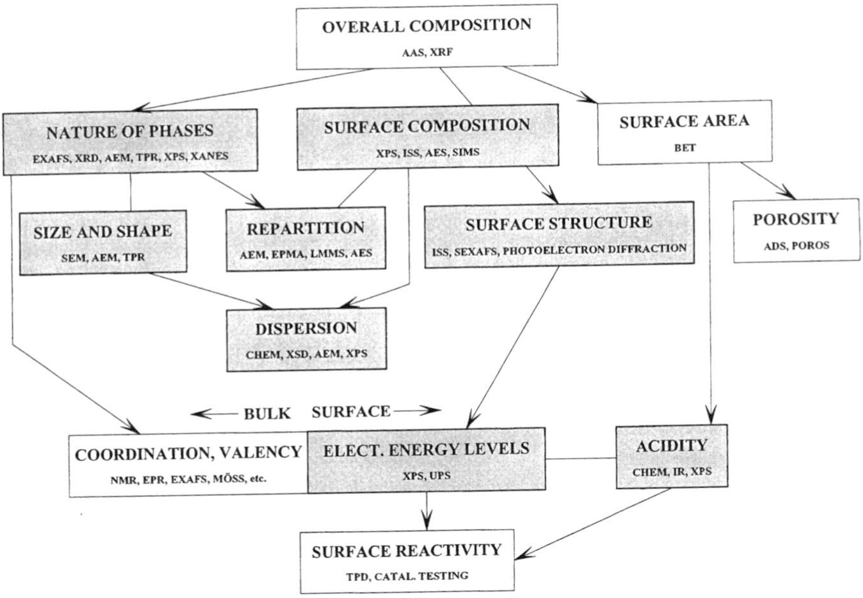

Figure 1. Scheme of information available from various techniques applied in catalyst characterization (redrawn from Ref. 1). Fields in which surface spectroscopies can contribute are highlighted.

Where appropriate, combinations of XPS with other characterization methods are described. The temperature-programmed and sorption techniques, which are important methods in the surface characterization of catalysts, have not been considered here in much detail because to do so would be beyond the scope of this handbook (but see Appendix 3). Easily accessible information on the state-of-the-art and on problems in this field can be found in the literature [2,3]. It should be mentioned that calibration and validation, necessary for successful application of the sorption methods, can be accomplished using surface spectroscopic tools.

For the "best practice" requirements in the application of surface analytical methods frequent reference is made below to the most recent standard terminologies, practices and guides, published in the relevant *Annual Book of ASTM Standards*. ASTM Committee E-42 on surface analysis has published standards for surface analysis over the past 20 years. On the international scene the ISO/TC 201 committee on surface chemical analysis was established in November 1991 to address the need for international standards. Subcommittees or working groups of ISO/TC 201 may request ASTM documents to be used as a starting point in the development of ISO standards. ISO defines the environment, scope, purpose, organization and work programs in a strategy policy statement for ISO/TC 201, which is in continual revision.

2. APPLICABILITY OF SURFACE SPECTROSCOPIES IN CATALYST CHARACTERIZATION

This section considers the applicability of the four principal surface analytical techniques that are usually applied to catalyst surface characterization. They are XPS (coexisting acronym is ESCA which also includes XAES), AES, SIMS, including its nondestructive static SIMS mode and the use of neutral atoms as primary particles in fast atom bombardment mass spectrometry (FAB-SIMS or FABMS), and last, ISS.

XPS is the most commonly used and most useful surface analysis method in catalyst characterization. It can be used for both qualitative and quantitative analysis for almost all kinds of catalysts used in heterogeneously catalyzed reactions. XPS studies of oxide, sulfide, fluoride, halide, etc., catalysts, supported metal catalysts, Raney or gauze metal catalysts and zeolite catalysts are all possible. The samples can be studied in any of the precursor, calcined, reduced, activated, deactivated, aged or poisoned states. Real industrial catalysts can be analyzed as well as fundamental model systems. Quantitative analysis is possible either with the help of empirical sensitivity factors or by "standard-free" methods. In the latter case appropriate theoretical models are used with photoionization cross-section tables, inelastic mean free path (λ) data and individual spectrometer transmission functions. However, there are also some features that restrict the application of XPS in catalyst characterization:

1. The analysis must be carried out under good high-vacuum conditions, at the worst, and preferably UHV.
2. X-ray damage has to be kept under control.
3. The analysis can suffer from line interference, which may occur easily with multicomponent catalyst samples.
4. Detection of minority constituents (<1 at.%) can be impossible due to lack of sensitivity.
5. Lateral resolution is limited at present to a few μm with the most advanced spectrometers.

The information depth in XPS can be specified as the thickness of material from which 95% of the measured signal is generated [4], and is of the order of a few nm, typically 3λ (see Chapter 4). An enhanced surface specificity is possible in the angle-resolved mode but only when smooth and planar surfaces are analyzed. A consequence of the magnitude of the information depth of XPS is that with high-surface-area samples the bulk, as well as inner and outer surfaces, contributes to the signals detected. Obviously, in that situation XPS is no longer a true surface analytical technique. For more detailed information about the application of XPS the reader is referred to Ref. 2. Note that a worldwide effort, under the heading of spectromicroscopy, is currently under way to achieve significantly enhanced lateral resolution in XPS [5]. The overall objective is a lateral resolution in the few nm range, which would match the demands of catalyst characterization.

The applicability of AES is limited in the characterization of real catalysts. This is due to the fact that, generally speaking, nonconducting samples cannot be analyzed. Thus only certain types of samples from the real world, e.g., Raney or metal gauze catalysts, are suitable for AES analysis. On the other hand, the applicability of AES is wide when fundamental models are studied in catalytically motivated surface science. Many aspects of gas and temperature treatments of metal or semiconductor surfaces, including segregation phenomena in multicomponent systems, have been studied extensively and successfully. AES may, in principle, provide chemical-state information with information depths very similar to those characteristic of XPS. However, the damage induced by the intense beam of energetic primary electrons must be carefully controlled. Another feature of AES, the possibility of applying it in a scanned mode (SAM) at sub-μm lateral resolution down to some tens of nm, makes it attractive in special cases. Niemantsverdriet [2] has considered the state of the art of AES in his handbook on spectroscopies applied to catalyst characterization.

SIMS is acceptable in catalyst characterization but has been used very much less often than has XPS. It can, in principle, be used for studying real catalysts including zeolites. Again there is a severe charging problem, which must be solved for each individual case with the appropriate means, e.g., an electron flood gun or the use of neutrals as primary particles (FAB-SIMS). This is one reason why SIMS is unlikely to be used as a tool for the routine analysis of technical catalysts. So it is not surprising that the most impressive SIMS applications have been made on model catalysts simulating selected features of real catalysts. In such system the samples can be "designed" in a way which allows optimal SIMS analysis. Finally, the SIMS method has contributed successfully to surface science studies of gases on single-crystal surfaces, including kinetics, motivated by catalysis science. There are important advantages of SIMS. The first one is the enhanced surface specificity with respect to XPS. Dedicated simulations have revealed that the secondary ions detected in SIMS originate exclusively from the first two monolayers (roughly 85% stem from the first layer [6]). The second advantage of SIMS is its unique elemental sensitivity, down to the ppm level for certain elements, e.g., the alkali metals. This feature is useful when traces of promoters or poisons are of interest. The third advantage is the possibility of elemental or (limited) chemical mapping of a catalyst surface with lateral resolution down to roughly 50 nm (imaging SIMS) by using secondary element or molecular ion species. Molecular fragments may carry chemical-state information provided SIMS is applied in its relatively nondamaging, static mode (SSIMS). The main disadvantage of SIMS is that the quantification of raw data is extremely difficult. Sputter yields and ionization probabilities are the main parameters defining secondary-ion yields in SIMS. Sputter yield data for multicomponent materials such as real catalysts scarcely exist. The ionization probability itself varies over five decades across the periodic table of elements. Furthermore, there is a huge matrix effect when the same element or a molecular fragment is sputtered from chemically different surface sites. On the other hand, once understood, this same extreme matrix effect can be exploited for purposes of qualitative analysis on model catalysts. The current state of affairs of SIMS in catalyst characterization has been reviewed [7].

A thorough overview of the advantages and disadvantages of the applicability of low-energy ion scattering (LEIS) spectroscopy, usually referred to as ISS, in catalysis research has been given in Ref. 8. ISS can be applied to real as well as to model catalyst samples. Moreover, it can make a valuable contribution in catalytically motivated surface science studies. When the surfaces of nonconductive catalyst samples are to be analyzed, charge compensation by flood gun is recommended by most workers in the field. The ISS method is extremely surface specific because its analytical data are defined almost exclusively by the outermost atomic layer. The main applicability of the method in catalyst characterization is in following the location or dispersion of elemental constituents at a catalyst's surface. This can be achieved, in the most favorable cases, in a semiquantitative manner by utilizing elemental intensity ratios. A shortcoming of ISS is that it cannot, generally speaking, provide chemical-state information. ISS is therefore used in catalyst studies most often with other surface analysis techniques, for instance with XPS, providing complementary information. The ISS/XPS combination is favored because in standard photoelectron spectrometers the energy analyzer, used for photoelectron kinetic energy analysis, can be easily switched to scattered ion energy measurement. An ion source suitable for ISS is usually available with standard spectrometer configurations.

3. SAMPLE DAMAGE

Without any doubt analysis is invalidated when the sample is damaged by the probe itself. X-ray, electron and ion induced sample damage is common during XPS, AES, SIMS or ISS analysis of catalysts, and can cause the spectrum to change as a function of time of exposure. The most prominent damage effect is the chemical reduction of sensitive species by heat or by particle or photon irradiation in vacuum. In particular, catalyst precursors such as precipitates, impregnated supports, metal salts and organometallic compounds, must be analyzed with utmost caution. However, there are also many examples in which catalysts in calcined states are also sensitive to irradiation. A well-known one is the reduction of Cu^{2+} species in calcined copper catalysts during XPS analysis. A detailed study of the damage effects observed with several Cu^{2+} compounds has been made in Ref. 9; in Fig. 2 are shown the results in an extreme case. Several approaches aimed at reducing this damage have been proposed and evaluated. First of all, the heat load on the sample surface should be minimized, which can be achieved either by cooling the X-ray housing or by using a monochromated X-ray beam. Additional electron irradiation of the sample to compensate for charging is necessary in the latter case. Cooling the sample holder can be useful, but the real problem is removing the heat from the specimen surface to the sample holder, where the cooling actually occurs. Condensation of residual gas constituents then often occurs. In any case it is always possible to minimize the irradiation time itself, especially when a sensitive state-of-the-art spectrometer and appropriate sample preparation are used. Compromises with respect to the optimum signal-to-noise ratio then have to be considered.

Recently, an XPS data base on polymers, another class of radiation-sensitive materials, has been published [10]. The authors sought an approximate guide to the damage rate which would be useful in the optimization of measurement conditions, including acquisition times. They defined a degradation index which gives the percentage damage after t = 500 min at a fixed X-ray source power. The X-ray degradation index of a sample is obtained from a graph of X_t/X_0 plotted versus time t, where X is some characteristic parameter to be obtained from sequentially recorded XPS spectra (for instance, an atomic ratio or the intensity

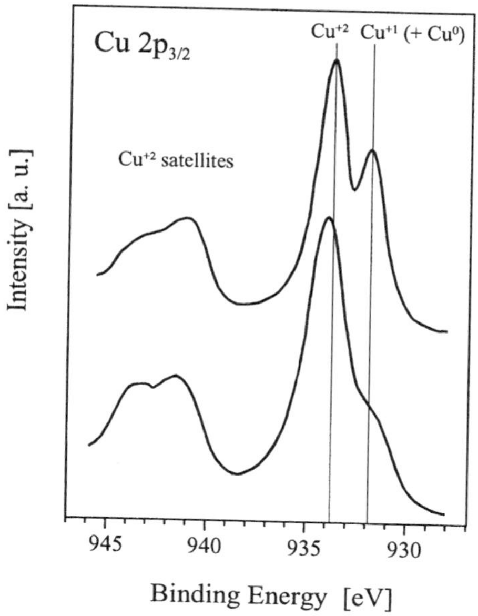

Figure 2. XP Cu $2p_{3/2}$ spectra from $CuCl_2$ partially reduced during 3 h X-ray irradiation in different spectrometers (upper spectrum: high thermal load; lower spectrum: reduced thermal load due to water-cooled X-ray housing) at equal X-ray doses of 8×10^{17} photons/mm^2). (From Ref. 9.)

of selected spectral features). This procedure can be applied easily to XPS analysis of catalysts and to the other surface spectroscopies. In an analysis of calcined Cu^{2+} catalysts the authors successfully used the $2p$ satellite/main peak intensity ratio as the decreasing X_t parameter in order to determine optimal times for the acquisition of spectra from a nearly nondamaged sample.

The conclusion is that, because sample damage cannot be fully ruled out, quotation of a damage index as defined above may improve significantly the traceable quality of analytical data. The requirements of the ASTM standard practice E1015-90 [11] could be met in this way.

4. SAMPLE PREPARATION

Industrial catalysts are, in most cases, pellets pressed from active component powders, often mixed with certain binders, for example technical grade graphite or kaolinite. In their normal form the pellets are unsuitable for surface analysis because of several experimental problems, amongst them surface charging and high rates of outgassing into the spectrometer vacuum system. Despite this they can in principle be analyzed by XPS. However, it is recommended that catalyst pellets are ground to powder before analysis. When binders are used, the samples are extremely heterogeneous and there is a risk of spectral interference between the active component features and those originating from the binder. Powders are also the usual form of laboratory-made catalysts, but the addition of binders should be avoided whenever possible. A very simple way of mounting a catalyst powder for analysis is by sticking it on a sample holder with the help of a double-sided adhesive film. Here the analyst must be aware of spectral interferences (typically C and Si signals) caused by the adhesive material itself. Improved signal-to-noise ratios, together with reduced sample damage risk (cf. Section 3), may be obtained when the powders are pressed into self-supporting thin wafers. For example, alumina-supported catalysts can be prepared in this way. However, it often occurs that binder-free catalyst powders do not give stable wafers. Under those conditions, the powders can be pressed as thin films into metal grids or soft metal foils, e.g., In or Pb [7,12]. When a study of heat treatment effects is intended, it has been found useful to press the powder into a thin but integral film on a low-volatility metal foil which assists the lateral homogeneity of the temperature distribution. Al or Au foils are highly suitable [13]. Another useful approach for the fixing of catalyst powders on a sampler holder is the use of a catalyst particle dispersion in a simple, high-purity organic solvent [14]. The sample holder must be covered by the dispersion. A stable layer of powder particles is obtained after evaporation of the solvent provided the right combination of solvent, particle concentration and sample holder metal has been found. Good results have been obtained with zeolite dispersions in isopropanol or high-purity water, deposited on Ni surfaces.

Finally, reference must be made to the ASTM *Standard Guide for Specimen Handling in AES, XPS and SIMS* (E1078-90) [15], which covers common recommendations and requirements for sample handling prior to, during and following surface analysis and is of basic importance for catalyst characterization, too.

Often a catalyst sample has to be analyzed in a condition which is sensitive to atmospheric exposure. For instance, samples taken from pilot or plant reactors can be stored in containers under a protective atmosphere and inserted into a spectrometer with the help of a glove box [12]. This sounds simple, but a fully controlled realization of this experimental concept is a great challenge for the analyst. Another approach relies on the preparation of a selected catalyst state within a special reaction chamber containing heat and gas treatment options, which allows transfer of the sample into the analysis chamber without any air exposure. Some surface analysis instrumentation manufacturers offer this optional chamber on their standard spectrometers. There are many papers on this topic, covering the whole range from basic research on intermediates in catalyzed reactions on single-crystal faces [16] to the activation of, for instance, commercial Cu-Zn-Al oxide methanol synthesis catalysts [13].

More recently it has been found that thin alumina or silica films on conducting Al or Si substrates, acting as model supports, offer excellent possibilities for the study of catalyst preparation by all surface analysis techniques currently available. Catalytically active components can be anchored on these supports with exactly the same wet chemical impregnation techniques and heat/gas treatments as applied normally in technical catalyst production [17]. Sophisticated procedures have allowed the preparation of well-defined γ-alumina grained layers on Al substrates [18]. One advantage of the approach is the strongly enhanced spectral, and therefore also chemical, resolution obtained for instance in XPS analysis. This is illustrated in Fig. 3.

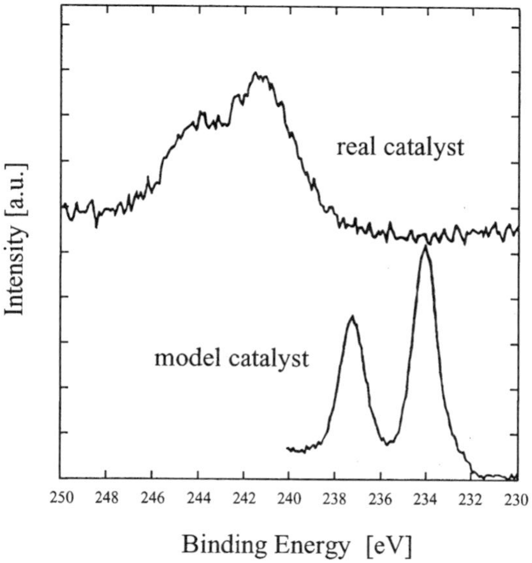

Figure 3. XP Mo 3*d* spectra of MoO_3 on (upper spectrum) commercial silica and (lower spectrum) a thin-film SiO_2 model support, recorded under the same conditions. The spectra are presented as recorded. The positive binding energy shift and the line broadening for the upper spectrum are due to charging. Clearly the signal-to-noise-ratio and the (chemical) resolution of the method are significantly improved in the lower spectrum. (From Ref. 2.)

5. CHARGING OF INSULATOR SURFACES BY THE PROBE

Irradiation of an insulator by photon, electron or ion beams results in charging of the specimen surface. Often catalysts are insulating samples. Real catalysts are often multiphase systems with complex morphology. In the course of analysis the surface of this type of specimen can become differentially charged. The phenomenon of lateral surface potential distributions on irradiated insulator surfaces has been verified recently for XPS by a spectromicroscopic approach [19,20]. Figure 4 demonstrates how individual spectra originating from the differentially charged regions of the same sample can be mutually shifted in energy. Merging them results in a broadened "total spectrum," as usually measured in a standard XPS analysis. As a consequence not only is the chemical resolution of the technique reduced, but any attempt to analyze the peak shape in terms of physical parameters would be irrelevant. That is the reason

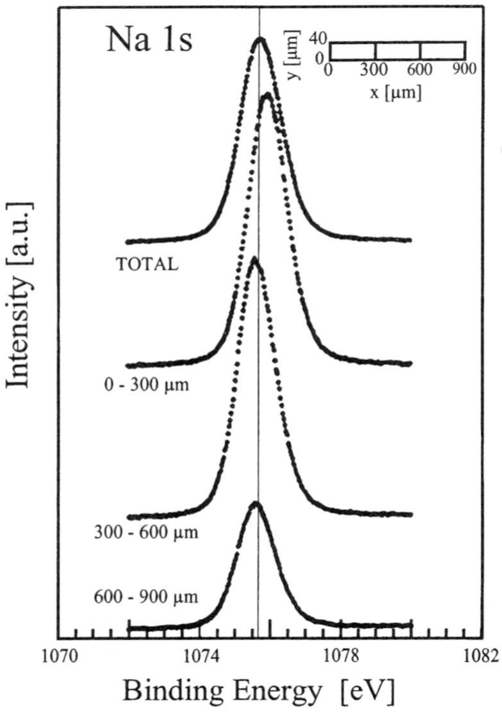

Figure 4. Na 1*s* XP spectra from an NaCl surface recorded from three adjacent 40 × 300 μm areas along a line, taken in the presence of a flood gun. The local potential defines peak parameters, e.g., the BE. (From Ref. 19.)

why the authors in Ref. 19 recommend that the surface analysis of insulators should always be performed in the laterally resolved mode.

In any case the charging problem is critical in chemical-state analysis of an insulating catalyst surface because reliable and reproducible results can be obtained only when the charging is properly under control; that is, stabilization, homogenization and compensation, or quantitative determination of the distribution of surface charge, must be established during analysis. In the literature a broad range of individual approaches to the handling of the charging problem can be found. Most solutions are based on the use of electron flood guns, often combined with decoration of part of the sample surface with conductive materials (grids, apertures, gold overlayers). However, there is as yet no unified approach. For XPS there is a standard guide to charge control and charge referencing techniques published by ASTM [21], which describes the state of the art. Recent advances in monochromatic XPS instrumentation, i.e., providing substantially enhanced photon flux density, require now the development of improved charge neutralization strategies. The manufacturers of surface analytical equipment have considered this problem and now provide redesigned electron flood guns with high output which meet the requirement of an enlarged effective emission area with a small energy spread of thermionic electrons [20]. Another successful approach is to use a well-stabilized flood gun built coaxially into the analyzer lens assembly.

6. CHEMICAL-STATE ANALYSIS WITH XPS BY FINGERPRINTING AND REFERENCE TO DATABASES OR TO CHEMICAL-STATE PLOTS

Obtaining chemical-state information about the elements present at a catalyst surface is the most important objective of qualitative catalyst characterization. It can be achieved, more or less successfully, by employing XPS on the basis of specific chemical shifts in the binding energies of photopeaks. These energies depend on the respective nearest-neighbor coordination spheres of the atoms under consideration (cf. Chapter 5). One approach is simply that of fingerprinting, which relies on reference binding energy data taken from the literature, from databases (the most prominent is the NIST *Standard Reference Data Base 1.0* [22]) and from the analysis of well-defined reference samples with properly calibrated instruments. Energy calibration should follow the recommendations given in Ref. 23 which are under consideration as an ISO standard.* An ideal reference is a single-phase sample with no differences between bulk properties (which can be determined by standard chemical and structure analytical methods) and those of the surface. Unfortunately, in most cases relevant

*An ASTM standard is in preparation, too.

to catalyst characterization, such ideal reference samples cannot be found. Often a specimen surface will undergo reaction with atmosphere or humidity, thus, for instance, oxide surfaces may become hydroxylated. An expensive way out of this problem is to prepare reference samples in vacuo when that is possible.

Once reliable BE reference data are available, a reliable method of comparison of them with data obtained from the catalyst surface under consideration must be found. But first the BE scale of the photoelectron spectrum must be corrected for charging. The earliest, and still the most commonly used, method of static charge correction of the BE scale of insulating catalyst specimens is to employ the C 1*s* peak binding energy of the adventitious carbon (aliphatic species), which is present on all samples of interest here. Usually it is fixed at a value, BE_{fix}, between 284.6 and 285.0 eV (there are different conventions!) and the whole photoelectron spectrum is then shifted by the binding-energy difference $BE_{fix} - BE_{measured}$ [21]. This convenient calibration method possesses several uncertainties, as pointed out in Refs. 20 and 24. During XPS analysis of real catalysts in particular it is often found to be impossible to apply the correction procedure adequately because large systematic errors may intrude due to the complexity of the C 1*s* spectra. The problem can be regarded as a special effect arising from recent developments in XPS equipment. Most of the new systems possess oil-free pumps. Thus the internal source of the uniform external reference material "adventitious carbon" has dried up. Under these conditions the nature of the contamination carbon, which is still observed, can be very different depending on the individual history of a sample. It can no longer be taken for granted that the aliphatic hydrocarbon species unequivocally dominates the C 1*s* spectrum. Figure 5 gives a typical example taken from Ref. 25. In that situation it is useful to look for other photopeaks which could be used for static charge referencing. In catalyst characterization support-related photopeaks are interesting candidates provided they can be proved to be sufficiently stable. For instance, Moretti [26] has reported results for an impregnated alumina catalyst where binding energies used for qualitative analysis were obtained by charge referencing with the Al 2*p* and Al 2*s* lines (at 74.5 and 119.3 eV, respectively). In selected cases the corresponding Si photopeaks could also be used for charge referencing in silica-supported catalysts. Another example is the use of the Zn $2p_{3/2}$ photopeak as a static charge reference in the study of different CuO-ZnO impregnated and coprecipitated catalysts [27]. The possibilities and limitations of these approaches have been discussed to some extent in Ref. 25, where it was concluded that the errors in the resulting binding and kinetic energies obtained after referencing cannot be smaller than ± 0.2 eV.

An alternative, but also not universally used, method for BE referencing is to use the photoemission from well-defined noble metal particles deposited on a sample surface to provide an external static charge reference [20,28,29]. Suitable Au and Pd particles can be obtained from a colloidal dispersion. Figure 6 gives

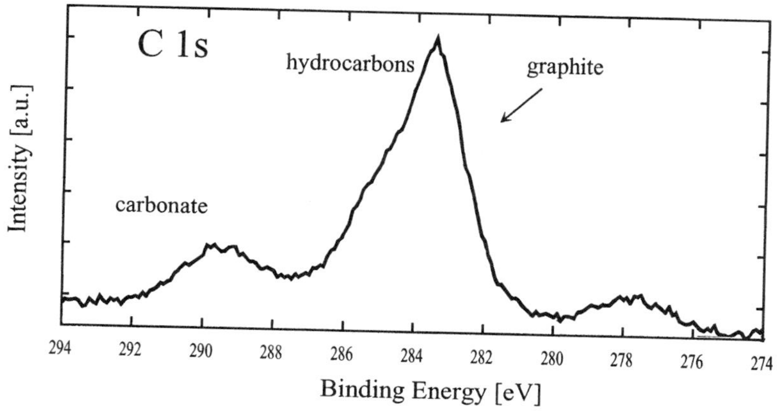

Figure 5. Complex XP C 1*s* spectrum recorded from a calcined Cu-Zn-Al oxide/graphite-binder methanol synthesis catalyst; the spectrum contains carbonate, hydrocarbon and graphite contributions. Obviously, determination of the aliphatic hydrocarbon reference C 1*s* binding energy by fitting this spectrum would not yield an unambiguous result. (From Ref. 25.)

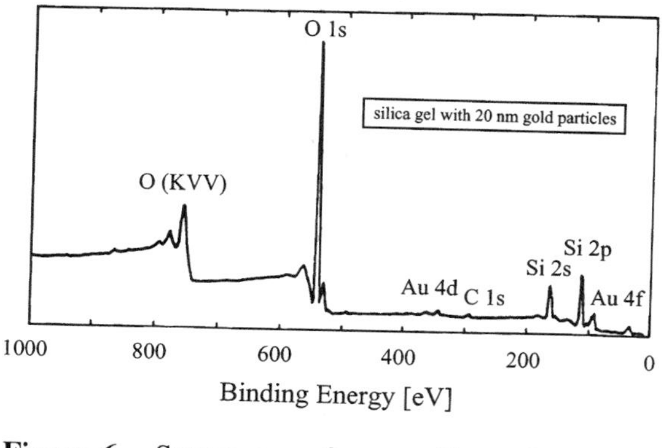

Figure 6. Survey scan from a silica gel sample (pressed powder pellet) with 20 nm diameter gold particles deposited on it. The high-resolution spectrum of the Au 4*f* doublet allows a static charge correction to be applied. (From Ref. 29.)

an example in which 20 nm diameter Au particles were deposited onto a silica gel support to provide an external Au $4f_{7/2}$ charge reference. The main advantage of this approach is that the metal particles are characterized by a well-defined particle size distribution. Moreover, the respective photopeaks are sharp, and the Au $4f_{7/2}$ and Pd $3d_{5/2}$ binding energies are accepted as references by the whole community [23]. The disadvantages of methods based on *in situ* evaporation of Au onto a specimen surface, which results in a rather broad Au particle size distribution, accompanied by particle size effects on the shape and the BE of the Au photopeak, are thus avoided. In addition, this external reference allows confirmation or otherwise of the positions of possible internal static charge reference lines, among them the adventitious carbon C 1*s* reference [25,29]. Finally it must be remembered that the simple fingerprint approach outlined above for qualitative analysis of a catalyst sample allows differentiation between coexisting species characterized by binding-energy differences of ≥0.5 eV, at a rough estimate, when the spectra are compared with those from high-quality reference samples under optimal conditions (or from a reliable database).

A more sophisticated approach to qualitative analysis is to use the photopeak BEs plotted together with the KEs of X-ray-excited Auger peaks and the modified Auger parameter α^* in a so-called chemical-state plot. Details of the principles of the chemical-state plots have been reviewed [30]. Briefly, the BE of the most intense photopeak of the element of interest is plotted on the abscissa while the KE of a prominent Auger transition is plotted on the ordinate. α^* is defined as the sum of the KE of the Auger electron and the BE of the photoelectron. With the use of α^* charge corrections are unnecessary because they cancel as a result of the summation. Moreover, the basic principles underlying the concept of the Auger parameter imply that detailed considerations of the physical chemistry of a sample are possible. For instance, differences in the final-state extra-atomic relaxation energy can be measured for an element in different chemical states. This energy is related to core hole screening and, consequently, to polarazibility and local geometry surrounding a photoionized atom. Usually an α^* grid in chemical-state plots is drawn as a family of parallel lines with unity slope. All points on any of these lines have the same α^*. One advantage of presenting data in this way is that each well-defined chemical state occupies a unique position on the two-dimensional grid, thus giving rise to the possibility of fingerprinting based on more than one parameter. However, to reach that objective, BE and KE data must be corrected carefully for charging.

Another advantage is that, where chemical shifts in core BEs are too small for unambiguous fingerprinting, α^* often has greater chemical sensitivity because the chemical shifts measured in X-ray excited Auger spectra are usually larger than those of the photopeaks. There are many examples of the successful application of this method of using the spectral data in both catalyst characterization and catalysis-related studies. Some instructive examples include a study of precipitation, calcination and reduction steps for Cu-Zn-Al-oxide methanol synthesis catalysts [13,25] and the determination of the coordination of Al and Si in aluminosilicates including zeolites [31]. Figure 7 presents the results of another study in which γ-alumina was analyzed before and after activation for catalyzed halogen exchange reactions [32]. Activation of γ-alumina with $CHClF_2$ at 525 K produced a single Al surface species whose position on the Al chemical plot (a) was shifted toward the aluminum hydroxyfluoride positions, but whose α^* value remained the same as before activation. In the F chemical-state plot (b) the position of the activated γ-alumina was a long way from those of the Al-fluoride or hydroxyfluoride reference samples. The conclusion was that this Al-F surface could be regarded as a precursor of an Al hydroxyfluoride; i.e., at a certain number of sites on the γ-alumina surface OH^- had exchanged with F^- during activation. Activation in $CHClF_2$ at 675 K resulted in two coexisting species, indicated by the two identical symbols in Fig. 7. For one of these species the positions on the plots were very similar to those observed for the single species found after activation at 525 K. For the other the positions on the plots were at the same α^* values as in Al-fluorides or hydroxyfluorides but with BEs substantially higher than in AlF_3. The conclusion reached in Ref. 32 was that the latter sample consisted probably of a disordered (i.e., small particles either with departures from octahedral coordination or under the strong influence of the support) high F/OH ratio Al hydroxyfluoride species coexisting with a slightly fluorinated γ-alumina surface and chlorinated and fluorinated coke.

Activation in CH_3-$CClF_2$ at 525 K again resulted in single Al and and F species combined to form an amorphous Al hydroxyfluoride phase undetectable by XRD analysis. Only this activation procedure provided the full catalytic activity obtained with β-AlF_3, the "reference catalyst," which correlates well with the fact that γ-alumina activated in CH_3-$CClF_2$ at 525K can be seen to occupy positions on the two chemical-state plots very close to those of β-AlF_3.

Another method of chemical-state analysis by fingerprinting, which relies on the evaluation of loss features on the high-BE sides of core-level peaks in XP spectra, can also be applied to catalyst characterization. These features, often referred to as satellites or shakeup peaks, are observed with high intensities in the XP spectra of compounds (oxides, halides, fluorides, etc.) of catalytically active transition and rare earth metals, such as Fe, Co, Ni, Cr, Cu and La. Detailed discussions of the physical origins of the loss features have been published [33]. For Cu especially these satellites have often been exploited successfully to differentiate between Cu^+ and Cu^{2+} species (for a typical Cu^{2+} $2p$ spectrum, from $CuCl_2$, see Fig. 2; cf. also Ref. 13). In another example, loss features associated with iron oxides and hydroxide have been used [34]. Recently, satellites in Cr^{3+} $2p$ spectra recorded from Cr_2O_3, and α- and β-CrF_3 have been analyzed in detail, stimulated by investigations of chromia halogen exchange catalysts [35]. It was found there that by application of a well-established curve-fitting procedure, the analytical potential of the satellites for a fingerprinting approach could also be confirmed for Cr. Table 1 and Fig. 8 present the respective data and spectra. The surprising fact emerged that this satellite structure in the Cr^{3+} XP spectra had not been considered or even mentioned in

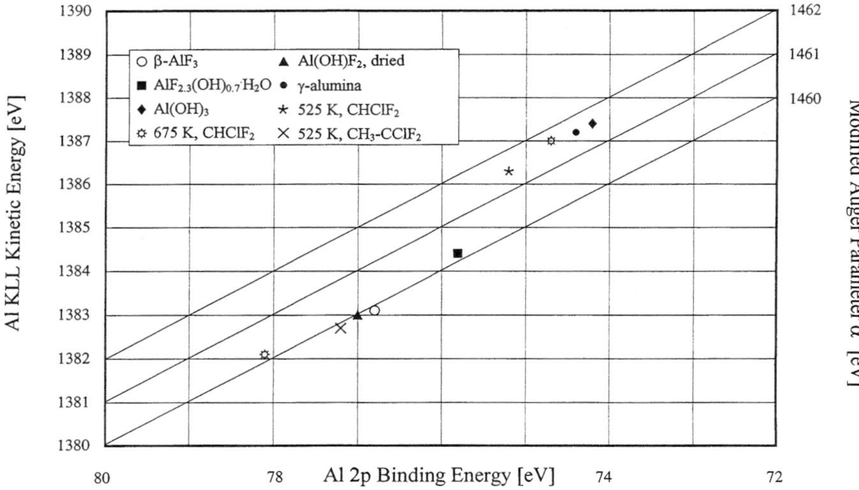

(a)

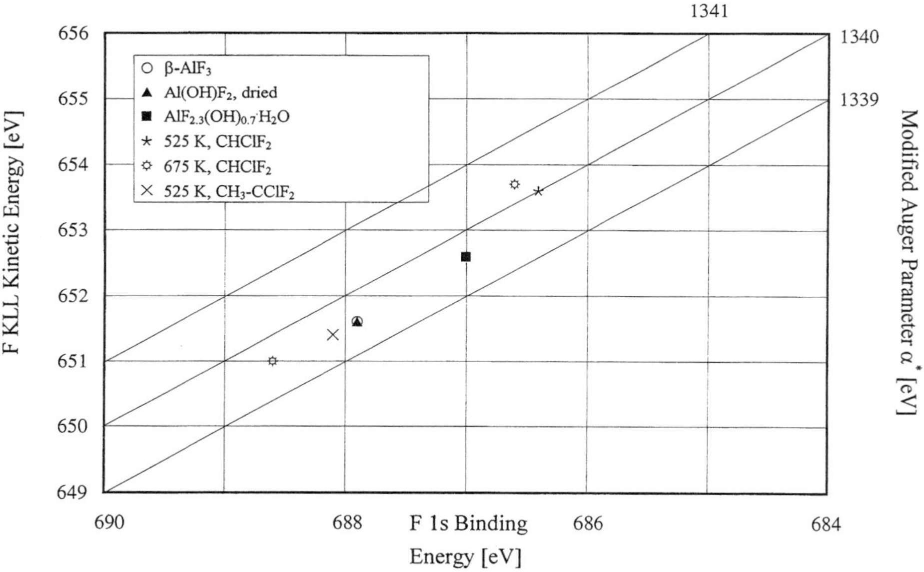

(b)

Figure 7a,b: Al (a) and F (b) chemical state plots for γ-alumina activated for catalyzed halogen exchange reactions, with $CHClF_2$ and CH_3-$CClF_2$ at different temperatures. Data recorded from reference samples (β-AlF_3, $Al(OH)F_2$, $AlF_{2.3}(OH)_{0.7} \cdot H_2O$ and $Al(OH)_3$) are included for comparison. (Redrawn from Ref. 32.)

Table 1. Spectroscopic Parameters for Cr^{3+} in Different Compounds

Sample	BE[a] Cr $2p_{3/2}$ (eV)	Cr $2p$ doublet spin-orbit split (eV)	Satellite to main peak distance (eV)	Satellite to main peak intensity ratio	Modified Auger parameter α^* (eV)
α-Cr_2O_3	576.3	9.4	11.1	0.19	1104.8
α-CrF_3	580.1	9.3	11.9	0.29	1105.5
β-CrF_3	579.8	9.1	12.0	0.34	1105.4

[a]Static charge reference: C 1*s* at 284.8 eV.
Source: From Ref. 35.

many earlier studies [35]. Such an omission was probably due to the superposition of the Cr $2p_{3/2}$ satellite and the Cr $2p_{1/2}$ main peak.

Finally, it must be emphasized that the usefulness of the fingerprint approaches discussed above is limited to extended solid phases. Complex problems occur when particle size effects have to be taken into account, such as in the important class of supported metal catalysts. Metal ions or metal clusters in a zeolite framework represent other difficult examples relevant to catalysis. The problem is that in a fingerprint analysis relying on chemical shifts in core-level and Auger electron energies, it is impossible to differentiate between an extended phase of an oxidized metal and highly dispersed zero-valent particles (clusters) of the same metal, because in both cases there are usually positive BE shifts with respect to the extended metallic phase. This is very important when the reduction states of supported metal catalysts and other systems (e.g., Cu-Zn-Al oxide catalysts) need to be studied. BE shifts in XPS are generally discussed by considering initial- and final-state effects (see Chapter 5). In the cases of small particles or clusters the size and nuclearity, respectively, determine the electron structure, giving rise to initial-state alterations. The reduced core-hole screening in metal clusters supported on insulating supports affects the final state. There are several studies in this field (for examples, see Refs. 36–39) which agree that the initial-state effect is the one that is predominant in determining the shift, because variations of the support give only small effects (cf. Fig. 9).

Returning to the problem of the surface chemical-state analysis of supported metal catalysts note that the particle size dependence of the BE should be known, or at least a reliable trend in it. There are different multitechnique ways of measuring it. The metal of interest can be evaporated or sputtered onto a flat substrate, for instance an oxidized Si wafer, a sapphire single crystal or pyrolytic graphite. Mean particle-size-related submonolayer coverages can then be estimated by quantitative XPS [36,37]. More directly, STM has been used to

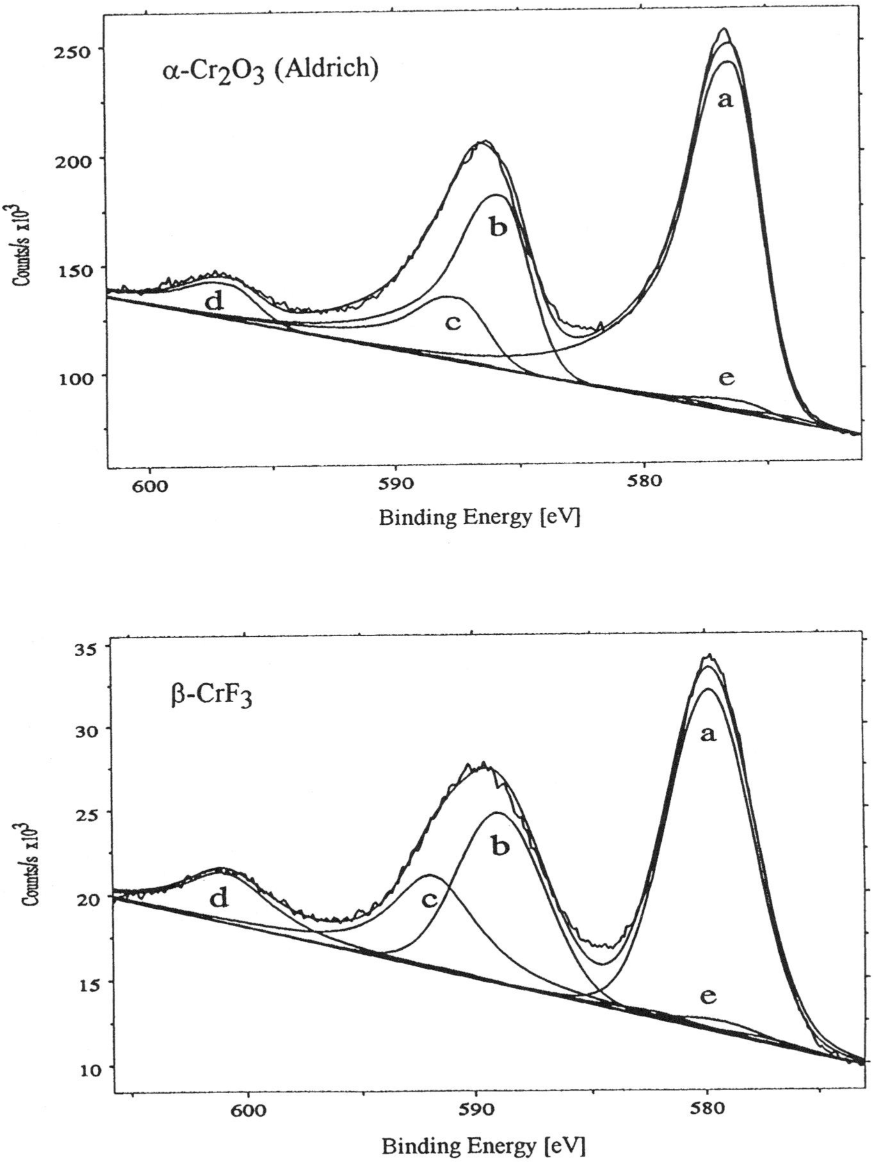

Figure 8. Cr 2p XP spectra of chromia and β-CrF_3. The curve fit shows that the apparent three-peak spectrum is actually a result of the superposition of a Cr 2$p_{3/2}$ and 2$p_{1/2}$ "main peak" doublet (a, b) with the corresponding doublet of satellites (c, d); (e) represents Al K$\alpha_{3,4}$ contributions. (From Ref. 35.)

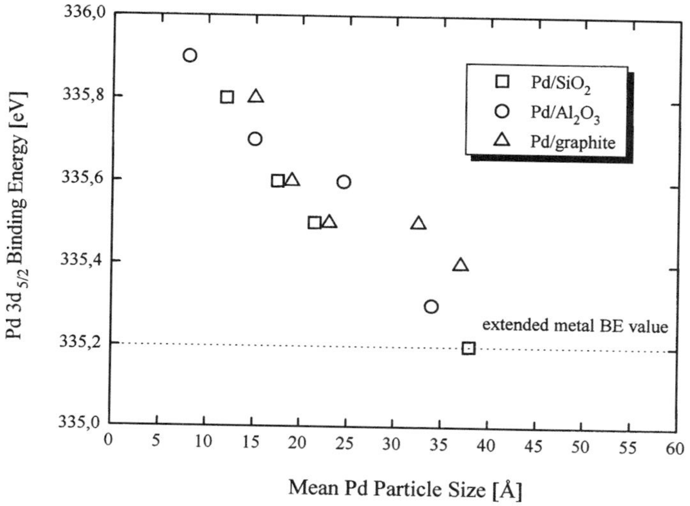

Figure 9. The particle size dependence of the Pd $3d_{5/2}$ binding energy measured after Pd dispersion by reduction of precursors on γ-alumina (200 m^2/g), silica (250 m^2/g) and wide-pore graphite-like carbon. The mean particle sizes were estimated by H_2-titration and electron microscopy. The dotted line represents the BE measured from a metal foil. (Redrawn from Ref. 38.)

measure a mean particle size for Pd particles supported on highly oriented pyrolytic graphite (HOPG) [40]. Alternatively, it has been found possible with a mass separation technique to collect single sized particles from a liquid metal ion source on an amorphous carbon substrate, which could then be analyzed by XPS [41]. However, the preferred approach is to prepare the dispersed metal in the same way as in catalyst preparation on typical supports, where the mean particle size can be estimated carefully by titration experiments or, when possible, directly by HRTEM. In this way a direct correlation between XPS BE and mean supported metal particle size can be obtained (cf., for instance, Ref. 38 and Fig. 9). Additional checks on the structural properties of the supported metal particles could be provided by EXAFS.

In the case of small clusters entrapped in zeolites or of zeolites with exchanged metal cations the complexity of the XPS data interpretation increases. Here the effects must be considered of the zeolite lattice field and of the coordination to ligands on a photoionized metal atom, which should have strong effects on the relaxation of the core hole, thus influencing the final state of the photoionized atom. The unusual chemical situation is demonstrated, for example, by the fact that intermediate Me^{1+} states of transition metals can be stabilized by a zeolite lattice field. However, the results from dedicated (and expensive) basic research confirm clearly the diagnostic potential of XPS (cf., e.g., Ref. 42). This includes redox and dehydration treatments, chemisorption, and migration effects considered to occur during temperature treatments. BE shift data from transition metals along with their valence-state changes in zeolites, useful for fingerprint analysis, have been collected in Ref. 43. Moreover, it has been shown for oxidized and reduced Cu-exchanged A-, X- and Y-zeolites that the Auger parameter as employed in a chemical-state plot is also very useful in chemical-state analysis of Cu in a zeolite [39,44].

7. CHEMICAL-STATE ANALYSIS WITH SSIMS BY FINGERPRINTING

An alternative, but more limited, way of obtaining chemical-state information about the elements present at a catalyst surface is the application of SSIMS (see also Chapter 6). Such information can be extracted only when it can be proved that the secondary ion-cluster emission is a direct result of fragmentation of the surface itself; i.e., direct cluster emission processes determine the spectra. Assuming that this requirement is fulfilled, an identified cluster ion can yield unique information about which elements were nearest neighbors on the surface of a catalyst. SSIMS applications in catalyst characterization are usually based on this assumption. The alternative process of cluster emission, that is, atomic recombination in the proximity of the sputtered surface, cannot lead to such straightforward interpretation of SSIMS data. Moreover, the unequivocal identification of the elemental composition of a cluster ion requires, in the case of complex multicomponent samples such as catalysts, high-resolution mass spectrometry with sufficient sensitivity, e.g., sector field or, a better choice since sensitivity can never be too high, time-of-flight (ToF) instruments.

The most important approach, similarly to that in XPS, is fingerprinting that relies on reference secondary-ion spectra taken from the literature or from databases (the most comprehensive can be found in the *Handbook of Static Secondary Ion Mass Spectrometry* or the Wiley Static SIMS Library [45]) or from the analysis of well-defined reference samples. Unfortunately, the number of inorganic compounds relevant to catalysis, whose spectra are given in the above database, is rather limited. One of the problems is that these materials often possess surfaces that are reactive to the environment, and they therefore suffer from surface corrosion. In addition, standard reagent-grade-quality chemical compound surfaces sufficiently free of organic contamination are difficult to find. Bearing in mind the high surface specificity of SSIMS, spectra may become distorted by both organic contaminants and corrosion by atmospheric components. When reporting the SSIMS spectra of reference samples for inclusion in the common fingerprint database, the recommendations of the relevant ASTM standard practice [46] should be considered carefully. As well as direct fingerprinting, logical deduction of the surface structure from the form of the fragmentation pattern and knowledge of the principles determining fragmentation processes can be helpful. In actual fact such applications depend on the state of the art in the interpretation of inorganic mass spectra, which must be extracted from the current literature.

A comprehensive review of the application of SIMS techniques to the analysis of the surface composition and structure to the study of catalyst preparation and the effects of promoters and poisons as well as to the study of reactivity has been given in Ref. 7. Various kinds of samples have been investigated, among them metal oxide catalysts, supported metal catalysts and zeolites in a variety of states. In most of the studies the occurrence of certain molecular ions or fragmentation patterns has been exploited successfully for surface chemical analysis. One SSIMS study of catalyst preparation, the formation of metallic Rh finely dispersed on alumina, which is an NO reduction component in automotive exhaust catalysis, is worth reporting here [17]. It shows once again the potential in the use of a thin-film alumina model support in providing optimal conditions for the application of surface spectroscopies. In that study the catalyst was prepared in the usual way by electrostatic adsorption of negatively charged $[RhCl_x(OH)_y(H_2O)_z]^{3(-x-y)+}$ complexes on the surface of an alumina film support which was positively charged in acid Rh chloride solution. Rh^{3+} precursors were decomposed by subsequent reduction in hydrogen, thus providing dispersed and catalytically active metallic Rh. Analysis of molecular secondary ions of the type $RhCl^-$ showed that a precursor-related RhCl species existed at the alumina surface after adsorption, which then decomposed at reduction temperatures below 200°C as verified by a decrease in $RhCl^-$ yields. Higher reduction temperatures were needed to remove chlorine from the support to which chloride ions are known to bind strongly. ClO^- secondary ions are regarded as diagnostic of this chlorine-support binding. Residual Cl is considered to have adverse effects on the catalytic properties of the activated sample, whereas high reduction temperatures may lead to Rh sintering, which is also undesirable. It may be concluded that measurement of the Cl^-, ClO^- and $RhCl^-$ secondary-ion yields and their correlation with activation process parameters (cf. Fig. 10) may help, when taken together with additional data obtained from other methods, to optimize the catalyst activation procedure itself.

Another application, using FABMS, in which the association of Pt and Sn on γ-alumina with a high surface area as measured by BET, was studied by monitoring $PtSn^+$ and $PtSnO^+$ secondary cluster ions, has been described in Ref. 47. The interpretation, supported by dedicated TPR studies, was that Sn, Pt and O are nearest neighbors at the surface of the catalyst, which is a prerequisite

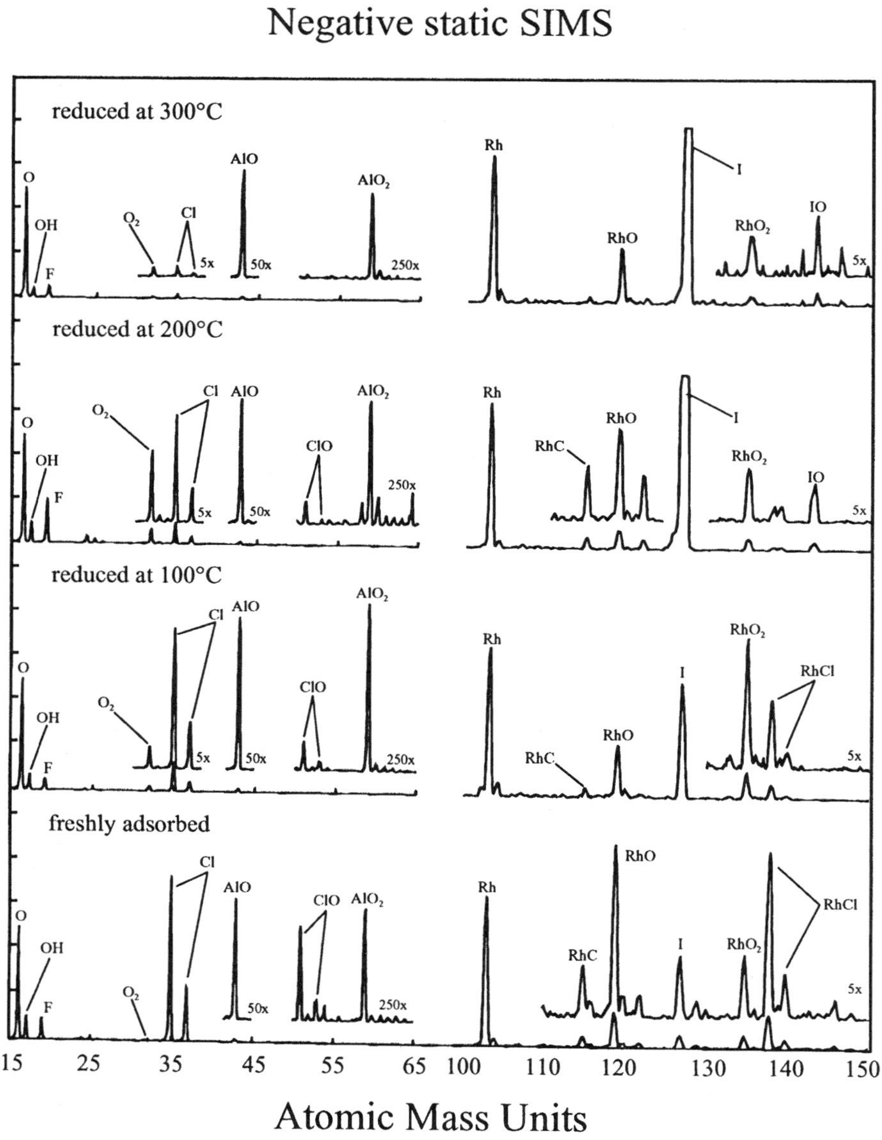

Figure 10. Negative SSIMS spectral sequences recorded from an Rh/thin film alumina model catalyst. The freshly adsorbed state and the states after reduction in 1 atm of H_2 at different temperatures were analyzed. Key secondary ions are Cl^-, ClO^-, indicating chlorine-support contact, and $RhCl^-$, indicating rhodium-chlorine contact. (From Ref. 17.)

for the formation of Pt-Sn alloy particles during reduction in hydrogen. The other interesting result was that there were many different support-related $Al_xO_yH_z^{\pm}$ cluster ions in the spectra but neither $MeAl^+$ nor $MeAl_xO_yH_z^+$ ions. The latter species would be expected if cluster-ion formation by recombination determined the mass spectrum. The study has shown that FABMS can provide information on alloying phenomena for supported multimetal catalysts.

The usefulness of SSIMS in the field of catalysis-related surface science is already well known. SSIMS studies of CO adsorption on single-crystal surfaces of Cu, Ru, Ni, Pd and Pt have been summarized briefly [45]. The observation that different CO adsorption structures (linearly bonded, bridged, triple bridged and mixed states) can be distinguished by the characteristic Me_xCO^+ ($x = 1, 2, 3$) secondary cluster-ion yield fractions in the SSIMS spectra, i.e., by fingerprinting, confirms that these rather delicate structures can be identified by the method.

8. MISCELLANEOUS

8.1. The Molecular Probe Approach: Assessment of Acid-Base Properties

The qualitative and (semi)quantitative determination of acid-base properties is of major interest in catalyst characterization. Measurements of these properties are normally made with IR spectroscopy, calorimetry or gravimetry using appropriate molecular probes such as H-bond acceptors, nitriles, pyridine, etc. However, in certain situations, such as when dealing with solids of low surface area, or when acid site concentrations are of interest, application of these techniques becomes difficult. XPS has proved to be an alternative approach which can also be used for a validation of the former, more common, methods. Chemical shifts (cf. Chapter 5) exhibited by key atoms in an adsorbed molecule can be used in such a way as to provide qualitative information about the character of the acid sites at the sample surface. The approach can be made quantitative by evaluation of the XPS intensities of the key elements, provided that an appropriate model has been developed for the analysis of the respective XPS intensity data [48,49]. As an example, the BE shifts in the N 1*s* spectra from adsorbed pyridine can be used semiquantitatively to differentiate between Lewis acid sites and weak and strong Brønsted acid sites, on various zeolites (H-ZSM-5, H-Y) [48,50]. The strength of the Lewis basicity of basic sites on alkali cation zeolites has also been estimated by measuring the N 1*s* BE shifts in chemisorbed pyrrole molecules. The shifts were correlated with the electron donor capability of lattice oxygen atoms adjacent to the alkali cations [51]. The approach is not restricted to zeolite analysis. Acid site concentrations normalized per aluminum atom have also been obtained with silica-alumina gels [49].

This molecular probe XPS approach has been extended successfully to the assessment of the acid-base properties of a completely different class of material, i.e., polymers. Workers in this field have used various molecular probes, e.g., trichloromethane, dimethyl sulfoxide and hexafluoroisopropanol, in which the key atoms whose core-level shifts were determined following adsorption were Cl, S, and F, respectively [52].

It is clear that there is further potential for the molecular probe approach. For instance, it is attractive to apply another type of molecular probe, developed for the quantitative determination of hydroxyl groups on polymer surfaces by XPS [53], to the characterization of catalyst supports such as silica-gel or alumina. The authors found that the derivatization of surface OH groups by trifluoroacidanhydride (TFAA) was specific and quantitative, thus allowing the total number of these groups on the surface to be estimated. However, it should be emphasized that such XPS analysis is rather sophisticated and should not be regarded as a routine method.

8.2. Alloying at Bimetallic Supported Catalysts

Bimetallic systems have long been of interest in catalysis. Adding a metallic modifier to the primary metal may provide ligand and ensemble effects or multifunctionality resulting in enhanced catalyst performance. The characterization of bimetal catalysts has been found to be a considerable challenge to surface analysis techniques, and multimethod approaches are usually required. The contribution of XPS has been that of the confirmation of the presence of zero-valent metals at the surface, which is a prerequisite for alloy particle or cluster formation on the catalyst support. Of course, particle size effects on the binding energies must be considered carefully (cf. Section 6). Moreover, there have been XPS studies on Pt/*sp*-metal alloys, which have revealed that alloying may also have characteristic effects on binding energies (shifts of up to ≈ 0.5 eV in Sn 3*d* and Pt 4*f* in Pt-Sn alloys), and on widths and asymmetries of the core-level lines of alloy constituents [54]. SSIMS can be used [47,55] by, for instance, looking for $A_xB_y^+$ cluster ions in the secondary spectra; they should appear when metal A atoms are adjacent to metal B atoms forming an alloy cluster (cf. Section 7).

A powerful but expensive tool for tackling this problem is the technique of EXAFS, but it requires that the constituents have high scattering cross sections [56]. From an analysis of EXAFS data the number of atoms of a particular kind at a particular distance from a chosen absorber atom, i.e., those in coordination spheres, can be evaluated. An EXAFS determination for each alloy constituent can provide information on the individual environment of atoms of that constituent on the support, thus indicating whether alloy formation has occurred.

For highly dispersed metals on high surface-area supports, EXAFS, which is a bulk analysis method, actually monitors phenomena at the catalyst surface. EXAFS is also a very powerful method for investigating promoter-support interactions (cf. Sharpe et al. [56]).

The interpretation of XPS, SSIMS and EXAFS data obtained from complex systems such as bimetallic catalysts ought to be supported by other more direct methods, e.g., XRD, Mössbauer spectroscopy or analytical electron microscopy (whenever possible) or by the indirect methods usually applied in catalyst characterization, e.g., chemisorption, TPR, etc. As an illustration there have been some successful multitechnique studies on Pt-Re and Pt-Sn reforming catalysts [57]. From XPS, EXAFS and L-edge X-ray absorption spectroscopy (XAS) it was found that, instead of substantial Pt-Re alloying, coexisting Pt metal and rhenium dioxide species were formed. The latter are suggested as providing enhanced stability of the catalyst against the sintering of Pt particles. Study of the Pt-Sn reforming catalyst by XPS, EXAFS, TPR, XRD and reaction measurements has revealed that the formation of bimetallic Pt-Sn entities depends on the conditions used in the preparation of the catalysts and on the nature of the support. Thus, Sn is present primarily as an Sn(II)-oxide species with a high affinity for alumina after high-temperature reduction of Pt-Sn/γ-Al_2O_3, whereas on silica alloy formation was found to occur under the same conditions.

8.3. In-depth Analysis

Depth profiling of a catalyst surface by ion etching combined with surface analysis (cf. Chapter 7) would doubtless be a very attractive tool in catalyst characterization. However, the range of application of this technique is rather restricted. It has, for instance, been used successfully in the analysis of Raney nickel catalysts [58]. The effects of segregation phenomena at Al/Ni alloy surfaces and the degree of reactivity of the aluminum on the development of the surface during leaching with NaOH to produce active hydrogenation catalysts were investigated by Auger depth profiling and XPS. This example is an isolated one and shows that such in-depth analyses are in principle not very different from those usually undertaken in the field of metallurgy (cf. Chapters 15 and 11) in which corrosion, passivation and segregation phenomena are studied.

In-depth analysis based on ion etching performed on standard multiphase/multicomponent catalysts is full of pitfalls and should be avoided. The notion that it should be possible to derive structural models for catalysts (e.g., layered promoter or "cherry" models, diffusion of promoter components into the bulk, etc.) from such profile data must be firmly rejected. An instructive example has been given [59] in which is reported a study of Ni-Mo/Al_2O_3 catalysts with ISS sputter depth profiling. The result was that the shape of the curve, relating Ni/Mo ISS intensity ratios to increasing bombardment time, was influenced strongly by the incident-ion energy. The reason is that multiphase/multicomponent samples such as catalysts may suffer from strong ion beam modification effects during sputtering. This has been documented for some years [60,61] and has been discussed in standard guides for specimen handling and depth profiling published by ASTM [15,62]. It is dangerous to believe that surface oxides, contaminant carbon or coke, as examples, could be removed from a catalyst surface by sputter cleaning in order to reveal the "real" sample surface. Sputtering of catalyst surfaces results invariably in the formation of chemically, compositionally and structurally altered surface layers. The most important chemical effect is the loss of high vapor pressure constituents, for instance oxygen, thus causing reduction. The composition of any particular phase may be altered by mixing, implantation, segregation and migration effects. Structural damage includes amorphization, recrystallization and redistribution. Reference 60 provides a data base of more than 100 entries consisting of binary and ternary metal compounds containing O, N, F, S and Cl, in which the surface composition had been altered by sputtering.

9. QUANTITATIVE SURFACE ANALYSIS OF CATALYSTS: COMPOSITION, DISPERSION AND COVERAGE

In most cases, XPS has been used to acquire quantitative data concerning catalyst surfaces. Quantitative data of interest consist of elemental concentration ratios, dispersion of the catalytically active component, or promoter, in the case of supported catalysts, and the coverage of the support by the promoter. For a given loading and support surface area dispersion and coverage are not independent parameters. Dispersion itself is defined as the ratio of surface promoter atoms to total number of promoter atoms. It is correlated with the catalyst activity.

There are two principal ways in which surface concentration data may be obtained by the evaluation of XPS intensities. One relies on a first-principle description of photoelectron emission from a solid surface, the other on empirically determined elemental sensitivity factors (cf. Chapter 4 and 5). Either approach can be used in its simplest form to estimate elemental surface concentration ratios from XPS intensity ratios for a catalyst surface, provided the sample is homogeneous and isotropic, i.e., all elements are uniformly distributed in the surface layer sampled by XPS. However, a heterogeneous catalyst is in fact just that—heterogeneous; it can be multiphase and of complex structure. Operative words here are porosity, inner and outer surface, texture, segregation, etc. Nevertheless the simple procedures have been used extensively in catalyst characterization studies for straightforward interpretation of XPS in-

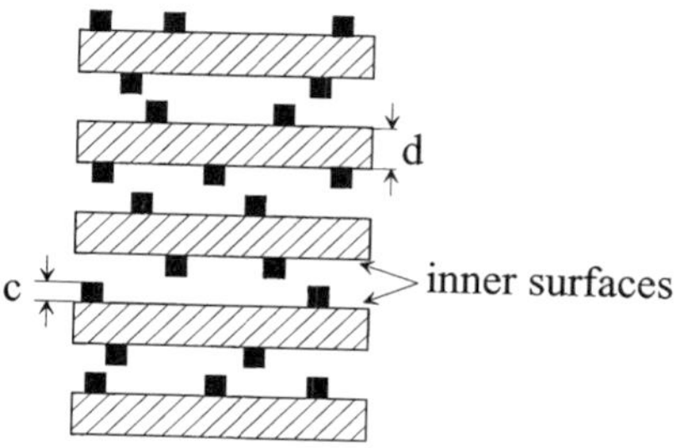

Figure 11. Stratified layer model for supported catalysts developed for the quantitative interpretation of XPS intensity data (cf. Ref. 63). The support is modeled as a stack of sheets of thickness d. This thickness is determined by the density ρ_s and the surface area A_s of the support by $d = 2/\rho_s A_s$. The promoter is represented by single-sized cubes with the dimension c. Within the model the surface coverage Θ can be calculated for a given promoter weight fraction x and density ρ_p by $\Theta = x/(1-x)A_s\rho_p c$.

tensity ratios, although, in many cases, it was not established that samples met the requirements of the procedure. Papers lacking this validation are, from both scientific and quality assurance points of view, substandard. For the limiting case of high surface-area (> 200 m^2/g), high dispersion supported catalysts, it has been shown [63] that XPS provides "bulk characteristic" promoter/support element concentration ratios because the average depth of information in the method means that the signal integrates over the inner pore system.

Taking into account dispersion and coverage, for the important class of high-area supported catalysts, appropriate quantitative models have been derived which are based on the idea that such samples can be modeled as stacks of support material sheets each covered by promoter particles (cf. Fig. 11 or Ref. 63). The surface area of the support and the density of the support material define the corresponding sheet thickness d. For example, a typical porous silica is characterized by d = 2.6 nm. When considering this stratified layer model it turns out that the simple expression

$$\left(\frac{I_p}{I_s}\right)_{\text{monodisperse promotor}} = \left(\frac{p}{s}\right)\frac{\sigma_p}{\sigma_s} \tag{1}$$

can be derived for the case of monodisperse (i.e., single atomic distribution) promoter particles. Thus, a measured XPS intensity ratio I_p/I_s is equal to the atomic ratio p/s of the chosen promoter and support elements multiplied by the factor σ_p/σ_s, the photoemission cross section ratio, provided that the difference in kinetic energies of the respective photoelectrons is small [63]. Crystallite growth or sintering phenomena of the promoter on the support can also be taken into account with the assumption of single-sized cubic promoter crystallites. In that case I_p/I_s depends also on the crystallite dimension c. The relation between the promoter/support XPS intensity ratios characteristic of the coarse-dispersed promoter particles and the monodisperse ones is

$$\frac{(I_p/I_s)_{\text{coarsely dispersed promoter}}}{(I_p/I_s)_{\text{monodisperse promoter}}} = \frac{1-e^{-\alpha}}{\alpha} \tag{2}$$

where $\alpha = c/\lambda$ is a dimensionless crystallite size. Hence it follows that the mean crystallite size of the promoter c can be determined from the respective XPS intensity ratios and Eq. (2) when λ, the inelastic mean free path length of the photoelectrons, originating from the promoter and traveling through the promoter, is known [63]. Once the mean crystallite size is known, the mean dispersion of the promoter can be estimated.

The stratified-layer model introduced by Kerkhof and Moulijn [63] has been developed further by considering random orientation of the stacks, which should be closer to the real situation [64]. The authors of Ref. 64 also analyzed the dependence of I_p/I_s on the promoter particle shape. The result was that, for genuinely randomly oriented samples, the XPS signal of a support phase, present as convex particles of equal size but of arbitrary shape, is determined by the surface/volume ratio of the support compound. This property is proportional to dispersion. Thus, in the extended model the experimental XPS I_p/I_s ratio is a direct measure of the dispersion. I_p/I_s can be converted into dimensions characteristic of some likely geometries of particles on the support (e.g., layers with a uniform thickness d, spheres with diameter $3d$, or half-spheres with radius $2.25d$ [64]), together with an appropriate coverage. However, the following parameters must be known exactly: elemental concentrations and densities of promoter and support, photoemission cross sections characteristic of the respective promoter and support core levels, inelastic mean free path lengths of the respective photoelectrons, analyzer transmission and specific area of support. Once again it should be emphasized that application of the stratified-layer model by Kerkhof and Moulijn [63] as well as the randomly oriented layer model by Kuipers et al. [64] requires single-sized and uniformly distributed promoter particles. Because in many practical situations that is not true, Kaliaguine et al. [65] have developed another important extension accounting for two populations of crystallites on the outermost support sheet, characterized by different crystallite sizes and coverages. Relevant practical situations described by this model are, for instance, the repartition of support material between the outer surface and the pores of a support, and promoter segregation to the outer surface. Its application requires that information on the crystallite size and coverage for both populations be obtained by independent methods such as electron microscopy or XRD.

Finally it must be said that extreme care must be exercised in the interpretation of XPS intensity ratios recorded from catalysts. The accuracy of the determination of dispersion and coverage data in this way suffers from the general inhomogeneity of practical samples, from the effects of surface roughness arising from the fact that most samples are investigated in the form of powder, and from problems in estimating parameters such as inelastic mean free paths, material densities, etc. However, when the standard techniques for catalyst characterization, e.g., selective chemisorption, TEM, XRD line profile analysis, etc., are unable to estimate dispersion data or crystallite sizes, it can be useful to try XPS as an alternative means of troubleshooting. This might involve specialized extensions of the models presented in Refs. 63, 64 (cf., e.g., Refs. 66, 67) or the development of entirely new approaches to fit individual situations. Again, only detailed descriptions of the data reduction procedures used to reach this goal will meet the requirements of best practice in XPS summarized in Ref. 11.

Another, more direct way of estimating the coverage of a catalyst support by promoter species should be mentioned. This approach relies on the unique surface specificity of ISS [68]. In Ref. 68 coverage data obtained by ISS on a titania support (surface area ca. 50 m^2/g) with different WO_3 loadings have been reported (cf. Fig. 12). The coverage of the support by monolayer-like W species was monitored by following the changes in the Ti ISS intensity, characteristic of the support, normalized to the O ISS peak intensity. The surface coverage Θ was estimated from

$$\Theta = 1 - \frac{(\mathrm{Ti/O})_{\mathrm{cat}}}{(\mathrm{Ti/O})_{\mathrm{sup}}} \qquad (3)$$

where Ti/O are ISS peak intensity ratios of Ti and O measured without (subscript sup) and with (subscript cat) WO_3 loading. Details concerning the assumptions to be made in surface coverage measurement by ISS are discussed in Ref. 69. The main result is shown in Fig. 12b, in which surface coverages estimated by Eq. (3) are plotted. Up to loadings of 8–10 wt.% WO_3 a monolayer-like dispersion of W species was observed. The leveling off of the plot of coverage for higher loadings indicates multilayer formation. The advantage of direct ISS analysis of the support coverage is that it can be applied also to systems where a chemisorption approach suffers from the severe problem that a molecule has to be found which adsorbs, in this example, exclusively on tungsten oxide and not on the TiO_2 support.

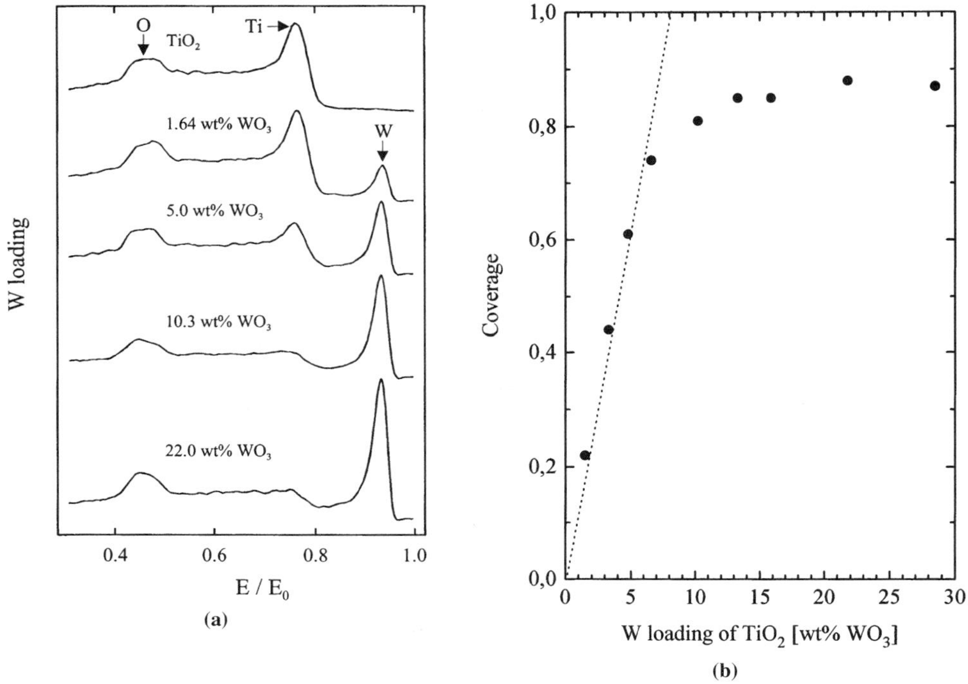

Figure 12a,b. ISS spectra (a) obtained for different WO_3 loadings of a 50 m^2/g titania support plotted together with the spectrum of the unloaded support. Surface coverage data determined by application of Eq. (3) are shown in (b). The dotted line represents theoretical coverages obtained by assuming monolayer-like dispersion of the W phase and 1:1 stoichiometry of the reaction between WO_3 and surface hydroxyls on the support. (Redrawn from Ref. 68.)

REFERENCES

1. B. Delmon, *Surf. Interface Anal. 9*, 195 (1986).
2. J. W. Niemantsverdriet, *Spectroscopy in Catalysis*, VCH, Weinheim, 1993, Chap. 3.
3. R. L. Austermann, D. R. Denley, D. W. Hart, P. B. Himelfarb, R. M. Irwin, M. Narayana, R. Szentirmay, S. C. Tang and R. C. Yeates, *Anal. Chem. 59*, 68R (1987).
4. *Standard Terminology Relating to Surface Analysis* (*E673-95c*), Annual Book of ASTM Standards, Vol. 03.06, American Society for Testing and Materials, West Conshohocken, 1996, p. 845.

5. G. Margaritondo, C. Coluzza and R. Sanjinés, Photoemission: From Hertz and Einstein to spectromicroscopy, in *Photoemission: From the Past to the Future* (C. Coluzza, R. Sanjinés and G. Margaritondo, eds.), Ecole Polytechnique Fédérale de Lausanne, 1992, pp. 44–95.
6. D. E. Harrison, Jr., *Rad. Effects 70*, 1 (1983).
7. H. J. Borg and J. W. Niemantsverdriet, Applications of secondary ion mass spectrometry in catalysis and surface chemistry, in *Catalysis: A Specialist Periodical Report*, Vol. 11 (J. J. Spivey and A. K. Agarwal, eds.), The Royal Society of Chemistry, Cambridge, 1994, pp. 1–50.
8. B. A. Horrell and D. L. Cocke, *Catal. Rev.-Sci. Eng. 29*, 447 (1987); H. Niehus, W. Heiland and E. Taglauer, *Surf. Sci. Rep. 17*, 213 (1993).
9. J. C. Klein, Ch. P. Li, D. M. Hercules and J. F. Black, *Appl. Spectrosc. 38*, 729 (1984).
10. G. Beamson and D. Briggs, *High Resolution XPS of Organic Polymers: The Scienta ESCA 300 Database*, Wiley, Chichester, 1992, Chap. 12.
11. *Standard Practice for Reporting Spectra in X-ray Photoelectron Spectroscopy (E1015-90)*, Annual Book of ASTM Standards, Vol. 03.06, American Society for Testing and Materials, Philadelphia, 1994, p. 770.
12. J. Grimblot, L. Gengembre and A. D'Huysser, *J. Electron Spectrosc. Relat. Phenom. 52*, 485 (1990).
13. B. Peplinski, W. E. S. Unger and I. Grohmann, *Appl. Surf. Sci. 62*, 115 (1992).
14. H. Shimada, N. Matsubayashi, M. Imamura, T. Sato and A. Nishijima, *Catal. Lett. 39*, 125 (1996).
15. *Standard Guide for Specimen Handling in Auger Electron Spectroscopy, X-ray Photoelectron Spectroscopy, and Secondary Ion Mass Spectrometry (E1078-90)*, Annual Book of ASTM Standards, Vol. 03.06, American Society for Testing and Materials, West Conshohocken, 1996, p. 884.
16. G. A. Somorjai, *Chemistry in Two Dimensions—Surfaces*, Cornell University Press, Ithaca, 1981.
17. G. B. Hoflund, D. A. Asbury and R. E. Gilbert, *Thin Solid Films 129*, 139 (1985); H. J. Borg, L. C. A. van den Oetelaar and J. W. Niemantsverdriet, *Catal. Lett. 17*, 81 (1993).
18. E. Ruckenstein and M. L. Malhotra, *J. Catal. 41*, 303 (1976); A. Jiménez-Gonzáles and D. Schmeisser, *J. Catal. 130*, 332 (1991); B. G. Frederick, G. Apai and T. N. Rhodin, *Surf. Sci. 250*, 59 (1991).
19. C. Coluzza, J. Almeida, T. Dell'Orto, F. Gozzo, H. Berger, D. Bouvet, M. Dutoit, S. Contarini and G. Magaritondo, *J. Appl. Phys. 76*, 3710 (1994).
20. U. Gelius, B. Wannberg, Y. Nakayama and P. Baltzer, ESCA studies of polymers and insulators using monochromatic X-ray excitation, in *Photoemission: From the Past to the Future* (C. Coluzza, R. Sanjinés and G. Margaritondo, eds.), Ecole Polytechnique Fédérale de Lausanne, 1992, pp. 143–211.

21. *Standard Guide to Charge Control and Charge Referencing Techniques in X-ray Photoelectron Spectroscopy (E1523-93)*, Annual Book of ASTM Standards, Vol. 03.06, American Society for Testing and Materials, West Conshohocken, 1996, p. 913.
22. C. D. Wagner and D. M. Bickham, *NIST Standard Reference Data Base 20: NIST X-ray Photoelectron Spectroscopy Database Version 1.0*, National Institute of Standards and Technology, Gaithersburg, MD, 1989. (Version 2.0 was announced to cover also chemical-state plots for many elements in J. R. Rumble, D. M. Bickham and C. J. Powell, *Surf. Interface Anal. 19*, 241 (1992).)
23. M. T. Anthony and M. P. Seah, *Surf. Interface Anal. 6*, 95 (1984).
24. P. Swift, *Surf. Interface Anal. 4*, 47 (1982); S. Kohiki and K. Oki, *J. Electron Spectrosc. Rel. Phenom. 33*, 375 (1984).
25. Th. Gross, A. Lippitz, W. E. S. Unger, A. Lehnert and G. Schmid, *Appl. Surf. Sci. 78*, 345 (1994).
26. G. Moretti, *Surf. Interface Anal. 17*, 745 (1991).
27. Y. Okamoto, K. Fukino, T. Imanaka and S. Teranishi, *J. Phys. Chem. 87*, 3747 (1983); G. Moretti, M. Lo Jacono, G. Fierro and G. Minelli, *Surf. Interface Anal. 9*, 246 (1986); G. Moretti, G. Fierro, M. Lo Jacono and P. Porta, *Surf. Interface Anal. 14*, 325 (1989).
28. Th. Gross, K. Richter, H. Sonntag and W. E. S. Unger, *J. Electron Spectrosc. Relat. Phenom. 48*, 7 (1989).
29. Th. Gross, M. Ramm, H. Sonntag, W. E. S. Unger, H. M. Weijers and E. H. Adem, *Surf. Interface Anal. 18*, 59 (1992).
30. C. D. Wagner and A. Joshi, *J. Electron Spectrosc. Relat. Phenom. 47*, 283 (1988).
31. C. D. Wagner, D. E. Passoja, H. F. Hillery, T. G. Kinisky, H. A. Six, W. T. Jansen and J. A. Taylor, *J. Vac. Sci. Technol. 21*, 933 (1982).
32. A. Hess, E. Kemnitz, A. Lippitz, W. E. S. Unger and D.-H. Menz, *J. Catal. 148*, 270 (1994).
33. C. S. Fadley, in *Electron Spectroscopy: Theory, Techniques and Applications*, Vol. 2 (C. R. Brundle and A. D. Baker, eds.), Academic Press, New York, 1978, pp. 1–156; D. D. Sarma, P. Vishnu Kamath and C. N. R. Rao, *Chem. Phys. 73*, 71 (1983); B. W. Veal and A. P. Paulikas, *Phys. Rev. B: Condensed Matter* 31, 5399 (1985).
34. K. Wandelt, *Surf. Sci. Rep. 2*, 1 (1982).
35. I. Grohmann, E. Kemnitz, A. Lippitz and W. E. S. Unger, *Surf. Interface Anal. 23*, 887 (1995).
36. S. Kohiki, *Appl. Surf Sci. 25*, 81 (1986); S. Kohiki and S. Ikeda, *Phys. Rev. B 34*, 3786 (1986).
37. I. Jirka, *Surf. Sci. 232*, 307 (1990).

38. L. V. Nosova, M. V. Stenin, Yu. N. Nogin and Yu. A. Ryndin, *Appl. Surf. Sci. 55*, 43 (1992).
39. G. Moretti and P. Porta, *Surf. Interface Anal. 20*, 675 (1993).
40. A. Sartre, M. Phaner, L. Porte and G. N. Sauvion, *Appl. Surf. Sci. 70/71*, 402 (1993).
41. S. B. DiCenzo, S. D. Berry and E. H. Hartford, *Phys. Rev. B. 38*, 8465 (1988).
42. W. Grünert, U. Sauerlandt, R. Schlögl and H. G. Karge, *J. Phys. Chem. 97*, 1413 (1993).
43. E. S. Shpiro, G. V. Antoshin and Kh. M. Minachev, in *Catalysis on Zeolites* (D. Kalló and Kh. M. Minachev, eds.), Academiai Kiadó, Budapest, 1988, pp. 43–93 and references cited therein.
44. B. A. Sexton, T. D. Smith and J. V. Sanders, *J. Electron Spectrosc. Relat. Phenom. 35*, 27 (1985).
45. D. Briggs, A. Brown and J. C. Vickerman, *Handbook of Static Secondary Ion Mass Spectrometry (SIMS)*, J. Wiley, Chichester, 1989, Library of Spectra, pp. 17–155. The Wiley Static SIMS Library, Project Directors J. Vickerman and D. Briggs, Wiley, Chichester, 1996.
46. *Standard Practice for Reporting Mass Spectral Data in Secondary Ion Mass Spectrometry* (SIMS) (*E1504-92*), Annual Book of ASTM Standards, Vol. 03.06, American Society for Testing and Materials, West Conshohocken, 1996, p. 908.
47. W. E. S. Unger, G. Lietz, H. Lieske and J. Völter, *Appl. Surf. Sci. 45*, 29 (1990).
48. C. Defossé and P. Canesson, *J. Chem. Soc. Faraday Trans. I 72*, 2565 (1976).
49. C. Defossé, P. Canesson, P. G. Rouxhet and B. Delmon, *J. Catal. 51*, 269 (1978).
50. R. Borade, A. Sayari, A. Adnot and S. Kaliaguine, *J. Phys. Chem. 94*, 5989 (1990)
51. M. Huang, A. Adnot and S. Kaliaguine, *J. Catal. 137*, 322 (1992).
52. J. F. Watts and M. M. Chehimi, *Int. J. Adhesion Adhesives 15*, 91 (1995); N. Shahidzadeh-Ahmadi, M. M. Chehimi, F. Arefi-Khonsari, N. Foulon-Belkacemi, J. Amouroux and M. Delamar, *Colloids Surf. A105*, 277 (1995).
53. A. Chilkoti and B. Rattner, Chemical derivatization methods for enhancing the analytical capabilities of X-ray photoelectron spectroscopy and static secondary ion mass spectrometry, in *Surface Characterization of Advanced Polymers* (L. Sabbatini and P. G. Zambonin, eds.), VCH, Weinheim, 1993, Chap. 6.
54. T. T. P. Cheung, *Surf. Sci. 177*, 493 (1986).
55. S. Y. Lai and J. C. Vickerman, *J. Catal. 90*, 337 (1984).

56. J. H. Sinfelt, G. H. Via, G. Meitzner and F. W. Lytle, Structure of bimetallic catalysts, in *Catalyst Characterization Science* (M. L. Deviney and J. L. Gland, eds.), ACS Symposium Series No. 288, American Chemical Society, Washington, DC, 1985, Chap. 22; L. R. Sharpe, W. E. Heinemann and R. C. Elder, *Chem. Rev. 90*, 705 (1990).
57. Pt-Re/Al_2O_3: J. H. Onuferko, D. R. Short and M. J. Kelley, *Appl. Surf. Sci. 19*, 227 (1984); G. Meitzner, G. H. Via, F. W. Lytle and J. H. Sinfelt, *J. Phys. Chem. 87*, 6354 (1987); D. Bazin, H. Dexpert and P. Lagarde, *Characterization of heterogeneous catalysts: The EXAFS tool*, Topics in Current Chemistry, Vol. 145, Springer-Verlag, Heidelberg, 1988, pp. 69–80; Pt-Sn/Al_2O_3 and Pt-Sn/SiO_2: B. A. Sexton, A. E. Hughes and K. Foger, *J. Catal. 88*, 466 (1984); G. Meitzner, G. H. Via, F. W. Lytle, S. C. Fung and J. H. Sinfelt, *J. Phys. Chem. 92*, 2925 (1988).
58. J. C. Klein and D. M. Hercules, *Anal. Chem. 53*; 754 (1981).
59. S. Kasztelan, J. Grimblot and J. Bonelle, *J. Chem. Phys. 80*, 793 (1983).
60. G. Betz and G. K. Wehner, Sputtering of multicomponent materials, in *Sputtering by Particle Bombardment—Physics and Applications* (R. Behrisch, ed.), Springer-Verlag, Berlin, 1981, Chap. 1.
61. R. Kelly, *Nucl. Instr. Methods 182/183*, 351 (1981); R. Kelly, in *Ion Bombardment Modification of Surfaces: Fundamentals and Applications* (O. Auciello and R. Kelly, eds.), Elsevier, Amsterdam, 1984, p. 79.
62. *Standard Guide for Depth Profiling in Auger Electron Spectroscopy (E1127-91)*, Annual Book of ASTM Standards Vol. 03.06, American Society for Testing and Materials, West Conshohocken, 1996, p. 892.
63. F. P. J. M. Kerkhof and J. A. Moulijn, *J. Phys. Chem. 83*, 1612 (1979).
64. H. P. C. E. Kuipers, H. C. E. van Leuven and W. M. Visser, *Surf. Interface Anal. 8*, 235 (1986).
65. S. Kaliaguine, A. Adnot and G. Lemay, *J. Phys. Chem. 91*, 2886 (1987).
66. I. Grohmann and Th. Gross, *J. Electron Spectrosc. Relat. Phenom. 53*, 99 (1990).
67. M. A. Stranick, M. Houalla and D. M. Hercules, *J. Catal. 103*, 151 (1987).
68. J. N. Fiedor, M. Houalla, A. Proctor and D. M. Hercules, *Surf. Interface Anal. 23*, 234 (1995).
69. M. A. Eberhadt, M. Houalla and D. M. Hercules, *Surf. Interface Anal. 20*, 766 (1993).

18

Adhesion Science and Technology

JOHN F. WATTS University of Surrey, Guildford, Surrey, England

1. INTRODUCTION

In the quarter of a century or so that has elapsed since the surface analysis methods XPS, AES and SIMS became commercially available they have moved from their original "home" in the laboratory of the surface chemist and surface physicist to become methods of applied surface analysis. The wide range of disciplines within which these methods have been successfully applied can be readily appreciated by the diverse contents of this book. Some areas, such as corrosion, catalysis and polymers, have featured in the literature since the early 1970s while other disciplines have come to surface analysis rather more recently.

Adhesion research has made use of surface analysis since the mid-1970s, but by the end of that decade there were, perhaps, only two or three research groups worldwide that had a strong presence in both adhesion and surface analysis. Surface analysis papers at adhesion conferences were few and far between, and those involved in surface analysis and adhesion tended to seek the company of likeminded analysts. By the end of the following decade the situation had changed dramatically and surface analysis was then seen to feature prominently in adhesion conferences and journals; there is now a biennial international conference on surface analysis and adhesion. This change was partly a result of the wider availability of surface analysis equipment, but also reflected a slight shift in emphasis of the adhesion community. It was no longer acceptable merely to report performance data for the plethora of systems available, but there was now a requirement to consider interfacial chemistry of the adhesion process and the locus of failure, in more detail, and to relate relevant parameters back to the performance of an adhesive joint or organic coating.

The adhesion phenomena that will be considered in this chapter are essentially those that occur between organic polymers and metal oxides, and will build on previous reviews of the 1980s [1,2]. Before any investigation of adhesion can commence a thorough appreciation of surface analysis in the fields of polymers and oxidation is an advantage. As adhesion loss is often associated with environmental degradation of the metal substrate, some experience in this field can also be valuable. The description of adhesion as a multidisciplinary, multifaceted subject also applies when considering the use of surface analysis in an adhesion investigation.

The role of surface analysis in adhesion research can be divided into three general areas: assessment of substrate surface characteristics prior to bonding, identification of the exact locus of failure of an adhesive joint or organic coating following some kind of mechanical or chemical perturbation, and analysis of interface chemistry with a view to gaining a better understanding of the adhesion process itself. From a chronological point of view the above sequence has been essentially the same as that in which the use of surface analysis has developed in adhesion, and is the approach that will be adopted in the main body of this chapter. In addition, the nature of interfacial interactions in adhesion and the contributions that surface analysis can make in this area will be considered, together with suggestions for the best course of action for the investigations of the very complex formulations that are now commonplace in adhesives and coatings.

The goal of almost all investigations of adhesion is to produce a "better" bond, which may be stronger, more durable, tougher or cheaper; there are many criteria that may be applied. The contribution of surface analysis investigations will be that of providing a better understanding of the system, thus enabling an improved prescription of substrate pretreatment, formulation chemistry, systems chemistry, etc., in order to achieve the elusive better bond. Throughout this chapter such considerations will be the focus of discussion and the relevance of the surface analytical data to the overall systems approach will be reviewed continually.

2. CHARACTERISTICS OF THE SOLID SUBSTRATE

Because the forces responsible for adhesion act over very short distances (ca. 0.5 nm) it is self-evident that any extraneous material between polymer and substrate may have a possibly deleterious effect on the level of adhesion achieved. Such material may be organic or inorganic, and the occurrence of both, and the manner in which adhesion may be affected, is well documented.

2.1. Organic Contamination

One of the earliest examples of the problem-solving capability of surface analysis is from the American automobile industry. In the early 1970s the industry was faced with impending federal legislation regarding improved corrosion resistance (against perforation) of automobile bodies. There was a body of evidence that poor performance in an accelerated corrosion test (the salt spray test) was related to the concentration of surface carbon on the steel sheet stock. An investigation by Hospadaruk et al. [3] provided a direct correlation between surface carbon, as measured by AES and a combustion method, and the time taken to arrive at failure in the salt spray test. Thus AES, together with XPS, rapidly became accepted methods for the assessment of surface contamination. This correlation with performance is illustrated by the data of Fig. 1, and provided a major plank in the case for clean steel for automobile use. In a later investigation Iezzi and Leidheiser used SAM to provide further details of the deleterious effect of the adventitious hydrocarbon layer by describing the manner in which it interfered with the formation of a uniform conversion coating [4].

Prior to painting, sheet steel is invariably treated with an inorganic conversion coating based on zinc phosphate, which accounts for the characteristic acicular surface morphology resistant to corrosion and providing enhanced coatings adhesion. The presence of patches of oily deposit was shown to interfere with the phosphate growth mechanism, resulting in areas devoid of the phosphate deposit; these areas subsequently showed inferior adhesion and corrosion resistance behavior [4].

The case for a low level of surface hydrocarbon is now well established, and XPS and AES both provide means by which the level, type and thickness of contamination can be readily evaluated. Nevertheless a great deal of work is still undertaken in this area with the aim of establishing the reasons for premature failure of an adhesive joint or coating. The organic contamination is often more complex than the "oily" or "greasy" nonspecific hydrocarbon that is tacitly implied. An example is the presence at the bonding surface of release agents based on fluorocarbons or poly(dimethyl siloxane). Compounds of this type are found on backing papers of adhesive tapes and preformed composite structures, and it is not unknown for them to remain on, or be transferred to, the substrate prior to bonding, leading to low strength in the subsequent bond. Analysis of the substrate failure surface by XPS or SIMS can lead to rapid and unequivocal identification of the contaminant. A good example of this problem is that of the adhesive bonding of glass or carbon-fiber-reinforced plastic components, where XPS has been able to identify the source of poor bond performance to be associated with residues from such backing, or release, papers

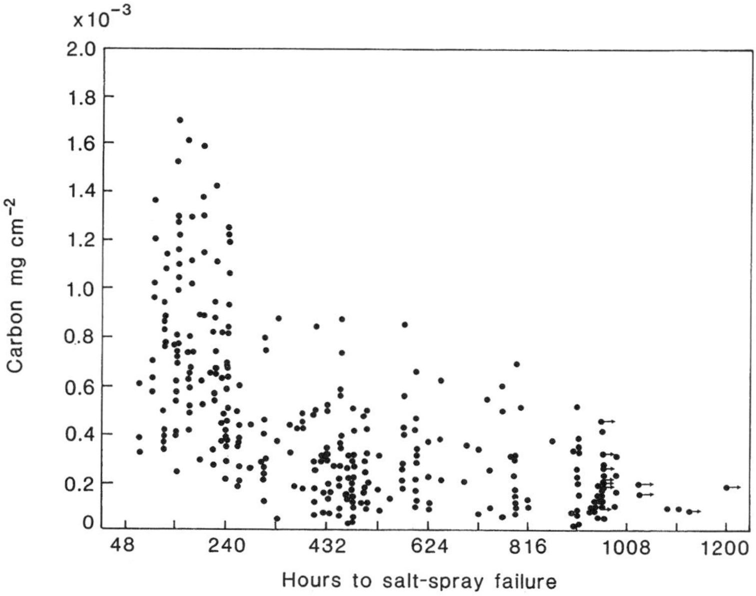

Figure 1. Correlation of steel surface cleanliness with coated panel salt spray performance. Carbon concentration was evaluated by combustion methods, but was soon to be superseded by AES. (From Ref. 3.)

[5,6]. The bonding surfaces may be protected by a release paper, whose purpose is to protect the surface from handling damage, but unfortunately removal of the paper often leaves behind a residue of the release agent, which is then responsible for poor bond properties. The solution to this dilemma is an additional treatment, following removal of the release paper, in order to ensure a clean surface devoid of fluorine- or silicon-containing products [5,6].

2.2. Oxide Films at Metal Surfaces

All metals, with the exceptions of the very noble ones, form oxide films on their surfaces on exposure to the atmosphere, and the compositions of these surface oxides are of critical importance to the adhesion scientist. In some cases it is possible to draw parallels between the outcomes of exposures in UHV clean conditions and ambient oxidation; this is certainly true for pure iron exposed to water vapor in UHV [7], which results in the formation of an FeOOH surface layer. Mild steel responds in the same manner, when cleaned by an abrasion process and passivated by air exposure. The early literature on the composition of metal surfaces following atmospheric exposure was reviewed by Castle in 1979 [8], and little seems to have been published on this topic during the intervening years.

The situation with alloys is often rather different since the oxide film may be formed by the oxidation of some, rather than all, of the constituent elements. For example, the surface oxide of stainless steel is a mixed Fe/Cr spinel enriched in chromium relative to the alloy composition, while that of cupro-nickel coinage metal is NiO. The composition of the surface phase is usually a function of equilibrium thermodynamics and process parameters such as oxygen partial pressure. There are well-documented cases in which a change in process parameters has led to the formation of a different surface phase and hence to an adverse effect on adhesion. Such a case is that of the annealing of some Al-Mg alloys; should the oxygen partial pressure in the closed-coil annealing process become too low, then the surface phase will be MgO instead of Al_2O_3. The resulting adhesion loss occurs because MgO is a brittle friable oxide which spalls readily from the parent metal, in marked contrast to Al_2O_3 which is a tenacious, strongly adhering oxide. Extension of the interpretation of such observations to other systems must, however, be treated with caution. The presence of a magnesium-rich surface phase associated with poor levels of adhesion does not necessarily mean that surface magnesium is always a predictor or diagnostic of poor adhesion. In the surface bonding of the above-mentioned Al-Mg alloys, etched alloy surfaces gave the expected poor performance and led to a cautionary warning [9]. Subsequent, more comprehensive, investigations [10] indicated that if the Mg were part, chemically, of the oxide film (rather than as a surface segregated layer of MgO), as occurs in the thick anodized film on Al-Mg alloys, then good bond performance could be achieved. This is illustrated graphically in Fig. 2, where the XPS spectrum (a) of commercial-purity aluminum with poor durability is compared with that (b) from an anodized Mg-containing Al alloy which, although having a substantial concentration of Mg in the spectrum, had excellent durability [10]. The commercial treatments used prior to the adhesive bonding of aluminum for structural purposes involve the electrochemical thickening of the oxide (i.e., anodizing) in acid media. This can lead to oxide films with thicknesses of around 1 μm, for which AES has been widely used in a depth profiling mode to monitor thickness and composition. An example shown in Fig. 3 illustrates Auger depth profiling through the interface region of an adhesive joint prepared on a very thin metallic aluminum substrate [11].

In general, it is now regarded as a requirement of any investigation of adhesion that the adherend surfaces are fully characterized by surface analysis *before* any adhesion tests are undertaken. This does, of course, beg the question concerning the inevitable hydrocarbon contamination. Such contamination will always be present in the spectrum recorded prior to bonding, the general opin-

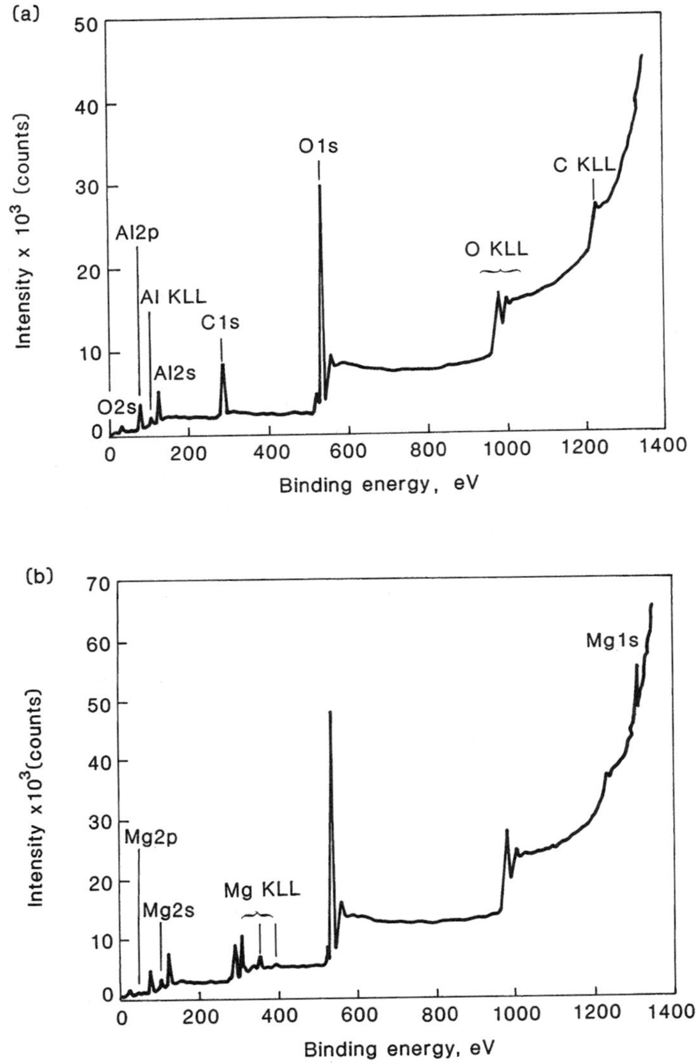

Figure 2. XPS survey spectra of aluminum alloys following chromic acid anodizing. (a) Commercial-purity aluminum which showed extremely poor durability, (b) Alloy BS L157 (A1–5Cu–1Mg–1Si–1Mn) which exhibited extremely good durability despite the high concentration of Mg in the surface analysis. (From Ref. 10.)

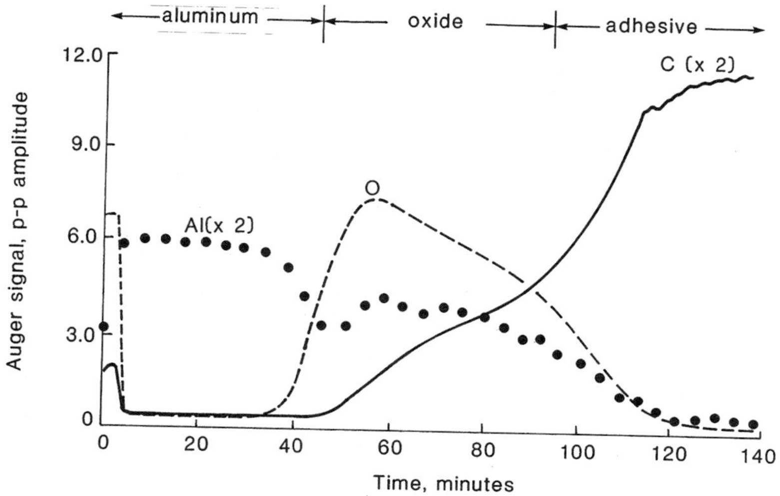

Figure 3. Sputter depth profile, by AES, through the interface region of an adhesive joint on a phosphoric acid anodized ultrathin Al substrate. (From Ref. 11.)

ion being that it will be displaced on application of the polymer phase via specific interactions with the substrate sufficient to overcome the weak dispersion forces between contaminant and metal oxide. Documented evidence of failure occurring in a preexistent contaminant layer is rare.

2.3. Carbon Fiber Composite Materials

In order to achieve strong interfacial bonds between the fibers and the resin matrices of composites, carbon fibers must be treated in either an electrochemical or a gas-phase process. XPS has been widely used to study the surface chemical modifications brought about by such treatments and the literature relating to these studies was reviewed by Castle and Watts in 1988 [12]. In essence the understanding of the surface properties has been reached by detailed peak fitting of the C 1*s* spectra from carbon fibers following electrochemical treatment and by the use of a labeling method known as chemical derivatization. The work relating to peak fitting of the C 1*s* spectra of carbon fibers is discussed at length in Chapter 14. In chemical derivatization a small molecule or ion is used to label a specific chemical group that is thought to be present on the material surface but cannot be detected uniquely by XPS. In the case of carbon fibers, Ba^+, Ag^+ and Mg^{2+} have all been used to good effect to label acidic sites (—COOH) that are present at very low concentrations at the fiber surface. A variation of the chemical derivatization methodology is the determination, by XPS, of adsorption isotherms, which were found, in a recent investigation, [13], to be of the Langmuir type, allowing monolayer coverage of the labelant (Ag^+ or Mg^{2+} ions), and hence the surface acidity, to be estimated. The source of nitrogen in the XPS spectra of carbon fibers has long been a topic of debate and it is only recently, with the advent of high-transmission, high-resolution spectrometers, that the source has been established unequivocally. On going from untreated to treated (in NH_4HCO_3 electrolyte) fibers, the binding energy of the N 1*s* peak shifts to a slightly higher value [14], reflecting a change in the nitrogen chemistry from that in the residue of the poly(acrylo nitrile) precursor to that in the ammonium ions adsorbed from the electrolyte. Another powerful tool, ToF-SIMS, has also been applied to the analysis of carbon fiber surfaces and shown to provide information complementary to that from XPS studies [15,16]. The thorny question of the level of electrochemically induced topography which cannot be distinguished by SEM, now seems to have been resolved by scanning probe microscopy (STM and AFM). Recent studies have shown conclusively the development of edge features and exposed basal planes following electrochemical treatment [17].

Although chemical analysis of the substrate is now a straightforward matter, recent research has concentrated on the extension of such analyses to other properties such as surface polarity and acido-basic characteristics. These are extremely important and may point the way to a predictive approach to bond integrity. They are discussed below in Section 5.

3. FAILURE ANALYSIS: IDENTIFICATION OF THE LOCUS OF FAILURE

In the use of surface analysis for the failure analysis of fracture surfaces there have been some spectacular successes over the years, and it could be argued that it was the success in this area that made such a strong impression on the adhesion community, leading to the adoption of surface analysis so enthusiastically. XPS in particular can be said to provide a chemical overview of the failure in a manner analogous to the overview of morphology provide by SEM fractography. Nowadays identification of the locus of failure of adhesive joints and organic coatings is a routine undertaking carried out with great success in many laboratories around the world. The whole question of whether interfacial failure really exists has been the subject of much scrutiny, and with recent advances in the application of ToF-SIMS to the problem it is clear that the situation is not as well defined as was hitherto imagined. In this section several types of failure analysis investigation will be reviewed, and in order to group them in a logical order they will be categorized according to substrate.

3.1. Adhesion to Brass

One of the first published investigations into practical adhesion phenomena in which surface analysis techniques were used was that into the adhesion of rubber to brass by van Ooij in the mid-1970s. Subsequently this particular research was extended and refined, and publications by the same author on the topic now cover more than two decades. The adhesion of vulcanized rubber to brass is of critical importance in the tire industry since the steel wires used for tire reinforcement are brass-plated to ensure good bonding to the rubber of the tire. The need to brass-plate the wires had been known for many years but it was not until van Ooij's benchmark publications of 1977 that the chemistry involved in this particular adhesion process became fully understood.

In order to be able to apply XPS in particular to the study of adhesion, van Ooij [18] used thin brass strips 1 cm wide rather than attempt to study the very fine wires themselves. When investigating the 70/30 brass alloy used in the tire industry, he found [19] that its surface consisted of a thin film of Cu(I) oxide on top of a thicker layer of copper and zinc oxide. When the alloy is heated, the zinc oxide grows in preference to the copper oxide, and in extreme cases leads to the solid-state reduction of the surface Cu_2O phase by the metallic zinc. The mechanism of adhesion of rubber to the brass substrate is expected to result in formation of both copper and zinc sulfides, by reaction between the elemental sulfur, which yields an active sulfide ion at the oxidized brass surface, and both the metallic copper, which is subsequently oxidised to a Cu(I) ion, and zinc ions, that are present in the inorganic film on the brass. In fact it is the Cu_xS phase which is responsible for the extent of adhesion between the two phases; if the concentration of Cu_xS is optimized at the interface, it can lead to improved adhesion as a result of the catalytic effect of the sulfide on vulcanization, but if in too high a concentration, the sulfide will lead to embrittlement of the mixed copper and zinc sulfide film. The situation which is thought to exist at the rubber/brass interface is illustrated schematically in Fig. 4. The consequences of this unequivocal link between alloy surface condition and adhesion performance were of great technological importance since tire manufacturers were then able to predict with good accuracy the expected adhesion that would be achieved between their own particular rubber formulation and brass-plated steel wires from any particular source. The potential existed, as in the steel cleanliness example from the United States of America, to use surface analysis in a quality assurance role.

In subsequent work [20–22] van Ooij developed the use of model compounds (in the case of natural rubber, squalene) to simulate the physical and chemical interactions at the interface of the brass/natural rubber system. Cuts in a tire carcass may lead to exposure of the steel/brass interface and the galvanic effects that give rise to dezincification may then become important [23]. The use of duplex Zn-Ni on Zn-Co alloys as replacements for brass has been proposed in order to

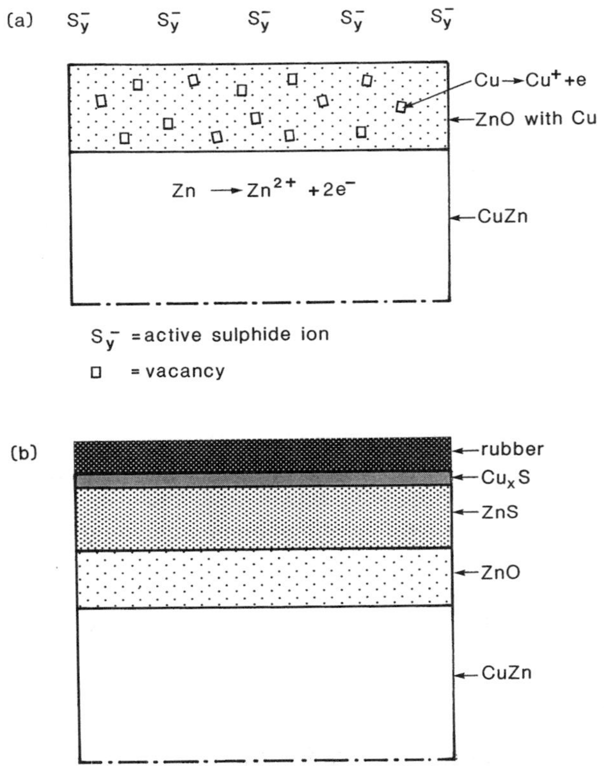

Figure 4. Schematic representation of rubber-to-brass adhesion, based on XPS analyses, from the work of van Ooij [22]. (a) The situation prior to vulcanization indicating the formation of the active sulfide ion from elemental sulfur at the brass surface; (b) the formation of Cu_xS following vulcanization.

circumvent the problems associated with the corrosion of brass [24]. The electrodeposited duplex layers have been optimized in terms of the process parameters and adhesion levels achievable with natural rubber and are found to be comparable in performance to the traditional steel/brass system. The advantages of such a system are that the corrosion problems are much reduced [25]. The important conclusion arising from the work of van Ooij is that by arriving at a complete understanding of a traditional system it is possible to predict exactly the developments needed for an improved system. Van Ooij has also pioneered the use of SSIMS in such adhesion studies and has published a comprehensive treatise on the SSIMS spectra of a series of synthetic rubbers [26].

In the application of protective organic coatings to brasses and bronzes there appear to have been relatively few studies that have made use of surface analysis. One such study is that of Castle et al. [27] into the localized protection of aluminum brass (Cu-76, Zn-22, Al-2) power station condenser tubes by a commercial acrylic lacquer. By careful examination, by XPS, of the failure surfaces produced by mechanical testing following exposure to hot saline solution for several days, it was possible to identify the failure mechanism as that associated with electrochemical activity within a developing crevice. In the study, by XPS or AES, of metal surfaces that are the result of electrochemical activity, it is often informative to report the ratio of characteristic ions from the aqueous exposure medium as a means of deducing the prior electrochemical history of the electrode. The excess of chloride to sodium ions in the surface analysis was indicative of the development of an anodic crevice, which is in marked contrast to the cathodic delamination case histories so often seen in coatings failure and quoted in the next section. The analyses on both sides of the failure path showed the same excess which became more marked with increasing coating thickness. At longer exposure times the anodic crevice conditions, which were solely a result of electrochemical activity within the crevice formed between the aluminum brass substrate and the detached acrylic coating, broke down, probably as a result of mass transport through the thickness of the polymer film. This example illustrates clearly the importance of taking into account corrosion phenomena when possible adhesion failure mechanisms are being considered. In this case the mineral layer present on the brass following power station operation was protective [28] and acted as a cathode to the brass anodic underfilm region that had been cleaned prior to the application of the acrylic lacquer. This situation is not usually encountered, the more common situation being that of the exposed metal acting as a local anode to the coated metal which becomes cathodic.

It is perhaps worthwhile at this point reviewing briefly the electrochemical processes that are associated with metallic corrosion. The reaction responsible for the degradation of the metal (Me) occurs at the anode, and for this reason the process is often described as anodic dissolution:

$$\mathrm{Me} \rightarrow \mathrm{Me}^{x+} + xe^{-} \tag{1}$$

The reaction has led to the production of an Me^{x+} ion and x electrons each with a single negative charge. As electrical neutrality must be maintained the electrons are consumed in the cathodic reaction that occurs on an adjacent portion of the electrode termed the cathode. In aqueous electrochemistry the reaction involves water and oxygen as well as the electrons:

$$2\mathrm{H_2O} + \mathrm{O_2} + 4e^{-} \rightarrow 4\mathrm{OH}^{-} \tag{2}$$

This reaction is referred to as the cathodic reduction of water (and oxygen) and leads to an increase in pH (alkalinity) in the environs of the cathode. In the aluminum brass example coatings failure occurred in the region of the anodic surface, while in the examples in the next section coatings detachment from the cathodic surface is described. The latter process is known as cathodic delamination or disbondment.

3.2. Adhesion of Organic Coatings to Steel

Because mild steel is such a technologically important substrate there have been many studies carried out on the adhesion to it of organic systems, predominantly in the form of coatings. The main difficulty that arises is that any exposure trial in an aqueous environment will lead to the production of copious amounts of corrosion product, i.e., rust, that can make any subsequent investigation of failure mechanism extremely difficult. Various authors have sought to circumvent this problem in different ways, including the separation of the anodic and cathodic sites by the use of a sacrificial anode such as zinc, or an impressed current method (frequently used in the cathodic protection of underground pipelines), or the use of aqueous environments containing low partial pressures of oxygen which, in turn, keep the level of anodic activity (rusting) at a modest level.

Knowledge accumulated from the investigation of failures associated with low-carbon steels (often referred to by the generic name of mild steel) has allowed the interfacial chemistry of the failure process to be well understood. Implicit in the mechanism of adhesion loss is the influence of corrosion, already mentioned, and the recognition that the electrochemical process of rusting will have a marked effect on durability of the coating or the adhesive joint. Thus, in a test in which the joint or coating is exposed to an aqueous or humid environment, it is important to consider the effect of such exposure on both substrate and polymer. In the case of exposure at the free corrosion potential (the potential that freely rusting steel will assume, also known as the rest potential), there is the possibility that postfailure corrosion will cloud the analytical issues; care must be taken not to confuse such postfailure activity with that responsible for failure in the first place. The first recognition of the need to study degradation in both phases was in an investigation into the adhesion of polybutadiene to steel exposed to a saline solution, either at rest potential or with an impressed cathodic potential [29,30]. It was shown by XPS that two characteristic types of failure occurred. At the exposed metal cathodically generated alkali brought about interfacial failure and subsequent blistering, which caused, for the specimen at rest potential, an extension of the anodic (rusting) site. In the case of cathodic polarization (to simulate cathodic protection) the surface remained very clean with a low level of carbon as shown by XPS (Fig. 5a). Outside this region downward diffusion of solvated cations (i.e., cations surrounded by a tightly bound sheath of water

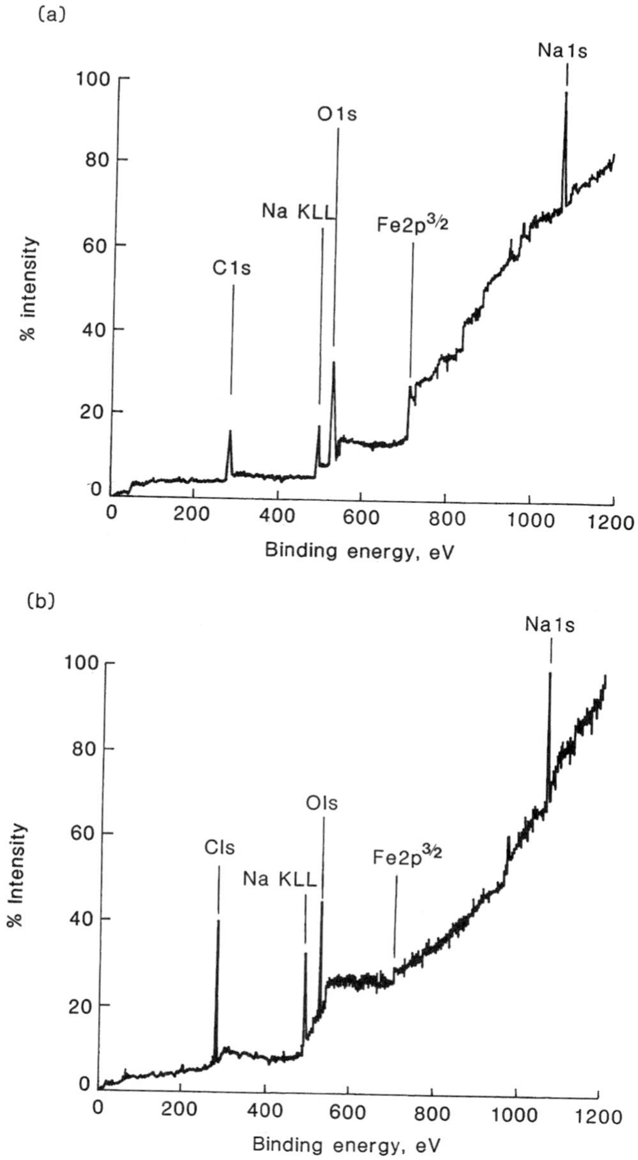

Figure 5. XPS survey spectra from polybutadiene/mild steel failure interfaces. (a) Blistered region representative of cathodic delamination and lateral diffusion of hydroxyl ions; (b) wet adhesion failure resulting from the downward diffusion of water molecules. (From Ref. 30.)

molecules), through the coating led to a failure process in which a thin (ca. 4 nm) layer of polymer remained adhering to the substrate (Fig. 5b). The latter process has also been invoked by other authors for polybutadiene [31] and for other coatings [32,33]. The failure mechanism was at the time ascribed to alkaline hydrolysis of the near interface polymer, and terms such as "active saponification" were often used. Some years later a more plausible explanation was offered [34], on the basis that the presence of alkali cations was not a necessary condition for such failure, and that it was the presence of water within the polymer that was responsible for failure. The strands of the argument can be brought together as follows: the damage to the polymer in this type of failure is reversible and shares all the characteristics of the phenomenon known in the paint industry as wet adhesion (in which tests are carried out in warm water). The explanation assumes that the locus of failure characterized by the XPS spectrum of Fig. 5b is the result of the downward diffusion of water molecules through the thickness of the paint film; the molecules accumulate near the oxide/polymer interface and lead to swelling and plasticization of the polymer. Such damage is reversible. The role of the solvated cations is to control the rate at which water molecules reach the proximity of the interface. It has been postulated that the thermodynamic activity of water molecules within the polymer phase may be all important in determining the kinetics of the process [35]. What little information there is on the activity of water in the solution to which coatings have been exposed does indeed support this hypothesis [36].

All the investigations referred to have used the approach of exposing samples to the chosen environment within the laboratory, peeling back the polymeric phase with tweezers or by a similar method, and then analyzing the interfacial metal and polymer surfaces. For true interfacial failure this means that the metal oxide surface, which has a very high surface free energy, will inevitably become covered with hydrocarbon contamination, as this leads to a reduction in the surface free energy and is the thermodynamically favored process (see, for example, the spectrum of Fig. 5a). The task for the analyst is then to attribute the carbon signal in the spectrum to either, or both, contamination and coating/adhesive residues. This is no easy task but may be tackled by using fine structure within the C 1*s* spectrum, angle-resolved XPS, quantitative surface analysis and molecular definition of the organic residue by ToF-SIMS. The alternative is to separate the polymer from the metal adherend within the spectrometer using either custom-made or commercial devices [37,38].

Although the assay of carbon can provide an indication as to whether the failure is truly interfacial or is cohesive within the polymer phase, much additional information can be revealed concerning polymer degradation chemistry by peak-fitting the C 1*s* spectrum. In the case of epoxy coatings, the formation of carboxylate species brought about by alkaline hydrolysis, as well as carbonate residues, has been identified by Dickie and colleagues [31–33] at interfacial failure surfaces. The latter feature is ascribed to the degradation of urea and urethane components of the paint, although other authors have attributed a similar observation to the presence of dissolved carbon dioxide in the electrolyte [57]. Using angle-resolved XPS, Watts et al. were able to determine the orientation of the polymeric fragments remaining on the steel surface and then to relate their observations to chain scission within the polymer [39]; the scission occurred adjacent to the residual epoxy groups, which always remain in the outermost region of the thin (2–3 nm) overlayer remaining on the metal substrate. On the interfacial polymer side of the failure they were below the surface.

The native oxide present on mild steel is often chemically modified prior to painting by application of a conversion coating intended to improve corrosion resistance. Such treatments are often used in the automobile and consumer goods industry, in which those based on zinc phosphate and chromate are commonplace. Dickie et al. [31] have studied the failure of coated prephosphated steel and have commented on the chemical degradation which appears to be a precursor to adhesion loss. It is well known that the underfilm alkalinity associated with cathodic delamination can bring about phosphate dissolution [40]. In the study of a simple phosphoric acid wash treatment Watts [41] noted that a surface iron phosphate phase was deposited to which adhesion was enhanced, in comparison with adhesion to the bare steel substrate. For chromate coatings the interfacial chemistry can be complex. Figure 6 illustrates fine structure within the Cr 2*p* spectrum from a chromate coating applied to steel in the as-coated condition and following cathodic delamination [42]. The decrease in hexavalent chromium is a result of the reduction reaction occurring at the cathode surface, which in turn leads to loss of adhesion. In this case the cathodic exposure conditions have, unusually, brought about degradation of the inorganic phase rather than that of the interfacial bonding.

The adhesive bonding of mild steel adherends has received scant attention apart from a very thorough investigation by Kinloch et al. [43] in the early days of the application of surface analysis to adhesion. This undoubtedly reflects the industrial view that welding and other joining methods are preferable to adhesive bonding of mild steel for structural applications. The situation is changing, however, and much has been published in recent years on the durability of adhesively bonded steel; surface studies will undoubtedly follow. Attempts have been made to develop inorganic thin film pretreatments for mild steel, and although such treatments show promise they are far from being commercially useful at the present time [44–46].

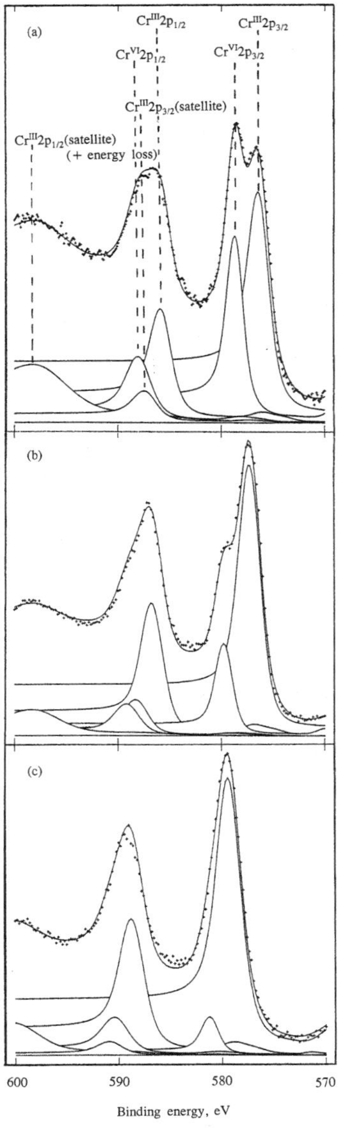

Figure 6. Cr $2p$ XPS spectra from the chromated steel/epoxy system. (a) Metal surface following chromate treatment; (b) interfacial metal surface following cathodic delamination. Note the reduction in the Cr^{VI} component when compared with the as-received substrate prior to coating. The underside of the polymer coating, (c), shows a very low concentration of Cr^{VI}. (From Ref. 42.)

In the characterization of failure interfaces on steel and other substrates, XPS has been the most useful analytical technique. While the strengths of XPS in such applications are well known and have been extensively reviewed [1,2], there is of course one major shortcoming with standard XPS, and that is the lack of spatial resolution. All the investigations described above have employed XPS in its standard (area integrating) form in which some several mm^2 contribute to the signal. This integration has its advantages, not least because there are then no problems concerning the provision of a representative analysis, but in the last decade there have been rapid advances in the development of small-area XPS (SAX, generally taken to mean XPS point analysis from a region 20–200 μm in diameter) and imaging XPS (iXPS, chemical maps at an image resolution of a few μm or better). The various ways in which SAX and iXPS can be achieved have recently been reviewed [48]; examples of their use in adhesion science are few as yet, but some have now appeared [47,49–51]. Recent work [47] using the Scienta ESCA 300 spectrometer has shown the power of the approach. This particular spectrometer makes use of a position-sensitive detector to record the electron intensity transmitted by a large (30-cm radius) concentric hemispherical analyzer. The output in the dispersive plane of the analyzer is essentially a small portion of the electron kinetic energy spectrum, while in the nondispersive plane the output axis represents distance in a linear direction along the specimen. This form of output is known as an *E*:*X* plot (representing energy and distance), the color-coded intensity representing the electron intensity at any energy and position at the specimen. Thus, this spectrometer produces a complex linescan output, the analysis area being $3500 \times 80\ \mu m^2$.

The adhesive bonding of mild steel with a commercial rubber-toughened epoxy adhesive was investigated using a lap shear geometry; specimens were exposed to water containing a low partial pressure of oxygen (to ensure that the level of rusting remained low). Following the aqueous exposure the joints were pulled in a mechanical testing machine and the fracture surfaces examined by iXPS, SAX and ToF-SIMS [47]. The failure surfaces consisted, visually, of a small central region of cohesively failed adhesive surrounded by regions of apparently interfacial failure. The *E*:*X* images indicated clearly that the oxide was thicker at the edge of the joint, but in the central region both Fe^{metal} and Fe^{oxide} contributions were apparent in the Fe 2*p* image, suggesting that the oxide was of about the same thickness as the air-formed film present when the joints were assembled. The SAX analyses showed a very low concentration of carbon at the interfacial metal surfaces, indicating that the failure mechanism, in this case, was interfacial in nature. The Na^+ assay showed that failure had been produced by cathodic delamination. A two-stage process was observed, the faster process giving rise to a zero-volume debond, in which the interfacial bonds are broken but there is no separation of polymer phase from the metal substrate,

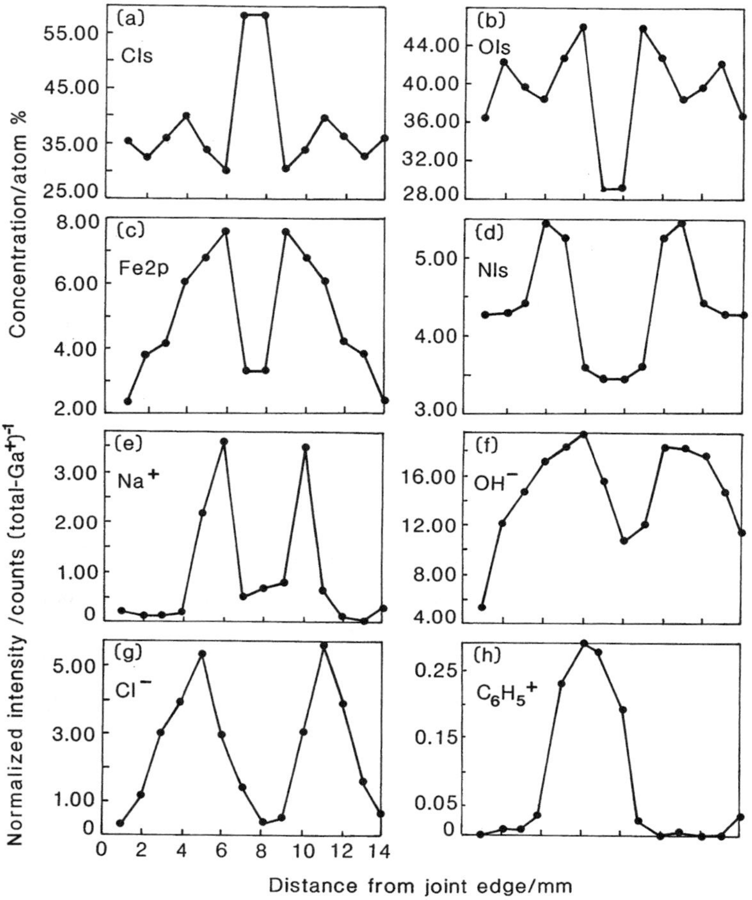

Figure 7. Linescans from a series of point analyses of the metal substrate of a failed adhesive joint by small-area XPS and ToF-SIMS; (a) C l*s*, (b) O l*s*, (c) Fe 2*p*, (d) N 1*s*, (e) $^{23}Na^+$, (f) $^{17}OH^-$, (g) $^{35}Cl^-$, (h) $^{77}C_6H_5^+$. (From Ref. 47.)

which was followed by a slower separation and crack opening stage, that led to the development of a wide crevice in which mass transport could easily take place. Small-area XPS was used in conjunction with ToF-SIMS to record elemental linescans across the metal and adhesive failure surfaces. Examples of these linescans from the metal surface are reproduced in Fig. 7 and merit a brief explanation. The "island" of epoxy adhesive remaining in the central region of the fracture surface is readily identified in the C 1*s* XPS linescan of Fig. 7a, the carbon concentration of 30–35 at .% being entirely consistent with an interfacial failure. The O 1*s* and Fe 2*p* data of Fig. 7b and c delineate the iron oxide on the exposed metal (i.e., where the O 1*s* intensity is higher than the adhesive, ca. 45 as against 30 at.%); the Fe 2*p* signal increases gradually as the crevice tip (adjacent to the intact adhesive) is approached, due to an increase in the Fe^{metal} contribution. The N 1*s* excursion at the crevice tip results from enhanced adsorption of nitrogenous species on the cathodic surface. The reasons for this are unclear. The presence of a sharp rise, in Fig. 7e, of Na^+ concentration (even though the specimens were exposed in ultrapure water) shows that cathodic conditions prevailed at the tip of the disbondment crevice formed between the detached adhesive and metallic substrate, and in Fig. 7f can be seen a corresponding rise in OH^- concentration in the same region. These observations identify explicitly the underfilm alkalinity, generated by cathodic activity, that is responsible for adhesion loss. The chloride ion (Fig. 7g) reaches a maximum slightly away from the crevice tip. The aromatic ion $C_6H_5^+$ is monitored in Fig. 7h and, as expected, identifies the exposed epoxy adhesive. It is interesting to note that the profiles in Fig. 7 peak at slightly different points within the failure region, demonstrating the role that the electrode potential within the crevice plays in the interfacial chemistry. The manner in which the maxima of the various species occur as a function of distance within the crevice, is analogous to that in conventional chromatography, where the distance traveled by the various "fronts" in, for example, thin-layer chromatography can be related to the chemistry of a particular species. It can be envisaged that the XPS and ToF-SIMS data of Fig. 7 describe the deposition of various chemical species at local conditions of pH and electrode potential within the crevice. Using this approach it should be feasible to map potential at a crevice tip by the appropriate ionic addition to the aqueous solution to which the tip is exposed.

The above example further emphasizes the fact that in order to understand the interfacial reactions responsible for adhesion loss the electrochemical response of the substrate must be understood. Such an understanding becomes even more essential when studying stainless steels where the polymer must interact with the air-formed passive film. In order to achieve the interaction many, mainly acidic, pretreatments have been developed on an empirical basis. Examples of the type of investigation required can be found in the work of Gettings and Kinloch, who investigated both the surface characteristics [52] and the nature of the failure surfaces [53] in some detail.

3.3. Zinc Surfaces

Although the adhesion of polymers to zinc itself is of relatively little interest, much work has been carried out on adhesion phenomena associated with zinc-coated steel such as hot-dipped galvanized steel (HDGS) or electrogalvanized steel (EGS). Such substrates are notoriously difficult to paint and are therefore often used in conjunction with a conversion coating of the type discussed in the previous section.

The interaction of a poly(vinyl chloride) (PVC) formulation with HDGS has been studied by Dickie et al. [54,55]. They showed that the PVC resin underwent extensive dehydrochlorination at the metal surface, an effect which could be reduced somewhat by the addition of a resin modifier such as dicyandiamide. Their conclusions were based on quantitative XPS analyses and careful binding energy determinations of the Cl $2p_{3/2}$ core level. The concentration of Cl at the interfacial failure surface was about half that within the bulk of the adhesive, and some 5% of this element within the interfacial region was present as an ionic compound, amine hydroxychloride, (Cl $2p_{3/2}$ = 197.7 eV compared with 199.7 eV for the adhesive component). Scant attention was paid to the zinc chemistry, which is a pity since Zn is particularly amenable to study by way of the Auger parameter, (the sum of the Zn $2p_{3/2}$ photoelectron and Zn $L_{2,3}M_{4,5}M_{4,5}$ Auger peak energies), and has been shown to be useful in the study of pretreatment processes [56]. The Auger parameter concept is particularly useful in the study of samples subject to electrostatic sample charging (such as inorganic pretreatment layers), since the increase in binding energy of the photoelectron peak is exactly offset by a reduction in the kinetic energy of the Auger peak. The chemical information available by this method is limited to Auger transitions that involve three core-like electrons, but in such cases (F, Mg, Al, Si, Cu, Zn, Ge, and so on) the chemical information available may be more certain than from the use of the photoelectron binding energy alone. For example the Zn Auger parameter values for ZnO and $ZnCl_2$ are 2009.8 and 2009.2 eV, respectively.

The adhesion of organic systems to zinc has also been studied by van Ooij et al. [50], who extended the coating variables to include pretreatment of the metal and the degree of cure of the organic system. Their work is important in that, perhaps for the first time, it indicated that the degree of cure of a fully formulated system might be detectable by careful consideration of the ToF-SIMS spectrum. They also reiterated the idea, already reported anecdotally by both Dickie et al. [33] and Watts and Castle [57], that certain of the coating constituents can be adsorbed preferentially at the metal oxide/polymer interface. With the increasing sophistication of analytical tools this idea has received ever greater support. Using an XPS elemental mapping method Haack et al. [49] have shown convincingly, by determining the lateral distribution of the elemental markers for cathodic and anodic reactions, that both anodic and cathodic areas of activity can be observed at the failure surface of adhesively bonded EGS, and they associated failure within the anodic regions to the poor cohesive strength of the zinc corrosion products. Indeed the readiness with which Zn forms so-called white rust seems to suggest that the words of caution regarding mild steel studies are just as relevant to Zn.

3.4. Aluminum Alloys

As might be expected it is in the aerospace industry that so much research into the pretreatment and adhesive bonding of aluminum and its alloys has been carried out. The early work was reviewed by Venables [58], who showed conclusively that for the successful adhesive bonding of aluminum adherends a complex pretreatment schedule must be followed that provided a characteristic microporous morphology. The introductory article by Davis [59] on surface analysis includes also many examples concerning adhesive bonding. The various methods of pretreatment have been reviewed by Olefjord [60] and Clearfield et al. [61]. Some of the initial studies of the failure of adhesively bonded aluminium were performed on failed Boeing wedge test specimens, in which failure mechanisms of the type illustrated in Fig. 8, ascribing failure to the hydration of the anodic oxide layer [58], were suggested. This suggestion led to the investigation of a number of hydration inhibitors that were thought might arrest the chemical degradation and thus improve joint durability. The assumption that such a hydration mechanism is the only cause of failure in aluminum adhesive joints has now been questioned, although recent electrochemical studies have shown unambiguously that subadhesive hydration can, and does, occur [62]. An important observation by Watts et al. was that in cyclic exposure trials of T-peel specimens the damage caused by immersion in water was partially reversible in the drying stage of the cycle [63]. This indicated a phenomenon similar to the wet adhesion process described earlier and ruled out gross hydration of the aluminum oxide phase which would have

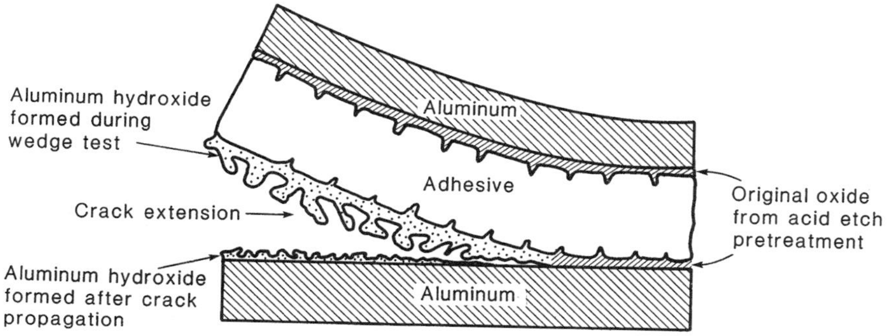

Figure 8. Proposed failure mechanism by oxide hydration of a wedge test joint exposed to an aqueous environment after Venables [58]. Controversy exists as to whether such hydration is the direct cause of failure or is a postfailure event.

involved a large increase in volume. The latter hydration, if it occurs, would not be reversible and the original bond integrity would not be retrievable, contrary to observation. Analysis of the failure surfaces by XPS showed that failure had occurred close to, but not at, the oxide/polymer interface [38]. It was assumed that a thin layer of adhesive remained on top of the porous oxide and filled the outer porous structure. As water and other aggressive species will diffuse in from the edge of an adhesive joint, failure will proceed from the edge toward the center in the manner illustrated for adhesively bonded steel in Fig. 8.

Care must be taken when selecting a sample for analysis to ensure that it is the corrosion or exposure induced failure surface that is chosen and that the central "island" of adhesive is not inadvertently included in the analysis. This does not pose a problem in the case of microanalysis methods such as AES or SIMS, but can be a very real concern in the case of XPS when used in its standard, large-area, mode. The situation is, however, simplified to a certain extent as the central region of adhesive that has failed in a cohesive manner can usually be identified visually (for example, a rubber-toughened epoxy may be black, which contrasts well with the metallic substrate). Figure 9a illustrates schematically the appearance of a failed T-peel test piece from the above investigation [38]; apparent interfacial failure at the specimen edge can be identified visually or with a low-power optical microscope, as one interfacial surface will be that of uncoated metal and the other will consist of the entire thickness of the adhesive. Small-area XPS spectra were recorded from both the edges of the joint and from the central region and are shown in Fig. 9b–d (page 803). The nitrogen peak (at a binding energy of ca. 400 eV) is diagnostic of the

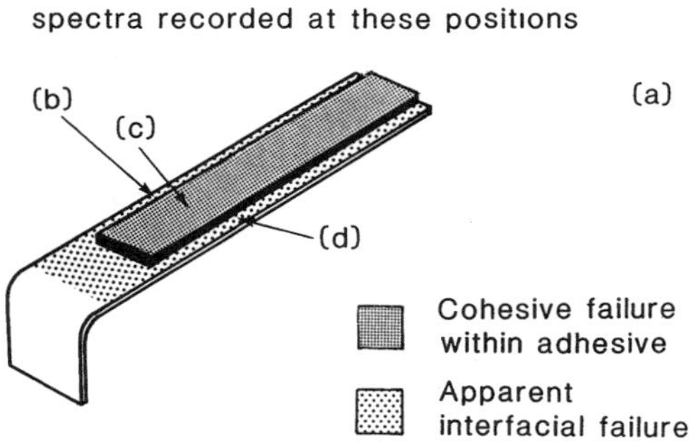

Figure 9. T-peel failure of adhesively bonded aluminum exposed to water at 50°C for 10 weeks; (a) schematic of joint.

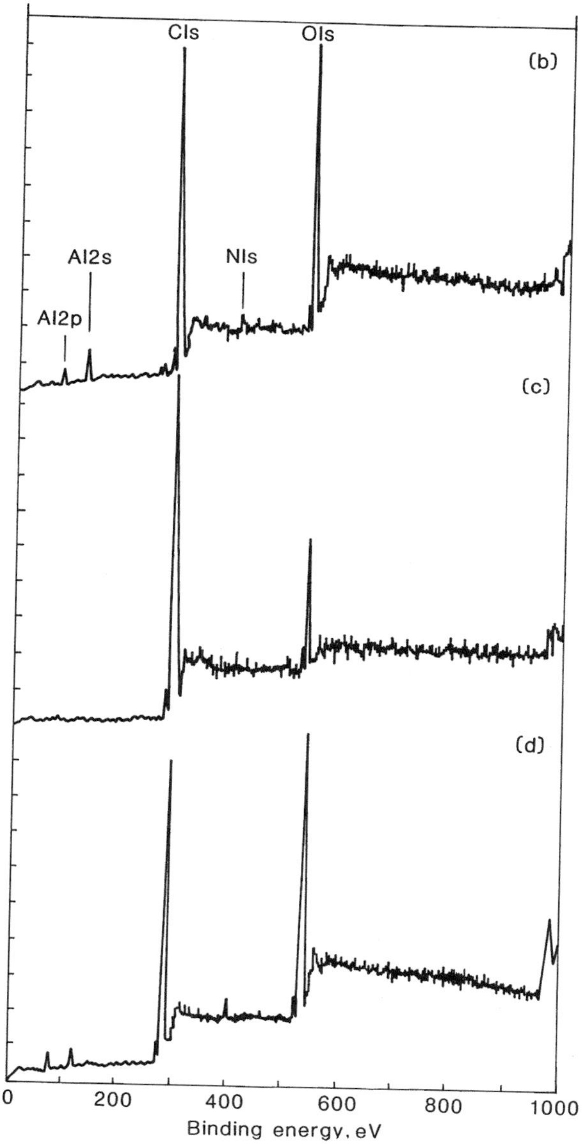

Figure 9, continued. Spectra recorded at positions b, c, and d of Fig. 9a: (b,d) small-area XPS analyses (250 μm) from the edge regions, and (c) the central cohesive region. (From Ref. 38.)

adhesive, which can also be recognized by the intense C 1*s* line. The two spectra recorded from the edge of the specimen also show Al 2*p* and 2*s* peaks, which, together with the C 1*s* line, indicate that rather than being interfacial in nature the failure is actually within the adhesive and that there is a thin layer of polymer remaining on the substrate side of the failed joint. Clearly, substrate hydration has not occurred, since one of the characteristics of a voluminous corrosion product, as far as surface analysis is concerned, is a low concentration of adventitious carbon. *In situ* XPS studies of miniature peel specimens (using Ar^+ profiling) have indicated a very deep transition zone between the metal oxide of the substrate and the polymeric adhesive. This observation has directly confirmed the existence of a extended mechanical interphase, where gross penetration of the adhesive into the porous pretreatment layer occurs, which had previously been hypothesized on the basis of morphological observations by SEM and TEM [64].

3.5. Composite Materials

Although the study of carbon fibers can be traced back to the early days of XPS, and much is now known about the surface chemistry and morphology induced by electrolytic treatments, there have been relatively few studies of composite fracture surfaces. This is perhaps not surprising as the failure surfaces are complex consisting of 6 μm fibers in a polymeric matrix. Attempts have been made to employ SAM to establish failure mechanisms [65,66], but it is generally accepted that ToF-SIMS will play the major role. Initial work with conventional SIMS was impressive and showed that the untreated fiber and matrix failed interfacially, while, on the other hand, an electrochemically oxidized fiber with its resin matrix failed predominantly cohesively [67]. In early studies using quadrupole SIMS the fiber was characterized by CN^- (present from the poly(acrylo nitrile) precursor) and the resin by Cl^- from the residual epichlorohydrin. Both ion signals were strong enough to provide good-quality maps, but subsequent work showed that the CN^- was often also present in the matrix (e.g., in the case of an amine cured epoxy), so this approach was clearly not universally applicable. Later studies have used ToF-SIMS [68,69] and the potential now exists for identifying major fragments of the resin at the fracture surface. ToF-SIMS analyses carried out on double-cantilever-beam test pieces of a carbon fiber composite fabricated from "pre-preg" material (fiber tows preimpregnated with partially cured resin that allows rapid and straightforward composite manufacture) showed that the fracture surfaces were decorated with poly(dimethyl siloxane) that had diffused from the outer surfaces (protected by a peel film) into the bulk during cure [70].

3.6. Ceramics

Most work associated with the adhesion of polymeric materials to ceramic substrates has been carried out in relation to the microelectronics industry. Buchwalter et al. [71–73] have shown the importance of the substrate surface condition in the level of adhesion achieved by a polyimide film to SiO_2, Al_2O_3 and MgO. Their work has revealed the importance of specific interactions of the acid-base type in adhesion, and formed part of the burgeoning body of data in this area. In essence the approach maintains that the forces of adhesion can be treated quantitatively, at a microscopic level, by considering interactions such as those between Lewis or Brönsted acids and bases. The basic tenor of this concept, of which the Lewis approach is the more widely applicable, is that one component in the adhesion couple can be regarded a the Lewis acid (electron acceptor) and the other as a Lewis base (electron donor), the specific interaction responsible for adhesion being the formation of a Lewis acid-base adduct or complex. The work of Buchwalter, cited above, has shown clearly how the extent of donor-acceptor forces at the polymer/ceramic interface can influence the locus of failure. The forces required to bring about failure are related to the types of polyimide precursors employed. For example, the peel strengths of polyimides on SiO_2 and Al_2O_3 are higher than on an MgO substrate, and the locus of failure is cohesive within the polymer. On MgO substrates the peel strength is higher for an ester-derived polyimide than for an acid-derived variety. The failure is of a mixed-mode type in the case of the acid polyimide and is ascribable to carboxylate salt formation and the degradation of MgO by the formation of strong acid-base interactions between the basic MgO and the polyamic acid precursor. With the more neutral ester polyimide the locus of failure was within a weak boundary layers of the polyimide itself [73]. The question of the acid-base characteristics of metal oxides and polymers is of crucial importance in adhesion and will be considered again later.

Oxide substrates in general and alumina in particular are not good electrical conductors and may give rise to a phenomenon known as vertical differential charging (VDC) in the resultant XPS spectrum. Such a phenomenon may be observed in the XPS analysis of insulating materials and results from a change in the electrostatic potential of the specimens as a function of the depth over which the analysis takes place (around 5 nm). This means that the resultant spectrum is actually the convolution of a series of spectra which are shifted a small amount, relative to each other, on the energy scale. In practice the result is a spectrum, recorded by the spectrometer, that is broader than anticipated for the particular spectrometer conditions employed (pass energy, analyzer slits, X-ray source) and shows very little, if any, fine structure, as this will have been destroyed by the convolution effect.

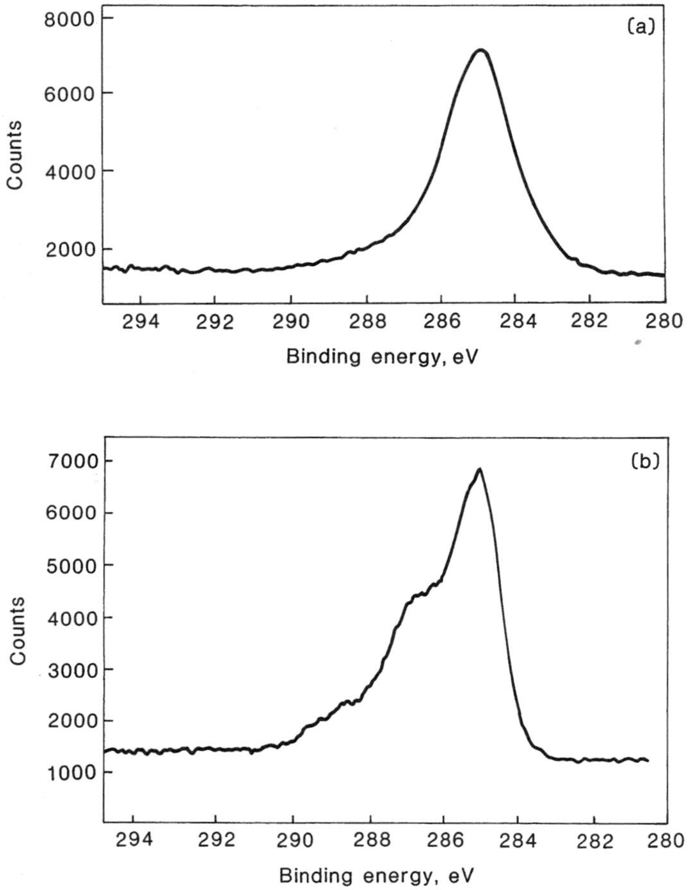

Figure 10. C 1*s* spectra from a failure surface of adhesively bonded alumina recorded with (a) achromatic Al K_α and (b) monochromatic Al K_α radiation. (From Ref. 74.)

In adhesion studies of polymer layers on ceramics particular care must be taken to ensure that the presence of VDC does not lead to an erroneous assignment of the failure characteristics. The potential difficulty of VDC is well illustrated by the work of Taylor and Watts as part of an investigation of the adhesion of a photocured resin to quartz, alumina and silicon [74]. The spectra of Fig. 10 were taken from the oxide failure surfaces of an adhesive joint that had failed following aqueous exposure [74]. Figure 10a shows the C 1*s* spectrum recorded using achromatic Al K_α radiation from a standard twin-anode assembly; the spectrum is rather broad and featureless, and there is a temptation to ascribe failure to a weak boundary layer of preexistent hydrocarbon contamination. When the spectrum was recorded again using monochromatic Al K_α radiation (Fig. 10b) with proper charge control (i.e., an electron flood gun) the C 1*s* spectrum then revealed much fine structure which could be related directly to polymer composition. The adhesive employed was an aromatic methacrylate material with an aliphatic reactive diluent added at a level of about 10 w/w%. The absence of a $\pi \rightarrow \pi^*$ shakeup satellite in the spectrum of Fig. 10b indicates that the locus of failure was associated with a segregated layer of the reactive diluent. This conclusion was confirmed subsequently by quantitative angular-resolved XPS [75], which enabled the overlayer thickness to be estimated at about 1–2 nm. Reformulation of the resin removed the segregated monolayer of reactive diluent and changed the nature of failure to a cohesive mode located within the aromatic adhesive. The reformulation also improved the durability of the joint [76]. The results of this work are important in that they provide a good illustration of vertical differential charging and also emphasize the problems associated with the examination of thin polymeric films on insulators. They also show how the conclusions from surface analysis can be fed back into the formulation process to enable a weak boundary layer to be removed, thus providing a product with superior properties.

3.7. Summary

In the analysis of surfaces produced by adhesive failure at interfaces, XPS is currently the method of choice, and with good reason. XPS can identify and quantify thin polymer overlayers and deduce their orientation, and it can monitor degradation in both polymer and metal oxide and the diffusion of active species within the disbondment crevice. The latter can in turn be related to electrode potential. Failure as a result of the ingress of water is a common occurrence, and such a failure is characterized by a thin (1–2 nm) overlayer of polymer remaining on the substrate. Failure as a result of electrochemical activity such as cathodic disbondment is much closer to the classical concept of interfacial separation. With the use of SAX and iXPS analysis can be performed on the μm scale, the usual criticism that XPS is only an area-integrating technique no longer applying. The two remaining shortcomings of XPS for adhesion studies are the poor detection limits for some elements and lack of molecular specificity. This leads to the conclusion that the complementary nature of SSIMS will make that technique extremely attractive for the study of adhesion processes, used in conjunction with XPS. Initial work has been very promising [50,76], and as all polymer coatings and adhesives are invariably multicomponent systems, future work will concentrate on the identification of those individual constituents of the formulation that have segregated to the polymer/metal interface, and on the correlation of such data with durability and other performance data. Many of the components of such systems will yield similar XPS spectra and so the additional provision of SSIMS analysis will enable their identification in an unambiguous manner. The definition of the exact locus of failure will, however, remain the preserve of XPS as a result of the ease with which quantitative analyses can be obtained, together with the determination of overlayer thicknesses.

Investigation of failure surfaces is often the starting point in an adhesion study and has resulted in many successful surface analysis investigations. There is, however, much additional information that is required to be able to interpret the surface analysis data successfully: a knowledge of substrate characteristics prior to coating or bonding, the manner in which the adherend behaves electrochemically, the formulation of the polymeric phase, and so on, are all important parameters. The forensic analysis of failure surfaces, undertaken with the aim of determining the cause of failure, can by itself sometimes lead to a fuller understanding of interfacial characteristics responsible for bonding, but, in general, that is not so. It is better to start with a knowledge of bonding characteristics and use that as a basis for understanding the exact mechanism of failure, although such a route is experimentally difficult. The analysis of buried interfaces between polymers and metal oxides is very challenging, and in the following section some of the ways in which this can be achieved are reviewed.

4. PROBING THE BURIED INTERFACE

Adhesion is a consequence of the chemical or physical interaction between two surfaces, one of which is a solid and the other a liquid, temporarily more mobile. As a consequence of the way in which adhesion is achieved practically, the interface or interphase region, where the bonds responsible for adhesion occur, is buried below many μm, or even mm, of solid substrate and solidified (generally crosslinked) polymer. The dimensions of the interphase region are likely to be of the order of nm at the most (unless an extended mechanical interphase is present of the type found in the adhesive bonding of anodized aluminum alloys) so that direct examination of interphase chemistry is best considered as an exercise in the analysis of a deeply buried interface. This situation is encountered frequently by those working in microelectronics and in corrosion and oxidation research. Removal of material is invariably accomplished with an energetic ion beam, either as part of a dynamic SIMS analysis or in conjunction with AES or (less often) XPS, in order to produce by sputtering a profile of composition with depth. Such an approach is, however, not appropriate in adhesion studies as the organic material will undergo gross degradation, and eventually graphitization, in the ion beam. Thus the study of interphase chemistry has led to the development of some rather ingenious ways in which the sample, often a real adhesive joint or a coated substrate, and containing a buried interface, can be prepared for surface analysis [77]. In the following section the more successful approaches that have been used will be described briefly.

Mechanical, chemical or electrochemical perturbations can sometimes provide useful results as described earlier, but rely on the interphase being the weakest link, which is not always so. Exposure of the metal substrate by causing the polymer to swell with *N*-methyl pyrrolidone (NMP) has been proposed and works well for some coatings [78]; indeed the use of the time taken to displace the polymer from the substrate by swelling has been suggested as a test for estimating the strength of adhesion of the coating/substrate system [79]. The method was used to provide the specimen for the illustration of Fig. 11 which shows the XPS spectra from a phosphated EGS substrate in the as-received condition and then after coating with a commercial radiation curing (by UV light) system and removal with NMP [80]. The substrate following coating removal is characterized by a modest F 1*s* peak, which is diagnostic of the anionic initiator, included in the formulation to initiate, in the presence of UV light, the curing reaction within the coating. The spectrum indicates clearly segregation of the initiator to the metal surface during application and immediately prior to cure, and such segregation must be taken into account before any adhesion mechanism can be postulated. There is of course a danger that air exposure will modify the interphase chemistry, particularly if reactive species are exposed, but such exposure can be avoided by the use of an argon glove box attached to the spectrometer [89].

The opposite approach is removal of the metal from the adhesive joint, as used extensively by Watts and Castle for the investigation of interfacial chemistry. In the case of ferrous alloys the metal substrate can be dissolved in a methanolic iodine or bromine solution, leaving the oxide film supported on the polymer or adhesive. The supported film can then be mounted for analysis and ion sputtering in order to obtain a depth profile through the oxide toward the oxide/polymer interface. Notwithstanding the danger of ion-beam-induced reduction the authors were able to show convincingly the formation of an Fe(II)-containing carboxylate interphase between polybutadiene and mild steel [30]. Identification of the role of the polymer as an agent in the reduction of Fe(III) oxide was not in itself surprising as the

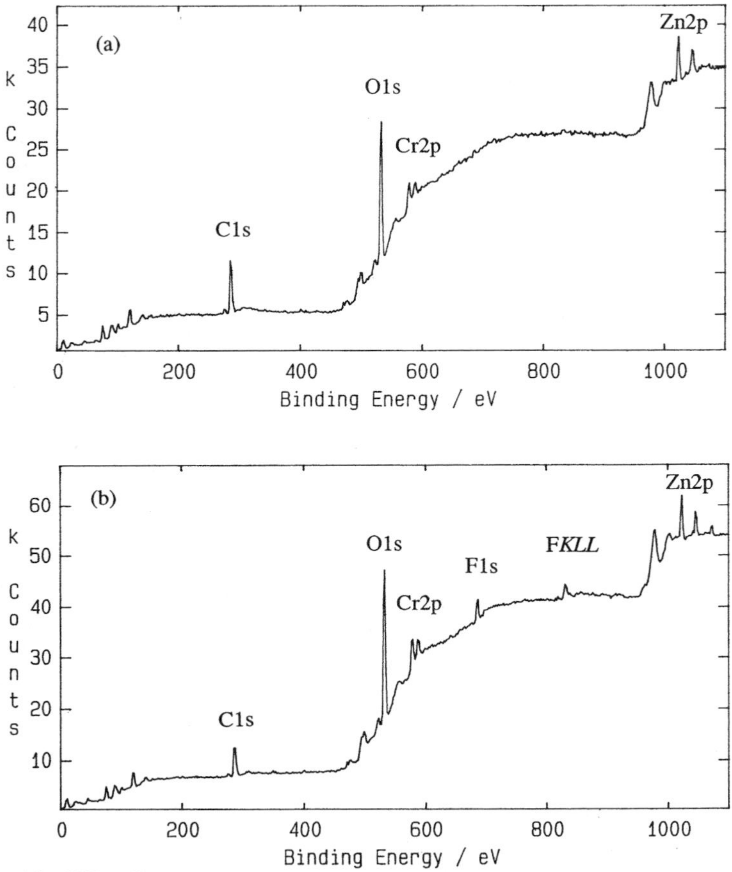

Figure 11. The use of NMP for the removal, by swelling, of an adherent polymer coating. XPS survey spectra of (a) uncoated substrate, (b) substrate after removal of radcure coating.

polymer was known to cure by an oxidative process, and the reduction effect was later confirmed by Mössbauer spectroscopy [81]. For aluminum alloys a similar stripping approach can be adopted using sodium hydroxide solution [82].

A more elegant way to the same end of buried interface analysis, but one that moves away from the real situation, is the use of thin-film model substrates. A methodology has been described [11] in which a thin film of aluminum (0.5–1 μm) was deposited on a secondary adherend and then anodized and bonded in the usual fashion. The thin film was then separated from its backing and the sandwich specimen of Al/Al_2O_3/adhesive presented for analysis by AES or XPS with the metal side uppermost. The depth profile shown in Fig. 3 is from this type of specimen.

Another modeling approach involves the deposition of an ultrathin film onto a solid substrate. If the film is thin enough, XPS can then be used to look through the film to observe interfacial chemistry directly. Such an approach was first tried in a study of the process of polymer metallization for Cr, Fe, Ni [83,84], Ti [85] and Al [86] and has recently been extended to the use of thin polymer films on metallic substrates [87]. The best results have been achieved using very dilute polymer solutions (0.01–0.025 w/w/%), polished substrates and very high resolution (monochromatic Al K_α) XPS [88]. The information that can be gleaned from such studies is impressive but relies on careful peal–fitting of the C1s spectra, as will be discussed in the section on acid-base interactions.

Mechanical sectioning at cryotemperatures in order to prevent polymer spreading and smearing can also be useful for interface analysis. The well-known ball-cratering technique used with AES for gross depth profiling has been developed with a cold stage to enable polymer films to be sectioned successfully [89]. However, electron-excited Auger analysis can yield chemical-state information for only a few elements, so the natural extension of the method was the development of a taper section stage for use with small-area XPS [77]. The design of such a stage was based on the use of an auxiliary HV chamber attached to the preparation chamber of the spectrometer. Taper sections of ca. 1° incline angle could be cut on the cooled specimen with a conventional tungsten carbide end mill. The specimen could then be transferred directly to the analysis position, where the angle of taper transformed to a depth of about 1 μm if a 100 μm small-area XPS facility were used. This has proved invaluable for the analysis of chemistry in polymer [89] and inorganic systems [56].

5. ORGANOSILANE ADHESION PROMOTERS

The use of organofunctional silanes as adhesion promoters is well documented [90], and much is now known about their mode of operation [91]. Surface analysis has played, and continues to play, an important role in the understanding of the action of these complex materials, which have been studied on glass, and on steel and other metals. There is now general agreement that the

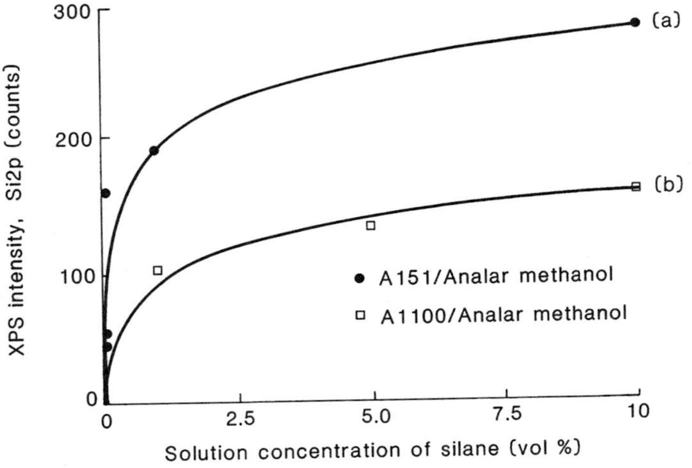

Figure 12. XPS adsorption isotherms for methanolic solutions of (a) vinyl and (b) amino ethoxy silanes on Fe. (From Ref. 92.)

enhancement of adhesion properties is the result of covalent bonding between the silane molecule and hydrated metal oxide, which is consistent with the theoretical models of the last three decades. However, one of the assumptions of these models which is being questioned is the existence of a uniform, toothbrush-like monolayer of molecules at the oxide surface. Such molecular organization has been observed in some cases but does not occur with all molecules of this type showing adhesion-enhancing properties.

Two important investigations dealing with silane organization on iron surfaces were published some two decades ago. Bailey and Castle demonstrated how XPS could be employed to monitor adsorption from the liquid phase by constructing adsorption isotherms of surface concentration (as determined by XPS) as functions of solution concentration [92]. The methodology was also used to assess the adsorption of Ag^+ and Mg^{2+} ions on carbon fiber surfaces, the surface concentration of acidic sites [13], and the adsorption of macromolecules onto the surface of an intrinsically conducting polymer, using ToF-SIMS instead of XPS [93]. In the earlier work Bailey and Castle studied the adsorption of two ethoxysilanes on iron and recorded the data reproduced in Fig. 12. There was enhanced adsorption of the vinyl triethoxysilane relative to the amino propyl triethoxysilane molecule. The isotherms were of the Temkin type, indicating chemisorption, which for the vinyl silane can be ascribed to a specific interaction between the silicon end of the hydrolyzed silane molecule and

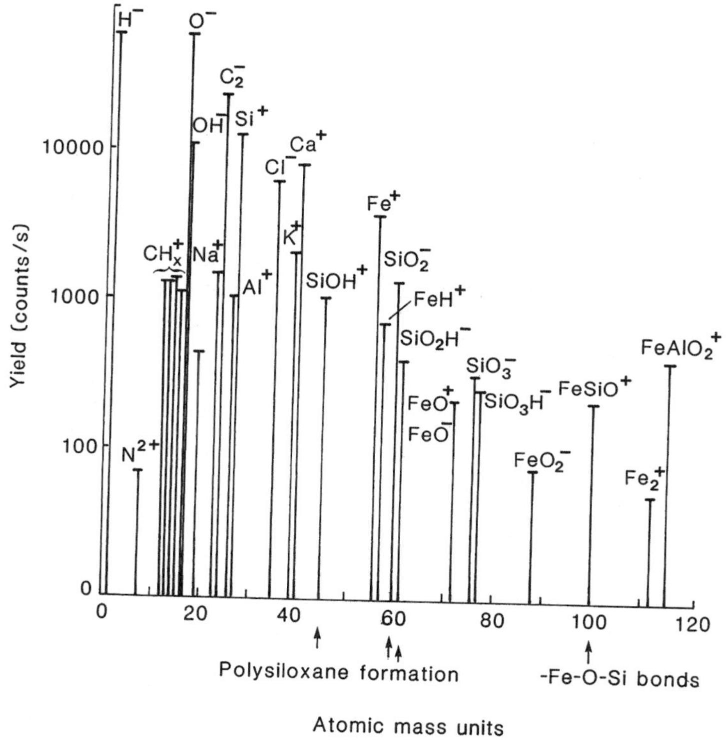

Figure 13. SIMS spectrum from an interfacial failure surface of a silane-primed adhesive joint showing good durability. The ion at $m/z = 100^+$ is thought to be diagnostic of an Fe—O—Si bond. (From Ref. 94.)

the hydrated iron oxide surface. The amino silane molecule was less strongly adsorbed, thought to be a result of the propensity of either end of the molecule (silane or amino) to interact with the metal surface, so the molecular orientation was less clear. As will be seen, the situation was resolved to a certain extent two decades later with the aid of computational chemistry.

The other important observation from the same era was made by Gettings and Kinloch using SSIMS [94]. When studying the fracture surfaces of adhesive joints fabricated using various silane-based primers, they were able to draw a direct correlation between good joint durability and the presence of an $m/z = 100^+$ ion in the SSIMS spectrum at the failure surface (Fig. 13). The ion was attributed to an Fe–O–Si covalent bond and was present only for the γ-glycidoxy propyl trimethoxysilane (GPS); for the first time improved durability could be ascribed to primary bond formation. Their observation was corroborated by other investigations [37]. The measurements of Gettings and Kinloch were made with a quadrupole mass analyzer, and it is therefore pertinent to ask whether the analysis conditions were truly static since those analyzers operate in a serial acquisition mode. Indeed, later ToF-SIMS work by van Ooij on layers (assumed to be only a monolayer in thickness) of silane on iron rather than fracture surfaces indicated that an ion such as that above may not be formed when silane films are applied to iron [95,96]. The question was resolved, to a large extent, by the ToF-SIMS investigation of Davis and Watts [97] in which the organization of GPS molecules on iron was studied. The layer deposited from 2% methanolic solution was shown to be poorly ordered and 1.7 nm thick. In the SIMS spectrum there was initially no peak at m/z = 100^+, but on gentle ion etching a well-defined peak at that mass appeared. Although alternative assignments for the peak exist (for example, $FeOC_2^+$ or SiO_3^+), they can be eliminated by consideration of the relative intensities of related ions [97]. Thus, it seems that both the conclusions from the earlier observations were correct; however, the assumption that the quadrupole analysis provided true static conditions is still not fully resolved.

Silane deposition and organization has also been studied on many other metal surfaces, including stainless steels [98], electrogalvanized steel [99], aluminum, titanium [100,101] and copper [102]. For about 10 years Boerio et al. have investigated the interaction of a variety of organosilanes with many metallic substrates. Their work is particularly significant in that they have combined XPS with vibrational spectroscopies such as reflection adsorption infrared spectroscopy (RAIRS) and surface-enhanced Raman spectroscopy (SERS) to provide enhanced chemical information. SERS is applicable only to highly Raman active compounds, so in the field of adhesion the only possible study has been that of the interaction of silver with organic materials [103,104]. Silanes are used widely in the size that is applied to glass fibers immediately after manufacture, and the interaction has been studied by Yates and West, using an Ag L_α source to record the Si 1*s*–Si KLL Auger parameter [105] (a method also used by Cave et al. to study the interaction of octadecyltrichlorosilane with aluminum [106]). The measurement of this Auger parameter is particularly useful as, in principle, it allows the polarizability of the oxygen atom in Si–O compounds to be determined, which, in turn, can be related to specific bond formation in a manner that could not be achieved merely by recording the Si 2*p* photoelectron spectrum. The most extensive investigation of the interaction of organo-silanes with glass fibers has been carried out by Jones et al. [107–109], who, on the basis of ToF-SIMS results, proposed the scheme in Fig. 14.

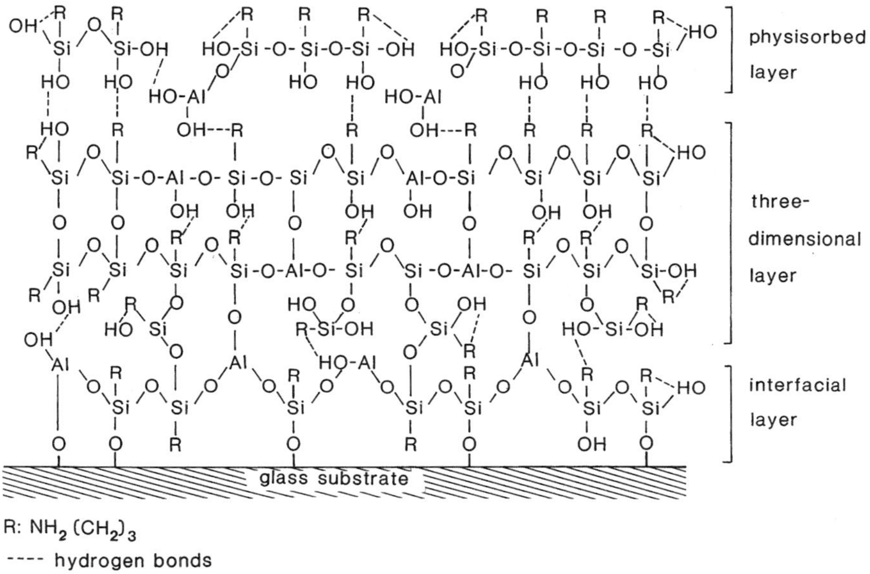

Figure 14. Schematic of the interaction of hydrolyzed γ-amino propyl triethoxy silane molecules with an E-glass fiber surface, based on the ToF-SIMS work of Wang and Jones [108].

6. ACID-BASE INTERACTIONS IN ADHESION

6.1. Evaluation of Acid-Base Interactions in Adhesion

It has become clear that in describing the nature of interfacial interactions responsible for adhesion it is incorrect merely to describe a "polar" contribution to the surface free energy and then arrive at an interface free-energy term using a geometric mean approach. A more rigorous description of such forces can be provided by adopting the donor-acceptor approach pioneered by Fowkes [111]. This is now widely accepted by the adhesion community, with many workers seeking to clarify and quantify the role that acid-base interactions have to play in adhesion phenomena. Progress up to the early 1990s has been summarized in a Festschrift in honor of F. M. Fowkes, which provides a comprehensive overview of the achievements to that time [112]. However, only recently have surface analysis methods been employed to determine the acid-base properties of metal oxides and polymers. The advantages of using surface-specific techniques are immediately obvious to those working in applied surface science; conventional methods such as flow microcalorimetry, FT-IRS, and the like rely generally on high-surface-area solid oxides, which means that either the film must be removed from the metal and ground to a powder or the polymer must be taken into solution. Such methods are perfectly acceptable in the study of homogeneous materials but cannot be used to measure concentration gradients, which might occur in surface treatment of a metal substrate, or the occurrence of segregation effects in coatings or adhesives. It is appropriate at this stage to review the three approaches to the quantification of acid-base forces, two of which are based on a formal thermochemical term, the exothermic enthalpy of acid base interaction ($-\Delta H_{AB}$), and the third being merely a monotonic scale indicating the strength of interaction. The acid-base contribution to the work of adhesion, W_{AB}, has been defined by Fowkes [111] as

$$W_{AB} = fN(-\Delta H_{AB}) \tag{3}$$

where N is the number of acid-base pairs available for bonding at the interface, and f is an entropy factor, specific to a particular system, which allows the conversion of enthalpy to free energy. Although originally taken as unity, f is now known to have values as low as 0.15.

The exothermic enthalpy of acid-base interaction (in kcal mol^{-1}) can be predicted from Drago's equation [113],

$$-\Delta H_{AB} = E_E E_B + C_A C_B \tag{4}$$

in which each Lewis acid or base is characterized by parameters E and C, where E is related to the susceptibility of a species of undergoing an electrostatic interaction and C to the susceptibility of taking part a covalent bond. The subscripts A and B refer to acidic and basic species, respectively. Equation (4) can be used to predict ΔH_{AB} for a polymer adsorbed on an inorganic substrate provided that the E and C parameters for these materials are known. The equation was devised originally to describe the adsorption of small molecules, and its wider application to the adsorption of macromolecules on solids has, up to now, neglected entropy considerations. It is hoped that future work will resolve this shortcoming in what is a very useful approach.

The enthalpy of formation of a Lewis acid-base adduct can also be estimated by using Gutmann's [114] donor and acceptor numbers, DN and AN, where DN characterizes the basicity and AN the acidity of Lewis species. The enthalpy is given (approximately) by the simple relationship

$$-\Delta H_{AB} = \frac{DN \cdot AN}{100} \tag{5}$$

The donor number concept, introduced some 30 years ago, is defined as the negative of the enthalpy of formation of the acid-base adduct formed between the base under investigation and a reference Lewis acid, antimony pentachloride;

$$DN = -\Delta H(\mathrm{SbCl_5{:}base}) \tag{6}$$

This was complemented some 10 years later by the addition of the acceptor number for the estimation of acidic properties, defined as the relative ^{31}P-NMR shift observed when triethylphosphine oxide was dissolved in the candidate acid. The scale was normalized by assigning an *AN* value of 0 to the NMR shift obtained with hexane, and of 100 to that obtained from the $SbCl_5{:}Et_3PO$ interaction in dilute 1,2-dichloroethane solution. The accuracy of the Drago and Gutmann methods has recently been appraised critically by Jensen [115].

An alternative to these thermochemical approaches has been suggested by Bolger [116], who defined a Δ parameter as the numerical difference between the isoelectric point of the inorganic surface (IEPS), i.e., the pH of an aqueous solution in which the solid surface would exist in a state of electrical neutrality, and the acid ionization constant of the polymer (pK_a), namely

$$\Delta \mathrm{pH} = \mathrm{pH_{surf}} - \mathrm{p}K_a \tag{7}$$

Using standard values of IEPS for acidic (SiO_2), basic (MgO) and neutral (Al_2O_3) surfaces, and pK_a values for some 30 organic compounds, Bolger has shown how the parameter provides an empirical index of acid-base interaction. He formulates the equation in such a way that the numerical value of Δ will always be large and positive if acid-base forces are maximized.

$$\Delta_\mathrm{A} = \mathrm{IEPS_{(B)}} - \mathrm{p}K_{a\,\mathrm{(A)}} \qquad \text{(organic acids)}$$

$$\Delta_\mathrm{B} = \mathrm{p}K_{a\,\mathrm{(B)}} - \mathrm{IEPS_{(A)}} \qquad \text{(organic bases)}$$

Bolger identified three regimes of acid-base interaction defined by the magnitude and sign of the parameter:

1. Δ is large and negative. Acid-base forces are zero or negligible.
2. Δ is small with sign either negative or positive. Acid-base forces are comparable to the dispersion force component between the two phases.
3. Δ is large and positive. Strong acid-base interactions, perhaps resulting in complete reaction between inorganic and organic phases.

6.2. The XPS Chemical Shift and Acid-Base Interactions

It has been pointed out by Chehimi [117], on the basis of XPS analyses by Burger and Fluck [118], that there is an excellent correlation between XPS chemical shifts and $-\Delta H_{AB}$ for a series of quick-frozen solutions of $SbCl_5$ with a series of Lewis bases in dichloroethane. As Drago's *E* and *C* parameters were

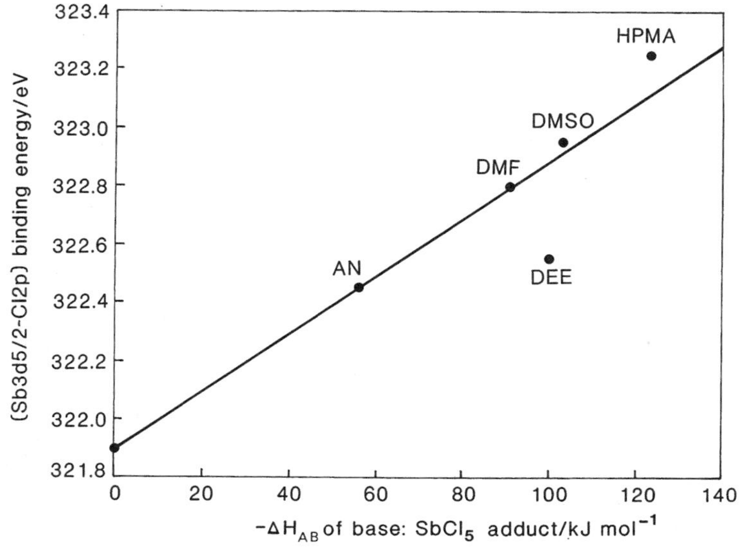

Figure 15. Correlation of XPS data with $-\Delta H_{AB}$ for $SbCl_5$: Lewis base complexes in quickly frozen solutions of dichloromethane. (From Ref. 117.) Key to data points: AN, acetonitrile; DEE, diethylether; DMF, dimethylformamide; DMSO, dimethylsulfoxide; HPMA, hexamethylphosphoramide.

available for all but one of the Lewis bases and for $SbCl_5$, the thermochemical term could be evaluated using Eq. (4). If these parametric data are plotted against the Sb $3d_{5/2}$ binding energy, referenced to the Cl $2p$ binding energy, the linear relationship shown in Fig. 15 is obtained. In the original work [118] an XPS binding energy value (Sb $3d_{5/2}$ – Cl $2p$) of 322.75 eV was obtained for the trimethylphosphate:$SbCl_5$ adduct in dichloroethane. The values of Drago's parameters for the base were not available, but by interpolation of the data of Fig. 15 the enthalpy of formation of this complex was estimated by Chehimi to be –89.9 kJ mol^{-1} [116]. It is this type of relationship that has been used in the determination of the acid-base properties of polymers in the solid state.

6.3 The Use of Vapor Phase Organic Probes for the Determination of $-\Delta H_{AB}$

The experimental procedures have been described in detail elsewhere [119,120]. In brief, FT-IRS was used to determine $-\Delta H_{AB}$ of polymer:solvent couples in very dilute solutions (ca. 0.02 mol dm^{-3}), using the method of Fowkes [121]. XPS measurements were made on thin films of the candidate polymers by exposing them to the organic vapor for a normalized vapor pressure time of 4 atm-min at 20°C. This exposure ensured that the volume of polymer probed by XPS (i.e., approximately the outer 5 nm) was saturated with organic vapor and that the chemical shift in the XPS line characteristic of the organic vapor was a measure of the extent of any acid-base interactions. The method is illustrated in Fig. 16 for a series of polymers exposed to trichloromethane, and it can be seen that there is indeed a linear correlation between the thermochemical term measured by FT-IRS and the XPS Cl 2*p* binding energy of the organic vapor used to interrogate the solid polymer. In order to use this method for unknowns it is necessary first to establish a calibration curve of the type in Fig. 16, for a particular solvent using a series of homopolymers, and then the magnitude of $-\Delta H_{AB}$ of the interaction between that solvent and a more complex polymeric system, such as a fully formulated coating or adhesive, can be interpolated in the manner illustrated for trimethylphosphate from the data of Fig. 15.

6.4. Quantitative Acid-Base Characteristics of the Polymer

Clearly $-\Delta H_{AB}$, as determined in the manner described above for a polymer with a particular solvent, is not an intrinsic property of the polymer. To obtain an intrinsic ΔH_{AB} it is necessary to record data for the candidate polymer with

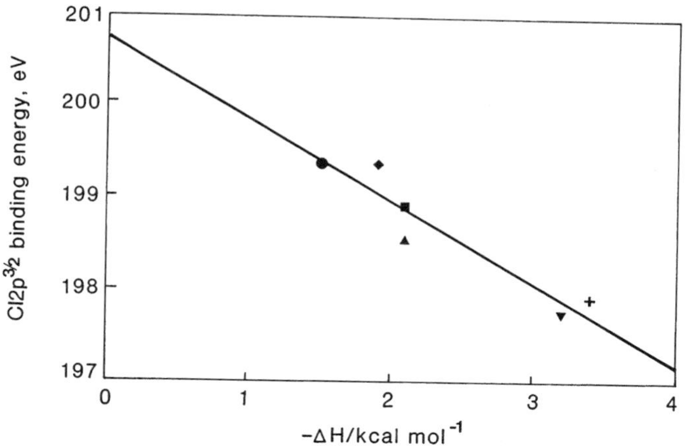

Figure 16. Correlation of Cl $2p_{3/2}$ binding energy with $-\Delta H_{AB}$ for trichloromethane absorbed in a series of polymers. (From Ref. 120.) Key to data points: + aromatic moisture cured urethane; ● poly(vinyl acetate); ▲ poly(methylmethacrylate); ▼ poly(ethylene oxide); ◆ poly(*n*-butylmethacrylate; ■ poly(cyclohexyl methacrylate).

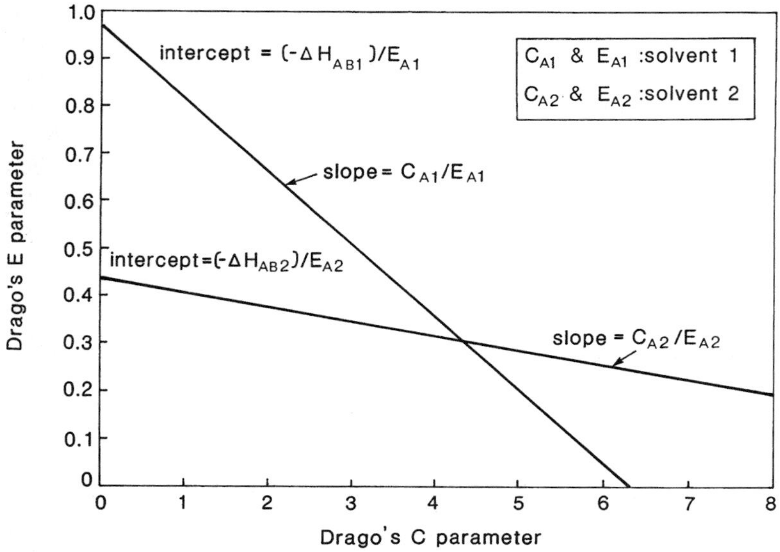

Figure 17. Graphical method for the estimation of the E and C parameters of a polymer using two solvents of known E and C values, and a knowledge of the exothermic acid-base interaction between the polymer and solvent 1 ($-\Delta H_{AB1}$) and solvent 2 ($-\Delta H_{AB2}$).

two solvents of known acid-base characteristics. The values of Drago's E and C parameters for the polymer can then be found by evaluation of Eq. (4) for the two acid-base adducts (the polymer with solvent 1 [E_{A1} and C_{A1}] and with solvent 2 [E_{A2} and C_{A2}]), which provide the thermochemical terms $-\Delta H_{AB1}$ and $-\Delta H_{AB2}$, respectively. The most convenient method of obtaining the solution is the graphical method of Fowkes [111], illustrated in Fig. 17.

With orthogonal axes representing the E and C parameters and by rearranging Eq. (4) it is possible to define two lines which have slopes of C_{A1}/E_{A1} and C_{A2}/E_{A2}, and intercepts with the ordinate of $-\Delta H_{AB1}/E_{A1}$ and $-\Delta H_{AB2}/E_{A2}$, for the respective interactions of the polymer with solvent 1 and solvent 2. The intercepts of the two lines provide numerical values of Drago's parameters for the polymer under consideration. For the hypothetical data of Fig. 17 the polymer characteristics are $E_B = 0.30$, and $C_B = 4.35$ (kcal mol^{-1})$^{0.5}$. In this manner it is possible to estimate Drago's parameters for systems which are not amenable to the more straightforward FT-IRS method. Virtually all commercial products fall into this category since segregation or depletion of minor components are known to occur at both the inorganic/organic interface and the free surface.

Such phenomena make the use of a bulk solution method such as FT-IRS uncertain for the elucidation of acid-base properties.

Other XPS approaches have been used to identify and rank such acid-base properties of polymers. Watts and Chehimi sought to develop a solid-state titration method in which candidate polymers were cast onto soda lime glass, the extent of uptake of the sodium ion, a very mobile Lewis acid, being indicative of the basicity of the polymer [122]. The same authors employed angular-resolved XPS to study the orientation of the carbonyl group of poly(methyl methacrylate), which is basic, when the polymer was cast onto acidic or basic substrates [123]. A pronounced orientation of the functional group on the acidic substrate was found. The study was extended using monochromatic XPS and the thin-film approach of Leadley and Watts, described in Section 4; subtle changes in the C 1*s* spectrum were observed, as shown in Fig. 18 [124], and ascribed to the nature of the interaction between PMMA and metal oxide surfaces. Schematics of the type of interactions proposed, on the basis of the high-resolution XPS data, are shown in Fig. 19 [124].

The thin-film approach has been shown to be very informative when dilute solutions of commercial polymer formulations are applied to metallic substrates, as the segregation of minor components can be observed readily [42,80]. The adsorption of amine molecules, as analogs of an epoxy adhesive, has been studied by XPS and SIMS [125–128] and the complementary nature of the two techniques demonstrated once again. The existence of a strong donor-acceptor interaction with anodized aluminum was postulated in the later work, and the integrity of the adsorption process appeared to be related to the existence and density of Brönsted sites on the Al_2O_3 surface [127].

All the work described above has established XPS firmly as a flexible method for the evaluation of the acid-base properties of homopolymers, but the technologically more important advantage of XPS is its ability to analyze thin modified or segregated layers not amenable to traditional forms of analysis [129]. The molecular probe technique has been used to good effect by Shahidzadeh et al. in the study of the plasma treatment of poly(propylene) film [130–132]. Their work identified the need to select basic probe molecules for the assessment of acidic surfaces, and dimethyl sulfoxide was shown to be a good choice, although the photoelectron cross section for sulfur is low, as it is for the Cl 2*p* core level used for trichloromethane; this puts an effective limit on the detectability of the molecular probe, i.e., the number (but not the strength) of the acid-base pairs detectable.

6.5. Acid-Base Properties of Inorganic Surfaces

The chemical specificity of XPS again provides in this area a powerful route to the determination of the acid-base properties of surfaces. Indeed there appears to be a direct linear correlation between certain XPS data and the isoelectric

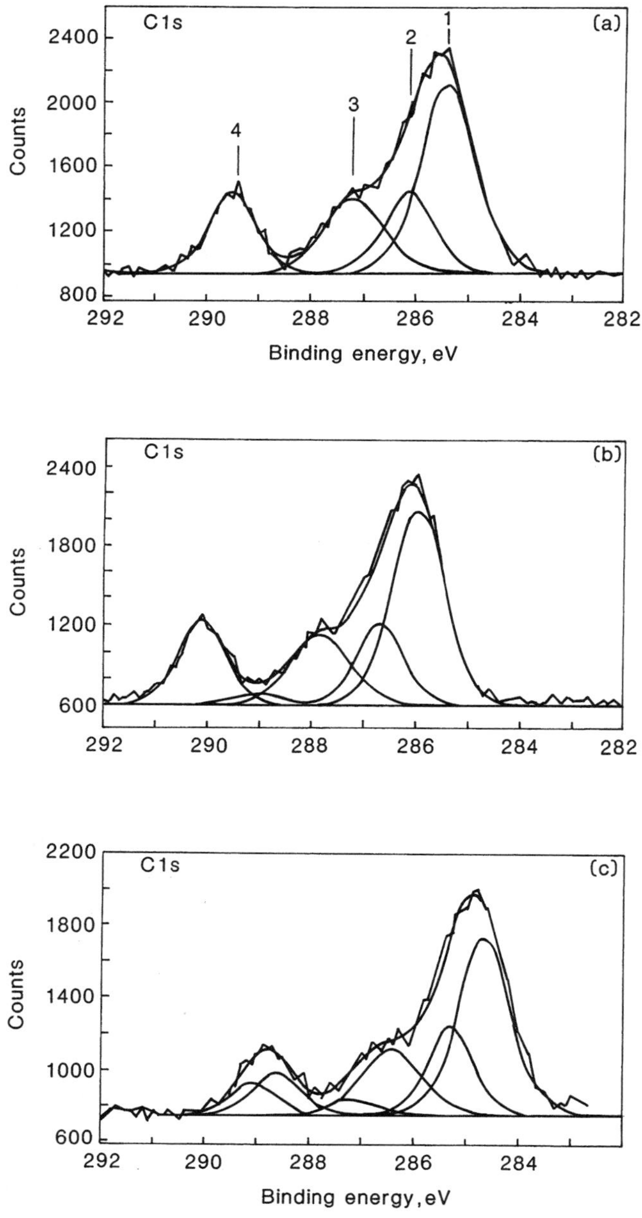

Figure 18. C 1*s* XPS spectra of thin films (ca. 2 nm) of PMMA on oxidized substrates of increasing basic character: (a) silicon (acidic), (b) aluminum (weakly basic), (c) nickel (strongly basic). (From Ref. 124.) The numbers of the components in (a) refer to the schematic of Fig. 19. Additional components in (b) and (c) are related to the specific interactions shown in Fig. 19.

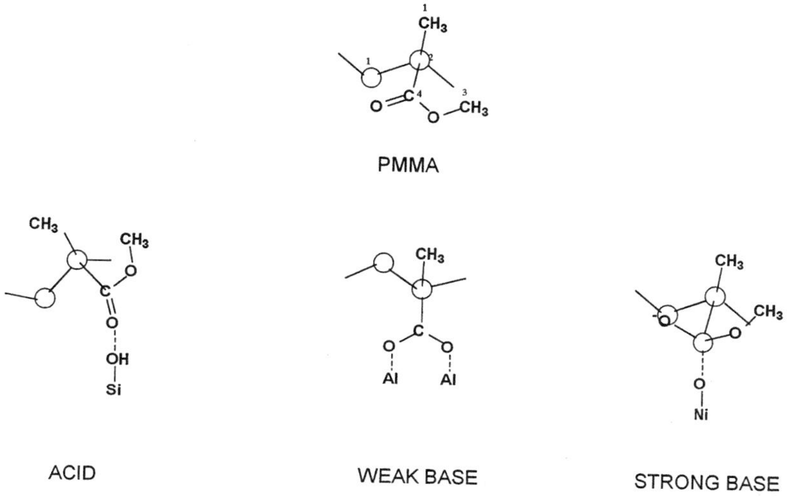

Figure 19. Schematic representations of the types of bonding thought to give rise to the spectra of Fig. 18.

point of the solid surface (IEPS), as pointed out by Delamar [133]. By summing the separation between the photoelectron peak from oxidized metal and the accepted value of the metal photoelectron peak, with the separation of the O 1*s* peak from that of an arbitrarily defined binding energy, and then plotting the summation against IEPS, a linear relationship was found. The Brönsted characteristics of metal surfaces were studied by Watts and Gibson [134], who extended the cation-exchange method of Beard and Simmons [135] to include both anion and cation exchange. The method is based on the fact that, in the presence of an aqueous medium, the hydroxyl groups on a hydrated metal surface may act as either acids or bases to produce cationic (MOH_2^+) or anionic (MOH^-), complexes, where M is the metal under consideration [134, 135]. The concentration of each of these species can then be determined by using a cationic (K^+) or anionic (PO_4^{2-}) exchange process. By carrying out the exchanges at a series of solution pH values, the acid equilibrium constants for the formation of the metal complexes can be determined. The IEPS is the value of solution pH at which the surface concentration of cationic complex ions ($MOH_2{}^+$) is equal to the concentration of anionic species (MOH^-), and it can be determined conveniently as the mean of the two constants determined by the ion-exchange procedures described above. The method has been criticized by Delamar [136], who suggested possible improvements to obtain data of greater reliability. Notwithstanding the shortcomings the method has been used to assess the acid-base properties of oxidized titanium and iron surfaces under both idealized [135] and real situations [134]. In a detailed investigation Kurbatov et al. [137,138] showed how slight changes in the hydroxylated surface layer present on iron surfaces could be effected by surface preparation protocols.

6.6 Concluding Remarks Regarding Donor-Acceptor Interaction

Application of the concepts of acceptor-donor interaction is an active area of research in adhesion science. In the last few years the foundations have been laid that will allow XPS, and to a lesser extent SSIMS, to play an important role in such investigations. Although information concerning the acid-base properties of polymer and inorganic surfaces is generally difficult to obtain, it is clear that the goal of achieving a predictive approach to adhesion and to the hydrolytic stability of adhesive joints and coated substrates is now much closer than a decade ago.

7. COMPUTATIONAL CHEMISTRY AND MOLECULAR MODELING

Although not strictly part of a chapter dealing with experimental surface analysis techniques for solving adhesion and related problems, discussion of the use of computer techniques for the modeling of adhesion phenomena is included, as it is an emerging approach which is often used in conjunction with surface analysis. Although computer chemistry methods based on quantum mechanics, e.g., self-consistent field (SCF) molecular orbital calculations, have been available for many years, it is the development of molecular simulation packages running on workstations that has provided the impetus in the applied fields of adhesion and catalysis. The difference between the two approaches is well known but is worth repeating. In the SCF approach *ab initio* calculations are made and all electron-electron interactions, as well as overlap integrals, are calculated. In order to simplify the computational process an approximation is often made in which only the valence electrons are considered, i.e., those electrons actually taking part in chemical interactions. A popular SCF method which employs this valence shell approximation is the complete-neglect-of-differential-overlap approach (CNDO), in which certain other electron-electron interactions are also neglected. The use of these methods in adhesion has been reviewed by Cain [139], and an elegant example of the application of SCF calculations to adhesion is the study, by Holubka et al., of acrylate and methacrylate esters on aluminum oxide [140]. The mechanism of the interaction of dicyandiamide (a widely used epoxy curing agent) has also been reported [141].

Molecular simulation methods, on the other hand, use semiempirical quantum chemical calculations and do not estimate the probability of chemical bond formation. They use essentially molecular dynamics, based on the integration of Newton's laws of motion, over many (several million) very short (femtosecond) time increments. The molecules under study are represented by spheres of known mass interconnected elastically and are defined by a force field which contains data relevant to the molecular structure such as atomic masses, bond lengths, bond angles, and so on. Other factors that affect molecular motion, such as temperature and secondary bonds must also be included. The first step is optimization of the geometry of the molecular structures by calculating the total energy as described by the appropriate energy expression. Molecular motion, for instance, of a molecule relative to a solid surface, can then be calculated for all the atomic masses present by imparting an initial velocity to them and evaluating the response after many time increments. Further information is available in standard texts on computational chemistry such as [142].

Molecular dynamics have been used by Sennett et al. to study the adsorption of epoxy analogs and of silanes on alumina and haematite surfaces [143,144]. Taylor et al. have shown that the failure of a photocured resin, described in Section 3, in which 1–2 nm of aliphatic reactive diluent remained on the inorganic substrate, was consistent with a monolayer of such material at the interface. The orientation of the molecule was 30–40° to the solid surface, which indicated a monolayer coverage of the same dimensions as that determined by angular-resolved XPS [75,76]. A similar approach has been used by Williams to study the orientation of organosilane molecules on EGS surfaces [145]. The limitations of molecular dynamics, so far as the application to adhesion is concerned, are associated with establishing the initial location of the molecule on the solid substrate, but can be overcome by using an adsorption routine in the manner described by Davis and Watts in a study of silane adsorption [46,146]. Using a commercial molecular dynamics suite of programs, marketed by Molecular Simulations Inc., under the name of Cerius^2 [147], those authors used the Sorption Package of Cerius^2 for the initial approach of the molecule to the inorganic surface, to determine the local energy minimum, which was then used as input data for molecular dynamics. The trajectory file associated with simulation was interrogated and the structure with the lowest total energy frame extracted. In this manner it was possible to study the adsorption characteristics of silane molecules on an inorganic substrate without the uncertainty regarding the initial interaction site. The adsorption of three organosilanes was investigated, and the lowest energy frame for the three molecules on a periodic FeOOH lattice is shown in Fig. 20 [146]. Two of these studies repeated the work of Bailey and Castle [92], and it is gratifying that the data for vinyl triethoxy silane (Fig. 20a) were in complete accord with those from XPS [92]. The result for amino propyl triethoxysilane (APS) was particularly informative in that the XPS data showed no preference for adsorption via either functional group, and this conclusion was confirmed and clarified by

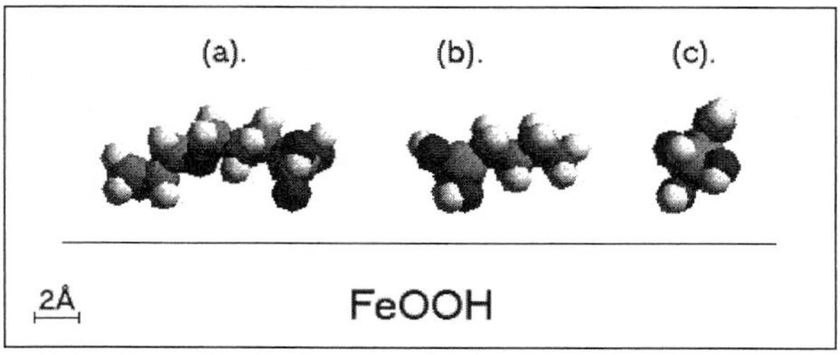

Figure 20. Molecular dynamics simulations of the positional arrangement of fully hydrolyzed organosilane molecules on an FeOOH substrate, obtained using the Cerius[2] Sorption package: (a) glycidoxypropyl trimethoxysilane; the silane head is at the right-hand end of the molecule; (b) γ-aminopropyl triethoxysilane: the silane head is at the left-hand end of the molecule; (c) vinyltriethoxy silane: the silane head of the molecule is adjacent to the FeOOH surface. The scale bar is 0.2 nm. This figure was produced using the original (color) computer graphics files obtained by Davis [146].

molecular dynamics, which indicated that either the amino or the silane ends of the molecule may approach the substrate, resulting in a parallel orientation for the molecule on the surface (Fig. 20b). For glycidoxy propyl trimethoxysilane the situation is known to be more complicated, since XPS and ToF-SIMS data showed that the overlayer was some 1.7 nm thick, implying a complex overlayer. The molecular dynamics simulation of Fig. 20c shows that either endgroup may interact with the substrate, as for APS, but that the layer is far from parallel. These examples indicate clearly that, although the molecular simulation routines have an important role to play in adhesion science, they must not be relied upon in isolation to provide unambiguous results. As is often the case calculations must be supported by experimental data. The attraction of the molecular simulation approach may lie more in its use in screening candidate materials for a comprehensive research program.

8. FUTURE PROSPECTS

In this chapter the development of the application of surface analytical techniques to adhesion problems, in particular that of XPS, over the past 20 years has been described. Now the use of XPS and SIMS in forensic analysis of interfacial failure surfaces is relatively straightforward and may provide direct information regarding the actual mechanism of adhesion. Recent extensions of the capability of XPS into the realm of high spatial resolution will ensure that the impetus in the use of XPS is maintained. The advent of ToF-SIMS a decade ago has had an extremely important impact on adhesion research, particularly with respect to the molecular identification of organic species at failure surfaces. Future work will see the development of theories capable of predicting segregation phenomena at interfaces in adhesive joints, coatings and related systems. Such work will be based on careful examination of the behavior of fully formulated (i.e., prototype or commercial) systems.

There has always been a tacit assumption that adhesion phenomena show very little lateral variation, which is probably an oversimplification. Adhesion scientists can learn something from corrosion investigations in this connection; SAM has been shown to be a powerful tool in understanding the breakdown of passivity around inclusions [148], and if used intelligently may well help in understanding adhesion mechanisms too. For example, the interactions of organic systems with heterogeneities in the microstructure may provide preferential sites for the initiation of failure. Imaging XPS should be applicable to such an example, but it is more likely that the scanned probe techniques of AFM and STM will be able to provide the required spatially resolved information.

The development of SPM and particularly AFM has not been ignored by the adhesion world. The analysis of failure surfaces by AFM is still in its infancy but it is already clear that AFM may sometimes be more useful than SEM. Thus AFM has been used successfully in the analysis of the failure surfaces of adhesively bonded galvanized steel [149] and in the study of the adsorption of polymers on conducting polymer substrates [150,151]. Some of the author's unpublished studies have shown that the corona treatment of polyolefines gives rise to a characteristic morphology and that failure interfaces from peel joints of polyolefine substrates can yield more useful information from AFM than from SEM investigations [152].

The study of interface chemistry by surface analysis following exposure by chemical dissolution or mechanical sectioning methods will remain important, probably in parallel with studies on model systems. The acid-base approach to adhesion is now widely accepted, and consequently the chemical information available from monochromated XPS will assume even greater importance. Together with this chemical approach will come an extension of the determination of adsorption isotherms by XPS or SSIMS to the minor constituents of commercial formulations, while molecular dynamics simulations will provide additional information.

The last few years have seen much activity in the closely associated areas of surface analysis and adhesion science, and there is no reason to doubt that the association will provide a great deal more exciting science for many years to come.

ACKNOWLEDGMENTS

It is a pleasure to thank those involved in adhesion research at the University of Surrey over the last two decades. In particular, the contributions of Professor Jim Castle, Dr. Mohamed Chehimi, Dr. Steve Davis, Dr. Stuart Leadley and Dr. Alison Taylor have been pivotal in the progress of much of the work described in this chapter. Thanks go to Steve Greaves and Andy Brown for skilful experimental assistance with XPS and ToF-SIMS, respectively, and to Sheila Rudman for help with the graphics.

REFERENCES

1. J. F. Watts, in *Organic Coatings I* (A. Wilson, H. Prosser and J. W. Nicholson, eds.) Applied Science, London, (1987), pp. 137–187.
2. J. F. Watts, *Surf. Interface Anal. 12*, 497 (1988).
3. V. Hospadaruk, J. Huff, R. W. Zurilla and H. T. Greenwood, SAE Technical Paper Series 780186, 1978.
4. R. A. Iezzi and H. Leidheiser Jr, *Corrosion 37*, 28 (1981).
5. B. M. Parker and R. Waghorne, *Composites 13*, 280 (1982).
6. A. J. Kinloch, G. K. A. Kadokian and J. F. Watts, *Phil. Trans. Roy Soc. A338*, 83 (1992).
7. M. W. Roberts and P. R. Wood, *J. Electron. Spectrosc. 11*, 431 (1977).
8. J.E. Castle, in *Corrosion Control by Coatings* (H. Leidheiser Jr., ed.), Science Press, Princeton, NJ, (1979), pp. 435–454.
9. A. J. Kinloch and N. R. Smart, *J. Adhesion 12*, 23 (1981).
10. P. Poole and J. F. Watts, *Int. J. Adhes. Adhes. 5*, 33 (1985).
11. J. S. Solomon, D. Hanlin and N. T. McDevitt, in *Adhesion and Adsorption of Polymers*, Part A (L.-H. Lee, ed.), Plenum, New York, 1980, pp. 103–122.
12. J. E. Castle and J. F. Watts, in *Interfaces in Polymer, Ceramic and Metal Matrix Composites* (H. Ishida, ed.), Elsevier Science, 1988, pp. 57–71.
13. C. A. Baillie, J. F. Watts and J. E. Castle, *J. Mater. Chem. 2*, 939 (1992).
14. M. R. Alexander and F. R. Jones, *Surf. Interface Anal 212*, 230 (1994).
15. M. R. Alexander and F. R. Jones, *Carbon 33*, 569 (1995).
16. M. R. Alexander and F. R. Jones, *Carbon 34*, 1093 (1996).
17. P. A. Zhdan, D. Grey and J. E. Castle, *Surf. Interface Anal 22*, 290 (1994).
18. W. J. van Ooij, *Surf. Sci 68*, 1 (1977).
19. W. J. van Ooij, *Surf. Tech. 6*, 1 (1977).
20. W. J. van Ooij, *Rubber Chem. Tech. 52*, 605 (1979).
21. W. J. van Ooij, *Rubber Chem. Tech. 57*, 421 (1984).

22. W. J. van Ooij and, M. E. F. Biemond, *Rubber Chem. Tech 57*, 686 (1984).
23. J. Giridhar, W. J. van Ooij and J. Hahn, *Kautsch. Gummi Kunstst. 44*, 348 (1991).
24. J. Giridhar and W. J. van Ooij, *Surf. Coat Tech. 52*, 17 (1992).
25. J. Giridhar and W. J. van Ooij, *Surf. Coat Tech. 53*, 35 (1992).
26. W. J. van Ooij and M. Nahmias, *Rubber Chem. Tech. 62*, 656 (1989).
27. J. E. Castle, Z. B. Luklinska and M. S. Parvizi, *J. Mater. Sci. 19*, 3217 (1984).
28. J. E. Castle, D. C. Epler and D. Peplow, *Corros. Sci. 16*, 145 (1976).
29. J. E. Castle and J. F. Watts, *Corrosion Control by Organic Coatings* (H. Leidheiser, ed.) NACE (Houston) (1981) pp. 78–86.
30. J. F. Watts and J. E. Castle, *J. Mater. Sci. 18*, 2987 (1983).
31. R. A. Dickie, J. S. Hammond and J. W. Holubka, *Ind. Eng. Chem. Prod. R & D 20*, 339 (1981).
32. J. W. Holubka, J. S. Hammond, J. E. deVries and R. A. Dickie, *J. Coat. Tech. 52*, 63 (1980).
33. J. S. Hammond, J. W. Holubka, J. E. deVries and R. A. Dickie, *Corros. Sci. 21*, 239 (1981).
34. J. E. Castle, J. F. Watts, P. J. Mills and S. A. Heinrich, in *Corrosion Protection by Organic Coatings* (M. W. Kendig and H. Leidheiser, Jr., eds.), The Electrochemical Society, Pennington, NJ, (1987), pp. 68–83.
35. J. E. Castle, Organic Coatings: 53rd International Meeting of Physical Chemistry, P.-C. Lacaze, (ed.), AIP Conf. Proc. 354, AIP Press, Woodbury, NY, 1995, pp. 432–447.
36. X. H. Jin, K. C. Tsay, A. Elbasir and J. D. Scantlebury in *Corrosion Protection by Organic Coatings* M. W. Kendig and H. Leidheiser, Jr. (eds.), The Electrochemical Society, Pennington, NJ, 1987, pp. 37–47.
37. R. Cayless and D. L. Perry, *J. Adhes. 26*, 113 (1988).
38. J. F. Watts, R. A. Blunden and T. J. Hall, *Surf. Interface Anal. 16*, 227 (1990).
39. J. F. Watts, J. E. Castle and S. J. Ludlam, *J. Mater. Sci. 21*, 2965 (1986).
40. T. R. Roberts, J. Kolts and J. H. Steele, SAE Technical Paper Series 800443, 1980.
41. J. F. Watts, *J. Mater. Sci. 19*, 3459 (1984).
42. M. Murase and J. F. Watts, submitted to *J. Mater. Chem.*
43. M. Gettings, F. S. Baker and A. J. Kinloch, *J. Appl. Polym. Sci. 21*, 2375 (1977).
44. R. A. Cayless and L. B. Hazell, European Patent 0331 284 A1, 1989.
45. S. J. Davis, J. F. Watts and L. B. Hazell, *Surf. Interface Anal. 21*, 460 (1994).

46. S. J. Davis, Ph.D. thesis, University of Surrey, 1995.
47. S. J. Davis and J. F. Watts, *J. Mater. Chem. 6*, 479 (1996).
48. J. F. Watts, in *Encylopaedia of Analytical Science* (R. McRae, ed.), Academic Press 1995, 5884.
49. L. P. Haack, M. A. Bolt, S. L. Kaberline, J. E. deVries and R. A. Dickie, *Surf. Interface Anal 20*, 115 (1993).
50. W. J. van Ooij, A. Sabata and A. D. Appelhans, *Surf. Interface Anal 17*, 403 (1991).
51. A. M. Taylor, J. F. Watts, H. Duncan and I. W. Fletcher, *J. Adhes. 46*, 145 (1994).
52. M. Gettings and A. J. Kinloch, *Surf. Interface Anal. 1*, 165 (1979).
53. M. Gettings and A. J. Kinloch, *Surf. Interface Anal. 1*, 189 (1979).
54. J. E. deVries, J. W. Holubka and R. A. Dickie, *J. Adhes. Sci. Tech. 3*, 189 (1989).
55. J. E. deVries, L. P. Haack, J. W. Holubka and R. A Dickie, *J. Adhes. Sci. Tech. 3*, 203 (1989).
56. Y. Yoshikawa and J. F. Watts, *Surf. Interface Anal. 20*, 379 (1993).
57. J. F. Watts and J. E. Castle, *J. Mater. Sci. 19*, 2259 (1984).
58. J. D. Venables, *J. Mater. Sci. 19*, 2431 (1984).
59. G. D. Davis, in *Adhesive Bonding* (L.-H. Lee, ed.), Plenum, New York, 1991, pp. 139–173.
60. I. Olefjord and L. Kozma, *Mat. Sci. Tech. 3*, 860 (1987).
61. H. M. Clearfield, D. K. McNamara and G. D. Davis, in *Adhesive Bonding* (L.-H. Lee, ed.) Plenum, New York, 1991, pp. 203–237.
62. G. D. Davis, P. L. Whisnant and J. D. Venables, *J. Adhes. Sci. Tech. 9*, 433 (1995).
63. J. F. Watts, J. E. Castle and T. J. Hall, *J. Mater. Sci. Lett. 7*, 176 (1988).
64. J. P. Sargent, *Int. J. Adhes. Adhes. 14*, 21 (1994).
65. C. Cazeneuve, J. E. Castle and J. F. Watts, *Interfacial Phenomena in Composite Materials '89* (F. R. Jones, ed.), Butterworth, 1989, pp. 88–96.
66. C. Cazeneuve, J. F. Watts and J. E. Castle, *J. Mat. Sci. 25*, 1902 (1990).
67. F. R. Jones, P. Denison, A. Brown, P. Humphrey and J. Harvey, *J. Mater. Sci. 23*, 2153 (1988).
68. M. J. Hearn and D. Briggs, *Surf. Interface Anal. 17*, 421 (1991).
69. S. Yumitori, D. Wang and F. R. Jones, *Composites 25*, 698 (1994).
70. J. F. Watts and P. E. Vickers, Fourth Int. Conf. on Adhesion and Surface Analysis, Loughborough Universtiy of Technology, UK, 1996, Paper P4.
71. L. P. Buchwalter, *J. Adhes. Sci. Tech. 1*, 341 (1987).
72. L. P. Buchwalter, *J. Adhes. Sci. Tech. 4*, 697 (1990).
73. T. S. Oh, L. P. Buchwalter and J. Kim, *J. Adhes. Sci. Tech. 4*, 303 (1990).
74. J. F. Watts and A. M. Taylor, *J. Adhes. 46*, 161 (1994).

75. A. M. Taylor, J. F. Watts, J. Bromley-Barratt and G. Beamson, *Surf. Interface Anal. 21*, 697 (1994).
76. A. M. Taylor, C. H. McLean, M. Charlton and J. F. Watts, *Surf. Interface Anal. 23*, 342 (1995).
77. J. E. Castle and J. F. Watts, *Adv. Mat. J. 1*, 16 (1990).
78. W. J. van Ooij, T. H. Tisser and M. E. F. Biedmond, *Surf. Interface Anal. 6*, 197 (1984).
79. W. J. van Ooij, R. A. Edwards, A. Sabata and J. Zappia, *J. Adhes. Sci. Tech. 7*, 897 (1993).
80. A Rodriguez, J. F. Watts and C. Lowe, Fourth Int. Conf. on Adhesion and Surface Analysis, Loughborough Universtiy of Technology, UK, 1996, Paper P1.
81. H. Leidheiser, S. Music and G. W. Simmons, *Nature 297*, 667 (1982).
82. Jiping Liu, Ph.D. thesis, University of Surrey, 1993.
83. J. M. Burkstrand, *Appl. Phys. Lett. 33*, 387 (1978).
84. J. M. Burkstrand, *J. Vac. Sci. Technol. 20*, 440 (1980).
85. F. S. Ohuchi and S. C. Freilich, *J. Vac. Sci. Technol. A4*, 1039 (1986).
86. P. Stoyanov, S. Akhter and J. M. White, *Surf. Interface Anal. 15*, 50 (1990).
87. M. M. Chehimi and J. F. Watts, *J. Adhes. Sci. Tech. 6*, 377 (1992).
88. S. R. Leadley and J. F, Watts, submitted to *J. Elec. Spec. 85*, 107 (1997).
89. A. N. MacInnes, Ph.D. thesis, University of Surrey, 1990.
90. E. P. Plueddeman, *Silane Coupling Agents*, Plenum, New York, 1992.
91. K. L. Mittal, *Silane and Other Coupling Agents*, VSP, Zeist, The Netherlands, 1992.
92. R. Bailey and J. E. Castle, *J. Mater. Sci. 12*, 2049 (1977).
93. M.-L. Abel, M. M. Chehimi, A. M. Brown, S. R. Leadley and J. F. Watts, *J. Mater. Chem. 5*, 845 (1995).
94. M. Gettings and A. J. Kinloch, *J. Mater. Sci. 12*, 2511 (1977).
95. W. J. van Ooij and A. Sabata, *J. Adhes. Sci. Tech. 5*, 843 (1991).
96. W. J. van Ooij and A. Sabata, *Surf. Interface Anal. 20*, 475 (1993).
97. S. J. Davis and J. F. Watts, *Int. J. Adhes. Adhes. 16*, 5 (1996).
98. W. J. van Ooij and A. Sabata, in *Polymer/Inorganic Interfaces*. (R. L. Opila, F. J. Boerio and A. W. Czandena, eds.), Materials Research Society, Pittsburgh, PA, 1993, pp. 155–160.
99. S. G. Hong and F. J. Boerio, *Surf. Interface Anal. 21*, 650 (1994).
100. F. J. Boerio, C. A. Gosselin, R. G. Dillingham and H. W. Liu. *J. Adhes. 13*, 159 (1981).
101. D. J. Ondrus and F. J. Boerio, *J. Coll. Interf. Sci. 124*, 349 (1988).
102. R Chen and F. J. Boerio, *J. Adhes. Sci. Tech. 4*, 453 (1990).
103. J. T. Young and F. J. Boerio, *Surf. Interface Anal. 20*, 341 (1993).
104. Y. M. Tsai, F. J. Boerio, W. J. van Ooij, D. K. Kim and T. Rau, *Surf. Interface Anal. 23*, 261 (1995).

105. K. Yates and R. H. West, *Surf. Interface Anal. 5*, 113 (1983).
106. N. G. Cave, A. J. Kinloch, S. C Mugford and J. F. Watts, *Surf. Interface Anal. 17*, 120 (1991).
107. D. Wang, F. R. Jones and P. Denison, *J. Mater. Sci. 27*, 36 (1992).
108. D. Wang and F. R. Jones, *Surf. Interface Anal. 20*, 457 (1993).
109. D. Wang and F. R. Jones, *Composites 50*, 215 (1994).
110. A. Zettlemoyer, in *Interface Conversion for Polymer Coatings* (P. Weiss and C. D. Cheever, eds.), Elsevier, New York, 1968, pp. 208–237.
111. F. W. Fowkes, *J. Adhes. Sci. Tech. 1*, 2 (1987).
112. K. L. Mittal and H. R. Anderson (eds.), *Acid-Base Interactions: Relevance to Adhesion Science and Technology*, VSP, Utrecht, The Netherlands, 1991.
113. R. S. Drago, G. C. Vogel and T. E. Needham, *J. Amer. Chem. Soc. 93*, 6014 (1971).
114. V. Gutmann, *The Donor-Acceptor Approach to Molecular Interactions*, Plenum, New York, 1978.
115. W. B. Jensen, *J. Adhes. Sci. Tech. 5*, 1 (1991).
116. J. C. Bolger, in *Adhesion Aspects of Polymeric Coatings* (K. L. Mittal, ed.), Plenum, New York, 1983, pp. 3–18.
117. M. M. Chehimi, *J. Mater. Sci. Lett. 10*, 908 (1991).
118. K. Burger and E. Fluck, *Inorg. Nucl. Chem. Lett. 10*, 171 (1974).
119. M. M. Chehimi, J. F. Watts, J. E. Castle and S. N. Jenkins, *J. Mater. Chem. 2*, 209 (1992).
120. M. M. Chehimi, J. F. Watts, W. K. Eldred, K. Fraoua and M. Simon, *J. Mater. Chem. 4*, 305 (1994).
121. F. M. Fowkes, D. O. Tischler, J. A. Wolfe, L. A. Lannigan, C. M. Ademu-John and M. J. Halliwell, *J. Polym. Sci, Polym. Chem. Edn. 22*, 547 (1984).
122. M. M. Chehimi and J. F. Watts, *J. Adhes. 41*, 81 (1993).
123. M. M. Chehimi and J. F. Watts, *J. Elec. Spectrosc. Relat. Phenom. 63*, 393 (1993).
124. S. R. Leadley and J. F. Watts, *J. Adhes. 60*, 175 (1997).
125. S. Affrossman, N. M. D. Brown, R. A. Pethrick, V. K. Sharma and R. J. Rurner, *Appl. Surf. Sci. 16*, 469 (1983).
126. S. Affrossman and S. M. MacDonald, *Langmuir 10*, 2257 (1994).
127. C. Fauquet, P. Dubot, L. Minel, M.-G. Barthes-Labtousse, M. Rei Vilar and M. Villante, *Appl. Surf. Sci. 81*, 435 (1994).
128. S. Affrossman and S. M. MacDonald, *Langmuir 12*, 2090 (1996).
129. M. M. Chehimi, Acid-base properties of polymers in the solid state, in *Handbook of Advanced Materials Testing* (N. P. and P. N. Cheremisinoff, eds.), Marcel Dekker, New York, 1996, chap. 33.

130. N. Shahidzadeh, Ph.D. thesis, Université Pierre et Marie Curie Paris VI, 1996.
131. M. Tatoulian, F. Arefi-Khonsari, N. Shahidzadeh-Ahmadi and J. Amoroux, *Int. J. Adhes. Adhes. 15*, 177 (1995).
132. N. Shahidzadeh-Ahmadi, M. M. Chehimi, F. Arefi-Khonsari, J. Amoroux and M. Delamar, *Plasmas Polyms. 1*, 27 (1996).
133. M. Delamar, *J. Elec. Spec. 53*, c11 (1990).
134. J. F. Watts and E. M. Gibson, *Int. J. Adhes. Adhes. 11*, 105 (1991).
135. G. W. Simmons and B. C. Beard, *J. Phys. Chem. 91*, 1143 (1987).
136. M. Delamar, *J. Elec. Spec. 67*, R1 (1994).
137. G. Kurbatov, E. Darque-Ceretti and M. Aucouturier, *Surf. Interface Anal. 18*, 811 (1992).
138. G. Kurbatov, E. Darque-Ceretti and M. Aucouturier, *Surf. Interface Anal. 20*, 402 (1993).
139. S. R. Cain, *J. Adhes. Sci. Tech. 4*, 333 (1990).
140. J. W. Holubka, R. A. Dickie and J. C. Cassatta, *J. Adhes. Sci. Tech. 6*, 243 (1992).
141. J. W. Holubka and J. C. Ball, *J. Adhes. Sci. Tech. 4*, 443 (1990).
142. *Reviews in Computational Chemistry* (K. B. Lipkowitz and D. B. Boyd, eds.), VCH New York, 1990.
143. M. S. Sennett, W. X. Zukas and S. E. Wentworth, *Comp. Polym. Sci. 2*, 124 (1992).
144. M. S. Sennett, S. E. Wentworth and A. J. Kinloch, *J. Adhes. 54*, 23 (1995).
145. B. Williams, 47th Chemists Conf., Scarborough UK, 1995, British Steel Research and Development Department.
146. S. J. Davis, Ph.D. thesis, University of Surrey, 1996
147. Molecular Simulation Inc., Cambridge, UK.
148. J. E. Castle, L. Sun and H. Yan, *Corros. Sci. 36*, 1093 (1994).
149. J. Jethwa, Ph.D. thesis, Imperial College of Science, Technology and Medicine, 1995.
150. M. M. Chehimi, M.-L. Abel, B. Saodi, M. Delamar, N. Jammul, J. F. Watts and P. A. Zhdan *Polimery 41*, 75 (1996).
151. M.-L. Abel, M. M. Chehimi, P. A. Zhdan and J. F. Watts, *Synth. Mets. 81*, 23, (1996).
152. P. A. Zhdan and J. F. Watts, unpublished work.

19

Archaeomaterials*

E. PAPARAZZO Istituto di Struttura della Materia del CNR, Rome, Italy

1. INTRODUCTION

Because SAM and XPS have made such decisive contributions to the fields of metallurgy and corrosion of "modern" materials [1–4], it is to be expected that those techniques might be able to solve specific problems in the surface and interface chemistry of ancient objects, such as the characterization of corrosion products formed after centuries-long burial in the soil, or the compositional analysis of the surfaces of artefacts before any restoration or conservation is attempted. Knowledge of the surface chemistry of ancient objects can contribute both to research into their history and provenance and to decisions about their correct postexcavation treatment.

As examples, two related types of ancient objects have been analyzed, both of Roman origin: a lead pipe, or *fistula*, and a series of leaded bronzes. Information about their surface chemical condition is essential, but, since their surfaces are highly heterogeneous, so is spatially resolved information.

*This work is dedicated to Pia Schizzano Proskauer.

2. CHOICE OF TECHNIQUES FOR THE STUDY OF ARCHAEOMATERIALS

2.1. Bulk Techniques

Analysis of metallic archaeomaterials is usually performed by a variety of conventional and well-established techniques. These include (i) atomic absorption spectrometry (AAS) and inductively coupled plasma atomic emission spectrometry (ICP); (ii) microscopical techniques spanning the range from optical microscopy to scanning electron microscopy (SEM), the latter often coupled with electron probe microanalysis (EPMA); and (iii) X-ray diffraction (XRD) measurements. While these techniques provide the kind of information that is essential to the understanding of bulk chemical composition and microstructure, they lack surface specificity. The outermost surface layers are of crucial importance in phenomena that have general relevance to archaeometry. It is the surface that is first exposed to weathering before other regions of the material are affected, and it is the surface that suffers initial chemical degradation induced by the environment. Ambient-induced oxidation starts at the surface and is followed by corrosion, which eventually propagates into the bulk. AAS, ICP and EPMA have high elemental sensitivities (about 1 ppm), but their sampling depths are greater than 1 μm. Although their analyses can be quantified easily, they cannot identify chemical states. Another important limitation of those techniques is their poor ability to detect elements with $Z < 13$. In particular, their elemental sensitivities are poor for oxygen and carbon, which are ubiquitous elements in the corrosion products of metallic materials.

2.2. Surface-Specific Techniques: XPS and SAM

Besides being surface specific, XPS and SAM possess the following advantages:

1. With the exception of H and He, all elements of the Periodic Table are detectable, including oxygen and carbon, and the elemental sensitivity is about 0.5 at %.
2. In XPS, chemical-state information is easily obtained, since most elements can be identified as in metallic (e.g., Cu^0) or oxidized form (e.g., Cu^{2+}). Information may even be available as to the distinction between the chemical states of a given element with the same oxidation state but in different chemical environments (e.g., the O 1*s* binding energy (BE) has different values according to whether the O^{2-} species is present in anhydrous oxides or in hydrates).
3. The relative accuracies in quantitative determination are typically ±5% in XPS and ±20–30% in SAM, which are comparable to those obtained with bulk techniques. Quantitative determination is often combined with Ar^+

etching, which removes surface material and extends the analysis beyond the outer surface.

4. SAM reveals the elemental lateral distribution with a spatial resolution that can be as good as 50 nm, which bridges the gap between SEM imaging (typically 5 nm) and EPMA mapping (ca. 1 μm).

In the present context both XPS and SAM can be regarded as "destructive" techniques because specimens of the appropriate dimensions (i.e., about 1 cm^2 in area, and 0.3–3 mm thick) have to be excised from the object under study. However, dissolution is not involved, nor does the sample require any chemical etching, as with AAS/ICP and SEM, respectively. The only step that must be taken prior to analysis is that of the removal of any gross incrustations from the surface. The reasons for this are as follows:

1. While being otherwise advantageous, the surface specificities of XPS and SAM would prevent analytical access to the region beneath the overlying material.
2. The incrustations, even if only a few mm thick, usually contain trapped moisture which might contaminate the analysis chamber and make evacuation to the 10^{-8} Pa range impossible. As will be shown, analysis of the original patina can be performed safely by SAM if the area of patina is reduced to ca. 1 mm^2 while the remaining area is scraped mechanically until a "bare" metallic surface is obtained.

3. ROMAN LEAD PIPE *FISTULA*

3.1. Description of Material and Specimen

Lead pipes, or *fistulae*, which were used for the reticulation of water, are among the most common artefacts of Roman metallurgy [5–8]. They were made from lead sheets that were rolled about a cylindrical mandrel to produce a pipe shape, with a gap left longitudinally; the gap edges were then closed to form a joint. Figure 1 shows the *fistula* analyzed here. It was provided by the "Antiquarium Comunale" of Rome (catalogue number 23822) and was found originally in the excavation carried out at the "Villa Palombara–Horti Lamiani," of the Esquiline Hill, Rome, in 1875 [6]. It bears the inscription STATIONIS PROPRIAE PRIVATAE DOMINI N(OSTRI) ALEXANDRI AUG(USTI) (literally, of the very private station of our emperor Alexander Augustus) and came from a residence of the emperor Alexander Severus, who held power from 222 to 235 A.D., which dates the *fistula* exactly. The main objectives of the analysis were determination of the surface elemental composition, and identification of the corrosion products and their lateral distribution. Particular emphasis was placed on

Figure 1. Side view of the *fistula* provided by courtesy of the 'Antiquarium Comunale' of Rome. (Reproduced from Ref. 8, by permission of Elsevier Science, Amsterdam.)

analysis of the joint, in order to ascertain whether chemical changes had occurred there relative to other regions, and on the microchemistry of the original patina.

A cross section of the *fistula*, approximately 2 cm wide, was cut, and samples for analysis were taken from the joint and from other regions of the section (hereafter "pipe") approximately 2 cm from the joint.

3.2. Results

Table 1 gives Auger and photoelectron energies and Pb Auger parameters for the *fistula* and comparison materials. Figure 2 shows atomic percentages from Auger point spectra recorded from three spots on the patina (1 to 3) in Fig. 3 on the as-received pipe, and from another three (4 to 6) on the surface from which the original patina had been removed by scraping. The principal element in the patina was C, which reached percentages three times higher than those measured in the

Table 1. Photoelectron BEs, Auger KEs, and Values of α^*_{Pb} (BE(Pb $4f_{7/2}$) + KE(Pb $N_6O_{4,5}O_{4,5}$)), Recorded from the Joint of the *fistula* and from Comparison Materials, Analyzed in the As-received (AR) Condition and after Being Subjected to Prolonged Ar^+ Etching (AE)[a]

Sample	Condition	Pb $4f_{7/2}$	Sn $3d_{5/2}$	O 1s	Pb N_6OO	α^*_{Pb}
Fistula	AR	136.8	486.5	530.2		
(joint)		138.0		532.1	92.1	230.1
	AE	136.8	486.4	530.3	96.5	233.3
		138.0	484.9	531.8	92.1	230.1
Pb	AR	136.8		530.1		
		138.2		531.8	92.0	230.2
	AE	136.8		530.1	96.3	233.1
		138.0		531.7	92.0	230.0
$Pb_{.66}Sn_{.33}$	AR	137.0	486.6	530.2		
		138.0		532.1	92.1	230.1
	AE	136.9	485.0	530.2	96.3	233.2
			486.6	531.8		
Sn	AR		484.9	530.3		
			486.5	531.5		
	AE		485.0			
SnO_2	AR		486.5	530.3		
	AE		486.3	530.2		

[a]All data are in eV, and both the accuracy and reproducibility are within ± 0.2 eV.
Source: Reproduced from Ref. 10, by permission of John Wiley & Sons, Ltd.

scraped regions. The C seemed to be mainly organic in origin (rather than as carbonate, found in the XPS spectra of the scraped material) given that the [C]/[Pb] ratio greatly exceeded that in $PbCO_3$. The lead was fully oxidized in the patina, i.e., as Pb^{2+}, since the Pb $N_6O_{45}O_{45}$ signal was at 92.6 eV [9].

Typical Auger and XPS spectra recorded from the joint and the pipe (both analyzed "bare") are shown in Figs. 4 and 5. Sn was present in the joint but not in the pipe. The surfaces of both regions contained Pb, O, Cl, and C, with the concentration of C nearly three times higher in the joint than in the pipe. Given the elemental detection limits of both Auger and XPS, any concentration of Sn below 0.5 at.% would not have been detected. Bulk chemical analysis with ICP indicated that there was some Sn in the pipe material, as an impurity, but at a concentration of < 0.3 at.% compared with more than 5 at.% in the joint [10]. Figures 6 and 7 show, in spectra (a), the respective Auger Sn MNN and XPS Sn $3d_{5/2}$ spectra recorded from the joint after 30 min Ar^+ etching at an estimated depth of analysis of 30 nm. For comparison, the same spectral features

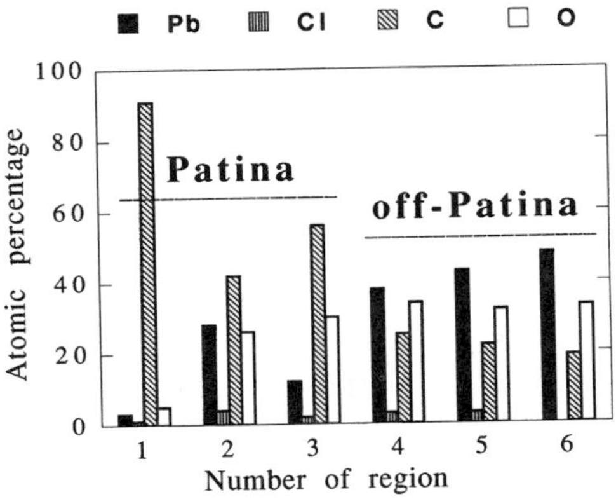

Figure 2. Atomic percentages obtained from Auger point spectra recorded from six different regions on the *fistula* surface; the patina side (regions 1 to 3) and the "bare" side (regions 4 to 6) (Fig. 3). (Reproduced from Ref. 10, by permission of John Wiley & Sons, Ltd.)

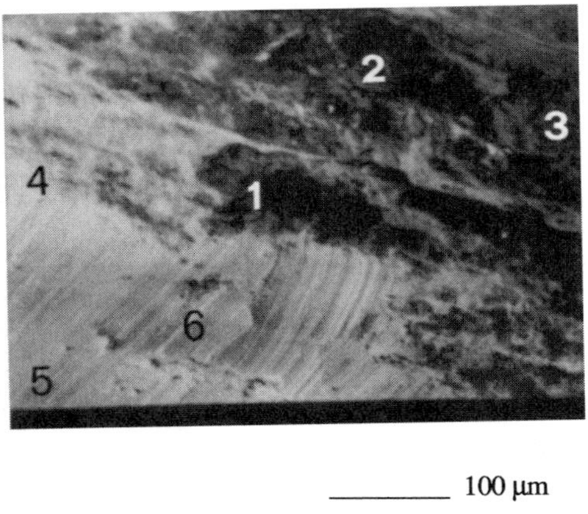

Figure 3. SEM image of the area from which the atomic percentages of Fig. 2 were measured in the locations numbered 1 to 6. (Reproduced from Ref. 10, by permission of John Wiley & Sons, Ltd.)

from SnO_2 (b) and Sn (c) are also shown. Both Auger and photoelectron spectra from the joint resemble those of SnO_2 more closely than those of Sn. Even allowing for Ar^+ etching effects, which might have modified to some (unknown) extent the genuine depth profile of the Sn chemistry in the joint, the relative amount of oxidized Sn there is much higher than in air-exposed Sn

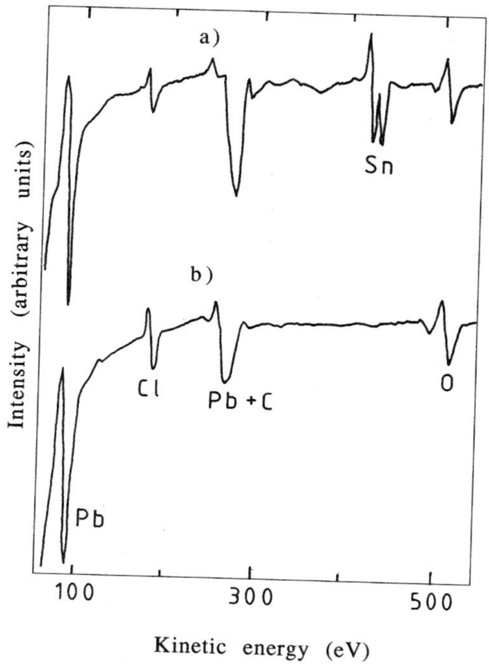

Figure 4. Typical Auger point analysis spectra recorded from (a) the joint and (b) the pipe. Both regions were analyzed in the as-received condition. (Reproduced from Ref. 8, by permission of Elsevier Science, Amsterdam.)

metal analyzed separately after the same etching treatment. The in-depth elemental distributions in the joint (a) and in the pipe (b) are shown in Figs. 8 and 9, using AES and XPS, respectively. Comparison of the two pairs of figures (a) and (b), i.e., analysis of areas typically of a few μm^2 with those typically of a few mm^2, reveals lateral inhomogeneities in the compositions of both joint and pipe, apart from the presence of Sn in the former. Thus, locally (AES) the Sn concentration reached values nearly five times those averaged over a large area (XPS). Also, although Cl was virtually undetectable in XPS, its local concentration was as much as 10 to 20 at.% in selected regions of the *fistula*.

Lateral inhomogeneity in the joint is also shown in Fig. 10, in which the elemental concentrations measured by AES point analyses from four different regions, denoted A to D in Fig. 11, are compared. Elemental Auger images recorded from the same area are displayed in Fig. 12. The Sn and O images correlate well, with the highest Sn content being associated with the highest O content (e.g., Fig. 11, region A). Cl was associated only with Pb, presumably

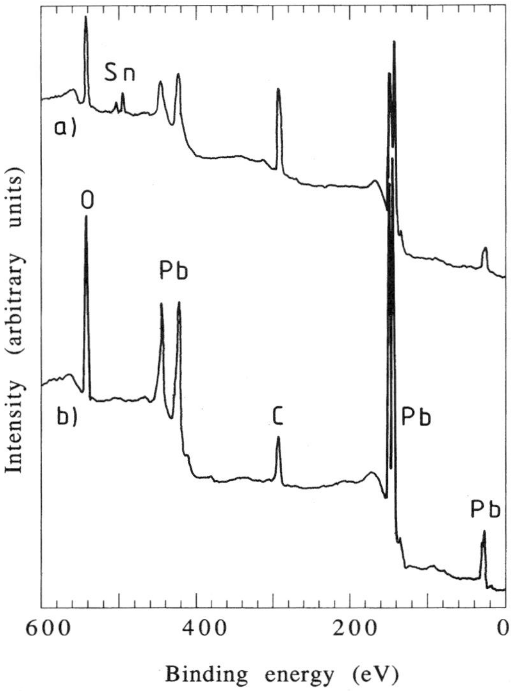

Figure 5. XPS spectra recorded (a) from the joint and (b) the pipe. Both regions were analyzed in the as-received condition. (Reproduced from Ref. 8, by permission of Elsevier Science, Amsterdam.)

as $PbCl_2$, since the Cl image is the negative of both O and Sn images; the highest Cl contents were associated with the lowest Sn content (Fig. 11, regions B and D).

Electron backscattering effects are expected to be a source of error in the Auger images of Fig. 12, as well as for all the Auger quantitative results. However, the virtual coincidence of the O and Sn Auger images recorded from the joint was confirmed by both XPS (Fig. 7, no electron backscattering) and Sn MNN spectra (Fig. 6), indicating unambiguously that most of the Sn was in a SnO_2-like form.

Figure 13 shows Pb $4f_{7/2}$ spectra from the joint and the pipe as a function of depth of analysis. Metallic Pb is characterized by a component at ca. 137 eV, and oxidized Pb (Pb^{2+}) by that at ca. 138.5 eV. It can be seen that the relative intensity of the metallic form increased with depth. The Pb^0/Pb^{2+} ratio would, of course, have been affected by Ar^+ etching, which produces an apparent sur-

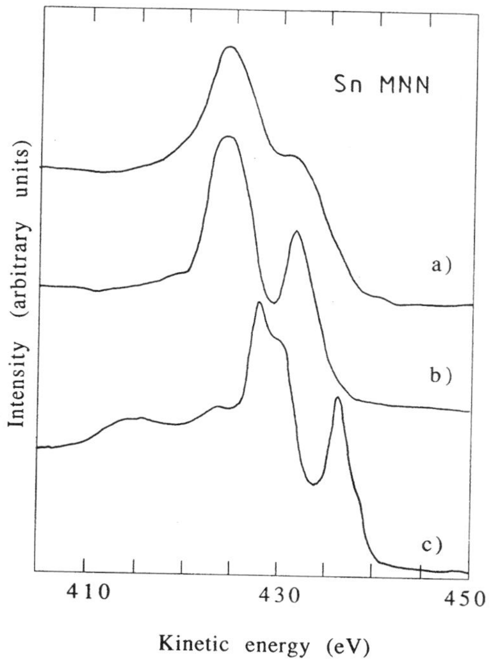

Figure 6. High-resolution Auger Sn MNN spectra, recorded in the integral mode for (a) the joint after 30 min Ar^+ etching, (b) SnO_2 and (c) Sn metal. (Reproduced from Ref. 8, by permission of Elsevier Science, Amsterdam.)

face enrichment in the metallic phase via the chemical reduction $Pb^{2+} \rightarrow Pb^0$ [11]. Note, however, that at all depths of analysis, i.e., after virtually identical etching treatments, the relative concentration of oxidized Pb was always higher for the pipe than for the joint. This suggests that there was a significant difference in the near-surface chemistry of Pb between the two regions of the *fistula*.

In the C 1*s* spectra, both inside and outside the joint, there were two components at ca. 285 and ca. 289 eV. The first can be ascribed to hydrocarbon residues and to graphitic contamination, and the second to carbonate. The O 1*s* spectra from the same areas were invariably complex. The two principal components were at ca. 530 eV, from anhydrous oxides, and at ca. 532.5 eV (high binding energy (HBE)), due to hydroxides and carbonates.

Depth profiles using XPS and milder etching conditions are shown in Figs. 14 and 15 for the joint and the pipe, respectively. The atomic ratios in the figures are (i) C (hydrocarbon)/C (carbonate), recorded until the carbonate component was undetectable; (ii) O (lattice)/O (HBE), where the HBE includes both

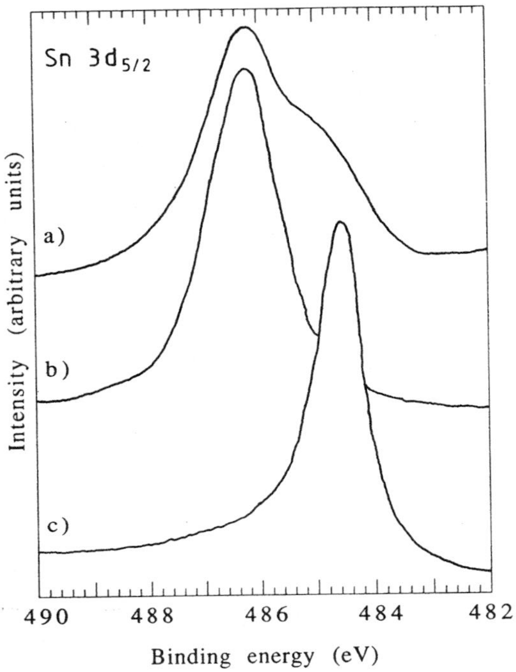

Figure 7. High-resolution XPS Sn $3d_{5/2}$ spectra for (a) the joint after 30-min Ar^+ etching, (b) SnO_2 and (c) Sn metal. (Reproduced from Ref. 8, by permission of Elsevier Science, Amsterdam.)

hydroxide and carbonate components; and (iii) Pb (metal)/Pb (oxidized). Carbonate and hydrated layers were obviously much thicker in the pipe than in the joint. Although the carbonates and hydroxides would have suffered ion beam damage to an unknown extent, the similarities of the materials and the substantial differences in the profiles suggest that the latter contain genuine information. Thus, although the surface removal rates under the etching conditions were unknown, it is reasonable to assume that the amount removed scales with etching time and that the apparent differences in the profiles are significant.

Because of the presence of so much Sn in the joint, it is clear that the Romans used soldering rather than welding to seal the pipe. There was also much less carbonate and hydroxide in the joint, indicating that the extent of Pb corrosion was less there than in the pipe. From this it is instructive to examine the effect of Sn content on the ambient oxidation of lead. Comparison measurements were therefore made on a "modern" Pb-Sn (~30% Sn in weight) alloy as used by today's

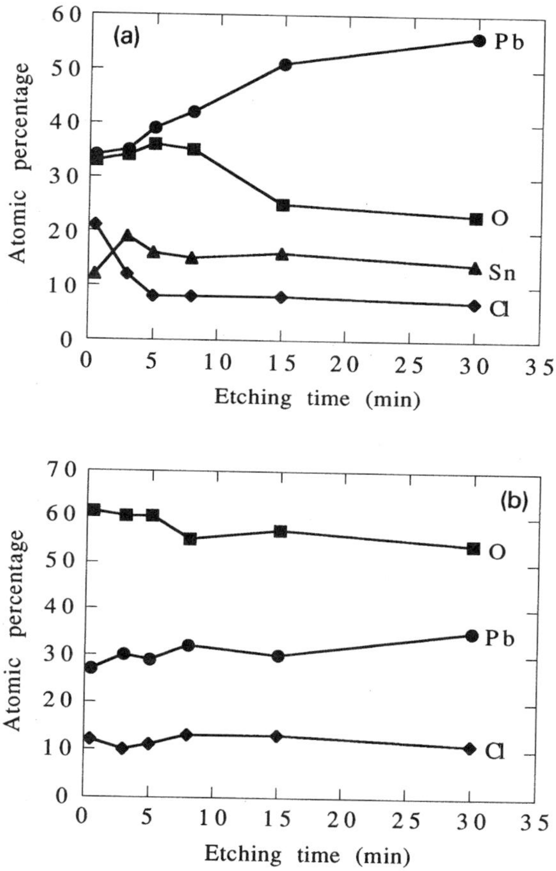

Figure 8. Elemental depth profiles recorded by AES for (a) the joint and (b) the pipe. (Reproduced from Ref. 8, by permission of Elsevier Science, Amsterdam.)

plumbers, on the individual pure metals, and on a Pb/Pb-Sn solder. It would also be useful to compare the interface chemistry of the materials used in modern plumbing technology with that of the materials used by the Roman *plumbarii*. In Fig. 16 metallic percentages, defined as $M(Pb,Sn)=100\times[M^0]/\{[M^0]+[M^{n+}]\}$, measured by XPS from the pure metals (open symbols) and from the alloy (full symbols) are plotted as a function of etching time. It can be seen that at all depths Pb is significantly less oxidized in the alloy, in that Pb is entirely metallic in the alloy profile after the fourth etch, whereas at the same depth in the pure metal it

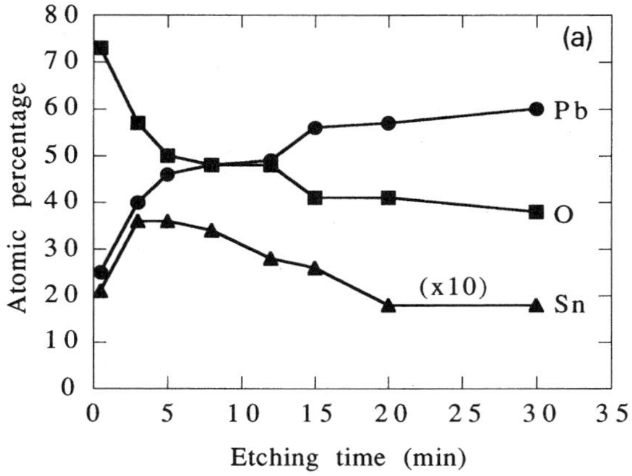

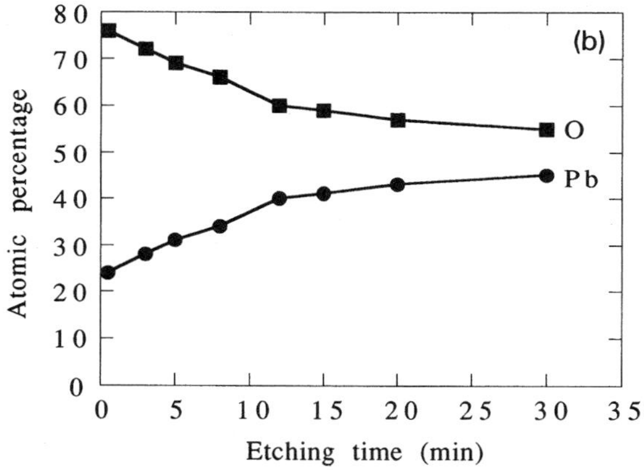

Figure 9. Elemental depth profiles recorded by XPS for (a) the joint (note the ×10 amplification factor for Sn) and (b) the pipe. (Reproduced from Ref. 8, by permission of Elsevier Science, Amsterdam.)

is still nearly 50% oxidized. Conversely, Sn shows the opposite trend, with much higher oxide content in the alloy than in the metal; even after 25 min etching of the alloy no metallic Sn had been detected.

In order to simulate the soldering procedure used by the Romans for sealing the joints of the *fistulae*, some Pb-Sn alloy was melted on a lead sheet first wetted with olive oil, thus producing an interface. Figure 17 shows an SEM image of this soldered interface, and Fig. 18 the C Auger image recorded over an en-

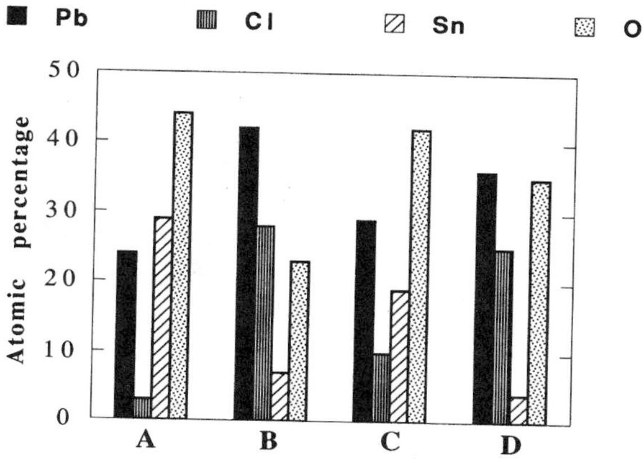

Figure 10. Lateral inhomogeneity in the elemental composition of the joint from Auger spectra recorded at four different locations, A to D, marked in Fig. 11. All four points were of area 1 × 1 μm². (Reproduced from Ref. 8, by permission of Elsevier Science, Amsterdam.)

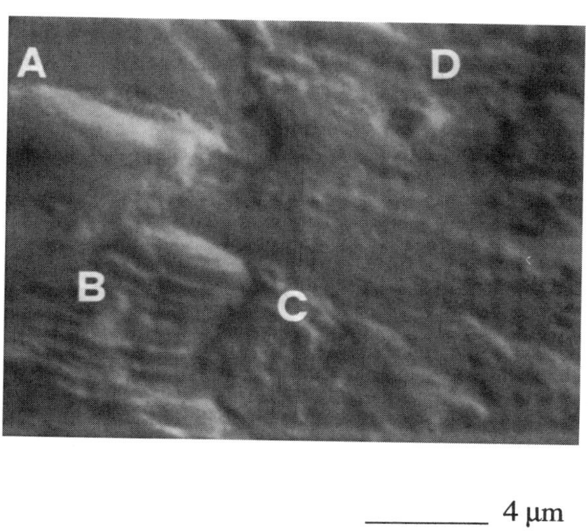

Figure 11. SEM image of the area from which the four Auger point spectra of Fig. 10 were recorded. (Reproduced from Ref. 8, by permission of Elsevier Science, Amsterdam.)

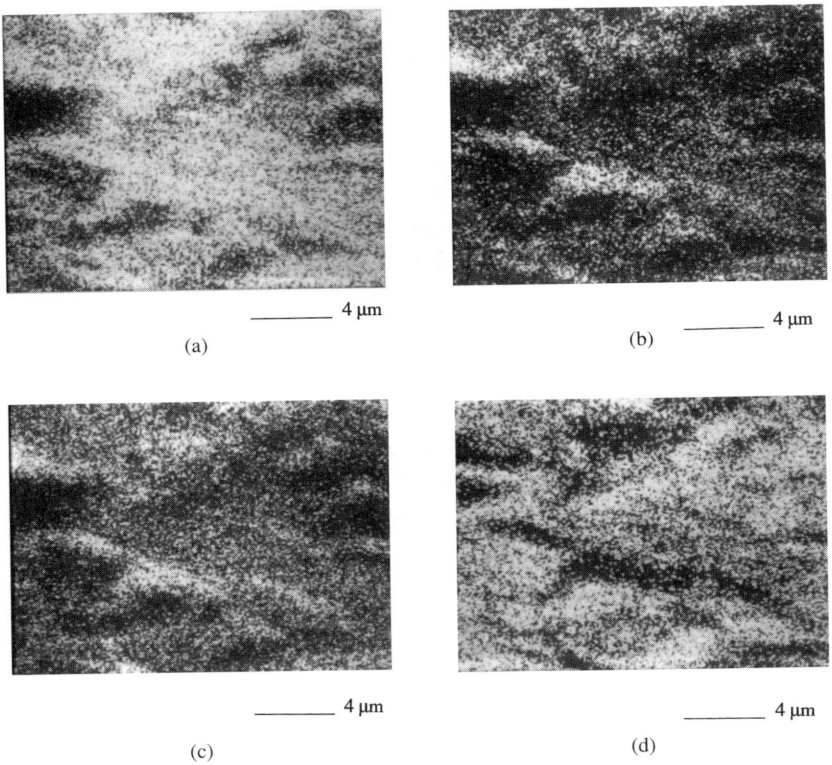

Figure 12. Auger maps ([peak-background]/background) recorded from the joint (in the area shown in Fig. 11) after 2 min Ar^+ etching, for the following elements: (a) Pb, (b) O, (c) Sn and (d) Cl. (Reproduced from Ref. 8, by permission of Elsevier Science, Amsterdam.)

larged area of $10 \times 10\ \mu m^2$. In Fig. 19 is shown the distribution of atomic percentages from Auger point spectra recorded along a line 200 μm long across the interface. Note the almost symmetric distribution of the carbon concentration on passing from one side of the interface to the other. Figures 20–22 show, respectively, electron-induced Pb $N_{6,7}O_{4,5}O_{4,5}$, Pb $M_5N_{6,7}N_{6,7}$ and Sn $M_{4,5}NN$ spectra from various materials including standards and the *fistula* in as-received and etched conditions. With these three transitions a useful extension of the depth sensitivity of XPS and SAM can be gained. In Fig. 23 the attenuation lengths (AL) characteristic of the Auger and (Al K_α-excited) photoemission transitions used here [12] are plotted. The range of AL extends from ca.

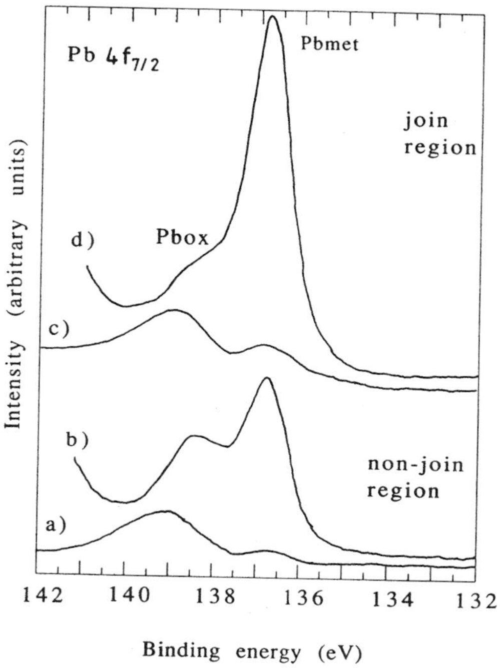

Figure 13. Evolution with depth of the Pb $4f_{7/2}$ spectra in the joint and in the pipe. Spectra (a) and (c) were recorded in the as-received condition, and spectra (b) and (d) after 12-min Ar^+ etching. (Reproduced from Ref. 8, by permission of Elsevier Science, Amsterdam.)

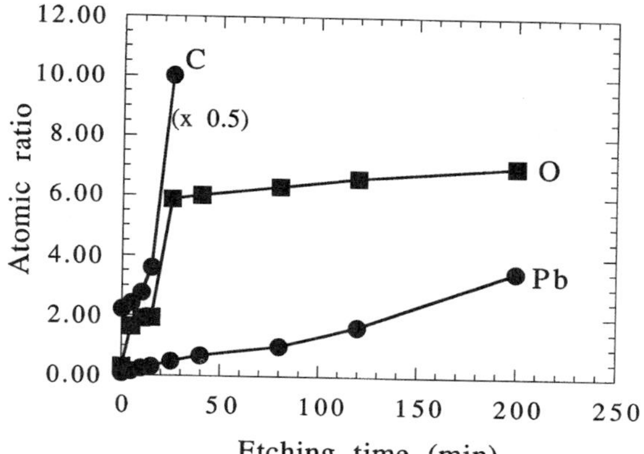

Figure 14. Depth profile of atomic ratios derived from XPS spectra in the joint of the *fistula*. The atomic ratios, derived from peak fittings in the Pb $4f_{7/2}$, O 1*s* and C 1*s* regions, are defined in the text. (Reproduced from Ref. 10, by permission of John Wiley & Sons, Ltd.)

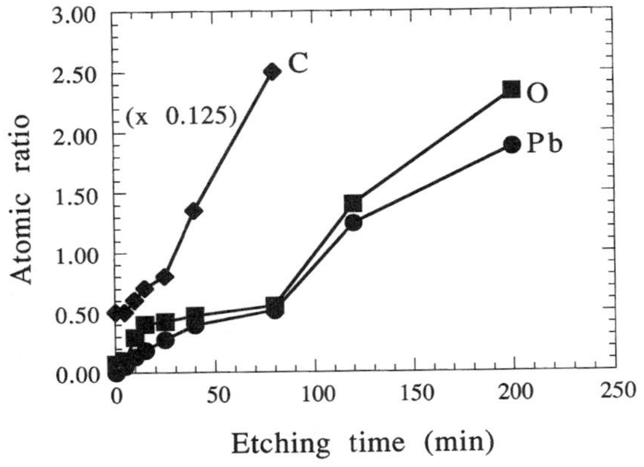

Figure 15. The analog of Fig. 14 for the pipe region of the *fistula*. (Reproduced from Ref. 10, by permission of John Wiley & Sons, Ltd.)

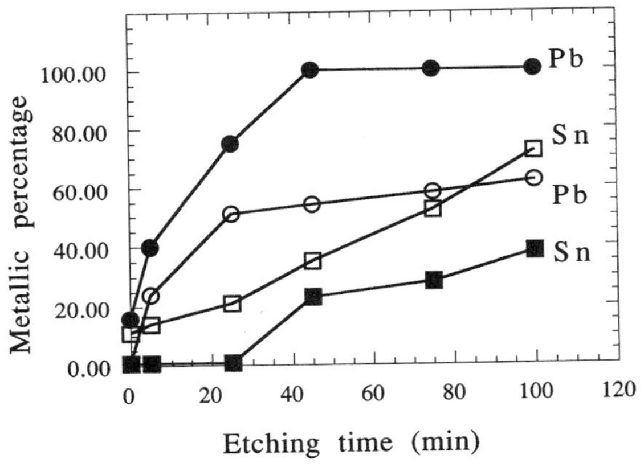

Figure 16. Evolution with depth of the percentages of Pb and Sn in the metallic states, defined as $M(Pb,Sn)=100\times([M^0]/\{[M^0]+[M^{n+}]\})$, measured from the individual metals (open symbols) and from the alloy (filled symbols). (Reproduced from Ref. 10, by permission of John Wiley & Sons, Ltd.)

——— 100 μm

Figure 17. SEM image of the solder interface between the Pb-Sn alloy (lower region) and Pb (upper region). (Reproduced from Ref. 10, by permission of John Wiley & Sons, Ltd.)

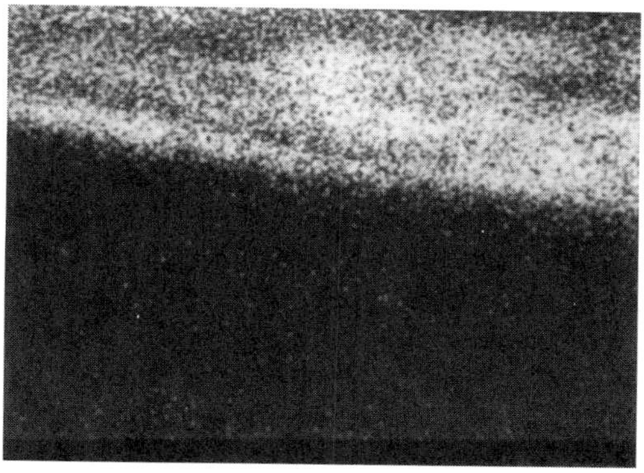

——— 2 μm

Figure 18. Auger image of C ([peak-background]/background) at the solder interface between the Pb-Sn alloy (black field) and Pb (white field). (Reproduced from Ref. 10, by permission of John Wiley & Sons, Ltd.)

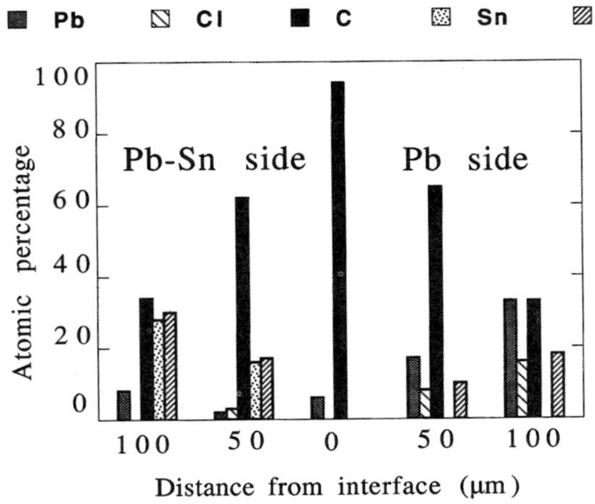

Figure 19. Distribution of atomic percentages from Auger point spectra across the solder interface between the Pb-Sn alloy and Pb. (Reproduced from Ref. 10, by permission of John Wiley & Sons, Ltd.)

0.5 to 2.6 nm, compared with the range of 1.7 to 2 nm available from the more "usual" XPS transitions.

3.3. Discussion

Direct evidence has been provided by XPS and SAM of both the nature and the elements present in the original patina, the bare joint and the bare pipe, and of their chemical states. The latter information compensates for the relatively poor elemental sensitivity, insufficient for the detection of those trace elements (e.g., Cu, Fe, Co, etc.) that had been detected by ICP analysis [10]. Gowland [13] reported in 1901 from wet chemical analysis that some Sn was present in the joint of a *fistula*. But neither the "old" analysis nor any modern ICP/AAS analysis has been able to identify the microchemistry of the Sn phases.

In the joint Sn was found mostly in the oxidized form, Sn^{4+}, but some Sn^{0} was present as well (Figs. 6 and 7). It was accompanied by a lower level of oxidized Pb and higher levels of both carbonates and hydrates, compared to the pipe (Figs 14 and 15). The carbonates and hydrates were consistent with the presence of hydrocerrusite, $2PbCO_3{\cdot}Pb(OH)_2$. From the results it seems that the Romans might have had some ability in limiting the occurrence of

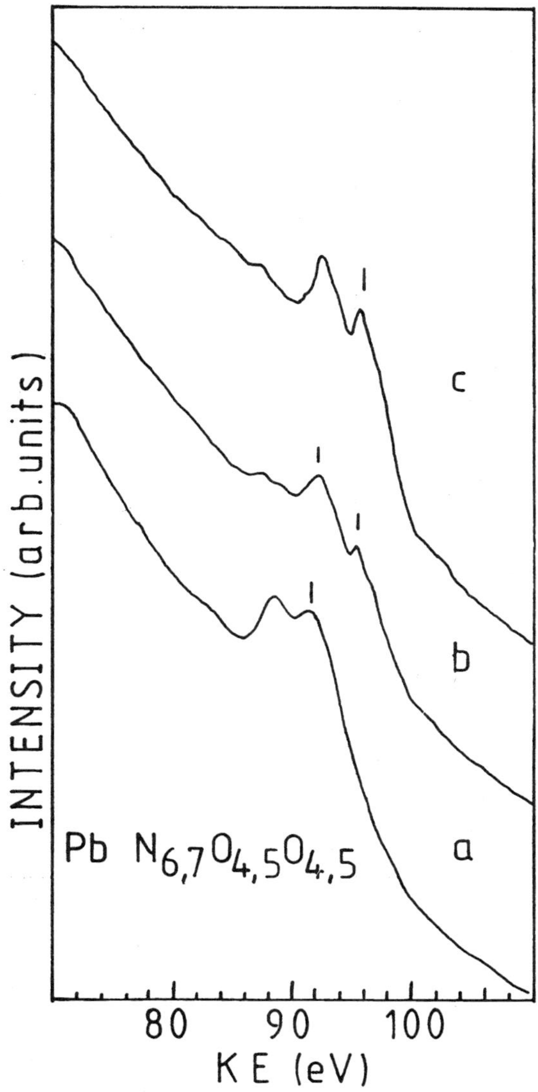

Figure 20. Electron-induced Pb $N_{6,7}O_{4,5}O_{4,5}$ spectra recorded from (a) as-received lead sheet; (b) Ar^+-etched lead sheet; (c) Ar^+-etched Pb-Sn alloy. The marks pinpoint the maxima of the $N_{6,7}O_{4,5}O_{4,5}$ component for the Pb^0 and Pb^{2+} states. (Reproduced from Ref. 10, by permission of John Wiley & Sons, Ltd.)

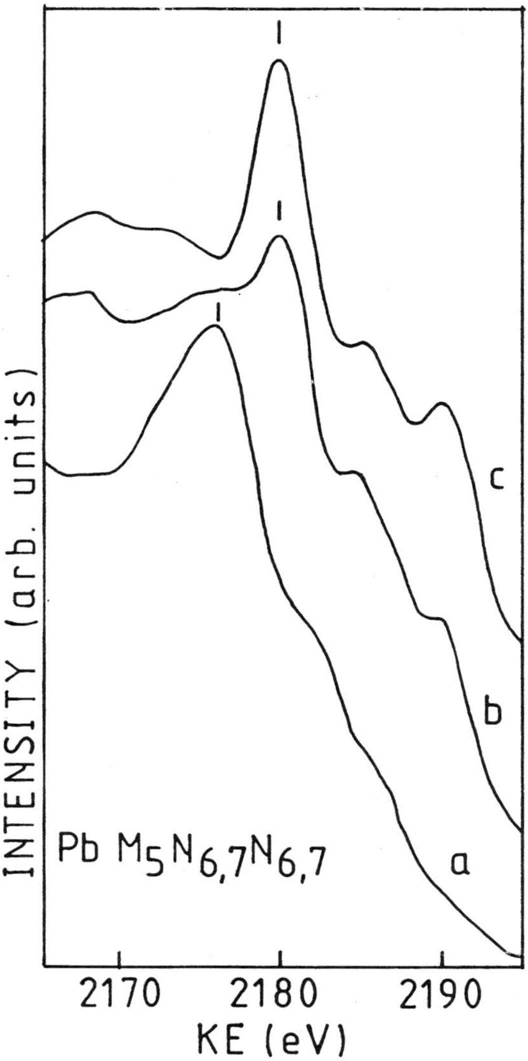

Figure 21. Electron-induced Pb $M_5N_{6,7}N_{6,7}$ spectra recorded from (a) the patina of the *fistula* in the as-received condition; (b) Ar^+ etched lead sheet; (c) Ar^+ etched Pb-Sn alloy. The marks at the maxima define the chemical shift between Pb^{2+} (a) and Pb^0 (b and c). (Reproduced from Ref. 10, by permission of John Wiley & Sons, Ltd.)

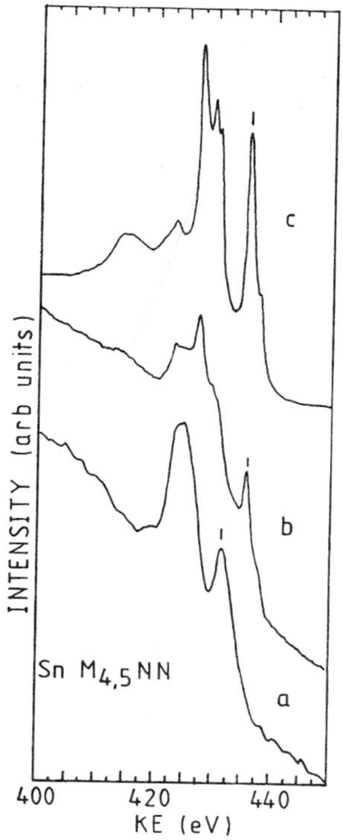

Figure 22. Electron-induced Sn $M_{4,5}NN$ spectra recorded from (a) SnO_2; (b) Ar^+-etched Pb-Sn alloy; (c) high-purity clean Sn metal. The marks pinpoint the maximum of the M_5NN component. (Reproduced from Ref. 10, by permission of John Wiley & Sons, Ltd.)

$2PbCO_3 \cdot Pb(OH)_2$ (a chemical compound of which they were aware, and which they called *cerussa*) during soldering.

The use of a Pb-Sn alloy for soldering Roman lead pipes is described by Pliny: "*..tertiarium . . . in quo duae sunt nigri portiones et tertia albi.*" (*tertiarium,* i.e., in which two parts are lead and the third tin). The reduction in the Sn content found in the joint (Fig. 19) must have been due to dilution of the alloy by the Pb of the pipe. The comparison measurements made on the Pb-Sn alloy confirmed that Sn plays a sacrificial role in limiting the oxidation of Pb, and provided a rationale for the reduced oxidation of the Pb in the joint.

Both point spectra and elemental imaging in SAM were able to provide much information about elemental, chemical, and compositional inhomogeneities across the surface of the pipe. Thus the SAM images of Fig. 12 re-

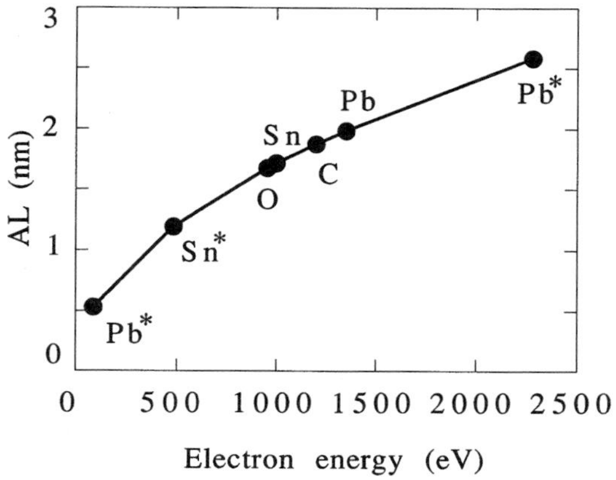

Figure 23. Attenuation lengths (AL) characteristic of the Auger and (Al K_{α}-excited) photemission transitions used here. An asterisk marks the ALs corresponding to the Auger transitions reported in Figs. 20–22. (Reproduced from Ref. 10, by permission of John Wiley & Sons, Ltd.)

veal, with sub-μm spatial resolution, local discontinuities in the chemical composition of the joint, leading to deviations from the ideal solder composition. In addition to SnO_2, both lead oxides and $PbCl_2$ were present. The microchemistry of the Pb-Sn/Pb interface after soldering (Fig. 18), an attempt to simulate ancient procedures [14], was also revealed. Although a detailed study was not made of the chemical role played by hydrocarbons in the process, their presence at the interface proved essential to successful soldering, for without the initial wetting with oil no joint could be made. Indeed, the Auger image in Fig. 18 shows, with ca. 0.5 μm spatial resolution, that there was intimate wetting of the lead by the alloy; the oil presumably minimized the oxidation of Pb during soldering [14]: "*Iungi non potest hoc ei sine oleo*" (joining is not possible of this (tin) to that (lead) without oil). Auger microanalysis also revealed that hydrocarbons were the main contaminants of the original patina (Fig. 2), presumably deriving from organic substances trapped there from the soil.

Use of less common Auger transitions adds further diagnostic value to the Auger analysis. Thus the chemical shift in the Pb $M_5N_{6,7}N_{6,7}$ transition ($Pb^0 \rightarrow Pb^{2+}$ = 3.7 eV) (Fig. 21), is more than twice that of the more commonly used Pb $4f_{7/2}$ XPS line [15]. The same applies to the Pb $N_{6,7}O_{4,5}O_{4,5}$ transition [9] (Fig. 20 and Table 2), where the chemical shift is nearly as large as the N_6–N_7 spin-orbit splitting. A substantial chemical shift between oxidized Sn and Sn^0 is also associated with the Sn $M_{4,5}NN$ transition (4.3 eV for the M_5NN component), compared with a value of 1.5 eV for the Sn $3d_{5/2}$ (Table 1 and Ref. 16), and is accompanied by a spectacular change in the line shape. Such line shape changes make it easy to see when a chemical change has occurred. On the other hand, when different chemical states coexist, curve fitting of Auger spectra is still too difficult in comparison with XPS, which probably accounts for its limited use. The extended AL range available using the above Auger transitions, plotted in Fig. 23, can be used to examine in-depth elemental distribution. For example, the high surface specificity of the Pb $N_6O_{4,5}O_{4,5}$ transition (AL = 0.53 nm) accounts for the absence of Pb^0 in both the *fistula* and the Pb sheet analyzed in the as-received condition (Table 1), whereas some metallic Pb was detected in the Pb $4f_{7/2}$ spectrum (AL = 1.98 nm) for the same two materials.

4. ROMAN LEADED BRONZES

Bronze was used by the Romans for the fabrication of objects of different sizes and uses ranging from small tools to huge statues [17–20]. The three leaded bronze samples (hereafter 1 to 3) and the copper sample (used for comparison purposes) analyzed here were provided by the "Museo della Civiltà Romana" (Rome). They date from the third century B.C. and were found in an excavation carried out in Carsòli (Italy) in 1908 [21].

4.1. Specification of the Problem

A leaded bronze is a more complex material than a lead *fistula*, as it is a Cu-Sn-Pb alloy. The two requirements are to relate the chemical states and concentrations of the elements present in the external layers to those in the bulk and to examine lateral inhomogeneity by comparison of XPS with SAM. Analysis of the microchemical composition of the original patina is also of interest.

4.2. Results

The bulk elemental compositions of the three bronzes as determined by ICP/EPMA are reported in Fig. 24. Trace elements at concentrations down to 0.01 wt.% [22] were also detected. Table 2 summarizes photoelectron BEs and Auger KEs for Cu, Sn and Pb, and the modified Auger parameter [23] for Cu, defined as α^*_{Cu} = BE(Cu $2p_{3/2}$) + KE(Cu L_3VV). Cu was present as Cu^+ at the surfaces (as-received condition) of all the bronzes, as indicated by α^*_{Cu} = 1849.1 ± 0.1 eV, and confirmed by the line shape of the Cu $2p_{3/2}$ spectrum (Fig. 25), which lacked the shake-up structure characteristic of Cu^{2+} species [24]. Neither Sn nor Pb showed any variation in valence state as a function of bulk content, being always in the Sn^{4+} and Pb^{2+} forms, respectively.

Although bronzes 1 and 3 contained similar bulk Pb contents, Pb was found by XPS on the surface only for bronze 1. Small quantities of chlorides and

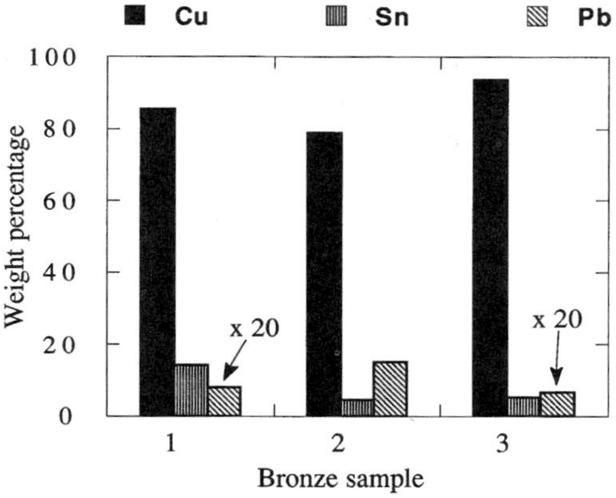

Figure 24. Weight percentages of Cu, Sn and Pb for the three bronzes. Note the ×20 amplification factor for Pb in bronzes 1 and 3.

Table 2. Photoelectron BEs, Auger KEs and Values of α^*_{Cu} (BE(Cu $2p_{3/2}$) + KE(Cu L_3 VV)) Measured from the Three Bronzes and from the Copper Sample as Functions of the Etching Time[a]

Sample	Etch (min)	Cu $2p_{3/2}$ (eV)	Sn $3d_{5/2}$ (eV)	Pb $4f_{7/2}$ (eV)	Cu L_3VV (eV)	α_{Cu} (eV)
1	0	932.5	486.5	138.1	916.5	1849.0
	9	932.6	484.9	136.9	918.5	1851.1
			486.6	138.0		
2	0	932.5	486.5	138.2	916.6	1849.1
	9	932.5	485.0	136.8	918.6	1851.1
			486.5	138.2	916.5	1849.0
3	0	932.5	486.5		916.6	1849.1
	9	932.6	484.9		918.6	1851.1
			486.6		916.6	1849.2
Cu	0	932.5			916.7	1849.2
	9	932.5			918.5	1851.0
					916.5	1849.0

[a]All data are in eV, and both the accuracy and reproducibility are within ± 0.2 eV.

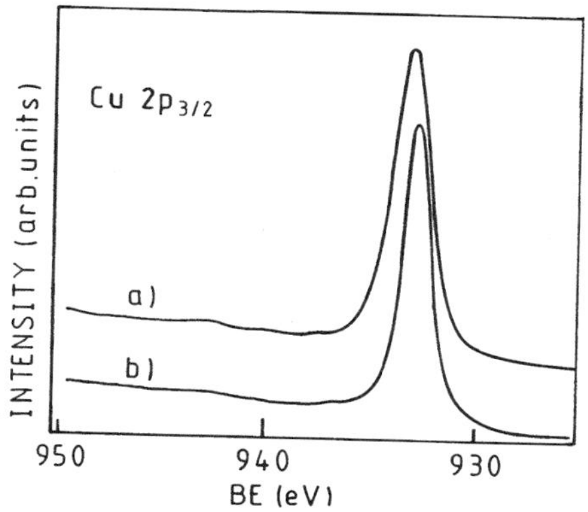

Figure 25. Typical Cu $2p_{3/2}$ spectra recorded from the bronzes: (a) in the as-received condition and (b) after Ar^+ etching.

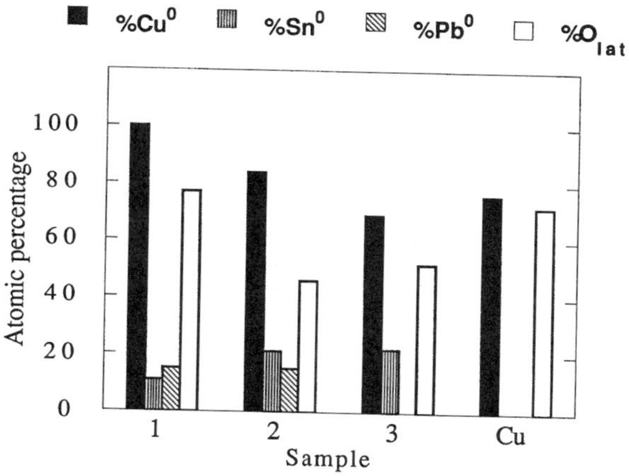

Figure 26. Percentages of the constituents in the metallic state, and of O, for the three bronzes and the copper sample after 9 min Ar^+ etching.

sulfates (ca. 0.4 at.%) were also detected. The surfaces of all samples were hydrated, since the O 1*s* spectra were dominated by a component at ca. 532.0 eV (also arising partly from carbonate identified by a minor component at ca. 289 eV in the C 1*s* spectrum), whereas that at ca. 530.3 eV, characteristic of anhydrous lattice oxides, was smaller.

After Ar^+ etching, metallic copper, $\alpha^*_{Cu} = 1851.1 \pm 0.1$ eV, as well as metallic Sn and Pb, were found. The partitioning between metallic and oxidized phases varied with bulk composition. Figure 26 shows percentages of the metallic state $\%M^{(0)}$ (M = Cu, Sn, Pb), defined as $100 \times I_M(0)/[I_M(0) + I_M ox]$, and percentages of lattice oxygen defined as $\%O_{1at} = 100 \times I_O 1at/[I_O 1at + I_O hydrate]$, measured after Ar^+ etching at an estimated depth of 9 nm. The XPS measurements showed that (i) bronze 1 with the highest bulk Sn content had the greatest surface concentration of $Cu^{(0)}$, the lowest $Sn^{(0)}$, and the highest O_{1at}, and (ii) on the same sample Cu was entirely metallic, whereas oxidized residues were still detectable on the "pure" material. Etching treatments four times as long were necessary to remove completely the oxidized species from the Cu standard. Artefacts associated with Ar^+ etching must have affected the quantification, but can be assumed to have caused only systematic errors since all samples received the same treatment. As a consequence, the differences found between them may be considered significant, as may those in the XPS profiles in Figs. 27–29.

Lateral elemental inhomogeneities at an estimated depth of 2 nm for bronze 1, recorded by SAM, are shown in Fig. 30. Eight separate regions, each ca. 1 μm^2 in size and distributed over an area of 50×50 μm^2, were analyzed. As discussed for the *fistula*, electron backscattering effects can be ruled out as the cause of the considerable variations in local composition. Cu was always the dominant metal element, and, except for regions G and H, its concentration was comparable to that of O. A similar composition was derived from XPS (no electron backscattering) for the same bronze after the same Ar^+ etching (Fig. 27). The eight points in Fig. 30 were selected from almost 200 analyzed to empha-

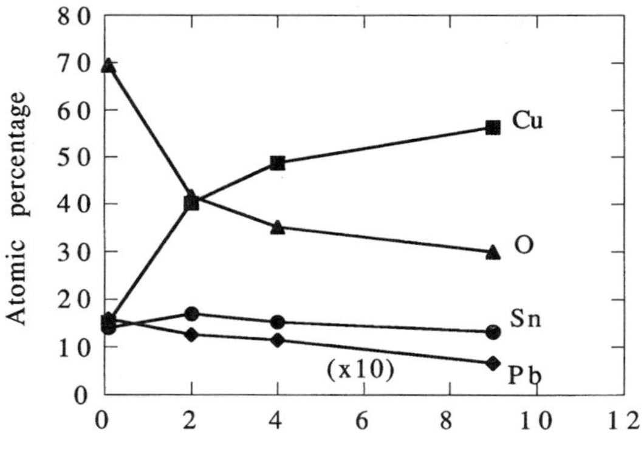

Figure 27. XPS depth profile of bronze 1. (Reproduced from Ref. 22 by permission of the American Institute of Physics.)

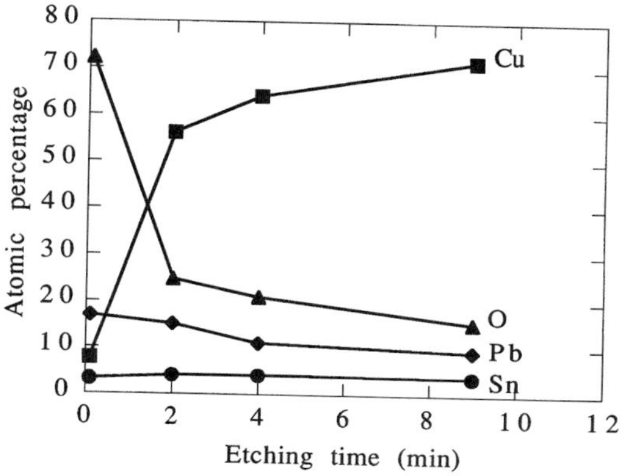

Figure 28. XPS depth profile of bronze 2. (Reproduced from Ref. 22 by permission of the American Institute of Physics.)

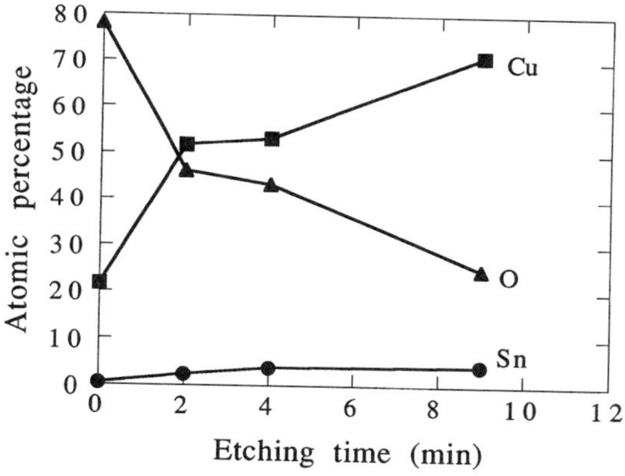

Figure 29. XPS depth profile of bronze 3. (Reproduced from Ref. 22 by permission of the American Institute of Physics.)

size not only lateral inhomogeneities but the occurrence of species undetected by XPS. In particular, Cl was found at some points (e.g., A, D, and F in the figure), at concentrations up to 10 at.%. The surface segregation of Pb in bronze 1 was also noticeable in some spots (e.g., regions B and D), where its concentration reached nearly 8 at.%.

Compositional variations across the interface between the original patina and the bare surface of bronze 3 were also explored. In the SEM image in Fig. 31 points 1, 2 and 3 correspond to the bare surface, interface, and patina, re-

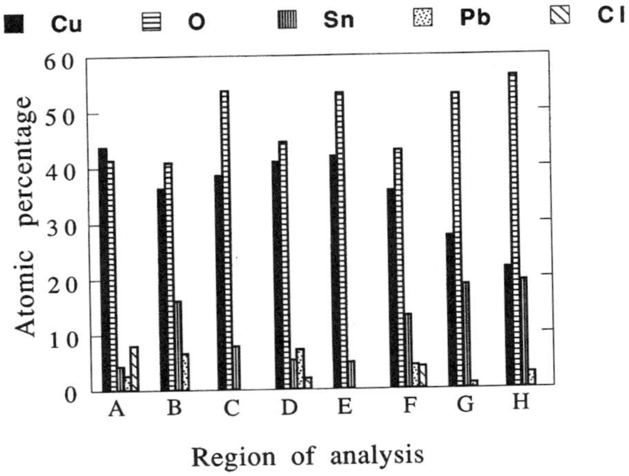

Figure 30. Atomic percentage distributions from Auger point spectra recorded from eight different regions of bronze 1. (Reproduced from Ref. 22 by permission of the American Institute of Physics.)

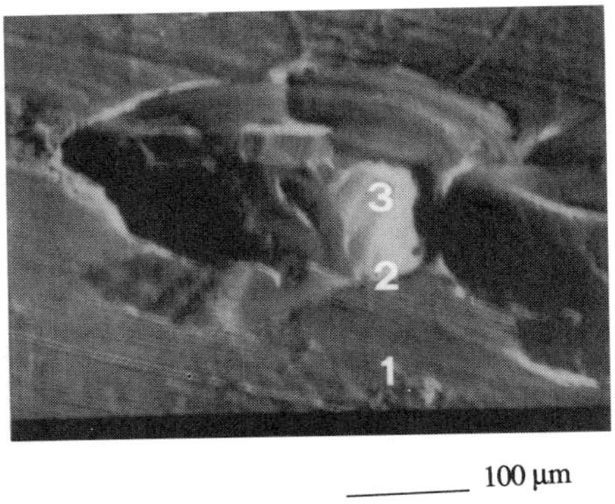

Figure 31. SEM image of bronze 3 showing the bare surface (point 1), the original patina (point 3), and the interface between the two (point 2). (Reproduced from Ref. 22 by permission of the American Institute of Physics.)

spectively, from which the elemental distributions in Fig. 32 were recorded by SAM. Although Sn was not present in the patina, its concentration reached ca. 3 at.% at the interface and ca. 10 at.% on the bare surface. S was enriched on passing from the bare surface to the patina side but was not observed at the interface. Cu increased progressively from the patina through the interface to

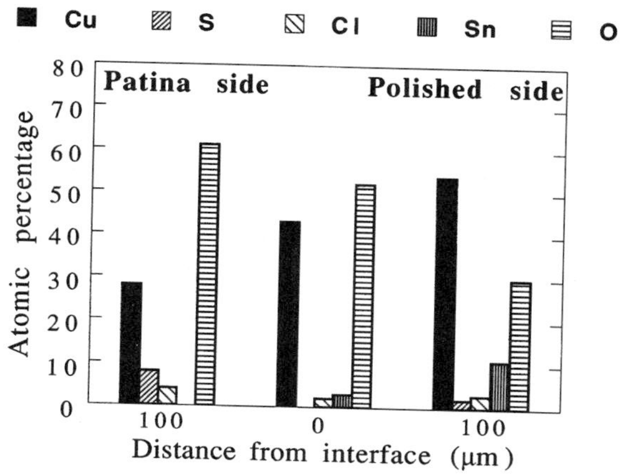

Figure 32. Atomic percentages from Auger point spectra recorded from different regions across the patina and the bare surface of bronze 3. (Reproduced from Ref. 22 by permission of the American Institute of Physics.)

the bare surface, and, unsurprisingly, the patina was more oxidized than the bare surface.

Ancient bronzes are very suitable for exploitation of the imaging capabilities of SAM because they are known to have inhomogeneous microchemistries [17–20]. Their analysis might therefore be useful for devising a sound "imaging" strategy. The questions to be answered are, what are the criteria for selection of an area for "good" imaging, and what is meant by "good" in any case? Obviously, good spatial resolution is necessary, but the principal criterion is whether or not the information in an Auger map is relevant to the material or the problem being analyzed. A good image will thus not only be relevant but will also have taken account of possible artefacts such as topography. For the bronzes any significant lateral inhomogeneities found for the alloying elements would be interesting. The first step must always be a thorough examination of surface morphology by SEM. Once an area is identified, Auger spectra at selected points and at higher spatial resolution must be recorded. If significant differences in elemental signal intensities between well-identified zones are found, there is a high probability of producing a good Auger map. Some examples of this procedure are given in the following.

An SEM micrograph recorded at a magnification of 10^4 from bronze 1 is shown in Fig. 33. Within this area, Auger point spectra recorded at a magnification of 5×10^4, in the KE range 40–600 eV, were of two types: the first was typical of regions high in Cu and low in Sn, whereas the second was the reverse. Auger maps for Cu and Sn were then recorded from the same area and

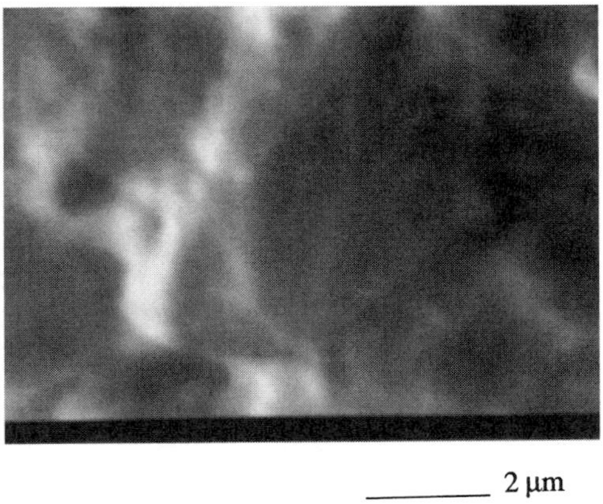

Figure 33. SEM image of an area on bronze 1. (Reproduced from Ref. 22 by permission of the American Institute of Physics.)

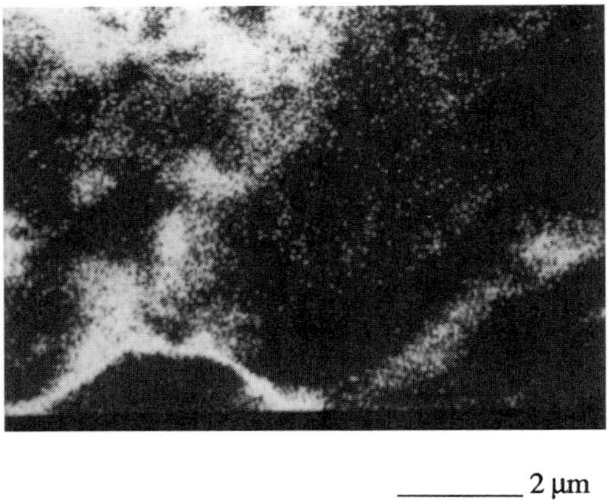

Figure 34. Auger image ([peak-background]/background) for Cu (using the $M_{2,3}VV$ signal) recorded from bronze 1 within the area displayed in Fig. 33. (Reproduced from Ref. 22 by permission of the American Institute of Physics.)

are given in Figs. 34 and 35, respectively. They are virtually complementary, with an estimated spatial resolution of 0.5 μm, whereas the Pb map, not shown, showed uniformity in Pb over the whole area. The same exercise was carried out for bronze 2 in Fig. 36. Again, Auger point spectra were of two kinds: one Cu-rich, the other Pb-rich. The lateral inhomogeneity was con-

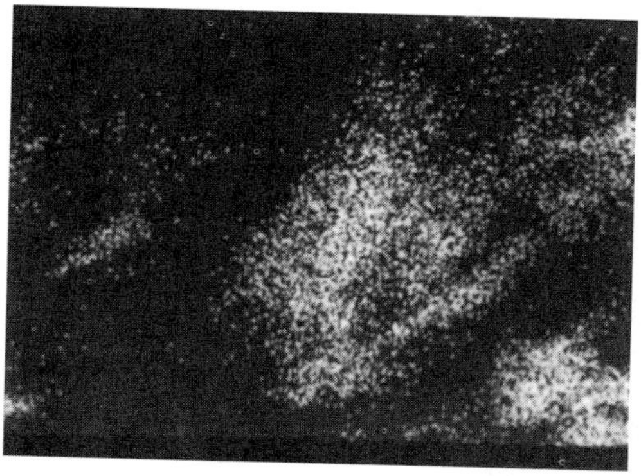

2 μm

Figure 35. Auger image ([peak-background]/background) for Sn (using the M_5NN signal) recorded from bronze 1 within the area displayed in Fig. 33. (Reproduced from Ref. 22 by permission of the American Institute of Physics.)

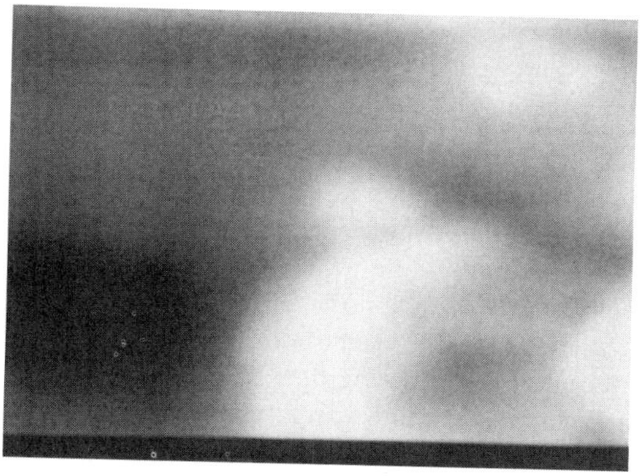

0.2 μm

Figure 36. SEM image of an area on bronze 2. (Reproduced from Ref. 22 by permission of the American Institute of Physics.)

firmed by the Auger images of Figs. 37 and 38, which show that the spatial distributions of Cu and of Pb, respectively, complemented each other, with an estimated lateral resolution of about 0.1 μm. In both pairs of Auger maps, topographic effects were minimized by using the [peak-background]/background correction procedure [25].

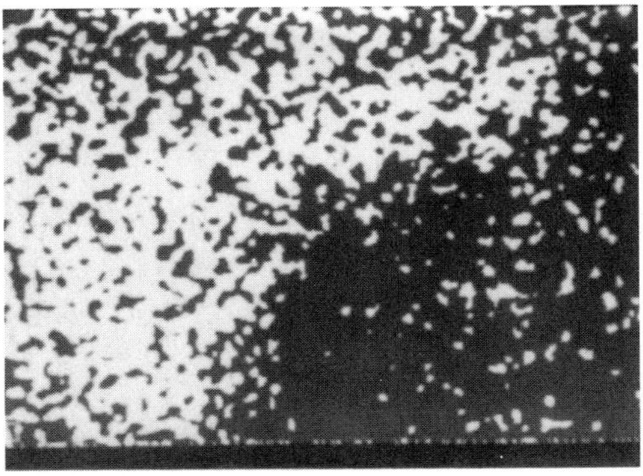

0.2 μm

Figure 37. Auger image ([peak-background]/background) recorded from bronze 2 for Cu (using the $M_{2,3}VV$ signal) within the area displayed in Fig. 36. (Reproduced from Ref. 22 by permission of the American Institute of Physics.)

A full evaluation of electron backscattering artefacts would require a detailed analysis, such as that of the MULSAM approach described by Prutton and coworkers [26–28], which is beyond the scope of this chapter. However, note that the differences in Auger point spectra and the corresponding Auger maps between the two bronzes must have arisen mainly from actual chemical inhomogeneities rather than from electron backscattering effects. Thus, in bronze 1 the concentration of Pb (a strong backscatterer) was almost uniform over the region in which analysis was performed, with local concentrations of 8 at.% at the most, as shown by Fig. 30. Furthermore, considering the BEs of the levels involved in the primary excitation events in Cu and Sn and the experimental conditions under which the measurements were performed (E_p = 10 keV, electron incidence angle of 45°), the difference in the backscattering terms between the two metals should be no more than 10% [29]. In bronze 2, although the apparent concentration of Cu may have been enhanced artificially by the underlying high backscatterer Pb, the opposite would not be true for Pb itself. In summary, the maps in Figs. 37 and 38 can be taken as true indications of the presence of two spatially separate domains of Cu and of Pb, although the Pb contrast in Fig. 38 might have been partially enhanced.

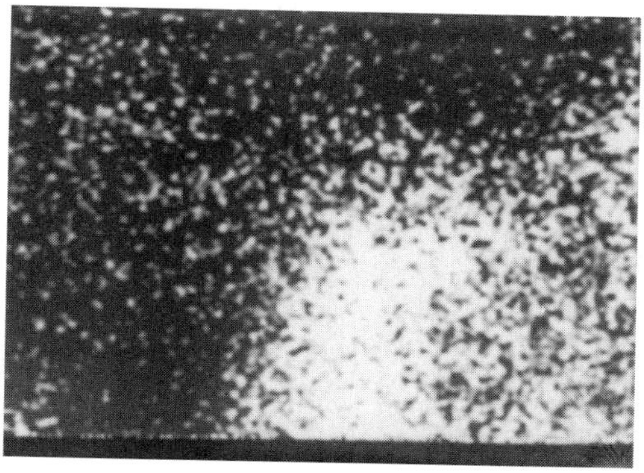

0.2 μm

Figure 38. Auger image ([peak-background]/background) recorded from bronze 2 for Pb (using the $N_{6,7}OO$ signal) within the area displayed in Fig. 36. (Reproduced from Ref. 22 by permission of the American Institute of Physics.)

4.3. Discussion

As for the *fistula*, XPS/SAM analyses have provided detailed information on the surface microchemistry of the bronzes. XPS BEs and the Auger parameter indicate that the samples were fully oxidized at the surface in the form of Cu^{+}, Sn^{4+}, and Pb^{2+} partially hydrated phases. There was also evidence for the presence of white lead, $2Pb(CO_3)_2 \cdot Pb(OH)_2$, at the surface of the high-lead bronze 2. At greater depths the proportion of metallic states increased in all samples. Sn appeared to have a beneficial effect in limiting the oxidation of Cu and the amounts of hydrated and carbonate phases; Sn in fact plays a sacrificial role in minimizing the surface corrosion of the bronzes. The elemental depth profiles in Figs. 27 to 29 reveal a reasonable correlation between surface and bulk concentrations (cf. Fig. 24 for the three constituents), except for the Pb surface enrichment in bronze 1. Hughes et al. have reported that lateral Pb segregation occurs in some leaded bronzes [30], and they interpreted that as deriving from a low cooling rate of the original casting which caused the lead to segregate ahead of the freezing front. In bronze 1, lead segregated to the first few nm of the surface almost entirely as a PbO-like phase, whose formation may have played an additional role in the phenomenon.

Auger point analysis revealed considerable lateral inhomogeneities in the microchemical compositions of the bronzes (Fig. 30), and was particularly useful in demonstrating the differences in elemental distribution across the interface between the bare bronze and the original patina (Fig. 32). Auger maps of the alloying metals revealed lateral segregations of Sn (Fig. 35) and Pb (Fig.

38) with lateral resolutions of about 0.5 and 0.1 μm, respectively. These microanalytical results are of particular methodological relevance because they represent an improvement over EPMA mapping and perhaps anticipate an ever wider use of SAM analysis in the study of the surface microchemistry of metallurgical archaeomaterials.

ACKNOWLEDGMENTS

Luciano Moretto is thanked for technical assistance, and thanks are also due to Massimo Brolatti for drawing the figures. Dr. Anna Mura Sommella, Director of the "Antiquarium Comunale" of Rome, and her co-workers, Dr. Carla Martini and Dr. Carla Salvetti, are thanked for giving permission to analyze the lead pipe or *fistula*, as well as for providing details on its provenance. Dr. Giuseppina Pisani-Sartorio, Head, Museo della Civiltà Romana of Rome, is thanked for making available the bronzes and the copper sample. Sincere gratitude goes to Dr. Clotilde D'Amato of the Museo della Civiltà Romana of Rome for her valuable help with the historical sources, for continued advice and encouragement, as well as for her introduction to the fascinating world of archaeology. Drs. A. Palmieri (ITABC-CNR, Monterotondo, Italy) and J.P. Northover (University of Oxford) are thanked for their collaboration with the ICP and EPMA analyses, respectively.

REFERENCES

1. I. F. Ferguson, *Auger Microprobe Analysis*, Adam Hilger, Bristol, 1989.
2. E. D. Hondros and M. P. Seah, *Physical Metallurgy* (R. W. Cahn and P. Haasen, eds.), Elsevier, Amsterdam, 1983, p. 856.
3. M. P. Seah, *Practical Surface Analysis* (D. Briggs and M. P. Seah, eds.), Vol. 1, 2nd ed. Wiley, Chichester, 1990, p. 311.
4. N. S. McIntyre and T. C. Chan, *Practical Surface Analysis*, (D. Briggs and M. P. Seah, eds), Vol. 1, 2nd ed., Wiley. Chichester, 1990, p. 485.
5. Marcus Vitruvius Pollio, *De Architectura*, VIII, 6, 1.
6. R. Lanciani, I Comentarii di Frontino intorno le acque e gli aquedotti (Italian), Salvinucci, Roma, 1880.
7. R. F. Tylecote, *Metallurgy in Archaeology*, E. Arnold, London, 1962, p. 94.
8. E. Paparazzo, *Appl. Surf. Sci. 74*, 61 (1994).
9. L. R. Pederson, *J. Electron Spectrosc. Relat. Phenom. 28*, 203 (1982).
10. E. Paparazzo, L. Moretto, C. D'Amato and A. Palmieri, *Surf. Interface Anal. 23*, 69 (1995).

11. K. S. Kim, W. E. Baitinger, J. W. Amy and N. Winograd, *J. Electron Spectrosc. Relat. Phenom. 5*, 351 (1974).
12. M. P. Seah and W. A. Dench, *Surf. Interface Anal. 1*, 2 (1979).
13. W. Gowland, *Archaeologia 57*, 81 (1901).
14. C. Plinius Secundus, *Naturalis Historia*, Vol. XXXIV, Einaudi, Turin, Italy p. 110.
15. C. D. Wagner and J. A. Taylor, *J. Electron Spectrosc. Relat. Phenom. 20*, 83 (1980).
16. A. W. C. Lin, N. R. Armstrong and T. Kuwana, *Anal. Chem. 49*, 1228 (1977).
17. M. Picon, S. Boucher and J. Condamin, *Gallia 26*, 245 (1968).
18. P. T. Craddock, *J. Archaeol. Sci. 4*, 103 (1977).
19. P. T. Craddock, *J. Archaeol. Sci. 5*, 1 (1978).
20. D. A. Scott, *Metallography and Microstructure of Ancient and Historic Metals*, The Getty Conservation Institute, Marina del Rey, CA, 1991.
21. A. Cederna, *Atti Accad. Naz. Lincei VIII Ser. (Ital.) 5*, 22 (1951).
22. E. Paparazzo, L. Moretto, J. P. Northover, C. D'Amato and A. Palmieri, *J. Vac. Sci. Technol. A13*, 1229 (1995).
23. J. C. Rivière, J. A. A. Crossley and B. A. Sexton, *J. Appl. Phys. 64*, 4585 (1988).
24. L. Alagna, E. Paparazzo, T. Prosperi and A. A. G. Tomlinson, *J. Chem. Res.* 352 (1982).
25. M. Prutton, L. A. Larson and H. Poppa, *J. Appl. Phys. 54*, 374 (1983).
26. M. Prutton, C. G. H. Walker, J. C. Greenwood, P. G. Kenny, J. C. Dee, I. R. Barkshire, R. H. Roberts and M. M. El Gomati, *Surf. Interface Anal. 17*, 71 (1991).
27. M. Prutton, I. R. Barkshire, M. M. El Gomati, J. C. Greenwood, P. G. Kenny and R. H. Roberts, *Surf. Interface Anal. 18*, 295 (1992).
28. I. R. Barkshire, M. Prutton, J. C. Greenwood and M. M. El Gomati, *Surf. Interface Anal. 20*, 984 (1993).
29. S. Ichimura, R. Shimizu and J. P. Langeron, *Surf. Sci. 124*, L49 (1983).
30. M. J. Hughes, J. P. Northover and B. E. P. Staniaszek, *Oxford J. Archaeol.* 359 (1982).

Appendix 1

Physical Constants and Conversion Factors

Table A1.1 Fundamental Physical Constants

Quantity	Symbol	Numerical value	SI units
Speed of light (in vacuum)	c	2.9979×10^{8}	$m\text{-}s^{-1}$
Elementary charge	e	1.6022×10^{-19}	C
Planck constant	h	6.6262×10^{-34}	J-s
Planck constant	$\hbar$	1.0546×10^{-34}	J-s
Avogadro constant	N_A	6.0221×10^{23}	mol^{-1}
Boltzmann constant	k_B	1.3807×10^{-23}	$J\text{-}K^{-1}$
Stefan-Boltzmann constant	σ	5.6703×10^{-14}	$W\text{-}m^{-2}\text{-}K^{-4}$
Atomic mass unit	amu	1.6606×10^{-27}	kg
Electron rest mass	m_e	9.1095×10^{-31}	kg
Proton rest mass	m_p	1.6273×10^{-27}	kg
Neutron rest mass	m_n	1.6750×10^{-27}	kg
Permittivity of vacuum	ε_o	8.8542×10^{-12}	$F\text{-}m^{-1}$
Permeability of vacuum	μ_o	1.2567×10^{-6}	$H\text{-}m^{-1}$
Fine-structure constant	α	7.2974×10^{-3}	
Bohr radius	a_o	5.2918×10^{-11}	m
Bohr magneton	μ_B	9.2741×10^{-24}	$J\text{-}T^{-1}$
Nuclear magneton	μ_N	5.0508×10^{-27}	$J\text{-}T^{-1}$

Table A1.2 Energy Conversion Factors and Equivalents

Quantity	Equals	Numerical value	Units
1 amu		931.50	MeV
1 m_ec^2		0.5110	MeV
1 m_pc^2		938.28	MeV
1 m_nc^2		939.57	MeV
1 eV		1.6022×10^{-19}	J
1 eV/h		2.4180×10^{14}	Hz
1 eV/hc		8.0655×10^{5}	m^{-1}
1 eV/k_B		1.1605×10^{4}	K
1 Rydberg		13.606	eV

Table A1.3 Pressure Conversion Factors

To get \ Multiply quantity by	kPa	Torr	Psi	mb	Atm
kPa	1	0.1333	6.895	10^{-1}	101.3
Torr	7.502	1	51.72	0.7502	760
Psi	0.145	1.934×10^{-2}	1	1.45×10^{-2}	14.7
mb	10	1.333	68.95	1	1013
Atm	9.869×10^{-4}	1.316×10^{-3}	6.8×10^{-2}	9.869×10^{-4}	1

Appendix 2

Data for the Elements and Isotopes

Table A2.1 Electronic Configuration of the Elements and Related Data

Element	Z	A^a (amu)	Structure[b]	Configuration	Density[c] $[(\times 10^3)\ kg/m^3]$
H	1	1.0079	HEX	$1s^1$	0.089
He	2	4.0026	HEX	$1s^2$	0.179
Li	3	6.941	BCC	$1s^2 2s^1$	0.53
Be	4	9.0122	HEX	$1s^2 2s^2$	1.85
B	5	10.81	TET	$1s^2 2s^2 2p^1$	2.34
C	6	12.01	DIA	$1s^2 2s^2 2p^2$	2.26
N	7	14.007	HEX	$1s^2 2s^2 2p^3$	1.03
O	8	15.999	CUB	$1s^2 2s^2 2p^4$	1.43
F	9	18.998	MCL	$1s^2 2s^2 2p^5$	1.97
Ne	10	20.18	FCC	$1s^2 2s^2 2p^6$	1.56
Na	11	22.989	BCC	$[Ne]3s^1$	0.97
Mg	12	25.305	HEX	$[Ne]3s^2$	1.74
Al	13	26.982	FCC	$[Ne]3s^2 3p^1$	2.70

Table A2.1 Continued

Element	Z	A[a] (amu)	Structure[b]	Configuration	Density[c] [($\times10^3$) kg/m^3]
Si	14	28.086	DIA	[Ne]$3s^2 3p^2$	2.33
P	15	30.974	CUB	[Ne]$3s^2 3p^3$	1.82
S	16	32.064	ORC	[Ne]$3s^2 3p^4$	2.07
Cl	17	35.453	ORC	[Ne]$3s^2 3p^5$	2.09
Ar	18	39.948	FCC	[Ne]$3s^2 3p^6$	1.78
K	19	39.09	BCC	[Ar]$4s^1$	0.86
Ca	20	40.08	FCC	[Ar]$4s^2$	1.54
Sc	21	44.956	HEX	[Ar]$3d^1 4s^2$	2.99
Ti	22	47.90	HEX	[Ar]$3d^2 4s^2$	4.51
V	23	50.942	BCC	[Ar]$3d^3 4s^2$	6.1
Cr	24	52.00	BCC	[Ar]$3d^5 4s^1$	7.19
Mn	25	54.938	CUB	[Ar]$3d^5 4s^2$	7.43
Fe	26	55.85	BCC	[Ar]$3d^6 4s^2$	7.86
Co	27	58.93	HEX	[Ar]$3d^7 4s^2$	8.9
Ni	28	58.71	FCC	[Ar]$3d^8 4s^2$	8.9
Cu	29	63.55	FCC	[Ar]$3d^{10} 4s^1$	8.96
Zn	30	65.38	HEX	[Ar]$3d^{10} 4s^2$	7.14
Ga	31	69.72	ORC	[Ar]$3d^{10} 4s^2 4p^1$	5.91
Ge	32	72.59	DIA	[Ar]$3d^{10} 4s^2 4p^2$	5.32
As	33	74.922	RHL	[Ar]$3d^{10} 4s^2 4p^3$	5.72
Se	34	78.96	HEX	[Ar]$3d^{10} 4s^2 4p^4$	4.79
Br	35	79.91	ORC	[Ar]$3d^{10} 4s^2 4p^5$	4.10
Kr	36	83.80	FCC	[Ar]$3d^{10} 4s^2 4p^6$	3.07
Rb	37	85.47	BCC	[Kr]$5s^1$	1.53
Sr	38	87.62	FCC	[Kr]$5s^2$	2.60
Y	39	88.91	HEX	[Kr]$4d^1 5s^2$	4.46
Zr	40	91.22	HEX	[Kr]$4d^2 5s^2$	6.49
Nb	41	92.91	BCC	[Kr]$4d^4 5s^1$	8.4
Mo	42	95.94	BCC	[Kr]$4d^5 5s^1$	10.2
Tc	43	98.91	HEX	[Kr]$4d^5 5s^2$	11.5
Ru	44	101.07	HEX	[Kr]$4d^7 5s^1$	12.2
Rh	45	102.90	FCC	[Kr]$4d^8 5s^1$	12.4
Pd	46	106.40	FCC	[Kr]$4d^{10} 5s^0$	12.0
Ag	47	107.87	FCC	[Kr]$4d^{10} 5s^1$	10.5
Cd	48	112.40	HEX	[Kr]$4d^{10} 5s^2$	8.65
In	49	114.82	TET	[Kr]$4d^{10} 5s^2 5p^1$	7.31
Sn	50	118.69	TET	[Kr]$4d^{10} 5s^2 5p^2$	7.30
Sb	51	121.75	RHL	[Kr]$4d^{10} 5s^2 5p^3$	6.62

Element	Z	A[a] (amu)	Structure[b]	Configuration	Density[c] [(×10³) kg/m³]
Te	52	127.60	HEX	$[Kr]4d^{10}5s^25p^4$	6.24
I	53	126.90	ORC	$[Kr]4d^{10}5s^25p^5$	4.94
Xe	54	131.30	FCC	$[Kr]4d^{10}5s^25p^6$	3.77
Cs	55	132.91	BCC	$[Xe]6s^1$	1.90
Ba	56	137.34	BCC	$[Xe]6s^2$	3.5
La	57	138.91	HEX	$[Xe]5d^16s^2$	6.17
Ce	58	140.12	FCC	$[Xe]4f^25d^06s^2$	6.77
Pr	59	140.91	HEX	$[Xe]4f^35d^06s^2$	6.77
Nd	60	144.24	HEX	$[Xe]4f^45d^06s^2$	7.00
Pm	61	145		$[Xe]4f^55d^06s^2$	
Sm	62	150.35	RHL	$[Xe]4f^65d^06s^2$	7.54
Eu	63	151.96	BCC	$[Xe]4f^75d^06s^2$	4.61
Gd	64	157.25	HEX	$[Xe]4f^75d^16s^2$	8.23
Tb	65	158.92	HEX	$[Xe]4f^95d^06s^2$	8.54
Dy	66	162.50	HEX	$[Xe]4f^{10}5d^06s^2$	8.78
Ho	67	164.93	HEX	$[Xe]4f^{11}5d^06s^2$	9.05
Er	68	167.26	HEX	$[Xe]4f^{12}5d^06s^2$	9.37
Tm	69	168.93	HEX	$[Xe]4f^{13}5d^06s^2$	9.31
Yb	70	173.04	FCC	$[Xe]4f^{14}5d^06s^2$	6.97
Lu	71	174.97	HEX	$[Xe]4f^{14}5d^16s^2$	9.84
Hf	72	178.49	HEX	$[Xe]4f^{14}5d^26s^2$	13.1
Ta	73	180.95	BCC	$[Xe]4f^{14}5d^36s^2$	16.6
W	74	183.85	BCC	$[Xe]4f^{14}5d^46s^2$	19.3
Re	75	186.2	HEX	$[Xe]4f^{14}5d^56s^2$	21.0
Os	76	190.20	HEX	$[Xe]4f^{14}5d^66s^2$	22.6
Ir	77	192.22	FCC	$[Xe]4f^{14}5d^76s^2$	22.5
Pt	78	195.09	FCC	$[Xe]4f^{14}5d^{10}6s^0$	21.4
Au	79	196.97	FCC	$[Xe]4f^{14}5d^{10}6s^1$	19.3
Hg	80	200.59	RHL	$[Xe]4f^{14}5d^{10}6s^2$	13.6
Tl	81	204.37	HEX	$[Xe]4f^{14}5d^{10}6s^26p^1$	11.85
Pb	82	207.19	FCC	$[Xe]4f^{14}5d^{10}6s^26p^2$	11.4
Bi	83	208.98	RHL	$[Xe]4f^{14}5d^{10}6s^26p^3$	9.8
Po	84	210	SC	$[Xe]4f^{14}5d^{10}6s^26p^4$	9.4
At	85	210		$[Xe]4f^{14}5d^{10}6s^26p^5$	
Rn	86	222	(FCC)	$[Xe]4f^{14}5d^{10}6s^26p^6$	(4.4)
Fr	87	223	(BCC)	$[Rn]7s^1$	
Ra	88	226		$[Rn]7s^2$	(5.0)
Ac	89	227	FCC	$[Rn]6d^17s^2$	10.1
Th	90	232.04	FCC	$[Rn]6d^27s^2$	11.7

Table A2.1 Continued

Element	Z	A[a] (amu)	Structure[b]	Configuration	Density[c] [(×10^3) kg/m^3]
Pa	91	231	TET	[Rn5$f^{2}6d^{1}7s^{2}$	15.4
U	92	238.03	ORC	[Rn}5$f^{3}6d^{1}7s^{2}$	19.07
Np	93	237.05	ORC	[Rn]5$f^{5}6d^{0}7s^{2}$	20.3
Pu	94	244	MCL	[Rn]5$f^{6}6d^{0}7s^{2}$	19.8
Am	95	243		[Rn]5$f^{7}6d^{0}7s^{2}$	11.8
Cm	96	247		[Rn]5$f^{7}6d^{1}7s^{2}$	
Bk	97	247		[Rn]5$f^{7}6d^{2}7s^{2}$	
Cf	98	251		[Rn]5$f^{9}6d^{1}7s^{2}$	

[a]The atomic mass number refers to the naturally occurring isotopic mixture.
[b]The structure refers to the most common phase: FCC = face-centered cubic; BCC = body-centered cubic; SC = simple cubic; CUB = cubic; TET = tetragonal; ORC = orthorhombic; HEX = hexagonal; DIA = diamond; RHL = rhombohedral; MCL = monoclinic.
[c]The density refers to the most common solid phase, generally at STP.

Table A2.2 Data for Stable Isotopes of the Elements[a]

Element	Z	Isotopic mass (amu)/relative abundance (%)		
H	1	1.00783	2.01411	
		99.985	0.015	
He	2	3.01605	4.0026	
		0.00015	100	
Li	3	6.01513	7.0160	
		7.42	92.58	
Be	4	9.01218		
		100		
B	5	10.0129	11.0093	
		19.6	80.4	
C	6	12.0000	13.0034	
		98.89	1.11	
N	7	14.0031	15.0001	
		99.63	0.37	
O	8	15.9949	16.9991	17.9992
		99.759	0.037	0.204
F	9	18.9984		
		100		
Ne	10	19.9924	20.9940	21.9914
		90.51	0.27	9.22
Na	11	22.9898		
		100		
Mg	12	23.9850	24.9858	25.9826
		78.99	10.00	11.01

Table A2.2 Continued

Element	Z	Isotopic mass (amu)/relative abundance (%)					
Al	13	26.9815					
		100					
Si	14	27.9769	28.9765	29.9738			
		92.23	4.67	3.09			
P	15	30.9738					
		100					
S	16	31.9721	32.9715	33.9679	35.9671		
		95.0	0.76	4.22	0.014		
Cl	17	34.9689	36.9671				
		75.77	24.23				
Ar	18	35.9676	37.9627	39.9624			
		0.337	0.063	99.60			
K	19	38.9637	39.964	40.9618			
		93.26	0.012	6.73			
Ca	20	39.9626	41.9586	42.9588	43.9555	45.9537	47.9524
		96.94	0.64	0.13	2.09	0.004	0.19
Sc	21	44.9559					
		100					
Ti	22	45.9526	46.9518	47.9480	48.9479	49.9448	
		8.0	7.5	73.7	5.5	5.3	
V	23	49.9472	50.9440				
		0.24	99.76				
Cr	24	49.9461	51.9405	52.9407	53.9389		
		4.35	83.79	9.50	2.36		

Mn	25	54.9381					
		100					
Fe	26	53.9396	55.9349	56.9354	57.9333		
		5.8	91.8	2.1	0.3		
Co	27	58.9332					
		100					
Ni	28	57.9353	59.9308	60.9310	61.9283	63.9280	
		68.27	26.10	1.13	3.59	0.91	
Cu	29	62.9296	64.9278				
		69.2	30.8				
Zn	30	63.9291	65.9260	66.9271	67.9249	69.9253	
		48.6	27.9	4.1	18.8	0.6	
Ga	31	69.9257	70.9249				
		60.4	39.6				
Ge	32	69.9243	71.9217	72.9234	73.9219	75.9214	
		20.5	27.4	7.8	36.5	7.8	
As	33	74.9216					
		100					
Se	34	73.9225	75.9192	76.9199	77.9173	79.9165	81.9167
		0.9	9.0	7.6	23.5	49.8	9.2
Br	35	78.9183	80.9163				
		50.3	49.7				
Kr	36	77.9204	79.9164	81.9135	82.9141	83.9115	85.9106
		0.35	2.25	11.6	11.5	57.0	17.3
Rb	37	84.9117	—				
		72.17	27.83				
Sr	38	83.9134	85.9094	86.9089	87.9056		
		0.5	9.9	7.0	82.6		

Table A2.2 Continued

Element	Z	Isotopic mass (amu)/relative abundance (%)									
Y	39	88.9059									
		100									
Zr	40	89.9043	90.9053	91.9046	93.9061	95.9082					
		51.4	11.2	17.1	17.8	2.8					
Nb	41	92.9060									
		100									
Mo	42	91.9063	93.9047	94.9058	95.9046	96.9058	97.9055	99.9076			
		14.8	9.3	15.9	16.7	9.6	24.1	9.6			
Tc	43	No natural stable isotope: 96.906 (2.6×10^6 y); 98.906 (6.0 h)									
Ru	44	95.9076	97.9055	98.9061	99.9030	100.9041	101.9037	103.9055			
		5.5	1.9	12.7	12.6	17.1	31.6	18.6			
Rh	45	102.9048									
		100									
Pd	46	101.9049	103.9036	104.9046	105.9032	107.9039	109.9045				
		1.0	11.0	22.2	27.3	26.7	11.8				
Ag	47	106.9051	108.9047								
		51.83	48.17								
Cd	48	105.907	107.9040	109.9030	110.9042	111.9028	112.9046	113.9036	115.9050		
		1.2	0.9	12.4	12.8	24.0	12.3	28.8	7.6		
In	49	112.9043	114.9041								
		4.3	95.7								
Sn	50	111.9040	113.9030	114.9035	115.9021	116.9031	117.9018	118.9034	119.9021	121.9034	123.9052
		1.0	0.7	0.4	14.7	7.7	24.3	8.6	32.4	4.6	5.6
Sb	51	120.9038	122.9041								
		57.3	42.7								

Te	52	119.9045	121.9030	122.9042	123.9028	124.9044	125.9032	127.9047	129.9067	
		1.0	2.45	0.9	4.6	7.0	18.7	31.7	34.5	
I	53	126.9044								
		100								
Xe	54	123.9061	125.9042	127.9035	128.9048	129.9035	130.9051	131.9042	133.9054	135.9072
		1.0	0.1	1.9	26.4	4.1	21.2	26.9	10.4	8.9
Cs	55	132.9051								
		100								
Ba	56	129.9062	131.9057	133.9043	134.9056	135.9044	136.9058	137.9050		
		0.1	1.0	2.4	6.6	7.9	11.2	71.7		
La	57	137.9068	138.9061							
		0.089	99.911							
Ce	58	135.907	137.9057	139.9053	141.9090					
		0.2	0.3	88.4	11.1					
Pr	59	140.9074								
		100								
Nd	60	141.9075	142.9096	143.9099	144.9122	145.9127	147.9165	149.9207		
		27.2	12.2	23.8	8.3	17.2	5.7	5.6		
Pm	61	No natural stable isotope: 144.913 (17.7 y); 146.915 (2.5 y)								
Sm	62	143.9117	146.9146	147.9146	148.9169	149.9070	151.9195	153.9220		
		3.1	15.1	11.3	13.9	7.4	26.6	22.6		
Eu	63	150.9196	152.9209							
		47.8	52.2							
Gd	64	151.9195	153.9207	154.9226	155.9221	156.9239	157.9241	159.9271		
		0.2	2.2	14.8	20.5	15.7	24.8	21.8		
Tb	65	158.9250								
		100								

Table A2.2 Continued

Element	Z	Isotopic mass (amu)/relative abundance (%)						
Dy	66	155.9238	157.9240	159.9248	160.9266	161.9265	162.9284	163.9288
		0.06	0.10	2.34	18.9	25.5	24.9	28.2
Ho	67	164.9303						
		100						
Er	68	161.9288	163.9293	165.9304	166.9320	167.9324	169.9355	
		0.1	1.6	33.4	22.9	27.0	15.0	
Tm	69	168.9344						
		100						
Yb	70	167.9339	169.9349	170.9365	171.9366	172.9383	173.9390	175.9427
		0.1	3.1	14.3	21.9	16.2	31.7	12.7
Lu	71	174.9409	—					
		97.4	2.6					
Hf	72	173.9403	175.9417	176.9435	177.9439	178.9460	179.9468	
		0.2	5.2	18.5	27.1	13.8	35.2	
Ta	73	179.9415	180.9480					
		0.012	99.988					
W	74	179.9470	181.9483	182.9503	183.9510	185.9543		
		0.1	26.3	14.3	30.7	28.6		
Re	75	184.9530	186.9560					
		37.4	62.6					
Os	76	183.9526	185.9539	186.9560	187.9560	188.9586	189.9586	191.9612
		0.02	1.58	1.6	13.3	16.1	26.4	41.0
Ir	77	190.9609	192.9633					
		37.3	62.7					
Pt	78	189.9600	191.9614	193.9628	194.9648	195.9650	197.9675	
		0.01	0.78	32.9	33.8	25.3	7.2	

Au	79	196.9666						
		100						
Hg	80	195.9658	197.9668	198.9683	199.9683	200.9703	201.9706	203.9735
		0.2	10.1	16.9	23.1	13.2	29.7	6.8
Tl	81	202.9723	204.9745					
		29.5	70.5					
Pb	82	203.973	205.9745	206.9759	207.9766			
		1.4	24.1	22.1	52.4			
Bi	83	208.9804						
		100						
Po	84	No natural stable isotope: 208.9825 (103 y)						
At	85	No natural stable isotope: 210.9875 (7.2 h)						
Rn	86	No natural stable isotope: 222.0175 (3.823 d)						
Fr	87	No natural stable isotope: 223.0198 (22 m)						
Ra	88	No natural stable isotope: 226.0254 (1620 y)						
Ac	89	No natural stable isotope: 227.0278 (21.2 y)						
Th	90	230.0311 (8×10^4 y); 232.0382 (1.41×10^{10} y)						
Pa	91	No natural stable isotope: 231.0359 (3.5×10^4 y)						
U	92	235.0439	238.0508					
		0.72	99.27					
Np	93	No natural stable isotope: 237.0480 (2.14×10^6 y)						
Pu	94	No natural stable isotope: 239.0522 (2.436×10^4 y)						
Am	95	No natural stable isotope: 241.0567 (458 y)						
Cm	96	No natural stable isotope: 245.065 (9.32×10^3 y)						
Bk	97	No natural stable isotope: 247.0702 (1.4×10^3 y)						
Cf	98	No natural stable isotope: 249.0748 (360 y)						

[a]Some masses are quoted for elements which have no stable naturally occurring isotopes. The relative abundances will be dependent on rate of production and/or the particulars of the decay chain.

Source: Adapted from *Handbook of Chemistry and Physics*, 64th ed. (R. C. Weast, ed.), CRC Press, Boca Raton, FL, 1983, pp.B-233–B-316.

Appendix 3
Less Commonly Used Techniques for Analysis of Surfaces and Interfaces

GAR B. HOFLUND University of Florida Gainesville, Florida

J. C. RIVIÈRE AEA Technology, Oxon, UK

1. ULTRAVIOLET PHOTOEMISSION SPECTROSCOPY (UPS)

UPS is a complex technique capable of being used in different ways to obtain different types of information. In its conventional form, UPS is rather similar to valence-band XPS (VBXPS) in that it provides information about the electronic structure near the Fermi level. However, there are several important differences between UPS and VBXPS. First, ultraviolet (UV) photons have very low energies compared to X-rays. Since the UV photons have energies not very different from those of the valence-band electrons, the cross sections for photoemission are much larger than for X-rays, resulting in much greater signal intensity in UPS. Second, X-rays promote VB electrons to empty states very far above the Fermi level, compared to the effect of UV photons. The density-of-states (DOS) just above the Fermi level has discrete energy levels for which the spacing may be chemically significant. Also, selection rules may play a role in forbidding excitation of electrons in a particular filled orbital to some unfilled orbitals. Thus, a UPS spectrum provides a joint density of states because the latter is determined by the electronic structures of both filled and unfilled levels near the Fermi level. This is not so in VBXPS because the valence electrons are promoted to states so far above the Fermi level that the spacings are very close. The DOS recorded in VBXPS therefore reflects the actual DOS of the filled valence-band region. Third, the low energies of the UV photons mean that the electronic structure cannot be probed very far below the Fermi level.

In most surface science laboratories, UV photons are produced using inert-gas plasma-discharge lamps. The discharge lines available are determined by the transition energies of the particular inert gas used in the lamp and lie between 8 and 45 eV. Due to the $\Delta\phi$ term in Eq. (1), Chapter 4, UPS can record the electronic structure as far as the energy of the photon, minus $\Delta\phi$, below the Fermi level. Although the pressure in the discharge region of a UV lamp is high (10 to 250 Pa) and there are no window materials available with high optical transmissions in the above energy range, the background pressure in the analysis chamber is maintained in the UHV region by a differential pumping system. In any case the inert gases, such as He, used in the lamp do not present a serious contamination problem if they are prepurified [1].

An early and elegant example of the type of information obtainable from UPS is given in Fig. 1 [2]. In (a) are shown the spectra from clean Ni(111) (full curve) and from the same surface after exposure to 2.4 L of benzene at room temperature (broken curve). The difference spectrum (i.e., "clean" subtracted from "adsorbed") is shown in (b). Another difference spectrum, this time for benzene adsorbed at 150 K, is in (c). For comparison purposes the benzene gas-phase UPS spectrum is given in (d), where the energy scale has been shifted to align the peaks with those in (c).

During adsorption at low temperature, benzene molecules interact only weakly (van der Waals adsorption) with Ni(111) and as a result the UPS difference spectrum (c) at 150 K looks very similar to the gas-phase spectrum. The species on the surface is basically unaltered condensed benzene. Comparison of difference spectrum (c) with that in (b), for room-temperature adsorption, reveals both that the benzene π orbitals have shifted with respect to the other orbitals, and that most of the nickel *d*-electron density has been lost. Clearly there has been a dissociative and strong interaction of the benzene with the Ni(111) surface since the electronic structure of the benzene molecule has been disrupted and the Ni *d*-electrons have been involved in the bonding. This type of change in electronic structure is indicative of chemisorption.

UPS can be performed in a more versatile manner by using the radiation from a synchrotron storage ring [3]. The technique is usually referred to as synchrotron radiation photoelectron spectroscopy (SRPS). When a synchrotron is operated as a storage ring, UV light is generated by the electrons as they are accelerated around the ring. The light is plane-polarized, and the UV photon energy can be varied continuously by changing monochromator settings. These

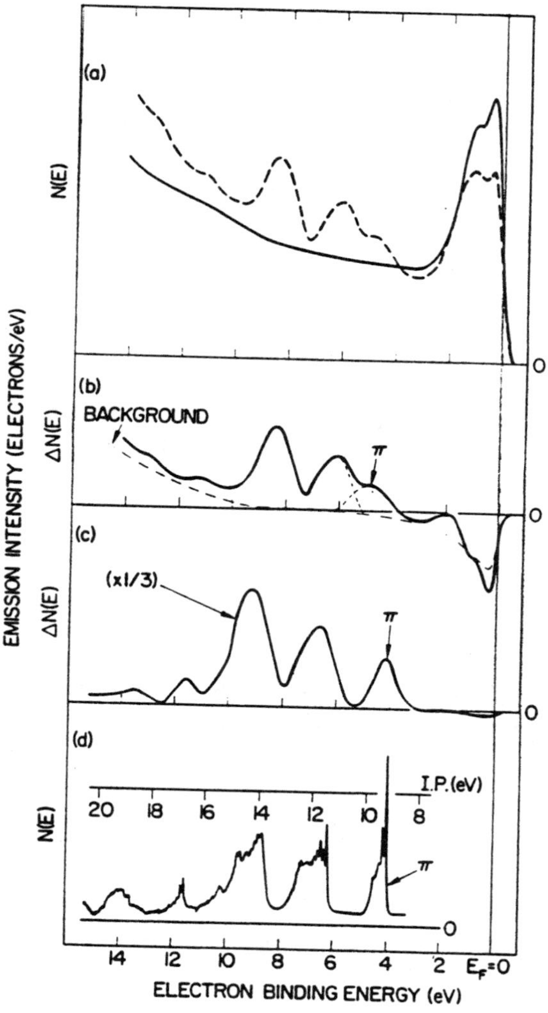

Figure 1. (a) Photoemission spectra *N(E)* for clean Ni(111), $\Phi = 5.4$ eV (full curve), and after exposure to 2.4 L of benzene at room temperature (broken curve). The energy of the unpolarized photons was 21.2 eV, and electrons were collected by the energy analyzer over a large solid angle. (b) Adsorbate-induced difference in emission $\Delta N(E)$. i.e., the difference between the broken and the full curves in (a). $\Delta\Phi = -1.4$ eV. (c) $\Delta N(E)$ for a condensed benzene layer formed at $T \approx 150$ K at a benzene pressure of 4.6×10^{-5} Pa. $\Delta\Phi = -1.6$ eV. (d) Gas-phase photoelectron spectra of benzene. The energy scale in (d) has been shifted so that the assigned levels line up with those in (c). Note that the π-level is shifted between (b) and (c). (From Ref. 2.)

features allow different types of UPS experiments to be performed which would not be possible using a single-line UPS lamp. For example, the initial DOS distribution can be obtained by scanning the photon energy and analyzer parameters so that the electrons are excited into just one final state. A similar experiment can be performed to map the final density of states using photoemission from only one initial state. In both cases selection rules can influence the spectra.

Another application of UPS is mapping of the band structure of single crystals using angular resolution (ARUPS) [4]. The method consists of recording a set of spectra for a range of polar angles at equal intervals, at a particular azimuthal angle, and then repeating the process for each of several other azimuthal angles. The polar angle is that made by the axis of the electron detector with the surface normal, and the azimuthal angle is that between the projection of the detector axis on the crystal surface and the vertical plane containing the polar angle. In a strict sense the method is valid only for free-electron-like metals, but it has had surprisingly general success. Features in the UPS spectra are seen to disperse with energy as the two angles are varied, and from the dispersions, when plotted in **k**-space, bands can be constructed along various crystal symmetry directions.

A more thorough method of mapping band structure, and one that has been used with semiconductors as well as metals, is to combine angular resolution with energy resolution. For the latter a source of variable photon energy, such as a synchrotron, is required. Since many spectra need to be acquired, taking a long time, it is vital in this type of work that the base vacuum conditions be very good (i.e., pressures $\leq 10^{-9}$ Pa), otherwise the surface condition of the crystal may change during the experiment.

Yet another variation on the angle-resolved theme is the determination of the orientation of adsorbates on single-crystal surfaces by recording the intensities of photoemission from various adsorbate molecular orbitals as a function of the angle of photon incidence at a fixed collection angle. Again, by accumulating many such spectra at a series of different collection angles, a very accurate picture of adsorbate arrangement can be obtained [3].

REFERENCES

1. G. B. Hoflund and R. P. Merrill, *J. Vac. Sci. Technol. A 1*, 1560 (1983).
2. J. E. Demuth and D. E. Eastman, *Phys. Rev. Lett. 32*, 1123 (1974).
3. B. Feuerbacher. B. Fitton and R. F. Willis (eds.), *Photoemission and the Electronic Properties of Surfaces*, Wiley, Chichester, 1978.
4. E. Jensen, R. A. Bartynski, T. Gustafson, E. W. Plummer, M. Y. Chou, M. L. Cohen and G. B. Hoflund, *Phys. Rev. B 30*, 5500 (1984).

2. ELECTRON ENERGY LOSS SPECTROSCOPY (ELS)

In a secondary electron distribution (e.g., in Fig. 23, Chapter 4), electron energy loss features will be found in the energy range some 50 to 100 eV below the elastic peak. They arise from inter- and intraband transitions as well as from bulk and surface plasmon losses. Their interpretation is often difficult so that in normal AES analysis they are usually neglected. Nevertheless, all the instrumentation required to collect AES data can be used quite conveniently to collect ELS data as well. It is necessary only to scan the KE region from the base of the elastic peak on the low-KE side to some chosen energy 50 or 100 eV lower. Care must be taken not to scan over any part of the elastic peak because its intensity is so relatively high that the electron multiplier in the electron analyzer will be damaged.

The advantage of ELS is that it can provide chemical-state information in many of those cases in which XPS cannot, e.g., when there are spectral features too closely spaced to resolve. Depth-sensitive ELS chemical-state information can also be obtained by varying the primary electron beam energy and the angles of incidence and collection [1]. These useful attributes can be illustrated by ELS studies of Sn metal [1], of Sn oxides [2] and of the oxidation of Sn metal [3]. Sn is an interesting case because the BEs of the Sn 3*d* lines for SnO and SnO_2 lie only 0.19 eV apart as described in Chapter 4 and shown in Fig. 4 of that chapter. ELS can distinguish the different Sn oxides because it samples both the filled electronic levels in the valence band and the empty levels just above the Fermi level; the ELS spectrum therefore reflects the joint density of states. ELS is thus highly sensitive to chemical state because the valence levels are involved directly in the chemical bonding, whereas the core levels are influenced only by the nature of the chemical bonding.

The predominant features in the ELS spectrum obtained from Sn metal and shown in Fig. 2 [1] consist of peaks due to single and multiple surface and bulk plasmon losses. A bulk plasmon corresponds to the collective oscillation of free electrons throughout the three-dimensional volume of a metal, and a surface plasmon corresponds to oscillations restricted to the two-dimensional near-surface region. A surface plasmon has an energy loss theoretically $2^{-1/2}$ times that of the bulk plasmon energy loss. For Sn metal the energy loss due to the bulk plasmon lies at 14 eV and that of the surface plasmon at 10 eV. Multiple loss features, such as a bulk-bulk loss at 28 eV, a bulk-surface loss at 24 eV, and a surface-surface loss at 20 eV, are also apparent in Fig. 2. The spectra in (a) and (b) were recorded using a primary beam energy of 600 eV but with different incident and collection angles. Spectrum (b) can be seen to be more surface specific than spectrum (a) because the surface plasmon is more prominent than the bulk plasmon, and the bulk-surface plasmon than the bulk-bulk plasmon. The effect of progressive reduction of the primary energy is shown in the se-

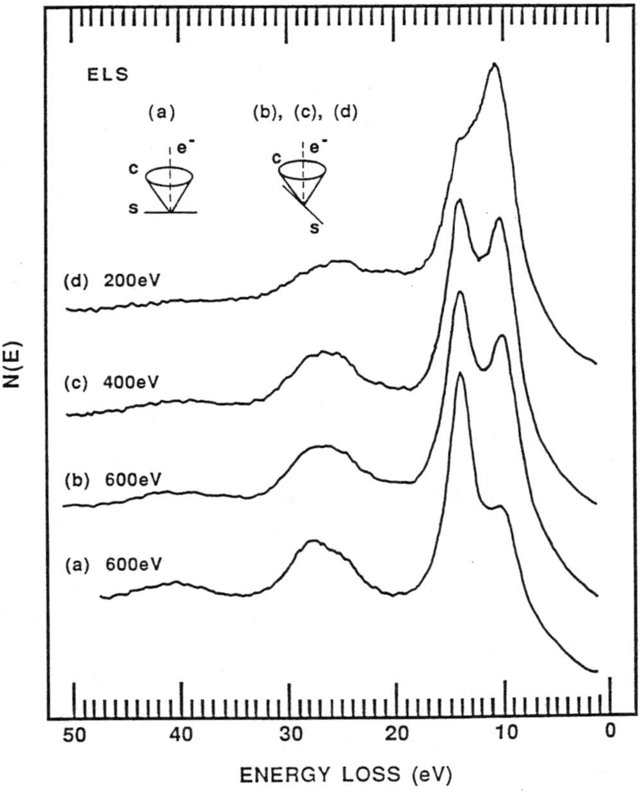

Figure 2. ELS spectra recorded from clean polycrystalline tin using primary beam energies of 200, 400 and 600 eV. The integral CMA electron gun was used, and all electrons emitted into the 360° conical section were collected. For (a) the sample was oriented so that the normal to its surface was along the CMA axis and for (b), (c) and (d) the sample was tilted at an angle of 45° to the CMA axis. The experimental configurations are indicated in the inserts. S is the sample, e is the primary electron beam from the CMA gun and C represents the solid angle of the collected electrons. (From Ref. 1.)

quence (b), (c) and (d). As the primary energy is reduced progressively from 600 to 200 eV, the surface plasmon features become more prominent with respect to the corresponding bulk plasmon features, indicating increased surface specificity.

The interpretation of ELS spectra from oxides is considerably more complex. In oxides the electrons are tightly bound, and there are no free electrons in metallic states. The ELS features therefore consist of losses due to the excitation of electrons from filled to empty levels. In order to understand the nature of oxide ELS features, comparison must be made with spectra from standards, as described by Powell [4], and the electronic structure of the oxides must be analyzed [2]. ELS spectra from a polycrystalline tin oxide surface which had been annealed at 750°C in vacuum are shown in Fig. 3 as a function of primary beam energy. The ELS features from an SnO standard lie at 9, 13.5 and 27 eV, while the predominant ELS feature from SnO_2 lies at 20 eV [2–4]. In addition to these oxides, another Sn oxide has been identified and referred to as a transitional oxide because

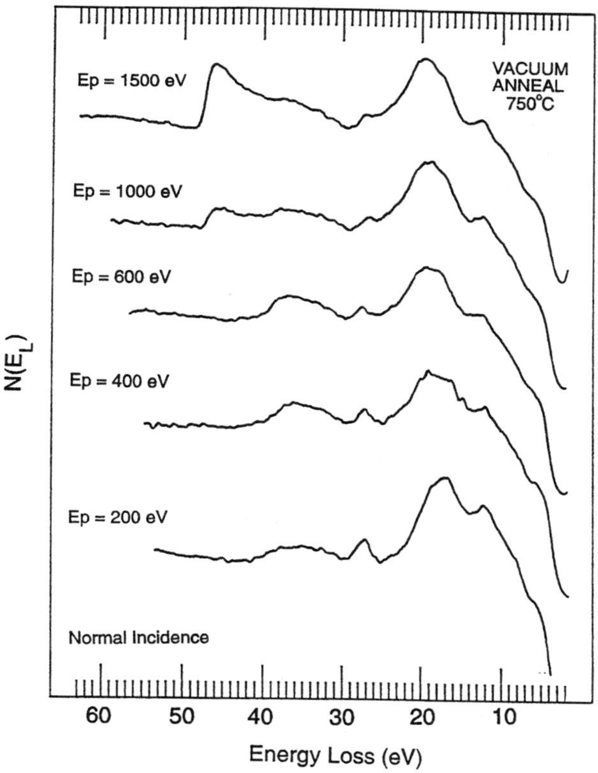

Figure 3. Variation with primary energy E_p of the *N(E)* ELS spectrum from an annealed tin oxide surface. Because of the relationship between KE and inelastic mean free path, the set of spectra corresponds to a depth profile of the annealed material. The growth of the 27-eV feature is due to increasing oxygen deficiency as the spectra become more surface specific. (From Ref. 2.)

it has an O-to-Sn ratio between those of SnO and SnO_2 and occurs when SnO_2 is reduced or SnO is oxidized. It has an ELS feature at 17 eV. As the primary-beam energy is increased from 200 to 1500 eV, the sampled depth increases and significant consequential changes are observed in the ELS spectra. The near-surface region (i.e., at 200 eV primary energy) consists mostly of SnO and the transitional oxide. With increasing primary energy and hence sampled depth the SnO features reduce in magnitude and a contribution from SnO_2 at 20 eV becomes apparent. The spectrum recorded at 1500 eV primary energy is due mostly to SnO_2. The depth-dependent ELS spectra thus indicate that the oxide surface is layered with SnO at the outer surface and the transitional oxide beneath the SnO but above the bulk SnO_2.

REFERENCES

1. G. B. Hoflund and G. R. Corallo, *Surf. Interface Anal. 13*, 33 (1988).
2. D. F. Cox and G. B. Hoflund, *Surf. Sci. 151*, 202 (1985).
3. G. B. Hoflund and G. R. Corallo, *Phys. Rev. B 46*, 7110 (1992).
4. R. A. Powell, *Appl. Surf. Sci. 2*, 397 (1979).

3. ELECTRON-STIMULATED DESORPTION (ESD)

There are numerous ways in which an electron beam striking a solid surface can alter the composition of the near-surface region, including thermal effects, electromigration, electron-stimulated desorption, deposition of contaminants from the filament, dissociation and deposition of background gas species and others. The severity or otherwise of each of these effects depends on the nature of the solid surface being examined and on the mode of operation of the particular electron beam technique. Generally, ESD is the most destructive process, and must always be taken into account when, for example, collecting AES data. ESD is a process which results in the emission of positive and negative ions, neutrals and metastables, when an electron beam strikes a solid even at primary beam energies as low as 10 eV. That the process can alter the composition of a surface drastically during Auger analysis is illustrated in Fig. 4 [1]. The Auger spectrum in (a) was recorded from a contaminated polycrystalline tin oxide surface, and that in (b) from the same surface after an exposure of 1 h to the primary electron beam (10 μA, 3 keV, 0.1 mm diameter spot). The large electron dose has caused the desorption of sulfur, chlorine, carbon, oxygen and other elements with consequential alteration of the surface composition. An ESD positive-ion spectrum obtained by mass-analyzing the emitted positive ions is shown in Fig. 5 [2].

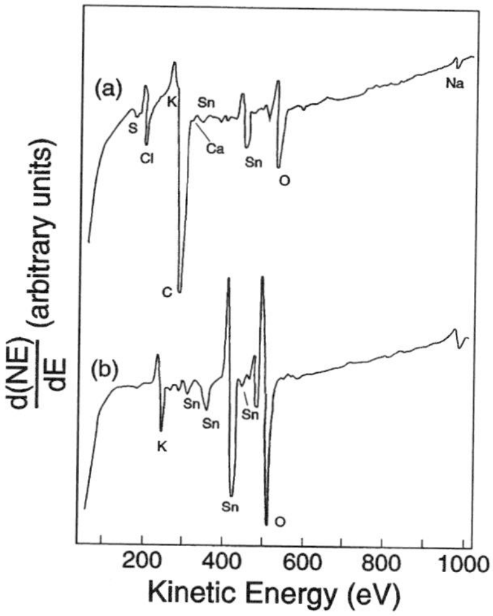

Figure 4. AES spectra recorded from a contaminated polycrystalline tin oxide surface (a) before and (b) after a large electron dose. The change in composition is due to ESD. (From Ref. 1.)

Clearly, the ESD effect must be minimized in analytical techniques using incident electron beams (e.g., AES), otherwise information will be lost. ESD is a complex process and can include various mechanisms, as described by Hoflund [3]. Many variables are involved in ESD. For a given surface the desorbing particle current, i_d, can be expressed as

$$i_d = i_d(E_p, Q_p, \Psi_p, \phi_p, E_d, \Psi_d, \phi_d, \theta, T, t) \tag{1}$$

where E_p is the primary electron beam energy, Q_p is the primary beam flux, ψ_p and ϕ_p are the polar and azimuthal angles of incidence of the primary beam. E_d is the kinetic energy of the desorbing particle, ψ_d and ϕ_d describe the direction of desorption. θ is the surface coverage, T is the surface temperature and t is the exposure time to the electron beam. At a constant primary beam energy, the ESD ion current is a linear function of the primary beam current as long as other processes such as desorption due to heating, surface diffusion, etc., are unimportant. This implies that each desorption process is an isolated event independent of other desorption processes. Consider a case in which an adsorbate exists in just one binding state on a given surface and that it desorbs by just one mechanism. Under these assumptions the total particle yield, I, can be written as the sum of the yields (particles/time) of neutrals and metastables, i^o, positive ions, i^+, and negative ions, i^-.

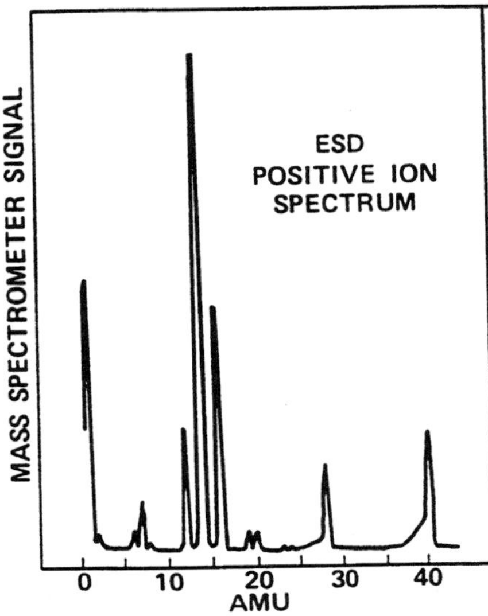

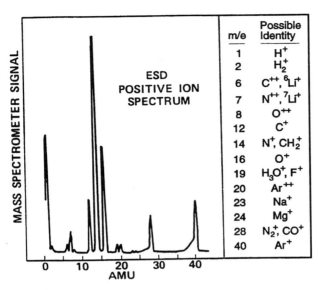

m/e	Possible Identity
1	H^+
2	H_2^+
6	C^{++}, $^6Li^+$
7	N^{++}, $^7Li^+$
8	O^{++}
12	C^+
14	N^+, CH_2^+
16	O^+
19	H_3O^+, F^+
20	Ar^{++}
23	Na^+
24	Mg^+
28	N_2^+, CO^+
40	Ar^+

Figure 5. ESD positive-ion spectrum recorded from a contaminated tin oxide surface using a 1000 eV electron beam. Unambiguous assignment of the peaks is not possible. For example the peak at *m/e* = 28 could be CO^+ or N_2^+ or a mixture, and that at *m/e* = 19 could be F^+ or H_3O^+. (From Ref. 2.)

$$I = i^{o} + i^{+} + i^{-} \tag{2}$$

A total cross section for ESD, Q_{total}, can be defined by the equation

$$\frac{I}{A\varepsilon} = \frac{I_p}{A\varepsilon} Q_{\text{total}}(N) \tag{3}$$

where I_p is the total electron beam current to the sample, A is the area of the sample irradiated, ε is the charge on an electron and (N) is the surface concentration of chemisorbed species. A non-steady-state material balance for the chemisorbed species yields

$$-\frac{d(N)}{dt} = \frac{I_p}{A\varepsilon} Q_{\text{total}}(N) \tag{4}$$

which integrates to

$$\frac{(N)}{(N)_0} = \exp\left(\frac{-I_p}{A\varepsilon} Q_{\text{total}} t\right) \tag{5}$$

where $(N)_0$ is the initial surface concentration of chemisorbed species. If (N) can be measured as a function of time using AES, ISS, or some other technique, then Q_{total} can be determined provided that I_p and A are known. Rather than measuring (N) as a function of time, Q_{total} may be determined by monitoring i^+ as a function of time since

$$\left(\frac{i^+}{i_0^+}\right) = \frac{(N)}{(N)_0} = \exp\left(\frac{-I_p}{A\varepsilon} Q_{total} t\right) \tag{6}$$

This discussion suggests several ways of minimizing the ESD effect in techniques such as AES. The first is to reduce the primary beam current and the total scanning time as much as possible while maintaining a reasonable signal-to-noise ratio. The second is to increase the irradiated area by defocusing the electron gun. Of course the latter is not possible in SAM, and ESD can be a very serious problem in that technique. The third, when focused electron beams are used, is to raster over a larger area on the sample in order to minimize the damage in any individual small area. The fourth, in connection particularly with AES, is to record spectra of the most ESD sensitive elements first by multiplexing, so that unwanted regions of the Auger spectrum are not scanned. The fifth is to use Eq. (6) to calculate $(N)_0$, if the total ESD cross section is known, by collecting (N) versus t data. The ESD problem has not been given much consideration in the design and operation of most commercially available electron beam techniques, so that each operator must assess how the problem affects the quality of the data for each type of sample analyzed. One of the most convenient ways of testing for ESD damage in AES is to record sequentially Auger spectra at regular intervals. If there are changes in the relative peak heights, then damage is occurring and steps must be taken to minimize it.

However, precisely because of the desorption effect, ESD is a powerful surface science technique in its own right and can provide important information complementary to other surface science techniques. For example, AES and ESD data could be collected simultaneously from the same surface by using the primary electron beam as the excitation source for both techniques. As is clear from Eq. (1), there are many parameters which can be varied in ESD to provide different types of information. One variation is simply to collect ESD ion mass spectra as shown in Fig. 5. ESD is highly sensitive to many elements and can often detect them with sensitivities greater than those of the electron spectroscopies. Most importantly, ESD is able to detect surface hydrogen which can be observed in AES and XPS only by its influence on peak shapes when present in large quantities. Furthermore, ESD is more surface specific than AES or XPS because the desorbing ions originate only within the outermost atomic layer. Ions created beneath the surface are neutralized before escaping or trapped within the solid. The ESD mass spectrum can be recorded with a quadrupole mass spectrometer with an energy prefilter similar to that used in

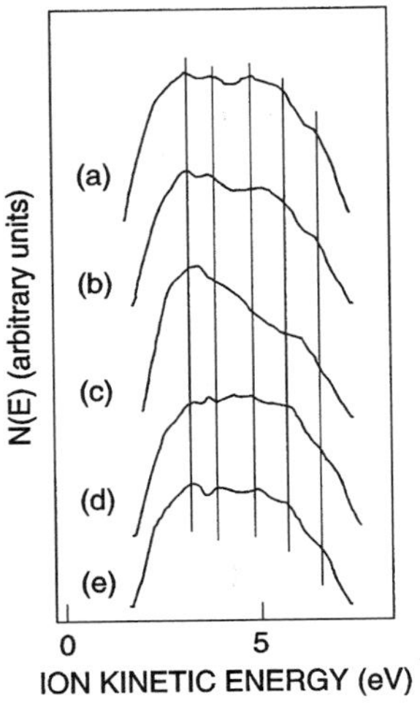

Figure 6. The upper portions of the H^+ energy distribution spectra recorded from a Si(100) surface. (a) the clean surface, (b) after exposure to 240 eV electrons for 30 min. (c) exposure as in (b) but with an additional exposure to 400 eV electrons for 25 min, (d) the beam-exposed surface in (c) after annealing at 400°C for 15 min. and (e) the beam-exposed surface in (c) after annealing at 950°C for 10 min. (From Ref. 6.)

static secondary-ion mass spectrometry (SSIMS) [4] or with a CMA operated as a ToF mass spectrometer [5].

Another variation is that of energy analysis of the desorbing ions, which allows various chemical states or bonding sites of a given species to be distinguished. This variation is referred to as electron-stimulated desorption/ion-energy distribution (ESDIED), and an example of desorption of H^+ from a clean Si(100) surface studied by ESDIED is shown in Fig. 6 [6]. AES, ISS and XPS indicated that this surface was clean; i.e., the adsorbed H was not detected. An important question frequently addressed in the semiconductor literature is that of the location of the bonding site of H. The data in Fig. 6 indicate that the H atoms bond at five different sites on Si(100). However, as in the technique of temperature-programmed desorption (TPD), it is not possible to specify just what the different bonding sites are. These data also indicate that the bulk material can be a source of surface hydrogen. With exposure to the electron beam, the surface hydrogen is depleted. During annealing hydrogen migrates from the bulk Si to the surface and repopulates the depleted states.

ESD can be performed in two other modes. One is a threshold-type experiment in which the primary beam energy is ramped and the yield of a given species monitored. Thresholds in the yields appear at various core-level and valence-level energies, providing information about bonding structure [7]. The second is electron-stimulated desorption/ion-angular distribution (ESDIAD). In this mode the angular distributions of the desorbing ions are measured to provide bonding site information on single-crystal surfaces [8–11].

REFERENCES

1. G. B. Hoflund, G. L. Woodson, D. F. Cox and H. A. Laitinen, *Thin Solid Films 78*, 357 (1981).
2. G. B. Hoflund, D. F. Cox, F. Ohuchi, P. H. Holloway and H. A. Laitinen, *Appl. Surf. Sci. 14*, 281 (1982–83).
3. G. B. Hoflund, *Scanning Electron Micros. IV*, 1391 (1985).
4. R. L. Ingelbert and J. F. Hennequin, in *Secondary Ion Mass Spectrometry SIMS III*, (A. Benninghoven, J. Giber, J. Laszlo, M. Riedel and H. W. Werner, eds.), Springer-Verlag, Berlin, 1982, p. 57.
5. M. M. Traum and D. P. Woodruff, *J. Vac. Sci. Technol. 17*, 1202 (1980).
6. C. F. Corallo and G. B. Hoflund, *Surf. Interface Anal. 12*, 297 (1988).
7. M. L. Knotek, *Surf. Sci. 91*, L17 (1980).
8. T. E. Madey, F. P. Netzer, J. E. Houston, D. M. Hanson and R. Stockbauer, in *Desorption Induced by Electronic Transitions DIET I*, (N. H. Tolk, M. M. Traum, J. C. Tully and T. E. Madey, eds.), Springer-Verlag, Berlin, 1983, p. 120.
9. T. E. Madey, J. T. Yates, Jr., A. M. Bradshaw and F. M. Hoffmann, *Surf. Sci. 89*, 370 (1979).
10. T. E. Madey, in *Inelastic Particle-Surface Collisions*, (E. Taglauer and W. Heiland, eds.), Springer-Verlag, Berlin, 1981, p. 80.
11. J. T. Yates, Jr., in *Surface Imaging and Visualization* (A. T. Hubbard ed.), CRC Press, Boca Raton, FL, 1995, p. 157.

4. VIBRATIONAL SPECTROSCOPIES

4.1. Infrared Techniques

Techniques using infrared radiation as the incident probe have been used widely in transmission for the analysis of solids, liquids and gases, and in reflectance for the analysis of liquids (both static and dynamic), pastes, powders and some solids. Some of the techniques have been adapted for use in an UHV environment to provide valuable information about surfaces, complementary to that from other techniques.

4.1.1. Attenuated Total Reflectance (ATR)

When infrared radiation passes through a prism made of certain infrared transmitting materials of high refractive index, it is totally internally reflected. This total internal reflection creates an evanescent (standing) wave at the crystal surface. If now another material (the sample) is brought into intimate contact with the totally reflecting surface of the crystal, the evanescent wave extends into the sample, and its intensity decays exponentially with distance from the surface of the sample. The reflected radiation is then attenuated in intensity at those wavelengths at which absorption occurs in the sample, so the absorption spectrum characteristic of the sample can be recorded as a function of wavelength. By reference to standard spectra analysis is therefore possible.

The most commonly used prism material is ZnSe, which has a refractive index of 2.4. According to the strength of absorption in the sample, prisms can be fabricated to allow the number of reflections to range from 1 to ca. 20. The angle of incidence of the infrared radiation can be varied within limits, which is very useful since such variation changes the depth of penetration of the evanescent wave into the sample, and hence the sampling depth of the technique. As an example, if the depth of penetration is taken as the distance over which the intensity of the evanescent wave decays to $1/e$ of that at the crystal surface, then for ZnSe that depth would be 2 μm at an incident angle of 45° and 1.1 μm at 60°. If the crystal material were Ge (with $n = 4.0$), then the depth would reduce to 0.5 μm at 60°. It is very important that the sample be in genuinely intimate contact with the crystal surface, which in the context of application in surface and interface analysis implies that the technique would be most useful for the analysis of thin films deposited in vacuo directly onto a substrate.

4.1.2. Reflection Absorption Infrared Spectroscopy (RAIRS)

In conventional infrared spectroscopy absorption spectra are measured in transmission, i.e., before and after passage of infrared radiation through the sample. Such a configuration is unsuited to studies of surfaces since the sample must be both infrared transparent and very thin, and in any case the contribution to the total absorption arising from changes at the surface would be too small to be measurable.

Thus infrared spectroscopy must be used in the reflectance mode in order to be surface sensitive. If infrared radiation is reflected specularly from an extremely good reflector on which there is an adsorbed layer, then the reflected light will show intensity losses at those frequencies that match vibrational frequencies at the surface. Such frequencies might be characteristic of the molecular structure within the adsorbed layer, modified or otherwise as a result of attachment to the surface, or of the molecule-surface system itself. Should the molecular structure not be distorted significantly by the surface (i.e., the original eigenmodes are intact), as in physisorption, then the reflected light spectrum will resemble that of the free molecule, which can then be identified. More interesting from the surface scientific point of view, however, are the effects on the molecule of interaction with the substrate, as in chemisorption, which will result in a reflected radiation spectrum that is more difficult to interpret but has a richer information content. The information from the technique might therefore be compositional but is more likely to be chemical.

In order that RAIRS can be used at all, there are several requirements. First, as stated above, the surface must be an extremely good reflector, which limits the application to those metals whose surfaces can be polished to a mirror finish, such as Cu, Ag, Au, Pt, Al and Ni. Second, the absorption cross-section of the adsorbed layer at the characteristic vibrational frequencies must be large enough that the resultant signal-to-noise ratio allows sufficiently good energy resolution to be used. Third, the theory of external reflectance IRS shows that plane-polarized radiation should be used and that the angle of incidence should be close to grazing. The optimum theoretical angle of incidence is 88°, and the angles used in practice are close to that, typically 85–86°.

Most of the RAIRS applications have been concerned with basic work on model catalyst systems, since many of the metals listed above as being suitable reflecting substrates are also by a fortuitous coincidence catalytically active. Also, many of the molecules of interest in catalytic processes possess large absorption cross-sections in the infrared. As a result much valuable insight has been gained into the mechanisms of surface interactions, particularly in catalysis but also in other areas.

Although RAIRS provides a great deal of chemical information about the effects of attachment of molecules to surfaces, and vice versa, it cannot normally give direct quantitative information about the amount of material present. It is normal, therefore, to use it in conjunction with one or more complementary techniques such as XPS or ISS so that quantification is available.

As with all infrared techniques, there are great advantages to be gained in using Fourier transform methods, and nowadays RAIRS should strictly speaking be called FT-RAIRS. The obvious advantages are not only much improved signal-to-noise and faster data collection but also the recording of all spectral features simultaneously, called multiplexing.

4.2. Electron Impact Technique

4.2.1. High-Resolution Electron Energy Loss Spectroscopy (HREELS)

When an electron of very low energy (2–5 eV) approaches a solid surface, it can be scattered inelastically by either of two mechanisms. It can interact with the dipole field associated with a particular surface vibrational mode, or it can be scattered from an atomic core at the surface. The former is called "dipole" scattering, and the latter "impact" scattering. The surface vibrations might be those of atoms of the solid itself or of an adsorbed molecule, and likewise the scattering cores might be those of the substrate or the molecule. In both cases measurement of the energy distribution of the scattered electrons reveals peaks arising from those electrons that have interacted with the surface vibrations and lost characteristic amounts of energy as a result. As for ELS (see Section 2), HREELS is a loss spectroscopy.

In HREELS nearly all measurements have involved dipole scattering, because of the effects of the so-called dipole selection rules. One of the principal effects is that because only the vibrations that induce dipole moments normal to the surface will be excited, the current of dipole scattered electrons will be peaked strongly along the direction of specular reflection. Another consequence is that when the intensities of loss peaks are calculated from dipole scattering theory, as proportions of the primary electron current, and as a function of the primary energy, sharp maxima are found in the range 2–6 eV. By contrast, scattering as a result of impact shows very little dependence on either scattering direction or primary energy. For optimum signal-to-noise, and thus optimum energy resolution, HREELS measurements are therefore performed in specular reflection and at primary energies between 2 and 5 eV.

Because both primary and scattered electron energies are so low (losses up to about 1 eV are usually measured), extreme precautions have to be taken in HREELS to restrict the energy spread in the primary beam to the absolute minimum achievable. Thus the entire surroundings of the experimental apparatus have to be shielded very carefully indeed from all magnetic and electrostatic fields, and no material can be used anywhere near the specimen or in the electron optics that might acquire a surface charge. Since HREELS is a loss spectroscopy, the energy resolution available in the spectra is dependent absolutely on the spread in the primary beam, hence the necessity for taking such measures. The best figure for the spread (i.e., the resolution) currently available is 0.5 meV, although at that resolution the count rate is still low. However, perfectly adequate count rates at energy resolutions not much inferior to the latter figure are achievable, as demonstrated in Fig. 7, from Ibach et al. [1]. There the HREELS spectrum from a film of cyclohexane condensed on the W(110) surface at 110 K is shown at an energy resolution of 0.88 meV. For comparison purposes the energetic positions of the vibrational modes in the gas phase are entered at the top of the figure. It is clear that the condensed cyclohexane is in the same molecular form as in the gas phase, i.e., no chemisorptive reaction with the W has occurred at 110 K, and that the energy resolution in HREELS is now good enough to resolve very closely spaced vibrational components.

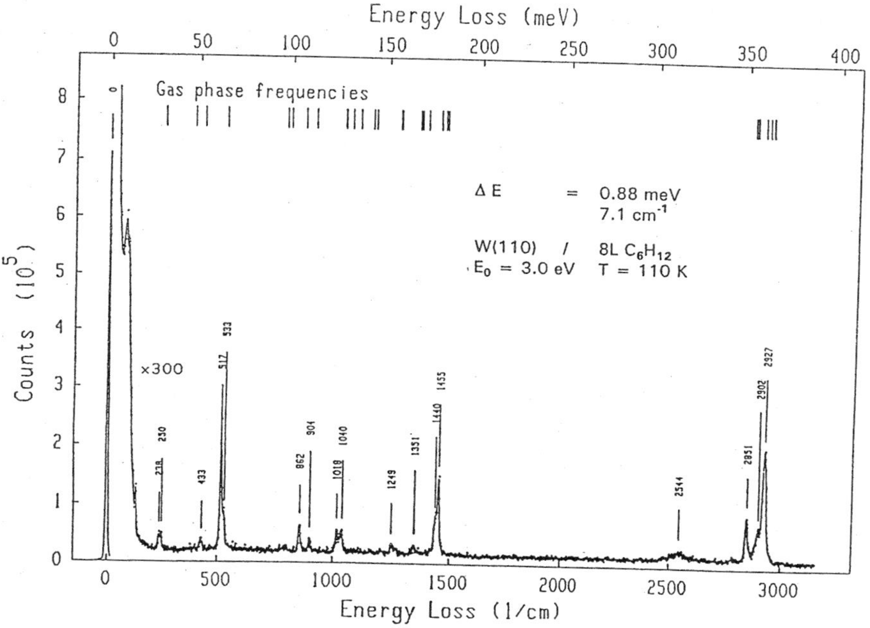

Figure 7. HREELS spectrum from a W(110) surface covered with a film of cyclohexane at 110 K. Primary electron energy is 3.0 eV, and energy resolution is 0.88 meV. For comparison the gas-phase vibrational frequencies are indicated at the top of the figure. (From Ref. 1.)

Since both HREELS and RAIRS are vibrational spectroscopies, and the same selection rules apply, their information contents must overlap. This is demonstrated in Fig. 8 [2], in which the HREELS and RAIRS spectra from a Cu(111) surface covered with about 10 molecular layers of cyclohexane at low temperature are shown. The vibrational spectra appear at the same energetic positions in both techniques, but it should be noted that whereas RAIRS has the advantage of better energy resolution HREELS is able to record spectra down to losses close to 0 cm^{-1}. For reasons of IR transmission of window materials, the cutoff in RAIRS is in the region 400–800 cm^{-1}.

The improvement in energy resolution in HREELS over the space of seven to eight years can be seen by comparing Figs. 7 and 8, since the spectrum in Fig. 8 was recorded at ΔE = 8.0 meV, but that in Fig. 7 at ΔE = 0.88 meV.

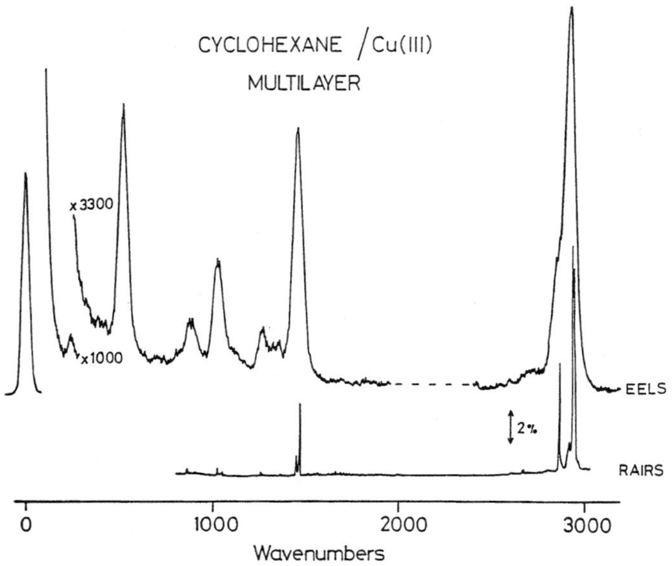

Figure 8. Comparison of the HREELS (upper) and RAIRS (lower) vibrational spectra from a multilayer of cyclohexane condensed on Cu(111) at low temperature. Energy resolution was 8.0 meV. RAIRS has superior energy resolution but suffers a frequency cutoff in the range 400–800 cm^{-1}. Note the difference in energy resolution between the HREELS spectra in this figure and in Fig. 7, demonstrating the progressive improvement in the HREELS capability with time. (From Ref. 2.)

REFERENCES

1. H. Ibach, M. Balden and S. Lehwald, *J. Chem. Soc. Faraday Trans.*, in press.
2. M. A. Chesters, *J. Electron Spectrosc. 38*, 123 (1986).

Appendix 4

Core-Level Binding Energies, Auger Kinetic Energies and Modified Auger Parameters for Some Chemical Elements in Various Compounds (all data are in electron-volts)[a]

Element	Compound	Photo-line	Auger-line	α*	Ref.
Carbon (*Z* = *6*)		C 1*s*			
	Carbide	ca. 281.8			Table 1 (Chap. 8)
	Aromatics	284.6			Table 2 (Chap. 5)
	Saturated hydrocarbon	285.0–285.3			Table 2 (Chap. 5)
	Amine	285.8–286.2			Table 2 (Chap. 5)
	Alcohol, phenol ester	286.1–286.5			Table 2 (Chap. 5)
	C–O in ester	286.7			Table 2 (Chap. 5)
	Nitrile	286.8			Table 3 (Chap. 5)
	Amide	287.5–288.0			Table 3 (Chap. 5)
	Carbonyl	287.6–287.9			Table 2 (Chap. 5)
	Urea	289.0			Table 3 (Chap. 5)
	C=O in ester	289.1			Table 2 (Chap. 5)
	COO	289.3			Table 2 (Chap. 5)
	Carbamate	289.5			Table 3 (Chap. 5)
	Carbonate	290.4–290.6			Table 2 (Chap. 5)
	π-π*	291.7			Table 2 (Chap. 5)

Element	Compound	Photo-line	Auger-line	α*	Ref.
Nitrogen ($Z = 7$)		N 1*s*			
	Nitride	397.3–398.5			Table 3 (Chap. 5)
	Nitrile	399.6			Table 3 (Chap. 5)
	Amine (primary)	399.3			Table 3 (Chap. 5)
	Amine (secondary)	400.2			Table 3 (Chap. 5)
	Amide	399.6–400.6			Table 3 (Chap. 5)
	Urea	400.6			Table 3 (Chap. 5)
	Carbamate	398.7			Table 3 (Chap. 5)
	Quat. salt	401.3			Table 3 (Chap. 5)
	NO	403.0			Table 3 (Chap. 5)
	NO_2^-	405.9			Table 3 (Chap. 5)
	NO_3^-	407.5			Table 3 (Chap. 5)
Oxygen ($Z = 8$)		O 1*s*	O $KL_{2,3}L_{2,3}$		
	TiO_2	529.8			Table 1 (Chap. 8)
	O–C	533.0			Table 1 (Chap. 8)
	SiO_2	533.3	506.5	1039.8	1
	SiOx[b]	533.1	506.7	1039.8	5
	SiO_2(Hydr)[c]	533.0	506.6	1039.6	7
	Al_2O_3	531.5	507.5	1039.0	1
	Cr_2O_3	530.3	511.8	1042.1	1
	FeO	530.5	512.2	1042.7	1
	Fe_2O_3	530.0	512.2	1042.2	1
	$FeAl_2O_4$	531.7	507.6	1039.3	1
	Fe_2SiO_4	532.5	509.0	1041.5	2
	$CoGa_2O_4$	531.3	511.3	1042.6	3
	$NiGa_2O_4$	531.2	509.8	1041.0	4
	$FeGa_2O_4$	531.2	510.1	1041.3	6
	Ga_2O_3	530.0	510.6	1040.6	6
	a-$Zr(HPO_4)_2{\cdot}H_2O$[d]	530.4	510.4	1040.8	8
	a-$Zr(HPO_4)_2{\cdot}H_2O$[d]	530.6	510.2	1040.8	8
Aluminum ($Z = 13$)		Al 2*p*	Al $KL_{2,3}L_{2,3}$		
	Al_2O_3	74.5	1388.2	1462.7	1
	$FeAl_2O_4$	74.5	1387.9	1462.4	1
Silicon ($Z = 14$)		Si 2*p*	Si $KL_{2,3}L_{2,3}$		
	SiO_2	103.8	1607.8	1711.6	9
	SiOx[b]	102.1	1610.7	1712.8	5
	SiO_2(Hydr)[c]	103.8	1607.6	1711.4	7
		Si 2*s*	Si $KL_{2,3}L_{2,3}$		
	SiO_2	154.8	1607.6	1762.4	5
	SiOx[b]	153.2	1610.7	1763.9	5
Titanium ($Z = 22$)		Ti 2*p*			
	Ti	453.9–454.3			Table 1 (Chap. 8)
	TiN	454.7			Table 1 (Chap. 8)
	TiN_xO_{1-x}	455.4			Table 1 (Chap. 8)
	TiN_xO_y	456.9–457.3			Table 1 (Chap. 8)
	TiN_{1+x}	456.9			Table 1 (Chap. 8)
	TiO	454.9–455.2			Table 1 (Chap. 8)
	Ti_2O_3	457.0–457.7			Table 1 (Chap. 8)
	TiO_2	457.9–459.3			Table 1 (Chap. 8)

Element	Compound	Photo-line	Auger-line	α*	Ref.
Chromium (*Z* = *24*)		Cr $2p_{3/2}$			
	α-Cr_2O_3	576.3		1104.8	Table 1 (Chap. 17)
	α-CrF_3	580.1		1105.5	Table 1 (Chap. 17)
	β-CrF_3	579.8		1105.4	Table 1 (Chap. 17)
Gallium (*Z* = *31*)		Ga $2p_{3/2}$	Ga $L_3M_{4,5}M_{4,5}$		
	$CoGa_2O_4$	1118.5	1063.1	2181.6	3
	$NiGa_2O_4$	1118.4	1062.7	2181.1	4
	$FeGa_2O_4$	1117.2	1062.7	2179.9	6
	Ga_2O_3	1117.1	1062.6	2179.7	6
Silver (*Z* = *47*)		Ag $3d_{5/2}$[10, 11]		α* [12]	
	Ag^0	368.3		726.1	
	Ag_2O	367.8		724.5	
	AgO	367.4		724.0	
Cadmium (*Z* = *48*)		Cd $3d_{5/2}$	Cd $M_4N_{4,5}N_{4,5}$		
	CdTe	405.2	382.2	787.4	13
	$Cd_{.5}Mn_{.5}Te$	404.7	381.4	786.1	13
Tellurium (*Z* = *52*)		Te $3d_{5/2}$	Te $M_4N_{4.5}N_{4.5}$		
	CdTe	572.7	490.7	1063.4	13
	MnTe	572.7	492.3	1065.0	13
	$Cd_{.5}Mn_{.5}Te$	572.7	491.5	1064.2	13
	$Cd_{.5}Mn_{.5}Te(ox)^e$	575.7	487.1	1062.8	13
Lead (*Z* = *82*)		Pb $4f_{7/2}$	Pb $N_6O_{4.5}O_{4.5}$		
	Pb	136.8	96.3	233.1	14
	$Pb_{.66}Sn_{.33}$	136.9	96.3	233.2	14

[a]Additional data can be found on the UK ESCA Users Group web site: http://www.surrey.ac.uk/MSE/ESCA/ESCA/home.html. Further information from Dr. K. R. Hallam, University of Bristol, Interface Analysis Centre, Oldbury House, 121, St. Michael's Hill, Bristol, BS2 8BS, UK, Fax: +44-117-925-5646, e-mailL k.r.hallam@bristol.ac.uk.
[b]Silicon suboxide species ($0 < x < 2$) formed after prolonged Ar^+ etching ($E_{Ar^+} \geq 6$ keV) of stoichiometric SiO_2 surfaces.
[c]Data obtained from aged silica showing sign of surface hydration.
[d]These data refer to the "anhydrous" component of oxygen.
[e]Oxidized Te component due to air exposure.

REFERENCES

1. E. Paparazzo, *J. Electron Spectrosc. Relat. Phenom. 43*, 97 (1987).
2. E. Paparazzo, *Vuoto 17*, 286 (1987).
3. F. Leccabue, C. Pelosi, E. Agostinelli, V. Fares, D. Fiorami and E. Paparazzo, *J. Cryst. Growth 79*, 410 (1986).
4. A. Pajaczowska, O. De Melo, F. Leccabue, C. Pelosi, D. Fiorani, A. M. Testa and E. Paparazzo, *J. Cryst. Growth 104*, 498 (1990).

5. E. Paparazzo, M. Fanfoni, C. Quaresima and P. Perfetti, *J. Vac. Sci. Technol. A8*, 2231 (1990).
6. F. Leccabue, R. Panizzeri, B. E. Watts, D. Fiorani, E. Agostinelli, A. Testa and E. Paparazzo, *J. Cryst. Growth 112*, 644 (1991).
7. E. Paparazzo, M. Fanfoni, E. Severini and S. Priori, *J. Vac. Sci. Technol. A10*, 2892 (1992).
8. E. Paparazzo, E. Severini, A. J. Lopez, P. M. Torres, P. O. Pastor, E. R. Castellon and A. A. G. Tomlinson, *J. Mater. Chem. 2*, 1175 (1992).
9. E. Paparazzo, *J. Phys. D: Appl. Phys. 20*, 1091 (1987).
10. J. F. Weaver and G. B. Hoflund, *J. Phys. Chem. 98*, 8519 (1994).
11. J. F. Weaver and G. B. Hoflund, *Chem. Mater. 6*, 1693 (1994).
12. C. D. Wagner, W. M. Riggs, L. E. Davis, J. F. Moulder and G. E. Muilenberg (eds.) *Handbook of X-ray Photoelectron Spectroscopy*, Perkin-Elmer Corp., Physical Electronics Division, Eden Prairie, MN, 1979.
13. O. De Melo, F. Leccabue, R. Panizzeri, C. Pelosi, G. Bocelli, G. Calestani, V. Sagredo, M. Chourio and E. Paparazzo, *J. Cryst. Growth 104*, 780 (1990).
14. E. Paparazzo, L. Moretto, C. D'Amato and A. Palmieri, *Surf. Interface Anal. 23*, 69 (1995).

Appendix 5

Documentary Standards in Surface Analysis: The Way of the Future?

S. J. HARRIS British Aerospace, Sowerby Research Centre, Filton, Bristol, United Kingdom

1. INTRODUCTION

Commercial surface analysis systems have been available since around 1970 [1]. Most of the early instruments were dedicated to longer-term fundamental research, even if they were located at industrial research centers. However, since the latter part of the 1980s, the most popular surface analytical techniques, Auger electron spectroscopy (AES), secondary-ion mass spectrometry (SIMS) and X-ray photoelectron spectroscopy (XPS), have gained a greater level of acceptance in industry due to their improved reliability. Surface analysis is now routinely used to solve complex industrial problems in both research and quality assurance environments. It has been specifically the move toward the use of these techniques in quality-assurance-type applications that has started to force the development of national/international documentary standards in order to formalize the methods of application of the techniques.

From an industrial point of view, documentary standards are essential for any laboratory which follows a quality assurance regime, since

1. *They should improve the reliability of the analytical results obtained.*
2. *They should reduce the level of skill required to perform routine analytical measurements.*
3. *Data can be transferred between different analytical laboratories, with a high degree of confidence if the laboratories all follow similar procedures.*

Given these perceived advantages, industry naturally prefers to use analytical methods which conform to predefined standards, particularly if critical measurements are to be made on key products. In fact the need for documentary standards in all types of analysis can only increase in the near future, since the number of companies adopting quality systems, such as ISO 9000, is increasing every year. A major reason for the adoption of this type of quality system is that companies which already have ISO 9000 certification tend to favor the use of suppliers/subcontractors who also have some form of accreditation. Therefore, companies which do not adopt a recognized quality system may find their business opportunities reduced in the near future. For better or for worse, conformance to an ISO 9000–type quality system may become essential.

The first set of documentary standards for surface analysis were written by ASTM committee E-42 in the early 1980s [1]. Currently there are 22 relevant ASTM standards for surface analysis [2], covering topics such as terminology, energy-scale calibration and the widths of interfaces in sputter depth profiling using SIMS. The ASTM standards are not used extensively outside the United States, and certainly within the United Kingdom they tend to be used as the basis for more descriptive local standards, containing the working practices of a given laboratory.

Under the NAMAS (National Measurement Accreditation Service) quality system, detailed procedures are written by a particular analytical laboratory which apply only to the instrumentation in that laboratory. Unfortunately, these in-house procedures are not made available to the wider analytical community. However, in 1991, ISO technical committee 201 on surface chemical analysis (TC201) was set up specifically to develop documentary standards for the most industrially developed surface analytical techniques. The standards written by this technical committee are targeted directly at the requirements of the **average** industrial user, and, since they are ISO standards, are available to any analyst upon request [1].

In the following sections, the structure of TC201, the various subcommittees, and the current ISO standards under development will be described. The work is very much a continuing process, and the list of fledgeling standards will grow rapidly over the next few years.

2. ISO TECHNICAL COMMITTEE 201 ON SURFACE CHEMICAL ANALYSIS

TC201 consists of experts from most of the countries which are actively involved in the development and application of surface analytical techniques. The main technical committee is divided into various subcommittees, which are led by secretariats based in Germany, Japan, Sweden, the United States and the United Kingdom. Other countries, such as Australia, Hungary and Korea, also make important contributions to the development of the ISO standards. The widespread of interests of these different countries ensure that the standards will be effective, focused, relevant and, perhaps most critically from an industrial point of view, practical! The scope and purpose of ISO TC201 have been defined in the strategic policy statement and are as follows [3]:

Scope

Standardisation in the field of surface chemical analysis in which beams of electrons, ions, neutral atoms or molecules, or photons are incident on the specimen material and scattered or emitted electrons, ions, neutral atoms or molecules, or photons are detected.

Purpose

(a) *To promote the harmonization of requirements concerning instrument specifications, instrument operations, specimen preparation, data acquisition, data processing, qualitative analysis, quantitative analysis and reporting of results*
(b) *To establish consistent terminology*
(c) *To develop recommended procedures and to promote the development of reference materials and reference data to ensure that surface analyses of the needed precision and accuracy can be made*

ISO standards are developed only when there is a clear industrial need, as well as an adequate technical and skill base, to enable the documentary standard to be produced within a realistic time scale. Hence the main technical committee and the subcommittees are drawn from a cross section of industrial and academic users who are well established in their respective fields.

2.1. Structure of ISO Technical Committee 201

The main ISO committee on surface analysis has seven subcommittees (SC) and two main working groups (WG). Also each subcommittee may have a num-

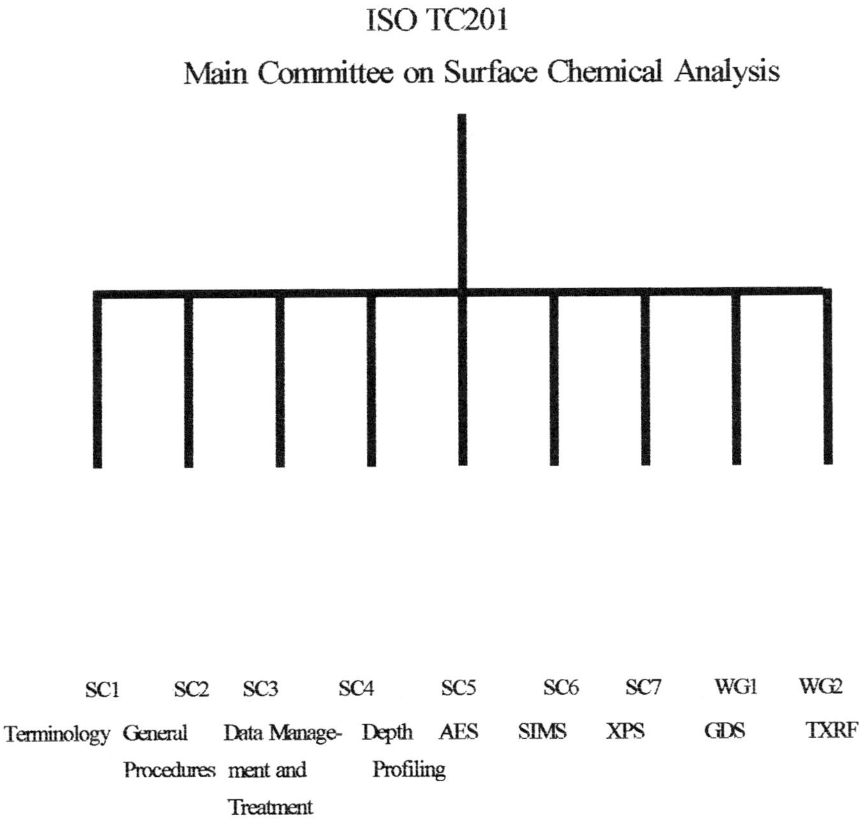

Figure 1. Structure of ISO TC201.

ber of working groups and study groups associated with it. The basic structure of ISO TC201 is shown in Fig. 1.

The basic structure was agreed upon in Japan in 1992, and the positions of the chairpersons of the main committee and of the subcommittees were subsequently confirmed at Catania in 1993. While the basic structure of the technical committee will remain relatively constant over the next few years, the working groups, study groups and *ad hoc* groups associated with the subcommittees will change, depending on the current requirements of a given subcommittee. The chairpersons, convenors, secretaries and the host countries involved in ISO TC201 are given in Table 1 [4].

The countries represented at the TC201 committee meetings fall into two distinct categories:

1. P, or participating, status, for those countries actively involved
2. O, or observer, status, for countries retaining a watching brief on the work

Table 2 lists the various O and P member countries affiliated with TC201 [4]. The individual TC201 subcommittees will have different O and P members, which are not listed in Table 2.

Over the years, other national and international bodies have developed procedures which apply to surface analytical methods. ISO documentary standards take a considerable time to write, and frequently it can take seven years from the initial proposal to the draft international standard stage [5]. Hence, to avoid replication and to increase the rate of development of the ISO standards, ISO TC201 has been particularly active in setting up links with many of these associations, and the present Category A liaisons are with the following bodies:

The International Union of Pure and Applied Chemistry (IUPAC)
The International Union for Vacuum Science, Technique and Applications (IUVSTA)
The Versailles Project on Advanced Materials and Standards (VAMAS)

These international bodies send representatives, who act in an advisory capacity to the TC201 committee meetings. Members from related ISO technical committees, such as TC202 on microbeam analysis, are also kept informed of the current activities of TC201, and there is an agreement with that particular ISO technical committee to cooperate in the areas of terminology, general procedures and data management and treatment [6]. From this description of the structure and composition of the TC, it is clear that a considerable commitment has already been made by the surface analysis community to the development of standards.

2.2. ISO Technical Committee 201 Subcommittees

ISO TC201 Subcommittee 1: Terminology

This is a particularly important subcommittee, since the work in this area affects the activities of all the other subcommittees. The scope of the subcommittee is to *standardize the definitions of the terms used in surface chemical analysis* [3]. At present eight countries are participating actively in the work of the subcommittee under the guidance of Dr. M. P. Seah from the United Kingdom.

Initially the subcommittee set up a working group led by Dr. Seah, which reviewed the existing terminology standards and the texts of ASTM, IUPAC and other relevant sources. When this had been completed, a second working group

Table 1 Chairpersons, Secretaries and Convenors for ISO TC201

Committee/ subcomm./ working gp.	Chairperson	Convenor	Secretary	Fax number	Host country
TC201 Main Committee	Dr. Cedric Powell NIST Fellow Maryland		Mr. Jiro Tokita Jap. Standards Assoc., Tokyo	+81-3-3582-2390	Japan
SC1 Terminology	Dr. Martin Seah NPL Teddington		Dr. John H. Thomas III 3M Corp. Minnesota	+1-612-733-0648	USA
SC2 Gen. Procedures	Dr. William Stickle Hewlett-Packard Oregon		Dr. Nigel G. Cave Motorola, Inc. Mesa, Arizona	+1-602-655-5013	USA
SC3 Data Manage. and Treatment	Dr. David Sykes ISST Loughborough Univ.		Dr. Albert Carley Dept. of Chem. Univ. of Wales Cardiff	+44-1222-874030	UK
SC4 Depth Profiling	Prof. S. Hofmann NRIM Tsukuba		Mr. Jiro Tokita Jap. Standards Assoc., Tokyo	+81-3-3582-2390	Germany

SC5 AES	Dr. Donald R. Baer Battelle Pacific Northwest Labs Washington		Dr. Robert F. Davis IBM Corp. New York	+1-914-892-6256	USA
SC6 SIMS	Dr. K. Tsunoyama Kawasaki Steel Corp., Chiba		Mr. Terumi Oda Osaka Science and Tech. Centre	+81-6-443-3767	Japan
SC7 XPS	Dr. John F. Watts Dept. Mater. Sci. and Eng. Surrey Univ.		Dr. Robert K. Wild Interface Anal. Centre, IAC Univ. of Bristol	+44-117-9255646	UK
WG1 Glow Discharge Spectroscopy		Dr. Arne Bengtson Inst. for metallforskning Stockholm		+46-8-723-0423	Sweden
WG2 Total Reflection X-ray Fluorescence Spectroscopy		Prof. Yohichi Gohshi Depl. Appl. Chem. Univ. of Tokyo		+81-3-3815-0053	Japan

Source: From Ref. 4.

Table 2 P and O Member Countries Affiliated to ISO TC201

Country	Representative	Title	Institution	Fax number
ISO/CS	Dr D Gaudillière	Dir. Technical Gp.	ISO Central Secretariat, Geneva	+41-22-733-34-30
Austria (P)	Mr P Waldhauser		Österreichisches Normungsinstitut, Wien	+43-1-213-00-650
China (P)	Mr Li Chuanqing	Dir. General	State Bureau of Technical Supervison, Beijing	+86-10-203-10-10
Germany (P)	Mr O Nather		DIN Deutsches Institut für Normung, Berlin	+49-30-26-01-1231
Italy (P)	Mr E Martinotti	Vice Pres. Delegue	Ente Nazionale Italiano di Unificazione, Milano	+39-2-70-02-41
Japan (P)	Mr Shoichi Saba	JISC President	c/o Standards Division, Tokyo	+81-3-3580-1418
Korea (P)	Mr Seung-Bae Lee	Director General	Nat. Institut. of Technology and Quality, Anyang City	+82-3-43-84-18-61
Russian Federation (P)	Prof. S F Bezverkhy	President	Committee of the Russian Federation for Standardisation, Moscow	+7-095-237-6032
Slovenia (P)	Mr Peter Palma	Director	Standards and Metrology Institute, Ljubljana	+386-61-314-88
Sweden(P)	Mr Svante Lundin		SIS-Standardiseringen i Sverige, Stockholm	+46-8-30-77-57
Switzerland (P)	Mr Carl-Arthur Eder		Swiss Association for Standardization, Zürich	+41-1-254-54-74
UK (P)	Mr Bernard Shelley		British Standards Institute, London	+44-181-996-74-00
	Dr Robert K Wild	Secretary National Committee	Interface Analysis Centre, Bristol	+44-117-92564
USA (P)	Ms Kayla Serotte		American Natl Standards Institute, NY	+1-212-398-0023
Algeria (O)	M Djenidi Bendaoud	Director General	Institut Algerien Normalisation, Alger	+213-2-61-09-71

Australia (O)	Mrs B Catteau	Intl Coordinator	Standards Australia, Strathfield NSW	+61-2-746-4766
Belgium (O)	M P Croon	Directeur General	Institut Belge de Normalisation, Bruxelles	+32-2-733-42-64
Egypt (O)	Dr A B El-Sebia	President	Egyptian Organisation for Standardisation and Quality Control, Cairo	+20-2-354-9720
Finland (O)			FIMET, Standards Department, Helsinki	+358-0-625-020
France (O)	Dr F Goldschmidt		Assoc. Française de Nomalisation Tout Europe, Paris	+33-1-42-91-56-56
Hungary (O)	Mr Gyorgy Ponyai	General Director	Magyar Szavanyugyi Testulet, Budapest	+36-1-218-5125
India (O)			Bureau of Indian Standards, New Delhi	+91-11-331-4062
Ireland (O)	Mr E Paterson	Director	National Standards Authority of Ireland, Dublin	+353-1-36-9821
Norway (O)	Mr E Braathen		Norges Standardiseringsforbund, Oslo	+47-22-04-92-11
Philippines (O)	Mr R V Navarrete	Director	Bureau of Product Standards, Manila	+63-2-890-4926
Poland (O)	Mr M Lukaszewicz	President	Polish Committee of Standardisation, Warszawa	+48-22-20-07-41
Romania (O)	Mr Mihail Ciocodica	General Director	Institutul Roman de Standardisation, Bucuresti	+40-1-210-0833
Singapore (O)	Mr Loh Wah Sing	Dep. Director	Institute of Standards and Industrial Research	+65-779-4359
South Africa (O)			Bureau of Standards, Pretoria	+27-12-344-1568
Turkey (O)	Mr Ahmet Kurter		Turk Standardlari Enstitusu, Ankara	+90-312-425-4399

Source: From Ref. 4.

was formed, which is currently developing a working draft of an ISO standard entitled *Surface Chemical Analysis—Vocabulary*. This draft standard is currently being reviewed by the members of the subcommittee and the nominated multinational experts from the P status countries. Table 3 lists the P status countries and the current list of nominated experts for this working group [7].

This particular document has already reached ISO working draft status. The relatively rapid development of such a long and complex document has been due partly to the existence of relevant material from other groups, such as ASTM and IUPAC, but also to the considerable efforts of its convenor. However, it will still be several years before the standard will be available for publication as a full international standard!

ISO TC201 Subcommittee 2: General Procedures

Subcommittee 2 is responsible for developing those standards that are relevant to two or more subcommittees, such as subcommittees 5 and 7 on AES and XPS, respectively. There are currently three working groups within this subcommittee: WG1 on specimen handling, WG2 on reference materials and WG3 on the reporting of results.

WG1 (on specimen handling) is developing two standards, namely *Specimen Handling and Preparation for Auger Electron Spectroscopy, X-ray Photoelectron Spectroscopy and Secondary-Ion Mass Spectrometry* and *Specimen Preparation, Mounting, and Analysis for Auger Electron Spectroscopy, X-ray Photoelectron Spectroscopy and Secondary-Ion Mass Spectrometry*. Although from their titles the standards appear to be similar, the first is specifically for the customer, namely the scientist who supplies the samples to the surface an-

Table 3 National Experts for Working Group 2, ISO TC201/SC1

Working group Convenor WG2	Participating countries in SC1	Nominated national experts for SC1/WG2
MP Seah (UK)	Germany	SJ Harris (UK)
	Japan	A Carley (UK)
	Russia	W Gries (Germany))
	Switzerland	R Opila (USA)
	UK	D Baer (USA)
	USA	K Bomben (USA)
	Hungary	C Powell (USA)
	Poland	R Shimizu (Japan)
		V Rakhovski (Russia)

alyst for examination, while the second will apply specifically to the surface analyst. These standards have been based on an existing ASTM standard and consequently are fairly well advanced. Table 4 lists the nominated national experts for the working group in SC2.

The work of WG 2 on reference materials has been slower, since any such ISO standards must be based on the application of existing reference materials. It should be noted that ISO does not itself develop new reference materials. Also, the documentary standards drafted in subcommittee 2 should in principle define the application of reference materials used by two or more surface analytical techniques. Given that such reference materials tend to be technique specific, the progress of this particular working group has been somewhat limited.

Table 4 National Experts for the Working Groups in ISO TC201/SC2

Working group convenor WG1	Participating countries in SC2	Nominated national experts For SC2/WG1
W Stickle (USA)	China Germany Japan UK USA Hungary	A Wirth (UK) D Sykes (UK) M P Seah (UK) R Bradley (UK) G Davis (USA) K Bomben (USA) R Opila (USA) S Ichimura (Japan) K Muira (Russia)
Working group convenor WG2		**Nominated national experts for SC2/WG2**
W Gries (Germany)		S J Harris (UK) T English (UK) K Tinkham (UK) F Stevie USA)
Working group convenor WG3		**Nominated national experts for SC3/WG2**
R Bradley (UK)		S J Harris (UK) R Bulpett (UK) A Chew (UK) S Gaarenstroom (USA) S Ichimura (Japan)

Despite these problems, WG2 currently has two items under development, *Procedure for Certification of Ion-Implanted Reference Materials for Analysis of Semiconductor Wafers* and *Ion-Implanted Dosimetry Calibration by Surface Analytical Techniques*.

WG2 has also commissioned surveys from the various P member countries on the current working practices of analysts within those countries with regards to reference materials. At present, a limited response has been obtained from the United Kingdom, while Japan is still conducting its survey. The results from the latter survey should help to formulate the future policy of WG2.

WG3 (on reporting of results) currently has no official ISO work items, but has put forward several discussion documents, which will become work items in the near future. Initially WG3 attempted to draft a general reporting of results document, but the latter was too complex, given the wide range of techniques considered. Hence, technique-specific structures have been developed which define all the key parameters that should be noted in instrument logbooks, on spectra and in final reports. At present, discussion documents exist only for the reporting of XPS, Auger and SIMS spectra, which may ultimately be developed by the individual subcommittees rather than by subcommittee 2.

ISO TC201 Subcommittee 3: Data Management and Treatment

This is an area of key importance, given the increasing refinement of instrument data systems and the reliance of the analyst upon them. The subcommittee work includes the standardization of databases and data transfer systems and the specification of the properties of algorithms used in surface analysis data systems [3]. At present there are two working groups in this particular subcommittee: WG1 on data transfer and storage, and WG2 on surface science data models. The subcommittee has also formed a study group which has surveyed the use of existing data algorithms. This latter group has effectively completed its initial work and is currently inactive.

WG1 has two approved work items, *A Standard Data Transfer Format for the Digital Exchange of Spectra, Maps and Profiles in Surface Chemical Analysis* and *Surface Chemical Analysis—Information Formats*. The former is by far the most advanced ISO standard currently in the TC201 portfolio. This is because the standard is based on the existing VAMAS transfer format, which has already gained fairly widespread acceptance in Europe and Japan. That particular work item is currently a registered draft international standard (DIS) and may be available within the next two years as a full ISO standard. Table 5 lists the nominated national experts for the working groups in ISO TC201/SC3.

Table 5 National Experts for Working Groups in ISO TC201/SC3

Working group convenor WG1	Participating countries in SC3	Nominated national experts for SC3/WG1
M P Seah (UK)	Germany	P Coxon (UK)
	Japan	A Carrick (UK)
	Korea	B Kisakurek (Turkey)
	Switzerland	K Yoshihara (Japan)
	UK	T Sekine (Japan)
	USA	D Watson (USA)
	Hungary	K Wagner (USA)
		H Goretski (Germany)
		H Störi (Germany)
Working group convenor WG2		**Nominated national experts for SC3/WG2**
S Garenstroom (USA)		P Coxon (UK)
		M P Seah (UK)
		B Kosakurek (Turkey)
		K Yoshihara (Japan)
		S Fukushima (Japan)

The progress within WG2 has perhaps been a little slower, given that there are no existing national standards to adopt or modify. A new work item has been proposed, entitled *A Data Dictionary for the Description of X-ray Photoelectron and Auger Electron Spectroscopy Data Records*; work is still in progress on the initial draft document.

ISO TC201 Subcommittee 4: Depth Profiling

Through the considerable efforts of the chairman of subcommittee 4, the work program of SC4 is well defined and should ultimately result in the development of a number of documentary standards. The scope of the subcommittee is the *standardization of instrument specification, instrument calibration, instrument operation, data acquisition, and data processing used to determine composition versus depth with surface analytical techniques [3]* With these objectives in mind, WG1 has defined the following areas which will require the development of ISO standards:

1. Destructive depth profiling, which includes sputtter and mechanical profiling in conjunction with surface analytical methods

2. Nondestructive depth profiling, which includes angle-resolved measurements, Rutherford backscattering spectroscopy (RBS), nuclear reaction analysis (NRA), peak intensity measurements from two or more peaks from the same element, and elemental intensity measurements as a function of incident electron energy [3]

At present only a draft document on the measurement procedure for determining sputtered depth in depth profiling has been developed, but this should become a full working draft in the near future.

WG2 (reference materials for depth profiling) has drafted a standards for system optimization using superlattices such as AlAs/GaAs. The working group is also considering a new work item on the optimization of sputter depth profiles using single-layer materials. The number of experts actively involved in the development of these standards is relatively large, and considerable progress should be made in this area in the next few years. The current composition of the two working groups in subcommittee 4 is given Table 6.

ISO TC201 Subcommittee 5: Auger Electron Spectroscopy (AES)

Due to the considerable overlap in the areas of interest of subcommittee 5 on Auger electron spectroscopy and subcommittee 7 on X-ray photoelectron spectroscopy, the two subcommittees have worked closely together and formed joint working groups. This has resulted in the drafting of a number of joint discussion documents which form the basis of working draft standards. For example, both subcommittees have worked together on documents which will define methods of spectral quantification. These include procedures on defining spectral intensity and on the use of relative sensitivity factors for quantitative analysis.

A survey defining the key topics for future standards has been conducted among the nominated national experts in each country. This survey is particularly important since realistically there is only a limited amount of manpower available for writing these standards, and priorities must therefore be established in order to make the most effective use of the resources. The results of this survey will help define the work plan for subcommittees 5 and 7 over the next few years. Table 7 lists the experts in WG1 and the study group.

ISO TC201 Subcommittee 6: Secondary Ion Mass Spectrometry (SIMS)

The subcommittee has focused initially on developing documentary standards for dynamic SIMS, specifically sputter depth profiling and the analysis of silicon-based materials. Although the title of the subcommittee suggests that it will only develop standards applicable to SIMS, procedures for sputtered neutral mass spectrometry (SNMS) and fast atom bombardment mass spectrometry (FABMS) will also be developed.

Table 6 National Experts for Working Groups in ISO TC201/SC4

Working group convenor WG1	Participating countries in SC4	Nominated national experts for SC4/WG1
S Hofmann (Germany)	Germany	D Skinner (UK)
	Japan	M Dowsett (UK)
	Korea	D Sykes (UK)
	Russia	M P Seah (UK)
	Switzerland	W H Gries (Germany)
	UK	K Kajiwara (Japan)
	USA	Dae-Won Moon (Korea)
	Hungary	H J Mathieu (Switz)
		A Torrisi (Italy)
		C Anderson (USA)
		J Geller (USA)
		D Simmons (USA)
		W Stickle (USA)
		A Zalar (IUVSTA)
Working Group ConvenorWG2		**Nominated national experts for SC4/WG2**
K Kajiwara (Japan)		M Dowsett (UK)
		D Sykes (UK)
		M P Seah (UK)
		S J Harris (UK)
		W H Gries (Germany)
		S Hofmann (Germany)
		Dae-Won Moon (Korea)
		H J Mathieu (Switz)
		R Payling (Australia)
		L Kövér (Hungary)
		C Anderson (USA)
		J Geller (USA)
		W Stickle (USA)

At present only one standard is under development, *Determination of Boron Content in Silicon—Secondary Ion Mass Spectrometric Method*. However, three other new standards may be started in the near future, namely

1. Reporting sputter depth profile data from SIMS
2. Reporting mass spectral data from SIMS
3. Determining SIMS relative sensitivity factors from ion-implanted external standards

Table 7 National Experts for Working Group and Study Group in ISO TC201/SC5

Working group convenor WG1	Participating countries in SC5	Nominated national experts for SC5/WG1
C Anderson (USA)	Germany	J F Watts (UK)
	Japan	M P Seah (UK)
	Korea	R Bulpett (UK)
	Russia	A Zarasiz (Turkey)
	UK	A Jablonski (Poland)
	USA	Chang-Y Hwang (Korea)
	Hungary	T Gross (Germany)
	Poland	W H Gries (Germany)
		Y Fukuda (Japan)
		N Usuki (Japan)
		L Kövér (Hungary)
		K Childs (USA)
		S Kreber (USA)
		R Opila (USA)
		N Turner (USA)
		H Wildman (USA)
		W Stickle (USA)
Study group for future standards joint with SC7		**Nominated national experts for the study group**
		S J Harris (UK)
		M P Seah (UK)
		R Bulpett (UK)
		S Evans (UK)
		T Gross (Germany)
		S Hoffman (Germany)
		W H Gries (Germany)
		T Sekine (Japan)
		A Tanaka (Japan)
		K Childs (USA)
		S Kreber (USA)
		C Anderson (USA)
		N Turner (USA)

Table 8 National Experts for Working Groups in ISO TC201/SC6

Working group convenor WG1	Participating countries in SC6	Nominated national experts for SC6/WG1
Y Homma (Japan)	China	Y Wang (China)
	Germany	M Dowsett (UK)
	Japan	A Chew (UK)
	Korea	A Kurter (Turkey)
	Russia	V Rakhovski (Russia)
	UK	D Kruger (Germany)
	USA	W H Gries (Germany)
		D Simmons (USA)
		S Bryan (USA)
		E-H Cirlin (USA)
		C Hintzman (USA)
		H Luftman (USA)
		F Stevie (USA)
		N Winograd (USA)
Working group convenor WG2		**Nominated national Experts for SC6/WG2**
D Simmons (USA)		A Chew (UK)
		K Tinkham (UK)
		C Liangzhen (China)
		N Tsuda (Japan)
		C Becker (USA)
		S Bryan (USA)
		E-H Cirlin (USA)
		F Stevie (USA)
		C Magee (USA)
		P Williams (USA)

These standards are likely to be based on existing ASTM standards and a discussion document submitted by working group 3, subcommittee 2, on the reporting of SIMS and SNMS data. The membership of both SIMS working groups is given in Table 8.

ISO TC201 Subcommittee 7: X-ray Photoelectron Spectroscopy (XPS)

As previously discussed, subcommittee 7 has worked closely with subcommittee 5 to develop standards applicable to AES and XPS. While subcommittee 5 has concentrated on data processing, WG1 of subcommittee 7 has started to develop standards which describe the specification of AES and XPS instruments. Currently there are two working drafts on instrument specification, one dealing with Auger spectrometers and the second with XPS-based instruments.

WG2 has already produced a working draft on the calibration of the energy scales in X-ray photoelectron spectrometers. This document is in part based on the earlier work of Seah [8] and an existing ASTM standard [2]. The experts involved in these working groups are listed in Table 9.

2.3. ISO TC201 Working Groups

The subcommittees associated with TC201 are for those areas/methods of analysis in which the level of international interest in the development of ISO standards analytical techniques was particularly high. Two working groups have also been created for glow discharge spectrometry (WG1) and total reflection X-ray fluorescence spectroscopy (WG2), which are directly affiliated to the main technical committee. Working groups were formed in these areas, not full subcommittees, since, although the science supporting these techniques is well developed and standards could be formulated, at present there was not a sufficient level of interest from the international community to support full subcommittees. Table 10 lists the national experts for these two working groups.

Currently each working group has one ISO approved work item. That in WG1 is *Glow Discharge Optical Emission Spectrometry (GDOES)—Introduction for Use*, and that in WG2 is *Surface Chemical Analysis—Measurement of Surface Elemental Contamination on Silicon Wafers by Total Reflection X-ray Fluorescence Spectroscopy (TXRF)*. Both documents are in the working draft stage and awaiting consideration for status as full committee draft standards. WG1 has also prepared a discussion document which is awaiting approval as a new work item, entitled *Procedure for the Determination of Thickness and Elemental Composition by GDOES of Zn-Based Metallic Coatings.*

3. CONCLUSIONS

The development of standards in surface analysis is the inevitable result of the gradual maturing of surface analytical methods, coupled with the increasing drive in industry toward the use of quality systems based on ISO 9001. The need for documentary standards in this field is clear, and the next 10 years will see the rapid development of the number of standards for all the analytical techniques in this area. The only thing that might restrict their development is the lack of available manpower to write them. Documentary standards take a con-

Table 9 National Experts for Working Group and Study Group in ISO TC201/SC7

Working group convenorWG1	Participating countries in SC7	Nominated national experts for SC7/WG1
J F Watts (UK)	Germany Japan Korea Russia Sweden UK USA Hungary Poland	S Evans (UK) M P Seah (UK) P Coxon (UK) R K Wild (UK) A Zarasiz (Turkey) Chang-Y Hwang (Korea) T Gross (Germany) Y Fukuda (Japan) N Usuki (Japan) N Crist (Japan) L Kövér (Hungary) B Beard (USA) R Davies (USA) J Fulgham (USA) S Kreber (USA) R Opila (USA) S Simko (USA) N Turner (USA) H Wallace (USA) W Stickle (USA)
Working group convenor WG2		**Nominated national experts for SC7/WG2**
M P Seah (UK)		C Powell (USA) C Anderson (USA) K Bomben (USA)
Study group for future standards joint with SC5		**Nominated national experts for the study group**
J F Watts (UK)		S J Harris (UK) P Coxon (UK) A Carrick (UK) S Evans (UK) T Gross (Germany) W H Gries (Germany) A Zarasiz (Turkey) Y Fukuda (Japan) N Usuki (Japan)

Table 10 National Experts for Working Group in ISO TC201

Working group convenor Glow Discharge Spectrometry WG1	Nominated national experts for WG1
A Bengtson (Sweden)	T English (UK) K Tinkham (UK) J Angell (Austria) A Quentmeier (Germany) K Hirokawa (Japan) Jin-Chun Woo (Korea) V I Rakhovski (Russia) L A Vasilyev (Russia) M R Winchester (USA) D Fang (USA) C Lazik (USA) R Marcus (USA) J Mitchell (USA)
Working group convenor Total Reflection X-ray Fluorescence Spectrometry WG2	**Nominated national experts for WG2**
Y Goshi (Japan)	S Biswas (UK) S Pahike (Germany) A Iida (Japan) K Kawai (Japan) Sung-Chul Kang (Korea) V I Rakhovski (Russia) L A Vasilyev (Russia) A Diebold (USA) R S Hockett (USA) A Fisher-Colbrie (USA) M Zaltiz (USA)

siderable time to conceive and develop, and require a substantial level of commitment in order to make it to full international standard status. Although a considerable amount has already been done, the surface analysis community needs to increase its level of commitment to this project to ensure that the appropriate standards are developed, and, more critically, are available when required. If a standard is too early, then it is likely to be badly defined and not easy to use, and if it is late it will not be used. The effort is required now!

REFERENCES

1. M. P. Seah, *VAM Bull., No. 13*, 16 (1995).
2. *Analytical Chemistry for Metals, Ores and Related Materials (II): E356 to latest: Molecular Spectroscopy; Surface Analysis*, Vol. 03.06 ASTM Standards, 1995.
3. ISO document number ISO/TC 201 N80, 1996-08-09.
4. ISO document number ISO/TC 201 N81, 1996-09-02 (N76 revised).
5. ISO document number ISO/TC 201 N82, 1996-09-02 (N77 Revised).
6. ISO document number ISO/TC 201 Ref. #2, 1992-07-01.
7. ISO document number ISO/TC 201 N69, 1995-10-27.
8. M. P. Seah and G. C. Smith, Appendix 1, in *Practical Surface Analysis*, Vol. 1, *Auger and X-ray Photoelectron Spectroscopy* (D. Briggs and M. P. Seah, eds.), Wiley, New York, 1990.

Index